Lecture Notes in Computer Science 3099

Commenced Publication in 1973
Founding and Former Series Editors:
Gerhard Goos, Juris Hartmanis, and Jan van Leeuwen

Editorial Board

Takeo Kanade
Carnegie Mellon University, Pittsburgh, PA, USA
Josef Kittler
University of Surrey, Guildford, UK
Jon M. Kleinberg
Cornell University, Ithaca, NY, USA
Friedemann Mattern
ETH Zurich, Switzerland
John C. Mitchell
Stanford University, CA, USA
Moni Naor
Weizmann Institute of Science, Rehovot, Israel
Oscar Nierstrasz
University of Bern, Switzerland
C. Pandu Rangan
Indian Institute of Technology, Madras, India
Bernhard Steffen
University of Dortmund, Germany
Madhu Sudan
Massachusetts Institute of Technology, MA, USA
Demetri Terzopoulos
New York University, NY, USA
Doug Tygar
University of California, Berkeley, CA, USA
Moshe Y. Vardi
Rice University, Houston, TX, USA
Gerhard Weikum
Max-Planck Institute of Computer Science, Saarbruecken, Germany

Springer
Berlin
Heidelberg
New York
Hong Kong
London
Milan
Paris
Tokyo

Jordi Cortadella Wolfgang Reisig (Eds.)

Applications and Theory of Petri Nets 2004

25th International Conference, ICATPN 2004
Bologna, Italy, June 21-25, 2004
Proceedings

Volume Editors

Jordi Cortadella
Universitat Politècnica de Catalunya, Software Department
Campus Nord, Jordi Girona Salgado 1-3, 08034 Barcelona, Spain
E-mail: jordi.cortadella@upc.es

Wolfgang Reisig
Humboldt-Universität zu Berlin, Institut für Informatik
Unter den Linden 6, 10099 Berlin, Germany
E-mail: reisig@informatik.hu-berlin.de

Library of Congress Control Number: 2004107502

CR Subject Classification (1998): F.1-3, C.1-2, G.2.2, D.2, D.4, J.4

ISSN 0302-9743
ISBN 3-540-22236-7 Springer-Verlag Berlin Heidelberg New York

This work is subject to copyright. All rights are reserved, whether the whole or part of the material is concerned, specifically the rights of translation, reprinting, re-use of illustrations, recitation, broadcasting, reproduction on microfilms or in any other way, and storage in data banks. Duplication of this publication or parts thereof is permitted only under the provisions of the German Copyright Law of September 9, 1965, in its current version, and permission for use must always be obtained from Springer-Verlag. Violations are liable to prosecution under the German Copyright Law.

Springer-Verlag is a part of Springer Science+Business Media

springeronline.com

© Springer-Verlag Berlin Heidelberg 2004
Printed in Germany

Typesetting: Camera-ready by author, data conversion by Olgun Computergrafik
Printed on acid-free paper SPIN: 11013112 06/3142 5 4 3 2 1 0

Preface

This volume contains the proceedings of the 25th International Conference on Application and Theory of Petri Nets (ICATPN 2004). The aim of the Petri net conferences is to create a forum for discussing progress in the application and theory of Petri nets.

Typically, the conferences have 100–150 participants, one third of these coming from industry, whereas the others are from universities and research institutions. The conferences always take place in the last week of June.

The conference and a number of other activities are coordinated by a steering committee with the following members: Wil van der Aalst (The Netherlands), Jonathan Billington (Australia), Jrg Desel (Germany), Susanna Donatelli (Italy), Serge Haddad (France), Kurt Jensen (Denmark), Maciej Koutny (United Kingdom), Sadatoshi Kumagai (Japan), Giorgio De Michelis (Italy), Tadao Murata (USA), Carl Adam Petri (Germany, Honorary Member), Wolfgang Reisig (Germany), Grzegorz Rozenberg (The Netherlands, Chairman) and Manuel Silva (Spain).

The 2004 conference was organized by the Department of Computer Science of the University of Bologna, Italy. We would like to thank the organizing committee, chaired by Roberto Gorrieri, for the effort invested in making the event successful. We are also grateful to the following sponsoring institutions and organizations: Associazione Italiana per l'Informatica ed il Calcolo Automatico (AICA), Microsoft Research, and Network Project & Solutions (NPS Group).

We received a total of 62 submissions from 26 different countries. The program committee finally selected 19 regular papers and 5 tool presentation papers. This volume comprises the papers that were accepted for presentation. Invited lectures were given by Gianfranco Ciardo, Roberto Gorrieri, Thomas A. Henzinger, Wojciech Penczek, Lucia Pomello and William H. Sanders. Their papers are also included in this volume.

Several tutorials and workshops were also organized within the conference, covering introductory and advanced aspects related to Petri nets. Detailed information can be found at the conference URL (`www.cs.unibo.it/atpn2004`).

We would like to thank all those who submitted papers to the Petri net conference. We are grateful to the program committee members and the referees for their valuable effort in reviewing and selecting the papers. We gratefully acknowledge Andrei Voronkov (University of Manchester) for his technical support with the PC management tool. Finally, we would like to mention the excellent cooperation with Springer-Verlag during the preparation of this volume.

April 2004 Jordi Cortadella and Wolfgang Reisig

Organizing Committee

Nadia Busi
Mario Bravetti
Roberto Gorrieri (Chair)
Roberto Lucchi
Gianluigi Zavattaro

Tools Demonstration

Gianluigi Zavattaro (Chair)

Program Committee

W. van der Aalst, The Netherlands
L. Bernardinello, Italy
D. Buchs, Switzerland
N. Busi, Italy
S. Christensen, Denmark
G. Ciardo, USA
J. Cortadella, Spain
(Co-chair, Applications)
J. Desel, Germany
X. He, USA
H. Klaudel, France
H.C.M. Kleijn, The Netherlands
M. Koutny, UK
S. Kumagai, Japan
J. Lilius, Finland
P. Moreaux, France
M. Mukund, India
C. Lakos, Australia
L. Recalde, Spain
W. Reisig, Germany
(Co-chair, Theory)
W. Sanders, USA
P.S. Thiagarajan, Singapore
E. Vicario, Italy
H. Voelzer, Germany
A. Yakovlev, UK

Referees

B. Adsul
A. Agostini
M. Alanen
A. Aldini
A. Alexander
A.K. Alves de Medeiros
M.A. Bednarczyk
M. Bernardo
G. Berthelot
F.S. de Boer
M. Bonsangue
A.M. Borzyszkowski
L. Brodo
F. Burns
A. Bystrov
J. Campos
F. Cardone
M.-Y. Chung
J.M. Colom
Z. Dai
D. Daly
C. Delamare
A. Dennunzio
S. Derisavi
R. Devillers
G. Di Marzo Serugendo
J. Ding
S. Donatelli
Z. Dong
C. Dutheillet
J. Engelfriet
R. Esser
J. Ezpeleta
J. Freiheit
D. de Frutos Escrig
G. Gallasch
S. Gaonkar
F. Garca-Valls
R. Gorrieri
M. Griffith
L. Groenewegen
S. Haar
S. Haddad
N. He
K. Heljanko

K. Hiraishi
H.J. Hoogeboom
A. Horvath
Y. Huang
J. Hulaas
D. Hurzeler
N. Husberg
G. Hutzler
R. Janicki
J.B. Jrgensen
G. Juhás
J. Júlvez
V. Khomenko
E. Kindler
L.M. Kristensen
V. Lam
C. Laneve
T. Latvala
R. Leporini
K. Lodaya
M. Loregian
R. Lorenz
L. Lucio
P. Madhusudan
O. Marroqun Alonso
F. Martinelli
C. Mascolo
J. Merseguer
V. Milijic
A.S. Miner
T. Miyamoto
L. Mo
K. Narayan Kumar
C. Neumair
A. Niaouris
A. Norta
A. Ohta
E. Pelz
O.-M. Penttinen
I. Petre
S. Peuker
G.M. Pinna
D. Poitrenaud
L. Pomello
F. Pommereau
J.-F. Pradat-Peyre
C.J. Prez-Jimnez
R. Ramanujam
M. Ribaudo
S. Roch
D. Rodrguez
A. Romanovsky
E. Roubtsova
L. Sassoli
K. Schmidt
R. Segala
S. Sendall
T. Shi
H. Shiizuka
D. D'Souza
J. Steggles
T. Suzuki
E. Teruel
C. Bui Thanh
N.A. Thomas
S. Tini
F. Tricas Garca
D. Tutsch
N. Uchihira
T. Ushio
A. Valmari
B. van Dongen
S. Vanit-Anunchai
D. Varacca
K. Varpaaniemi
E. Verbeek
T. Watanabe
T. Weijters
L. Wells
M. Westergaard
D. Xu
S. Yamaguchi
H. Yu
J. Yu
G. Zavattaro

Table of Contents

Invited papers

Full papers

Tool Papers

Positive Non-interference in Elementary and Trace Nets

Nadia Busi and Roberto Gorrieri

Dipartimento di Scienze dell'Informazione, Università di Bologna
Mura A. Zamboni, 7, 40127 Bologna, Italy

Abstract. Several notions of non-interference have been proposed in the literature to study the problem of confidentiality in concurrent systems. The common feature of these non-interference properties is that they are all defined as extensional properties based on some notion of behavioural equivalence on systems. Here we also address the problem of defining non-interference by looking at the structure of the net systems under investigation. We define *structural* non-interference properties based on the absence of particular places in the net. We characterize a structural property, called *PBNI+*, that is equivalent to the well-known behavioural property *SBNDC*. We start providing a characterization of *PBNI+*on contact-free Elementary Net Systems, then we extend the definition to cope with the richer class of Trace nets.

1 Introduction

Non-interference has been defined in the literature as an extensional property based on some observational semantics: the high part of a system does not interfere with the low part if whatever is done at the high level produces no visible effect on the low part of the system. The original notion of non-interference in [9] was defined, using trace semantics, for system programs that are deterministic. Generalized notions of non-interference were then designed to include (nondeterministic) labeled transition systems and finer notions of observational semantics such as bisimulation (see, e.g., [13, 7, 12, 14, 8]). Relevant properties in this class are the trace-based properties *SNNI* and *NDC*, as well as the bisimulation-based properties *BSNNI*, *BNDC* and *SBNDC* proposed by Focardi and Gorrieri some years ago [7, 8] on a CCS-like process algebra. In particular, *SNNI* states that a system R is secure if the two systems $R \setminus H$ (all the high level actions are prevented) and R/H (all the high level actions are permitted but are unobservable) are trace equivalent. *BNDC* intuitively states that a system R is secure if it is bisimilar to R in parallel with any high level process Π w.r.t. the low actions the two systems can perform. And *SBNDC* tells that a system R is secure if, whenever a high action h is performed, the two instances of the system before and after performing h are bisimilar from a low level point of view.

In the first part of the paper we show that these non–interference properties can be naturally defined also on Petri Nets; in particular – to keep the presentation as simple as possible – we use Elementary Nets [6]. The advantage of this

J. Cortadella and W. Reisig (Eds.): ICATPN 2004, LNCS 3099, pp. 1–16, 2004.
© Springer-Verlag Berlin Heidelberg 2004

proposal is the import in the Petri Net theory of security notions that makes possible the study of security problems. Technically, what we do is to introduce two operations on nets, namely parallel composition (with synchronization in TCSP-like style) and restriction, and suitable notions of observational equivalences on the low part of the system (low trace equivalence and low bisimulation); then, five security properties are defined and compared in a rather direct way. In particular, the two properties based on low trace semantics, namely *SNNI* and *NDC*, are equivalent. On the contrary, in the bisimulation case, *BSNNI* is weaker than *BNDC*, which turns out to be equivalent to *SBNDC*.

In this approach, the security property is based on the dynamics of systems; they are all defined by means of one (or more) equivalence check(s); hence, non-interference checking is as difficult as equivalence checking, a well-studied hard problem in concurrency theory.

In the second part of the paper we address the problem of defining statically non-interference for Elementary nets, by looking at the structure of the net systems under investigation:

- in order to better understand the causality and conflict among different system activites, hence grounding more firmly the intuition about what is an interference, and
- in order to find more efficiently checkable non-interference properties that are sufficient conditions for those that have already received some support in the literature.

We define structural non-interference properties based on the absence of particular places in the net. We identify two special classes of places: *causal places*, i.e., places for which there are an incoming high transition and an outgoing low transition; and, *conflict places*, i.e. places for which there are both low and high outgoing transitions. Intuitively, causal places represent potential source of interference (*hilo* flow for *high input – low output*), because the occurrence of the high transition is a prerequisite for the execution of the low transition. Similarly, conflict places represent potential source of interference (*holo* flow for *high output – low output*), because the occurrence of a low event tells us that a certain high transition will not occur.

We show that when causal and conflict places are absent, we get a property, called *Positive Place–Based Non–Interference* (*PBNI+* for short), which turns out to be equivalent to *SBNDC*. More precisely, the net N has no causal and no conflict places if and only if it satisfies *SBNDC*.

In the third part of the paper we extend the definition of *PBNI+* to cope with the richer class of Trace nets[1]. We provide an example showing how our property can be used to capture the information flows arising in a shared variable that can be accessed and modified by both high and low users.

The paper is organised as follows. In Section 2 we recall the basic definitions about transition systems and Elementary Nets. In Section 3 we recast the behavioural approach to non-interference properties, originally defined in a process algebraic setting, on Elementary Nets. The original structural property *PBNI+* for Elementary Nets is introduced in Section 4. In Section 5, after recalling the basic definitions about Trace Nets, we extend the definition of *PBNI+* to Trace Nets. Finally, some conclusive remarks are drawn.

2 Basic Definitions

Here we recall the basic definition about transition systems and elementary net systems we will use in the following.

2.1 Transition Systems

Definition 1. *A transition system is a triple $TS = (St, E, \rightarrow)$ where*

- *St is the set of states*
- *E is the set of events*
- *$\rightarrow \subseteq St \times E \times St$ is the transition relation.*

In the following we use $s \stackrel{e}{\rightarrow} s'$ to denote $(s, e, s') \in \rightarrow$.

Given a transition $s \stackrel{e}{\rightarrow} s'$, s is called the source, *s' the* target *and e the* label *of the transition.*

A rooted *transition system is a pair (TS, s_0) where $TS = (St, E, \rightarrow)$ is a transition system and $s_0 \in St$ is the* initial state.

2.2 Elementary Net Systems

Definition 2. *An* elementary net *is a tuple $N = (S, T, F)$, where*

- *S and T are the (finite) sets of* places *and* transitions, *such that $S \cap T = \emptyset$*
- *$F \subseteq (S \times T) \cup (T \times S)$ is the flow relation*

A set over the set S of places is called a *marking.* Given a marking m and a place s, if $s \in m$ then we say that the place s contains a token, otherwise we say that s is empty.

Let $x \in S \cup T$. The *preset* of x is the set ${}^\bullet x = \{y \mid F(y, x)\}$. The *postset* of x is the set $x^\bullet = \{y \mid F(x, y)\}$. The preset and postset functions are generalized in the obvious way to set of elements: if $X \subseteq S \cup T$ then ${}^\bullet X = \bigcup_{x \in X} {}^\bullet x$ and $X^\bullet = \bigcup_{x \in X} x^\bullet$. A transition t is enabled at marking m if ${}^\bullet t \subseteq m$ and $t^\bullet \cap m = \emptyset$. The firing (execution) of a transition t enabled at m produces the marking $m' = (m \setminus {}^\bullet t) \cup t^\bullet$. This is usually written as $m[t\rangle m'$. With the notation $m[t\rangle$ we mean that there exists m' such that $m[t\rangle m'$.

An *elementary net system* is a pair (N, m_0), where N is a net and m_0 is a marking of N, called *initial marking.* With abuse of notation, we use (S, T, F, m_0) to denote the net system $((S, T, F), m_0)$.

The set of *markings reachable from* m, denoted by $[m\rangle$, is defined as the least set of markings such that

- $m \in [m\rangle$
- if $m' \in [m\rangle$ and there exists a transition t such that $m'[t\rangle m''$ then $m'' \in [m\rangle$.

The set of *firing sequences* is defined inductively as follows:

- m_0 is a firing sequence;
- if $m_0[t_1\rangle m_1 \dots [t_n\rangle m_n$ is a firing sequence and $m_n[t_{n+1}\rangle m_{n+1}$ then $m_0[t_1\rangle m_1 \dots [t_n\rangle m_n[t_{n+1}\rangle m_{n+1}$ is a firing sequence.

Given a firing sequence $m_0[t_1\rangle m_1 \dots [t_n\rangle m_n$, we call $t_1 \dots t_n$ a *transition sequence*. The set of transition sequences of a net N is denoted by $TS(N)$. We use σ to range over $TS(N)$. Let $\sigma = t_1 \dots t_n$; we use $m[\sigma\rangle m_n$ as an abbreviation for $m[t_1\rangle m_1 \dots [t_n\rangle m_n$.

The *marking graph* of a net N is

$$MG(N) = ([m_0\rangle, T, \{(m, t, m') \mid m \in [m_0\rangle \wedge t \in T \wedge m[t\rangle m'\})$$

A net is *simple* if the following condition holds for all $x, y \in S \cup T$: if ${}^\bullet x = {}^\bullet y$ and $x^\bullet = y^\bullet$ then $x = y$.

A marking m contains a *contact* if there exists a transition $t \in T$ such that ${}^\bullet t \subseteq m$ and $not(m[t\rangle)$. A net system is *contact–free* if no marking in $[m_0\rangle$ contains a contact. A net system is *reduced* if each transition can occur at least one time: for all $t \in T$ there exists $m \in [m_0\rangle$ such that $m[t\rangle$.

In the following we consider contact-free net systems that are simple and reduced.

3 A Behavioural Approach to Non-interference for Elementary Nets

In this section we recall from [5] some basic properties, initially proposed in a process algebraic setting by Focardi and Gorrieri [7,8]. Our aim is to analyse systems that can perform two kinds of actions: high level actions, representing the interaction of the system with high level users, and low level actions, representing the interaction with low level users. We want to verify if the interplay between the high user and the high part of the system can affect the view of the system as observed by a low user. We assume that the low user knows the structure of the system, and we check if, in spite of this, he is not able to infer the behavior of the high user by observing the low view of the execution of the system.

Hence, we consider nets whose set of transitions is partitioned into two subsets: the set H of high level transitions and the set L of low level transitions. To emphasize this partition we use the following notation. Let L and H be two disjoint sets: with (S, L, H, F, m_0) we denote the net system $(S, L \cup H, F, m_0)$.

The non-interference properties we are going to introduce are based on some notion of *low* observability of a system, i.e., what can be observed of a system from the point of view of low users. The low view of a transition sequence is nothing but the subsequence where high level transitions are discarded.

Definition 3. *Let $N = (S, L, H, F, m_0)$ be a net system. The* low view *of a transition sequence of N is defined as follows:*

$\Lambda_N(\varepsilon) = \varepsilon$

$\Lambda_N(\sigma t) = \begin{cases} \Lambda_N(\sigma)t & \text{if } t \in L \\ \Lambda_N(\sigma) & \text{otherwise} \end{cases}$

The definition of Λ_N *is extended in the obvious way to sets of transitions sequences:* $\Lambda_N(\Sigma) = \{\Lambda_N(\sigma) \mid \sigma \in \Sigma\}$ *for* $\Sigma \subseteq (L \cup H)^*$.

Definition 4. *Let* N_1 *and* N_2 *be two net systems. We say that* N_1 *is* low-view trace equivalent *to* N_2, *denoted by* $N_1 \overset{\Lambda}{\approx}_{tr} N_2$, *iff* $\Lambda_{N_1}(TS(N_1)) = \Lambda_{N_2}(TS(N_2))$.

We define the operations of parallel composition (in TCSP-like style) and restriction on nets, that will be useful for defining some non-interference properties.

Definition 5. *Let* $N_1 = (S_1, L_1, H_1, F_1, m_{0,1})$ *and* $N_2 = (S_2, L_2, H_2, F_2, m_{0,2})$ *be two net systems such that* $S_1 \cap S_2 = \emptyset$ *and* $(L_1 \cup L_2) \cap (H_1 \cup H_2) = \emptyset$. *The parallel composition of* N_1 *and* N_2 *is the net system*

$N_1 \mid N_2 = (S_1 \cup S_2, L_1 \cup L_2, H_1 \cup H_2, F_1 \cup F_2, m_{0,1} \cup m_{0,2})$

Definition 6. *Let* $N = (S, L, H, F, m_0)$ *be a safe net system and let* U *be a set of transitions. The restriction on* U *is defined as* $N \backslash U = (S, L', H', F', m_0)$, *where*

$L' = L \setminus U$

$H' = H \setminus U$

$F' = F \setminus (S \times U \cup U \times S)$

Strong Nondeterministic Non-Interference (*SNNI* for short) is a trace-based property, that intuitively says that a system is secure if what the low-level part can see does not depend on what the high-level part can do.

Definition 7. *Let* $N = (S, L, H, F, m_0)$ *be a net system. We say that* N *is SNNI iff* $N \overset{\Lambda}{\approx}_{tr} N \backslash H$.

The intuition is that, from the low point of view, the system where the high level transitions are prevented should offer the same traces as the system where the high level transitions can be freely performed. In essence, a low-level user cannot infer, by observing the low view of the system, that some high-level activity has occurred.

As a matter of fact, this non-interference property captures the information flows from high to low, while admits flows from low to high. For instance, the net N' of Figure 1 is *SNNI* while the net N'' is not *SNNI*.

An alternative notion of non-interference, called *Nondeducibility on Composition* (*NDC* for short), says that the low view of a system N in isolation is not to be altered when considering each potential interaction of N with the high users of the external environment.

Definition 8. *Let* $N = (S, L, H, F, m_0)$ *be a net system. We say that* N *is a high-level net if* $L = \emptyset$.

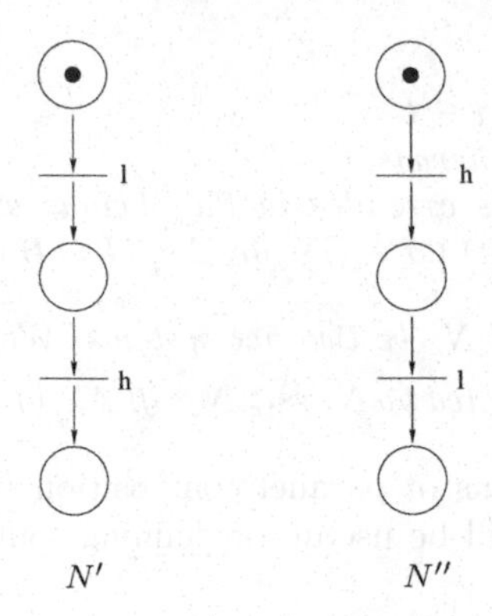

Fig. 1. The net system N' is *SNNI* while N'' is not *SNNI*.

Definition 9. *Let $N = (S, L, H, F, m_0)$ be a net system. N is NDC iff for all high-level nets $K = (S_K, \emptyset, H_K, F_K, m_{0,K})$: $N \backslash H \overset{\Lambda}{\approx}_{tr} (N \mid K) \backslash (H \setminus H_K)$.*

The left-hand term represents the low view of the system N in isolation, while the right-hand term expresses the low view of N interacting with the high environment K (note that the activities resulting from such interactions are invisible by the definition of low view equivalence). *NDC* is a very intutive property: whatever high level system K is interacting with N, the low effect is unobservable. However, it is difficult to check this property because of the universal quantification over high systems. Luckily enough, we will then prove that *SNNI* and *NDC* are actually the same non-interference property.

Theorem 1. *Let $N = (S, L, H, F, m_0)$ be a net system. N is SNNI if and only if N is NDC.*

The two properties above are based on (low) trace semantics. It is well-known [8] that bisimulation semantics is more appropriate than trace semantics because it captures also some indirect information flows due to, e.g., deadlocks. For this reason, we now consider non-interference properties based on bisimulation. To this aim, we first need to introduce a notion of low–view bisimulation.

Definition 10. *Let $N_1 = (S_1, L_1, H_1, F_1, m_{0,1})$ and $N_2 = (S_2, L_2, H_2, F_2, m_{0,2})$ be two net systems. A* low–view bisimulation *from N_1 to N_2 is a relation on $\mathcal{M}(S_1) \times \mathcal{M}(S_2)$ such that if $(m_1, m_2) \in R$ then for all $t \in \bigcup_{i=1,2} L_i \cup H_i$:*

- *if $m_1[t\rangle m_1'$ then there exist σ, m_2' such that $m_2[\sigma\rangle m_2'$, $\Lambda_{N_1}(t) = \Lambda_{N_2}(\sigma)$ and $(m_1', m_2') \in R$*
- *if $m_2[t\rangle m_2'$ then there exist σ, m_1' such that $m_1[\sigma\rangle m_1'$, $\Lambda_{N_2}(t) = \Lambda_{N_1}(\sigma)$ and $(m_1', m_2') \in R$*

If $N_1 = N_2$ we say that R is a low–view bisimulation on N_1.

We say that N_1 is low–view bisimilar *to N_2, denoted by $N_1 \overset{\Lambda}{\approx}_{bis} N_2$, if there exists a low–view bisimulation R from N_1 to N_2 such that $(m_{0,1}, m_{0,2}) \in R$.*

The first obvious variation on the theme is to define the bisimulation based version of *SNNI*, yielding *BSNNI*.

Definition 11. *Let $N = (S, L, H, F, m_0)$ be a net system. We say that N is BSNNI iff $N \overset{\Lambda}{\approx}_{bis} N \backslash H$.*

Obviously, *BSNNI* $\subseteq$ *SNNI*. The converse is not true: the net N in Figure 2 is *SNNI* but not *BSNNI*. Note that *SNNI* misses to capture the indirect information flow present in this net: if the low transition l cannot be performed, the low user can infer that the high transition h has been performed, hence deducing one piece of high knowledge.

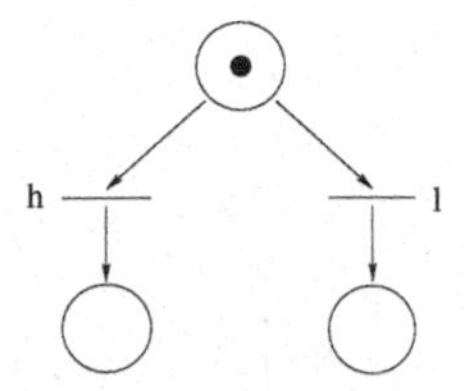

Fig. 2. A net system that is *SNNI* but not *BSNNI*.

Similarly, *BNDC* can be defined from *NDC*, yielding a rather appealing security property, which is finer than *BSNNI*.

Definition 12. *Let $N = (S, L, H, F, m_0)$ be a net system. N is BNDC iff for all high-level nets $K = (S_K, \emptyset, H_K, F_K, m_{0,K})$: $N \backslash H \overset{\Lambda}{\approx}_{bis} (N \mid K) \backslash (H \setminus H_K)$.*

Theorem 2. *Let $N = (S, L, H, F, m_0)$ be a net system. If N is BNDC then N is BSNNI.*

Unfortunately, the converse is not true: Figure 3 reports a net that is *BSNNI* but not *BNDC*; the reason why can be easily grasped by looking at their respective marking graphs in Figure 4.

BNDC is quite appealing but, because of the universal quantification on all possible high level systems, it is difficult to check. The next property, called *Strong Bisimulation Non Deducibility on Composition* (*SBNDC* for short), is actually an alternative characterization of *BNDC* which is easily checkable.

Definition 13. *Let $N = (S, L, H, F, m_0)$ be a net system. N is SBNDC iff for all markings $m \in [m_0\rangle$ and for all $h \in H$ the following holds:*

if $m[h\rangle m'$ then there exists a low–view bisimulation R on $N \backslash H$ such that $(m, m') \in R$.

Theorem 3. *Let $N = (S, L, H, F, m_0)$ be a net system. N is BNDC if and only if N is SBNDC.*

The theorem above holds because we are in an unlabeled setting: transitions are not labeled. In [7, 8] it is proved that – for the Security Process Algebra – *SBNDC* is strictly finer than *BNDC*.

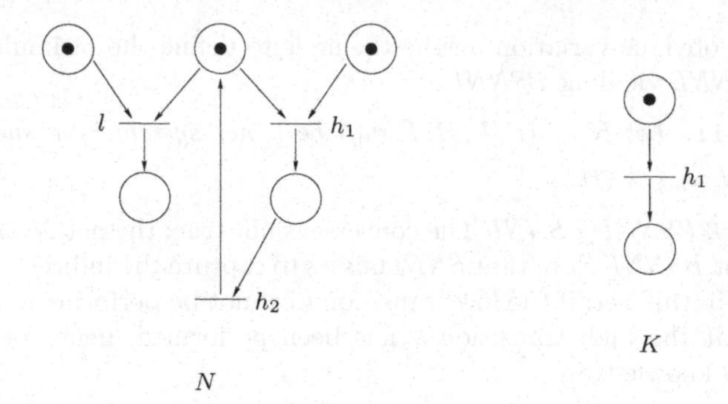

Fig. 3. A net system that is *BSNNI* but not *BNDC*.

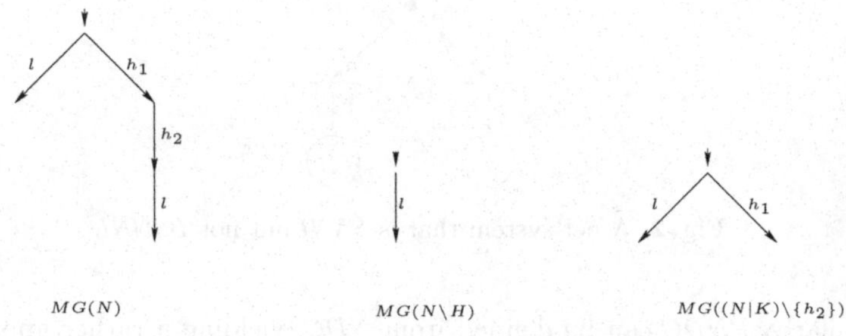

Fig. 4. The marking graphs of the net systems N, $N\backslash H$ and $(N \mid K)\backslash\{h_2\}$.

4 Positive Place-Based Non-interference in Elementary Nets

In [4, 5] we defined two notions of non-interference, namely, *PBNI* and *RBNI*, aiming at capturing any kind of information flow from high users to low users. Those notions capture both positive and negative informations on the high behaviour of the system. More precisely, a positive information flow arises when the occurrence of a high level transition can be deduced from the low level behaviour of the system, whereas a negative information is concerned with the fact that a high level transition has not occurred.

In this paper we provide a characterisation of positive information flows, i.e., we consider a system secure if it is not possible to deduce that some high level action has been performed by observing the low level behaviour.

To this aim, we define the *PBNI+* property based on the absence of some kinds of places in a net system. Consider a net system $N = (S, L, H, F, m_0)$.

Consider a low level transition l of the net: if l can fire, then we know that the places in the preset of l are marked before the firing of l; moreover, we know that such places become unmarked after the firing of l. If there exists a high level action h that produces a token in a place s in the preset of l (see the system N_1 in Figure 5), then the low level user can infer that h has occurred if he can perform the low level action l. We note that there exists a causal dependency

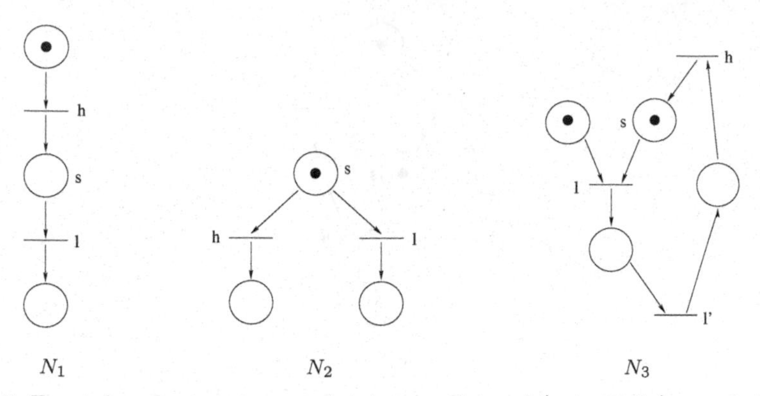

Fig. 5. Examples of net systems containing conflict and (potentially) causal places.

between the transitions h and l, because the firing of h produces a token that is consumed by l. Consider now the situation illustrated in the system N_2 of Figure 5: in this case, place s is in the preset of both l and h, i.e., l and h are competing for the use of the resource represented by the token in s. Aware of the existence of such a place, a low user knows that the high-level action h has been performed, if he is not able to perform the low-level action l. Place s represents a conflict between transitions l and h, because the firing of h prevents l from firing.

Our idea is to consider a net system secure if it does not contain places of the kinds illustrated above.

In order to avoid the definition of a security notion that is too strong, and that prevents systems that do not reveal information on the high-level actions that have been performed, we need to refine the concepts illustrated above. Consider the net system N_3 reported in Figure 5. Although s is a potentially causal place, the net system has to be considered secure, as the (unique) possible firing of l is at the initial marking, hence it is not caused by h. For s to be a source of information on the occurrence of h, there must exists a firing sequence where l consumes a token produced by h. In other words, s is an active causal place if there exists a path in $MG(N_3)$ connecting (an occurrence of) h to (an occurrence of) l, such that the transitions occurring after h and before l do not produce tokens in s.

Regarding conflicts, consider the net system N_4 reported in Figure 6. At first sight, the net N_4 could appear not secure because of the presence of the conflict place s. However, we note that the occurrence of h has no effect on the low behaviour of the system, as the possibility to fire l has already been ruled out by the firing of transition l'. Hence, for s to be a source of information on the occurrence of h, there must exist a reachable marking where the firing of h rules out the possibility to fire (immediately or after some other transitions) l. In other words, s is an active conflict place if there exists a path in $MG(N_4)$ connecting the source of (an occurrence of) h to (an occurrence of) l, such that the transitions occurring in the path do not produce tokens in s.

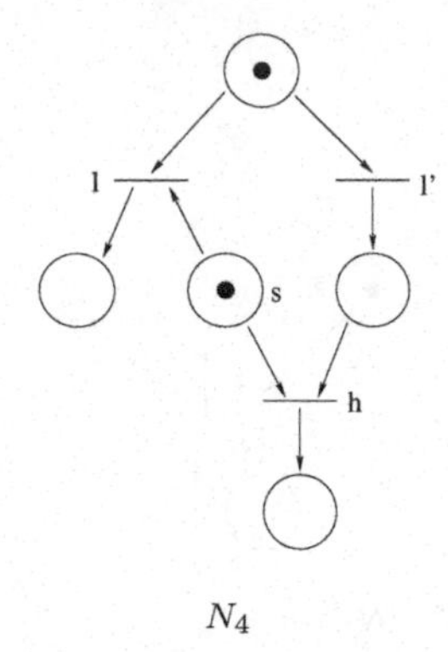

N_4

Fig. 6. A net system containing a potentially conflict place but no active conflict places.

Definition 14. *Let $N = (S, L, H, F, m_0)$ be an elementary net system. Let s be a place of N such that $s^\bullet \cap L \neq \emptyset$.*

The place $s \in S$ is a potentially causal place *if ${}^\bullet s \cap H \neq \emptyset$.*

A potentially causal place s is an active causal place *if the following condition holds: there exist $l \in s^\bullet \cap L$, $h \in {}^\bullet s \cap H$, $m \in [m_0\rangle$ and a transition sequence σ such that $m[h\sigma l\rangle$ and $s \notin t^\bullet$ for all $t \in \sigma$.*

The place $s \in S$ is a potentially conflict place *if $s^\bullet \cap H \neq \emptyset$.*

A potentially conflict place is a active conflict place *if the following condition holds: there exist $l \in s^\bullet \cap L$, $h \in s^\bullet \cap H$, $m \in [m_0\rangle$ and a transition sequence σ such that $m[h\rangle$, $m[\sigma l\rangle$ and $s \notin t^\bullet$ for all $t \in \sigma$.*

Definition 15. *Let $N = (S, L, H, F, m_0)$ be an elementary net system. We say that N is PBNI+ (*positive Place Based Non-Interference*) if, for all $s \in S$, s is neither an active causal place nor an active conflict place.*

We have that the absence of both causal and conflict places is a necessary and sufficient condition for *SBNDC*.

Theorem 4. *Let $N = (S, L, H, F, m_0)$ be an elementary net system. If N is PBNI+ then N is SBNDC.*

Theorem 5. *Let $N = (S, L, H, F, m_0)$ be an elementary net system. If N is SBNDC then N is PBNI+.*

Corollary 1. *Let $N = (S, L, H, F, m_0)$ be an elementary net system. Then N is PBNI+ iff N is BNDC.*

5 Non-interference in Trace Nets

In this section we extend the definition of *PBNI+* to cope with the richer class of Trace nets [1] and we show that the results presented in the previous section for elementary nets continue to hold also in this setting. Finally, we provide an example to show how our property can be used to capture the information flows arising in a shared variable that can be accessed and modified by both high and low users.

5.1 Trace Nets

Trace nets [1] are an extension of elementary nets: besides the classical flow arcs, also arcs permitting to test for presence/absence of tokens in a place, as well as arcs permitting to fill/empty a place regardless of its previous contents, are added.

Definition 16. *A* trace net *is a tuple* $N = (S, T, W)$, *where*

- S *and* T *are the (finite) sets of* places *and* transitions, *such that* $S \cap T = \emptyset$
- $W : (S \times T) \to \{in, out, nop, read, inhib, set, reset\}$ *is the flow function, such that* $\forall t \in T \exists s \in S : W(s,t) \neq nop$.

The arcs of kind *in* and *out* correspond to the flow arcs of elementary nets: more precisely, a flow arc from a place s to a transition t is represented in trace nets by setting $W(s,t) = in$, whereas a flow arc from t to s is represented by setting $W(s,t) = out$. The arcs of kind *read* and *inhib* permit to test a condition on a place, without altering its contents. A read (resp. inhibitor) arc from s to t requires that s contains a token (resp. no tokens) for t to fire. The arcs of kind *set* and *reset* permit to set the contents of the place to a given value, independently of the previous contents of the place. A set (resp. reset) arc from s to t sets the number of tokens in place s to 1 (resp. 0) when t fires. Finally, an arc of kind *nop* denotes the absence of any kind of relation among the contents of s and the firing of t.

We adopt the graphical convention proposed in [2], and depicted in Figure 7 to draw trace nets: input (resp. output) arcs are represented as directed segments with an arrow on the transition (resp. place) side; read (resp. inhibitor) arcs are represented as segments with a small black (resp. white) circle on the transition side; set (resp. reset) arcs are represented as segment with a small black (resp. white) circle on the place side.

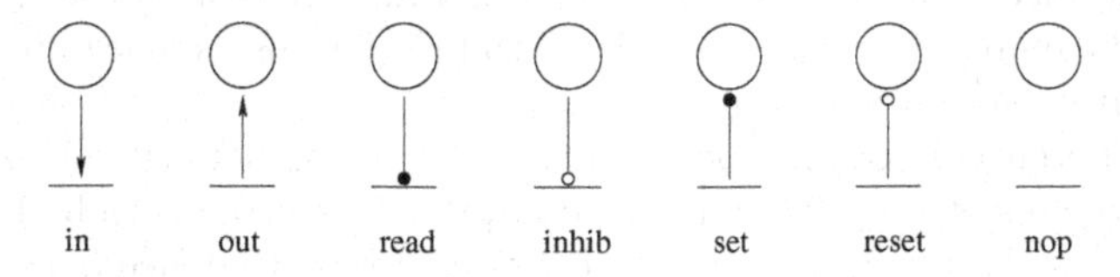

Fig. 7. Graphic conventions for drawing trace nets.

To lighten the definitions of enabling of a transition and of the firing rule, we introduce the following auxiliary relations. Intuitively, t *test*1 s (resp. t *test*0 s)

holds if it is necessary that s contains one (resp. zero) tokens for t to fire. On the other hand, $t\ set1\ s$ (resp. $t\ set0\ s$) holds if, after the firing of t, s contains one (resp. zero) tokens.

Given$s \in S$ and $t \in T$, we define the following relations:
$t\ test1\ s$ iff $W(s,t) \in \{in, read\}$
$t\ test0\ s$ iff $W(s,t) \in \{out, inhib\}$
$t\ set1\ s$ iff $W(s,t) \in \{out, set\}$
$t\ set0\ s$ iff $W(s,t) \in \{in, reset\}$

A transition t is enabled at marking m if $\{s \mid t\ test1\ s\} \subseteq m$ and $\{s \mid t\ test0\ s\} \cap m = \emptyset$.

The firing (execution) of a transition t enabled at m produces the marking $m' = \{s \in S \mid t\ set1\ s\} \cup \{s \in m \mid W(s,t) = nop\}$. This is usually written as $m[t\rangle m'$.

A *trace net system* is a pair (N, m_0), where N is a trace net and m_0 is a marking of N, called *initial marking.* With abuse of notation, we use (S, T, W, m_0) to denote the trace net system $((S, T, W), m_0)$.

The definitions of reachable markings, firing sequences, marking graph and reduced system given in Section 2 for elementary nets apply also for trace nets.

In the following we consider trace net systems that are reduced.

5.2 Positive Place Based Non-interference for Trace Nets

As already done for elementary nets, we consider trace nets whose set of transitions is partitioned into two subsets: the set H of high level transitions and the set L of low level transitions. Also for trace nets, given two disjoint sets L and H, with (S, L, H, W, m_0) we denote the trace net system $(S, L \cup H, W, m_0)$.

We extend the definitions of causal an conflict places for Elementary nets of Section 4 to Trace nets.

An extension of the definition of potentially causal place for contact free elementary nets to trace nets leads to the following: a place s is a potentially causal place if there exist a high transition h that puts a token in s (either through an output or a set arc) and a low transition l that needs place s to be full to fire (because s and l are connected by a read or an input arc). However, as contact freeness no longer holds for trace nets, we have to take into account also causal dependencies arising from the fact that a high transition h removes a token contained in a place s (either through an input or a reset arc) that is required to be empty for a low transition l to fire (because s and l are connected by an output or an inhibitor arc).

Also for potentially conflict places, two kinds of conflicts can arise. Similarly to elementary nets, s is a potentially conflict place if there exists a high transition h that removes a token from s and a low transition l that needs place s to be full to fire. Morevoer, s has to be considered a potentially conflict place also if there exists a high transition h that produces a token in S and a low transition l that needs place s empty to fire.

Definition 17. *Let $N = (S, L, H, W, m_0)$ be a trace net system.*

The place $s \in S$ is a potentially causal place *if there exist $h \in H$, $l \in L$ and $X \in \{0,1\}$ such that h setX s and l testX s.*

A potentially causal place s is an active causal place *if the following condition holds: there exist $h \in H$, $l \in L$ and $X \in \{0,1\}$ such that:*

- *h setX s*
- *l testX s*
- *there exists a marking $m \in [m_0\rangle$ and a transition sequence σ such that*
  - *$m[h\sigma l\rangle$*
  - *$s \in m$ iff $X = 0$*
  - *for all $t \in \sigma$: $\neg(t$ set1 $s)$ and $\neg(t$ set0 $s)$*

The place $s \in S$ is a potentially conflict place *if there exist $h \in H$, $l \in L$ and $X \in \{0,1\}$ such that h setX s and l test$(1-X)$ s.*

A potentially conflict place is a active conflict place *if the following condition holds: there exist $h \in H$, $l \in L$ and $X \in \{0,1\}$ such that:*

- *h setX s*
- *l test$(1-X)$ s*
- *there exists a marking $m \in [m_0\rangle$ and a transition sequence σ such that*
  - *$m[h\rangle$ and $m[\sigma l\rangle$*
  - *$s \in m$ iff $X = 0$*
  - *for all $t \in \sigma$: $\neg(t$ set1 $s)$ and $\neg(t$ set0 $s)$*

Definition 18. *Let $N = (S, L, H, F, m_0)$ be a trace net system. We say that N is PBNI+ (*positive Place Based Non-Interference*) if, for all $s \in S$, s is neither an active causal place nor an active conflict place.*

The definitions of parallel composition and restriction presented in Section 3 are extended to trace nets in the obvious way:

Definition 19. *Let $N_1 = (S_1, L_1, H_1, W_1, m_{0,1})$ and $N_2=(S_2, L_2, H_2, W_2, m_{0,2})$ be two trace net systems such that $S_1 \cap S_2 = \emptyset$ and $(L_1 \cup L_2) \cap (H_1 \cup H_2) = \emptyset$. The parallel composition of N_1 and N_2 is the trace net system*
$N_1 \mid N_2 = (S_1 \cup S_2, L_1 \cup L_2, H_1 \cup H_2, W_1 \cup W_2, m_{0,1} \oplus m_{0,2})$

Definition 20. *Let $N = (S, L, H, W, m_0)$ be a trace net system and let U be a set of transitions. The restriction on U is defined as $N \backslash U = (S, L', H', W', m_0)$, where*
$L' = L \setminus U$
$H' = H \setminus U$
$W' = W \setminus (S \times U)$

The results presented in Section 3 continue to hold also for trace nets:

Theorem 6. *Let $N = (S, L, H, W, m_0)$ be a trace net system. N is BNDC if and only if N is SBNDC.*

Theorem 7. *Let $N = (S, L, H, W, m_0)$ be a trace net system. N is PBNI+ if and only if N is SBNDC.*

5.3 Example: Binary Memory Cell

In this section we recast the example of a binary memory cell proposed in [3] in our framework.

A binary memory cell can contain a binary value, i.e., either 0 or 1. The memory cell is accessible to both high and low users, that can read and write a value in the cell. The trace net representing the binary cell is reported in Figure 8.

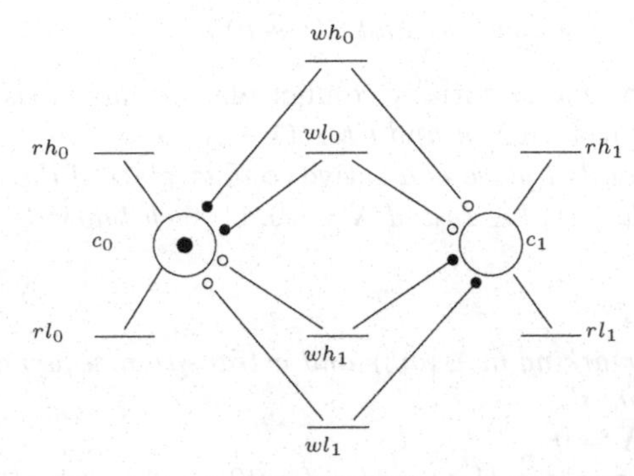

Fig. 8. The trace net modeling a binary memory cell.

A token in place c_0 (resp. c_1) represents the fact that the current value contained in the cell is 0 (resp. 1). Each operation is modeled by two transitions, one for each binary value that can be contained in the cell. For example, transition wh_0 (resp. wh_1) is performed by a high level user that writes the value 0 (resp. 1) in the cell. A write operation of e.g. value 0 in the cell is represented by a transition that sets the contents of place c_0 (i.e., puts one token in place c_0 regardless of its previous contents), and resets the contents of place c_1 (i.e., removes the possible token present in place c_1). A read operation of e.g. value 0 is represented by a transition with a read arc on place c_0, i.e., a transition that can happen only if place c_0 contains a tokens.

As already pointed out in [3], the binary memory cell depicted in Figure 8 is completely insecure, as a high level user can send confidential information to a low level user through the binary cell. In fact, the binary cell is not *PBNI+* because of the existence of (at least) the active causal place c_1. Note that c_1 is a potentially causal place, because the high transition wh_1 has a set arc on c_1, and the low transition rl_1 has a read arc on c_1. Moreover, if we consider the firing sequence $\{c_0\}[wh_1\rangle\{c_1\}[rl_1\rangle\{c_1\}$ also the conditions for the potentially causal place c_1 to be an active causal place are fulfilled.

In order to avoid the flow of information from the high user to the low user, we can either forbid all the read operations performed by a low user or forbid all the write operations performed by a high user, thus obtaining the trace nets depicted in Figures 9 and 10.

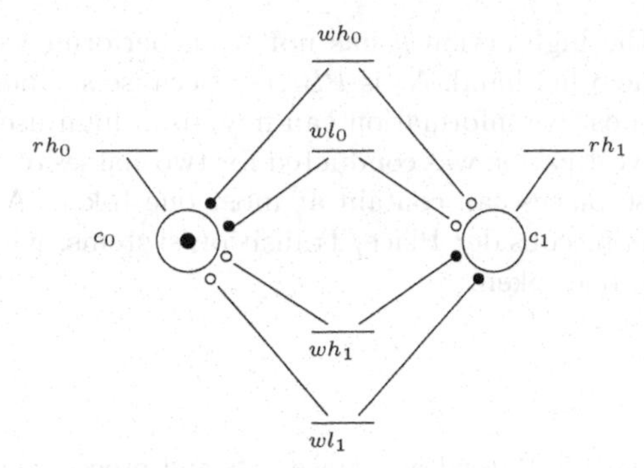

Fig. 9. The trace net obtained by removing the low level read operations.

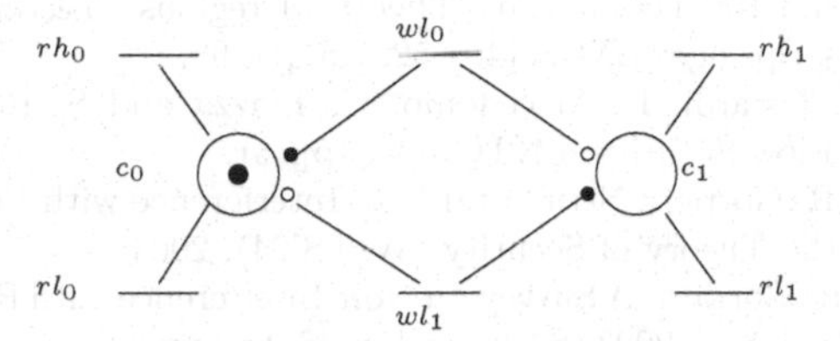

Fig. 10. The trace net obtained by removing the high level write operationsl.

The trace nets obtained in this way do not contain any (potential) causal or conflict place.

6 Conclusion

A structural non-interference property *PBNI+* is proposed to firm more strongly the intuition about the nature of interferences and to obtain more efficiently checkable property. We start defining *PBNI+* for the simpler class of contact free elementary net systems, then the definition is extended to the richer class of trace net systems. We showed that *PBNI+* turns out to be equivalent to the behavioural property *SBNDC*, initially proposed in a process algebraic setting [7, 8].

The property *PBNI+* is structural because no notion of observational equivalence is considered in its definition; however, to be precise, the definition of *PBNI+* requires a limited exploration of the state space (marking graph), hence it is in some sense a *behavioural* property.

The main difference of *PBNI+* w.r.t. the non-interference properties *PBNI* and *RBNI*, introduced in [4, 5], is that *PBNI+* captures only flows of positive information from high users to low users, whereas *RBNI* (and in some cases also *PBNI*) capture both positive and negative flows. Consider for example the net system N_4 in Figure 6: the net N_4 does not satisfy *PBNI* and *RBNI*, because of the presence of the conflict place s. Indeed, there exists a negative information flow from high users to low users: if the low action l is performed, the low user knows that the high action h has not been performed (and will never be performed). On the other hand, N_4 is *PBNI+*, because s is not an active conflict place. Indeed, no positive information can flow from high users to low users.

The current investigation was conducted for two classes of "safe" net systems, i.e., for nets whose places can contain at most one token. A natural extension of this approach is to consider Place/Transition systems, where each place can contain more than one token.

References

1. E. Badouel and Ph. Darondeau. Trace nets and process automata. *Acta Informatica*, 32, 647–679, 1995.
2. E. Badouel and Ph. Darondeau. Theory of regions. *Lectures on Petri Nets I: Basic Models*, Springer LNCS 1491:529:586, 1998.
3. A. Bossi, R. Focardi, D. Macedonio, C. Piazza and S. Rossi. Unwinding in Information Flow Security. ENTCS, to appear.
4. N. Busi and R. Gorrieri. Structural Non-Interference with Petri Nets. Workshop on Issues in the Theory of Security (WITS'04), 2004.
5. N. Busi and R. Gorrieri. A Survey on Non-Interference with Petri Nets. Advanced Course on Petri Nets 2003, Springer LNCS, to appear.
6. J.Engelfriet and G. Rozenberg. Elementary Net Systems *Lectures on Petri Nets I: Basic Models*, Springer LNCS 1491, 1998.
7. R. Focardi, R. Gorrieri, *"A Classification of Security Properties"*, Journal of Computer Security 3(1):5-33, 1995
8. R. Focardi, R. Gorrieri, *"Classification of Security Properties (Part I: Information Flow)"*, Foundations of Security Analysis and Design - Tutorial Lectures (R. Focardi and R. Gorrieri, Eds.), Springer LNCS 2171:331-396, 2001
9. J.A. Goguen, J. Meseguer, *"Security Policy and Security Models"*, Proc. of Symposium on Security and Privacy, IEEE CS Press, pp. 11-20, 1982
10. C. A. Petri, *Kommunikation mit Automaten*, PhD Thesis, Institut für Instrumentelle Mathematik, Bonn, Germany, 1962.
11. W. Reisig, "Petri Nets: An Introduction", EATCS Monographs in Computer Science, Springer, 1985.
12. A.W. Roscoe, *"CSP and Determinism in Security Modelling"*, Proc. of IEEE Symposium on Security and Privacy, IEEE CS Press, pp. 114-127, 1995
13. P.Y.A. Ryan, *"Mathematical Models of Computer Security"*, Foundations of Security Analysis and Design - Tutorial Lectures (R. Focardi and R. Gorrieri, Eds.), Springer LNCS 2171:1-62, 2001
14. P.Y.A. Ryan, S. Schneider, *"Process Algebra and Noninterference"*, Proc. of 12th Computer Security Foundations Workshop, IEEE CS Press, pp. 214-227, 1999

Reachability Set Generation for Petri Nets: Can Brute Force Be Smart?*

Gianfranco Ciardo

Department of Computer Science and Engineering
University of California, Riverside
Riverside, CA, 92521, USA
ciardo@cs.ucr.edu

Abstract. Generating the reachability set is one of the most commonly required step when analyzing the logical or stochastic behavior of a system modeled with Petri nets. Traditional "explicit" algorithms that explore the reachability graph of a Petri net require memory and time at least proportional to the number of reachable markings, thus they are applicable only to fairly small systems in practice. Symbolic "implicit" algorithms, typically implemented using binary decision diagrams, have been successfully employed in much larger systems, but much of the work to date is based on breadth-first search techniques best suited for synchronous hardware verification. Here, instead, we describe recently-introduced data structures and algorithms particularly targeted to Petri nets and similar asynchronous models, and show why they are enormously more efficient for this application. We conclude with some directions for future research.

1 Introduction

Petri nets [29, 32] are an excellent formalism to model a large class of discrete-state systems that exhibit a large amount of asynchronous behavior, yet have the capability to occasionally synchronize some of their activities. Such systems, often called *globally-asynchronous locally-synchronous*, arise in flexible manufacturing, communication protocols, asynchronous circuit design, embedded flight controllers, and, of course, distributed software, to name a few examples.

Once a system has been modeled with a Petri net, its logical behavior can be studied with a variety of techniques, such as *invariant analysis*, *reachability graph generation*, and *simulation or token game*.

Invariant analysis [18, 29] is always applicable to ordinary Petri nets and is often quite efficient, since it directly operates on the incidence matrix of the net, but it has limitations. For example, it can prove that a particular marking is not reachable from the initial marking, but it can never prove that it is reachable. One of its most useful applications is perhaps the investigation of boundedness

* Work supported in part by the National Aeronautics and Space Administration under grant and NAC-1-02095 and by the National Science Foundation under grants CCR-0219745 and ACI-0203971.

J. Cortadella and W. Reisig (Eds.): ICATPN 2004, LNCS 3099, pp. 17–34, 2004.
© Springer-Verlag Berlin Heidelberg 2004

properties: a place is bounded if it is covered by a place-invariant. However, if a Petri net has inhibitor arcs, invariant analysis does not take them into account, so that a place might still be bounded even if it is not covered by any place-invariant. With self-modifying nets [38], the situation is even worse, since only limited place-invariant algorithms have been proposed for this class [7].

At the opposite extreme, the token game can be played on any discrete-state Petri net, regardless of the extensions one might use, but it is akin to testing. Simulation is an essentially incomplete type of analysis that, while helpful in understanding the system under study, is quite time-consuming and cannot be relied upon when attempting to discover all undesirable behaviors.

Only reachability analysis can provide complete answers, since it does generate *all* the possible markings and behaviors of the net. Of course, even this statement needs to be tempered by practical considerations. For ordinary Petri nets, we might not be able to store the entire reachability graph of the net, thus we might fail to provide an answer. For Turing-equivalent extensions of Petri nets, even just being able to tell whether a net is bounded, thus whether it makes sense to attempt its reachability set exploration, is an undecidable problem since the halting problem can be reduced to it [1, 23]. In practice, we are limited to hoping that the reachability set is not only bounded, but small enough to fit in the memory of our computer. Such a brute-force approach makes nevertheless sense, and many research efforts have been directed at finding efficient algorithms and data structures to generate and store the reachability set. While these works cannot change the theoretical nature of the limitations, they can change the practical impact of the approach, by making it applicable to larger and more realistic systems.

In this paper, we present our approach to scale-up the applicability of reachability set generation through the use of *decision diagram* and *Kronecker operator* techniques. This approach reflects work spanning several years, starting from our ICATPN 1999 [28] and 2000 [9] papers, up to the *saturation* algorithm [10, 11] and its applications to verification [14, 15]. The "smart" in the title is a reference to our software tool SMART [8], which implements all the algorithms and data structures we discuss, and was used to obtain the results we present. The rest of paper proceeds as follows. Section 2 contains background material on explicit and symbolic reachability set generation. Section 3 discusses our data structures targeted to Petri nets and the saturation algorithm. Finally, Section 4 outlines some of our ongoing research to further improve the efficiency and applicability of this approach.

2 Traditional Breadth-First Reachability Set Generation

This section describes our extended class of Petri nets, defines their reachability set, and summarizes traditional explicit and BDD-based breadth-first algorithms for the generation of their reachability set. In contrast to the algorithms and data structures described in Section 3, though, these algorithms are not particularly targeted to Petri nets, in the sense that they do not exploit the structural properties of the net.

2.1 Our Extended Petri Net Class

We consider an extended class of Petri nets, *self-modifying nets with inhibitor arcs*, described by a tuple of the form $(\mathcal{P}, \mathcal{T}, \mathbf{F}^-, \mathbf{F}^+, \mathbf{F}^\circ, \mathbf{i}^{init})$ where

- $\mathcal{P}$ and $\mathcal{T}$ are sets of *places* and *transitions* satisfying $\mathcal{P} \cap \mathcal{T} = \emptyset$ and $\mathcal{P} \cup \mathcal{T} \neq \emptyset$.
- $\mathbf{F}^-: \mathcal{P} \times \mathcal{T} \times \mathbb{N}^{|\mathcal{P}|} \rightarrow \mathbb{N}$, $\mathbf{F}^+: \mathcal{P} \times \mathcal{T} \times \mathbb{N}^{|\mathcal{P}|} \rightarrow \mathbb{N}$, and $\mathbf{F}^\circ: \mathcal{P} \times \mathcal{T} \times \mathbb{N}^{|\mathcal{P}|} \rightarrow \mathbb{N} \cup \{\infty\}$ are $|\mathcal{P}| \times |\mathcal{T}|$ *incidence* matrices; $\mathbf{F}^-_{p,t}$, $\mathbf{F}^+_{p,t}$, and $\mathbf{F}^\circ_{p,t}$ are the *marking-dependent* [7, 38, 39] cardinalities of the corresponding *input*, *output*, and *inhibitor* arcs.
- $\mathbf{i}^{init} : \mathcal{P} \rightarrow \mathbb{N}$ is a *marking*; $\mathbf{i}^{init}_p$ is the initial number of *tokens* in place p.

The evolution of the marking $\mathbf{i} : \mathcal{P} \rightarrow \mathbb{N}$ is governed by the same rules as for ordinary Petri nets, keeping in mind that the cardinality of any arc is evaluated in the current marking, i.e., prior to the firing of any transition:

Enabling rule: A transition t is *enabled* in $\mathbf{i}$ if, for all places p, the input arc is satisfied, $\mathbf{i}_p \geq \mathbf{F}^-_{p,t}(\mathbf{i})$, and the inhibitor arc is not, $\mathbf{i}_p < \mathbf{F}^\circ_{p,t}(\mathbf{i})$.

Firing rule: The firing of an enabled transition t in marking $\mathbf{i}$ leads to marking $\mathbf{j}$, where, for all places p, $\mathbf{j}_p = \mathbf{i}_p - \mathbf{F}^-_{p,t}(\mathbf{i}) + \mathbf{F}^+_{p,t}(\mathbf{i})$. We write $\mathcal{N}_t(\mathbf{i}) = \{\mathbf{j}\}$, to stress that, for general discrete-state formalisms, the *next-state function* $\mathcal{N}_t : \widehat{\mathcal{S}} \rightarrow 2^{\widehat{\mathcal{S}}}$ applied to a single state $\mathbf{i}$ returns a set of states, not necessarily a single state[1]. A notational advantage of a set-valued $\mathcal{N}_t$ is that we can write $\mathcal{N}_t(\mathbf{i}) = \emptyset$ to indicate that t is not enabled in marking $\mathbf{i}$.

The class of ordinary Petri nets is obtained by requiring that all the arc cardinalities be constant instead of marking-dependent, and inhibitor arcs are disallowed by requiring $\mathbf{F}^\circ_{p,t}(\mathbf{i})$ to be ∞ for all p, t, and $\mathbf{i}$. We stress that the extensions we introduced are extremely useful in practice, as they allow us to have much more compact models. Indeed, once we allow these extensions, one-to-one translations from other formalisms into Petri nets often become possible; this may constitute a low-cost first step to a conceptual framework where many different solution techniques are applicable to arbitrary formalisms, an approach taken in both Möbius [19] and SMART [8].

2.2 The Reachability Set

Using just our knowledge about the number of places in the Petri net, we can define the *potential reachability set* as $\widehat{\mathcal{S}} = \mathbb{N}^{|\mathcal{P}|}$. In other words, all we know about the markings of the net is that they are tuples of $|\mathcal{P}|$ natural numbers. Given a Petri net, its *(actual) reachability set* $\mathcal{S}$ is instead the set of markings

[1] Indeed, this is already the case for *generalized stochastic* Petri nets [2], if we restrict $\mathcal{S}$ to be the *tangible* markings. The firing of a *timed* transition t in a tangible marking $\mathbf{i}$ might lead to a *vanishing* marking $\mathbf{j}$, which can then enable sequences of *immediate* transitions. In this case, $\mathcal{N}_t(\mathbf{i})$ is the set of tangible markings reachable from $\mathbf{j}$ through the firing of immediate transitions alone. An analogous situation arises in ordinary Petri nets if we choose to partition $\mathcal{T}$ into *internal*, or *invisible*, transitions and *external*, or *observable*, transitions.

that can be actually reached from the initial marking through any number of transition firings. Let $\mathcal{N}_t(\mathcal{X}) = \cup_{\mathbf{i}\in\mathcal{X}}\mathcal{N}_t(\mathbf{i})$, for any $\mathcal{X} \subseteq \widehat{\mathcal{S}}$; let $\mathcal{N}_{T'} = \bigcup_{t\in T'}\mathcal{N}_t$, for any $T' \subseteq T$; finally, denote by "*" the reflexive and transitive closure of a relation. Then, $\mathcal{S} = \{\mathcal{N}_T^*(\mathbf{i}^{init})\}$.

It is well known that Petri nets with inhibitor arcs are Turing-equivalent [1], thus there is no algorithm to determine whether $\mathcal{S} \subseteq \widehat{\mathcal{S}}$ is finite for our class of nets. The reachability set algorithms we consider, then, are forced to assume that the reachability set is indeed finite prior to attempting its generation.

2.3 Explicit Reachability Set Generation

A trivial algorithm to generate the reachability set simply maintains a set of *old* markings $\mathcal{O}$, initially empty, and a set of *unexplored* markings $\mathcal{U}$, initially equal to $\{\mathbf{i}^{init}\}$. Iteratively, each unexplored marking $\mathbf{i} \in \mathcal{U}$ is moved to $\mathcal{O}$, after having added to $\mathcal{U}$ any marking $\mathbf{j} \in \mathcal{N}_T(\mathbf{i})$ not yet in $\mathcal{O} \cup \mathcal{U}$, and the process ends when $\mathcal{U}$ becomes empty. If $\mathcal{U}$ is maintained as a queue, the reachability set is explored in *breadth-first order*, that is, if a marking $\mathbf{i}$ is added to $\mathcal{U}$ before another marking $\mathbf{j}$, then the distance of $\mathbf{i}$ from $\mathbf{i}^{init}$ is at most that of $\mathbf{j}$.

Unfortunately, this *explicit* approach requires time at least proportional to $|\mathcal{S}| \cdot |\mathcal{P}|$, while the memory requirements depend on the encoding technique used for the markings. If a single hash table or tree is used to store and search $\mathcal{O}$ and $\mathcal{U}$, the memory requirements are proportional to the number of elements, $|\mathcal{S}|$, times the size of each element, $|\mathcal{P}|$; using more sophisticated (but still explicit) multi-level data-structures, the requirements can be proportional to just $|\mathcal{S}|$ for most Petri nets in practice [12]. In any case, such an approach is feasible only when $|\mathcal{S}|$ is of the order of 10^7 or 10^8, given current workstation capabilities.

To avoid these limitations, researchers turned to *implicit*, or *symbolic*, algorithms. In particular, encodings based on *binary decision diagrams* (BDDs) [4], which had been used with enormous success in the areas of *symbolic model checking* and *formal verification of digital hardware*, were proposed and implemented for Petri nets as well. The next section is based on the approach used in [30].

2.4 BDD-Based Reachability Set Generation

A BDD on the variables $(x_K, \ldots, x_1)$, or K-level BDD, is a directed acyclic graph *rooted* at a node r, such that each non-terminal node is labeled with one of the K levels and has two outgoing edges labeled with the values 0 and 1, respectively, while the terminal nodes, labeled with level 0, can only be the special nodes 0 or 1, corresponding to the boolean functions 0 and 1. Let $lvl(p)$ be the level of node p, and let $p[0]$ and $p[1]$ be the nodes pointed by the 0-edge and the 1-edge of node p, respectively. The BDD is *ordered* [4] if, for any two nodes p and q encountered, in order, along a path from the root, $lvl(p) > lvl(q)$. The BDD is *reduced* [4] if it does not contain any *redundant* node p (i.e., such that $p[0] = p[1]$) nor any *duplicate* nodes p and q (i.e., such that $lvl(p) = lvl(q)$, $p[0] = q[0]$, and $p[1] = q[1]$). We assume reduced and ordered BDDs from now on.

```
BfSsGen(i^init, N_T)
1. S ← {i^init};                    • known markings
2. U ← {i^init};                    • unexplored known markings
3. while U ≠ ∅ do                   • there are still unexplored markings
4.   X ← N_T(U);                    • possibly new markings
5.   U ← X \ S;                     • truly new markings
6.   S ← S ∪ U;
7. return S;
```

Fig. 1. A BDD-based breadth-first generation algorithm.

A BDD encodes a function $f_r : \{0,1\}^K \to \{0,1\}$, according to the following recursive definition for the function encoded by a node p:

$$f_p = \begin{cases} \overline{x}_{lvl(p)} \cdot f_{p[0]} + x_{lvl(p)} \cdot f_{p[1]} & \text{if } lvl(p) > 0 \\ p & \text{if } lvl(p) = 0 \end{cases}.$$

A fundamental property of BDDs is that they are *canonical*: if we encode multiple boolean functions over the same set of K boolean variables using BDDs with *shared nodes* to avoid duplicate nodes, two functions are identical iff they have the same root.

Given a *safe* Petri net, we can store any set of markings $\mathcal{X} \subseteq \widehat{\mathcal{S}} = \{0,1\}^{|\mathcal{P}|}$ by encoding its indicator function in a $|\mathcal{P}|$-level BDD rooted at r,

$$f_r(\mathbf{i}_{|\mathcal{P}|}, \ldots, \mathbf{i}_1) = 1 \Leftrightarrow (\mathbf{i}_{|\mathcal{P}|}, \ldots, \mathbf{i}_1) \in \mathcal{X}.$$

Analogously, we can store any relation over $\widehat{\mathcal{S}}$, thus any function from $\widehat{\mathcal{S}}$ to $2^{\widehat{\mathcal{S}}}$, such as $\mathcal{N}_\mathcal{T}$, in a $2|\mathcal{P}|$-level BDD. Then, we can generate the reachability set $\mathcal{S}$ using the BDD-based algorithm of Fig. 1. Starting from the set containing only the initial marking, we compute the set of new markings reachable in one step, and iterate until the set $\mathcal{U}$ of unexplored markings is empty. Of course, $\mathcal{S}$, $\mathcal{U}$, $\mathcal{X}$, and $\mathcal{N}_\mathcal{T}$ are encoded by BDDs.

Instead of encoding $\mathcal{N}_\mathcal{T}$ as a single BDD representing the *transition relation*, we can encode it using $4|\mathcal{T}|$ very simple boolean functions [31]:

- $APM_t = \prod_{p:\mathbf{F}^-{p,t}=1} (\mathbf{i}_p = 1)$ (all predecessor places of t are marked)
- $NPM_t = \prod_{p:\mathbf{F}^-{p,t}=1} (\mathbf{i}_p = 0)$ (no predecessor place of t is marked)
- $ASM_t = \prod_{p:\mathbf{F}^+{p,t}=1} (\mathbf{i}_p = 1)$ (all successor places of t are marked)
- $NSM_t = \prod_{p:\mathbf{F}^+{p,t}=1} (\mathbf{i}_p = 0)$ (no successor place of t is marked)

Then, the *topological image computation* for a transition t can be expressed as

$$\mathcal{N}_t(\mathcal{U}) = (((\mathcal{U} \div APM_t) \cdot NPM_t) \div NSM_t) \cdot ASM_t$$

where "$\div$" indicates the *cofactor* operator[2] and "$\cdot$" indicates boolean conjunction. This approach, which inherently uses a *partitioned* representation for $\mathcal{N}_\mathcal{T}$

[2] Given a boolean function f over $(x_K, \ldots, x_1)$ and a literal $x_k = v_k$, with $K \geq k \geq 1$ and $v_k \in \{0,1\}$, the cofactor $f \div x_k$ is defined as $f(x_K, \ldots, x_{k+1}, v_k, x_{k-1}, \ldots, x_1)$; the extension to multiple literals, $f \div (x_{k_c} = v_{k_c}, \ldots, x_{k_1} = v_{k_1})$, is recursively defined as $f(x_K, \ldots, x_{k_c+1}, v_{k_c}, x_{k_c-1}, \ldots, x_1) \div (x_{k_{c-1}} = v_{k_{c-1}}, \ldots, x_{k_1} = v_{k_1})$.

as already proposed for symbolic model checking [24], also opens new possibilities with respect to the exploration order. To exactly recreate the effect of statement 4 in Fig. 1, we must substitute it with:

$\mathcal{X} \leftarrow \emptyset$;
for each $t \in \mathcal{T}$ do
  $\mathcal{X} \leftarrow \mathcal{X} \cup \mathcal{N}_t(\mathcal{U})$;

However, if we don't mind straying a little from a strict breadth-first order in our reachability set exploration, we can use *chaining* [34], to (heuristically) accelerate the algorithm. This is achieved by substituting statements 4 and 5 in Fig. 1 with

for each $t \in \mathcal{T}$ do
  $\mathcal{U} \leftarrow \mathcal{U} \cup \mathcal{N}_t(\mathcal{U})$;
$\mathcal{U} \leftarrow \mathcal{U} \setminus \mathcal{S}$;

Without chaining, the *number of iterations* required to generate $\mathcal{S}$ equals the maximum distance of any marking from $\mathbf{i}^{init}$ plus one, to detect convergence; this is the *sequential depth* of the net. With chaining, the number of iterations at most equals the sequential depth, but it is usually smaller in practice, and can in principle be reduced by a factor of up to $|\mathcal{T}|$. Intuitively, chaining works better if the order in which the for-loop enumerates the transitions is such that the firing of t is attempted only after the firing of any transition whose output places are input places of t has been attempted as well. Of course, this is possible only if the net is acyclic; in the cyclic case, we must somehow "break the cycles". However, decision diagrams are complex data-structures, and the cost of each iteration depends on the *size of the BDDs* that are being manipulated. It is possible, although uncommon, that the size of the BDD representing the currently-known $\mathcal{S}$ at each iteration is substantially larger when the algorithm employs chaining, so that the overall runtime increases even if the number of iterations is reduced. Unfortunately, the evolution of the size of the BDD for $\mathcal{S}$ is not well understood at the moment, so the best we can do is employ heuristics and learn from experiments on practical models.

3 Exploiting the Petri Net Structure

In the previous section, we saw how BDDs can be used to generate the reachability set of a safe Petri net. However, except for partitioning $\mathcal{N}_\mathcal{T}$ into $\cup_{t\in\mathcal{T}}\mathcal{N}_t$ and for ordering the transitions according to their input/output places when chaining is used, nothing we have discussed is specific to Petri nets. In this section, we present instead data-structures and algorithms that are particularly targeted to the reachability set generation of arbitrary Petri nets and similar *asynchronous* models. The main differences are the use of *multi-way decision diagrams* to encode sets of markings, instead of BDDs, the use of *Kronecker operators* on sparse boolean matrices to encode the next-state function, again instead of BDDs, and the use of a new *iteration strategy*. Together, these ideas allows us to exploit the *locality of effect* inherently present in Petri nets and related models of computation, and to reduce by many orders of magnitude both runtime and memory consumption with respect to breadth-first BDD-based methods.

3.1 Multi-way Decision Diagrams

If a net is not safe, BDD-based techniques can still be used, but multiple boolean variables are needed for each place [31], complicating the encoding of $\mathcal{S}$ and $\mathcal{N}_{\mathcal{T}}$. An alternative is to explicitly allow non-boolean variables in the decision diagram. Consider a *structured* set $\widehat{\mathcal{S}} = \mathcal{S}_K \times \cdots \times \mathcal{S}_1$, where each set $\mathcal{S}_k$ is of the form $\{0, 1, \ldots, n_k - 1\}$, for $K \geq k \geq 1$. A *quasi-reduced ordered multi-way decision diagram* (MDD)[3] over $\widehat{\mathcal{S}}$ is a directed acyclic edge-labeled multi-graph where:

- Nodes are organized into K *levels* numbered from K down to 1, plus one additional level numbered 0. We write $\langle k|p\rangle$ to denote a node at level k, where p is a unique index for that level.
- Level K contains a single *non-terminal* node $\langle K|r\rangle$, the *root*, whereas levels $K-1$ through 1 contain one or more non-terminal nodes.
- Level 0 can contain only the *terminal* nodes, $\langle 0|0\rangle$ and $\langle 0|1\rangle$.
- A non-terminal node $\langle k|p\rangle$ has n_k arcs pointing to nodes at level $k-1$. If the ith arc, for $i \in \mathcal{S}_k$, is to node $\langle k-1|q\rangle$, we write $\langle k|p\rangle[i] = q$.
- Duplicate nodes are not allowed (but redundant nodes, where all arcs point to the same node, are allowed).

We can extend our arc notation to paths. The index of the node at level $l-1$ reached from a node $\langle k|p\rangle$ through a sequence $(i_k, \ldots, i_l) \in \mathcal{S}_k \times \cdots \times \mathcal{S}_l$, for $K \geq k > l \geq 1$, is recursively defined as

$$\langle k|p\rangle[i_k, i_{k-1}, \ldots, i_l] = \langle k-1|\langle k|p\rangle[i_k]\rangle[i_{k-1}, \ldots, i_l].$$

An MDD can encode subsets of *tuples* in $\widehat{\mathcal{S}}$ using a similar idea as for BDDs. The k-tuples encoded by, or "below", $\langle k|p\rangle$ are

$$\mathcal{B}(\langle k|p\rangle) = \{\beta \in \mathcal{S}_k \times \cdots \times \mathcal{S}_1 : \langle k|p\rangle[\beta] = 1\},$$

while the $(K-k)$-tuples reaching, or "above", $\langle k|p\rangle$ are

$$\mathcal{A}(\langle k|p\rangle) = \{\alpha \in \mathcal{S}_K \times \cdots \times \mathcal{S}_{k+1} : \langle K|r\rangle[\alpha] = p\}.$$

For any node $\langle k|p\rangle$ on a path from the root $\langle K|r\rangle$ to $\langle 0|1\rangle$, the set of K-tuples $\mathcal{A}(\langle k|p\rangle) \times \mathcal{B}(\langle k|p\rangle)$ is a subset of the set of K-tuples encoded by the MDD, which we can write as $\mathcal{B}(\langle K|r\rangle)$ or $\mathcal{A}(\langle 0|1\rangle)$. By convention, we reserve the indices 0 and 1 at any level for nodes encoding the empty and full sets, that is, $\mathcal{B}(\langle k|0\rangle) = \emptyset$ and $\mathcal{B}(\langle k|1\rangle) = \mathcal{S}_k \times \cdots \times \mathcal{S}_1$. An example of an MDD with $K = 4$ is in Fig. 2, which shows (a) the composition of the sets $\mathcal{S}_k$; (b) the MDD; (c) a more readable representation of the MDD, where we omit nodes $\langle k|0\rangle$, for any k, and node $\langle 0|1\rangle$, as well as their incoming arcs, and we draw only the nonzero entries of the remaining nodes; (d) the set of 4-tuples encoded by the MDD.

[3] MDDs were introduced in [25, 26], in the fully-reduced version. However, they were implemented on top of a BDD library, not natively as we do in our tool SMART.

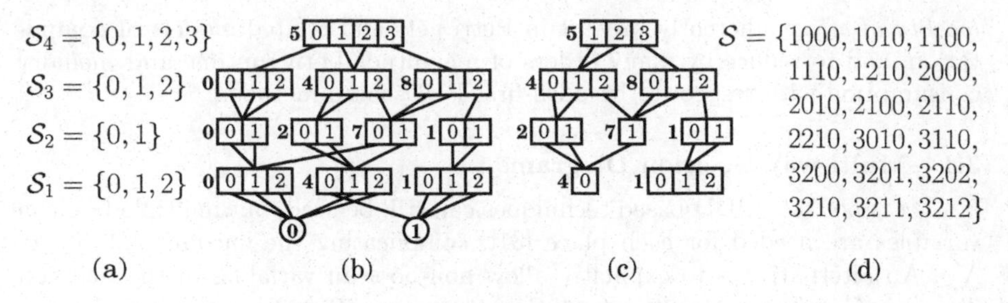

$\mathcal{S}_4 = \{0,1,2,3\}$
$\mathcal{S}_3 = \{0,1,2\}$
$\mathcal{S}_2 = \{0,1\}$
$\mathcal{S}_1 = \{0,1,2\}$
$\mathcal{S} = \{1000, 1010, 1100, 1110, 1210, 2000, 2010, 2100, 2110, 2210, 3010, 3110, 3200, 3201, 3202, 3210, 3211, 3212\}$
(a)
(b)
(c)
(d)

Fig. 2. An example of an MDD and the set of 4-tuples it encodes.

To encode the reachability set of an arbitrary Petri net, we could then use an MDD over $\mathcal{S}_{|\mathcal{P}|} \times \cdots \times \mathcal{S}_1$, where $\mathcal{S}_p = \{0, 1, \ldots, m_p\}$ and m_p is the maximum number of tokens place p may ever contain. However, this simplistic approach would have several drawbacks:

- $\mathcal{S}_p$ might have "holes", that is, even if the bound m_p for the number of tokens in p is achieved in some reachable marking, there might not be a reachable marking with exactly c tokens in place p, for some $0 \leq c < m_p$.
- Given the possible presence of *invariants* in the Petri net, some of the $|\mathcal{P}|$ components of the marking might be completely determined from the value of other components, thus the corresponding levels would waste space without carrying information.
- Such a fine choice for the variables of the MDD might compromise our ability of using the Kronecker encoding described in the next section, if the net contains marking-dependent arc cardinalities. In any case, it might be more efficient to have fewer variables, i.e., levels in the MDD.
- The value of m_p might not be known in advance.

Thus, we use instead K *indexing* functions to map submarkings into natural numbers. Given a Petri net, we partition its places into K subsets $\mathcal{P}_K, \ldots, \mathcal{P}_1$, so that a marking is a collection of the corresponding K submarkings, $(\mathbf{i}_K, \ldots, \mathbf{i}_1)$, with $\mathbf{i}_k \in \mathbb{N}^{|\mathcal{P}_k|}$, for $K \geq k \geq 1$, and define K partial functions $\psi_k : \mathbb{N}^{|\mathcal{P}_k|} \to \mathbb{N}$. In practice, each ψ_k only needs to map the *known* submarkings for $\mathcal{P}_k$, $\mathcal{S}_k$, to the range $\{0, \ldots, n_k - 1\}$ of natural numbers, where $n_k = |\mathcal{S}_k|$. We *dynamically* do so as follows:

- Initially, $\mathcal{S}_k = \{\mathbf{i}_k^{init}\}$ and $\psi_k(\mathbf{i}_k^{init}) = 0$, that is, we map the only known submarking for $\mathcal{P}_k$, the initial submarking, to the first natural number.
- Every time we discover a new submarking $\mathbf{i}_k$ for $\mathcal{P}_k$, we add it to $\mathcal{S}_k$, and define $\psi_k(\mathbf{i}_k) = |\mathcal{S}_k| - 1$.
- For any $\mathbf{i}_k \in \mathbb{N}^{|\mathcal{P}_k|} \setminus \mathcal{S}_k$, that is, any other tuple that is not a submarking discovered so far, we let $\psi_k(\mathbf{i}_k)$ be undefined; its index won't be needed.

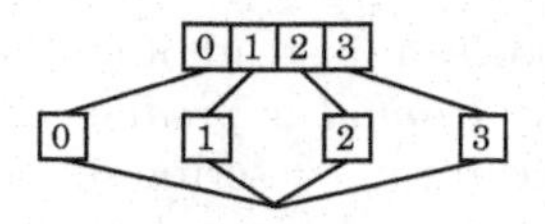

Fig. 3. An identity transformation for a generic (i_k, j_k) pair with $n_k = 4$.

Using these K indexing functions, which can be incrementally defined *during* reachability set generation [11], a marking can be identified with a K-tuple:

$$(\mathbf{i}_K, ..., \mathbf{i}_1) \in \mathbb{N}^{|\mathcal{P}_K|} \times \cdots \times \mathbb{N}^{|\mathcal{P}_1|}$$
$$\Longleftrightarrow$$
$$(\psi_K(\mathbf{i}_K), ...\psi_1(\mathbf{i}_1)) = (i_k, ..., i_1) \in \widehat{\mathcal{S}} = \{0, ..., n_K - 1\} \times \cdots \times \{0, ..., n_1 - 1\}.$$

From now on, we let i_k be the index corresponding to the submarking $\mathbf{i}_k$, i.e., $i_k = \psi_k(\mathbf{i}_k)$, and we use $\mathcal{S}_k$ to mean indifferently the set of known submarkings for $\mathcal{P}_k$, or the set of their indices obtained through ψ_k.

This approach gives us complete flexibility for the choice of the MDD variables, since we can have as little as a single variable (when $K = 1$, $\mathcal{S}_1 = \mathcal{S}$ and we perform an explicit reachability set generation), or as many as $|\mathcal{P}|$ variables, so that each place corresponds to a different level of the MDD. We can even generalize our notation to the case $K > |\mathcal{P}|$, which would then allows us to use multiple variables, although not necessarily binary, per place.

3.2 Kronecker Representation of the Transition Relation

The idea of a BDD encoding of the next-sate function $\mathcal{N}_{\mathcal{T}}$ is clearly still applicable when MDDs are used instead of BDDs. We can simply define an MDD over $(\mathcal{S}_K \times \mathcal{S}_K) \times \cdots \times (\mathcal{S}_1 \times \mathcal{S}_1)$, rooted at $\langle\langle K|s\rangle\rangle$, so that $\langle\langle K|s\rangle\rangle[(i_K, j_K), ..., (i_1, j_1)] = 1$ iff $(\mathbf{j}_K, ..., \mathbf{j}_1) \in \mathcal{N}_{\mathcal{T}}(\mathbf{i}_K, ..., \mathbf{i}_1)$, where, to avoid confusion, we use a double bracket notation for the nodes of "$2K$-level" MDDs. However, this widely employed approach suffers from an inherent problem: an inappropriate definition of what constitutes a redundant node in the decision diagram representing the transition relation of an asynchronous system. For formalisms such as Petri nets, where an event (transition) usually affects only a few state components (places), most of the tuples $((i_K, j_K), \ldots, (i_1, j_1))$ encoded by the MDD for $\mathcal{N}_{\mathcal{T}}$ have many (i_k, j_k) pairs where $i_k = j_k$, that is, the kth component (submarking) remains unchanged. This *identity transformation*, though, is not efficiently encoded by ordinary decision diagrams, as shown in Fig. 3.

We then adopt a representation of $\mathcal{N}_{\mathcal{T}}$ inspired by the *Kronecker encoding of the infinitesimal generator of a Markov chain*, an idea championed by Plateau [33] that has received much attention in the last decade from researchers in the area of stochastic Petri nets [5, 6, 21]. When the partition of the places in the net is *Kronecker consistent*, we can write

$$\mathcal{N}_{\mathcal{T}} = \bigcup_{t \in \mathcal{T}} \mathcal{N}_{K,t} \times \cdots \times \mathcal{N}_{1,t} \qquad \text{or} \qquad \mathbf{N} = \sum_{t \in \mathcal{T}} \bigotimes_{K \geq k \geq 1} \mathbf{N}_{k,t},$$

if $\mathcal{N}_{\mathcal{T}}$ is seen as a $|\widehat{\mathcal{S}}| \times |\widehat{\mathcal{S}}|$ boolean (instead of real, as for Markov chains) matrix $\mathbf{N}$, where "$\otimes$" indicates the *Kronecker product* operator [20] and the $K \cdot |\mathcal{T}|$ boolean matrices $\mathbf{N}_{k,t}$, of size $n_k \times n_k$, describe the (local) effect of each transition t on each submarking $\mathbf{i}_k$. In particular, t is *locally enabled* by submarking $\mathbf{i}_k$ iff not all entries of the $i_k{}^{\text{th}}$ row of $\mathbf{N}_{k,t}$ are zero, written $\mathbf{N}_{k,t}[i_k, \cdot] \neq \mathbf{0}$, and it is *(globally) enabled* in marking $(\mathbf{i}_K, \ldots, \mathbf{i}_1)$ iff it is locally enabled in each submarking $\mathbf{i}_k$. The identity transformations mentioned above simply correspond to having $\mathbf{N}_{k,t} = \mathbf{I}$; in this case, t and k are said to be *independent*, since the enabling of t is not affected by $\mathbf{i}_k$, and the firing of t does not change $\mathbf{i}_k$ [9, 28]. By exploiting *event locality*, i.e., knowledge of the (k, t) pairs for which $\mathbf{N}_{k,t} \neq \mathbf{I}$, we can not only reduce the storage requirements, as there is no need to store identity matrices, but also enormously reduce the computational cost, as we describe in the next section.

We conclude this section by observing that, for Petri nets, two important properties hold, further strengthening the case for our approach:

- The $\mathbf{N}_{k,t}$ matrices that do need to be stored are extremely sparse. For ordinary Petri nets, each row contains at most one nonzero entry, since the effect of firing a particular transition t is deterministic.
- For ordinary Petri nets, the Kronecker consistency requirement is satisfied *for any partition of* $\mathcal{P}$. This remains true even if we allow inhibitor arcs and a limited form of marking dependency where the cardinality of an input, output, or inhibitor arc between p and t depends at most on the number of tokens in p (as a special case, this includes *reset arcs*, which were proposed [3] as a mechanism to empty a p when t fires).

3.3 The Saturation Algorithm

We are now able to describe an iteration strategy that greatly reduces the computational costs of reachability set generation by exploiting event locality. Let $\mathit{Top}(t)$ and $\mathit{Bot}(t)$ be the highest and lowest levels on which t depends:

$$\mathit{Top}(t) = \max\{k : \mathbf{N}_{k,t} \neq \mathbf{I}\} \qquad \text{and} \qquad \mathit{Bot}(t) = \min\{k : \mathbf{N}_{k,t} \neq \mathbf{I}\}.$$

Each transition depends on at least one level (if not, it can be eliminated), so

$$\forall t \in \mathcal{T}, \quad K \geq \mathit{Top}(t) \geq \mathit{Bot}(t) \geq 1,$$

and we can partition $\mathcal{T}$ into (up to) K classes,

$$\mathcal{T}_k = \{t \in \mathcal{T} \ : \ \mathit{Top}(t) = k\}, \qquad \text{for } K \geq k \geq 1$$

where some $\mathcal{T}_k$ might be empty. Then, we define the restriction of $\mathcal{N}_{\mathcal{T}}$ to the transitions that depend only levels k or below:

$$\mathcal{N}_{\leq k} = \bigcup_{1 \leq l \leq k} \mathcal{N}_{\mathcal{T}_l} = \bigcup_{t:\, \mathit{Top}(t) \leq k} \mathcal{N}_t,$$

and use it to define the main concept behind the saturation algorithm [10].

Definition 1. *An MDD node $\langle k|p\rangle$ at level k is said to be* saturated *if it represents a fixed point with respect to the firing of any transition that does not affect any level above k, that is, $\mathcal{B}(\langle k|p\rangle) = \mathcal{N}^*_{\leq k}(\mathcal{B}(\langle k|p\rangle))$.*

Clearly, any MDD node reachable on a path from a saturated MDD node must also be saturated and, since $\mathcal{N}_{\leq K} \equiv \mathcal{N}_{\mathcal{T}}$, we have $\mathcal{B}(\langle K|r\rangle) = \mathcal{N}^*_{\mathcal{T}}(\mathcal{B}(\langle K|r\rangle))$ when the root $\langle K|r\rangle$ of the MDD is saturated. Thus, if the MDD initially encodes the K-tuple corresponding to the initial marking and no other tuple, $\mathcal{B}(\langle K|r\rangle) = \{(i^{init}_K, ..., i^{init}_1)\}$, we have $\mathcal{B}(\langle K|r\rangle) = \mathcal{S}$ once $\langle K|r\rangle$ is saturated.

The saturation mechanism differs from the traditional node manipulation of breadth-first BDD algorithms in several fundamental ways:

- To saturate a node $\langle k|p\rangle$ whose descendants are already saturated, we attempt the firing of any transition $t \in \mathcal{T}_k$. A traditional approach, even with chaining, would build the MDD corresponding to $\mathcal{N}_t(\mathcal{S})$ and then union it with $\mathcal{S}$ (of course, $\mathcal{S}$ is the *currently known* reachability set). However, saturation *modifies $\langle k|p\rangle$ in place* [9] because, due to locality, we can say that,
  - If marking $\mathbf{i} \in \mathcal{S}$ corresponds to a path going through $\langle k|p\rangle$, i.e., $\mathbf{i} \in \mathcal{A}(\langle k|p\rangle) \times \mathcal{B}(\langle k|p\rangle)$, and if $\{\mathbf{j}\} = \mathcal{N}_t(\mathbf{i})$, then $\mathbf{i}_l$ and $\mathbf{j}_l$ can differ only if the level l satisfies $k \geq l \geq Bot(t)$.
  - In addition, if there is another marking $\mathbf{i}' \in \mathcal{A}(\langle k|p\rangle) \times \mathcal{B}(\langle k|p\rangle)$, such that $\mathbf{i}'_l = \mathbf{i}_l$ for $k \geq l \geq Bot(t)$, then $\{\mathbf{j}'\} = \mathcal{N}_t(\mathbf{i}')$, that is, t is enabled also in $\mathbf{i}'$ and, furthermore, it must be that $\mathbf{j}'_l = \mathbf{j}_l$ for $k \geq l \geq Bot(t)$, and $\mathbf{j}'_l = \mathbf{i}_l$ for $l > k$ or $l < Bot(t)$.

  In other words, we can modify $\langle k|p\rangle$ to take into account the firing of t without even having to know what paths $\mathcal{A}(\langle k|p\rangle)$ lead to this node. This is much less computationally expensive than building the MDD for $\mathcal{N}_t(\mathcal{S})$ and can greatly reduce the number of MDD nodes that are temporarily generated and then discarded soon thereafter because they become disconnected.
- The reachable markings are not found in breadth-first order. Indeed, there is no concept of a global iteration: once the root is saturated, the entire state space $\mathcal{S}$ has been saturated.
- As with all decision diagram manipulations, we employ a *unique table* to detect duplicate nodes, and *operation caches* to speed-up computation (in particular a *union cache* and a *firing cache*). However, only saturated nodes are ever inserted in the unique table or referenced in the operation caches. Again, this implies that far fewer nodes and cache entries are allocated; these large memory savings translate into large execution time savings as well.

The pseudocode for the saturation algorithm is shown in Fig. 4. The functions listed perform the following operations:

Generate() builds an MDD rooted at $\langle K|r\rangle$ encoding $\mathcal{N}^*_{\mathcal{T}}(\mathbf{i}^{init})$, and returns r, that is, it builds the MDD encoding $\mathcal{S}$.

Saturate(k, p) updates $\langle k|p\rangle$ in-place, to encode $\mathcal{N}^*_{\leq k}(\mathcal{B}(\langle k|p\rangle))$, and returns whether the node changed.

Fire(t, k, p) updates $\langle k|p\rangle$, also in-place, to encode $\mathcal{N}^*_{\leq k-1}(\mathcal{N}^*_t(\mathcal{B}(\langle k|p\rangle)))$, and returns whether the node changed; it is always called with $k = Top(t)$.

```
Generate()
 1. p ← 1;
 2. for k = 1 to K do
 3.   S_k ← {0};                    • ψ_k(i_k^init) = 0
 4.   C_k ← ∅;   • the confirmed indices
 5.   Confirm(k, 0);
 6.   r ← NewNode(k);
 7.   ⟨k|r⟩[0] ← p;
 8.   Saturate(k, r);
 9.   CheckIn(k, r);
10.   p ← r;
11. return r;
```

```
Saturate(k, p)
 1. chng ← true;
 2. while chng do
 3.   chng ← false;
 4.   foreach t ∈ T_k do
 5.     chng ← chng ∨ Fire(t, k, p);
```

```
Fire(t, k, p)
 1. chng ← false;
 2. L ← {i ∈ S_k: ⟨k|p⟩[i] ≠ 0 ∧ N_{k,t}[i,·] ≠ 0};
 3. while L ≠ ∅ do
 4.   pick and remove i from L;
 5.   f ← RecFire(t, k−1, ⟨k|p⟩[i]);
 6.   if f ≠ 0 then
 7.     foreach j s.t. N_{k,t}[i, j] = 1 do
 8.       u ← Union(k−1, f, ⟨k|p⟩[j]);
 9.       if u ≠ ⟨k|p⟩[j] then
10.         if j ∉ C_k then Confirm(k, j);
11.         ⟨k|p⟩[j] ← u;
12.         chng ← true;
13.         if N_{k,t}[j, ·] ≠ 0 then
14.           L ← L ∪ {j};
15. return chng;
```

```
Confirm(k, i)
 1. C_k ← C_k ∪ {i};
 2. i_k ← ψ_k^{-1}(i);
 3. foreach t s.t. N_{k,t} ≠ I do
 4.   foreach j_k ∈ N_{k,t}(i_k) do
 5.     j ← ψ_k(j_k);
 6.     if j = "undefined" then
 7.       ψ_k(j_k) ← |S_k|;
 8.       j ← |S_k|;
 9.       S_k ← S_k ∪ {j};
10.     N_{k,t}[i, j] ← 1;
```

```
RecFire(t, l, q)
 1. if l < Bot(t) then return q;
 2. if Cached(FIRE, l, t, q, s) then
 3.   return s;
 4. s ← NewNode(l); chng ← false;
 5. L ← {i ∈ S_l: ⟨l|q⟩[i] ≠ 0 ∧ N_{l,t}[i,·] ≠ 0};
 6. while L ≠ ∅ do
 7.   pick and remove i from L;
 8.   f ← RecFire(t, l−1, ⟨l|q⟩[i]);
 9.   if f ≠ 0 then
10.     foreach j s.t. N_{l,t}[i, j] = 1 do
11.       u ← Union(l−1, f, ⟨l|s⟩[j]);
12.       if u ≠ ⟨l|s⟩[j] then
13.         if j ∉ C_l then Confirm(l, j);
14.         ⟨l|s⟩[j] ← u;
15.         chng ← true;
16. if chng then
17.   Saturate(l, s);
18. CheckIn(l, s);
19. PutInCache(FIRE, l, t, q, s);
20. return s;
```

```
Union(k, p, q)
 1. if p = q then return p;
 2. else if p = 1 or q = 1 then return 1;
 3. else if p = 0 then return q;
 4. else if q = 0 then return p;
 5. else if k = 0 then return p ∨ q;
 6. if Cached(UNION, k, p, q, u) then
 7.   return u;
 8. u ← NewNode(k);
 9. for i = 0 to n_k − 1 do
10.   Arc(k,u,i,Union(k−1,⟨k|p⟩[i],⟨k|q⟩[i]));
11. u ← CheckIn(k, u);
12. PutInCache(UNION, k, p, q, u);
13. return u;
```

```
Arc(k, p, i, q)
 1. s ← ⟨k|p⟩[i];   • remember old node
 2. if q = s then return
 3. ⟨k|p⟩[i] ← q;
 4. ⟨k − 1|q⟩.inc ← ⟨k − 1|q⟩.inc + 1;
 5. if ⟨k − 1|s⟩.inc = 1 then
 6.   DeleteNode(k − 1, s);
 7. else
 8.   ⟨k − 1|s⟩.inc ← ⟨k − 1|s⟩.inc − 1;
```

Fig. 4. *Generate*, *Saturate*, *Fire*, *RecFire*, *Confirm*, *Union*, and *Arc*.

RecFire(t, l, q) builds an MDD rooted at $\langle l|s\rangle$ encoding $\mathcal{N}^*_{\leq l}(\mathcal{N}_t(\mathcal{B}(\langle l|q\rangle)))$, and returns s; it is always called with $l < Top(t)$.

Confirm(k, i) builds row $\mathbf{N}_{k,t}[i, \cdot]$ for each transition t that depends on level k, by examining the places $\mathcal{P}_k$ of the Petri net and the arcs connecting them to t. If it discovers new submarkings $\mathbf{j}_k$, it adds them to $\mathcal{S}_k$ and assigns them the next available index. At any point in time, $\mathbf{N}_{k,t}$ has rows indexed by *confirmed* submarking indices $\mathcal{C}_k$, but column indices refer to $\mathcal{S}_k$.

Union(k, p, q) return an index u such that $\mathcal{B}(\langle k|u\rangle) = \mathcal{B}(\langle k|p\rangle) \cup \mathcal{B}(\langle k|q\rangle)$; it is always called on saturated nodes, thus the result is also saturated, by definition.

Arc(k, p, i, q) sets the arc $\langle k|p\rangle[i]$ to q, and updates the *incoming arc counts* for the node previously pointed by $\langle k|p\rangle[i]$ and for q.

In addition, the following functions are used without giving their pseudocode:

CheckIn(k, p) searches the level-k unique table to determine whether $\langle k|p\rangle$ is a duplicate of an existing node $\langle k|q\rangle$; if it is, it deletes $\langle k|p\rangle$ and returns q; otherwise it inserts $\langle k|p\rangle$ in the table and returns p.

PutInCache$(UNION, k, p, q, u)$ and *PutInCache*$(FIRE, l, t, q, s)$ insert in the level-k cache the result of an operation.

Cached$(UNION, k, p, q, u)$ and *Cached*$(FIRE, l, t, q, s)$ attempt to retrieve from the level-k cache the result of an operation, and return *true* if it is found.

NewNode(k) allocates a new MDD node $\langle k|p\rangle$ at level k, sets all its arcs to 0, and returns p.

DeleteNode(k, p) deletes the MDD node $\langle k|p\rangle$.

3.4 Results

To illustrate the effectiveness of the saturation algorithm, we recall some of the results from [11]. Table 1 compares the final and peak memory and the runtime of our tool SMART [8] and of a traditional symbolic breadth-first implementation in NuSMV [17], a model checking tool built on top of the CUDD library [36]. The examples of Petri nets used for the study are the dining philosophers problem [31], a slotted ring system [31], a round robin mutual exclusion [22], and a flexible manufacturing system (FMS) [16]. The first three models are safe Petri nets, the parameter N affects the height of the MDD but not the size of the local state spaces (except for $\mathcal{S}_1$ in the round robin model, which grows linearly in N). The FMS has instead a 19-level MDD, and N affects the size of the nodes. The enormous advantage of our data structures and of the saturation algorithm is clearly visible in terms of both memory and runtime reduction.

4 What's Ahead

While saturation offers large runtime and memory savings, many opportunities for further improvements still need to be explored on several fronts. In the following sections, we survey some research activities in which we are currently involved.

Table 1. Generation of the state space: SMART vs. NuSMV.

N	Reachable states	Final memory (KB)		Peak memory (KB)		Time (sec)	
		SMART	NuSMV	SMART	NuSMV	SMART	NuSMV
Dining Philosophers: $K=N$, $\lvert\mathcal{S}_k\rvert=34$ for all k							
20	3.46×10^{12}	4	4,178	5	4,192	0.01	0.4
50	2.23×10^{31}	11	8,847	14	8,863	0.03	13.1
100	4.97×10^{62}	24	8,891	28	15,256	0.06	990.8
200	2.47×10^{125}	48	21,618	57	59,423	0.15	18,129.3
5,000	6.53×10^{3134}	1,210	—	1,445	—	65.55	—
Slotted Ring Network: $K=N$, $\lvert\mathcal{S}_k\rvert=15$ for all k							
5	5.39×10^{4}	1	502	5	507	0.01	0.1
10	8.29×10^{9}	5	4,332	28	8,863	0.06	6.1
15	1.46×10^{15}	10	771	80	11,054	0.18	2,853.1
100	2.60×10^{105}	434	—	15,753	—	41.72	—
Round Robin Mutual Exclusion: $K=N+1$, $\lvert\mathcal{S}_k\rvert=10$ for all k except $\lvert\mathcal{S}_1\rvert=N+1$							
10	2.30×10^{4}	5	917	6	932	0.01	0.2
20	4.72×10^{7}	18	5,980	20	5,985	0.04	1.4
30	7.25×10^{10}	37	2,222	41	8,716	0.09	5.6
100	2.85×10^{32}	357	13,789	372	21,814	2.11	2,836.5
150	4.82×10^{47}	784	—	807	—	7.04	—
FMS: $K=19$, $\lvert\mathcal{S}_k\rvert=N+1$ for all k except $\lvert\mathcal{S}_{17}\rvert=4$, $\lvert\mathcal{S}_{12}\rvert=3$, $\lvert\mathcal{S}_7\rvert=2$							
5	1.92×10^{4}	5	2,113	6	2,126	0.01	1.0
10	2.50×10^{9}	16	1,152	26	8,928	0.02	41.6
25	8.54×10^{13}	86	17,045	163	152,253	0.16	17,321.9
150	4.84×10^{23}	6,291	—	16,140	—	18.50	—

4.1 Beyond Kronecker Consistency

The saturation algorithm as presented in [10, 11] relies on the assumption that the model partition allows $\mathcal{N}_\mathcal{T}$ to be decomposed in a Kronecker consistent way, that is, $\mathcal{N}_\mathcal{T} = \bigcup_{t\in\mathcal{T}} \mathcal{N}_{K,t} \times \cdots \times \mathcal{N}_{1,t}$.

For ordinary Petri nets, this is not a restriction at all while, for more generals formalisms, such as Petri nets with marking-dependent arc cardinalities, this requirement forces us to either split events into multiple events or merge submodels into larger submodels. Either way, this can cause a large increase in the computational requirements, since it brings us closer to an explicit approach.

To see the practical importance of this limitation, one can simply consider the semantic of an *assignment* in a procedural programming language, which is easily modeled using marking-dependent arc cardinalities, see Fig. 5. The effect of the assignment statement on the left is correctly captured by transition t in the Petri net on the right. However, if we partition the Petri net so that places a, b, and c are in classes by themselves, we have no way of expressing the effect of $\mathcal{N}_t$ on a without knowing the number of tokens in b and c as well. In other words, any Kronecker consistent partition must assign a, b, and c to the same class, unless we split t into $\lvert\mathcal{S}_a \times \mathcal{S}_b \times \mathcal{S}_c\rvert$ cases.

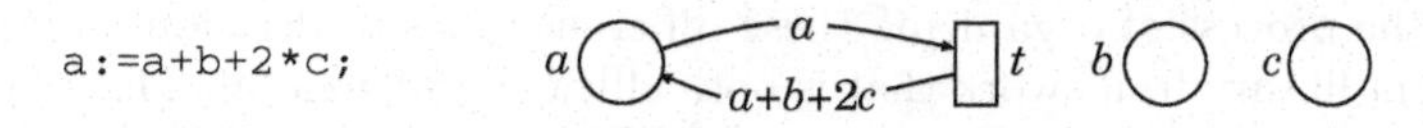

Fig. 5. An ordinary assignment and its extended Petri net model.

We are currently working on a different, MDD-based, encoding of the next-state function $\mathcal{N}_{\mathcal{T}}$ which, while retaining the efficiency of a Kronecker representation when the partition is indeed Kronecker consistent, "gracefully adapts" to the presence of non-Kronecker consistent behavior in the models and it still allows us to perform a saturation-type iteration, even when the sets $\mathcal{S}_k$ are not known a-priori.

4.2 Ordering and Partitioning Heuristics Based on the Net Structure

It is well known that the *order* of the variables in a BDD can greatly affect the efficiency of BDD operations and that finding an optimal ordering is an NP-complete problem [37]. *Dynamic* variable reordering heuristics have been proposed [35], but they can be quite expensive, thus they are usually triggered only when the size of the BDD exceeds some threshold. When using MDDs instead of BDDs, the optimization problem has even more parameters, since we need to decide not just the order of the variables (the order of the K subnets) but the definition of the variables themselves (the composition of each $\mathcal{P}_k$, for $K \geq k \geq 1$).

We are currently working on a heuristic that, using the structure of the Petri net and our concept of locality, attempts to *statically* find a good order that reduces the cost of firing a transition, both in terms of the actual computation for the recursive *Fire* calls, and in terms of the change in the size of the MDD it causes. Preliminary results are quite promising.

4.3 Implementation and Libraries

The research results described in this paper have been implemented in the tool SMART [8]. Begun over a decade ago as a tool for the explicit solution of Markov models, such as GSPNs, SMART has evolved in several directions. Of particular relevance here are Kronecker techniques for the encoding of the infinitesimal generator of a continuous-time Markov chain [5] and of the next-state function [28]; *matrix diagrams* [13, 27] as an alternative data structure to store the matrices $\mathbf{N}_{k,t}$; and, of course, the use of saturation to generate the reachability set [10, 11], to compute the *distance function* of each marking from the initial marking [14], and to carry on CTL model checking on a Petri net model [15].

As a consequence of implementing these ideas in SMART, we have, over the years, refined our data structures and algorithms for the manipulation of decision diagrams in a variety of versions: boolean-valued vs. integer-valued vs. real-valued; edge valued vs. terminal-valued; quasi-reduced vs. fully-reduced. We are now in the process of organizing these different types of decision diagrams into a coherent library framework that is both efficient and general. This new library will be at the core of a new version of SMART. It will also be freely distributed to researchers who want to use its raw capabilities in their own tools.

5 Conclusion

We have presented an exhaustive brute force, but nevertheless smart (or is that SMART?) approach to the symbolic generation of the reachability set of an extended class of Petri nets. Using the idea of *saturation*, the size of the decision diagram encoding the state space grows much less than with conventional breadth first iteration strategies, and this, in turn, results in must faster execution times.

Of particular relevance to Petri net researchers is the fact that saturation, and the concept of *event locality* on which saturation relies, were inspired by structural properties that are inherent in Petri nets and similar asynchronous formalisms.

Thanks to these new algorithms, the logical behavior of realistic Petri net models with very large reachability sets is now more feasible than ever before.

References

1. T. Agerwala. A complete model for representing the coordination of asynchronous processes. Hopkins Computer Research Report 32, Johns Hopkins University, Baltimore, Maryland, July 1974.
2. M. Ajmone Marsan, G. Balbo, G. Conte, S. Donatelli, and G. Franceschinis. *Modelling with Generalized Stochastic Petri Nets*. John Wiley & Sons, New York, 1995.
3. T. Araki and T. Kasami. Some decision problems related to the reachability problem for Petri nets. *Theoretical Computer Science*, 3:85–104, 1977.
4. R. E. Bryant. Symbolic boolean manipulation with ordered binary-decision diagrams. *ACM Comp. Surv.*, 24(3):293–318, 1992.
5. P. Buchholz, G. Ciardo, S. Donatelli, and P. Kemper. Complexity of memory-efficient Kronecker operations with applications to the solution of Markov models. *INFORMS J. Comp.*, 12(3):203–222, 2000.
6. P. Buchholz and P. Kemper. Numerical analysis of stochastic marked graphs. In *Proc. 6th Int. Workshop on Petri Nets and Performance Models (PNPM'95)*, pages 32–41, Durham, NC, Oct. 1995. IEEE Comp. Soc. Press.
7. G. Ciardo. Petri nets with marking-dependent arc multiplicity: properties and analysis. In R. Valette, editor, *Proc. 15th Int. Conf. on Applications and Theory of Petri Nets*, LNCS 815, pages 179–198, Zaragoza, Spain, June 1994. Springer-Verlag.
8. G. Ciardo, R. L. Jones, A. S. Miner, and R. Siminiceanu. Logical and stochastic modeling with SMART. In P. Kemper and W. H. Sanders, editors, *Proc. Modelling Techniques and Tools for Computer Performance Evaluation*, LNCS 2794, pages 78–97, Urbana, IL, USA, Sept. 2003. Springer-Verlag.
9. G. Ciardo, G. Luettgen, and R. Siminiceanu. Efficient symbolic state-space construction for asynchronous systems. In M. Nielsen and D. Simpson, editors, *Proc. 21th Int. Conf. on Applications and Theory of Petri Nets*, LNCS 1825, pages 103–122, Aarhus, Denmark, June 2000. Springer-Verlag.

10. G. Ciardo, G. Luettgen, and R. Siminiceanu. Saturation: An efficient iteration strategy for symbolic state space generation. In T. Margaria and W. Yi, editors, *Proc. Tools and Algorithms for the Construction and Analysis of Systems (TACAS)*, LNCS 2031, pages 328–342, Genova, Italy, Apr. 2001. Springer-Verlag.
11. G. Ciardo, R. Marmorstein, and R. Siminiceanu. Saturation unbound. In H. Garavel and J. Hatcliff, editors, *Proc. Tools and Algorithms for the Construction and Analysis of Systems (TACAS)*, LNCS 2619, pages 379–393, Warsaw, Poland, Apr. 2003. Springer-Verlag.
12. G. Ciardo and A. S. Miner. Storage alternatives for large structured state spaces. In R. Marie, B. Plateau, M. Calzarossa, and G. Rubino, editors, *Proc. 9th Int. Conf. on Modelling Techniques and Tools for Computer Performance Evaluation*, LNCS 1245, pages 44–57, St. Malo, France, June 1997. Springer-Verlag.
13. G. Ciardo and A. S. Miner. A data structure for the efficient Kronecker solution of GSPNs. In P. Buchholz, editor, *Proc. 8th Int. Workshop on Petri Nets and Performance Models (PNPM'99)*, pages 22–31, Zaragoza, Spain, Sept. 1999. IEEE Comp. Soc. Press.
14. G. Ciardo and R. Siminiceanu. Using edge-valued decision diagrams for symbolic generation of shortest paths. In M. D. Aagaard and J. W. O'Leary, editors, *Proc. Fourth International Conference on Formal Methods in Computer-Aided Design (FMCAD)*, LNCS 2517, pages 256–273, Portland, OR, USA, Nov. 2002. Springer-Verlag.
15. G. Ciardo and R. Siminiceanu. Structural symbolic CTL model checking of asynchronous systems. In W. Hunt, Jr. and F. Somenzi, editors, *Computer Aided Verification (CAV'03)*, volume 2725 of *LNCS*, pages 40–53, Boulder, CO, USA, July 2003. Springer-Verlag.
16. G. Ciardo and K. S. Trivedi. A decomposition approach for stochastic reward net models. *Perf. Eval.*, 18(1):37–59, 1993.
17. A. Cimatti, E. Clarke, F. Giunchiglia, and M. Roveri. NuSMV: A new symbolic model verifier. In *CAV '99*, LNCS 1633, pages 495–499. Springer-Verlag, 1999.
18. J. M. Colom and M. Silva. Convex geometry and semiflows in P/T nets: A comparative study of algorithms for the computation of minimal p-semiflows. In *10th Int. Conf. on Application and Theory of Petri Nets*, pages 74–95, Bonn, Germany, 1989.
19. D. Daly, D. D. Deavours, J. M. Doyle, P. G. Webster, and W. H. Sanders. Möbius: An Extensible Tool for Performance and Dependability Modeling. In B. R. Haverkort, H. C. Bohnenkamp, and C. U. Smith, editors, *Proc. 11th Int. Conf. on Modelling Techniques and Tools for Computer Performance Evaluation, Schaumburg, IL*, LNCS 1786, pages 332–336. Springer-Verlag, Mar. 2000.
20. M. Davio. Kronecker products and shuffle algebra. *IEEE Trans. Comp.*, C-30:116–125, Feb. 1981.
21. S. Donatelli. Superposed generalized stochastic Petri nets: definition and efficient solution. In R. Valette, editor, *Proc. 15th Int. Conf. on Applications and Theory of Petri Nets*, LNCS 815, pages 258–277, Zaragoza, Spain, June 1994. Springer-Verlag.
22. S. Graf, B. Steffen, and G. Lüttgen. Compositional minimisation of finite state systems using interface specifications. *Formal Asp. of Comp.*, 8(5):607–616, 1996.
23. M. Hack. Decidability questions for Petri nets. Technical Report 161, Laboratory for Computer Science, Massachusetts Institute of Technology, Cambridge, MA, June 1976.

24. J.R. Burch, E.M. Clarke, and D.E. Long. Symbolic model checking with partitioned transition relations. In A. Halaas and P.B. Denyer, editors, *Int. Conference on Very Large Scale Integration*, pages 49–58, Edinburgh, Scotland, Aug. 1991. IFIP Transactions, North-Holland.
25. T. Kam. *State Minimization of Finite State Machines using Implicit Techniques*. PhD thesis, University of California at Berkeley, 1995.
26. T. Kam, T. Villa, R. Brayton, and A. Sangiovanni-Vincentelli. Multi-valued decision diagrams: theory and applications. *Multiple-Valued Logic*, 4(1–2):9–62, 1998.
27. A. S. Miner. Efficient Solution of GSPNs using canonical matrix diagrams. In *Proc. 9th Int. Workshop on Petri Nets and Performance Models*, pages 101–110, Aachen, Germany, Sept. 2001.
28. A. S. Miner and G. Ciardo. Efficient reachability set generation and storage using decision diagrams. In H. Kleijn and S. Donatelli, editors, *Proc. 20th Int. Conf. on Applications and Theory of Petri Nets*, LNCS 1639, pages 6–25, Williamsburg, VA, USA, June 1999. Springer-Verlag.
29. T. Murata. Petri Nets: properties, analysis and applications. *Proc. of the IEEE*, 77(4):541–579, Apr. 1989.
30. E. Pastor, J. Cortadella, and O. Roig. Symbolic analysis of bounded Petri nets. *IEEE Trans. Comp.*, 50(5):432–448, May 2001.
31. E. Pastor, O. Roig, J. Cortadella, and R. Badia. Petri net analysis using boolean manipulation. In R. Valette, editor, *Proc. 15th Int. Conf. on Applications and Theory of Petri Nets*, LNCS 815, pages 416–435, Zaragoza, Spain, June 1994. Springer-Verlag.
32. C. Petri. *Kommunikation mit Automaten*. PhD thesis, University of Bonn, Bonn, West Germany, 1962.
33. B. Plateau. On the stochastic structure of parallelism and synchronisation models for distributed algorithms. In *Proc. ACM SIGMETRICS*, pages 147–153, Austin, TX, USA, May 1985.
34. O. Roig, J. Cortadella, and E. Pastor. Verification of asynchronous circuits by BDD-based model checking of Petri nets. In G. De Michelis and M. Diaz, editors, *Proc. 16th Int. Conf. on Applications and Theory of Petri Nets*, LNCS 935, pages 374–391, Turin, Italy, June 1995. Springer-Verlag.
35. R. Rudell. Dynamic variable ordering for ordered binary decision diagrams. In *Proc. IEEE Int. Conf. on Computer-Aided Design*, pages 139–144, Nov. 1993.
36. F. Somenzi. CUDD: CU Decision Diagram Package, Release 2.3.1. http://vlsi.colorado.edu/~fabio/CUDD/cuddIntro.html.
37. S. Tani, K. Hamaguchi, and S. Yajima. The complexity of the optimal variable ordering problems of shared binary decision diagrams. In *4th International Symposium on Algorithms and Computation (ISAAC)*, 1993.
38. R. Valk. On the computational power of extended Petri nets. In *7th Symp. on Mathematical Foundations of Computer Science*, LNCS 64, pages 527–535. Springer-Verlag, 1978.
39. R. Valk. Generalizations of Petri nets. In *Mathematical foundations of computer science*, LNCS 118, pages 140–155. Springer-Verlag, 1981.

Embedded Software: Better Models, Better Code

Thomas A. Henzinger

EPFL and University of California, Berkeley

Embedded software is increasingly deployed in safety-critical applications, from medical implants to drive-by-wire technology. This calls for a rigorous design and verification methodology. The main difference between embedded and traditional software is that in the embedded case non-functional aspects, such as reactivity with respect to physical processes, resource usage, and timing, are integral to the correct behavior of the software. Yet traditional software description mechanisms – including specification, modeling, and programming languages – rarely address these non-functional aspects; indeed, the systematic abstraction of real time and other physical constraints in models of sequential and concurrent computation, from Turing machines to software threads, has been one of the great success stories in computer science. For the principled design of embedded software, instead, we have to recombine computation and physicality.

A central challenge is to achieve this recombination at the modeling level: suitable models should be neither too abstract (this makes the generation of efficient code difficult) nor too close to the execution platform (this makes code verification and code reuse difficult). For example, many traditional real-time programming models are based on *priorities* [1]. These models are arguably not sufficiently abstract, and the resulting code is often unpredictable with respect to both timing (jitter) and functionality (data races). Some newer, elegant programming models are based on the *synchrony assumption*, which postulates that all computation is infinitely faster than the physical environment [2]. However, in highly resource-constrained embedded contexts this assumption may be too abstract, and producing efficient code for distributed platforms is difficult. We advocate a third modeling paradigm, based on the *LET* (*logical execution time*) *assumption*, which is less abstract than synchronous models but more abstract than priority-based models.

The starting point of LET is that the model specifies not only the functionality of embedded software, but also its timing: when inputs (sensors) are read, and when outputs (actuators) are written. It is then the model compiler's job to ensure that the specified timing is exhibited by the generated code, in the same way in which a classical compiler ensures that the functional properties of a high-level program are implemented by the generated code. In other words, the software engineer specifies the reactive behavior of embedded software, and the model compiler checks its schedulability. For this, the compiler needs information about the platform, such as worst-case execution times, but once the schedulability check succeeds, the compiler produces *deterministic* code whose observable behavior is independent of execution-time variations and scheduling

J. Cortadella and W. Reisig (Eds.): ICATPN 2004, LNCS 3099, pp. 35–36, 2004.
© Springer-Verlag Berlin Heidelberg 2004

decisions. The key is that in the generated code, the output of a software task is made available to actuators and to other tasks only when specified in the model, at the end of the LET, not when the task completes its actual execution on the processor (which is usually earlier). In other words, the task's specified *logical* execution time, not its varying *physical* execution time, determines program behavior. This minimizes jitter and removes the possibility of data races, and therefore ensures reproducible behavior in embedded software. Furthermore, the model can be recompiled on different platforms and the code can be composed with other software, without changing its functionality nor its timing. In this way, LET-based modeling leads to predictable, portable, and composable real-time code.

LET semantics can be realized in many different modeling languages, such as dataflow-based or Petri net-based formalisms. We have primarily worked with a textual syntax called Giotto: a Giotto program consists of a collection of sequential modes, each of which invokes a set of concurrent, periodic LET tasks. This model is geared toward high-performance control applications, and we have demonstrated that code generated automatically from Giotto is sufficiently efficient to control a model helicopter in flight.

The work reported here is joint with Arkadeb Ghosal, Ben Horowitz, Christoph Kirsch, Rupak Majumdar, Slobodan Matic, and Marco Sanvido [3–7].

References

1. Alan Burns and Andy J. Wellings. *Real-Time Systems and Programming Languages*. Addison-Wesley, 2001.
2. Nicolas Halbwachs. *Synchronous Programming of Reactive Systems*. Kluwer Academic Publishers, 1993.
3. Thomas A. Henzinger, Benjamin Horowitz, and Christoph M. Kirsch, Giotto: a time-triggered language for embedded programming. *Proceedings of the IEEE* **91**, 2003, pp. 84–99.
4. Thomas A. Henzinger and Christoph M. Kirsch, The Embedded Machine: Predictable, portable real-time code. *Proceedings of the International Conference on Programming Language Design and Implementation* (PLDI), pp. 315–326. ACM Press, 2002.
5. Thomas A. Henzinger, Christoph M. Kirsch, Rupak Majumdar, and Slobodan Matic. Time-safety checking for embedded programs. *Proceedings of the Second International Workshop on Embedded Software* (EMSOFT), Lecture Notes in Computer Science 2491, pp. 76–92. Springer-Verlag, 2002.
6. Thomas A. Henzinger, Christoph M. Kirsch, and Slobodan Matic. Schedule-carrying code. *Proceedings of the Third International Workshop on Embedded Software* (EMSOFT), Lecture Notes in Computer Science 2855, pp. 241–256. Springer-Verlag, 2003.
7. Arkadeb Ghosal, Thomas A. Henzinger, Christoph M. Kirsch, and Marco A.A. Sanvido. Event-driven programming with logical execution times. *Proceedings of the 7th International Workshop on Hybrid Systems: Computation and Control* (HSCC), Lecture Notes in Computer Science 2993, pp. 357–371. Springer-Verlag, 2004.

Specification and Model Checking of Temporal Properties in Time Petri Nets and Timed Automata*

Wojciech Penczek[1,2] and Agata Półrola[3]

[1] Institute of Computer Science, PAS, Ordona 21, 01-237 Warsaw, Poland
[2] Institute of Informatics, Podlasie Academy, Sienkiewicza 51, 08-110 Siedlce, Poland
penczek@ipipan.waw.pl
[3] Faculty of Mathematics, University of Lodz, Banacha 22, 90-238 Lodz, Poland
polrola@math.uni.lodz.pl

Abstract. The paper surveys some of the most recent approaches to verification of properties, expressible in some timed and untimed temporal logics (LTL, CTL, TCTL), for real-time systems represented by time Petri nets (TPN's) and timed automata (TA). Firstly, various structural translations from TPN's to TA are discussed. Secondly, model abstraction methods, based on state class approaches for TPN's, and on partition refinement for TA, are given. Next, SAT-based verification techniques, like bounded and unbounded model checking, are discussed. The main focus is on bounded model checking for TCTL and for reachability properties. The paper ends with a comparison of experimental results for several time Petri nets, obtained using the above solutions, i.e., either model abstractions for TPN's, or a translation of a net to a timed automaton and then verification methods for TA. The experiments have been performed using some available tools for TA and TPN's.

1 Introduction

Model checking provides for a promising set of techniques for hardware and software verification. Essentially, in this formalism verifying that a property follows from a system specification amounts to checking whether or not a temporal formula is valid on a model representing all possible computations of the system.

Recently, the interest in automated verification is moving towards concurrent real-time systems. Several models of such systems are usually considered in the literature, but timed automata (TA) [6] and time Petri nets (TPN's) [51] belong to the most widely used. For the above systems, one is, usually, interested in checking reachability or temporal properties that are most frequently expressed in either a standard temporal logic like LTL and CTL*, or in a timed extension of CTL, called TCTL [3].

However, practical applicability of model checking methods is strongly limited by the *state explosion problem*. For real-time systems, the problem occurs

* Partly supported by the State Committee for Scientific Research under the grant No. 3T11C 00426 and a special grant supporting ALFEBIITE.

J. Cortadella and W. Reisig (Eds.): ICATPN 2004, LNCS 3099, pp. 37–76, 2004.
© Springer-Verlag Berlin Heidelberg 2004

with a particular strength, which follows from infinity of the time domain. Therefore, existing verification techniques frequently apply symbolic representations of state spaces using either operations on Difference Bound Matrices [34], variations of Boolean Decision Diagrams [11, 84, 84], or SAT-related algorithms. The latter can exploit either a sequence of translations starting from timed automata and TCTL, going via (quantified) separation logic to quantified propositional logic and further to propositional logic [10, 57, 73] or a direct translation from timed automata and TCTL to propositional logic [63, 86, 92]. Finite state spaces, preserving properties to be checked, are usually built using detailed region approach or (possibly minimal) abstract models based on state classes or regions. Algorithms for generating such models have been defined for time Petri nets [14, 17, 35, 48, 58, 61, 82, 89], as well as for timed automata [3, 4, 21, 32, 65, 79].

It seems that in spite of the same underlying timed structure, model checking methods for time Petri nets and timed automata have been defined independently of each other. However, several attempts to combine the two approaches have been already made, concerning both a structural translation of one model to the other [28, 37, 42, 49, 64, 74] or an adaptation of existing verification methods [35, 58, 82].

The goal of this paper is to report on a recent progress in building abstract state spaces and application of symbolic methods for verification of both the time Petri nets and timed automata, either directly or indirectly via a translation from the former formalism to the latter. To this aim, firstly, various structural translations from TPN's to TA are discussed (see Section 3) to apply timed automata specific methods to time Petri nets. Temporal specification languages are introduced in Section 4. Model abstraction methods based on state classes approaches for TPN's and on partition refinement for TA are given in Section 5. Next, SAT-based verification techniques, like bounded (BMC) and unbounded model checking (UMC), are discussed in Section 6.

The idea behind BMC consists in translating the model checking problem for an existential fragment of some temporal logic (like ECTL or TECTL) on a fraction of a model into a test of propositional satisfiability, for which refined tools already exist [56]. Unlike BMC, UMC [50, 73] deals with unrestricted temporal logics checked on complete models at the price of a decrease of efficiency. In this paper the main focus is on SAT-related methods of BMC for TCTL [63] as well as for reachability and unreachability [86].

Each section of our paper is accompanied with pointers to the literature, where descriptions of complementary verification methods can be found. The paper ends with a comparison of experimental results for several time Petri nets, obtained using solutions discussed in the paper, i.e., either model abstraction techniques for TPN's, or a translation of a net to a timed automaton and then verification methods for TA. The experiments, described in Section 8, have been performed using the tools: Tina [16], Kronos [91], and VerICS [31].

2 Main Models of Real-Time Systems

We consider two main models of real-time systems: Petri nets with time and timed automata. First we define Petri nets, discuss their time extensions, and provide a definition of time Petri nets. Our attention is focused on a special kind of TPN's - distributed time Petri nets, which are then considered in our case studies.

The following abbreviations are used in definitions of both TA and TPN's. Let $\mathbb{R}$ ($\mathbb{R}_+$) denote the set of non-negative (positive) reals, $\mathbb{N}$ ($\mathbb{N}_+$) - the set of (positive) naturals, and $\mathbb{Q}_+$ - the set of non-negative rational numbers,

2.1 Petri Nets with Time

We start with the standard notion of Petri nets.

Definition 1. *A Petri net is a four-element tuple* $\mathcal{P} = (P, T, F, m_0)$, *where* $P = \{p_1, \ldots, p_{n_P}\}$ *is a finite set of* places, $T = \{t_1, \ldots, t_{n_T}\}$ *is a finite set of* transitions, $F : (P \times T) \cup (T \times P) \to \mathbb{N}$ *is the* flow function, *and* $m_0 : P \to \mathbb{N}$ *is the* initial marking *of* $\mathcal{P}$.

For each transition $t \in T$ we define its *preset* $\bullet t = \{p \in P \mid F(p,t) > 0\}$ and its *postset* $t\bullet = \{p \in P \mid F(t,p) > 0\}$. Moreover, in order to simplify some consequent notions, we consider only the nets, for which $\bullet t$ and $t\bullet$ are non-empty, for all transitions t. We use the following auxiliary notations and definitions:

- a *marking* of $\mathcal{P}$ is any function $m : P \to \mathbb{N}$;
- a transition $t \in T$ is *enabled* at m ($m[t\rangle$ for short) if $(\forall p \in \bullet t)\ m(p) \geq F(p,t)$, and *leads from marking* m *to* m', where for each $p \in P$, $m'(p) = m(p) - F(p,t) + F(t,p)$. The marking m' is denoted by $m[t\rangle$ as well, if this does not lead to misunderstanding;
- $en(m) = \{t \in T \mid m[t\rangle\}$ - the set of transitions enabled at m;
- for $t \in en(m)$, $newly\_en(m,t) = \{u \in T \mid u \in en(m[t\rangle) \wedge u \notin en(m')$ with $m'(p) = m(p) - F(p,t)$ for each $p \in P\}$ - the set of transitions newly enabled after firing t;
- a marking m is *reachable* if there exists a sequence of transitions $t^1, \ldots, t^l \in T$ and a sequence of markings $m^0, \ldots, m^l$ such that $m^0 = m_0$, $m^l = m$, and $t^i \in en(m^{i-1})$, $m^i = m^{i-1}[t^i\rangle$ for each $i \in \{1, \ldots, l\}$; the set of all the reachable markings of $\mathcal{P}$ is denoted by $RM_{\mathcal{P}}$;
- a net $\mathcal{P}$ is said to be *bounded* if all its reachable markings are bounded;
- two transitions $t_1, t_2 \in T$ are *concurrently enabled in* m if $t_1 \in en(m)$ and $t_2 \in en(m')$ with $m'(p) = m(p) - F(p,t_1)$ for each $p \in P$;
- a net $\mathcal{P}$ is *sequential* if none of its reachable markings concurrently enables two transitions;
- a net $\mathcal{P}$ is *ordinary* if the flow function F maps onto $\{0,1\}$;
- a net $\mathcal{P}$ is *1-safe* if it is ordinary and $m(p) \leq 1$, for each $p \in P$ and each $m \in RM_{\mathcal{P}}$.

Intuitively, Petri nets are directed weighted graphs with two types of nodes: places (representing conditions) and transitions (representing events), whose arcs correspond to these elements in the domain of the flow function, for which the value of this function is positive. The arcs are assigned positive weights according to the values of F.

The theory of Petri nets provides a general framework for modelling distributed and concurrent systems. Since for many of them timing dependencies play an important role, a variety of extensions of the main formalism, enabling to reason about temporal properties, has been introduced. In what follows, we present a brief survey of the approaches, based on [23, 71].

Petri nets with timing dependencies can be classified according to the way of specifying timing constraints (these can be timing intervals [51, 83] or single numbers [68]), or elements of the net with which these constraints are associated (places [27], transitions [51, 68] or arcs [1, 38, 83]). The next criterion is the interpretation of the timing constraints. When associated with a *transition*, the constraint can be viewed as its *firing time* (a transition consumes the input tokens when becomes enabled, but does not create the output ones until the delay time associated with it has elapsed [68]), *holding time* (when the transition fires, the actions of removing and creating tokens are done instantaneously, but the tokens created are not available to enable new transitions until they have been in their output place for the time specified as the duration time of the transition which created them [81]), or *enabling time* (a transition is forced to be enabled for a specified period of time before it can fire, and tokens are removed and created in the same instant [51]). A time associated with a *place* usually refers to the period the tokens must spend in the place before becoming available to enable a transition [27]. A timing interval on an *input arc* usually expresses the conditions under which tokens can potentially leave the place using this arc [38], whereas a timing interval on an *output arc* denotes the time when tokens produced on this arc become available [83]. Nets can be also classified according to *firing rules*: the *weak firing rule* means that the time which passes between enabling of the transition and its firing is not determined [80], the *strong earliest firing rule* requires the transition to be fired as soon as it is enabled and the appropriate timing conditions are met [38], whereas the *strong latest firing rule* means that the transition can be fired in a specified period of time, but no later than after certain time from its enabling, unless it becomes disabled by firing of another (conflicting) one [51]. The best known timed extensions of Petri nets are *timed Petri nets* by Ramchandani [68] and *time Petri nets* by Merlin and Farber [51]. In this paper, we focus on the latter.

Definition 2. *A time Petri net (TPN, for short) is a six-element tuple* $\mathcal{N} = (P, T, F, m_0, Eft, Lft)$, *where* (P, T, F, m_0) *is a Petri net, and* $Eft : T \to \mathbb{N}$, $Lft : T \to \mathbb{N} \cup \{\infty\}$ *are functions describing the* earliest *and the* latest firing times *of the transitions, where* $Eft(t) \leq Lft(t)$ *for each* $t \in T$.

A *concrete state* σ of $\mathcal{N}$ is an ordered pair $(m, clock)$, where m is a marking, and $clock : T \to \mathbb{R}$ is a function which for each transition $t \in en(m)$ gives the time elapsed since t became enabled most recently. By $clock + \delta$ we denote the function given by $(clock + \delta)(t) = clock(t) + \delta$ for all $t \in T$. Moreover, let $(m, clock) + \delta$ denote $(m, clock + \delta)$. The *concrete state space* of $\mathcal{N}$ is a structure $F_c(\mathcal{N}) = (\Sigma, \sigma^0, \to)$, where Σ is the set of all the concrete states of $\mathcal{N}$, $\sigma^0 = (m_0, clock_0)$ with $clock_0(t) = 0$ for each $t \in T$ is the initial state, and a timed consecution relation $\rightarrow \subseteq \Sigma \times (T \cup \mathbb{R}) \times \Sigma$ is defined by action- and time-successors as follows:

- for $\delta \in \mathbb{R}$, $(m, clock) \xrightarrow{\delta} (m, clock + \delta)$ iff $(clock + \delta)(t) \leq Lft(t)$ for all $t \in en(m)$ (*time successor*),
- for $t \in T$, $(m, clock) \xrightarrow{t} (m_1, clock_1)$, iff $t \in en(m)$, $Eft(t) \leq clock(t) \leq Lft(t)$, $m_1 = m[t\rangle$, and for all $u \in T$ we have $clock_1(u) = 0$ for $u \in newly\_en(m,t)$, and $clock_1(u) = clock(u)$ otherwise (*action successor*).

A σ_0-*run* ρ of $\mathcal{N}$ is a maximal sequence of concrete states $\rho = \sigma_0 \xrightarrow{\delta_0} \sigma_0 + \delta_0 \xrightarrow{t_0} \sigma_1 \xrightarrow{\delta_1} \sigma_1 + \delta_1 \xrightarrow{t_1} \sigma_2 \xrightarrow{\delta_2} \ldots$, where $t_i \in T$ and $\delta_i \in \mathbb{R}$, for all $i \in \mathbb{N}$. A state $\sigma \in \Sigma$ is *reachable* if there exists a σ^0-run ρ and $i \in \mathbb{N}$ such that $\sigma = \sigma_i + \delta_i$.

The set of all the reachable states of $\mathcal{N}$ is denoted by $Reach_{\mathcal{N}}$. A marking m is reachable if there is a state $(m, clock) \in Reach_{\mathcal{N}}$ for some function $clock$. A run ρ is said to be *progressive* iff $\Sigma_{i \in \mathbb{N}} \delta_i$ is unbounded.

The above notion of run can be used for interpreting untimed branching time temporal logics [58] or for checking reachability. Alternatively, runs can be defined such that each two consecutive time and action steps are combined [89][1]. However, in order to give semantics over dense paths for timed temporal logics, we assume that $\delta_i > 0$ for all $i \in \mathbb{N}$. This will become clear in Section 4.2.

In what follows, we consider only nets whose all the runs are infinite[2] and progressive by restricting the nets to contain neither two consecutive transitions whose earliest firing times are both equal to 0 nor one such a transition, which is a loop[3]. The set of all the σ-runs of $\mathcal{N}$, with $\delta_i > 0$ for all $i \in \mathbb{N}$, is denoted by $f_{\mathcal{N}}(\sigma)$.

In order to reason about systems represented by TPN's, we define, for a given set of propositional variables PV, a valuation function $V_{\mathcal{N}} : P \rightarrow 2^{PV}$, which assigns propositions to the places[4] of $\mathcal{N}$. Let $V_c : \Sigma \rightarrow 2^{PV}$ be a valuation function extending $V_{\mathcal{N}}$ such that $V_c((m, \cdot)) = \bigcup_{p \in m} V_{\mathcal{N}}(p)$, i.e., V_c assigns the same propositions to the states with the same markings. The structure $M_c(\mathcal{N}) = (F_c(\mathcal{N}), V_c)$ is called a *concrete model* of $\mathcal{N}$. Notice that for the same concrete model, we can interpret timed and untimed logics using appropriate notions of runs.

A time Petri net $\mathcal{N} = (P, T, F, m_0, Eft, Lft)$ is said to be sequential if the net $\mathcal{P} = (P, T, F, m_0)$ is so, and $\mathcal{N}$ is *1-safe* if for each reachable marking m of $\mathcal{N}$, we have $m(p) \leq 1$ for each $p \in P$. Unless otherwise stated, in what follows 1-safe TPN's are considered only. Moreover, we define a notion of a *distributed time Petri net*, which is an adaptation of the one from [41].

Definition 3. *Let I be a finite set of indices, and let $\mathfrak{N} = \{N^i \mid i \in I\}$, with $N^i = (P^i, T^i, F^i, Eft^i, Lft^i, m_0^i)$ be a family of 1-safe, sequential time Petri*

[1] This semantics is claimed to be equivalent w.r.t. CTL model checking.

[2] This can be checked by applying algorithms looking for deadlocks.

[3] This restriction is only for assuring an existence of a translation to timed automata, for which similar restrictions are assumed.

[4] Usually, there is one-to-one correspondence between the propositions and the places.

nets (called processes*), indexed I, with pairwise disjoint sets P^i of places, and satisfying the condition* $(\forall i_1, i_2 \in I)(\forall t \in T^{i_1} \cap T^{i_2})$ $(Eft^{i_1}(t) = Eft^{i_2}(t) \wedge Lft^{i_1}(t) = Lft^{i_2}(t))$*. A* distributed time Petri net $\mathcal{N} = (P, T, F, Eft, Lft, m_0)$ *is the union of the* processes N^i, *i.e.*, $P = \bigcup_{i \in I} P^i$, $T = \bigcup_{i \in I} T^i$, $F = \bigcup_{i \in I} F^i$, $Eft = \bigcup_{i \in I} Eft^i$, $Lft = \bigcup_{i \in I} Lft^i$ *and* $m_0 = \bigcup_{i \in I} m_0^i$.

It is easy to notice that a distributed net is 1-safe. The interpretation of such a net is a collection of sequential, non-deterministic processes with communication capabilities (via joint transitions). In what follows, we consider distributed nets whose all the processes are *state machines* (i.e., for each $t \in T^i$, $|\bullet t| = |t \bullet| = 1$), which implies that in any reachable marking m of $\mathcal{N}$, there is exactly one place p of each process with $m(p) = 1$. It is important to mention that each distributed net can be transformed to an equivalent net satisfying this requirement [43].

Example 1. Examples of (distributed) time Petri net are shown in Fig. 1. Each of the nets $5a$, $5b$, and $5c$ in the left-hand side consists of two disjoint processes with the sets of places $P^1 = \{p_1, p_2, p_3, p_4\}$ and $P^2 = \{p_6, p_7\}$, whereas the net on the right is composed of three communicating processes with the sets of places: $P^i = \{idle_i, trying_i, enter_i, critical_i\}$ with $i = 1, 2$, and $P^3 = \{place0, place1, place2\}$.

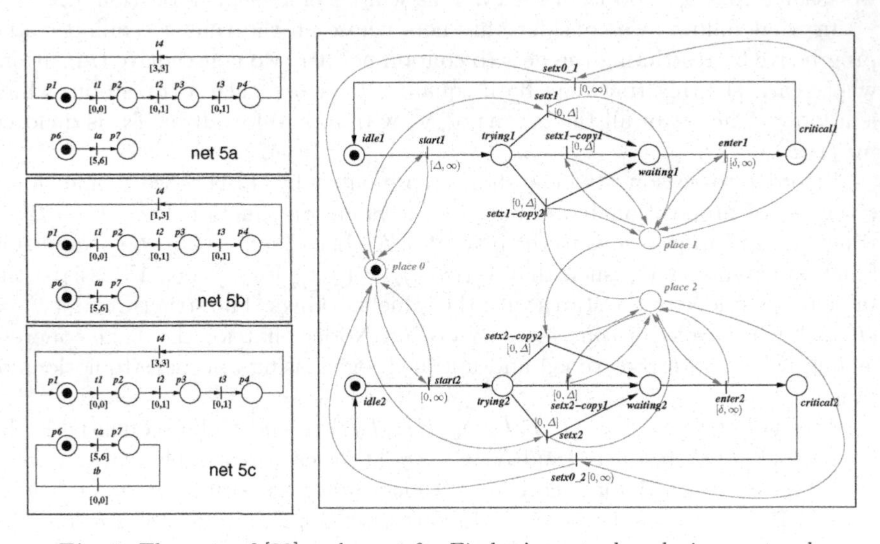

Fig. 1. The nets of [89] and a net for Fischer's mutual exclusion protocol

2.2 Timed Automata

In this section we briefly recall the concept of timed automata, which was introduced by Alur and Dill [6]. Timed automata are extensions of finite state automata with constraints on timing behaviour. The underlying finite state automata are augmented with a set of real time variables. We start with formalizing the above notions. Let $\mathcal{X} = \{x_1, \ldots, x_{n_\mathcal{X}}\}$ be a finite set of real-valued variables, called *clocks*. The set of *clock constraints* over $\mathcal{X}$ is defined by the following grammar:

$$\mathfrak{cc} := true \mid x_i \sim c \mid x_i - x_j \sim c \mid \mathfrak{cc} \wedge \mathfrak{cc},$$

where $x_i, x_j \in \mathcal{X}$, $c \in \mathbb{N}$, and $\sim \in \{\leq, <, =, >, \geq\}$. The set of all the clock constraints over $\mathcal{X}$ is denoted by $\mathcal{C}_\mathcal{X}^\ominus$, whereas its restriction, where differences of clocks are not allowed, is denoted by $\mathcal{C}_\mathcal{X}$. A *clock valuation* on $\mathcal{X}$ is a $n_\mathcal{X}$-tuple $v \in \mathbb{R}^{n_\mathcal{X}}$. For simplicity, we assume a fixed ordering on $\mathcal{X}$. The value of the clock x_i in v can be then denoted by $v(x_i)$ or $v(i)$, depending on the context. For a valuation v and $\delta \in \mathbb{R}$, $v + \delta$ denotes the valuation v' s.t. for all $x \in \mathcal{X}$, $v'(x) = v(x) + \delta$. Moreover, for a subset of clocks $X \subseteq \mathcal{X}$, $v[X := 0]$ denotes the valuation v' such that for all $x \in X$, $v'(x) = 0$ and for all $x \in \mathcal{X} \setminus X$, $v'(x) = v(x)$. Let $v \in \mathbb{R}^{n_\mathcal{X}}$, the satisfaction relation $\models$ for a clock constraint $\mathfrak{cc} \in \mathcal{C}_\mathcal{X}^\ominus$ is defined inductively as follows:

- $v \models true$,
- $v \models (x_i \sim c)$ iff $v(x_i) \sim c$,
- $v \models (x_i - x_j \sim c)$ iff $v(x_i) - v(x_j) \sim c$, and
- $v \models (\mathfrak{cc} \wedge \mathfrak{cc}')$ iff $v \models \mathfrak{cc}$ and $v \models \mathfrak{cc}'$.

For a constraint $\mathfrak{cc} \in \mathcal{C}_\mathcal{X}^\ominus$, let $[\![\mathfrak{cc}]\!]$ denote the set of all the clock valuations satisfying $\mathfrak{cc}$, i.e., $[\![\mathfrak{cc}]\!] = \{v \in \mathbb{R}^{n_\mathcal{X}} \mid v \models \mathfrak{cc}\}$. By a *(time) zone* in $\mathbb{R}^{n_\mathcal{X}}$ we mean each convex polyhedron $Z \subseteq \mathbb{R}^{n_\mathcal{X}}$ defined by a clock constraint, i.e., $Z = [\![\mathfrak{cc}]\!]$ for some $\mathfrak{cc} \in \mathcal{C}_\mathcal{X}^\ominus$ (for simplicity, we identify the zones with the clock constraints which define them). The set of all the zones for $\mathcal{X}$ is denoted by $Z(n_\mathcal{X})$.

Given $v, v' \in \mathbb{R}^{n_\mathcal{X}}$ and $Z, Z' \in Z(n_\mathcal{X})$, we define the following operations:

- $Z \setminus Z'$ is a set of disjoint zones s.t. $\{Z'\} \cup (Z \setminus Z')$ is a partition of Z;
- $Z \nearrow := \{v' \in \mathbb{R}^n \mid (\exists v \in Z)\ v \leq v'\}$, where $v \leq v'$ iff $\exists \delta \in \mathbb{R}$ s.t. $v' = v + \delta$;
- $Z[X := 0] = \{v[X := 0] \mid v \in Z\}$.

Notice that the operations $\nearrow$, $Z[X := 0]$, and the standard intersection preserve zones. These results and the implementation of $Z \setminus Z'$ can be found in [4, 79].

Definition 4. *A* timed automaton *(TA, for short) is a six-element tuple $\mathcal{A} = (A, L, l^0, E, \mathcal{X}, \mathcal{I})$, where A is a finite set of actions, L is a finite set of locations, $l^0 \in L$ is an initial location, $\mathcal{X}$ is a finite set of clocks, $E \subseteq L \times A \times \mathcal{C}_\mathcal{X}^\ominus \times 2^\mathcal{X} \times L$ is a transition relation. Each element e of E is denoted by $l \xrightarrow{a,\mathfrak{cc},X} l'$, which represents a transition from the location l to the location l', executing the action a, with the set $X \subseteq \mathcal{X}$ of clocks to be reset, and with the clock constraint $\mathfrak{cc}$ defining the enabling condition for e. The function $\mathcal{I} : L \rightarrow \mathcal{C}_\mathcal{X}^\ominus$, called a* location invariant, *assigns to each location $l \in L$ a clock constraint defining the conditions under which $\mathcal{A}$ can stay in l.*

If the enabling conditions and the values of the location invariant are in the set $\mathcal{C}_\mathcal{X}$ only, then the automaton is called *diagonal-free*. In what follows, we consider diagonal-free automata. In order to reason about systems represented by timed automata, for a set of propositional variables PV, define a valuation function $V_{\mathcal{A}} : L \to 2^{PV}$, which assigns propositions to the locations. Usually, we consider networks of timed automata, where the automata are composed by synchronization over labels. The reader is referred to [86] for a formal definition of composition.

Example 2. In Fig. 2, a network of TA for Fischer's mutual exclusion protocol with two processes is depicted[5]. The protocol is parameterised by the number of processes involved. In the general case, the network consists of n automata of processes, together with one automaton modelling a global variable X, used to coordinate the processes' access to their critical sections, which means that if the automaton of a process and the automaton of the variable X contain actions with a common label a, then the process is allowed to execute this action if and only if the automaton for X executes an action labelled by a as well.

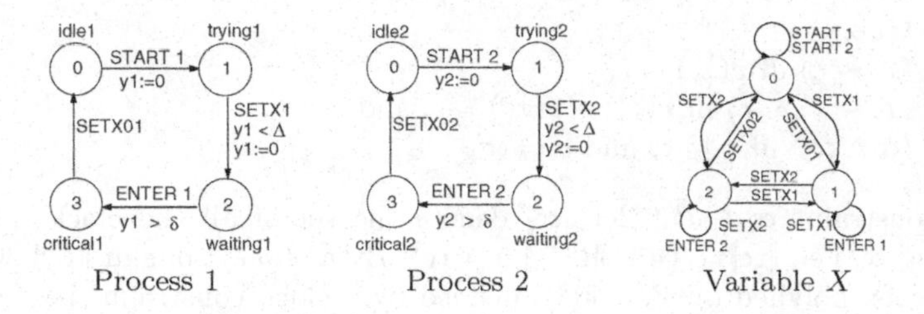

Fig. 2. Fischer's Mutual Exclusion Protocol for two processes

A *concrete state* of $\mathcal{A}$ is a pair (l, v), where $l \in L$ and $v \in \mathbb{R}^{n_\mathcal{X}}$ is a valuation. The *concrete (dense) state space* of $\mathcal{A}$ is a structure $F_c(\mathcal{A}) = (Q, q^0, \to)$, where $Q = L \times \mathbb{R}^{n_\mathcal{X}}$ is the set of all the concrete states, $q^0 = (l^0, v^0)$ with $v^0(x) = 0$ for all $x \in \mathcal{X}$ is the initial state, and $\to \subseteq Q \times (A \cup \mathbb{R}) \times Q$ is the transition relation, defined by action- and time-successors as follows:

- for $a \in A$, $(l, v) \xrightarrow{a} (l', v')$ iff $(\exists \mathfrak{cc} \in \mathcal{C}_\mathcal{X})(\exists X \subseteq \mathcal{X})$ such that $l \xrightarrow{a,\mathfrak{cc},X} l' \in E$, $v \in [\![\mathfrak{cc}]\!]$, $v' = v[X := 0]$ and $v' \in [\![\mathcal{I}(l')]\!]$ (*action successor*),
- for $\delta \in \mathbb{R}$, $(l, v) \xrightarrow{\delta} (l, v + \delta)$ iff $v + \delta \in [\![\mathcal{I}(l)]\!]$ (*time successor*).

For $(l, v) \in Q$, let $(l, v) + \delta$ denote $(l, v + \delta)$. A *q-run* ρ of $\mathcal{A}$ is a maximal sequence of concrete states: $q_0 \xrightarrow{\delta_0} q_0 + \delta_0 \xrightarrow{a_0} q_1 \xrightarrow{\delta_1} q_1 + \delta_1 \xrightarrow{a_1} q_2 \xrightarrow{\delta_2} \ldots$, where $q_0 = q \in Q$, $a_i \in A$ and $\delta_i \in \mathbb{R}$, for each $i \geq \mathbb{N}$. A state $q' \in Q$ is *reachable* if there exists a q^0-run $\rho = q_0 \xrightarrow{\delta_0} q_0 + \delta_0 \xrightarrow{a_0} q_1 \xrightarrow{\delta_1} \ldots$ and $i \in \mathbb{N}$ such that $q' = q_i + \delta_i$. The set of all the reachable states of $\mathcal{A}$ will be denoted by $Reach_\mathcal{A}$. A run ρ is said to be *progressive* iff $\Sigma_{i \in \mathbb{N}} \delta_i$ is unbounded. A timed automaton is *progressive* iff all its runs are progressive. For simplicity of the presentation,

[5] Two clocks are denoted by y_1 and y_2, whereas Δ and δ are parameters.

we consider only progressive timed automata. Note that progressiveness can be checked using sufficient conditions of [79].

Like for time Petri nets, the above notion of run can be easily used for interpreting untimed temporal logics or for checking reachability. However, in order to give semantics over dense paths corresponding to runs (see Section 4.2) for timed temporal logics, we assume that $\delta_i > 0$ for all $i \in \mathbb{N}$ in a run. So, by $f_{\mathcal{A}}(q)$ we denote the set of all such q-runs of $\mathcal{A}$.

The structure $M_c(\mathcal{A}) = (F_c(\mathcal{A}), V_c)$ is called a *concrete model* for $\mathcal{A}$, for a valuation function $V_c : Q \to 2^{PV}$ extending $V_{\mathcal{A}}$ such that $V_c(l, v) = V_{\mathcal{A}}(l)$ (i.e., V_c assigns the same propositions to the states with the same locations).

3 From TPN's to TA

There are two approaches to verifying properties of time Petri nets. Either specific algorithms are used or nets are translated to timed automata in order to exploit verification methods designed for automata[6]. Therefore, we first consider translations from TPN's to TA, and then review the most recent verification methods for both the formalisms.

Several methods of translating time Petri nets to timed automata have been already developed. However, in most cases translations produce automata, which extend timed automata. Some of the existing approaches are sketched below.

The most natural translation consists in defining an automaton whose locations correspond to the markings of the net, whereas the actions and the clocks - to its transitions. The invariant of a marking m expresses that no transition can be disabled by passage of time, whereas the enabling condition of an action labelled by $t \in T$ guarantees that the time passed since t became enabled is between $Eft(t)$ and $Lft(t)$. Executing an action labelled by t at m resets the clocks corresponding to the transitions in $newly\_en(m, t)$. This approach was used to define *detailed region graphs* for TPN's [58, 82]. Its formal description can be also found in [64].

Sifakis and Yovine [74] presented a translation of a subclass of *time stream Petri nets* [72] to automata whose invariants can be disjunctions of clock constraints. In these nets, each of the arcs (p, t) is assigned a timing interval which specifies when tokens in the place p become available to fire the transition t. An enabled transition t can be fired if (*) all the places in $\bullet t$ have been marked at least as long as the lower bounds of the corresponding intervals, and (**) there is at least one place in $\bullet t$ marked no longer than the upper bound of the corresponding interval. (1-safe) time Petri nets can be seen as a subclass of these nets. In order to translate a net $\mathcal{N}$ to an automaton, we define a location for each marking of $\mathcal{N}$, and associate a clock with each of its places. The actions are labelled by the transitions of $\mathcal{N}$. Executing an action labelled by $t \in T$ resets the clocks of the places in $t\bullet$, whereas its enabling condition corresponds to (*) given above. The invariant of a marking m states that (**) holds for each $t \in en(m)$.

[6] Usually, the concrete state spaces of both the models are required to be bisimilar.

In [25, 36], translations of (general) TPN's to automata equipped with shared variables and urgency modelling mechanisms are provided. The method of [25] generates a set of TA containing an automaton $\mathcal{A}_t$ with one clock for each $t \in T$, and a supervisor automaton $\mathcal{A}_s$. The locations of $\mathcal{A}_t$ correspond to the possible states of t (being enabled, disabled, and to its firing). The automaton $\mathcal{A}_s$ with *committed locations*[7] forces the other automata to change synchronously their states when a transition of the net is fired. Shared variables are used to model the number of tokens in the places. In the approach of [36], the transitions are classified according to their number of input, output and inhibitor places, and one automaton with the locations *disabled* and *enabled* is built for each of the classes obtained. Similarly as in [25], an additional automaton (with an *urgent transition*[8]) ensures a synchronous behaviour of the whole system when a transition is fired, whereas shared variables store the marking of the net.

Lime and Roux [49] proposed a method of translating a (general) time Petri net to a timed automaton, using an algorithm of building a *state class graph* G [14] (see Sec. 5.1). The nodes of G are *state classes*, i.e., pairs (m, I), where m is a marking and I is a set of inequalities. The translation leads to an automaton whose locations are the nodes of G, and the edges, labelled by transitions of the net, correspond to its successor relation. Every class (m, I) is assigned a set of clocks, each of which corresponds to all these transitions in $en(m)$ which became enabled at the same time. Executing an action labelled by t resets some clocks (associated with newly enabled transitions). It is also possible to assign a value of one clock to another (this goes beyond the standard definition of TA). Invariants and enabling conditions describe, respectively, when the net can be in a given marking, or when a transition can be fired at a class. Since the number of state classes is finite only if the net is bounded [14], the authors provide a condition for an on-line checking of unboundedness to ensure stopping the computation.

The paper [28] shows a method of translating extended 1-safe TPN's with finite values of the function Lft (called PRES+ models) to a set of extended timed automata. When applied to a non-extended net, the translation gives a network of standard TA. The set of transitions of the net is divided into disjoint *clusters*, i.e., sequences of transitions in which firing of one of them enables the next one. An automaton with one clock is built for each such a cluster.

Another translation [64] applies to distributed nets only. It is based on a different approach to the concrete semantics of the net, which is bisimulation equivalent[9] to that defined in Sec. 2.1. In this semantics, each process is assigned a "clock" measuring the time elapsed since its place became marked most recently. The automaton resulting from the translation uses the above clocks, whereas its locations are defined as the markings of the net, and its edges are labelled by the transitions. Firing of a transition resets the clocks related to the processes involved. The enabling conditions and invariants are possibly disjunctions of clock constraints, but can be modified to be zones (see [64] for details).

[7] A committed location is a location which has to be leaved as soon as it is entered.
[8] A transition which has to be taken as soon as it is enabled.
[9] The concrete state spaces are bisimilar, see [64] for a proof.

Besides translations from TPN's to TA, some methods of translating subclasses of TA to time Petri nets also exist [37], but because of lack of space, we do not discuss these approaches in this paper.

4 Main Formalisms for Expressing Properties

Properties of timed systems are usually expressed using temporal logics. In this section we look at the logics that are most commonly used. We start with untimed formalisms, which are later extended with time constraints.

4.1 Untimed Temporal Logics: LTL, CTL, CTL*

All the untimed logics we consider are subsets of CTL*. Therefore, we give syntax and semantics for CTL* only. The other logics are defined as its restrictions.

Syntax and Semantics of CTL*. Let $PV_L = \{\wp_1, \wp_2 \ldots\}$ be a set of propositional variables. The language of CTL* is given as the set of all the state formulas φ_s, defined using path formulas φ_p, by the following grammar:

$$\varphi_s := \wp \mid \neg\wp \mid \varphi_s \wedge \varphi_s \mid \varphi_s \vee \varphi_s \mid \mathrm{A}\varphi_p \mid \mathrm{E}\varphi_p$$
$$\varphi_p := \varphi_s \mid \varphi_p \wedge \varphi_p \mid \varphi_p \vee \varphi_p \mid \mathrm{X}\varphi_p \mid \varphi_p\mathrm{U}\varphi_p \mid \varphi_p\mathrm{R}\varphi_p$$

A ('for all paths') i E ('there exists a path') are path quantifiers, whereas X ('neXt'), U ('Until'), and R ('Release') are state operators. The formula $\varphi_p\mathrm{U}\psi_p$ expresses that ψ_p eventually occurs and that φ_p holds continuously until then. The operator R is dual to U. Derived path operators are: $\mathrm{G}\varphi_p \stackrel{def}{=} (\wp \wedge \neg\wp)\mathrm{R}\varphi_p$ and $\mathrm{F}\varphi_p \stackrel{def}{=} (\wp \vee \neg\wp)\mathrm{U}\varphi_p$. Sublogics of CTL* are defined below:

CTL (Computation Tree Logic): the temporal formulas are restricted to positive boolean combinations of: $\mathrm{A}(\varphi\mathrm{U}\psi), \mathrm{A}(\varphi\mathrm{R}\psi)$, $\mathrm{AX}\varphi$, and $\mathrm{E}(\varphi\mathrm{U}\psi), \mathrm{E}(\varphi\mathrm{R}\psi)$, $\mathrm{EX}\varphi$ only.

ACTL (Universal Computation Tree Logic): the temporal formulas are restricted to positive boolean combinations of: $\mathrm{A}(\varphi\mathrm{U}\psi), \mathrm{A}(\varphi\mathrm{R}\psi)$, and $\mathrm{AX}\varphi$ only.

ECTL (Existential Computation Tree Logic): the temporal formulas are restricted to positive boolean combinations of: $\mathrm{E}(\varphi\mathrm{U}\psi), \mathrm{E}(\varphi\mathrm{R}\psi)$, and $\mathrm{EX}\varphi$ only.

LTL (Linear Time Logic): the formulas of the form $\mathrm{A}\varphi$ are allowed only, where φ does not contain the path quantifiers A, E.

$\mathbf{L}_{-\mathrm{X}}$ denotes logic L without the next step operator X.

For example, $\mathrm{AFG}(\wp_1 \vee \wp_2)$ is an LTL formula, whereas $\mathrm{AFAG}(\wp_1 \vee \wp_2)$ is an ACTL formula. Each of the above logics can be extended by time constraints (defined later). Semantics of CTL* is defined over standard Kripke models.

Definition 5. *A* model *is a tuple $M = ((S, s^0, \rightarrow), V)$, where S is a set of states, $s^0 \in S$ is the initial state, $\rightarrow \subseteq S \times S$ is a total successor relation, and $V : S \rightarrow 2^{PV}$ is a valuation function.*

Let $|M|$ denote the number of states of M. For $s_0 \in S$ a path $\pi = (s_0, s_1, \ldots)$ is an infinite sequence of states in S starting at s_0, where $s_i \rightarrow s_{i+1}$ for all $i \geq 0$, and $\pi_i = (s_i, s_{i+1}, \ldots)$ is the i-th suffix of π.

- $M, s \models \wp$ iff $\wp \in V(s)$, $M, s \models \neg\wp$ iff $\wp \notin V(s)$,
- $M, x \models \varphi \vee \psi$ iff $M, x \models \varphi$ or $M, x \models \psi$, for $x \in \{s, \pi\}$,
- $M, x \models \varphi \wedge \psi$ iff $M, x \models \varphi$ and $M, x \models \psi$, for $x \in \{s, \pi\}$,
- $M, s \models \mathrm{A}\varphi$ iff $M, \pi \models \varphi$ for each path π starting at s,
- $M, s \models \mathrm{E}\varphi$ iff $M, \pi \models \varphi$ for some path π starting at s,
- $M, \pi \models \varphi$ iff $M, s_0 \models \varphi$, for a state formula φ,
- $M, \pi \models \mathrm{X}\varphi$ iff $M, \pi_1 \models \varphi$,
- $M, \pi \models \psi\mathrm{U}\varphi$ iff $(\exists j \geq 0)\ \big(M, \pi_j \models \varphi$ and $(\forall 0 \leq i < j),\ M, \pi_i \models \psi\big)$,
- $M, \pi \models \psi\mathrm{R}\varphi$ iff $(\forall j \geq 0)\ \big(M, \pi_j \models \varphi$ or $(\exists 0 \leq i < j),\ M, \pi_i \models \psi\big)$.

We adopt the initialized notion of validity in a model: $M \models \varphi$ iff $M, s^0 \models \varphi$.

For timed systems, untimed temporal logics are typically interpreted over concrete (or abstract) models, where a path is defined to correspond to a run that includes unrestricted time-successor steps (i.e., with $\delta_i \in \mathbb{R}$). The *model checking* problem for CTL* over timed systems is defined as follows: given a CTL* formula φ and a timed system $\mathcal{T}$ (i.e., a TPN $\mathcal{N}$ or a TA $\mathcal{A}$) together with a valuation function $V_{\mathcal{T}}$, determine whether $M_c(\mathcal{T}) \models \varphi$.

Besides untimed properties of timed systems, which are directly expressed using the above-defined temporal logics, *reachability* in these systems is usually checked. Given a propositional formula p, the reachability problem for timed automata (time Petri nets) consists in testing whether there is a reachable state satisfying p in $M_c(\mathcal{A})$ ($M_c(\mathcal{N})$, resp.). This problem can be obviously translated to the model checking problem for the CTL* formula EFp. However, in spite of that, several efficient solutions aimed at reachability checking only exist as well.

4.2 Timed Temporal Logics

Timed temporal logics can be interpreted over either discrete or dense models of time [9]. We consider the latter option. Since the model checking problem for TCTL* is undecidable [2], we focus on TCTL [3] and its subsets: TACTL and TECTL, defined analogously to the corresponding fragments of CTL.

The logic TCTL is an extension of $\mathrm{CTL}_{-\mathrm{X}}$ obtained by subscribing the modalities with time intervals specifying time restrictions on formulas. Formally, syntax of TCTL is defined inductively by the following grammar:

$$\varphi := \wp \mid \neg\wp \mid \varphi \vee \varphi \mid \varphi \wedge \varphi \mid \mathrm{A}(\varphi \mathrm{U}_I \varphi) \mid \mathrm{E}(\varphi \mathrm{U}_I \varphi) \ \mid \mathrm{A}(\varphi \mathrm{R}_I \varphi) \mid \mathrm{E}(\varphi \mathrm{R}_I \varphi)$$

where $\wp \in PV_L$ and I is an interval in $\mathbb{R}$ with integer bounds of the form $[n, n']$, $[n, n')$, $(n, n']$, (n, n'), (n, ∞), and $[n, \infty)$, for $n, n' \in \mathbb{N}$.

For example, $\mathrm{A}((\wp \wedge \neg\wp)\mathrm{R}_{[0,\infty)}(\wp_1 \Rightarrow \mathrm{A}((\wp \vee \neg\wp)\ \mathrm{U}_{[0,5]}\ \wp_2)))$ expresses that for all the runs, always when $\wp_1$ holds, $\wp_2$ holds within 5 units of time.

Semantics of TCTL over TA and TPN's. Let $\mathcal{T}$ be a timed system, i.e., a TPN $\mathcal{N}$ or a TA $\mathcal{A}$, $M_c(\mathcal{T}) = (F_c(\mathcal{T}), V_c)$ be its concrete model, and $\rho = s_0 \stackrel{\delta_0}{\rightarrow} s_0 + \delta_0 \stackrel{b_0}{\rightarrow} s_1 \stackrel{\delta_1}{\rightarrow} s_1 + \delta_1 \stackrel{b_1}{\rightarrow} s_2 \stackrel{\delta_2}{\rightarrow} \ldots$ be an s_0-run of $\mathcal{T}$, where $\delta_i \in \mathbb{R}_+$ for $i \in \mathbb{N}$. Recall that $f_\mathcal{T}(s)$, for a concrete state s, denotes the set of all the progressive runs of $\mathcal{T}$ starting at s (where the time steps are labelled with $\delta > 0$ only). In order to interpret TCTL formulas along a run, we introduce the notion of a *dense path corresponding to* ρ, denoted by π_ρ, which is a mapping from $\mathbb{R}$ to a set of states[10], given by $\pi_\rho(r) = s_i + \delta$ for $r = \Sigma_{j=0}^{i}\delta_j + \delta$, with $i \geq 0$ and $0 \leq \delta < \delta_i$. Next, we define semantics of TCTL formulas in the following way:

$$
\begin{array}{ll}
s_0 \models \wp & \text{iff } \wp \in V_c(s_0), \\
s_0 \models \neg\wp & \text{iff } \wp \notin V_c(s_0), \\
s_0 \models \varphi \vee \psi & \text{iff } s_0 \models \varphi \text{ or } s_0 \models \psi, \\
s_0 \models \varphi \wedge \psi & \text{iff } s_0 \models \varphi \text{ and } s_0 \models \psi, \\
s_0 \models \mathrm{A}(\varphi \mathrm{U}_I \psi) & \text{iff } (\forall\, \rho \in f_\mathcal{T}(s_0))(\exists r \in I)\big[\pi_\rho(r) \models \psi \wedge (\forall r' < r)\, \pi_\rho(r') \models \varphi\big], \\
s_0 \models \mathrm{E}(\varphi \mathrm{U}_I \psi) & \text{iff } (\exists\, \rho \in f_\mathcal{T}(s_0))(\exists r \in I)\big[\pi_\rho(r) \models \psi \wedge (\forall r' < r)\, \pi_\rho(r') \models \varphi\big], \\
s_0 \models \mathrm{A}(\varphi \mathrm{R}_I \psi) & \text{iff } (\forall\, \rho \in f_\mathcal{T}(s_0))(\forall r \in I)\big[\pi_\rho(r) \models \psi \vee (\exists r' < r)\, \pi_\rho(r') \models \varphi\big], \\
s_0 \models \mathrm{E}(\varphi \mathrm{R}_I \psi) & \text{iff } (\exists\, \rho \in f_\mathcal{T}(s_0))(\forall r \in I)\big[\pi_\rho(r) \models \psi \vee (\exists r' < r)\, \pi_\rho(r') \models \varphi\big].
\end{array}
$$

Again, we adopt the initialized notion of validity in a model: $M_c(\mathcal{T}) \models \varphi$ iff $M_c(\mathcal{T}), s^0 \models \varphi$, where s^0 is the initial state in $M_c(\mathcal{T})$.

It is important to mention that there is an alternative semantics for sublogics of $\mathrm{CTL}_{-\mathrm{X}}$, which can be interpreted over dense models by assuming that semantics of each untimed modality O is like semantics of the corresponding TCTL modality $\mathrm{O}_{[0,\infty)}$. The *model checking* problem for TCTL over a timed system $\mathcal{T}$ is defined as usual, i.e., given a TCTL formula φ and a timed system together with a valuation function $V_\mathcal{T}$, determine whether $M_c(\mathcal{T}) \models \varphi$.

5 Verification Methods for TPN's and TA

Basic methods of verification of timed systems consist in building their finite (possibly minimal) *abstract models* that preserve properties of interest, and then in running verification algorithms on these models, for instance, a version of the *state-labelling algorithm* of [3], or an algorithm which checks an equivalence of behaviours between the model and its specification.

Let $M_c(\mathcal{T}) = ((S, s^0, \rightarrow), V_c)$ be a concrete model of a timed system $\mathcal{T}$, where $\rightarrow$ is a B-labelled transition relation for a given set of labels B. By an abstract model we mean a structure $M_a = ((W, w^0, \rightarrow), V)$, where elements of W, called *abstract states*, are sets of concrete states of S, and satisfy at least the following conditions: $s^0 \in w^0$; $V(w) = V_c(s)$ for each $w \in W$ and $s \in w$, and for any $w_1, w_2 \in W$ and each $b \in B$

EE) $w_1 \stackrel{b}{\rightarrow} w_2$ if there are $s_1 \in w_1$ and $s_2 \in w_2$ s.t. $s_1 \stackrel{b}{\rightarrow} s_2$.

---

[10] This can be defined thanks to the assumption about $\delta > 0$.

Condition EE guarantees that abstract states are related if they contain related representatives. Other conditions on $\rightarrow$ depend on the properties to be preserved. Some of them are listed below:

EA) $w_1 \xrightarrow{b} w_2$ iff $(\exists s_1 \in w_1)(\forall s_2 \in w_2)\ s_1 \xrightarrow{b} s_2$;
AE) $w_1 \xrightarrow{b} w_2$ iff $(\forall s_1 \in w_1)(\exists s_2 \in w_2)\ s_1 \xrightarrow{b} s_2$;
U) $w_1 \xrightarrow{b} w_2$ iff $(\forall s_1 \in w_1^{cor})(\exists s_2 \in w_2^{cor})\ s_1 \xrightarrow{b} s_2$,
where $w^{cor} \subseteq w$ for each $w \in W$, and $s^0 \in (w^0)^{cor}$.

Condition EA restricts the abstract transition relation to the pairs of states, where all the representatives of the successor state have a related representative in the predecessor state. This condition is put on abstract models to preserve LTL. Condition AE, which specifies the symmetric property w.r.t. the successor and the predecessor of each pair of states in the abstract transition relation, is known as a *bisimulation condition.* So, it is used to ensure preservation of CTL* or TCTL. Condition U is a weakening of AE, which puts the same restriction, but only on a subset of each abstract state. It is know as a *simulation condition.* Similarly, U is applied to preserve ACTL* or TACTL. Next, we present the main approaches to generating various kinds of abstract models for timed systems.

5.1 State Classes Approaches

Methods of building abstract models for time Petri nets are usually based on the *state classes approach,* which consists in representing an abstract state by a marking and a set of linear inequalities. Many algorithms for building such models, for various restrictions on the definition of TPN's and approaches to their concrete semantics, are known in the literature. Below, we present the main solutions. For each of them, we give either the method for obtaining the model from some earlier-defined underlying structure, or a full description, i.e., a definition of the concrete states applied, a notion of a state class, and a condition on firability of a transition at a class together with a method of computing its successor resulting from this firing. Moreover, in order to obtain a finite abstract model, an equivalence relation on state classes is provided. In all the descriptions, we use the following notation: given a set of inequalities I, $sol(I)$ denotes the set of its solutions, and $var(I)$ - the set of all the variables appearing in I.

State Class Graph [14, 15]. The method of building this basic model of the state classes approach was defined for general TPN's. Their concrete states are represented by pairs $\sigma^F = (m, f)$, where m is a marking, and f is a *firing interval function* which assigns to each $t \in en(m)$ the timing interval in which t is (individually) allowed to fire. A *state class* of a TPN is a pair $C = (m, I)$, where m is a marking, and I is a set of inequalities built over the set $var(I) = \{v_i \mid t_i \in en(m)\}$. All the possible values of a variable v_i in the set $sol(I)$ (called the *firing domain* of C) form the timing interval, relative to the time C was entered, in which t_i can be fired. The state class is thus a set of concrete states whose values of firing interval functions are included in the firing domain.

The initial class of the graph is given by $(m_0, \{\text{“}Eft(t_i) \leq v_i \leq Lft(t_i)\text{”} \mid t_i \in en(m_0)\})$. A transition $t_i \in en(m)$ is *firable* at a class $C = (m, I)$ if the set of inequalities $I_1 = I \cup \{\text{“}v_i \leq v_j\text{”} \mid t_j \in en(m)\}$ is consistent (i.e., $sol(I_1) \neq \emptyset$), which intuitively means that t_i can fire earlier than any other enabled transition. The class $C' = (m', I')$, resulting from firing t_i, satisfies $m' = m[t_i\rangle$, whereas I' is obtained from I by the following four steps: (1) the set I is substituted by I_1 (i.e., the firablity condition for t_i is added to I), (2) all the variables in I are substituted by new ones reflecting the fact of firing t_i at the time given by v_i which relates the values of the variables to the time the class C' was entered, (3) the variables corresponding to the transitions disabled by firing of t_i are eliminated, and (4) the system is extended by the set of inequalities $\{\text{“}Eft(t_l) \leq v_l \leq Lft(t_l)\text{”} \mid t_l \in newly\_en(m, t)\}$.

Two classes are considered as equivalent if their markings and firing domains are equal. A TPN is of a bounded number of state classes if and only if it is bounded. Although the boundedness problem for TPN's is undecidable, in [14, 15] sufficient conditions for checking unboundedness while generating the state class graph are given. Since all the state classes satisfy the condition EA, the model preserves LTL formulas.

Geometric Region Graph [89]. The information carried by firing domains of the classes of the state class graph is not sufficient to check their *atomicity*[11]. To this aim, some additional information about the histories of firings is needed. Therefore, as a step towards building a model preserving CTL* properties, a modification of the state class graph for 1-safe nets with finite values of the function Lft, called *geometric region graph*, has been introduced. Its construction exploits the notion of concrete states given in Sec. 2.1. State classes are triples $C = (m, I, \eta)$, where I is a set of inequalities, and m is a marking obtained by firing from m_0 the sequence $\eta \in T^*$ of transitions, satisfying the timing constraints represented by I. The variables in $var(I)$ represent absolute (i.e., counted since the net started) firing times of the transitions in η (different firings of the same transition are then distinguished, and $v_i^j \in var(I)$ corresponds to j-th firing of $t_i \in T$). Unlike the construction of [14, 15], I can be seen as describing the history of states of a given class rather than these states as such.

The initial state class is given by $C^0 = (m_0, \emptyset, \epsilon)$, where ϵ is the empty sequence of transitions. Firing of $t_i \in en(m)$ at a class $C = (m, I, \eta)$ is described in terms of the *parents* of the enabled transitions, i.e., the transitions is η which most recently made the transitions in $en(m)$ enabled[12]. If t_i appears $k-1$ times in η, then it is firable at C iff the set of inequalities $I_1 = I \cup \{\text{“}parent(v_i^k, C) + Eft(t) \leq parent(v_j^l, C) + Lft(t_j)\text{”} \mid t_j \in en(m)$ and t_j appears $l-1$ times in $\eta\}$ is consistent[13], which means that t_i can be fired earlier than any other enabled transition. In the class $C' = (m, I', \eta')$, obtained by firing t_i at C,

[11] A class which satisfies the condition AE w.r.t. all its successors is called *atomic*.

[12] As the parents of the transitions enabled in m_0, a fictitious transition ν, which denotes the start of the net and fires at the time 0, is assumed.

[13] $parent(v_i, C)$ denotes the variable corresponding to the most recent appearance of the parent of t_i in η.

$m' = m[t\rangle$, $\eta = \eta t$, and the set of inequalities I' is equal to $I_1 \cup \{$"$Eft(t_i) \leq t_i^k - parent(t_i^k, C) \leq Lft(t_i)$"$\}$. This describes timing conditions which need to hold for firing of t_i at C. The classes are equivalent if their markings are equal, the enabled transitions have the same parents, and these parents could be fired at the same absolute times. All the classes of the graph satisfy the condition EA.

Atomic State Class Graph [89]. In order to build a model which preserves CTL* properties, the geometric state class graph needs to be refined until all its classes are atomic. The model obtained this way is called *atomic state class graph.* If a class $C = (m, I, \eta)$ is not atomic, then there is some inequality ξ non-redundant[14] in I and such that satisfaction of ξ is necessary for the concrete states in C to have descendants in a successor C' of C. The class C is then split into $C_1 = (m, I \cup \xi, \eta)$ and $C_2 = (m, I \cup \neg\xi, \eta)$. Their descendants are computed from copies of those of C, by modifying their sets of inequalities adding ξ and $\neg\xi$, respectively. As a result, the abstract states satisfy both AE and EA.

Pseudo-atomic State Class Graph [61]. The atomic state class graph's construction was further modified to generate *pseudo-atomic class graphs* which preserve the universal fragment of CTL* (i.e., ACTL*). Instead of AE, all the classes of these graphs satisfy the weaker condition U. The models are built in a way similar to atomic state class graphs, besides the fact that in some cases instead of splitting the classes only their *cor*s are refined.

Strong State Class Graph [17]. The paper by Berthomieu and Vernadat presents another approach to building models which are then refined to a CTL*-preserving structure, applicable[15] to the general class of TPN's. The solution combines, in some sense, these of [89] and [14, 15]. The definition of concrete states is taken from [14]. The authors define a *strong state class graph*, whose classes are of the form $C = (m, I)$, where I is a set of inequalities built over the set of variables corresponding to the transitions in $en(m)$, similarly as in [14, 15]. However, the value of a variable v_i corresponding to a transition $t_i \in en(m)$ gives the time elapsed since t_i was last enabled, which, in turn, corresponds in some sense to Yoneda's approach. The set I enables to compute a firing domain, and in consequence to define the set of concrete states that belong to the class.

The initial class of the graph is given by $C^0 = (m_0, \{$"$0 \leq v_i \leq 0$" $\mid t_i \in en(m_0)\})$. Firability of a transition t at a class as well as the set of inequalities of the successor are defined in terms of the times elapsed since the transitions were enabled, using additional temporary variables denoting possible firing times of t (see [17] for details). Two classes (m, I) and (m', I') are considered as equivalent if $m = m'$ and $sol(I) = sol(I')$. Similarly as in the previous approaches, all the classes in the strong state class graph satisfy the condition EA.

Strong Atomic State Class Graph [17]. In order to obtain a *strong atomic state class graph* preserving CTL* properties, the above model is refined in a

---

[14] A inequality ξ is non-redundant in the set of inequalities I if the solution sets of both $I \cup \xi$ and $I \cup \neg\xi$ are non-empty.

[15] However, transitions with infinite latest firing times require a special treatment.

way similar to the one of [89], i.e., its classes are partitioned until all of them are atomic. However, unlike [89], the classes satisfy AE but not necessarily EA, since an inequality added to the set I of a class C, which is partitioned, is not propagated to the descendants of C.

The Graphs of [48, 90]. In [90], a construction of a state class graph, applicable to 1-safe nets with finite values of the function Lft, was described. The method aimed at verification of formulas of a logic TNL (*Timed Temporal Logic for Nets*). Since building of these graphs is formula-guided, we do not provide it here, but focus on a similar approach proposed by Lilius for 1-safe nets with possibly infinite values of Lft, aimed at reachability checking using partial order reductions[16] [48]. It is based on the notion of concrete states given in Sec. 2.1. A state class is a pair (m, I), where m is a marking, and I is a set of inequalities describing the constraints on possible firing times of the transitions in $en(m)$. Values of the variables in $var(I)$ represent absolute times. There are two kinds of variables in $var(I)$: these which correspond to possible firing times of transitions in $en(m)$, and those which represent firing times of some of the transitions that are disabled in m and have been fired earlier. The latter are kept to determine the latest possible firing time of these $t \in en(m)$ which can become enabled by firings of different transitions. The set of inequalities of the initial class expresses that the differences between the possible firing times of each $t \in en(m_0)$ and the start of the net are between $Eft(t)$ and $Lft(t)$. Given a class (m, I), the set of inequalities of its successor $(m[t\rangle, I')$ obtained by firing $t \in en(m)$ is computed similarly as in Yoneda's approach, besides the fact that the condition saying that t can be fired earlier than any other $t' \in en(m)$ is not added to I' (see [90] and [48] for a detailed description and comparison). As a result, the solutions of I do not necessarily determine a feasible clock assignment as it was in [90]. However, there is a run of the net corresponding to each path in the graph. Applying an equivalence of classes ensures finiteness of the model, but, unlike the case of [90], it is not of a form of a tree. This is useful for partial order reductions.

The Reachability Graph of [22]. A completely different approach to state classes, aimed at computing reachability graphs of (general) time Petri nets, was proposed in [22]. It is based on a notion of *firing points* of transitions (defined as the instant when a given transition fires). The class produced by the $(n-1)$-th firing point (the points are numbered along sequences of transitions fired from m_0) is a triple (m, TS, TE), where m is a marking, $TS : \mathbb{N} \rightarrow en(m)$ gives the enabling order of transitions (the transitions of $TS(0)$ become enabled before those of $TS(1)$ etc.), whereas $TE : \mathbb{N} \times \mathbb{N} \rightarrow \mathbb{Q}_+ \cup \{\infty\}$ is a function which for a given pair i, j returns the minimal or the maximal time elapsed between the i-th and the j-th firing point at the point $n-1$. The resulting graph is finite iff the net is bounded.

5.2 Detailed Region Graph Approach

Next, we define abstract models for timed automata, which can be used for enumerative and symbolic verification. We start with a detailed region graph

[16] Firing transitions in different orders can lead to the same state.

approach. The main reason for introducing this approach is that our SAT-related methods (see Section 6) are based on a propositional encoding of detailed regions.

Given a timed automaton $\mathcal{A}$ and a TCTL formula φ. Let $\mathcal{C}_{\mathcal{A}} \subseteq \mathcal{C}_{\mathcal{X}}$ be a non-empty set containing all the clock constrains occurring in any enabling condition or in a state invariant of $\mathcal{A}$. Moreover, let $c_{max}(\varphi)$ be the largest constant appearing in $\mathcal{C}_{\mathcal{A}}$ and in any time interval in φ[17]. For $\delta \in \mathbb{R}$, $frac(\delta)$ denotes the fractional part of δ, and $\lfloor \delta \rfloor$ denotes its integral part.

Definition 6 (Equivalence of clock valuations). *For two clock valuations $v, v' \in \mathbb{R}^{n_{\mathcal{X}}}$, $v \simeq_{\mathcal{C}_{\mathcal{A}},\varphi} v'$ iff for all $x, x' \in \mathcal{X}$ the following conditions are met:*

1. *$v(x) > c_{max}(\varphi)$ iff $v'(x) > c_{max}(\varphi)$,*
2. *if $v(x) \leq c_{max}(\varphi)$ and $v(x') \leq c_{max}(\varphi)$ then*
   a.) $\lfloor v(x) \rfloor = \lfloor v'(x) \rfloor$,
   b.) $frac(v(x)) = 0$ iff $frac(v'(x)) = 0$, and
   c.) $frac(v(x)) \leq frac(v(x'))$ iff $frac(v'(x)) \leq frac(v'(x'))$.

Lemma 1 (Preserving TCTL, **[3]).**
Let $\mathcal{A} = (A, L, l^0, E, \mathcal{X}, \mathcal{I})$ be a timed automaton, $V_{\mathcal{A}}$ be a valuation function for $\mathcal{A}$, and $M_c(\mathcal{A})$ be the concrete model of $\mathcal{A}$. Moreover, let $l \in L$, and $v, v' \in \mathbb{R}^{n_{\mathcal{X}}}$ with $v \simeq_{\mathcal{C}_{\mathcal{A}},\varphi} v'$. Then, we have $M_c(\mathcal{A}), (l, v) \models \varphi$ iff $M_c(\mathcal{A}), (l, v') \models \varphi$.

The equivalence classes of the relation $\simeq_{\mathcal{C}_{\mathcal{A}},\varphi}$ are called *detailed zones*. The set of all the detailed zones is denoted by $DZ(n_{\mathcal{X}})$. A *detailed region* is a pair (l, Z), where $l \in L$ and $Z \in DZ(n_{\mathcal{X}})$.

The action and time successor relation in the set of the detailed regions can be defined via representatives (see [3, 63]), which gives us a finite abstract model, called the *detailed region graph* (DRG) preserving TCTL. In Section 6.2, we show how to encode detailed regions together with the transition relation in a symbolic way to accomplish bounded model checking.

5.3 Partition Refinement for TA

Partition refinement (minimization) is an algorithmic method of constructing abstract models for timed automata *on-the-fly*, i.e., without building concrete models first. Typically, the algorithm starts from an initial *partition* Π_0 of the state space Q of $\mathcal{A}$ (i.e., a set of disjoint *classes* the union of which equals Q), which respects (at least) the propositions of interest. The partition is then successively refined until it (or its reachable part, depending on the algorithm) becomes *stable*, i.e., satisfies the conditions required on the model to be generated. Let $M = (G, V)$ with $G = (W, w^0, \rightarrow)$ be a model generated by the above algorithm, where the elements of W are (reachable) classes of the stable partition Π, w^0 is the class containing the state q^0 of $\mathcal{A}$, and the successor relation is induced by that on Π. The classes of partitions are usually represented by regions, defined as follows. Given a timed automaton $\mathcal{A}$, let $l \in L$ and $Z \in Z(n_{\mathcal{X}})$.

---

[17] Obviously, ∞ is not considered as a constant.

A *region* $R \subseteq L \times \mathbb{R}^{n_{\mathcal{X}}}$ is a set of concrete states $R = \{(l, v) \mid v \in Z\}$, denoted by (l, Z). For regions $R = (l, Z)$ and $R' = (l, Z')$, we define their difference $R \setminus R' = \{(l, Z'') \mid Z'' \in Z \setminus Z'\}$. This operation potentially returns a set of regions, and is of exponential complexity in the number of clocks, which can cause some inefficiency of partition refinement algorithms.

Bisimulating Models. Three main minimization algorithms were introduced in [20, 59, 47]. They are aimed at generating *bisimulating models*, i.e., models whose classes satisfy the condition AE. The set of labels B usually corresponds to the set of actions of the automaton augmented with one additional label τ which denotes passing some time[18].

An adaptation of the method of [47] to the case of timed automata was presented in [87]. The algorithm starts from an initial partition Π_0 which respects the invariants and enabling conditions of $\mathcal{A}$ (i.e., for each $Y \in \Pi$, each $e = l' \xrightarrow{a,cc,X} l'' \in E$ and each $l \in L$ we have $Y \cap [\![\mathcal{I}(l)]\!], Y \cap [\![cc]\!] \in \{Y, \emptyset\}$), and stabilizes only the reachable classes. To ensure this, each reachable class is *marked* by a representative that is guaranteed to be a reachable state of $\mathcal{A}$ (initially, only the initial class is marked by q^0). If there exists a transition from a representative of the reachable class Y_1 to a state in a class Y_2, then Y_2 becomes reachable, and is marked by one of its known reachable states. If Y_1 is not stable, then it is partitioned into a class Y_1' (easily computed using the inverse images of the successors), which contains a representative r_{Y_1} and all the concrete states of Y_1 which have their successors in the same classes as r_{Y_1}, and $Y_1 \setminus Y_1'$ (which possibly is a set of classes). The concrete states known as reachable in $Y_1 \setminus Y_1'$ (if exist) are chosen as the representatives of (parts of) $Y_1 \setminus Y_1'$. Partitioning of Y_1 can make its predecessors unstable. The paper proposes also a solution to the problem of computing differences of regions. It is based on using a *forest structure*, consisting of trees corresponding to the locations of $\mathcal{A}$, which keep the history of partitionings.

The algorithm of [20] was applied to timed automata in many papers, and served to building bisimulating models for various time-abstracted (*ta*-) bisimulation relations. These are *strong ta-bisimulation* abstracting away the exact amount of time passed between two states, and *delay* and *observational* ta-bisimulations, which additionally consider as equivalent the states obtained by executing an action a and these obtained by passing some time and then executing a, or, respectively, passing some time, executing a, and then passing some time again (see [79] for details). Its generic pseudo-code is presented in Fig. 3. The algorithm starts from an initial class containing the state q^0, and then successively searches and refines reachable classes. A class $Y_1 \in \Pi$ which is unstable w.r.t. its successor Y_2 is split into Y_1' which contains all the concrete states which have successors in Y_2, and $Y_1 \setminus Y_1'$ (this can possibly be a set of regions). Partitioning of a class can result in unstability of its predecessors.

In [5], the algorithm of Fig. 3 was applied to generating abstract models for delay ta-bisimulation. The work [4] describes its implementation for the case of

[18] The actions of $\mathcal{A}$ can be taken as labels only, if a *discrete model* is to be built.

```
1.   Π := Π_0; reachable := {initial_class}; stable := ∅;
2.   while (∃X ∈ reachable \ stable) do
3.       C_X := Split(X, Π);
4.       if (C_X = {X}) then
5.           stable := stable ∪ {X};
6.           reachable := reachable ∪ the_successors_of_X_in_Π;
7.       else
8.           Y_X := {Y ∈ Π | Y has been split };
9.           reachable := reachable \ Y_X ∪ {Y ∈ C_X | initial_state ∈ Y};
10.          stable := stable \ the_predecessors_of_elements_of_Y_X_in_Π;
11.          Π := (Π \ Y_X) ∪ C_X;
12.      end if;
13.  end do;
```

Fig. 3. A generic minimization algorithm

strong time-abstracted bisimulation. A solution to the problem of computing differences of classes while generating a strong time-abstracted bisimulating model was presented in [79]. Similarly to that of [87], it consists in starting from an initial partition of the state space respecting the invariants and enabling conditions of $\mathcal{A}$. An unstable class Y is refined simultaneously w.r.t. all its time-, or action a-successors (on a given action a). Since their inverse images form a partition of Y, differences do not need to be computed.

The algorithm of Fig. 3 was modified to build other kinds of abstract models, preserving more restricted classes of properties. The solutions differ in the definitions of the partition of Q, the stability condition and the function *Split* used to refine unstable classes. Below, we sketch the main of them.

Simulating Models. The paper [32] shows a technique of building *simulating models*, preserving ACTL* properties. Comparing with bisimulating models, the condition on the successor relation between classes is relaxed, i.e., for each two reachable classes the condition U needs to be satisfied. The algorithm operates on a *cor*-partition $\Pi \subseteq 2^Q \times 2^Q$, consisting of pairs (Y, Y^{cor}) whose first elements form a partition of Q. Unstability of a class w.r.t. its successor can result in splitting one or both the classes, or modifying their *cors*. The problem of computing differences for these models (for any ta-simulation) has not been solved so far.

Pseudo-bisimulating Models. The paper [65] introduces *pseudo-bisimulating models*, preserving reachability properties. The main idea behind the definition consists in relaxing the condition on the transition relation on bisimulating models, formulated for all the predecessors of each abstract state, such that it applies only to one of them, reachable from the beginning state in the minimal number of steps. In this case, the algorithm deals with a *d-partition* $\Pi \subseteq 2^Q \times (\mathbb{N} \cup \{\infty\})$, which is a set of pairs $(Y, dpt(Y))$, whose first elements build a partition of Q. Unstability of a class w.r.t. its successor $(Y, dpt(Y))$ results in refining a predecessor of $(Y, dpt(Y))$ which is assumed to be reachable in the minimal number of steps. The models are usually built for the strong ta-bisimulation, using the method of avoiding computing differences given in [79].

Pseudo-simulating Models. Both the solutions for simulating and pseudo-bisimulating models were combined, resulting in a definition of reachability-preserving *pseudo-simulating models* [66].

Other Minimization Techniques. Besides of the methods based on the minimization algorithms presented above, some other solutions exist. The paper [76] describes a partition refinement technique, operating on a product of the specifications of a system and a property, and exploiting splitting histories, whereas [44] presents a method of building reachability-preserving abstract models, exploiting (timed and untimed) histories of concrete states.

5.4 Forward-Reachability Graphs for TA

Verification based on reachability analysis is usually performed on an abstract model known as *simulation graph* or *forward-reachability graph*. The nodes of this graph can be defined as (not necessarily convex) sets of detailed regions [21] or as regions [30, 46, 88]. Usually, the latter approach is used, which follows from a convenient representation of zones by Difference Bound Matrices [34]. In this case, the simulation graph can be defined as the smallest graph $G = (W, w^0, \to)$ such that

- $w^0 = (l^0, Z^0)$ with $Z^0 = v^0 \nearrow \cap [\![\mathcal{I}(l^0)]\!]$;
- for any $a \in A$ such that $e : l \xrightarrow{a,cc,X} l' \in$ E, and any $w = (l, Z) \in W$, if $Succ_a((l, Z)) \neq \emptyset$, then $w' = Succ_a((l, Z)) \in W$ and $w \xrightarrow{a} w'$, for $Succ_a((l, Z)) = (l', ((Z \cap [\![cc]\!])[X := 0]) \nearrow \cap [\![\mathcal{I}(l')]\!])$.

The graph is usually generated using a forward-reachability algorithm which, starting from w^0, successively computes all the successors $Succ_a(w)$ for all $w \in W$ generated in earlier steps. The process can be terminated if a state satisfying a given property is reached before the whole graph is built. This is called *on-the-fly* reachability analysis. The simulation graph of [21] is finite, since the number of detailed regions is so. However, when generated in the above-presented manner, finiteness of the graph needs to be ensured. This is done by applying an *extrapolation abstraction* whose general idea is based on the fact that for any constraint of $\mathcal{A}$ involving a clock x, the exact value of $v(x)$ is insignificant if greater than the maximal value this clock is compared with (see [13, 30] for details). Other abstractions, enabling to reduce the size of the graph still enabling reachability checking, were presented in [30, 46, 88]. There are also many solutions aimed at reducing the memory usage while generating models [12, 30, 46].

5.5 On-the-Fly Verification for TA

When the methods described above are used for verifying temporal properties, they require, in principle, to build a model first, and then to check a formula over this model. This can obviously make verification infeasible, especially when the size of the model is prohibitive. Therefore, there are approaches which offer on-the-fly solutions, i.e., a formula is checked over a model while its construction. Bouajjanni et al [21] define an algorithm, which starts with building a simulation graph for a TA. Then, cycles of this graph are refined. This process is guided by a formula. If a stable cycle is found, then this means that the formula holds and at that point the verification ends.

Another solution has been suggested by Dickhofer and Wilke [33] and Henzinger et al. [45]. The idea follows the standard approach to automata-theoretic model checking for CTL. So, first an automaton accepting all the models for a TCTL formula is built and the product of this automaton with the automaton corresponding to the detailed region graph is constructed while its non-emptiness is checked [33]. The method of [45] is slightly different as the product is constructed without building the automaton for a formula first.

6 Symbolic Data Structures and Verification

To store and operate on abstract models usually Difference Bound Matrices [34] are used for state classes of TPN's [35] or regions of TA [34]. To our knowledge, there are very few approaches to BDD- or SAT-based verification of TPN's [19], which mainly consist in describing the state space of a net in a way which enables to use an existing symbolic tool. However, such approaches to verification of untimed Petri nets are known [26, 29, 52, 53, 60, 62, 77], but a discussion of them goes beyond the scope of this paper. Therefore, in what follows we consider symbolic data structures used for verification of TA. To this aim we list BDD-based methods and focus on the most recent SAT-related techniques for TA.

6.1 BDD- and SAT-Based Methods for TA – Overview

BDD-Based Methods. A standard symbolic approach to representation of state spaces and model checking of untimed systems is based on Binary Decision Diagrams (BDD's) [24]. A similar approach is to apply BDD's for encoding discretizations of TA using the so-called Numeric Decision Diagrams [9]. Discretizations of TA can be also implemented using propositional formulas (see Section 6.2).

Another approach follows the solution suggested by Henzinger et al. [40], where the characteristic function of a set of states is a formula in separation logic (SL)[19]. SL formulas can be represented using Difference Decision Diagrams (DDD's) [54, 55]. A DDD is a data structure using separation predicates with the ordering of predicates induced by the ordering of the clocks. A similar approach is taken in [73], where a translation from quantified SL to quantified boolean logic [78] is exploited.

One can use also Clock Difference Diagrams (CDD's) to symbolically represent unions of regions [11]. Each node of a CDD is labelled with the difference

[19] SL extends the propositional logic by clock constraints.

of clock variables, whereas the outgoing edges with intervals bounding this difference. Alternatively, model checkers are based on data structures called Clock Restriction Diagrams (CRD) [85]. CRD is like CDD except for the fact that for each node the outgoing edges are labelled with an upper bound rather than with the interval of the corresponding difference of clock variables.

Another recent symbolic approach has been motivated by a dramatic increase in efficiency of SAT-solvers, i.e., algorithms solving the satisfiability problem for propositional formulas [93].

SAT-Based Methods. The main idea of SAT-based methods consists in translating the model checking problem for a temporal logic to the problem of satisfiability of a formula in propositional or separation logic. This formula is typically obtained by combining an encoding of the model and of the temporal property. In principle, there are two different approaches. In the first one, a model checking problem for TCTL (or reachability properties) is translated to a formula in separation logic [10, 57, 75] or quantified separation logic [73] and then either solved by MathSAT[20] or translated further to propositional logic and solved by SAT-solver. The second approach exploits a translation of the model checking problem from TCTL to CTL and then further to a propositional formula [63].

On the other hand, the approaches to SAT-based symbolic verification can be viewed as unbounded (UMC) or bounded (BMC). UMC [73] is for unrestricted TCTL (or timed μ-calculus) on the whole model, whereas BMC [57, 63] applies to an existential fragment of TCTL (i.e., TECTL) on a part of the model. However, it is possible to use the bounded approach for verifying universal properties as well, which is shown for unreachability properties in [92]. In what follows, we focus on BMC for TECTL as well as for unreachability properties of TA.

6.2 BMC for Timed Automata

BMC consists in translating the model checking problem of an existential TCTL formula (i.e., containing only existential quantifiers) to the problem of satisfiability of a propositional formula. This translation is based on bounded semantics satisfaction, which, instead of using possibly infinite paths, is limited to finite prefixes only. Moreover, it is known that the translation of the existential path quantifier can be restricted to finitely many computations [62].

In this section we describe how to apply BMC to TECTL. The main idea of our method consists in translating the TECTL model checking problem to the model checking problem for a branching time logic [3, 79] and then in applying BMC for this logic [62].

We start with showing a discretization of TA and a translation from TCTL to slightly modified CTL (called $\mathrm{CTL_y}$) on such discretized models. Then, we present BMC for $\mathrm{CTL_y}$, and a BMC method of checking unreachability for TA.

[20] MathSAT is a solver checking satisfiability of SL.

Discretization of TA. We define a discretized model for a timed automaton, which is based on the discretization of [92]. The idea behind this method is to represent detailed zones of a timed automaton by one or more (but finitely many) specially chosen representatives.

Let $\mathcal{A} = (A, L, l^0, E, \mathcal{X}, \mathcal{I})$ be a timed automaton with $n_{\mathcal{X}}$ clocks, $V_{\mathcal{A}}$ be a valuation function, and φ be a TCTL formula. As before, let $M_c(\mathcal{A}) = (F_c(\mathcal{A}), V_c)$ be the concrete model for $\mathcal{A}$. We choose the discretization step $\Delta = 1/d$, where d is a fixed even number[21] greater than $2n_{\mathcal{X}}$. The *discretized clock space* is defined as $\mathbb{D}^{n_{\mathcal{X}}}$, where $\mathbb{D} = \{k\Delta \mid 0 \leq k \leq 2c_{max}(\varphi) + 2\}$. This means that the clocks cannot go beyond $2c_{max}(\varphi) + 2$, which follows from the fact that for evaluating the TCTL formula φ over diagonal-free timed automata we do not need to distinguish between clock valuations above $c_{max}(\varphi) + 1$. Similarly, the maximal values of time delays can be restricted to $c_{max}(\varphi) + 1$, since otherwise they would make the values of clocks greater than $c_{max}(\varphi) + 1$. Thus, the set of values that can change a valuation in a detailed zone, is defined as $\mathbb{E} = \{k\Delta \mid 0 \leq k\Delta < c_{max}(\varphi) + 1\}$[22]. To make sure that the above two definitions can be applied we will guarantee that before any time transition, the value of every clock does not exceed $c_{max}(\varphi) + 1$.

Next, we define the set $\mathbb{U}$ of valuations that are used to 'properly' represent detailed zones in the discretized model, i.e., we take a subset of the valuations v of $\mathbb{D}^{n_{\mathcal{X}}}$ that preserve time delays by insisting that either the values of all the clocks in v are only even or only odd multiplications of Δ. To preserve action successors we will later use 'adjust' transitions. The set $\mathbb{U}$ is defined as follows:

$$\mathbb{U} = \{u \in \mathbb{D}^{n_{\mathcal{X}}} \mid (\forall x \in \mathcal{X})(\exists k \in \mathbb{N}) u(x) = 2k\Delta \vee (\forall x \in \mathcal{X})(\exists k \in \mathbb{N}) u(x) = (2k+1)\Delta\}$$

Now, we are ready to define a *discretized model* for $\mathcal{A}$, that is later used for checking reachability and unreachability of a propositional formula p, which, as we already mentioned, is expressed by the formula CTL $\varphi = \mathrm{EF}p$ interpreted over runs with unrestricted time-successor steps. Note that in this case $c_{max}(\varphi)$ depends only on $\mathcal{A}$.

Definition 7 (Discretized Model). *The* discretized model *for $\mathcal{A}$ is a structure $DM(\mathcal{A}) = ((S, s^0, \rightarrow_d), V_d)$, where $S = L \times \mathbb{D}^{n_{\mathcal{X}}}$, $s^0 = (l^0, v^0)$ is the initial state, the labelled transition relation $\rightarrow_d \subseteq S \times (A \cup \mathbb{E} \cup \{\epsilon\}) \times S$ is defined as*

1. *$(l, v) \xrightarrow{a}_d (l', v')$ iff $(l, v) \xrightarrow{a} (l', v')$ in $F_c(\mathcal{A})$, for $a \in A$, (action transition)*
2. *$(l, v) \xrightarrow{\delta}_d (l, v')$ iff $(\forall x \in \mathcal{X})(v(x) \leq c_{max}(\varphi) + 1)$, $v' = v + \delta$ and $v' \in [\![\mathcal{I}(l)]\!]$, for $\delta \in \mathbb{E}$ (time delay transition),*
3. *$(l, v) \xrightarrow{\epsilon}_d (l, v')$ iff $v' \in \mathbb{U}^{n_{\mathcal{X}}}$, $(\forall x \in \mathcal{X})(v'(x) \leq c_{max}(\varphi) + 1)$, and $v \simeq_{C_{\mathcal{A},\varphi}} v'$ (adjust transition).*

and the valuation function $V_d : S \rightarrow 2^{PV}$ is given by $V_d((l, v)) = V_{\mathcal{A}}(l)$.

[21] A good choice for d is the minimal such a number, which equals to 2^l for some l.
[22] By $\mathbb{E}_+$ we denote $\mathbb{E} \setminus \{0\}$.

Notice that the transitions in $DM(\mathcal{A})$ are labelled with actions of A, time delays of $\mathbb{E}$, or the epsilon label $\epsilon \notin A \cup \mathbb{E}$. The first two types of labels correspond exactly to labels used in the concrete model. The adjust transitions are used for moving within detailed zones to a valuation in $\mathbb{U}^{n_\mathcal{X}}$. The reason for defining action and adjust transitions separately consists in increasing efficiency of the implementation for checking reachability in timed automata.

However, for verification of more complex TCTL formulas φ we need to put some restrictions on $DM(\mathcal{A})$ by defining the restricted discretized (r-discretized) model $DM_R(\mathcal{A}) = ((S', s^0, \rightarrow'_d), V'_d)$, where $S' = L \times \mathbb{U}^{n_\mathcal{X}}$, $V'_d = V_d \cap S'$, and the transition relation $\rightarrow'_d \subseteq S' \times (A \cup \{\tau\}) \times S'$ is given as

1. $(l, v) \xrightarrow{\tau}{}'_d (l, v')$ iff $(l, v) \xrightarrow{\delta}_d (l, v')$ for some $\delta \in \mathbb{E}_+$, and if $(l, v) \xrightarrow{\delta'}_d (l, v'') \xrightarrow{\delta''}_d (l, v')$ with $\delta', \delta'' \in \mathbb{E}$ and $(l, v'') \in S$, then $v \simeq_{\mathcal{C}_{\mathcal{A},\varphi}} v''$ or $v' \simeq_{\mathcal{C}_{\mathcal{A},\varphi}} v''$, and if $v \simeq_{\mathcal{C}_{\mathcal{A},\varphi}} v'$, then $v \simeq_{\mathcal{C}_{\mathcal{A},\varphi}} v' + \delta''$ for each $\delta'' \in \mathbb{E}_+$ (time successor),
2. $(l, v) \xrightarrow{a}{}'_d (l', v')$ iff (l, v) is not boundary[23] and $((l, v) \xrightarrow{a}_d; \xrightarrow{\epsilon}_d (l', v')$ or $(l, v) \xrightarrow{\tau}{}'_d; \xrightarrow{a}_d; \xrightarrow{\epsilon}_d (l', v'))$, for $a \in A$ (action successor).

Intuitively, time successor corresponds to a move by time delay transition with the smallest δ to another region (if not final), whereas action successor corresponds to a move by action transition (adjusted by ϵ-transition to be in $\mathbb{U}^{n_\mathcal{X}}$), taken from non-boundary regions, possibly preceded by the time successor step. Since our definition is based on the notion of the detailed region graph [3], it is easy to notice that $DM_R(\mathcal{A})$ is its discretization, and as such, it can be used for checking the TCTL formula φ.

Translation from TCTL to CTL. Rather than showing BMC directly for TECTL over timed automata, which is quite a complex task, we first discuss a translation from TCTL to a slightly modified CTL ($\mathrm{CTL_y}$) and then BMC for $\mathrm{ECTL_y}$. In general, the model checking problem for TCTL can be translated to the model checking problem for a fair version of CTL [3]. However, since we have assumed that we deal with progressive timed automata only, we can define a translation to the $\mathrm{CTL_y}$ model checking problem [79].

The idea is as follows. Given a timed automaton $\mathcal{A}$, a valuation function $V_\mathcal{A}$, and a TCTL formula φ. First, we extend $\mathcal{A}$ with a new clock, action, and transitions to obtain an automaton $\mathcal{A}_\varphi$. The aim of the new transitions is to reset the new clock, which corresponds to all the timing intervals appearing in φ. These transitions are used to start the runs over which subformulas of φ are checked. Then, we construct the r-discretized model for $\mathcal{A}_\varphi$ and augment its valuation function. Finally, we translate the TECTL formula φ to an $\mathrm{ECTL_y}$ formula $\psi = \mathrm{cr}(\varphi)$ such that model checking of φ over the r-discretized model of $\mathcal{A}$ can be reduced to model checking of ψ over the r-discretized model of $\mathcal{A}_\varphi$.

Formally, let $\mathcal{X}$ be the set of clocks of $\mathcal{A}$, and $\{I_1, \ldots, I_r\}$ be a set of the non-trivial intervals appearing in φ. The automaton $\mathcal{A}_\varphi$ extends $\mathcal{A}$ such that

[23] A state (l, v) is boundary if for any $\delta \in \mathbb{E}_+$, $\neg(v \simeq_{\mathcal{C}_{\mathcal{A},\varphi}} v + \delta)$

- the set of clocks $\mathcal{X}' = \mathcal{X} \cup \{y\}$,
- the set of actions $A' = A \cup \{a_y\}$,
- the transition relation $E' \subseteq L \times A' \times \mathcal{C}_{\mathcal{X}'} \times 2^{\mathcal{X}'} \times L$ is defined as follows $E' = E \cup \{l \stackrel{a_y, true, \{y\}}{\longrightarrow} l \mid l \in L\}$.

Let $DM_R(\mathcal{A}_\varphi) = ((S, s^0, \rightarrow_d), V_d)$ be the r-discretized model for $\mathcal{A}_\varphi$. Denote by $\rightarrow_\mathcal{A}$ the part of $\rightarrow_d$, where transitions are labelled with elements of $A \cup \{\tau\}$, and by $\rightarrow_y$ the transitions that reset the clock y, i.e., labelled with a_y.

Next, we extend the set of propositional variables PV to PV' and the valuation function V_d to V. By $\wp_{y \in I_i}$ we denote a new proposition for every interval I_i appearing in φ, and by PV_φ the set of the new propositions. The proposition $\wp_{y \in I_i}$ is true at a state (l, v) of $DM_R(\mathcal{A}_\varphi)$ if $v(y) \in I_i$. Let V_φ be a function labelling each state of $DM_R(\mathcal{A}_\varphi)$ with the set of propositions from PV_φ true at that state and labelling each boundary state with $\wp_b$. Next, set $PV' = PV \cup PV_\varphi \cup \{\wp_b\}$ and define the valuation function $V : S \rightarrow 2^{PV'}$ as $V = V_d \cup V_\varphi$.

In order to translate a TCTL formula φ to the corresponding CTL formula ψ we need to modify the language of CTL to $\mathrm{CTL_y}$ by reinterpreting the next-time operator, denoted now by X_y. This language is interpreted over r-discretized models for $\mathcal{A}_\varphi$, defined above, where we assume that r is the number of non-trivial intervals appearing in φ. The modality X_y is interpreted only over the new transitions that reset the new clock y, whereas the other operators are interpreted over all the transitions except for the new ones. Formally, for $\wp \in PV'$, the set of $\mathrm{CTL_y}$ formulas is defined inductively as follows:

$$\alpha := \wp \mid \neg\wp \mid \alpha \wedge \alpha \mid \alpha \vee \alpha \mid \mathrm{X}_y\alpha \mid \mathrm{E}(\alpha \mathrm{U} \alpha) \mid \mathrm{E}(\alpha \mathrm{R} \alpha) \mid \mathrm{A}(\alpha \mathrm{R} \alpha) \mid \mathrm{A}(\alpha \mathrm{U} \alpha).$$

A *path* in $DM_R(\mathcal{A}_\varphi)$ is a maximal sequence $\pi = (s_0, s_1, \ldots)$ of states such that $s_i \rightarrow_\mathcal{A} s_{i+1}$ for each $i \in \mathbb{N}$. The relation $\models$ is defined like in Section 4.1 for all the $\mathrm{CTL_y}$ formulas, except for X_y, which is given as follows: $(l, v) \models \mathrm{X}_y\alpha$ iff $(l, v[\{y\} := 0)] \models \alpha$. Next, the TCTL formula φ is translated inductively to the $\mathrm{CTL_y}$ formula $\mathrm{cr}(\varphi)$ as follows:

- $\mathrm{cr}(\wp) = \wp$ for $\wp \in PV'$,
- $cr(\neg\wp) = \neg cr(\wp)$,
- $\mathrm{cr}(\alpha \vee \beta) = \mathrm{cr}(\alpha) \vee \mathrm{cr}(\beta)$,
- $\mathrm{cr}(\alpha \wedge \beta) = \mathrm{cr}(\alpha) \wedge \mathrm{cr}(\beta)$,
- $\mathrm{cr}(\mathrm{O}(\alpha \mathrm{U}_{I_i} \beta)) = \mathrm{X}_y\big(\mathrm{O}(\mathrm{cr}(\alpha)\mathrm{U}(\mathrm{cr}(\beta) \wedge \wp_{y \in I_i} \wedge (\wp_b \vee \mathrm{cr}(\alpha))))\big)$,
- $\mathrm{cr}(\mathrm{O}(\alpha \mathrm{R}_{I_i} \beta)) = \mathrm{X}_y\big(\mathrm{O}(\mathrm{cr}(\alpha)\mathrm{R}(\neg\wp_{y \in I_i} \vee (\mathrm{cr}(\beta) \wedge (\wp_b \vee \mathrm{cr}(\alpha)))))\big)$, for $\mathrm{O} \in \{\mathrm{E}, \mathrm{A}\}$.

It is easy to show that the validity of the TCTL formula φ over the concrete model of $\mathcal{A}$ is equivalent to the validity of the corresponding $\mathrm{CTL_y}$ formula $\mathrm{cr}(\varphi)$ over the r-discretized model of $\mathcal{A}_\varphi$ with the extended valuation function [3].

Next, we show a BMC method for $\mathrm{ECTL_y}$ over r-discretized models for TA. Since we have defined a translation from TCTL to $\mathrm{CTL_y}$, we obtain a BMC method for TECTL.

BMC for ECTL$_y$. In this section we present a SAT-based approach to ECTL$_y$ model checking over r-discretized models for timed automata, to which we refer as to models from now on. We start with giving a *bounded semantics* for ECTL$_y$ in order to define the *bounded model checking problem* and to translate it subsequently into a satisfiability problem [62].

Let φ be a TECTL formula, $\psi = \mathrm{cr}(\varphi)$, and $M = ((S, s^0, \rightarrow_d), V)$ be a r-discretized model for $\mathcal{A}_\varphi$ with the extended valuation function.

We start with some auxiliary definitions. For $k \in \mathbb{N}_+$ a *k-path* in M is finite sequence of $k+1$ states $\pi = (s_0, s_1, \ldots, s_k)$ such that $(s_i, s_{i+1}) \in \rightarrow_{\mathcal{A}}$ for each $0 \leq i \leq k$. For a k-path $\pi = (s_0, s_1, \ldots, s_k)$, let $\pi(i) = s_i$ for each $i \leq k$. By $\Pi_k(s)$ we denote the set of all the k-paths starting at s. This is a convenient way of representing a k-bounded subtree rooted at s of the tree resulting from unwinding the model M from s.

Definition 8 (k-model). *A k-model for M is a structure $M_k = ((S, s^0, P_k), V)$, where P_k is the set of all the k-paths of M, i.e., $P_k = \bigcup_{s\in S} \Pi_k(s)$.*

Define a function $loop : P_k \rightarrow 2^{\mathbb{N}}$ as: $loop(\pi) = \{l \mid 0 \leq l \leq k \wedge \pi(k) \rightarrow_{\mathcal{A}} \pi(l)\}$. Satisfaction of the temporal operator R on a k-path π in the bounded case can depend on whether or not π represents a path[24], i.e., $loop(\pi) \neq \emptyset$,

The main reason for reformulating the semantics of the modalities in the following definition in terms of elements of k-paths rather than elements of S or Π is to restrict the semantics to a part of the model.

Definition 9 (k-bounded semantics for ECTL$_y$). *Let M_k be a k-model and α, β be ECTL$_y$ subformulas of ψ. $M_k, s \models \alpha$ denotes that α is true at the state s of M_k. M_k is omitted if it is clear from the context. The relation $\models$ is defined inductively as follows:*

$$
\begin{array}{ll}
s \models \wp & \textit{iff } \wp \in V(s) \\
s \models \neg\wp & \textit{iff } \wp \notin V(s), \\
s \models \alpha \wedge \beta & \textit{iff } s \models \alpha \textit{ and } s \models \beta, \\
s \models \alpha \vee \beta & \textit{iff } s \models \alpha \textit{ or } s \models \beta, \\
s \models \mathrm{X}_y\alpha & \textit{iff } \exists s' \in S\left(s \rightarrow_y s' \textit{ and } s' \models \alpha\right), \\
s \models \mathrm{E}(\alpha \mathrm{U} \beta) & \textit{iff } \exists \pi \in \Pi_k(s)\Big(\exists_{0\leq j \leq k}\Big(\pi(j) \models \beta \textit{ and } \forall_{0 \leq i < j}\, \pi(i) \models \alpha\Big)\Big), \\
s \models \mathrm{E}(\alpha \mathrm{R} \beta) & \textit{iff } \exists \pi \in \Pi_k(s)\Big(\Big(\exists_{0\leq j \leq k}\Big(\pi(j) \models \alpha \textit{ and } \forall_{0 \leq i < j}\, \pi(i) \models \beta\Big)\Big) \textit{ or } \\
& \quad \Big(\forall_{0 \leq j \leq k}\, \pi(j) \models \beta \textit{ and } loop(\pi) \neq \emptyset\Big)\Big).
\end{array}
$$

Next, we describe how the model checking problem ($M \models \psi$) can be reduced to the bounded model checking problem ($M_k \models \psi$). In this setting we can prove that in some circumstances satisfiability in the $|M|$-bounded semantics is equivalent to the unbounded one.

Theorem 1. *Let $M = ((S, s^0, \rightarrow_d), V)$ be a model, ψ be an ECTL$_y$ formula and $k = |M|$. Then, $M, s^0 \models \psi$ iff $M_k, s^0 \models \psi$.*

---

[24] Note that a path is infinite by definition.

The rationale behind the method is that for particular examples checking satisfiability of a formula can be done on a small fragment of the model.

Next, we show how to translate the model checking problem for $\mathrm{ECTL_y}$ on a k-model to a problem of satisfiability of some propositional formula. Our method is based on [62], but we use the operator ER rather than less expressive EG. Proofs of correctness of our approach are extensions of the corresponding proofs in [62]. We assume the following definition of a submodel.

Definition 10. *Let $M_k = ((S, s^0, P_k), V)$ be a k-model of M. A structure $M'_k = ((S', s^0, P'_k), V')$ is a* submodel *of M_k if $P'_k \subseteq P_k$, $S' = States(P'_k)$, and $V' = V|_{S'}$, where $States(P'_k) = \{s \in S \mid (\exists \pi \in P'_k)(\exists i \leq k)\; \pi(i) = s\}$.*

The bounded semantics of $\mathrm{ECTL_y}$ over submodels M'_k is defined as for M_k (see Def. 9). Our present aim is to give a bound for the number of k-paths in M'_k such that the validity of ψ in M_k is equivalent to the validity of ψ in M'_k. Let $\mathrm{F_{ECTL}}_y$ be a set of the formulas of $\mathrm{ECTL_y}$.

Definition 11. *Define a function $f_k : \mathrm{F_{ECTL}}_y \to \mathbb{N}$ as follows:*

- $f_k(\wp) = f_k(\neg\wp) = 0$, *where* $\wp \in PV$,
- $f_k(\alpha \vee \beta) = max\{f_k(\alpha), f_k(\beta)\}$,
- $f_k(\alpha \wedge \beta) = f_k(\alpha) + f_k(\beta)$,
- $f_k(\mathrm{X}_y\alpha) = f_k(\alpha)$,
- $f_k(\mathrm{E}(\alpha \mathrm{U} \beta)) = k \cdot f_k(\alpha) + f_k(\beta) + 1$,
- $f_k(\mathrm{E}(\alpha \mathrm{R} \beta)) = k \cdot f_k(\beta) + f_k(\alpha) + 1$.

The function f_k determines the number of k-paths of a submodel M'_k sufficient for checking an $\mathrm{ECTL_y}$ formula.

The main idea is that we can check ψ over M_k by checking the satisfiability of a propositional formula $[M, \psi]_k = [M^{\psi, s^0}]_k \wedge [\psi]_{M_k}$, where the first conjunct represents (part of) the model under consideration and the second a number of constraints that must be satisfied on M_k for ψ to be satisfied.

Once this translation is defined, checking satisfiability of an $\mathrm{ECTL_y}$ formula can be done by means of a SAT-checker. Although from a theoretical point of view the complexity of this operation is no easier, in practice the efficiency of modern SAT-checkers makes the process worthwhile in many instances.

We now give details of this translation. We begin with the encoding of the transitions in the model under consideration. Since the set of states S of our model is finite, every element of S can be encoded as a bit vector of a length depending on the number of locations in L, the size of the set $\mathbb{D}$ and $c_{max}(\varphi)$. We do not give details of this encoding here. The interested reader is referred to [63, 92]. Each state s can be represented by a vector $w = (w[1], \ldots, w[l])$ (called a *global state variable*), where each $w[i]$ for $i = 1, \ldots, l$ is a propositional variable (called state variable). Notice that we distinguish between states s encoded as sequences of 0's and 1's and their representations in terms of propositional variables $w[i]$. A finite sequence $(w_0, \ldots, w_k)$ of global state variables is called a *symbolic k-path*.

In general we shall need to consider not just one but a number of symbolic k-paths. This number depends on the formula ψ under investigation, and it is returned as the value $f_k(\psi)$ of the function f_k.

To construct $[M, \psi]_k$, we first define a propositional formula $[M^{\psi,s^0}]_k$ that constrains the $f_k(\psi)$ symbolic k-paths to be valid k-paths of M_k. The j-th symbolic k-path is denoted as $w_{0,j}, \ldots, w_{k,j}$, where $w_{i,j}$ are global state variables for $1 \leq j \leq f_k(\psi)$, $0 \leq i \leq k$. Let PV_s be a set of state variables, $\mathcal{FORM}$ be a set of propositional formulas over PV_s, and let $lit : \{0,1\} \times PV_s \to \mathcal{FORM}$ be a function defined as follows: $lit(0, \wp) = \neg\wp$ and $lit(1, \wp) = \wp$. Furthermore, let w, v be global state variables. We define the following propositional formulas:

- $I_s(w) := \bigwedge_{i=1}^{l} lit(s[i], w[i])$ (encodes the state s of the model M, i.e., $s[i] = 1$ is encoded by $w[i]$, and $s[i] = 0$ is encoded by $\neg w[i]$).
- $\wp(w)$ is a formula over w, which is true for a valuation s_w of w iff $\wp \in V(s_w)$, where $\wp \in PV'$ (see page 62),
- $H(w, v) := \bigwedge_{i=1}^{l} w[i] \Leftrightarrow v[i]$ (equality of the two state encodings),
- $\mathcal{R}(w, v)$ is a formula over w, v, which is true for two valuations s_w of w and s_v of v iff $s_w \to_{\mathcal{A}} s_v$ (encodes the transition relation of the paths),
- $R_y(w, v)$ is a formula over w, v, which is true for two valuations s_w of w and s_v of v iff $s_w \to_y s_v$ (encodes the transitions resetting the clock y),
- $L_{k,j}(l) := \mathcal{R}(w_{k,j}, w_{l,j})$, (encodes a backward loop from the k-th state to the l-th state in the symbolic k-path j, for $0 \leq l \leq k$).

The translation of $[M^{\psi,s^0}]_k$, representing the transitions in the k-model is given by the following definition.

Definition 12 (Encoding of Transition Relation). *Let* $M_k = ((S, s^0, P_k), V)$ *be the k-model of M, and ψ be an* $\mathrm{ECTL_y}$ *formula. The propositional formula* $[M^{\psi,s^0}]_k$ *is defined as follows:*

$$[M^{\psi,s^0}]_k := I_{s^0}(w_{0,0}) \wedge \bigwedge_{j=1}^{f_k(\psi)} \bigwedge_{i=0}^{k-1} \mathcal{R}(w_{i,j}, w_{i+1,j})$$

where $w_{0,0}$, and $w_{i,j}$ for $0 \leq i \leq k$ and $1 \leq j \leq f_k(\psi)$ are global state variables. $[M^{\psi,s^0}]_k$ constrains the $f_k(\psi)$ symbolic k-paths to be valid k-paths in M_k.

The next step of our algorithm is to translate an $\mathrm{ECTL_y}$ formula ψ into a propositional formula. We use $[\alpha]_k^{[m,n]}$ to denote the translation of an $\mathrm{ECTL_y}$ subformula α of ψ at $w_{m,n}$ to a propositional formula, where $w_{m,n}$ are global state variables with $0 \leq m \leq k$ for $1 \leq n \leq f_k(\psi)$ (which correspond to $f_k(\psi)$ symbolic paths), and with $m = 0$ for $f_k(\psi)+1 \leq n \leq f_k(\psi)+r$ (which correspond to r global state variables[25] for representing states, where the clock y is reset). Note that the index n denotes the number of a symbolic path, whereas the index m the position at that path.

[25] Recall that r is the number of non-trivial intervals in φ, where $\psi = \mathrm{cr}(\varphi)$.

$$[\wp]_k^{[m,n]} := \wp(w_{m,n}),$$
$$[\neg\wp]_k^{[m,n]} := \neg\wp(w_{m,n}),$$
$$[\alpha \wedge \beta]_k^{[m,n]} := [\alpha]_k^{[m,n]} \wedge [\beta]_k^{[m,n]},$$
$$[\alpha \vee \beta]_k^{[m,n]} := [\alpha]_k^{[m,n]} \vee [\beta]_k^{[m,n]},$$
$$[\mathrm{X}_y\alpha]_k^{[m,n]} := \bigvee_{1\le j\le r}\left(R_y(w_{m,n}, w_{0,f_k(\psi)+j}) \wedge [\alpha]_k^{[0,f_k(\psi)+j]}\right),$$
$$[\mathrm{E}(\alpha \mathrm{U}\beta)]_k^{[m,n]} := \bigvee_{1\le i\le f_k(\psi)}\left(H(w_{m,n}, w_{0,i}) \wedge \bigvee_{j=0}^{k}\left([\beta]_k^{[j,i]} \wedge \bigwedge_{l=0}^{j-1}[\alpha]_k^{[l,i]}\right)\right),$$
$$[\mathrm{E}(\alpha \mathrm{R}\beta)]_k^{[m,n]} := \bigvee_{1\le i\le f_k(\psi)}\left(H(w_{m,n}, w_{0,i}) \wedge \left(\bigvee_{j=0}^{k}\left([\alpha]_k^{[j,i]} \wedge \bigwedge_{l=0}^{j-1}[\beta]_k^{[l,i]}\right) \vee \right.\right.$$
$$\left.\left.\bigwedge_{j=0}^{k}[\beta]_k^{[j,i]} \wedge \bigvee_{l=0}^{k} L_{k,i}(l)\right)\right).$$

Given the translations above, we can now check ψ over M_k by checking satisfiability of the propositional formula $[M^{\psi,s^0}]_k \wedge [\psi]_k^{[0,0]}$. The translation presented above can be shown to be correct and complete.

Theorem 2. *Let M be a model, M_k be a k-model of M, and ψ be an* $\mathrm{ECTL_y}$ *formula. Then, $M \models_k \psi$ iff $[\psi]_{M_k} \wedge [M^{\psi,s^0}]_k$ is satisfiable.*

We have all ingredients in place to give the algorithm for BMC of TECTL.

Definition 13. BMC *algorithm for* TECTL*:*

1. *Let φ be a* TECTL *formula and $\mathcal{A}$ be a timed automaton.*
2. *Let $M = ((S, s^0, \rightarrow_d), V)$ be the r-discretized model with the extended valuation function for $\mathcal{A}_\varphi$.*
3. *Let $\psi = \mathrm{cr}(\varphi)$ be the* $\mathrm{ECTL_y}$ *formula.*
4. *Set $k := 1$.*
5. *Select the k-model M_k.*
6. *Select the submodels M_k' of M_k with $|P_k'| \le f_k(\psi)$.*
7. *Encode the transition relation of all M_k' by a propositional formula $[M^{\psi,s^0}]_k$.*
8. *Translate ψ over all M_k' into a propositional formula $[\psi]_{M_k}$.*
9. *Check the satisfiability of $[M,\psi]_k := [M^{\psi,s^0}]_k \wedge [\psi]_{M_k}$.*
10. *If $[M,\psi]_k$ is satisfiable, then return $M_c(\mathcal{A}) \models \varphi$.*
11. *Set $k := k + 1$.*
12. *If $k = |M| + 1$, then return $M_c(\mathcal{A}) \not\models \varphi$ else go to 5.*

Checking Reachability with BMC. Reachability of a propositional formula p in a timed automaton $\mathcal{A}$ can be specified by the ECTL formula E*F*p. So, in principle, reachability can be verified using the above approach over $DM_R(\mathcal{A})$ slightly modified to incorporate zero time successor steps. It turns out, however, that a slight change in the technique can dramatically influence efficiency of the method in this case.

First of all, we consider k-paths over $DM(\mathcal{A})$, rather than over $DM_R(\mathcal{A})$, which means that action and adjust transitions are not combined, and the delay transition relation is transitive. Secondly, we use the notion of a *special k-path*, which satisfies the following conditions.

- It begins with the initial state.
- The first transition is a time delay one.
- Each time delay transition is directly followed by an action one.
- Each action transition is directly followed by an adjust one.
- Each adjust transition is directly followed by a time delay one.
- The above three rules do not apply only to the last transition of a special k-path.

Obviously, it is sufficient to use only one symbolic path to encode all the special k-paths. Thus, the reachability problem is translated to conjunction of the encoding of the symbolic path and the encoding of the propositional property p at the last state of that path. If this conjunction is satisfiable, then p is reachable. In Section 8 we discuss experimental results obtained using this method.

Checking Unreachability with BMC. Unreachability of a propositional formula p in a timed automaton $\mathcal{A}$ means that the ACTL formula AG$\neg p$ holds in $\mathcal{A}$. Again, for verification, we could check the ECTL formula EFp over a slightly modified $DM_R(\mathcal{A})$, but, in this case, we have to prove that this formula does not hold in the model. This is, obviously, one of the major problems with BMC, as in the worst case the algorithm needs to reach the upper bound for k, i.e., the size of the model.

There is, however, another approach to checking unreachability, which in many cases (the method is not complete) gives striking results. The idea is to use a SAT-solver to find a minimal (possible) k such that if $\neg p$ holds at all the k-paths, then it means that p is unreachable. The method described below finds such a k, if each path at which p holds only at the final state is finite. To this aim, a *free special k-path* is defined. It satisfies all the conditions on a special k-path except for the first one, i.e., it does not need to start at the initial state.

In addition we require that for a special k-path the following conditions hold:

- p holds only at the last state if the last transition is an action one,
- p holds only at the last two states if the last transition is an adjust one,
- p holds only at the last three states if the last transition is a time delay one.

Notice that if p holds in our model, then it holds at a path of length restricted by the length of a longest special k-path.

So, using one symbolic k-path, we encode all the free special k-paths in order to find the length of a longest one satisfying the above three conditions. If the above encoding is unsatisfiable for some $k = k_0$, then it means that we have found a longest free path. It is known that we can look for such a k by running the algorithm for the values of k satisfying $k \ mod \ 3 = 2$ only. Then, when we find k_0 for which the encoding is unsatisfiable, we can run the check for reachability of p up to $k = k_0 - 3$. If the reachability algorithm does not find the formula satisfiable for such a k, then it means that p is indeed unreachable.

Unfortunately, the above method is not complete, it fails when there are loops in the unreachable part of the state space involving states satisfying $\neg p$, from which a state satisfying p is reachable.

A solution to make the method complete by encoding that a free special path is loop-free i.e., no state repeats at the path, turns out to be ineffective in practice [92]. In Section 8 we discuss experimental results obtained using this method.

7 Existing Tools

Tools for TPN's. Some of the existing tools for Petri nets with time are listed below:

- **Tina** [16] - a toolbox for analysis of (time) Petri nets. It constructs state class graphs [14, 15] and performs LTL or reachability verification. In addition, Tina builds atomic state class graphs [17] to be used for verification of CTL formulas.
- **Romeo** [70] - a tool for time Petri nets analysis, which provides several methods for translating TPN's to TA [25, 49] and computation of state class graphs [35].
- **INA** (Integrated Net Analyser) - a Petri net analysis tool, supporting place/transition nets and coloured Petri nets with time and priorities. Among others, INA provides verification by analysis of paths for TPN's [67].
- **CPN Tools** [69] (a replacement for **Design/CPN**) - a software package for modelling and analysis of both timed and untimed Coloured Petri Nets, enabling their simulation, generating occurrence (reachability) graph, and analysis by place invariants.

Tools for TA. There are many tools using the approaches considered in this paper. Below, we list some of them and give pointers to the literature, where more detailed descriptions can be found.

- **Cospan** - a tool for verifying the behaviour of designs written in the industry standard design languages VHDL and Verilog. It implements an automata-based approach to model checking including an on-the-fly enumerative search (using zones in the timed case), as well as symbolic search using BDDs. A detailed description of timed verification can be found in [8].
- **Kronos** [91] is a tool which performs verification of TCTL using forward or backward analysis, and behavioural analysis, which consists in building the smallest finite quotient of a timed model (using minimization), and then checking whether the minimal model of the system simulates that of the specification. DBM's are used for representing zones.
- **UppAal2k** (a successor of UppAal) - a tool for modelling, simulation and verification of timed systems, appropriate for systems which can be described by a collection of non-deterministic processes with finite control structure and real-valued clocks, communicating through channels or shared variables. Forward reachability analysis, deadlock detection and verification of properties expressible in a subset of TCTL are available. Many optimizations are implemented, e.g. application of Clock Difference Diagrams (CDD's) to represent unions of regions [11].

- **Red** is a fully symbolic model checker based on data structures called Clock Restriction Diagrams (CRD) [85]. It supports TCTL model checking and backward reachability analysis.
- **HyTech** [39] - an automatic tool for the analysis of embedded systems. Real-time requirements are specified in the logic TCTL and its modification - ICTL (*Integrator Computation Tree Logic*), used to specify safety, liveness, time-bounded and duration requirements of hybrid automata. Verification is performed by a successive approximation of the set of states satisfying the formula to be checked, by iterating boolean operations and weakest-precondition operations on regions (see [7]).
- **Rabbit** [18] - a tool for BDD-based verification of real-time systems, developed for an extension of TA, called Cottbus Timed Automata, and providing reachability analysis.
- **VerICS** [31] - implements partition refinement algorithms and SAT-based BMC for verifying TCTL and reachability for timed automata and Estelle programs.

8 Experimental Results for Verifying TPN's and TA

In this section we compare experimental results for the four TPN's of Figure 1, and for the TA of Figure 2 modelling Fischer's mutual exclusion algorithm.

In the first table (Fig. 4) we give the sizes of several abstract models for all the nets, obtained using either state classes approaches (Yoneda's implementation, Tina), or minimization algorithms (VerICS, Kronos) applied to translations[26] to timed automata as well as directly to the TA of Figure 2. The experiments were performed on a PC (Intel Pentium III 640MHz), with the assumed limit on the execution time (30 min) and memory required (128 MB RAM + 128 MB of swap space under Linux). For the Fischer's protocol, we give the sizes of models for 3 processes as well as for the maximal number of processes a model could be generated for. The abbreviation *nsp* means that the tool does not support verification of time Petri nets, where Lft of a transition is equal to ∞.

The models generated can be divided into three groups, the first of which contains the structures to be used for reachability checking: (strong) state class graphs (SCG and SSCG, resp.), geometric region graphs, forward-reachability graphs obtained using *inclusion* [30] and extrapolation abstractions (*forw -ax - ai*), and pseudo-simulating models generated for the semantics which collects together time- and transition steps (*ps- discrete*). The next group includes various kinds of bisimulating models built for the above semantics: (strong) atomic state class graphs (SASCG and *atomic*, resp.), and models denoted by *bis. discrete*. In a separate class are bisimulating models for the dense semantics considered in this paper (i.e., *bis. dense*).

Notice that the sizes of models for the nets 5a, 5b, and 5c are comparable for different approaches and there is no tool (approach), which would outperform the other ones w.r.t. the sizes of all the types of models. For MUTEX of 3

[26] Processes-as-clocks translations of [64] are used.

		Net 5a		Net 5b		Net 5c		Mutex Δ=1, δ=2			Mutex Δ=2, δ=1		
		states	edges	states	edges	states	edges	noP	states	edges	noP	states	edges
OBTAINED BY TPN - SPECIFIC METHODS													
Tina	SASCG	36	61	62	163	80	204	3	65	96	3	152	240
								9	81035	280170	7	73600	200704
Tina	SCG	18	26	34	58	50	76	3	65	96	3	152	240
								9	81035	280170	7	73600	200704
Tina	SSCG	21	29	39	63	60	93	3	65	96	3	152	240
								9	81035	280170	7	73600	200704
impl.[89]	atomic	53	95	64	179	168	363	3	nsp	nsp	3	nsp	nsp
impl.[89]	geometric	16	25	32	57	105	170	3	nsp	nsp	3	nsp	nsp
OBTAINED BY TPN $\rightarrow$TA TRANSLATIONS													
VerICS	bis. dense	54	80	135	230	186	323	3	77	108	3	200	312
								3	77	108	3	200	312
VerICS	bis. discr.	26	47	46	135	80	204	3	65	96	3	152	240
								3	65	96	3	152	240
VerICS	ps- discr.	21	34	13	22	53	121	3	65	96	3	152	204
								3	65	96	3	152	204
Kronos	bis. dense	51	77	134	229	185	321	3	77	108	3	200	312
								5	807	1590	4	1008	1856
Kronos	forw -ai -ax	37	42	37	42	26	40	3	214	321	3	613	1084
								5	33451	62223	4	12850	27848

Fig. 4. Experimental results for the nets in Fig. 1

		TPN$\rightarrow$TA				TA			
Parameters	**NoP**	variables	**clauses**	**sec**	**MB**	variables	**clauses**	**sec**	**MB**
$\Delta = 1$, $\delta = 2$	8	61530	176319	10890.1	61.31	36461	103228	2326.3	34.5
$\Delta = 2$, $\delta = 1$	8	22552	64442	8.0	21.9	13357	37666	0.7	20.5
$\Delta = 2$, $\delta = 1$	10	29918	86002	14.5	23.5	17283	49034	1.1	20.2
$\Delta = 2$, $\delta = 1$	50	378203	1118763	99.7	100.5	156941	459722	21.8	31.7
$\Delta = 2$, $\delta = 1$	104	1411156	4200809	1397.6	577.9	528136	1562194	218.8	75.9
$\Delta = 2$, $\delta = 1$	310	-	-	-	-	3873940	11557290	21723.6	648.3

Fig. 5. BMC of VerICS for Fischer's protocol modelled by TPN and TA

processes, all the algorithms give nearly the same results, but only Tina can generate models for MUTEX of 9 processes within the assumed time limit.

In the second table (Fig. 5) we display the results of applying the BMC algorithm (for checking reachability and unreachability) to timed systems modelling Fischer's mutual exclusion. We verify that either mutual exclusion is violated for $\Delta = 2$ and $\delta = 1$, or is preserved for $\Delta = 1$ and $\delta = 2$. BMC is applied either directly to the TA, or to the timed automaton resulting from the translation of the TPN. For the case of $\Delta = 2$ and $\delta = 1$ we provide the time and memory resources needed to confirm satisfiability of the property on a special path of length $k = 17$, whereas for $\Delta = 1$ and $\delta = 2$ the time given is sum of the times required to check that a special free path of length $k = 44$ is the longest possible one and then to check satisfiability on the special path of that length. These experiments were performed on a PC (AMD Athlon XP 1800 - 1544MHz).

Notice that the BMC could verify MUTEX modelled by the TPN or the TA for $\Delta = 1, \delta = 2$ of 8 processes, and respectively of 104 and 310 processes, where the mutual exclusion was violated ($\Delta = 2$ and $\delta = 1$).

The above results show that verifying time Petri nets via translations to timed automata can sometimes give better results than using specific methods for nets. This is especially the case, when BMC is used for verification. Therefore, it seems interesting to investigate different combinations of specific methods for both timed automata and time Petri nets like for example BMC applied to models based on state class graphs rather that on r-discretized region graphs.

Acknowledgements

Many thanks to B. Berthomieu, O. H. Roux, M. Szreter, and B. Woźna for their comments that greatly helped to improve this paper. The authors are especially grateful to A. Zbrzezny for providing experimental results of BMC.

References

1. P. A. Abdulla and A. Nylén. Timed Petri Nets and BQOs. In *Proc. of ICATPN'01*, volume 2075 of *LNCS*, pages 53–70. Springer-Verlag, 2001.
2. R. Alur, C. Courcoubetis, and D. Dill. Model checking for real-time systems. In *Proc. of LICS'90*, pages 414–425. IEEE, 1990.
3. R. Alur, C. Courcoubetis, and D. Dill. Model checking in dense real-time. *Information and Computation*, 104(1):2–34, 1993.
4. R. Alur, C. Courcoubetis, D. Dill, N. Halbwachs, and H. Wong-Toi. An implementation of three algorithms for timing verification based on automata emptiness. In *Proc. of RTSS'92*, pages 157–166. IEEE Comp. Soc. Press, 1992.
5. R. Alur, C. Courcoubetis, D. Dill, N. Halbwachs, and H. Wong-Toi. Minimization of timed transition systems. In *Proc. of CONCUR'92*, volume 630 of *LNCS*, pages 340–354. Springer-Verlag, 1992.
6. R. Alur and D. Dill. Automata for modelling real-time systems. In *Proc. of ICALP'90*, volume 443 of *LNCS*, pages 322–335. Springer-Verlag, 1990.
7. R. Alur, T. Henzinger, and P. Ho. Automatic symbolic verification of embedded systems. *IEEE Trans. on Software Eng.*, 22(3):181–201, 1996.
8. R. Alur and R. Kurshan. Timing analysis in COSPAN. In *Hybrid Systems III*, volume 1066 of *LNCS*, pages 220–231. Springer-Verlag, 1996.
9. E. Asarin, M. Bozga, A. Kerbrat, O. Maler, A. Pnueli, and A. Rasse. Data-structures for the verification of Timed Automata. In *Proc. of HART'97*, volume 1201 of *LNCS*, pages 346–360. Springer-Verlag, 1997.
10. G. Audemard, A. Cimatti, A. Kornilowicz, and R. Sebastiani. Bounded model checking for timed systems. In *Proc. of FORTE'02*, volume 2529 of *LNCS*, pages 243–259. Springer-Verlag, 2002.
11. G. Behrmann, K. Larsen, J. Pearson, C. Weise, and W. Yi. Efficient timed reachability analysis using Clock Difference Diagrams. In *Proc. of CAV'99*, volume 1633 of *LNCS*, pages 341–353. Springer-Verlag, 1999.
12. J. Bengtsson. *Clocks, DBMs and States in Timed Systems.* PhD thesis, Dept. of Information Technology, Uppsala University, 2002.
13. J. Bengtsson and W. Yi. On clock difference constraints and termination in reachability analysis in Timed Automata. In *Proc. of ICFEM'03*, volume 2885 of *LNCS*, pages 491–503. Springer-Verlag, 2003.

14. B. Berthomieu and M. Diaz. Modeling and verification of time dependent systems using Time Petri Nets. *IEEE Trans. on Software Eng.*, 17(3):259–273, 1991.
15. B. Berthomieu and M. Menasche. An enumerative approach for analyzing Time Petri Nets. In *Proc. of the IFIP 9th World Computer Congress*, volume 9 of *Information Processing*, pages 41–46. North Holland/ IFIP, September 1983.
16. B. Berthomieu, P-O. Ribet, and F. Vernadat. The tool TINA - construction of abstract state spaces for Petri nets and Time Petri Nets. *International Journal of Production Research*, 2004. to appear.
17. B. Berthomieu and F. Vernadat. State class constructions for branching analysis of Time Petri Nets. In *Proc. of TACAS'03*, volume 2619 of *LNCS*, pages 442–457. Springer-Verlag, 2003.
18. D. Beyer. Rabbit: Verification of real-time systems. In *Proc. of the Workshop on Real-Time Tools (RT-TOOLS'01)*, pages 13–21, 2001.
19. A. Bobbio and A. Horváth. Model checking time Petri nets using NuSMV. In *Proc. of the 5th Int. Workshop on Performability Modeling of Computer and Communication Systems (PMCCS5)*, pages 100–104, September 2001.
20. A. Bouajjani, J-C. Fernandez, N. Halbwachs, P. Raymond, and C. Ratel. Minimal state graph generation. *Science of Computer Programming*, 18:247–269, 1992.
21. A. Bouajjani, S. Tripakis, and S. Yovine. On-the-fly symbolic model checking for real-time systems. In *Proc. of RTSS'97*, pages 232–243. IEEE Comp. Soc. Press, 1997.
22. H. Boucheneb and G. Berthelot. Towards a simplified building of Time Petri Nets reachability graph. In *Proc. of the 5th Int. Workshop on Petri Nets and Performance Models*, pages 46–55, October 1993.
23. F. Bowden. Modelling time in Petri nets. In *Proc. of the 2nd Australia-Japan Workshop on Stochastic Models (STOMOD'96)*, July 1996.
24. R. Bryant. Graph-based algorithms for boolean function manipulation. *IEEE Transaction on Computers*, 35(8):677–691, 1986.
25. F. Cassez and O. H. Roux. Traduction structurelle des Réseaux de Petri Temporels vers le Automates Temporisés. In *Proc. of 4ieme Colloque Francophone sur la Modélisation des Systémes Réactifs (MSR'03)*. Hermes Science, October 2003.
26. G. Ciardo, G. Lüttgen, and R. Simniceanu. Efficient symbolic state-space construction for asynchronous systems. In *Proc. of ICATPN'00*, volume 1825 of *LNCS*, pages 103–122. Springer-Verlag, 2000.
27. J. Coolahan and N. Roussopoulos. Timing requirements for time-driven systems using augmented Petri nets. *IEEE Trans. on Software Eng.*, SE-9(5):603–616, 1983.
28. L. A. Cortés, P. Eles, and Z. Peng. Verification of real-time embedded systems using Petri net models and Timed Automata. In *Proc. of the 8th Int. Conf. on Real-Time Computing Systems and Applications (RTCSA'02)*, pages 191–199, March 2002.
29. J-M. Couvreur, E. Encrenaz, E. Paviot-Adet, D. Pointrenaud, and P-A. Wacrenier. Data Decision Diagrams for Petri net analysis. In *Proc. of ICATPN'02*, volume 2360 of *LNCS*, pages 101–120. Springer-Verlag, 2002.
30. C. Daws and S. Tripakis. Model checking of real-time reachability properties using abstractions. In *Proc. of TACAS'98*, volume 1384 of *LNCS*, pages 313–329. Springer-Verlag, 1998.
31. P. Dembiński, A. Janowska, P. Janowski, W. Penczek, A. Półrola, M. Szreter, B. Woźna, and A. Zbrzezny. VerICS: A tool for verifying Timed Automata and Estelle specifications. In *Proc. of TACAS'03*, volume 2619 of *LNCS*, pages 278–283. Springer-Verlag, 2003.
32. P. Dembiński, W. Penczek, and A. Półrola. Verification of Timed Automata based on similarity. *Fundamenta Informaticae*, 51(1-2):59–89, 2002.

33. M. Dickhofer and T. Wilke. Timed Alternating Tree Automata: The automata-theoretic solution to the TCTL model checking problem. In *Proc. of ICALP'98*, volume 1664 of *LNCS*, pages 281–290. Springer-Verlag, 1998.
34. D. Dill. Timing assumptions and verification of finite state concurrent systems. In *Automatic Verification Methods for Finite-State Systems*, volume 407 of *LNCS*, pages 197 – 212. Springer-Verlag, 1989.
35. G. Gardey, O. H. Roux, and O. F. Roux. Using zone graph method for computing the state space of a Time Petri Net. In *Proc. of FORMATS'03*, volume 2791 of *LNCS*. Springer-Verlag, 2004.
36. Z. Gu and K. Shin. Analysis of event-driven real-time systems with Time Petri Nets. In *Proc. of DIPES'02*, volume 219 of *IFIP Conference Proceedings*, pages 31–40. Kluwer, 2002.
37. S. Haar, L. Kaiser, F. Simonot-Lion, and J. Toussaint. On equivalence between Timed State Machines and Time Petri Nets. Technical Report RR-4049, INRIA Rhône-Alpes, 655, avenue de l'Europe, 38330 Montbonnot-St-Martin, November 2000.
38. H-M. Hanisch. Analysis of place/transition nets with timed arcs and its application to batch process control. In *Proc. of ICATPN'93*, volume 691 of *LNCS*, pages 282–299. Springer-Verlag, 1993.
39. T. Henzinger and P. Ho. HyTech: The Cornell hybrid technology tool. In *Hybrid Systems II*, volume 999 of *LNCS*, pages 265–293. Springer-Verlag, 1995.
40. T. Henzinger, X. Nicollin, J. Sifakis, and S. Yovine. Symbolic model checking for real-time systems. *Information and Computation*, 111(2):193–224, 1994.
41. M. Huhn, P. Niebert, and F. Wallner. Verification based on local states. In *Proc. of TACAS'98*, volume 1384 of *LNCS*, pages 36–51. Springer-Verlag, 1998.
42. H. Hulgaard and S. M. Burns. Efficient timing analysis of a class of Petri Nets. In *Proc. of CAV'95*, volume 939 of *LNCS*, pages 923–936. Springer-Verlag, 1995.
43. R. Janicki. Nets, sequential components and concurrency relations. *Theoretical Computer Science*, 29:87–121, 1984.
44. I. Kang and I. Lee. An efficient state space generation for the analysis of real-time systems. In *Proc. of Int. Symposium on Software Testing and Analysis*, 1996.
45. O. Kupferman, T. A. Henzinger, and M. Y. Vardi. A space-efficient on-the-fly algorithm for real-time model checking. In *Proc. of CONCUR'96*, volume 1119 of *LNCS*, pages 514–529. Springer-Verlag, 1996.
46. K. G. Larsen, F. Larsson, P. Pettersson, and W. Yi. Efficient verification of real-time systems: Compact data structures and state-space reduction. In *Proc. of RTSS'97*, pages 14–24. IEEE Comp. Soc. Press, 1997.
47. D. Lee and M. Yannakakis. On-line minimization of transition systems. In *Proc. of the 24th ACM Symp. on the Theory of Computing*, pages 264–274, May 1992.
48. J. Lilius. Efficient state space search for Time Petri Nets. In *Proc. of MFCS Workshop on Concurrency, Brno'98*, volume 18 of *ENTCS*. Elsevier Science Publishers, 1999.
49. D. Lime and O. H. Roux. State class timed automaton of a time Petri net. In *Proc. of the 10th Int. Workshop on Petri Nets and Performance Models (PNPM'03)*. IEEE Comp. Soc. Press, September 2003.
50. K. L. McMillan. Applying SAT methods in unbounded symbolic model checking. In *Proc. of CAV'02*, volume 2404 of *LNCS*, pages 250–264. Springer-Verlag, 2002.
51. P. Merlin and D. J. Farber. Recoverability of communication protocols – implication of a theoretical study. *IEEE Trans. on Communications*, 24(9):1036–1043, 1976.

52. A. Miner and G. Ciardo. Efficient reachability set generation and storage using decision diagrams. In *Proc. of ICATPN'99*, volume 1639 of *LNCS*, pages 6–25. Springer-Verlag, 1999.
53. P. Molinaro, D. Roux, and O. Delfieu. Improving the calculus of the marking graph of Petri net with BDD like structure. In *Proc. of the 2nd IEEE Int. Conf. on Systems, Man and Cybernetics (SMC'02)*. IEEE Comp. Soc. Press, October 2002.
54. J. Møller, J. Lichtenberg, H. Andersen, and H. Hulgaard. Difference Decision Diagrams. In *Proc. of CSL'99*, volume 1683 of *LNCS*, pages 111–125. Springer-Verlag, 1999.
55. J. Møller, J. Lichtenberg, H. Andersen, and H. Hulgaard. Fully symbolic model checking of timed systems using Difference Decision Diagrams. In *Proc. of FLoC'99*, volume 23(2) of *ENTCS*, 1999.
56. M. Moskewicz, C. Madigan, Y. Zhao, L. Zhang, and S. Malik. Chaff: Engineering an efficient SAT solver. In *Proc. of the 38th Design Automation Conference (DAC'01)*, pages 530–535, June 2001.
57. P. Niebert, M. Mahfoudh, E. Asarin, M. Bozga, O. Maler, and N. Jain. Verification of Timed Automata via satisfiability checking. In *Proc. of FTRTFT'02*, volume 2469 of *LNCS*, pages 226–243. Springer-Verlag, 2002.
58. Y. Okawa and T. Yoneda. Symbolic CTL model checking of Time Petri Nets. *Electronics and Communications in Japan, Scripta Technica*, 80(4):11–20, 1997.
59. R. Paige and R. Tarjan. Three partition refinement algorithms. *SIAM Journal on Computing*, 16(6):973–989, 1987.
60. E. Pastor, J. Cortadella, and O. Roig. Symbolic Petri net analysis using boolean manipulation. Technical Report RR-97-08, UP/DAC, Univerisitat Politécnica de Catalunya, February 1997.
61. W. Penczek and A. Półrola. Abstractions and partial order reductions for checking branching properties of Time Petri Nets. In *Proc. of ICATPN'01*, volume 2075 of *LNCS*, pages 323–342. Springer-Verlag, 2001.
62. W. Penczek, B. Woźna, and A. Zbrzezny. Bounded model checking for the universal fragment of CTL. *Fundamenta Informaticae*, 51(1-2):135–156, 2002.
63. W. Penczek, B. Woźna, and A. Zbrzezny. Towards bounded model checking for the universal fragment of TCTL. In *Proc. of FTRTFT'02*, volume 2469 of *LNCS*, pages 265–288. Springer-Verlag, 2002.
64. A. Półrola and W. Penczek. Minimization algorithms for Time Petri Nets. *Fundamenta Informaticae*, 2004. to appear.
65. A. Półrola, W. Penczek, and M. Szreter. Reachability analysis for Timed Automata using partitioning algorithms. *Fundamenta Informaticae*, 55(2):203–221, 2003.
66. A. Półrola, W. Penczek, and M. Szreter. Towards efficient partition refinement for checking reachability in Timed Automata. In *Proc. of FORMATS'03*, volume 2791 of *LNCS*. Springer-Verlag, 2004.
67. L. Popova and S. Marek. TINA - a tool for analyzing paths in TPNs. In *Proc. of the Int. Workshop on Concurrency, Specification and Programming (CS&P'02)*, volume 110 of *Informatik-Berichte*, pages 195–196. Humboldt University, 1998.
68. C. Ramchandani. Analysis of asynchronous concurrent systems by timed Petri nets. Technical Report MAC-TR-120, Massachusets Institute of Technology, February 1974.
69. A. Ratzer, L. Wells, H. Lassen, M. Laursen, J. Qvortrup, M. Stissing, M. Westergaard, S. Christensen, and K. Jensen. CPN Tools for editing, simulating, and analyzing Coloured Petri Nets. In *Proc. of ICATPN'03*, volume 2679 of *LNCS*, pages 450–462. Springer-Verlag, 2003.

70. Romeo: A tool for Time Petri Net analysis. http://www.irccyn.ec-nantes.fr/irccyn/d/en/equipes/TempsReel/logs, 2000.
71. S. Samolej and T. Szmuc. Modelowanie systemów czasu rzeczywistego z zastosowaniem czasowych sieci Petriego. In *Mat. IX Konf. Systemy Czasu Rzeczywistego (SCR'02)*, pages 45–54. Instytut Informatyki Politechniki Śląskiej, 2002. In Polish.
72. P. Sénac, M. Diaz, and P. de Saqui Sannes. Toward a formal specification of multimedia scenarios. *Annals of Telecommunications*, 49(5-6):297–314, 1994.
73. S. Seshia and R. Bryant. Unbounded, fully symbolic model checking of Timed Automata using boolean methods. In *Proc. of CAV'03*, volume 2725 of *LNCS*, pages 154–166. Springer-Verlag, 2003.
74. J. Sifakis and S. Yovine. Compositional specification of timed systems. In *Proc. of STACS'96*, volume 1046 of *LNCS*, pages 347–359. Springer-Verlag, 1996.
75. M. Sorea. Bounded model checking for Timed Automata. In *Proc. of MTCS'02*, volume 68(5) of *ENTCS*. Elsevier Science Publishers, 2002.
76. R. L. Spelberg, H. Toetenel, and M. Ammerlaan. Partition refinement in real-time model checking. In *Proc. of FTRTFT'98*, volume 1486 of *LNCS*, pages 143–157. Springer-Verlag, 1998.
77. K. Strehl and L. Thiele. Interval diagram techniques for symbolic model checking of Petri nets. In *Proc. of DATE'99*, LNCS, pages 756–757. IEEE Comp. Soc. Press, 1999.
78. O. Strichman, S. Seshia, and R. Bryant. Deciding separation formulas with SAT. In *Proc. of CAV'02*, volume 2404 of *LNCS*, pages 209–222. Springer-Verlag, 2002.
79. S. Tripakis and S. Yovine. Analysis of timed systems using time-abstracting bisimulations. *Formal Methods in System Design*, 18(1):25–68, 2001.
80. J. Tsai, S. Yang, and Y. Chang. Timing constraint Petri nets and their application to schedulability analysis of real-time system specifications. *IEEE Trans. on Software Eng.*, 21(1):32–49, 1995.
81. W. van der Aalst. Interval timed coloured Petri nets and their analysis. In *Proc. of ICATPN'93*, volume 961 of *LNCS*, pages 452–472. Springer-Verlag, 1993.
82. I. B. Virbitskaite and E. A. Pokozy. A partial order method for the verification of Time Petri Nets. In *Fundamental of Computation Theory*, volume 1684 of *LNCS*, pages 547–558. Springer-Verlag, 1999.
83. B. Walter. Timed Petri nets for modelling and analysing protocols with real-time characteristics. In *Proc. of the 3rd IFIP Workshop on Protocol Specification, Testing, and Verification*, pages 149–159. North Holland, 1983.
84. F. Wang. Region Encoding Diagram for fully symbolic verification of real-time systems. In *Proc. of the 24th Int. Computer Software and Applications Conf. (COMPSAC'00)*, pages 509–515. IEEE Comp. Soc. Press, October 2000.
85. F. Wang. Verification of Timed Automata with BDD-like data structures. In *Proc. of VMCAI'03*, volume 2575 of *LNCS*, pages 189–205. Springer-Verlag, 2003.
86. B. Woźna, A. Zbrzezny, and W. Penczek. Checking reachability properties for Timed Automata via SAT. *Fundamenta Informaticae*, 55(2):223–241, 2003.
87. M. Yannakakis and D. Lee. An efficient algorithm for minimizing real-time transition systems. In *Proc. of CAV'93*, volume 697 of *LNCS*, pages 210–224. Springer-Verlag, 1993.
88. W. Yi, P. Pettersson, and M. Daniels. Automatic verification of real-time communicating systems by constraint-solving. In *Proc. of the 7th IFIP WG6.1 Int. Conf. on Formal Description Techniques (FORTE'94)*, volume 6 of *IFIP Conference Proceedings*, pages 243–258. Chapman & Hall, 1994.

89. T. Yoneda and H. Ryuba. CTL model checking of Time Petri Nets using geometric regions. *IEICE Trans. Inf. and Syst.*, 3:1–10, 1998.
90. T. Yoneda and B. H. Schlingloff. Efficient verification of parallel real-time systems. *Formal Methods in System Design*, 11(2):197–215, 1997.
91. S. Yovine. KRONOS: A verification tool for real-time systems. *Springer International Journal of Software Tools for Technology Transfer*, 1(1/2):123–133, 1997.
92. A. Zbrzezny. Improvements in SAT-based reachability analysis for Timed Automata. *Fundamenta Informaticae*, 2004. to appear.
93. L. Zhang, C. Madigan, M. Moskewicz, and S. Malik. Efficient conflict driven learning in a boolean satisfiability solver. In *Proc. of Int. Conf. on Computer-Aided Design (ICCAD'01)*, pages 279–285, 2001.

Formal Tools for Modular System Development*

Lucia Pomello and Luca Bernardinello

Dipartimento di Informatica, Sistemistica e Comunicazione
Università degli Studi di Milano – Bicocca
via Bicocca degli Arcimboldi 8, I-20126 Milano, Italy
{pomello,bernardinello}@disco.unimib.it

Abstract. The dualities event–condition and local state–global state in basic net theory are exploited in order to equip system designers with formal tools supporting modular system development. In the framework of categories with suitable morphisms of Elementary Net systems, Elementary Transition systems and Orthomodular posets of system local states (regions), some operations of composition and refinement are discussed as tools for modularity and abstraction.

1 Introduction

In the development of distributed systems a central role is played by formal tools supporting various aspects of modularity such as compositionality, refinement and abstraction.

In this context we consider as theoretical model Elementary Net systems, with the classical interpretation of Petri nets, in which conditions correspond to local assertions on the state of a system, while events represent local state transformations, i.e. events change the truth value of a subset of local assertions [26].

In this model the duality state–transition has multiple aspects which can be differently exploited for modularity.

Considering the duality *event–condition* and focussing on *events*, it is natural to adopt the design paradigm, almost independently introduced in the late 1970s by Tony Hoare with CSP [20] and Robin Milner with CCS [22], in which a concurrent/distributed system is modelled by the parallel composition of components, which interact with each other through *synchronous communication.*

In this frame, EN systems can be constructed as the composition of state machines, each one modelling a sequential component, through the superposition, possibly after suitable splitting, of events modelling the same synchronous communication action.

Furthermore, each EN system can be decomposed, possibly after complementation of some conditions, into monomarked state machine components.

These decomposition and composition operations have been characterized in different contexts in the literature. Let us just mention [9], [18] and the ones more directly concerning our work: [12], [5], and recently [3], where in particular the corresponding operation on a category of asynchronous systems is defined as rigid product and seen as a pullback.

* Partially supported by MIUR, CNR and IPIPAN

J. Cortadella and W. Reisig (Eds.): ICATPN 2004, LNCS 3099, pp. 77–96, 2004.
© Springer-Verlag Berlin Heidelberg 2004

In the conceptual framework introduced by Milner [23] to deal with concurrency, in which the *interaction* of a system with its environment, rather than its input-output function, is essential, an important role is played by the notion of an *observer*. This allows one to distinguish operationally between *internal*, i.e., non observable, system behaviour, and *external*, i.e., observable, system behaviour. Then two systems are equivalent if one cannot distinguish between the patterns of their interactions with all possible environments (i.e., all possible observers).

An equivalence notion based on action (interaction) observation supports *organizational refinement/abstraction*, allowing one to consider a system at a particular level of abstraction as the parallel composition of a number of interacting components, and to refine a component of the system as the parallel composition of a number of interacting subcomponents, hence getting a more structured model.

Following this approach, various equivalences based on action observation have been proposed in the literature for different models of concurrent systems, while other notions have been introduced specifically for net based systems. Some of these notions have been considered for event-labelled EN systems, on the basis of interleaving, step and partial order semantics, and have been compared with respect to their power of identifying EN systems. See [27] for a survey.

The need of considering equivalences based on partial order, instead of notions based on arbitrary interleaving, derives from rejecting the axiom of *action atomicity*, specific to CCS and related models, in order to satisfy the requirement of having the possibility of respecifying, at a subsequent later level in the development process, an action as the composition of more elementary ones.

This fact is explained by the following simple example [11]: the parallel composition of two actions x and y is equivalent, considering interleaving semantics, to the alternative choice between the sequences xy and yx. However, if x is refined by a sequence, let's say x_1x_2, then the parallel composition of x_1x_2 and y does not exhibite any more the same execution sequences as the system obtained by the same action refinement starting from the alternative choice between xy and yx. The first system exhibits also the sequence x_1yx_2, which is not a behaviour sequence of the second one.

From this consideration, researches have been developed investigating *refinement preserving equivalence* notions, based on partial order semantics. Net theory is founded on partial order semantics, and therefore some solutions have been proposed in this frame. For a survey of such notions on Petri Nets see [10], and, among others, [19], in which other models of concurrency are considered.

Coming back to the duality *event–condition*, the fact that Petri nets explicitly represent *conditions* suggested to apply the notion of observability also to them [13]. This leads to the opportunity, in system development, to abstract, through the corresponding equivalence relation, from the level of action description, and to preserve instead the transformations of the observable states.

State transformations in Petri nets are *local*, involving in any event occurrence only those local states which are preconditions or postconditions of the event itself.

In order to be consistent with this aspect of locality, the behaviour of an EN system should be characterized in terms of the local state transformations it goes through during its evolution. In this respect the notion of case graph is not sufficient since it considers global state transformations only, disregarding the actual local change of conditions.

In order to capture the locality in system evolution, the system state space has been characterized by a relational algebraic structure, called Local State Transformation (LST) algebra [28], in which both global and local states as well as local state transformations are explicitly taken into account. An LST algebra can be derived from the case graph by adding to the set of cases all the subsets of conditions obtained by closure with respect to set difference, and to the set of transitions all the local state transformations which are consistent with the transition rule. LST algebras allow a finer distinction between systems than the one induced by considering their case graphs: two EN systems have isomorphic LST algebras iff their reductions to simple, pure and 1-live systems yield isomorphic EN systems.

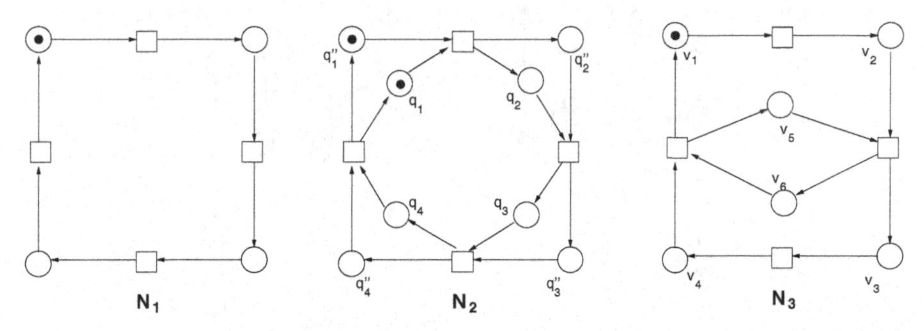

Fig. 1. Equivalence and LST algebra.

Let us consider for example the three EN systems N_1, N_2 and N_3 given in Figure 1. They have isomorphic case graphs. However, N_1 and N_2 have a cyclic behaviour in which each state transformation is *global*. There is no reason to distinguish between q_i and q_i'' (i=1,2), because they have the same extension, i.e.: whenever q_i is true also q_i'' is true, and as soon as q_i becomes false, also q_i'' becomes false. N_1 and N_2 are therefore equivalent with respect to the (*local* and *global*) state transformations they perform: their LST algebras are in fact isomorphic. In N_3 some state transformations are global, while others are local; precisely the case $\{v_1, v_5\}$ is transformed into the case $\{v_2, v_5\}$ by means of a local transformation, as well as $\{v_3, v_6\}$ into $\{v_4, v_6\}$; whereas the transformations of $\{v_2, v_5\}$ into $\{v_3, v_6\}$ and of $\{v_4, v_6\}$ into $\{v_1, v_5\}$ are global (they involve all components). For this reason N_3 is not considered equivalent, w.r.t. (*local* and *global*) state transformations, to the systems N_1 and N_2; the LST algebra of N_3 is not isomorphic to the algebra of N_1.

A class of injective morphisms between LST algebras has been introduced, allowing the comparison of system models at different level of abstraction in action description, together with a notion of state observability and of preorder and equivalence based on state observability, which generalize ideas introduced in [13].

These notions have been discussed in the framework of system development as conceptual tools supporting an incremental design based on *extension* and *expansion* of system requirements: a specification can be *extended*, through the notion of morphisms, by adding to a given model concurrent components or alternative components (i.e.: adding state transformations which can be performed in alternative to the given ones) or components in sequence, and it can be *expanded*, through the notion of state observability and related equivalence, by implementing some state transformations as more structured transformations.

For example, if we consider the EN systems N_1, N_2 and N_3 given in Figure 2 with the correspondences between conditions suggested by the labels, we can interpret the behaviour specified by N_1 as 'extended' by N_2, and the behaviour specified by N_2 as 'expanded' by N_3. The idea of using morphisms for supporting abstraction is not at all

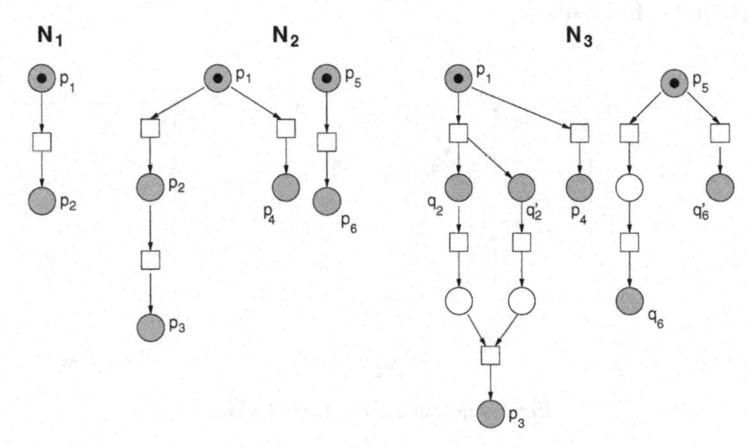

Fig. 2. Extension and expansion.

new in net theory. Starting with the definition proposed by Petri [26], different families of net morphisms have been proposed for this purpose in the literature: [14], [31], [32], [21], [24] are some of them. A comparison of morphisms for LST algebras with these notions has been sketched in [28].

In LST algebras each global state is a collection of local states (system cases are defined by sets of conditions), and this is the crucial point which allows one to distinguish N_1 and N_3 given in Figure 1 above. These two systems would be undistinguishable, but for implementation considerations, starting from a specification of the system behaviour given in terms of transition systems in which each global state is unstructured.

The duality *global state–local state* in net system state transformations has been investigated by A. Ehrenfeucht and G. Rozenberg introducing the notion of *region* for transition systems in [15].

A region of a transition system modelling a concurrent system behaviour is a set of states such that all occurrences of a given event have the same crossing relation with respect to the region itself, and this property holds for all events. It can be interpreted as a locally observable property.

In this way, regions allow one to extract the locality of state transformations from the specification of the behaviour of an EN system given in terms of a transition system, that is, in terms of global transformations. Regions are the basis of the synthesis procedure defined in [15] for obtaining, from an Elementary Transition (ET) system, the model of an EN system with case graph isomorphic to the given transition system. In particular, this synthesis procedure yields, among all the EN systems with isomorphic case graphs, the one with maximal set of conditions.

Obviously in this approach it is not possible to distinguish between the EN systems N_1 and N_3 of Figure 1 above: they have isomorphic case graphs.

In [24] the correspondence between ET systems and EN systems has been lifted to respect behaviour preserving morphisms, G-morphisms and N-morphisms, respectively, showing two functors which link the two corresponding categories and form an adjunction. Starting from the works in [15] and [24] a *theory of region* has been developed, see [1] for a survey.

In [4], [16] and subsequently in [8], the algebraic properties of the set of regions associated to an EN system have been studied. It turns out that such a set is a prime and coherent *orthomodular poset*, for simplicity called orthomodular poset in the following. Orthomodular posets can be seen as sets of partially overlapping Boolean algebras, and have been studied in the framework of algebraic models of quantum logic [29].

If we consider a system as composed by interacting sequential components, then the algebraic model of the observable properties of each component is a Boolean algebra, and the collection of interacting sequential components is represented by a particular family of overlapping Boolean algebras. Such a family is an orthomodular poset.

The maximal sets of mutually consistent observable properties, with respect to the structure of sequential components given by the family of Boolean algebras, can be interpreted as system states. These maximal sets are characterized by the notion of *prime filters*. Prime filters corresponds to states in transition systems.

In [8] the mutual relations among orthomodular posets, transition systems and net systems have been studied. In particular, a synthesis procedure is defined which associates, to a given prime and coherent orthomodular poset, a CE transition system, i.e., a transition system modelling the behaviour of a Condition Event system [17]. This is realized by considering as states of the CE transition systems prime filters of the orthomodular poset, and as events the order symmetric differences of prime filters. In this way, the new transition system contains all possible global states and all the possible state transformations, and this is due to the fact that orthomodular posets do not contain any information on the specific events to be considered.

As a result, an orthomodular poset is the abstraction of several CE transition systems, while a CE transition system is the abstraction of several CE net systems. (To an orthomodular poset more than one CE transition system can be associated; in turn, to a CE transition system more than one CE net system can be associated.) That is, the process associating an orthomodular poset to a net is, intuitively, a process of abstraction. Viceversa, the process of synthesis of a net from an orthomodular poset will be an implementation process.

The correspondence between prime and coherent orthomodular posets and CE transition systems has been lifted to respect G-morphisms and morphisms in orthomodular posets, respectively, showing two functors which link the two corresponding categories and form an adjunction [8].

We propose to consider *abstraction* and *composition* notions supported by the logic and algebraic structure of system local properties. For instance, one can observe some regions (local properties or conditions) and disregard some others, which are summarized (which are contained in set terms, and implied in logical terms) by the observed ones. This leads to the possibility of different perspectives in system observation, with the ability to model *different point of views* of the same system, in which some parts (or components) are more detailed, while others are seen at a more abstract level. This corresponds to the assumption that the model of a part of a system includes a detailed specification of the corresponding local component, and a, generally coarser, specification of the rest of the system, embodying the assumptions made by this part on the behaviour of the rest of the system.

Different models, each representing a different observation from a different perspective, of the same system, can be composed by identifying those regions representing the same observable property in the different models. The result of the composition is a new model comprising the details of both operands.

In other words, the two operands can be seen as different refinements of a common abstract view. The composite model describes the system at the highest level of detail comprising the two given operands, while respecting the same abstract view.

In order not to introduce any contradiction in the observation of those local properties, which – observed in the different views as the same property – have been identified and glued together in the new model, for each of these conditions, the events changing their value in the same way in the two operands have to be identified too; this is obtained by pairwise event superimposition.

This approach is quite new in the literature. The composition by identification of conditions has been widely used both in a theoretical frame and in applications, however the resulting net is usually obtained simply by 'place fusion' of the two net operands, without imposing any subsequent identification of events, see for example [18].

In [6] this new notion of composition of EN systems has been introduced and studied in the case in which only one property and its complement are common to the two components, and the operation has been discussed in the frame of modular design and as a tool to generalize the refinement operation proposed in [25].

In [7] the corresponding operation on ET systems is given and it is discussed an application to modular synthesis of EN systems.

In [3] it is shown that this operation is not a pullback in any category of transition/asynchronous systems with *flat* morphisms. Nevertheless, it is proved that the construction turns out to be a pullback in the richer category with *synchronising* morphisms, a generalization of morphisms studied in [2].

In the formal part of this paper, we give the definitions of the composition operation for EN systems and ET systems via two morphisms from the operands to a system modelling a common abstraction of the two operands, without imposing particular restrictions on the structure of this last system, and where the notions of morphisms are N-morphisms and G-morphisms, respectively, with some additional requirements, hence generalizing the compositions defined in [7]. Subsequently, the relations among the resulting composed systems in the two different contexts are discussed.

In particular, even if in the considered categories the operation is not a limit, like a pullback, however, the composed systems are related to their operands and, by morphism composition, to the common abstract system by morphisms respecting the additional requirements.

In the case of EN systems, the restrictions on N-morphisms here considered imply that if there is an N-morphism from a system Q to Q', then the subnet generated by the set of conditions of Q which are related to conditions in Q' is, after suitable simplification, isomorphic to Q'.

From this consideration it follows that the resulting net contains its operands, as well as the common abstraction, as subnets, up to simplification, because some synchronization events can have multiple instances. It may also happen that some conditions, other than the shared one, are forced to be identified.

This fact leads to an interpretation of the composition operation, which is slightly different from the previously discussed one. Each operand can be seen as made of the actual, local, component, and of an *interface* to the rest of the system. This interface is modelled by what we called the 'common abstraction' of the two operands. In other words, the two parts decide to 'agree' on interacting with each other via a common interface which defines the protocol of their interactions.

The here presented compositions support also *incremental design* in the following sense. Suppose to redesign each operand in a less abstract way, by means of a new system, which still fulfils the requirements specified by the more abstract operand, i.e., there is a morphism from the new to the old operand. Then the composition of these new less abstract operands yields a new system, which is less abstract than the old result, there is a morphism from the new to the old result.

In the following we present the composition operation on a few examples, with hints to the applications to modular synthesis and to refinement.

For what concerns the *synthesis* of EN systems, our proposal is motivated by the consideration that the applicability of the solution given in [15] is limited by two factors: first, the algorithm is NP-complete [1] with respect to the size of the transition system; second, the number of global states grows rapidly in relation with the number of system components.

We propose a technique for designing distributed systems in a modular way, blending the synthesis algorithm previously mentioned with the composition operation on net systems. This technique can be split into three main steps.

First, different aspects of the system behaviour, corresponding to different views of the system, are specified by means of different ET systems, in such a way that each specification conforms to a common abstraction.

In the second step, instead of composing these partial specifications into a unique model, and then synthesizing this global model, each behavioural specification is synthesized – with the traditional technique – into an EN system.

Finally, these different EN systems are composed, preserving the common abstraction. The behaviour of the resulting EN system coincides with the transition system which we would obtain by directly composing the partial specifications.

For example, let us consider the ET system A_I, shown in Figure 3, as a very abstract view of the system behaviour to be modelled, and let us imagine that this behaviour has

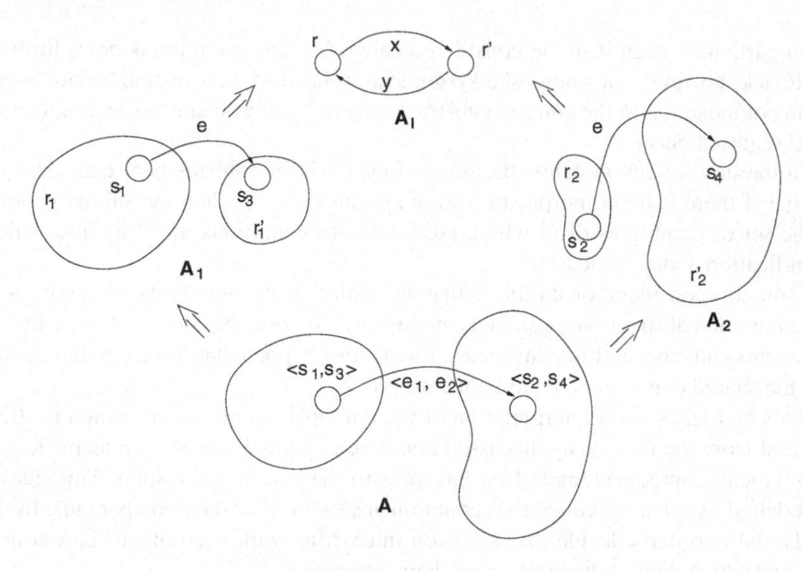

Fig. 3. Composition of transition systems.

been specified in a more detailed way from two different points of view obtaining the ET systems A_1 and A_2, which, both for space reasons and for being more general, are only sketched in Figure 3. Each morphism f_i, $i = 1, 2$, defines a point of view in the new specification, stating which properties (regions) are identified in the common abstraction, i.e. inducing a partition of the states of A_i into the two regions r_i, r'_i, which correspond through f_i to the non trivial regions r and r' of A_I.

Suppose we want to compose A_1 and A_2 into a new transition system A inglobing the two points of view, respecting the common abstraction A_I. That means that the two morphsms, f_1 and f_2, drive the construction of the new system.

A state in the composite system is a pair of states, one from each operand. The two 'local' states forming a composite state must agree on the properties (regions) that have been identified by the two morphisms; hence, in this case, the new state belongs to $(r_1 \times r_2) \cup (r'_1 \times r'_2)$.

Whenever $s_1 \xrightarrow{e} s_3$ is a transition in A_1, which does not cross the border of r_1, we will have a corresponding transition $(s_1, s_2) \xrightarrow{e} (s_3, s_2)$ in the composite system, for each pair of states whose first component is s_1 in the first state and s_3 in the second state of the pair, and similarly for transitions in A_2 which do not cross the border of r_2.

Transitions crossing the border of r_1 and r_2 must be treated differently. Any crossing from r_1 to r'_1 must be observed simultaneously with a crossing from r_2 to r'_2. Therefore, if $s_1 \xrightarrow{e_1} s_3$ is a transition in A_1 with $s_1 \in r_1$ and $s_3 \in r'_1$, and $s_2 \xrightarrow{e_2} s_4$ is a transition in A_2 with $s_2 \in r_2$ and $s_4 \in r'_2$, there will be a transition $(s_1, s_2) \xrightarrow{(e_1,e_2)} (s_3, s_4)$ in A. A similar construction applies to transitions crossing r_1 and r_2 in the other direction.

The result of the operation in this example is sketched in Figure 3. The two canonical projections of the set of states of A are actually morphisms of ET systems.

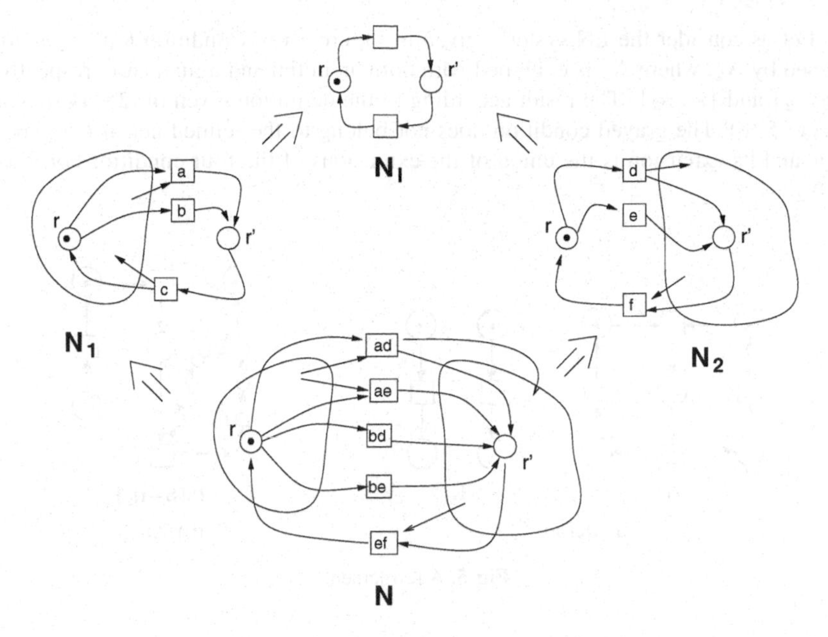

Fig. 4. Composition of net systems.

Coming back to the approach to design summarized above, instead of composing A_1 and A_2, we can consider A_1 and A_2 as behavioural specifications of a system from two perspectives. These specifications can be synthesized by means of the procedure introduced in [15], yielding two EN systems, respectively N_1 and N_2 as sketched in Figure 4. And now, we can compose these implementations.

In this case, we identify those conditions which are the image under the morphisms (β_1, η_1) and (β_2, η_2) of the same condition in N_I, the EN system obtained by synthesizing the abstract behaviour A_I.

As in the case of transition systems, this identification implies synchronizing events whose occurrence affects corresponding conditions in the two net systems.

As shown in the Figure, one event, a, must synchronize either with d or with e. This is realized by events (a, d) and (a, e) in the resulting EN system N.

The EN system N, just obtained by composing N_1 and N_2, is not the one we would get by synthesizing A; however the case graph of N is isomorphic to A, as stated later on in Theorem 4.

The consideration that the conditions of a net can be partially ordered, according to their extensions leads to a second application, where a condition is refined into a net. This is done so that, in the resulting net system, the extension of the refined condition coincides with the union of the extensions of all conditions making up the refining net system. This is very similar to – and actually generalizes – the refinement operation defined in [25]. A formal comparison with it is presented in [6]. Here, we show an example taken from [25], and reformulate it in our terms.

Let us consider the EN systems given in Figure 5 (a). Condition b in N_1 is to be refined by N_2, where N_2 is equipped with both an initial and a final case, respectively $\{s_1, s_2\}$ and $\{s_3, s_4\}$. The result according to the definition given in [25] is shown in Figure 5 (b). The grayed condition does not belong to the refined net: it corresponds to b, and its extension is the union of the extensions of the four conditions originally in N_2.

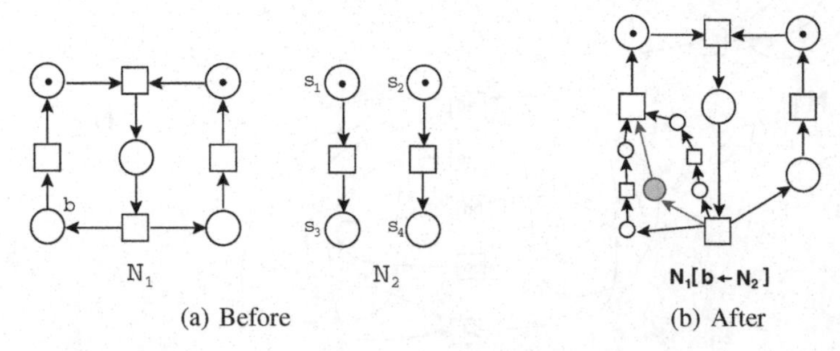

(a) Before (b) After

Fig. 5. A refinement.

In order to view this refinement as a composition of two EN systems, with respect to a third EN system representing the common abstraction, we have to construct this third EN system N_I and to apply a transformation to N_2.

In general, N_I is obtained by the subnet of N_1 generated by the condition b together with its complement b'. This subnet might not be simple w.r.t. events; in this case N_I is the subnet after event simplification, and its initial case is $\{b\}$ if b belongs to the initial case of N_1, $\{b'\}$ otherwise. In this specific case, N_I is as in Figure 6(a), N_2 is transformed by juxtaposing to it N_I and connecting the initial conditions of N_2 with an entering arc from the pre-event of b, if any, and the final conditions with an outgoing arc to the post-event of b, if any. In this way the extension of b equals the union of the extensions of the original conditions. N_2 modified is given in Figure 6(b).

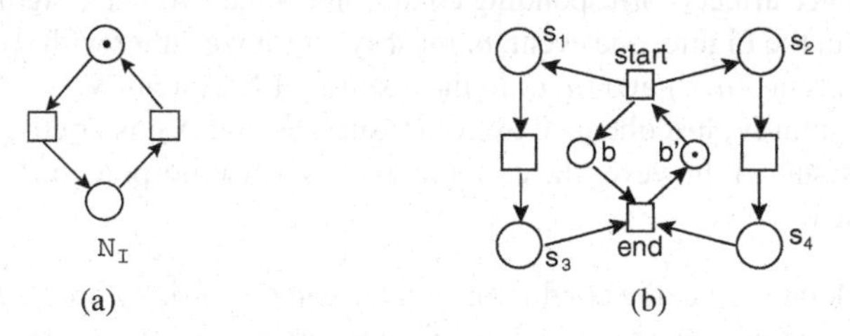

(a) (b)

Fig. 6. Refinement of a condition.

Obviously, there are two N-morphisms from N_1 and N_2 modified, respectively, to N_I and then we can apply our operation, getting the same net as before, except for the presence of b, and possibly b'; however, the behaviour is the same.

The ideas and results we have presented constitute a framework for the design and development of complex systems. In order to effectively exploit this formal framework we need to develop software tools implementing the operations and the verification of relations and properties above introduced. This is one of our short term plans.

On the theoretical side we intend to move in different directions. The composition operation can be defined and studied on orthomodular posets. Different notions of morphisms will be investigated in order to represent other relations between systems. N-morphisms and G-morphisms relate systems while maintaining the atomicity level of corresponding events. We are interested in morphisms supporting the refinement of events in the same line as morphisms between LST algebras. In this way, we would get a more general notion of local state observability.

We are interested in studying a composition operation on the basis of these new morphisms, and in investigating equivalence notions based on local state observability that could be congruences with respect to the operation.

We also aim at generalizing the notion of local state observability and the composition operation to other net classes such as Place Transition Nets.

In the following, after introducing some basic definitions, we summarize results concerning the relations among ETS, ENS, and orthomodular posets. Then we give the formal definitions of the composition operations, both for EN systems and ET systems, and state the main properties they satisfy.

2 Formal Definitions

2.1 Basic Definitions

In this section, we consider Elementary Net (EN) systems [30] and recall basic definitions of Elementary Transition (ET) systems and Orthomodular posets of regions, toghether with, respectively, N-morphisms, G-morphisms and O-morphisms, as introduced in [24], and in [8].

Definition 1. *Let $N_i = (B_i, E_i, F_i, c_{0i})$ be an EN system, for $i = 1, 2$. An N-morphism from N_1 to N_2 is a pair (β, η), where $\beta \subseteq B_1 \times B_2$, and $\eta : E_1 \rightharpoonup E_2$ is a partial function, such that:*

1. *$\beta^{-1} : B_2 \rightharpoonup B_1$ is a partial function;*
2. *if $\eta(e_1)$ is undefined, then $\beta(\mathrm{pre}_1(e_1)) = \emptyset = \beta(\mathrm{post}_1(e_1))$;*
3. *if $\eta(e_1) = e_2$, then $\beta(\mathrm{pre}_1(e_1)) = \mathrm{pre}_2(e_2)$ and $\beta(\mathrm{post}_1(e_1)) = \mathrm{post}_2(e_2)$;*
4. *$\forall (b_1, b_2) \in \beta : [b_1 \in c_{01} \Leftrightarrow b_2 \in c_{02}]$.*

Elementary Net systems together with N-morphisms form a category, denoted ENS.

An ET system is a transition system isomorphic to the case graph of an EN system and its definition is based on the notion of *region*.

Definition 2. A transition system *is a structure $A = (S, E, T, s_0)$, where S is a non-empty finite set of states, E is a finite alphabet of actions or labels, $T \subseteq S \times E \times S$ is the set of labelled edges or transitions, and s_0 is the initial state, satisfying the following properties:*

1. *each state is reachable from the initial state;*
2. $\forall (s, e, s') \in T : s \neq s'$;
3. $\forall (s, e_1, s_1), (s, e_2, s_2) \in T : [s_1 = s_2 \Rightarrow e_1 = e_2]$;
4. $\forall e \in E : \exists (s, e, s') \in T$.

Definition 3. *Let* $A = (S, E, T, s_0)$ *be a transition system. A set of states* $r \subseteq S$ *is a region of* A *iff* $\forall e \in E$, $\forall (s_1, e, s_1'), (s_2, e, s_2') \in T$ *we have:*

$$(s_1 \in r \wedge s_1' \notin r) \Rightarrow (s_2 \in r \wedge s_2' \notin r) \text{ and } (s_1 \notin r \wedge s_1' \in r) \Rightarrow (s_2 \notin r \wedge s_2' \in r).$$

The set of regions of an ET system A_i will be denoted by R_i. The set of regions containing a state s will be denoted by R_s.

If r is a region, then the set difference $S \setminus r$ is also a region, where S is the set of states. We will refer to $S \setminus r$ as the *complement* of r and will denote it by r'.

Definition 4. *A transition system* $A = (S, E, T, s_0)$ *is an* Elementary Transition system *iff it satisfies the following separation axioms:*

A1. $\forall s, s' \in S : R_s = R_{s'} \Rightarrow s = s'$
A2. $\forall s \in S, \forall e \in E : \text{pre}(e) \subseteq R_s \Rightarrow \exists s' \in S \ (s, e, s') \in T$

Definition 5. *Let* $A_i = (S_i, E_i, T_i, s_{0i})$ *be Elementary Transition systems, for* $i = 1, 2$. *A G-morphism from* A_1 *to* A_2 *is a total function* $f : S_1 \rightarrow S_2$ *such that:*

1. $f(s_{01}) = s_{02}$
2. $\forall (s_i, e_j, s_k) \in T_1 : [\, f(s_i) = f(s_k) \text{ or } \exists e_h \in E_2 : (f(s_i), e_h, f(s_k)) \in T_2 \,]$
3. *if* $(s_i, e_j, s_k) \in T_1$ *and* $(f(s_i), e_h, f(s_k)) \in T_2$, *then* $(f(s_m), e_h, f(s_n)) \in T_2$ *for all* $(s_m, e_j, s_n) \in T_1$

Each G-morphism $f : A_1 \rightarrow A_2$ uniquely determines a partial function $g : E_1 \rightharpoonup E_2$ defined by

$$g(e_1) = \begin{cases} e_2 & \text{if } (f(s), e_2, f(s')) \in T_2 \text{ for some } (s, e_1, s') \in T_1 \\ \textit{undefined} & \text{otherwise} \end{cases}$$

A basic property of G-morphism is that they preserve regions in the following sense.

Proposition 1. *Let* $f : A_1 \rightarrow A_2$ *be a G-morphism. Suppose* $r \subseteq S_2$ *is a region in* A_2. *Then* $f^{-1}(r)$ *is a region in* A_1.

Elementary Transition systems together with G-morphisms form a category, denoted ETS.

Orthomodular posets can be considered as a generalization of Boolean algebras, where meet ($\wedge$) and join ($\vee$) are partial operations, while each element has a complement.

Definition 6. *An* orthomodular poset $P = (P, \leq, 0, 1, (\,.\,)')$ *is a partially ordered set* $(P, \leq)$, *equipped with a minimum and a maximum element, respectively denoted by 0 and 1, and with a unary complement operation* $(\,.\,)' : P \rightarrow P$, *which satisfy the following conditions:* $\forall x, y \in P$

1. $(x')' = x$;
2. $x \leq y \Rightarrow y' \leq x'$;
3. $x \leq y \Rightarrow y = x \vee (y \wedge x')$;
4. $x \leq y' \Rightarrow x \vee y \in P$;
5. $x \wedge x' = 0$.

Elements of P are called *observations*. The third condition above is known as the *orthomodular law*. In an orthomodular poset two elements x and y are said *orthogonal*, denoted $x \perp y$, iff $x \leq y'$; while x and y are said *compatible*, denoted $x \ \$ \ y$, iff there exist pairwise orthogonal $x_0, y_0, z \in P$ such that: $x = x_0 \vee z$ and $y = y_0 \vee z$.

Notation $\$A$ is used to denote that observations in A are pairwise compatible.

Definition 7. *An orthomodular poset* $P = (P, \leq, 0, 1, (\,.\,)')$ *is* coherent *when* $\forall x, y, z \in P$ *such that* $\$\{x, y, z\}$, *it holds:* $(x \vee y)\$z$.

As shown in [29], an orthomodular poset P is coherent if and only if every pairwise compatible subset of it (every clique of \$ in P) admits an enlargement to a Boolean subalgebra of P.

A coherent orthomodular poset can be represented as a *transitive partial Boolean algebra*, i.e.: a family of partially overlapping Boolean algebras. As discussed in [8], if the elements of an orthomodular poset are the observable properties of a concurrent system, then each Boolean algebra models the observable properties of a system component.

We now present *filters* in coherent orthomodular posets as a generalization of filters in Boolean algebras. Filters, and in particular *prime filters*, play an important role since they correspond to states in ET systems.

Definition 8. *Let* $P = (P, \leq, 0, 1, (\,.\,)')$ *be an orthomodular poset;* $f \subseteq P$ *is a* filter *in* P *if* $\forall x, y \in P$*:*

1. $f \neq \emptyset$;
2. $(x \in f$ and $x \leq y) \Rightarrow y \in f$;
3. $(x, y \in f$ and $x\$y) \Rightarrow x \wedge y \in f$.

A filter f is said to be *proper* if $f \neq P$, so $0 \notin f$ and $1 \in f$. A proper filter f is *prime* if, $\forall x, y \in P$:

4. $(x\$y$ and $x \vee y \in f) \Rightarrow (x \in f$ or $y \in f)$.

Definition 9. *Let* P_1 *and* P_2 *be orthomodular posets. An O-morphism from* P_1 *to* P_2 *is a map* $\beta : P_1 \rightarrow P_2$ *such that:* $\forall x, y \in P_1$

$$\beta(0) = 0;$$
$$\beta(x') = (\beta(x))';$$
$$x \perp y \Rightarrow \beta(x \vee y) = \beta(x) \vee \beta(y).$$

O-morphisms preserve order, orthogonality and compatibility. Moreover, the inverse image of a filter with respect to an O-morphism is again a filter, as stated in the following.

Proposition 2. *Let P_1 and P_2 be orthomodular posets, and β a morphism from P_1 to P_2. Let $f \subset P_2$ be a filter of P_2. Then $\beta^{-1}(f)$ is a filter of P_1. If f is prime, then $\beta^{-1}(f)$ is prime.*

Definition 10. *Let $P = (P, \leq, 0, 1, (.)')$ be an orthomodular poset and PF(P) denote the set of prime filters of P. P is* prime *iff $\forall x, y \in P : x \neq y \Rightarrow (\exists f \in FP(P)$ such that $x \in f \Leftrightarrow y \notin f)$.*

Finite, prime and coherent orthomodular posets together with O-morphisms, as defined above, form a category, denoted PCOP [8].

2.2 Relations among Models

In [24] the authors define a pair of functors between the categories of Elementary Net systems and of Elementary Transition systems, proving that they form an adjunction. The first functor coincides with the computation of the case graph of an EN system; the second one gives a procedure of synthesis which, given an ET system, builds an EN system whose case graph is isomorphic to the transition system. Among all EN systems with isomorphic case graphs, the synthesis procedure yields the system with maximal set of conditions.

The set of regions of an ET system can be partially ordered by set inclusion. The result is a coherent prime orthomodular poset, where each state of the ETS corresponds to a prime filter.

Proposition 3. *Let $A = (S, E, T, s_0)$ be an Elementary Transition system. Let $R(A)$ be the set of all regions of A. Then $(R(A), \subseteq, \emptyset, S, (S \setminus .))$ is a coherent prime orthomodular poset. Let $s \in S$, and $F_s = \{r \in R(A) \mid s \in r\}$; then F_s is a prime filter.*

The correspondence between states and filters suggests a canonical way to construct a transition system, starting from an abstract coherent prime orthomodular poset: take all prime filters as states, and ordered symmetric differences as labels of transitions.

In this way, the transition system contains all global states and all state transformations consistent with the given structure of local properties.

Orthomodular posets do not contain any information neither on properties initially valid nor on the specific events to be considered.

In Condition Event systems a class of forward and backward reachable cases is considered instead of a particular initial case. In [8] the correspondence between the categories of prime and coherent orthomodular posets and of Condition Event transition systems has been formalized by means of two functors which form an adjunction.

2.3 Composing Elementary Net Systems

We now define the composite of two Elementary Net systems, N_1 and N_2, with respect to a third Elementary Net system N_I. The composition is driven by a pair of N-morphisms, (β_1, η_1) and (β_2, η_2), respectively from N_1 to N_I, and from N_2 to N_I.

The morphisms are subject to the following restrictions: $\beta_i^{-1} : B_I \to B_i$ is total and injective; $\eta_i : E_i \rightharpoonup E_I$ is surjective. In the following, we will use the term $\hat{\text{N}}$-morphism to denote an N-morphism satisfying these restrictions.

The following proposition is easily established.

Proposition 4. *Let (N, c_{in}) and (N', c'_{in}) be EN systems and (β, η) be an N-morphism from (N, c_{in}) to (N', c'_{in}), such that $\beta_i^{-1} : B' \to B$ is total and injective, and $\eta : E \rightharpoonup E'$ is surjective. Let N'' be the subnet of N generated by the set of conditions $B'' = \beta_i^{-1}(B')$, and let $\mathrm{simp}\,(N'')$ be obtained from N'' by event simplification. Then $\mathrm{simp}\,(N'')$ is isomorphic to N'.*

In the following, let D_i denote the domain of the binary relation β_i, and G_i denote the domain of the partial function η_i:

$$D_i = \{b \in B_i \mid \beta_i(b) \neq \emptyset\} \qquad G_i = \mathrm{dom}(\eta_i)$$

Definition 11. *Consider the following diagram in the category of Elementary Net systems:*

$$N_2 \xrightarrow{\eta_2, \beta_2} N_I \xleftarrow{\eta_1, \beta_1} N_1$$

with the additional restrictions that η_i is surjective, and β_i^{-1} is total injective, for $i = 1, 2$.

We define $N_1 \langle N_I \rangle N_2 = N = (B, E, F, c_{in})$ as follows:

1. $B = (B_1 \setminus D_1) \cup (B_2 \setminus D_2) \cup B_I$,
2. $E = (E_1 \setminus G_1) \cup (E_2 \setminus G_2) \cup E_{sync}$, *where* $E_{sync} = \{\langle e_1, e_2 \rangle\} \mid e_1 \in G_1, e_2 \in G_2, \eta_1(e_1) = \eta_2(e_2)$,
3. *F is defined by the following clauses:*
   (a) $\forall b \in (B_i \setminus D_i), \forall e \in (E_i \setminus G_i), i = 1, 2$

   $$(b, e) \in F \Leftrightarrow (b, e) \in F_i \qquad (e, b) \in F \Leftrightarrow (e, b) \in F_i$$

   (b) $\forall b \in (B_i \setminus D_i), \forall e \in G_1, \forall e_2 \in E_2$

   $$(b, \langle e, e_2 \rangle) \in F \Leftrightarrow \langle e, e_2 \rangle \in E, (b, e) \in F_1$$
   $$(\langle e, e_2 \rangle, b) \in F \Leftrightarrow \langle e, e_2 \rangle \in E, (e, b) \in F_1$$

   (c) $\forall b \in (B_i \setminus D_i), \forall e \in G_2, \forall e_1 \in E_1$

   $$(b, \langle e_1, e \rangle) \in F \Leftrightarrow \langle e_1, e \rangle \in E, (b, e) \in F_2$$
   $$(\langle e_1, e \rangle, b) \in F \Leftrightarrow \langle e_1, e \rangle \in E, (e, b) \in F_2$$

   (d) $\forall b \in B_I, \forall e = \langle e_1, e_2 \rangle \in E_{sync}$

   $$(b, e) \in F \Leftrightarrow (\beta_1^{-1}(b), e_1) \in F_1, (\beta_2^{-1}(b), e_2) \in F_2$$
   $$(e, b) \in F \Leftrightarrow (e_1, \beta_1^{-1}(b)) \in F_1, (e_2, \beta_2^{-1}(b)) \in F_2$$

4. $c_{in} = (c_{in}^1 \setminus D_1) \cup (c_{in}^2 \setminus D_2) \cup c_{in}^I$.

From this construction it follows immediately that $N = N_1 \langle N_I \rangle N_2$ as defined above is an Elementary Net system. Moreover, the net system N maps onto N_1 and N_2.

Definition 12. *Define $\gamma_i \subseteq B \times B_i$ and $\delta_i : E \longrightarrow E_i$ as follows:*

- $\gamma_1 = \{(b,b) \mid b \in B_1 \setminus D_1\} \cup \{(b, \beta_1^{-1}(b)) \mid b \in B_I\}$,
- $\gamma_2 = \{(b,b) \mid b \in B_2 \setminus D_2\} \cup \{(b, \beta_2^{-1}(b)) \mid b \in B_I\}$,
- $\forall e \in E_1 \setminus G_1 : \ \delta_1(e) = e, \delta_2(e) = undefined$,
- $\forall e \in E_2 \setminus G_2 : \ \delta_1(e) = undefined, \delta_2(e) = e$,
- $\forall \langle e_1, e_2 \rangle \in E : \delta_1(\langle e_1, e_2 \rangle) = e_1, \delta_2(\langle e_1, e_2 \rangle) = e_2$.

Theorem 1. *The pair (γ_i, δ_i) is an $\hat{\text{N}}$-morphism from $N = N_1 \langle N_I \rangle N_2$ to $N_i, i = 1, 2$. and the following diagram commutes.*

$$\begin{array}{ccc} N_I & \xleftarrow{\eta_1, \beta_1} & N_1 \\ \uparrow {\scriptstyle \eta_2, \beta_2} & & \uparrow {\scriptstyle \gamma_1, \delta_1} \\ N_2 & \xleftarrow{\gamma_2, \delta_2} & N \end{array}$$

From the previous commutative diagram and from Proposition 4 it follows that N contains N_1, N_2 and N_I as subnets. However, as discussed in [3], the operation is not a pullback in the category of ENS.

The next result states that the composition respects N-morphisms. Let us start from the diagram above and suppose N_1 is redesigned, in a less abstract way, by means of a new EN system N_1', which still fulfils the requirements specified by N_1, i.e.: there is an $\hat{\text{N}}$-morphism $(\beta_i', \eta_i') : N_i' \rightarrow N_i$, and suppose the same holds for N_2 with respect to N_2'. Then, by composition of morphisms, N_1' and N_2' turn out to be composable with respect to N_I. Moreover, their composition $N' = N_1' \langle N_I \rangle N_2'$ maps uniquely, by an $\hat{\text{N}}$-morphism, to N, the composite of the less detailed models N_1 and N_2. This fact is formalized in the following proposition.

Proposition 5. *Let $N = N_1 \langle N_I \rangle N_2$, and, for $i = 1, 2$, N_i' be EN systems such that there exist $\hat{\text{N}}$-morphisms $(\beta_i', \eta_i') : N_i' \rightarrow N_i$. Then N_1' and N_2' are composable w.r.t. N_I, yielding $N' = N_1' \langle N_I \rangle N_2'$ and there is a unique $\hat{\text{N}}$-morphism $(\beta, \eta) : N' \rightarrow N$ such that the following diagram commutes.*

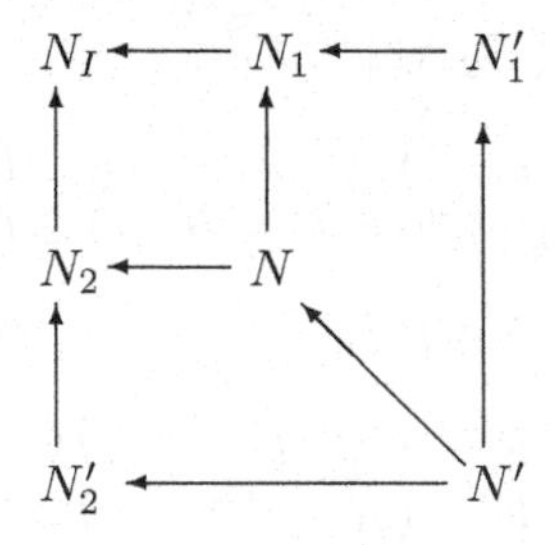

2.4 Composing Elementary Transition Systems

We now define the composition of two Elementary Transition systems, A_1 and A_2, with respect to a third Elementary Transition system A_I. The composition is driven by a pair of G-morphisms, f_1 and f_2, respectively from A_1 to A_I, and from A_2 to A_I.

The morphisms are subject to the following restrictions, for $i = 1, 2$: both $f_i : S_i \rightarrow S_I$ and $g_i : E_i \rightharpoonup E_I$, the partial functions derived from f_i, are surjective. In the following, such morphisms will be called $\hat{\text{G}}$-morphisms.

Let $L_i, i = 1, 2$ be the set of events which are in E_i and not in the domain of the partial function g_i; and let H be the set of pairs of events $\langle e_1, e_2 \rangle$ which are mapped to the same event in A_I by g_1 and g_2:

$$L_i = \{e \in E_i \mid \exists (s_1, e, s_2) \in T_i \wedge f_i(s_1) = f_i(s_2)\}$$
$$H = \{\langle e_1, e_2 \rangle \mid g_1(e_1) = g_2(e_2)\}$$

Definition 13. *Consider the following diagram in the category of Elementary Transition systems with the above restrictions on the morphisms f_1 and f_2.*

$$A_2 \xrightarrow{f_2} A_I \xleftarrow{f_1} A_1$$

We define $A_1 \langle A_I \rangle A_2 = A = (S, E, T, s_{in})$ as follows:

1. $S = \{(s_1, s_2) \in S_1 \times S_2 \mid f_1(s_1) = f_2(s_2)\}$,
2. $E = L_1 \cup L_2 \cup H$,
3. *T is such that $((s_1, s_2), e, (s_3, s_4)) \in T$ iff one of the following conditions is true:*
   (a) $(s_1, e_1, s_3) \in T_1 \wedge (s_2, e_2, s_4) \in T_2 \wedge \langle e_1, e_2 \rangle \in H$,
   (b) $(s_1, e, s_3) \in T_1, s_2 = s_4 \wedge e \in L_1$,
   (c) $s_1 = s_3, (s_2, e, s_4) \in T_2 \wedge e \in L_2$.
4. $s_{in} = (s_{in}^1, s_{in}^2)$.

It is immediate to prove that A is a transition system. We will sketch the proof that its reachable part is an ETS.

Theorem 2. *Let $\zeta_i \subseteq S \times S_i, i = 1, 2$, be the projection of an element of S into S_i, $i = 1, 2$, i.e.:*

$$\zeta_i = \{((s_1, s_2), s_i) \mid s_i \in S_i\}$$

Then ζ_i is a $\hat{\text{G}}$-morphism from $A = A_1 \langle A_I \rangle A_2$ to $A_i, i = 1, 2$, and the following diagram commutes.

$$\begin{array}{ccc} A_I & \xleftarrow{f_1} & A_1 \\ \uparrow f_2 & & \uparrow \zeta_1 \\ A_2 & \xleftarrow{\zeta_2} & A \end{array}$$

Strictly speaking, G-morphisms are defined for ETS, however, their definition and properties hold for more general transition systems.

Theorem 3. *The reachable part of $A = A_1 \langle A_I \rangle A_2$ as defined above is an Elementary Transition system.*

The theorem can be proved by first showing that A satisfies the axioms of separation, and then showing that the reachable part of a transition system satisfying the axioms of separation is elementary. The first part is easily established by using the properties of separation in A_1 and A_2, and a fundamental property of G-morphisms, namely that the inverse image of a region is a region.

As for ENS, also in this case the operation does not coincide with a pullback construction in the category of ETS. However, in [3] it is proved that a similar construction turns out to be a pullback in the richer category of asynchronous systems with *synchronising* morphisms, a generalization of morphisms studied in [2].

As well as the composition operation for EN systems respects N-morphisms, the composition for ET systems respects G-morphisms with the above restrictions: the same commutative diagram as that obtained in Proposition 5 holds for ET systems.

2.5 Relations between the Composition of ENS and the Composition of ETS

Let $A = A_1\langle A_I\rangle A_2$ be the ET system obtained by composing A_1 and A_2 with respect to A_I. Let $N(A_i)$, $i = 1, 2$, and $N(A_I)$ be the saturated EN systems obtained by applying to A_i and A_I, respectively, the synthesis procedure described in [15]. Then we get the following.

Theorem 4. *$N(A_1)$ and $N(A_2)$ can be composed with respect to $N(A_I)$, and the case graph of $N = N(A_1)\langle N(A_I)\rangle N(A_2)$ is isomorphic to A.*

This theorem can be proved by showing that all conditions in N correspond, by construction, to regions of A, and that these regions are enough to satisfy the axioms of separation for A.

This result is summarized in the following commutative diagrams, in which the arrows represent morphisms in the appropriate categories.

$$\begin{array}{ccc} A_I & \xleftarrow{h_1} & A_1 \\ h_2 \uparrow & & \uparrow g_1 \\ A_2 & \xleftarrow{g_2} & A \end{array} \qquad \begin{array}{ccc} N(A_I) & \xleftarrow{N(h_1)} & N(A_1) \\ N(h_2) \uparrow & & \uparrow n_1 \\ N(A_2) & \xleftarrow{n_2} & N \end{array}$$

References

1. Badouel E., Darondeau Ph.: Theory of Regions, Reisig W., Rozenberg G. (Eds.), Lectures on Petri Nets I: Basic Models, Advances in Petri Nets 1998, LNCS **1491**, Springer-Verlag, pp.529-586, 1998.
2. Bednarczyk, M. A.: Categories of asynchronous systems. PhD Thesis, University of Sussex, 1-88, 1988.
3. Bednarczyk M.A., Bernardinello L., Caillaud B., Pawlowski W., Pomello L.: Modular System Development with Pullbacks. van der Aalst W., Best E. (eds.), Application and Theory of Petri Nets 2003, LNCS **2679**, Springer-Verlag, pp. 140–160, 2003.
4. Bernardinello L.: Propriétés algébriques et combinatoires des régions dans les graphes, et leurs application à la synthèse de réseaux, PhD thesis, Université de Rennes I, 1998.

5. Bernardinello L., De Cindio F.: A Survey of Basic Net Models and Modular Net Classes. Rozenberg G. (ed.), Advances in Petri Nets 1992, LNCS **609**, Springer-Verlag, pp.: 304–351, 1992.
6. L. Bernardinello, C. Ferigato, L. Pomello, Composing Net Systems by identification of conditions, DISCO Internal Report, `http://www.mcd.disco.unimib.it/pub/compcond.ps`, 2001.
7. Bernardinello L., Ferigato C., Pomello L.: Towards modular synthesis of EN systems. Caillaud B. et al. (eds.) Synthesis and Control of Discrete Event Systems, Kluwer Academic Publishers, pp.103–113, 2002.
8. Bernardinello L., Ferigato C., Pomello L.: An Algebraic Model of Observable Properties in Distributed Systems. TCS **290**, pp.:637–668, 2003.
9. Best E., Devillers R., Koutny M.: Petri Net Algebra. EATCS Monographs on Theoretical Computer Science, Springer Verlag, 2001.
10. Brawer W., , Gold R., Vogler W.: Behaviour and equivalences Preserving Refinements of Petri Nets, Rozenberg G. (ed.), Advances in Petri Nets 1992, LNCS **483**, Springer-Verlag, pp. 1–46, 1991.
11. Castellano L., De Michelis G., Pomello L.: Concurrency vs Interleaving: an instructive example. EATCS Bull., **31**, pp. 12–15, 1987.
12. De Cindio F., De Michelis G., Pomello L., Simone C.: Superposed Automata Nets, Reisig (ed.), Application and Theory of Petri Nets 1982, IFB **52**, Springer, pp. 269–279, 1982.
13. De Cindio F., De Michelis G., Pomello L., Simone C.: A State Transformation Equivalence for Concurrent Systems: Exhibited Functionality Equivalence. Concurrency 88, LNCS **335**, Springer-Verlag, pp: 222–236, 1988.
14. Desel J., Merceron A.: Vicinity Respecting Net Morphisms. Rozenberg G. (ed.), Advances in Petri Nets 1990, LNCS **483**, Springer-Verlag, pp.: 165–185, 1990.
15. Ehrenfeucht A., Rozenberg G.: Partial (set) 2 structures, I & II. Acta Informatica, **27**, 4, pp.: 315–368, 1990.
16. Ferigato C.: Note su alcune proprietá algebriche, logiche e topologiche della concorrenza. Ph.D. Thesis, Universitá degli Studi di Milano e Torino, 1996.
17. Genrich H.J., Lautenbach K., Thiagarajan P.S.: Elements of General Net Theory, Goss G., Hartmanis J. (eds.) Net Theory and Applications, Advanced Course on General Net theory of Processes and Systems, LNCS **84** Springer Verlag, pp.21–164, 1980.
18. Girault C., Valk R.: Petri Nets for Systems Engineering, A Guide to Modelling, verification, and applications Springer Verlag, 2003
19. van Glabbeck R.: Comparative Concurrency Semantics and Refinement of Actions. PhD. Thesis, Centrum voor Wiskunde en Informatica, Amsterdam, 1990.
20. Hoare C.A.R.: Communicating Sequential Processes, Communication of the ACM, Vol. **21**, N. 8, pp. 666–671, 1978.
21. Meseguer J., Montanari U.: Petri Nets are Monoids, Information and Computation **88**, 105-155, Academic Press, 1990.
22. Milner R.: A Calculus of Communicating Systems, LNCS **92**, Springer-Verlag, 1980.
23. Milner R.: Elements of Interaction, Turing Award Lecture. Communications of the ACM **36**, N.1, pp.: 78–89, 1993.
24. Nielsen M., Rozenberg G., Thiagarajan P.S.: Elementary transition systems. Theoretical Computer Science, **96**, 1, pp.: 3–32, 1992.
25. Nielsen M., Rozenberg G., Thiagarajan P.S.: Elementary transition systems and refinement. Acta Informatica, **29**, pp.: 555–578, 1992.
26. Petri C.A.: Concepts of Net theory. Mathematical Foundations of Computer Science. Proc. Symp. and Summer School, High Tatras, pp.: 137–146,1973.

27. Pomello L., Rozenberg G., Simone C.: A Survey of Equivalence Notions for Net Based Systems. Rozenberg G. (ed.), Advances in Petri nets 1992, LNCS **609**, Springer-Verlag, pp.: 410–472, 1992.
28. Pomello L., Simone C.: An Algebraic Characterization of Elementary Net System (Observable) State Space, Formal Aspects of Computing (1992) **4**, pp. 612–637, 1992.
29. P. Pták, S. Pulmannová, Orthomodular Structures as Quantum Logics. Kluwer Academic Publishers, 1991.
30. Rozenberg G., Engelfriet J.: Elementary Net Systems, Reisig W., Rozenberg G. (eds.), Lectures on Petri Nets I: Basic Models, Advances in Petri nets, LNCS **1491**, Springer-Verlag, pp.:12–121, 1998.
31. Smith E. Reisig W.: The Semantics of a Net is a Net - an Exercise in General Net Theory. Concurrency and Nets, Voss K., Genrich H.J., Rosenberg G. (eds.), Springer Verlag, pp.: 461–479, 1987.
32. Winskel G.: Petri Nets, Algebras, Morphisms and Compositionality, Inf. & Comput., **72**, No. 3, pp. 197-238, 1987.

Stochastic Methods for Dependability, Performability, and Security Evaluation

William H. Sanders

Coordinated Science Laboratory and
Department of Electrical and Computer Engineering
University of Illinois at Urbana-Champaign
whs@uiuc.edu

Abstract. Stochastic methods are commonly used for dependability evaluation. In the mid 1970's, stochastic evaluation was proposed for combined performance/dependability evaluation, called performability evaluation. Extending reliability evaluation to include performance related behaviors presented new challenges, most notably due to the large difference in time scale of performance- and dependability-related events. Stochastic Petri nets, invented shortly thereafter, played an integral part in the development of performability evaluation methods. Most recently, stochastic evaluation has been proposed to quantify the security or survivability that a system provides, taking into account malicious attacks on the system. As with performability evaluation, attempts to stochastically evaluate system security and survivability met new challenges, and researchers have attempted to use stochastic net models to quantify the security and survivability a system will provide.

This invited presentation will survey the challenges, advances, and future research directions in the use of stochastic evaluation for dependability, performability, and security evaluation, paying particular attention to methods that make use of stochastic Petri nets and extensions. More specifically, I describe the challenges that were encountered as stochastic evaluation of successively more complex system properties was attempted, and show how net representations, together with new stochastic methods, enable their evaluation. In doing so, I show the relationship between methods for evaluating dependability, performability, and security, paying particular attention to issues that remain in creating a methodology for stochastically quantifying the security and survivability that a system provides to an end user.

J. Cortadella and W. Reisig (Eds.): ICATPN 2004, LNCS 3099, p. 97, 2004.
© Springer-Verlag Berlin Heidelberg 2004

Composition of Temporal Logic Specifications

Adrianna Alexander

Humboldt-Universität zu Berlin
Institut für Informatik
Unter den Linden 6, 10099 Berlin, Germany
alexander@informatik.hu-berlin.de

Abstract. The Temporal Logic of Distributed Actions (TLDA) is a new temporal logic designed for the specification and verification of distributed systems. TLDA supports a modular design of systems: Subsystems can be specified and verified separately and then be integrated into one system. The properties of the subsystems will be preserved in this process. TLDA can be syntactically viewed as an extension of TLA. We propose a different semantical model based on partial order which increases the expressiveness of the logic.

Keywords: temporal logic, specification, verification, compositional semantics, partial order

Introduction

The majority of existing specification techniques describes a system in an abstract formalism, e.g. as a transition system, a Petri net or a state chart. From such a model the behaviour of the system is usually derived, i.e. the set of all possible courses of actions (or states) occurring in the system. To analyse the systems means to reason about its behaviour in terms of a temporal logic. Thus, a system is described in an abstract formalism and its properties are specified as logical formulas.

Another way is to specify a system itself as a logical formula. We call such a formula a *specification* of a system. This formula generates a set of models, that is the set of actions (or states) courses of the system, in which the specification holds. This idea originates from Pnueli [19], and was a bit later applied, among others, by Lamport [15]. His Temporal Logic of Actions (TLA, [16, 1, 5]) is based on this idea. This approach has some considerable advantages:

A *property* of a system is also represented as a logical formula. Thus, no formal distinction is made between a system and a property. Hence, proving that a system possess a property is reduced to proving a logical *implication*.

Furthermore, compositional reasoning is eased significantly: A large system is composed of smaller components. Properties of the composed system should be derivable from the properties of its components. The components are represented as logical formulas. It can be shown that *parallel composition* of the components can basically be represented by *conjunction* of the formulas representing the components.

J. Cortadella and W. Reisig (Eds.): ICATPN 2004, LNCS 3099, pp. 98–116, 2004.
© Springer-Verlag Berlin Heidelberg 2004

These valuable features of TLA are retained in our logic, called *Temporal Logic of Distributed Actions*(TLDA for short). *Distributed* refers to the behavioural model of TLDA which is different from the model of TLA.

System behaviour is usually described by help of *variables*: A variable updates its value in the course of time. TLA describes a system execution as a *stuttering sequence*, e.g. an *interleaving* of variable updates in which each update can be repeated finitely often. TLDA, in contrast, describes a system execution as a *partially ordered* set of variable updates which will be called a *run*. This allows us to distinguish between runs describing *concurrent* and *nondeterministic* variable updates. Moreover, it can be determined, whether an update of a variable does not change the value of the variable in a run or whether the variable is not updated at all (see [8] for examples of such properties).

We assume that every system and, consequently, every run of a system have infinitely many variables, but we will explicitly describe updates of a finite subset of them only. These variables will be called *system variables*. The values of all other variables may change arbitrarily. Therefore, a set of all runs of a system will always be infinite. This set will be called the *behaviour* of the system.

Updates of variables occur either *concurrently* or are forced to occur *coincidently* in a run at a given time. This principle resembles the main idea of a Petri net. As a matter of fact, runs can be viewed as a particular class of occurrence nets (as defined in [10, 21] for instance).

The paper is organised as follows: We start in Sect.1 with an example of a system consisting of three components: a producer, consumer and buffer. In order to specify the system, we introduce first the basics of the logic TLDA (Sect. 2). We focus on modular design of systems: We show how to specify (Sect. 3) and verify (Sect. 4 and 5) the system components using TLDA. In Sect. 6 we integrate the components to one system and show that the properties of the components will be preserved by the composition.

1 Introductory Example

We are going to describe a behaviour of a *producer-consumer* system with a *buffer* having 1-item capacity: The producer spontaneously produces an item and retains it as long as the buffer is not empty. After the buffer gets empty, the item can be delivered to the buffer and the producer is ready to produce a new item. If there is an item in the buffer, the consumer can remove it from the buffer. Afterwards, he eventually consumes the item and is then ready to remove a new item from the buffer.

Instead of treating the entire system, we decompose it first into several subsystems. This is not always a simple task, but in this case we can well consider the producer-consumer system as consisting of three interacting components: a *producer*, a *consumer* and a *buffer* (cp. [21] for example). Now we describe the behaviour of the single components.

The producer alternates between two states: "ready to produce" (*rp* for short) and "ready to deliver" (*rd*). Suppose that the producer is initially in the state "ready to produce". We assume a variable *prod* which can take the values *rp* or *rd* to display the current state of the producer. Hence, the sequence $rp, rd, rp, \ldots$ of values of variable *prod* plus the set of updates of *prod*, each update explicitly depicted as a box, form a run of the producer system:

$$prod\colon\ rp \rightarrow\square\rightarrow rd \rightarrow\square\rightarrow rp \rightarrow\square\rightarrow rd \ldots \tag{1.1}$$

Likewise, for the consumer system we assume a variable *cons* to display the consumer's states "ready to remove" (rr) and "ready to consume" (rc). We fix "ready to remove" as the initial state of the consumer. Finally, for the buffer system we assume a variable *buff* ranging over $\{0, 1\}$ to denote the buffer's states "empty" and "filled". Suppose, the buffer is initially empty. Hence

$$cons\colon\ rr \rightarrow\square\rightarrow rc \rightarrow\square\rightarrow rr \rightarrow\square\rightarrow rc \ldots \tag{1.2}$$

is a run of the consumer system, and

$$buff\colon\ 0 \rightarrow\square\rightarrow 1 \rightarrow\square\rightarrow 0 \rightarrow\square\rightarrow 1 \ldots \tag{1.3}$$

is a run of the buffer system. Recall that a run actually comprises infinitely many variables. For example, a consumer's run contains also the variables *prod* and *buff* which can change arbitrarily. Consequently, there are infinitely many consumer's runs but in each of them variable *cons* will be updated like in (1.2). We will always graphically outline only a finite part of a run concerning updates of the system variables as done in (1.1)-(1.3).

We are now going to describe the behaviour of the producer-consumer system with the 1-item buffer. In a run of the entire system, variable *prod*, for example, will be updated from rp to rd as in (1.1). The same holds for *cons* and *buff*. Thus, a run of the entire system has to be a run of the producer, consumer and buffer system at the same time. This intuitively justifies the definition that the behaviour of the composed system is the *intersection* of the behaviours of its components.

The interaction between the components will be represented by forcing some updates of the subsystem's variables to occur coincidently. For example, delivering an item to the (necessarily empty) buffer by the producer immediately entails filling of the buffer. Thus, the *prod*'s update from rd to rp has to occur coincidently with the *buff*'s update from 0 to 1. Such a coincident update in a Petri net is called a *transition*. We adopt this naming in a very similar context (see Sect. 2.1 for the definition). A run of the producer-consumer system with a 1-item buffer, σ_1, is shown in (1.4).

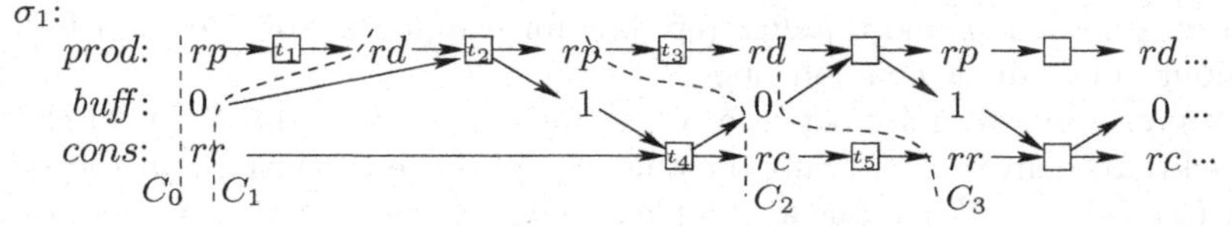

σ_1:

prod: rp → t_1 → rd → t_2 → rp → t_3 → rd → □ → rp → □ → rd ...
buff: 0 → t_2 → 1 → t_4 → 0 → □ → 1 → □ → 0 ...
cons: rr → t_4 → rc → t_5 → rr → □ → rc ...
C_0 C_1 C_2 C_3 (1.4)

2 The Logic TLDA

In this section, we briefly introduce the representation of the semantic model of TLDA, and then we define the syntax and semantics of our logic and introduce some basic notions used in the paper.

2.1 The Semantic Model

The semantic model of a TLDA formula is a *run* as already intuitively introduced and exemplified above in (1.1)–(1.4). The notation of a run resembles that of an occurrence net known from Petri net's theory (see [10] for instance). A run consists of a *history* of each variable and a set of *transitions.*

History and Transitions. Let *Var* be an infinite set of variables and *Val* an infinite set of values. A history of a variable $x \in Var$ is the sequence of its updates, i.e. a finite or infinite sequence $H(x) = x_0\ x_1\ x_2\ \ldots$ of values x_i from *Val*. x_i is the *local state* of x *at index* i. We abbreviate $H(x)$ to H_x. As an example, $H_{prod} = rp\ rd\ rp\ rd \ldots$ is the history of variable *prod*, and $H_{prod}(0) = rp$, $H_{prod}(1) = rd$ etc. The histories of the variables constitute a history of the run:

Let Val^+ and Val^ω denote the set of all non-empty, finite and infinite sequences of values, respectively, and let $Val^\infty \triangleq Val^+ \cup Val^\omega$. Then

$$H : Var \rightarrow Val^\infty$$

is a *history.*

A transition is an update of one variable or a synchronised update of several variables; technically a mapping:

$$t : V \rightarrow \mathbb{N}_0$$

where $\emptyset \neq V \subseteq Var$ is finite. $V = dom(t)$ includes all variables that are *involved* in transition t, $t(x)$ denotes the index of the involved update of x. In the run σ_1, transition t_1 is an update of variable *prod*. Thus, *prod* is the only variable involved in t_1. In contrast, transition t_2 depicts a synchronised update of variables *prod* and *buff*. Hence, *prod* and *buff* are involved in t_2.

The transitions in a run are (partially) ordered by a transitive successor relation $\prec$. Immediate succession is indicated by arrows in a run. Transitions in a run which are not related by $\prec$ are called *concurrent.* For instance, t_3 and t_4 as well as t_3 and t_5 are concurrent in σ_1. We require that for every transition t there is a finite number of transitions t_i with $t_i \prec t$. This completes the notions required for the definition of a run:

Run. Let H be a history, let T be a set of transitions. $\sigma = (H, T)$ is a run iff

- For every variable $x \in Var$ and for all i with $0 \leq i < l(H_x) - 1$, there exists exactly one transition $t \in T$ with $t(x) = i$.
- For all $t \in T$ and for all variables $x \in dom(t)$ holds: $0 \leq t(x) < l(H_x) - 1$.
- The relation $\prec$ on T is irreflexive.

Runs (1.1)–(1.4) fulfill these properties.

Since in a run σ the relation $\prec$ on T is transitive and irreflexive by definition, $\prec$ constitutes a *partial order* on the transition set of σ.

Cuts and Steps. $\prec$ can canonically be generalized to the local states of variables. A set of local states which are not related by $\prec$ forms a *cut*.

Formally, a mapping

$$C : Var \rightarrow \mathbb{N}_0$$

is called a *cut* in σ iff for each $t \in T$ and all $x, y \in dom(t)$ holds:

$$\text{if } t(x) < C(x), \text{ then } \ t(y) < C(y).$$

For instance, C_0 with $C_0(prod) = C_0(cons) = C_0(buff) = 0$, as well as C_1 with $C_1(prod) = 1$ and $C_1(cons) = C_1(buff) = 0$ are cuts in σ_1.

We say that a transition $t \in T$ *occurs at* a cut C in σ if t updates variables at the local states belonging to C, i.e. if $t(x) = C(x)$ for each $x \in dom(t)$.

From the definition of a cut arises an important observation: *Every two transitions that occur at C are concurrent in σ.*

Let U_C be the set of transitions that occur at C. From a given cut C the *successor* cut C' of C is reached by occurrence of all transitions from U_C. For example, from the *initial cut* C_0 in σ_1 the cut C_1 is reached by occurrence of t_1. The cut C_3 is reached from C_2 by occurrence of two transitions t_3 and t_5.

A cut C and its successor cut C', together with the transition set U_C form a *step* S_C in σ. Thus, S_C will canonically be defined by the cut C. Note that not every cut of a run σ can be reached by taking such *maximal* steps from C_0. Hence, the cuts do not constitute the run; they rather may be conceived as *observations* of the run.

This completes the set of structural notions required in the sequel.

2.2 Syntax of TLDA

Vocabulary. A vocabulary of TLDA is given by the following sets: a set of function symbols $\mathcal{F}$, a set of predicate symbols $\mathcal{P}$ (the symbol for equality $=$ is one of them in particular), a set of special symbols and a set of variables.

Additionally, TLDA expressions can include brackets, which we use in order to overwrite the binding priorities or just to increase readability.

Each predicate symbol and each function symbol has an arity. Constants can be thought of as 0-arity functions.

The set of special symbols consists of the standard boolean connectives $\neg$ and $\wedge$, the quantifiers $\exists$ and $\tilde{\exists}$ and the temporal operator $\Box$. The sets $\mathcal{P}$, $\mathcal{F}$ and the set of special symbols should be pairwise disjoint.

An infinite set of variables Var_{all} is partitioned into infinite disjoint sets of:

- rigid variables $Var_{rigid} = \{\mathsf{m}, \mathsf{n}, \ldots\}$,
- flexible variables $Var = \{x, y, \ldots\}$,
- primed flexible variables $Var' \triangleq \{x' | x \in Var\} = \{x', y', \ldots\}$,
- $\sim$-variables $\widetilde{Var} \triangleq \{\widetilde{a} | \emptyset \neq a \subseteq Var\} = \{\widetilde{\{x\}}, \widetilde{\{y\}}, \widetilde{\{x, y\}}, \ldots\}$.

$\widetilde{\{x\}}$ and $\widetilde{\{x, y\}}$ are abbreviated to $\widetilde{x}$ and $\widetilde{xy}$, respectively.

Terms. The terms of our logic, like in classical predicate logic, are made up of variables and functions applied to them. TLDA-terms over $\mathcal{F}$ are inductively defined as follows:

- Any variable from $Var_{rigid} \cup Var \cup Var'$ is a term.
- If $t_1, t_2, \ldots, t_n$ are terms and $f \in \mathcal{F}$ has arity n, then $f(t_1, t_2, \ldots, t_n)$ is a term.

Examples of terms are: y, x', $2 + \mathsf{n}$, $(x - 3 * y') + z$, where n is a rigid variable.

Formulas. Based on terms, we can continue to define the formulas of our logic. Formulas are divided into two classes: *step formulas* and *run formulas.*

The set of step formulas over $\mathcal{F}$ and $\mathcal{P}$ is inductively defined as follows:

- If P is a predicate taking n arguments, $n \geq 1$, and if $t_1, t_2, \ldots, t_n$ are terms over $\mathcal{F}$, then $P(t_1, t_2, \ldots, t_n)$ is a step formula.
- Any variable from the set $\widetilde{Var}$ is a step formula.
- If F and G are step formulas, then so are $\neg F$, $F \wedge G$ and $F \Rightarrow G$.
- If $x \in Var_{rigid} \cup Var \cup Var'$ and if F is a step formula, then so is $\exists x\, F$.
- If F is a step formula, then so is $\tilde{\exists}\, F$.

Examples of step formulas are: $x' \leq 1$, $y = (2 + \mathsf{n})$, $\neg\tilde{z} \wedge (\mathsf{m} \geq 0) \Rightarrow (z = 15)$, $\exists \mathsf{m}\, (\mathsf{m} = x^2)$, $\exists y'\, (\widetilde{xy} \Rightarrow (y' = x))$, n and m are rigid variables.

Note that we use some conventional arithmetical and logical abbreviations, including boolean abbreviations *true* (for $P \vee \neg P$), *false* (for $\neg$*true*), $\Leftrightarrow$, $\equiv$, as well as $\Diamond$ (for $\neg\Box\neg$).

The set of run formulas over $\mathcal{F}$ and $\mathcal{P}$ is inductively defined as follows:

- Any step formulas F is a run formula.
- If F and G are run formulas, then so are $\neg F$, $F \wedge G$ and $F \Rightarrow G$.
- If F is a run formula, then so is $\Box F$.
- If $x \in Var_{rigid}$ and if F is a run formula, then so is $\exists x\, F$.

$\Box(z < \mathsf{m})$, $\Box(\neg\tilde{z} \Rightarrow (x' = x + 5))$, $\Box(\exists \mathsf{m}\, (\mathsf{m}^2 = y))$ are examples of run formulas, m is the only rigid variable occurring in them.

2.3 Semantics of TLDA

Now we explain briefly the difference between the sets of variables introduced above.

Rigid variables stand for an unknown but fixed value and are commonly used in any kind of classical logic.

Flexible variables will mostly be called *program* variables. They are intended to describe changes in our systems: to every program variable a sequence of values, i.e. its history will be assigned. The histories represent an execution of a system. Nevertheless, every program variable has exactly one value in a particular cut of a system run. A value of this variable in the successor cut of a system run will be described by a corresponding primed program variable.

The partition of variables into rigid and flexible variables is a well known idea (see for example [18]) and primed variables have also been used before for describing values of variables in a successor state ([16, 18]).

The $\sim$-variables are independent from the values assigned to the program variables in a cut of a system run: The $\sim$-variables can only take boolean values and provide information about the synchronization of variable updates: They describe, which program variables are involved in the transitions of a step in a system run. Some subsequent examples will clarify this concept.

The semantics of the logic resembles those for other temporal logics. We assume a non-empty set *Val* of concrete values, called the *universe*, we interpret each function symbol in $\mathcal{F}$ as a concrete function on *Val*, and each predicate symbol in $\mathcal{P}$ as a predicate over *Val*. Formally, the *interpretation* I of $(\mathcal{F}, \mathcal{P})$ consists of the following set of data:

- a non-empty set *Val* (the truth values *true* and *false* are in particular elements of *Val*),
- for each n-ary $f \in \mathcal{F}$ a function $f^I : Val^n \to Val$, and
- for each $P \subseteq \mathcal{P}$ with n arguments a subset $P^I \subseteq Val^n$.

Evaluating Terms. Let r be a mapping $r : Var_{rigid} \to Val$ which associates with every rigid variable m a value $r(m)$ of the universe. The values of all program variables and primed program variables depend on a run.

Terms will be evaluated in steps of a run. So let $\sigma = (H, T)$ be a run and let S_C be a step of σ. We interpret a term in the set *Val* by replacing all variables with their values, as follows: Each rigid variable is replaced with its value according to the mapping r. Each program variable $x \in Var$ is replaced with the value $H_x(C(x))$, i.e. with the value assigned to x at the $C(x)$th index in its history H_x. This is intuitively the value of x in the cut C. Each primed program variable $x' \in Var'$ is replaced with $H_x(C'(x))$, i.e. with the value of x in the successor cut C'.

Formally, to compute the value of a term in S_C under the interpretation I and with respect to r we inductively define a (standard) mapping r_S as follows:

$$\begin{aligned} r_C(m) &= r(m) && \text{if } m \in Var_{rigid}, \\ r_C(x) &= H_x(C(x)) && \text{if } x \in Var, \\ r_C(x') &= H_x(C'(x)) && \text{if } x' \in Var', \text{ and} \\ r_C(f(t_1, \ldots, t_n)) &= f^I(r_C(t_1), \ldots, r_C(t_n)) && \text{if } t_1, \ldots, t_n \text{ are terms.} \end{aligned}$$

Evaluating Step Formulas. A model of a step formula consists of a step S_C of a run $\sigma = (H, T)$ and an interpretation I of $(\mathcal{F}, \mathcal{P})$ (for convenience, we will write simply S_C in place of (S, I) when a model of a step formula is concerned). Let r be a valuation mapping of rigid variables.

We define the notion $S_C \models_r \psi$ of ψ holding in S with respect to r for each step formula ψ by structural induction on ψ:

- $S_C \Vdash_r P(t_1, \ldots, t_n)$ iff $(r_S(t_1), \ldots, r_S(t_n)) \in P^I$.
- $S_C \Vdash_r \widetilde{a}$ iff $a \subseteq dom(t)$ for any $t \in U_C$,
  i.e. we replace $\widetilde{a}$ by the boolean value *true*, if a is a subset of the variables involved in a transition t occurring at C, and by *false* otherwise.
- $S_C \Vdash_r \neg F$, as well as $F \wedge G$ and $F \Rightarrow G$ are standard.
- $S_C \Vdash_r \exists x\, F$ for $x \in Var_{rigid}$ iff $S_C \Vdash_{r[x \leftarrow a]} F$ for some $a \in Val$, where $r[x \leftarrow a](x) = a$ and $r[x \leftarrow a](k) = r(k)$ for all $k \in Var_{rigid} \setminus \{x\}$.
- $S_C \Vdash_r \exists x\, F$ for $x \in Var$ iff there is a run $\sigma^* = (H^*, T)$ with $H_y(C(y)) = H^*_y(C(y))$ for all $y \in Var \setminus \{x\}$ and $H_x(C'(x)) = H^*_x(C'(x))$ for all $x \in Var$ so that $S^*_C \Vdash_r F$,
  i.e. F can be made *true* in S_C by setting a new value in the x-History H_x at index $C(x)$ without changing the value in H_x at $C'(x)$ and without changing the histories of all other variables y at $C(y)$ and $C'(y)$. Note that the transition set T remains the same in σ^*. Hence, it can be shown that C is a cut also in σ^* and S^*_C is therefore a step in σ^*. Thus, only the value of x can be changed, whereas the values of x' and all other variables, inclusive $\sim$-variables, do not change. We analogously define:
- $S_C \Vdash_r \exists x'\, F$ for $x' \in Var'$ iff there is a run $\sigma^* = (H^*, T)$ with $H_x(C(x)) = H^*_x(C(x))$ for all $x \in Var$ and $H_y(C'(y)) = H^*_y(C'(y))$ for all $y \in Var \setminus \{x\}$ so that $S^*_C \Vdash_r F$.
- $S_C \Vdash_r \tilde{\exists} F$ iff there is a run $\sigma^* = (H^*, T^*)$ with $H_x(C(x)) = H^*_x(C(x))$ for all $x \in Var$ and C is a cut in σ^* so that $S^*_C \Vdash_r F$.
  Note that $\tilde{\exists}$ does not quantifies a variable, but a step formula F. Thus, $\tilde{\exists} F$ is *true* in S_C if we can take a new step from the cut C, in which F holds, i.e. if we can make F *true* without changing the values of (unprimed) program variables.

Since the $\sim$-variables are essential for the logic, we give here some examples of their interpretation:

Let S_{C_1} and S_{C_2} be steps of σ_1 in (1.4). In S_{C_1} holds $\widetilde{prod\, buff}$, because there is apparently a transition t_2 occurring at C_1 in which both *prod* and *buff* are involved together. Consequently, it holds in particular that *prod* is involved in this transition, so $\widetilde{prod}$ is *true* in S_{C_1}, too. The same holds for variable *buff*.

In S_{C_2} hold $\widetilde{prod}$ and $\widetilde{cons}$, because of t_3 and t_5 in which *prod* and *cons* are involved, respectively. However, $\widetilde{prod\, cons}$ does *not* hold in S_{C_2}, since *prod* and *cons* are not involved together in a transition.

Evaluating Run Formulas. Now we extend the semantics to run formulas.

A model of a run formula is a pair (σ, C) consisting of a run $\sigma = (H, T)$ and a cut C of σ, and an interpretation I for $(\mathcal{F}, \mathcal{P})$ (we will write for convenience simply (σ, C) instead of (σ, C, I)). Analogously to step formulas, we define now the notion $(\sigma, C) \models_r \psi$ of ψ holding in (σ, C) with respect to r for each run formula ψ by structural induction on ψ:

- $(\sigma, C) \models_r F$ for a step formula F iff $S_C \Vdash_r F$.
- $(\sigma, C) \models_r \neg F$, as well as $F \wedge G$, $F \Rightarrow G$ and $\exists x F$ for $x \in Var_{rigid}$ are standard.
- $(\sigma, C) \models_r \Box F$ iff $(\sigma, C^*) \models_r F$ for every cut C^* of σ with $C^*(x) \geq C(x)$ for all $x \in Var$.

Notations. We usually omit an explicit denotation of the mapping r and write simply $(\sigma, C) \models \psi$ for $(\sigma, C) \models_r \psi$. Furthermore, if a run formula ψ holds in σ at the initial cut C_0, i.e. $(\sigma, C_0) \models \psi$, we write $\sigma \models \psi$.

The semantic model of a TLDA formula Φ is a run of a system. The set of all models of Φ will be denoted by $\Sigma(\Phi)$. Hence, $\Sigma(\Phi)$ is the behaviour of the system specified by Φ.

Now, let C_i and C_j be cuts of a run. $C_i \geq C_j$ abbreviates $C_i(x) \geq C_j(x)$ for all $x \in Var$.

The set of all variables except for the rigid occurring in a formula Φ will be denoted by $VA(\Phi)$. Φ will be called a *cut formula* iff no primed or $\sim$-variables occur in Φ, i.e. $VA(\Phi) \subseteq Var$. A formula Φ', obtained from a cut formula Φ by replacing each variable x by the primed variable x', will be called a *primed cut formula.* Hence, $VA(\Phi') \subseteq Var'$. From these definitions follows

Proposition 1 *Let σ be a run, C a cut of σ, Φ a cut formula and Φ' a primed cut formula.*
If $(\sigma, C) \models \Phi'$, then $(\sigma, C') \models \Phi$.

Finally, $V(\Phi) \triangleq \{x \in Var | \; x, \, x' \text{ occurs in } \Phi \text{ or } \widetilde{a} \text{ occurs in } \Phi \text{ and } x \in a\}$ denotes the set of Var-variables occurring in Φ.

3 Component Specification

With the logic of Section 2 we are now ready to specify the producer-consumer system of Section 1. We start with specifying the three subsystems. In Section 6 we show how to compose these specifications to obtain a specification of the entire system.

To begin we specify the producer system. Initially, the producer is "ready to produce". Hence, the value of variable *prod* equals *rp*. *Producing* an item will be viewed as an update of *prod* from *rp* to *rd*. Analogously, the update from *rd* to *rp* should represent *delivering* of an item. This will be specified by the following formulas *Produce* and *Deliver*:

$$\begin{aligned} Produce &\triangleq prod = rp \wedge prod' = rd \\ Deliver &\triangleq prod = rd \wedge prod' = rp \end{aligned}$$

Both formulas are to be alternatively applied only in system steps which involve variable *prod*. Hence the producer specification:

$$\begin{aligned} Producer \triangleq\; & prod = rp \wedge \\ & \Box\,(\widetilde{prod} \Rightarrow Produce \vee Deliver) \end{aligned}$$

We similarly specify the consumer system. We assume the consumer is initially "ready to remove" and can either *remove* an item from the buffer or *consume* an item. Thus,

$$\begin{aligned}
\mathit{Consumer} &\triangleq \mathit{cons} = \mathit{rr} \wedge \\
&\quad \Box\,(\widetilde{\mathit{cons}} \Rightarrow \mathit{Remove} \vee \mathit{Consume}), \text{ with} \\
\mathit{Remove} &\triangleq \mathit{cons} = \mathit{rr} \wedge \mathit{cons}' = \mathit{rc}, \text{ and} \\
\mathit{Consume} &\triangleq \mathit{cons} = \mathit{rc} \wedge \mathit{cons}' = \mathit{rr}
\end{aligned}$$

Finally, the buffer system. We need two additional conditions to specify that *emptying* of the buffer is only possible if the buffer contains an item at all and *filling* of the buffer can happen only if the buffer is empty.

$$\begin{aligned}
\mathit{Buffer} &\triangleq \mathit{buff} = 0 \wedge \\
&\quad \Box\,(\widetilde{\mathit{buff}} \Rightarrow \mathit{Filling} \vee \mathit{Emptying}) \wedge \\
&\quad \Box\,(\mathit{Filling} \Rightarrow \mathit{buff} = 0) \wedge \\
&\quad \Box\,(\mathit{Emptying} \Rightarrow \mathit{buff} > 0), \text{ with} \\
\mathit{Filling} &\triangleq \mathit{buff}' = \mathit{buff} + 1 \\
\mathit{Emptying} &\triangleq \mathit{buff}' = \mathit{buff} - 1
\end{aligned}$$

The specifications of the producer, consumer and buffer allow so far that each of those systems eventually gets stuck. To rule out such a behaviour we need a progress assumption. Progress will subsequently be considered in Sect. 5.3.

4 Component Verification: Safety

Based on the component specifications presented in Sect. 3, we are now able to deduce some (safety) properties of the systems. A property in our logic is a set of runs specified by a formula P. Thus, we do not distinguish between a behaviour of a system and a property of a system. We say that a property P *holds* for a system specified by Φ iff the behaviour of the system is contained in the property (see [9, 19] for instance). In the logic, $\Sigma(\Phi) \subseteq \Sigma(P)$ holds iff $\Phi \Rightarrow P$ holds. Hence, proving that a property P holds in a system specified by Φ amounts to proving the formula $\Phi \Rightarrow P$, or in other words: to proving P under the assumption Φ.

In this section we will exemplify a proof of a simple safety property holding in the producer system. We will derive this property from the system specification in a syntactic way using some inference rules of our natural deduction proof calculus. We present below only a part of the calculus, that is the rules necessary for the sample proofs.

4.1 Proof Rules

We denote an inference rule in the usual way: the set of premises of the rule is written above the line, the conclusion stands below the line. Sometimes we write a double line to denote that a rule can be applied in both directions: we may infer the lower formula from the upper one, as usual, but we may as well infer the upper one from the lower one.

Figure 1 shows some selected rules of the propositional calculus. They are all well known and do not need any explanation.

$$\frac{\Phi, \Psi}{\Phi \wedge \Psi} \wedge i \qquad \frac{\Phi \wedge \Psi}{\Phi} \wedge e_1 \qquad \frac{\Phi \wedge \Psi}{\Psi} \wedge e_2$$

$$\frac{\Phi, \Phi \Rightarrow \Psi}{\Psi} mp \qquad \frac{\Phi, \neg \Phi}{false} \neg e \qquad \frac{false}{\Phi} \bot e$$

$$\frac{\begin{matrix}\Phi \\ \vdots \\ \Psi\end{matrix}}{\Phi \Rightarrow \Psi} \rightarrow i \qquad \frac{\begin{matrix} & \Phi & \Psi \\ & \vdots & \vdots \\ \Phi \vee \Psi, & \chi, & \chi\end{matrix}}{\chi} \vee e$$

$$\frac{\neg \Phi \vee \Psi}{\Phi \Rightarrow \Psi} \leftrightarrow r \qquad \frac{\Phi \wedge (\Psi \vee \Omega)}{(\Phi \wedge \Psi) \vee (\Phi \wedge \Omega)} distr$$

Fig. 1. Selected propositional rules

There are some fundamental rules for $\sim$-variables in our logic. Here we present only one of them, claiming that if a program variable x is *not involved* in any transition of a step S_C of a run σ (i.e. $(\sigma, C) \models \neg \widetilde{x}$ holds), then the value of x may not change in that step. Hence the rule $\sim r$:

$$\frac{\neg \widetilde{x}}{x' = x} \sim r$$

Now we present three temporal rules which will later be used in our proofs. These rules will be supplemented by two other temporal rules for proving liveness properties in Sect. 5.2.

$$\frac{I \wedge \Phi \Rightarrow I', \ \ _{VA(I)=\{x\} \subseteq Var}}{I \wedge \Box \Phi \Rightarrow \Box I} inv \qquad \frac{\boxed{\begin{matrix}\vdots \\ \Phi\end{matrix}}}{\Box \Phi} \Box i \qquad \frac{\Box \Phi}{\boxed{\begin{matrix}\vdots \\ \Phi \\ \vdots\end{matrix}}} \Box e$$

The rule *inv* is a simplified version of an invariant rule and has the following meaning: Let I be a cut formula containing only one variable. (Note, that this assumption made for the rule stands above the line in a smaller font.) If $I \wedge \Phi$ holding in a run σ at a cut C implies that I holds in the successor cut C' of

C, too, which is equivalent with I' holds at C, then we can prove that $I \wedge \Box \Phi$ implies $\Box I$. Thus, I is an invariant.

For reasoning about $\Box$-formulas we introduce proof boxes. We adopt this notation with some slight modifications from [12]. By definition in Sect. 2.3, a formula $\Box \Phi$ holds in a run σ at a cut C iff Φ holds at all cuts C_i with $C_i \geq C$. Hence, we could work on Φ in one of those cuts, say C_j, to obtain a new formula, for example, Ψ. Since we have shown Ψ in an arbitrary cut $C_j \geq C$, we may deduce $\Box\Psi$ at the cut C. This is the intuitive meaning of both rules above: Reasoning in a concrete but arbitrary cut reachable from a given cut C as a starting point is marked by going into a box. The rule $\Box e$ says that if at any point in a proof we have $\Box\Phi$, we could open a box and put Φ into it. The rule $\Box i$ says that if we are in a box and could infer a formula Φ, then we could leave the box and deduce $\Box \Phi$.

4.2 Proving Safety Properties

Using the proof rules from the previous section we are now prepared for proving safety properties holding in a system. As an example, we prove that the property $\Box\, (\mathit{prod} \in \{\mathit{rp}, \mathit{rd}\})$ holds in the producer system. Hence, we show that we can derive this formula from the specification *Producer* of the producer system, thus we prove $\mathit{Producer} \Rightarrow \Box\, (\mathit{prod} \in \{\mathit{rp}, \mathit{rd}\})$.

In a proof we number all lines and we write on the right side a justification for each line: the name of the applied rule and the numbers of lines used for it. The scope of a temporary assumption (made in order to apply rule $\rightarrow i$ or $\vee e$) will be demarcated by an indentation.

Hence, the proof for $\mathit{Producer} \Rightarrow \Box\, (\mathit{prod} \in \{\mathit{rp}, \mathit{rd}\})$ is shown in Fig. 2.

We can prove in a very similar way that, for example, $\Box\, (\mathit{cons} \in \{\mathit{rr}, \mathit{rc}\})$ and $\Box\, (0 \leq \mathit{buff} \leq 1)$ hold in the consumer and buffer system, respectively.

5 Component Verification: Liveness

In this section we first introduce two new operators for proving liveness properties: a *leads-to* operator $\rhd$ and an *entails* operator $\blacktriangleright$. Secondly, we add to our proof calculus a few rules for proving liveness properties. We introduce subsequently the notion of *progress* and *fairness*. Finally, we show a sample proof of a liveness property holding in the buffer system.

5.1 Liveness Operators

We usually want to prove liveness properties having the following form: Whenever Φ holds once in a system, eventually Ψ holds, where Φ and Ψ each specify a property holding in a global state of a system. As an example, for the buffer-system we would like to prove: "Whenever the buffer is empty, it becomes eventually filled." This sort of properties will be expressed with help of the *leads-to* operator $\rhd$.

1.	$prod = rp \wedge \Box\,(\widetilde{prod} \Rightarrow (Produce \vee Deliver))$	given
2.	$prod \in \{rp, rd\} \wedge (\widetilde{prod} \Rightarrow (Produce \vee Deliver))$	assumption
2.1	$\widetilde{prod} \Rightarrow (Produce \vee Deliver)$	$\wedge e_2$ 2
2.2	$\neg\widetilde{prod} \vee Produce \vee Deliver$	$\leftrightarrow r$ 2.1
2.3	$prod \in \{rp, rd\}$	$\wedge e_1$ 2
2.4	$prod \in \{rp, rd\} \wedge (\neg\widetilde{prod} \vee Produce \vee Deliver)$	$\wedge i$ 2.2, 2.3
2.5	$prod \in \{rp, rd\} \wedge \neg\widetilde{prod}) \vee (prod \in \{rp, rd\} \wedge Produce)$ $\vee(prod \in \{rp, rd\} \wedge Deliver)$	twice *distr* 2.4
2.6	$prod \in \{rp, rd\} \wedge \neg\widetilde{prod}$	assumption
2.6.1	$\neg\widetilde{prod}$	$\wedge e_2$ 2.6
2.6.2	$prod' = prod$	$\sim r$ 2.6.1
2.6.3	$prod' \in \{rp, rd\}$	$\wedge e_1$ 2.6 and 2.6.2
2.7	$prod \in \{rp, rd\} \wedge Produce$	assumption
2.7.1	$prod = rp \wedge prod' = rd$	$\wedge e_2$ 2.7 and def. *Produce*
2.7.2	$prod' \in \{rp, rd\}$	$\wedge e_2$ 2.7.1 and math
2.8	$prod \in \{rp, rd\} \wedge Deliver$	assumption
2.8.1	$prod = rd \wedge prod' = rp$	$\wedge e_2$ 2.8 and def. *Deliver*
2.8.2	$prod' \in \{rp, rd\}$	$\wedge e_2$ 2.8.1 and math
2.9	$prod' \in \{rp, rd\}$	twice $\vee e$ 2.5–2.8
3.	$prod \in \{rp, rd\} \wedge (\widetilde{prod} \Rightarrow (Produce \vee Deliver))$ $\Rightarrow prod' \in \{rp, rd\}$	$\rightarrow i$ 2, 2.9
4.	$prod \in \{rp, rd\} \wedge \Box\,(\widetilde{prod} \Rightarrow (Produce \vee Deliver))$ $\Rightarrow \Box\,(prod \in \{rp, rd\})$	*inv* 3
5.	$\Box\,(prod \in \{rp, rd\})$	*mp* 1, 4 and math

Fig. 2. Proof of $\Box\,(prod \in \{rp, rd\})$

Definition 1 *Let σ be a run, let C be a cut of σ and let Φ and Ψ be cut formulas. $(\sigma, C) \models \Phi \rhd \Psi$ iff for all cuts C_i, $C_i \geq C$ holds: if $(\sigma, C_i) \models \Phi$, then there is a cut C_j, $C_j \geq C_i$, such that $(\sigma, C_j) \models \Psi$.*

From this definition follows immediately

Proposition 2 $(\sigma, C) \models \Phi \rhd \Psi$ *iff* $(\sigma, C) \models \Box\,(\Phi \Rightarrow \Diamond\Psi)$.

Now we define a one-step operator $\blacktriangleright$(*entails*). $\Phi \blacktriangleright \Psi$ intuitively means that if Φ holds in a run σ at a cut C_i and all variables occurring in Φ are involved in transitions that occur at C_i, then Ψ holds at the successor cut C_i' of C_i. Hence the definition

Definition 2 *Let σ be a run, let C be a cut of σ and let Φ and Ψ be cut formulas. $(\sigma, C) \models \Phi \blacktriangleright \Psi$ iff for all cuts C_i, $C_i \geq C$ holds: if $(\sigma, C_i) \models \Phi \wedge \bigwedge_{x \in V(\Phi)} \widetilde{x}$, then $(\sigma, C_i') \models \Psi$.*

5.2 Proof Rules for Liveness

So far both operators are defined semantically. With help of the rule $\blacktriangleright r_1$ the formula $\Phi \blacktriangleright \Psi$ can be syntactically derived in case Φ and Ψ are cut formulas.

This rule follows directly from Definition 2 and Proposition 1.

$$\frac{\Box\,(\Phi \wedge \bigwedge_{x\in V(\Phi)} \widetilde{x} \Rightarrow \Psi')}{\Phi \blacktriangleright \Psi} \blacktriangleright r_1$$

Finally, we give a rule for inference of a *leads-to* formula $\Phi \rhd \Psi$ for cut formulas Φ and Ψ: If Φ entails Ψ in a run σ, and the run never gets stuck in a cut at which the formula Φ holds, and if always *all* variables occurring in Φ can be involved in a step of σ, then $\Phi \rhd \Psi$ holds in σ. Formally

$$\frac{\Phi \blacktriangleright \Psi, \quad \Phi \rhd \bigvee_{x\in V(\Phi)} \widetilde{x}, \quad \Box\,(\bigvee_{x\in V(\Phi)} \widetilde{x} \Rightarrow \bigwedge_{x\in V(\Phi)} \widetilde{x})}{\Phi \rhd \Psi} \blacktriangleright r_2$$

Obviously, the rule $\blacktriangleright r_2$ is circular. But the second premise can in most cases be directly inferred from a system specification. More precisely, the progress requirement for a specification formula sharing some variables with Φ usually implies the formula $\Phi \rhd \bigvee_{x\in V(\Phi)} \widetilde{x}$ (see the subsequent section for definition and an example). Some more *leads-to* rules are

$$\frac{\Box\,(\Phi \Rightarrow \Psi)}{\Phi \rhd \Psi} \rhd l_1 \qquad \frac{\Box\Diamond\,\Phi}{\Psi \rhd \Phi} \rhd l_2 \qquad \frac{\Phi \rhd \Psi,\ \Psi \rhd \Omega}{\Phi \rhd \Omega} \rhd l_3$$

5.3 Progress and Fairness

If an update U described by a step formula Φ does not take place anymore in a run of a system, there are two possible explanations for this: Either the run (or a part of it) gets stuck in a step, in which U could possibly occur. This often indicates a failure in a system. Or another update involving the same variables always occurs in the run instead of U. These two concepts are clearly separated in our logic. In the first case we say, that the run ignores *progress* of the step formula Φ specifying U. In the second case we say that the run ignores *fairness* of Φ. The notion of progress originates from Petri nets theory (cf. [14, 20, 21]).

In order to define progress and fairness we introduce first for a step formula Φ the notion of *step variables*, denoted by $St(\Phi)$:

$$St(\Phi) \triangleq \{x \in Var \mid x' \text{ occurs in } \Phi \text{ or } \widetilde{a} \text{ occurs in } \Phi \text{ and } x \in a\}$$

and, secondly, the notion of $Enabled(\Phi)$:

$$Enabled(\Phi) \triangleq \widetilde{\exists}\Phi$$

By definition, Φ *is enabled* in a step S_C of a run σ iff the step formula $Enabled(\Phi)$ is true in S_C. $Enabled(\Phi)$ is true in S_C if it is possible to take a *new* step S_C^* from the cut C such that the formula Φ becomes true in S_C^*.

Based on these notions we define now fairness of a step formula Φ.

$$fair(\Phi) \triangleq \Box\Diamond\Phi \vee \Box\Diamond\neg Enabled(\Phi)$$

(Note that this is the common definition of *weak* fairness. *Strong* fairness of Φ will also be defined in the standard way. To simplify matters we omit here this definition.) We say that a run σ respects fairness of Φ iff $fair(\Phi)$ holds in σ.

Let $St(\Phi) = \{x_1, \ldots, x_n\}$ be the set of step variables of Φ.

$$progress(\Phi) \triangleq \Box\Diamond\,(\bigwedge_{1 \leq i \leq n} \neg Enabled(\Phi \land \widetilde{x_i})) \lor \Box\Diamond \bigvee_{1 \leq i \leq n} \widetilde{x_i}$$

A run σ respects progress of Φ iff $progress(\Phi)$ holds in σ.

We now revise our introductory example. Consider again the specification *Producer* of the producer system in Sect. 3. Assume a run σ_p in which $prod = rp$ initially holds and in which variable *prod* is never involved in a step of the run. We can easily state that σ_p is a proper run of the producer system, since the formula *Producer* holds in σ_p. Hence, the specification *Producer* of the producer system allows, for example, that the producer gets stuck in the state "ready to produce" and will never produce an item. We can prevent this by an additional requirement that every run of the producer respects progress of the specification formula *Producer*. Thus,

$$Producer_{live} \triangleq Producer \land progress(Producer)$$

specifies a producer system that never gets stuck. Since $St(Producer) = \{prod\}$, we obtain

$$progress(Producer) \triangleq \Box\Diamond\,(\neg Enabled(Producer \land \widetilde{prod})) \lor \Box\Diamond\widetilde{prod}$$

The formula $Producer \land \widetilde{prod}$ is continuously enabled in every run, hence $\Box\Diamond\,\neg Enabled(Producer \land \widetilde{prod})$ is always *false*. Thus, $\Box\Diamond\widetilde{prod}$ is the progress requirement for the producer system. Analogously, $\Box\Diamond\widetilde{cons}$ and $\Box\Diamond\widetilde{buff}$ ensure progress for the consumer and buffer system, respectively. Hence, the formulas

$$Consumer_{live} \triangleq Consumer \land \Box\Diamond\widetilde{cons} \text{ and}$$
$$Buffer_{live} \triangleq Buffer \land \Box\Diamond\widetilde{buff}$$

are the final specifications of the consumer and the buffer system, respectively.

5.4 Verifying Liveness Properties

As an example, Fig. 3 shows the proof of the property $(buff = 0) \rhd (buff = 1)$ holding in the buffer system. Some other liveness properties that can be analogously proven are $(prod = rp) \rhd (prod = rd)$ for the producer and $Consume \rhd \neg Enabled(Consume)$ for the consumer, for instance.

6 Parallel Composition

In this section we focus on *parallel composition* of systems and their specification in TLDA. Let S_1 and S_2 be systems specified by formulas Φ_1 and Φ_2, respectively. Parallel composition of S_1 and S_2 is defined as the intersection of their behaviours $\Sigma(\Phi_1) \cap \Sigma(\Phi_2)$. From this definition follows immediately by logical reasoning that

1.	$buff = 0 \wedge \Box(\widetilde{buff} \Rightarrow (Filling \vee Emptying)) \wedge \Box(Filling \Rightarrow buff = 0)$	
	$\wedge \Box(Emptying \Rightarrow buff > 0) \wedge \Box\Diamond\widetilde{buff}$	given
2.	$\Box(\widetilde{buff} \Rightarrow (Filling \vee Emptying))$	$\wedge e_2, \wedge e_1$ 1
3.	$\Box(Emptying \Rightarrow buff > 0)$	$\wedge e_1, \wedge e_2$ 1
4.	$\Box\Diamond\widetilde{buff}$	$\wedge e_2$
5.	$\widetilde{buff} \Rightarrow (Filling \vee Emptying)$	$\Box e$ 2
6.	$Emptying \Rightarrow buff > 0$	$\Box e$ 3
7.	$(buff = 0) \wedge \widetilde{buff}$	assumption
7.1	$Filling \vee Emptying$	*mp* 5, 7
7.2	$Filling$	assumption
7.2.1	$buff' = buff + 1$	def. *Filling* 7.2
7.2.2	$buff' = 1$	math 7, 7.2.1
7.3	$Emptying$	assumption
7.3.1	$buff > 0$	*mp* 6, 7.3
7.3.2	$false$	$\neg e$ 7, 7.3.1
7.3.3	$buff' = 1$	$\bot e$ 7.3.2
7.4	$buff' = 1$	$\vee e$ 7–7.3
8.	$(buff = 0) \wedge \widetilde{buff} \Rightarrow (buff' = 1)$	$\rightarrow i$ 7–7.4
9.	$\Box((buff = 0) \wedge \widetilde{buff} \Rightarrow (buff' = 1))$	$\Box i$ 8
10.	$(buff = 0) \blacktriangleright (buff = 1)$	$\blacktriangleright r_1$ 9
11.	$(buff = 0) \triangleright \widetilde{buff}$	$\triangleright l_2$ 4
12.	$(buff = 0) \triangleright (buff = 1)$	$\blacktriangleright r_2$ 10, 11

Fig. 3. Proof of $(buff = 0) \triangleright (buff = 1)$

the specification formula of the composed system is the *conjunction* $\Phi_1 \wedge \Phi_2$ of the specification formulas of the components. The idea of composition as conjunction has been suggested in [6, 7, 3, 4, 17]. Works on compositional semantics based on partial order are [11, 13].

We revise now our introductory example. Using the component specifications from Sect. 3 we demonstrate how to specify the composition of the producer, consumer and buffer system, in order to obtain a producer-concumer system (with a buffer) as described in Sect. 1.

By definition, the parallel composition of the three subsystems will be specified by conjunction $Comp \triangleq Producer_{live} \wedge Consumer_{live} \wedge Buffer_{live}$. We easily convince ourselves that every run of the producer-consumer system, as for example σ_1 in (1.4), is a model of *Comp*. But *Comp* additionally evolves models that are not proper runs of the producer-concumer system. (6.1) shows an example.

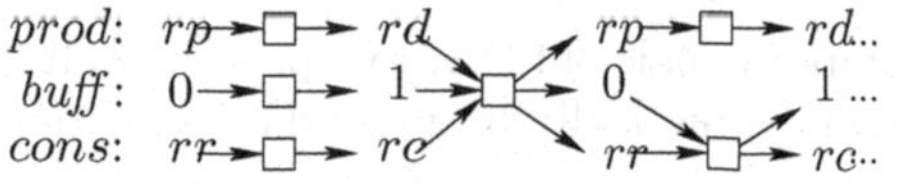

prod: rp → □ → rd ... rp → □ → rd...
buff: 0 → □ → 1 → □ → 0 ... 1 ...
cons: rr → □ → rc ... rr → □ → rc... (6.1)

The unwanted models do not properly *synchronise* the updates of *prod*, *cons* and *buff*. Hence, we strive for a further formula, *Sync*, to express additional constraints on the models. The final formula will then be the conjunction of *Producer*, *Consumer*, *Buffer* and *Sync*.

The formula *Sync* talks about synchronised updates of *prod*, *cons* and *buff* by help of the $\sim$-variables. Three properties are required: Firstly, delivery of an item into the buffer by the producer immediately causes filling the buffer, whereby the updates of *prod* and *buff* are synchronised. Secondly, removing an item from the buffer by the consumer is to equate with emptying the buffer, since the buffer has capacity of one item. The updates of *cons* and *buff* have to be synchronised in this case. And finally, the updates of *prod* and *cons* are never synchronised.

$$
\begin{aligned}
Sync \triangleq\ & \Box \neg \widetilde{prod\,cons} \\
& \wedge \Box(Deliver \Leftrightarrow Filling \Leftrightarrow \widetilde{prod\,buff}) \\
& \wedge \Box(Remove \Leftrightarrow Emptying \Leftrightarrow \widetilde{cons\,buff})
\end{aligned}
$$

Hence, the following formula *ProdCons* specifies the producer-consumer system:

$$ProdCons \triangleq Producer_{live} \wedge Consumer_{live} \wedge Buffer_{live} \wedge Sync$$

Obviously, the specification *ProdCons* of the composed system implies the specification formulas of its components, for instance $Producer_{live}$. Consequently, every property P holding in the producer system holds in the producer-consumer system, too, since $ProdCons \Rightarrow P$ can be easily inferred from $ProdCons \Rightarrow Producer_{live}$ and $Producer_{live} \Rightarrow P$. Hence, compositional reasoning is strongly supported by the specification method using TLDA.

There are several ways in which systems can be composed with each other. Sometimes system components are intended to run really in parallel. Such components have always disjoint system variables. They are specified in TLDA by formulas allowing an *arbitrary synchronisation* with others. These formulas have usually the form of an implication and describe how a given system variable changes *if* it is involved in a transition of the system. Examples of such formulas are *Deliver*, *Remove* and *Filling*. Thus, *Comp* specifies a producer, consumer and buffer system running in parallel.

But (6.1) shows that this is *not* the expected producer-consumer system, since a producer-consumer system requires a proper synchronisation of its system variables. We solve this problem by adding an appropriate synchronisation formula *Sync*. This can be viewed as a parallel composition with a third component. This demonstrates a particular specification method which properly works in the majority of cases.

7 Conclusion

We suggest a new temporal logic, TLDA, for specifying and verifying distributed systems. The logic can syntactically be viewed as an extension of TLA. TLDA, however, is interpreted on Petri nets-like partial order semantics. This allows us to specify some properties, as for example concurrency vs. nondeterminism, that can well be expressed in a partial order based formalism, but not in an interleaving based formalism like TLA (see also [8]).

Furthermore, we have shown that TLDA supports a modular design of systems: subsystems can be specified and verified separately and then be integrated into one system. The properties of the subsystems will be preserved in this process, i.e. they hold in the composed system, too.

Acknowledgement

I would like to thank especially one anonymous referee for detailed comments and very helpful suggestions.

References

1. M. Abadi. An axiomatization of Lamport's Temporal Logic of Actions. In J.Baeten and J.Klop, editors, *Proc. of Concur'90, Theories of Concurrency: Unification and Extension*, volume 458 of *LNCS*, pages 57–69, 1990.
2. M. Abadi and L. Lamport. Composing specifications. *ACM Transactions on Programming Languages and Systems*, 15(1):73–132, January 1993.
3. M. Abadi and L. Lamport. Decomposing specifications of concurrent systems. In E.-R. Olderog, editor, *Proc. of the Working Conference on Programming Concepts, Methods and Calculi (PROCOMET '94)*, volume A-56 of *IFIP Transactions*, pages 327–340. North-Holland, 1994.
4. M. Abadi and L. Lamport. Conjoining specifications. *ACM Transactions on Programming Languages and Systems*, 17(3):507–534, May 1995.
5. M. Abadi and S. Merz. On TLA as a logic. In M.Broy, editor, *Deductive Program Design*, NATO ASI series F. Springer-Verlag, 1996.
6. M. Abadi and G.D. Plotkin. A logical view of composition and refinement. In *Proc. of the 18th Ann. ACM Symposium on Principles of Programming Languages*, pages 323–332, 1991.
7. M. Abadi and G.D. Plotkin. A logical view of composition. *Theoretical Computer Science*, 114(1):3–30, 1993.
8. A. Alexander and W. Reisig. Logic of involved variables - system specification with Temporal Logic of Distributed Actions. In *Proc. of the 3rd International Conference on Aplication of Concurrency to System Design (ACSD'03)*, pages 167–176, Guimaraes, Portugal, 2003.
9. B. Alpern and F.B. Schneider. Defining liveness. *Information Processing Letters*, 21(4):181–185, October 1985.
10. E. Best and C. Fernandez. Nonsequential processes – a Petri net view. In W. Brauer, G. Rozenberg, and A. Salomaa, editors, *EATCS Monographs on Theoretical Computer Science*, volume 13. Springer-Verlag, 1988.
11. D. Gomm, E. Kindler, B. Paech, and R. Walter. Compositional liveness properties of EN-systems. In M.A. Marsan, editor, *Applications and Theory of Petri Nets 1993*, 14th *International Conference*, volume 691 of *LNCS*, pages 262–281. Springer-Verlag, June 1993.

12. M. Huth and M. Ryan. *Logic in Computer Science: Modelling and reasoning about systems.* Cambridge University Press, 2000.
13. E. Kindler. A compositional partial order semantics for Petri net components. In P. Azéma and G. Balbo, editors, *Application and Theory of Petri Nets 1997, 18th International Conference*, volume 1248 of *LNCS*, pages 235–252. Springer-Verlag, June 1997.
14. E. Kindler and R. Walter. Message passing mutex. In J. Desel, editor, *Structures in Concurrency Theory*, Workshops in Computing, pages 205–219. Springer-Verlag, may 1995.
15. L. Lamport. "Sometime" is sometimes "not never". In *Proc. of the 7th ACM Symposium on Principles of Programming Languages*, pages 174–185, January 1980.
16. L. Lamport. The Temporal Logic of Actions. *ACM Transactions on Programming Languages and Systems*, 16(3):872–923, May 1994.
17. L. Lamport. Composition: A way to make proofs harder. In A.Pnueli W.P.de Roever, H.Langmaack, editor, *Compositionality: The Significant Difference, International Symposium, COMPOS'97*, volume 1536 of *LNCS*, pages 402–423, September 1997.
18. Z. Manna and A. Pnueli. *The temporal logic of Reactive and Concurrent Systems: Specification.* Springer, 1992.
19. A. Pnueli. The temporal semantics of concurrent programs. *Theoretical Computer Science*, 13(1):45–61, 1981.
20. W. Reisig. Interleaved progress concurrent progress and local progress. In D.A. Peled, V.R. Pratt, and G.J. Holzmann, editors, *Partial Order Methods in Verification*, volume 29 of *DIMACS Series in Discrete Mathematics and Theoretical Computer Science*, pages 99–115. AMS, 1997.
21. W. Reisig. *Elements of Distributed Algorithms: Modeling and Analysis with Petri Nets.* Springer, 1998.

On the Use of Coloured Petri Nets for Object-Oriented Design

João Paulo Barros[1,2,*] and Luís Gomes[1]

[1] Universidade Nova de Lisboa / UNINOVA, Portugal
[2] Instituto Politécnico de Beja, Portugal
{jpb,lugo}@uninova.pt

Abstract. Behaviour specification in object-oriented design clearly benefits from the use of a formal, or semi-formal, visual specification language. This is attested by the adoption of a statecharts based notation by the Unified Modelling Language specification, and also by the several object-inspired Petri net classes. This paper defines a class of high-level nets, named Composable Coloured Petri nets, allowing the use of Coloured Petri nets in object-oriented design, namely for the specification of synchronous and asynchronous communication among objects, and the three most common abstractions: generalisation, classification, and composition. Starting from Coloured Petri nets, the paper shows how those abstractions can be modelled based on node fusion and with minimally intrusive syntax additions. Node fusions take two forms: one for modelling message passing, abstracting the interactions between objects, and another for modelling generalisation and composition, abstracting the system static structure.

Keywords: object-oriented design, net composition, coloured Petri nets, UML class diagrams.

1 Introduction

The original paper on statecharts by Harel [1] acknowledges Petri nets as "one of the best known and best understood solutions" for the behavioural description of complex reactive systems but lacking a "satisfactory hierarchical decomposition" mechanism. Also, the 1989 paper by Huber, Jensen and Shapiro [2] states that "In the literature there is almost no work on hierarchies in Petri Nets". Fortunately, this is no longer true. Especially after the paper by Huber, Jensen and Shapiro, numerous proposals for Petri nets hierarchical structuring and abstraction mechanisms were defined and implemented (e.g. [3–10]). Also, several extensions to Petri nets, with sophisticated abstraction constructs, were inspired by object-oriented concepts [11]. One can view these extensions as bringing object-oriented concepts to Petri nets with the goal of increasing the available abstraction levels and variety. Most of the proposals achieve this through additions to the well-known graphical and textual notations, and consequentially, they diverge, sometimes significantly from the well-known Petri nets

* Work partially supported by a PRODEP III grant (Concurso 2/5.3/ PRODEP/2001, ref. 188.011/01).

J. Cortadella and W. Reisig (Eds.): ICATPN 2004, LNCS 3099, pp. 117–136, 2004.
© Springer-Verlag Berlin Heidelberg 2004

syntax, semantics, or both. Unfortunately, the connection to the common object-oriented concepts and notations, most notably the Unified Modelling Language (UML)[12], is not always evident.

This paper also mixes Petri nets with object-oriented modelling. Our intention is to close the gap between high-level Petri nets and object-oriented notation and concepts, especially the ones present in UML. We remain close to Coloured Petri Nets (CPNs) [8] and rely, exclusively, on the addition of two distinct abstractions, both based on node fusion. With this, we strive for a minimal number of modifications to the usual CPN graphical and textual syntaxes.

We propose a class of nets named *Composable Coloured Petri nets* (CCPNs). They differ from other "Petri nets with objects" proposals in the following points:

1. They are strongly based on CPNs.
2. They totally avoid modifications to the CPNs graphical notation.
3. They imply minimal additions to the CPNs textual notations.
4. They use a composition operator, for the specification of generalisation and composition, and a different abstraction for message passing modelling.
5. They are reducible to a CPN.

Regarding semantics, Composable Coloured Petri nets are shown to be higher level representations for Coloured Petri nets [8] and, by consequence, to place/transition nets (e.g. [13, 8]). This makes available to CCPNs all the known theoretical and practical work on CPNs. From another perspective, this paper's contribution can be seen as an alternative to class behavioural specification by statecharts [1] as presently adopted in the Unified Modelling Language (UML) [12].

The minimal modifications to the known CPNs syntax and the "syntactic sugar" aspect of CCPNs also mean that the proposed higher-level structuring mechanisms must rely on well-known Petri nets concepts, namely transition fusion and place fusion. The former maps quite naturally to synchronous request invocation (or synchronous message passing), and the latter to asynchronous requests. Node fusions are also used to model two important abstractions: composition and (a form of) generalisation. Classification (an abstraction for a set of instances sharing a data structure and a behaviour) is supported by coloured tokens representing objects as pairs containing a class identifier (typically implicit), an instance identifier, and an optional data structure.

The following section defines CCPNs and Sect. 3 three shows how they can be translated to a CPN model. Section 4 defines a composition operation named net addition, which can be used for better abstracting and composing the static net structure. Section 5 shows how CCPNs, together with the addition operation, can be used to model class and object attributes, message passing (through requests), classification, generalisation, composition, and object life cycles. More concisely, templates, navigability, visibility, and multiple invocations, are also discussed. Section 6 presents an illustrative example inspired by the well-known readers-writers and producer-consumer problems. Section 7 reviews related approaches and, the last section, concludes with some pointers to future implementation work.

2 Composable Coloured Petri Nets

Inspired by the UML specification [12], we first define an object system as a set of classes and a set of requests (also defined as classes) and assume the existence of requests *create/destroy* for the dynamic creation/destruction of objects:

Definition 1 (Object System). *An Object System is a tuple* $OS = (\mathcal{O}, \mathcal{R}, V, C, class, attrib, Obj, \mathcal{E}, ppq)$ *verifying the requirements below:*

1. $\mathcal{O} = \{O_1, O_2, \ldots, O_n\}$ *is a finite set of classes (non-request classes).*
2. $\mathcal{R} = \{R_1, R_2, \ldots, R_m\}$ *is a finite set of request classes, such that* $\mathcal{O} \cap \mathcal{R} = \emptyset$.
3. V *is a set of variables.*
4. C *is a set of constants.*
5. *class is a function defined from variables and constants into non-request classes:* $class : (V \cup C) \rightarrow \mathcal{O}$.
6. *attrib is an attribute function defined from classes into tuples of non-request variables and constants:* $attrib : (\mathcal{O} \cup \mathcal{R}) \rightarrow (V \cup C)^*$
7. $Obj \subseteq \mathcal{O} \times \mathbb{N}_0 \times (V \cup C)^*$ *is a set of objects, where* $\forall (c, i, data) \in Obj, attrib(c) = data$.
8. $\{create, destroy\} \subseteq \mathcal{R} \wedge attrib(create) = attrib(destroy) = \emptyset$
9. $\mathcal{E} \subseteq \{e | e \in \{SEND, RECV\} \times \mathcal{O} \times \mathcal{R} \times (V \cup C)^*\}$ *is a finite set of events where:* $\forall (sr, c, r, pl) \in \mathcal{E}, attrib(e) = attrib(r) = pl$ *and* $\forall e_1^S = (SEND, c_1, r, pl_1), e_2^R = (RECV, c_2, r, pl_2) \in \mathcal{E}, \forall 1 \leq i \leq |attrib(c_1)|$, $class(attrib_i(e_1^S)) = class(attrib_i(e_2^S)) \wedge attrib_i(e_2^R) \notin C$ *(the attributes are compatible); we also say that* $class(e_1^S) = c_1$ *and* $class(e_2^R) = c_2$.
10. *ppq is a parameter passing qualifier function defined from event attributes into one of three qualifiers:* $ppq : attrib(\mathcal{E}) \rightarrow \{IN, OUT, INOUT\}$.

Notice that we distinguish between request *senders* and request *receivers* by the association of qualifiers *SEND* and *RECV* to the sender and receiver event, respectively. We omit the instance number in a request or event specification. This implies that, by default, we are implicitly referring to class level requests. If needed, the instance number should be part of the request attributes (preferably as the first one, as usually found in object-oriented programming languages implementations).

We also define the *attrib* function on events. For a given event e, we also use $attrib_i^a(e)$ for the ith actual attribute, and $attrib_i^f(e)$ for the ith formal attribute. We write $e^S \underset{req}{=} e^R$ for two events $e^S = (SEND, c_1, req, pl_1) \in \mathcal{E}^S$ and $e^R = (RECV, c_2, req, pl_2) \in \mathcal{E}^R$ having the same request req and compatible attributes. Additionally, for a request $r \in \mathcal{R}$ we also use $attrib_{IN}(r) = \{a \in attrib(r) | ppq(a) = IN\}$, as well as $attrib_{OUT}(r)$ and $attrib_{INOUT}(r)$ with the obvious analogous semantics.

As already stated, this paper proposes a way to specify class behaviour in an object system by a composition operation and a Petri nets class strongly based on CPNs. Compared to place/transition nets, CPNs provide tokens as data types. The tokens' data values can be manipulated by expressions associated to arcs. A transition firing is dependent on the existence of specific data values. These are named transition bindings. The handling of data values captures symmetries in the model and, consequently, typically allows a significant reduction in model size, compared to low-level nets. Additionally, it brings an increased flexibility for the modeller as the model complexity can be split between the graphical net structure and the net inscriptions (data values, expressions, functions, etc.). Next we present the CPN definition (Def. 2) as in [8] where further details can also be found. The definition assumes the existence of an inscription language allowing the specification of algebraic expressions, variables, and functions. Functions *Type* and *Var* apply in that context: *Type* returns the type(s) of variable(s) or function(s); *Var* returns the set of variables present in an attribute (or parameter) list, an expression, or a function. The subscript $_{MS}$ denotes multisets (bags), and $\mathbb{B} = \{true, false\}$.

Definition 2 (Coloured Petri Net). *A Coloured Petri net is a tuple* $N = (\Sigma, P, T, A, N, C, G, E, I)$ *satisfying the following requirements:*

1. *Σ is a finite set of non-empty types called colour sets (or simply* colours*).*
2. *P is a finite set of places.*
3. *T is a finite set of transitions.*
4. *A is a set of arcs such that: $P \cap T = A \cap P = A \cap T = \emptyset$.*
5. *N is a node function defined from A into $(P \times T \cup T \times P)$*
6. *C is a colour function mapping each place $p \in P$ into a colour from Σ.*
7. *G is a guard function. It is defined from T into expressions such that*
$$\forall t \in T, Type(G(t)) = \mathbb{B} \wedge Type(Var(G(t))) \subseteq \Sigma$$
8. *E is an arc expression function. It is defined from A into expressions such that: $\forall a \in A, Type(E(a)) = C(p(a))_{MS} \wedge Type(Var(E(a))) \subseteq \Sigma$ where $p(a)$ is the place connected to a* [1]*.*
9. *I is an initialisation function mapping each place $p \in P$ into a multiset over $C(p)$.*

Each arc has an associated function with free variables. Each of these variables' type belongs to the set of net types denoted by Σ. On each transition t firing, all variables in the respective input arcs are bound according to the token values in the transition input places. Next we classify the transition variables:

Definition 3 (Transition Variables). *Let* $N = (\Sigma, P, T, A, N, C, G, E, I)$ *be a CPN. For each transition $t \in T$ we use the following definitions:*

1. *Each variable occurring in, at least, one input arc expression is a transition input variable: $VarIn(t) = \{v | \exists p \in P, \exists a \in A, N(a) = (p,t) \wedge v \in Var(E(a))\}$.*
2. *Each variable occurring in, at least, one output arcs expression is a transition output variable: $VarOut(t) = \{v | \exists p \in P, \exists a \in A, N(a) = (t,p) \wedge v \in Var(E(a))\}$.*

---
[1] We will also use $t(a)$ for the transition connected to arc a.

3. *Each variable occurring in at least one input or output arc expression is a transition variable:* $Var(t) = VarIn(t) \cup VarOut(t)$.

A transition binding is one set of bound transition variables. It is defined as a function b, with $Var(t)$ as domain, that assigns a value to each transition variable. Additionally, the values in a transition binding must allow the respective transition guard to evaluate to *true*.

In a given object system, each class is defined by a CPN where some transitions can have associated events and some places can be fused. A CCPN can be seen as a way to interconnect a set of CPNs by a particular form of transition fusion, allowing synchronous communication, and also by place fusion, allowing asynchronous communication. We always assume that each CPN provides a distinct scope for the respective variables. In other words, two variables in two different classes are always distinct variables, unless related by pass-by-name in events as presented later.

Next we define CCPNs (Def. 4)[2]. The paragraphs following the definition explain it and should be read in parallel with it. Each CCPN contains a set of CPNs. This set is denoted $\mathcal{S}$. Generically, we will use X_s to denote "X of $s \in \mathcal{S}$", and $X_{\mathcal{S}}$ to denote the set $\bigcup_{s \in \mathcal{S}} X_s$. For example: P_s denotes the set of places in the CPN $s \in \mathcal{S}$; $P_{\mathcal{S}}$ stands for the set of all places in all CPNs in $\mathcal{S}$.

Definition 4 (Composable Coloured Petri Net). *Let* $OS = (\mathcal{O}, \mathcal{R}, V, C, class, attrib, Obj, \mathcal{E}, ppq)$ *be an object system. A CCPN is a tuple* $N = (\mathcal{S}, nc, PF, TE)$ *defining the behaviour of a set of classes* $\{O_1, \ldots, O_n\} \subseteq \mathcal{O}$ *satisfying the following requirements:*

1. $\mathcal{S}$ *is a set of CPNs such that:*
   (a) $\Sigma_{\mathcal{S}} \subseteq \mathcal{O}^*$
   (b) nc *is a net class function from nets in* $\mathcal{S}$ *to non-request classes in OS:* $nc : \mathcal{S} \to \mathcal{O}$.
   (c) *The sets of net elements are pairwise disjoint:*

$$\forall s_1, s_2 \in \mathcal{S}, s_1 \neq s_2 \Rightarrow (P_{s_1} \cup T_{s_1} \cup A_{s_1}) \cap (P_{s_2} \cup T_{s_2} \cup A_{s_2}) = \emptyset$$

   (d) $Var(T_{\mathcal{S}}) \subseteq V$.
2. PF *is a partition of the set of all places in* $\mathcal{S}$ *(2a,2b,2c,2d):*
   (a) $PF \subseteq \mathcal{P}(P_{\mathcal{S}})$
   (b) $PFE \in PF \Rightarrow PFE \neq \emptyset$
   (c) $\forall PFE_1, PFE_2 \in PF, (PFE_1 \neq PFE_2) \Rightarrow PFE_1 \cap PFE_2 = \emptyset$
   (d) $\forall p \in P_{\mathcal{S}}, \exists PFE \in PF, p \in PFE$
   (e) $\forall PFE \in PF, \forall p_1, p_2 \in PFE,\ C(p_1) = C(p_2) \wedge I(p_1) = I(p_2)$

---

[2] In the following definitions the notation is similar to the one used by Christensen and Petrucci for the definition of Modular Coloured Petri Nets (MCPNs)[7]. This should allow the reader an easier comparison between CCPNs and MCPNs. Section 7 presents further comparison.

3. TE is a partial function from transitions to events (3a), such that:
(a) $TE : T_{\mathcal{S}} \hookrightarrow \mathcal{E}$
(b) Event attributes must satisfy the following:

$$\exists t^S, t^R \in dom(TE), t^S \neq t^R \wedge e^S = TE(t^S) \Rightarrow e^R = TE(t^R) \wedge e^S \underset{req}{=} e^R \wedge$$
$$attrib_{IN}(e^S) \subseteq (VarIn(t^S) \cup C) \wedge attrib_{OUT}(e^S) \subseteq VarOut(t^S) \wedge$$
$$attrib_{OUT}(e^R) \subseteq VarIn(t^R) \wedge attrib_{INOUT}(e^R) \subseteq VarIn(t^R) \wedge$$
$$attrib_{INOUT}(e^S) \subseteq (VarIn(t^S) \cap VarOut(t^S))$$

1. Each colour is a list of object system non-request classes (1a). Clearly, if all operations on an object can be called independently from its internal state, behaviour modelling is not necessary. For this reason, CCPNs do not enforce all classes to be specified by a Petri net as this would be extremely counter-productive. In fact, excessive behavioural modelling has been pointed out as one of the leading causes of analysis paralysis [14]. Differently, each and every net corresponds to one non-request class in the object system(1b). Finally, all net nodes and arcs are considered to be disjoint (1c). All transition variables are object system variables (1d). The latter can also occur in events.
2. Place fusion sets are disjoint place sets, possibly across different CPNs, which are considered different occurrences of one single place. Each set in the PF partition on $P_{\mathcal{S}}$ is called a place fusion set. All places, in a given place fusion set, have the same associated colour and the same initialisation expressions (2e). We denote by $[p]$ the fusion set containing place $p \in P_{\mathcal{S}}$.
3. Some transitions can have an associated event (3a). For each *SEND* event in some transition, the corresponding *RECV* event must be associated to one different transition (3b). Event attributes (which include request parameters) must be included in proper subsets of the respective SEND and RECV transition variables, according to their parameter passing qualifiers. We talk about a *SEND transition* when a transition has one associated *SEND* event, and about a *RECV transition* when the transition has one *RECV* event.

The following section details how CCPNs can be translatable to CPNs.

3 From CCPNs to CPN

This section defines CCPNs semantics by defining their translation to a single CPN. As CPNs have a well documented (and implemented) semantics [8], this translation implicitly defines the CCPNs semantics. As each CCPN contains a set of CPNs the translation to a single CPN is centred in the two ways to interconnect CPNs, namely the set of place fusion sets (PF) and the association of events to transitions through the partial function TE.

From the net model perspective, each transition t with $TE(t) = \emptyset$, and each pair of one *SEND* event and one *RECV* event, define a synchrony group:

Definition 5 (Synchrony group). *Let $N = (\mathcal{S}, nc, PF, TE)$ be a CCPN. A synchrony group is either a single transition $t \in T_{\mathcal{S}}$, with no associated events or a triple containing one SEND transition, the non-request class in the SEND event, and the RECV transition. The set of all synchrony groups is denoted by SG:*

$$SG = \{t \in T_{\mathcal{S}} | TE(t) = \emptyset\} \cup \{(t^S, c^S, t^R) \in T_{\mathcal{S}} \times \mathcal{O} \times T_{\mathcal{S}} | \\ c^S = class(TE(t^S)) \wedge \exists s \in \mathcal{S}, t^R \in T_s \wedge nc(s) = c^S, TE(t^S) \underset{req}{=} TE(t^R)\}$$

A synchrony group is said to be trivial if it contains a single transition in the CCPNs. Accordingly, the set $\{t \in T_{\mathcal{S}} | TE(t) = \emptyset\}$ is called the set of trivial synchrony groups. Each synchrony group will correspond to a single transition in the equivalent CPN, with all the input and output arcs of the original transitions.

The possibility of manipulating the event class parameter (c^S) as a transition variable allows the specification of polymorphic request invocations, through translation to a disjunctive transition fusion. An example is presented in Sect. 6 (Fig. 9).

Although synchrony groups present some similarity to synchronous channels of Modular Coloured Petri nets [6,7], the synchrony groups rationale comes from the need to model method invocation, as usually found in object-oriented languages and defined in the UML specification. Furthermore, there is no concept of a channel, as found also in OCP-nets [15], as there are no channel type function or channel expressions. Event variables are a subset of the transition variables with the restrictions defined in point 3b) in Def. 4. As such, synchrony groups can be seen as a fusion of high level transitions but on a specific binding. This is imposed by the SEND transition, on the *IN* formal parameters of the receiving transition; and by the RECV transition, on the *OUT* actual parameters of the sending transition. As for transitions, we refer to the guard of a synchrony group *sg* as $G(sg)$, which is the conjunction of the guards of the *sg* transitions and a test on the event non-request class parameter (see point 5 in Def. 6).

Definition 6 (Synchrony group binding). *Let $sg = (t^S, c^S, t^R)$ be a non-trivial synchrony group and let $e^S = TE(t^S)$ and $e^R = TE(t^R)$. A binding b of sg is a function defined on the transitions variables of the synchrony group, $Var(sg) = Var(t^S) \cup Var(t^R)$, such that:*

1. $Var(t^S) \cap Var(t^R) = \emptyset$
2. $\forall 1 \leq i \leq |attrib^a(e^S)|, class(attrib^a_i(e^S)) = class(attrib^f_i(e^R)) \wedge$ $$attrib^a_i(e^S) \mapsto attrib^f_i(e^R)$$
3. $\forall v \in Var(sg), b(v) \in Type(v)$
4. $G(t^S)\langle b\rangle \wedge G(t^R)\langle b\rangle$
5. $class(e^S) \in VarIn(t^S) \Rightarrow \exists c \in Type(class(e^S)), class(e^S) = c$

Fig. 1 exemplifies the formation of a synchrony group from a request with attributes. In the synchrony group, the variables "*a*" and "*b*" replace, respectively, the variables "*c*" and "*d*" "by name": "*a*" and "*b*" are the request actual parameters, "*c*" and "*d*" are the request formal parameter. The parameter passing is

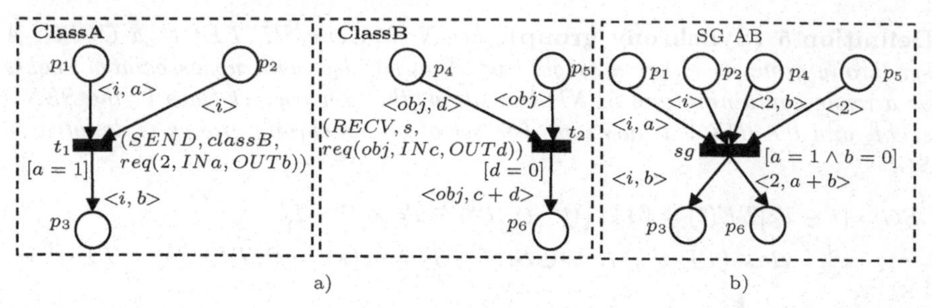

Fig. 1. a)Synchronous request with parameters, and b) respective synchrony group.

an Algol-like "pass by name". Note that the object identifier ("2" in the example) is also part of the request and, consequently, also of the receiving transition event parameters. On one hand, the class identifier (*ClassB* in Fig. 1) is used to specify the net (class) where the corresponding RECV transition resides (the one that should be considered as fused with the current SEND transition); on the other hand, on the RECV event it identifies the class where the respective SEND resides ("s" parameter in the example). Condition 5 in Def. 6 is used in Fig. 9 to model polymorphism.

We can now define the equivalent Coloured Petri net:

Definition 7. *For a given CCPNs, $N = (\mathcal{S}, nc, PF, TE)$ we define an equivalent coloured Petri net $CPN = (\Sigma', P', T', A', N', C', G', E', I')$ where*

1. $\Sigma' = \Sigma_{\mathcal{S}}$
2. $P' = PF$
3. $T' = SG$
4. $A' = \{(a, sg) \in A_{\mathcal{S}} \times SG \mid t(a) \in sg\}$
5. $\forall a = (a, sg) \in A', N'(a, sg) = N([p(a)], sg) \vee N'(sg, a) = N(sg, [p(a)])$
6. $\forall p' \in P', C'(p') = C(p')$
7. $\forall sg \in T', G'(sg) = G(sg)$
8. $\forall a' = (a, sg) \in A', E'(a') = E(a)$
9. $\forall p' \in P', I'(p') = I(p')$

4 Net Addition

Here we define a net composition operation (e.g. [2, 8, 16, 10]) based on the known concept of net compositions by node fusion. This operation has two aims: to provide an intuitive and uniform abstraction for generalisation and composition; to be clearly separated from the other node fusions already provided by CCPNs. For this reason, net addition should always be seen as applicable to CCPNs: in a translation to a CPN, the implicit node fusions in CCPNs (*TE* and *PF*) would be applied next. Although the net addition definition (Def. 8) is quite long, the intuition is very obvious: the interface nodes (IN_A and IN_B) are fused. Markings, in interface places, are added. They must be compatible in the sense that both colours are equal or one "includes" the other (it includes all values in the other's domain). All the other nodes are part of the resulting net and all arcs, and respective arc inscriptions, are preserved.

Definition 8. *Let $N = (\mathcal{S}, nc, PF, TE)$ be a CCPN, and let $N_A, N_B \in \mathcal{S}$ be two CPNs. We write $N_C = N_A(IN_A) \oplus N_B(IN_B)$, where $(P_A \cup T_A) \cap (P_B \cup T_B) = \emptyset$, $|IN_A| = |IN_B|$, and $|IN_A| > 0 \Rightarrow IN_A = (in_{A_1}, \ldots, in_{A_n}) \wedge IN_B = (in_{B_1}, \ldots, in_{B_n})$ where, together, IN_A and IN_B define pairs of node fusions named INF: $INF_{AB} = \{(in_{A_1}, in_{B_1}), \ldots (in_{A_n}, in_{B_n})\}$ satisfying $\forall 1 \le i \le n, (in_{A_i}, in_{B_i}) \in P_A \times P_B \vee (in_{A_i}, in_{B_i}) \in T_A \times T_B$. We also use the following definitions:*

1. $P(INF_{AB}) = \{(p_A, p_B) \in INF_{AB} | \{p_A, p_B\} \subseteq (P_A \cup P_B)\}$
2. $T(INF_{AB}) = \{(t_A, t_B) \in INF_{AB} | \{t_A, t_B\} \subseteq (T_A \cup T_B)\}$

The resulting N_C is defined as the following CPN:

$$
\begin{aligned}
&\Sigma_C = \Sigma_A \cup \Sigma_B \\
&P_C = (P_A \backslash P(IN_A)) \cup (P_B \backslash P(IN_B)) \cup P(INF_{AB}) \\
&T_C = (T_A \backslash T(IN_A)) \cup (T_B \backslash T(IN_B)) \cup T(INF_{AB}) \\
&A_C = A_A \cup A_B \\
&N_C = \{a \in (A_A \cup A_B) | p(a) \notin (IN_A \cup IN_B) \wedge t(a) \notin (IN_A \cup IN_B)\} \cup \\
&\qquad \{(a, ((p_A, p_B), (t_A, t_B))) \in ((A_A \cup A_B) \times (P_A \times P_B) \times (T_A \times T_B)) \mid \\
&\qquad\quad (a, (p_A, t_A)) \in N_A \vee (a, (p_B, t_B)) \in N_B\} \cup \\
&\qquad \{(a, ((t_A, t_B), (p_A, p_B))) \in ((A_A \cup A_B) \times (T_A \times T_B) \times (P_A \times P_B)) \mid \\
&\qquad\quad (a, (t_A, p_A)) \in N_A \vee (a, (t_B, p_B)) \in N_B\} \\
&C_C = \{(p, c) \in (C_A \cup C_B) | p \notin (IN_A \cup IN_B)\} \cup \\
&\qquad \{((p_A, p_B), c) | (p_A, c) \in C_A \wedge C_A \supseteq C_B \wedge (p_A, p_B) \in INF_{AB}\} \\
&G_C = \{(t, e) \in (G_A \cup G_B) | t \notin (IN_A \cup IN_B)\} \cup \\
&\qquad \{((t_A, t_B), e_A \wedge e_B) | (t_A, e_A) \in G_A \wedge (t_B, e_B) \in G_B \wedge (t_A, t_B) \in INF_{AB}\} \\
&E_C = E_A \cup E_B \\
&I_C = \{(p, m) \in (I_A \cup I_B) | p \notin (IN_A \cup IN_B)\} \cup \\
&\qquad \{((p_A, p_B), m_A + m_B) \mid \\
&\qquad\quad (p_A, m_A) \in I_A \wedge (p_B, m_B) \in I_B \wedge (p_A, p_B) \in INF_{AB}\}
\end{aligned}
$$

5 Object-Oriented Design and Composable CPNs

In this section we present and discuss how several OOD concepts are supported by CCPNs.

5.1 Attributes

Instance attributes and class level attributes are specified in lists inside tokens. Typically, tokens include the instance number, the instance attributes, or both.

The instance numbers inside place tokens allow the modelling of all class instances by a single Petri net. The Petri net becomes the orthogonal superposition of all object behaviours. This means the Petri net model is the object system model: there is no need to think of multiple diagrams, one for each object of the same class. Additionally, the concurrent execution of different object parts (intra-object concurrency) is straightforwardly modelled by Petri nets inherent capabilities, namely the graphical visualisation of forks and joins. Class level attributes can be modelled by a place (named CLA in Fig. 2) containing only class level attributes. A set of *SEND* and *RECV* transitions can then be used to access (and possibly modify) the class level attributes. These transitions effectively control the class level attributes visibility and model class level requests.

5.2 Requests

As in structured programming with procedure invocation, requests allow the abstraction of the inner details of a complex operation provided by a class or object. We see them as a fundamental abstraction when modelling object-oriented systems.

A request is "an initial message from one object to another" [12, pp. 2-311]. The UML specification gives the following definition for signals and operations:

> "Several kinds of requests exist between instances (e.g., sending a signal and invoking an operation). The former is used to trigger a reaction in the receiver in an asynchronous way and without a reply, while the latter applies an operation to an instance, which can be either done synchronously or asynchronously and may require a reply from the receiver to the sender. Other kinds of requests are used for example to create a new instance or to delete an already existing instance." [12, pp. 2-110].

As already presented, instance creation and destruction can be synchronous or asynchronous.

Asynchronous. Signals, as defined in the UML specification, seem to imply some form of event broadcast. Its non-local effect brings fundamental difficulties in their modelling by Petri nets. To avoid extensions to Petri nets, and also to CPNs, we propose the use of shared memory as the mean for signal communication. Shared memory is modelled by place fusion. Each signal is modelled by a place fusion set. Each place belonging to a place fusion set should be annotated with the set of places in the same fusion set or, alternatively, all should be annotated with the identifier of the "real" place in the set.

The UML specification defines a signal as "The specification of an asynchronous stimulus communicated between instances." [12, p. Glossary-13]. A restricted interpretation for this definition sees a signal as an asynchronous request between two objects, where one object knows the other. This interpretation is readily modelled by a place fusion between the place where the request is put and the place from where the request is retrieved. Place fusion sets also support a form of signal broadcast where one object sends a set of asynchronous requests to a known number of receiving objects. In the simplest approach, these are then left on their own: e.g. one receiving object can retrieve all the requests.

Synchronous. Synchronous requests are specified by synchrony groups. These support synchronisation between several processes, as usually found in process algebras (see Fig. 1). According to the UML specification, if the request is synchronous the receiver must have a well-defined reply point. CCPNs can enforce this by a pair of requests: from sender to receiver (method activation) and from receiver to sender (method return). This second synchronisation is also the only way for the requestor to "(...)suspend execution until the activity invoked by the request reaches a well defined point and sends a reply message back to the requestor." [12, pp. 2-311]. Fig. 2a illustrates this double synchronisation. Notice that the c and x parameters identify the calling object instance. This allows the mentioned eventual reply to the requestor by a second synchrony group.

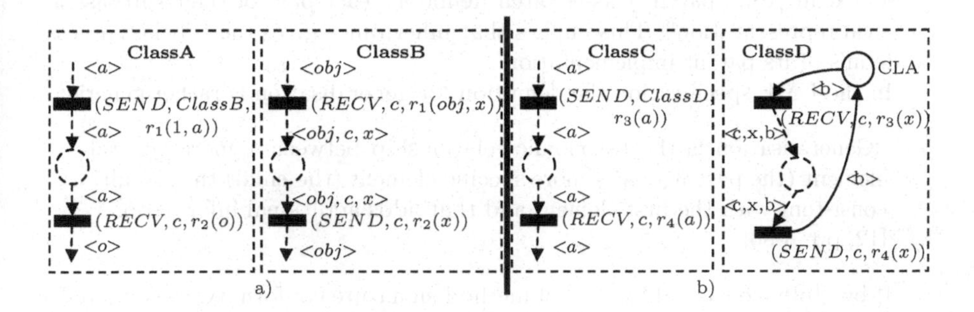

Fig. 2. a)Double synchronisation for non-atomic operation invocation. b)Double synchronisation for class level operation invocation; notice the use of the class level attributes place (CLA) accessed by two transitions with associated class level requests, and without an instance identifier in the request. Due to space limitations, the attribute qualifiers (*IN* in all attributes) were omitted.

If the method does not return data and its execution is considered atomic with no need for a reply request, it is enough to associate a code segment, in the target programming language, to the *RECV* transition. This is also supported by the CPN Tools and Design/CPN tools [17, 18] and can be helpful for code generation purposes.

Class level methods are implicitly modelled if an event has no instance number and accesses the *CLA* place in the respective CPN.

5.3 Classification

Classification provides the association between a set of related instances and the respective class. It is also extremely intuitive, as we are used to point sets of similar things (usually named objects) as belonging to a unique class. The similarity encompasses behaviour (or provided functionality) and structure: e.g. the set of human beings, or the set of doors. Nets easily model classification: the net structure enforces a common behaviour among the several tokens (representing objects) flowing along it. Additionally, high-level net (as CPNs) can also model exceptional behaviour by the use of guards, arc expressions, or both. Basically, each net is a class, and each object is associated to specific tokens.

5.4 Generalisation

Generalisation is a widely used abstraction as it relates classes in an intuitive manner. A good example is the work on Taxonomy by Carl Linnaeus, which allowed biologists to relate a large set of animal species classes through generalisations. Yet, generalisation, usually referred as inheritance, has several already well-known disadvantages (e.g. [19]). Basically, they result from the poor abstraction provided by class hierarchies as class inheritance "break encapsulation". As presented in [19], "parent classes often define at least part of their subclasses' physical representation". This implies that inheritance gives the subclass access to details of its parent implementation.

In the UML specification the definition for generalisation is rather generic:

> "Generalisation is the taxonomic relationship between a more general element (the parent) and a more specific element (the child) that is fully consistent with the first element and that adds additional information." [12, pp. 3-86]

If by "fully consistent" we mean method signature conformity, as commonly found in mainstream programming languages (e.g. C++, Java) and if we want net B to inherit from net A, we just need to extend net A with elements from net B, while maintaining the *RECV* events in net A. This extension is implemented through the defined net addition. This is the approach we propose here. Yet, it should be clear that method signature conformity is not enough to ensure type substitutability as the specialised class can modify the behaviour for a specific method of the general class. This has already been pointed out (e.g. [20]), in the context of distributed systems. Behavioural conformity becomes a much more important issue [21] and it has already been extensively studied, also in the Petri net context [22, 23]. Nevertheless, we believe, net addition can be a sufficiently "low-level" operation to support other inheritance types between nets.

5.5 Composition

Composition is another useful abstraction for object-oriented modelling and, arguably, a fundamental one [24]. On one hand it corresponds to yet another intuitive way to organise objects: objects inside objects. On the other hand it is to classes what methods (and the associated procedural abstraction) are to code.

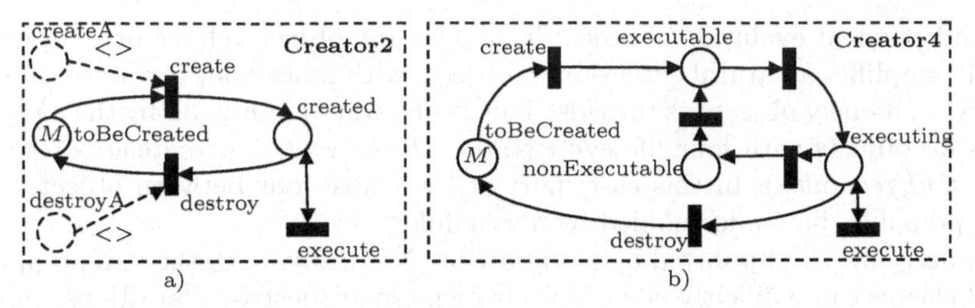

Fig. 3. a)Class *Creator2* modelling object creation and destruction; b) Class *Creator4* modelling a more detailed life-cycle. M stands for the marking $<1, D_1> + \cdots + <n, D_n>$ where n is the bound on the number of instances. Places *createA* and *destroyA* are used only for object asynchronous creation and destruction. All arcs with no inscription have the associated expression $<i, D_i>$.

We consider two types of composition: one where the component is created, and destructed, together with the composite object; other where the component object can be created after the composite creation and can also be destructed before the composite destruction. The first is sometimes called strong composition or just composition, and the second is sometimes called *aggregation* or even *association*. Yet, the latter two terms are sometimes not well-defined even in the UML specification [24].

We propose to model composition by net addition where the fusion nodes include the create and destroy transitions or associated places for asynchronous creation and destruction (see Fig. 3).

5.6 Object Creation and Destruction

The associated object system *OS* definition contains the create and destroy requests. They are responsible for the respective creation and destruction of object identifiers inside the associated Petri net class. The instance numbers of the objects, initially available to be created, are deposited in a place (*toBeCreated* in Fig. 3). The create transition removes tokens from place *toBeCreated* and deposits them in place *created*. These tokens can only be subtracted by the corresponding *destroy* transition. This gives rise to a simple complementary place pattern, where one place contains the set of creatable objects for a given class and its complementary place contains the already created objects (Fig. 3a). The upper bound on the number of class instances (objects) allows the generation of the complete state space (if its dimension allows it).

Considering a class C. The creation/destruction patterns (exemplified by classes *Creator2* and *Creator4* in Fig. 3) can be seen as a separate class C_{life} modelling the class objects life-cycle. The behaviour model is them specified by another class (C_{behav}). The former class is an obvious generalisation as many classes of objects can exhibit the same life cycle. Therefore the C_{behav} class inherits from C_{life}. With this we get a logical separation between the objects life cycle definition and its behaviour model. This intra-class modularity simplifies the independent evolution of object life-cycles and object behaviours. For example, it simplifies a natural extension to objects with more than two states in their life cycle, namely objects as threads. This is illustrated in Fig. 3b by the *Creator4* class for objects with four life cycle states: (*toBeCreated*, *executable*, *executing*, and *nonExecutable*). In this case, part of the transitions between object states will, probably, be made enabled by a scheduler.

Inheritance, is supported by net addition. Naturally, both the parent and the child classes can explicitly contain their own create/destroy transitions. In that case those should be fused, by considering them as interface nodes. Yet, we prefer to avoid the structural repetition: if one class inherits from another containing a create/destroy pattern, than it is enough to fuse all the relevant *SEND* and *RECV* transitions in the child class with an *execute* transition that tests if the intended object is in the correct state, or (if any object will do) simply tests if some object is ready (see Fig. 3). This is used in the example to be presented.

By using net addition, class inheritance is supported by a construction at the class level. This means the structural level in the Petri net. To avoid the use of the same abstraction for modelling the static part and the inter-object object interactions we defend the use of net addition, also for object composition. Yet, if the create/destroy pattern is complex enough, the SEND/RECV mechanism can be used. Anyway, the choice between composition (aggregation) and generalisation (inheritance) still remains unclear and even subjective as illustrated, for example, in chapter 24 of [25].

Finally, as previously noted, object creation can also be made asynchronously. In that case the create transition is no longer an interface node transition but an internal transition depending on an additional place which is made an interface node (dotted places and arcs in Fig. 3a). This fused place enables the internal create transition. A similar technique can be used for object destruction.

5.7 Other Concepts: Templates, Navigability, Visibility, and Multiple Invocation

Other common object-oriented concepts can be readily added to the presented CCPNs and net addition constructs. Here we briefly discuss templates and multiple invocations.

Class templates can be easily supported by CCPNs, through the addition of a notation for template parameters. Navigability results from use of the SEND/REQUEST events.

Visibility already exists at the level of *public* and *private* nodes: sets in the PF partition with more than one place, contain public places; private places are unitary places. Additional visibility would require more qualifiers (e.g. *protected*) eventually also applying to net addition fusion.

Multiple invocation is conjunctive transition fusion. This can be readily supported if one allows the specification of multiple classes in each *SEND* event.

6 Example

Here we briefly present the modelling of an example system by a CCPN. More specifically we present an annotated UML class diagram and the set of annotated CPNs in the CCPN.

The system contains seven active objects: one producer, three readers, one writer, one consumer of type A, and one consumer of type B. It also contains a passive object: the database, which provides methods to access the stored data and a way to control the access. Finally, we also model a start-up object responsible for the creation and destruction of all other objects.

First we present a UML class diagram (Fig. 4). The diagram presents generalisation, composition, and association relations. All are annotated with the node fusions supporting them. The system behaviour is the following: the *DBSystem*

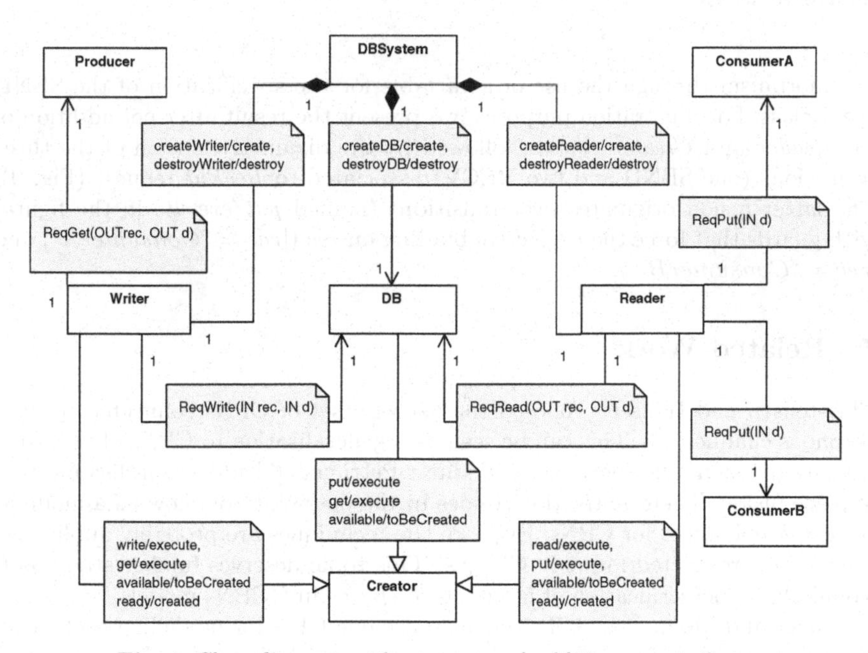

Fig. 4. Class diagram with request and addition annotations.

singleton object (Fig. 8),containing its own create/destroy pattern, knows the intended number of *Readers*, *Writer*, and *DB* to create (by composition). Those objects already have their own create/destroy patterns inherited from the *Creator* class (Fig. 5). Then the writer singleton (Fig. 6) asks the *Producer* (Fig. 7) for the data and the recipient class. After, the writer asks the *DB* object to put those registers (data plus recipient class name) in its database. The *Reader* objects ask the DB for registers (Fig. 9), which are then passed to the respective type of consumer: *ConsumerA* or *ConsumerB* (Fig. 7). Note here, the use of

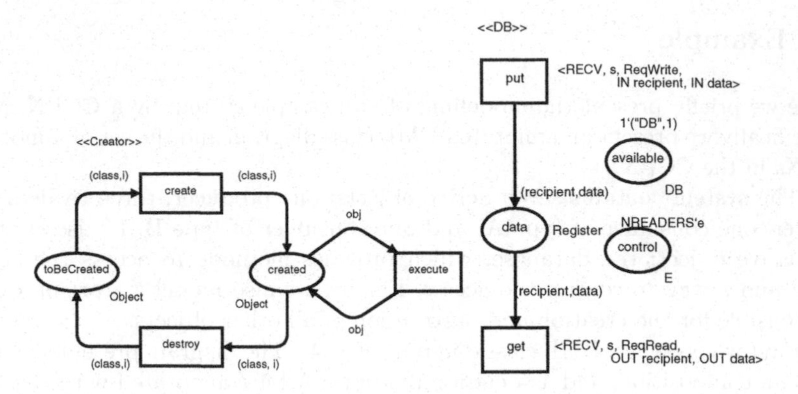

Fig. 5. Class *Creator* modelling object creation and destruction; and class *DB* modelling data access.

polymorphism through the use of a variable for the specification of the SEND event class. For illustration purposes, we present the result after net addition of the *Reader* and *Creator* classes, followed by the disjunctive fusion of the three transitions (one SEND and two RECV) associated to *ReqPut* request (Fig. 9). The latter fusion originates two transitions (named *put/execute* in the figure) with guards that force the respective binding on *rec* ($[rec = \text{“}ConsumerA\text{”}]$ and $[rec = \text{“}ConsumerB\text{”}]$).

7 Related Work

Christensen and Hansen proposed the use of a set of CPNs connected by synchronous channels [6]. This can be seen as a generalisation to CPNs of the usual transition fusion in low level nets. Modular Petri nets [7] add a non-disjoint form of place fusion. Their motivation resides in the possibility of allowing a modular state space analysis for CPNs. Even so, the techniques are probably applicable, even if in a restricted form, to CCPNs. This topic deserves further study as it would allow the verification of larger state spaces in CCPNs models.

Maier and Moldt [15] OCP-Nets also use one CPN for modelling each class. They force the use of two transition fusion pairs, similarly to the one presented in Fig. 2, to model method activation and return. The transition fusion is strongly based on synchronous channels. They propose the use of subtyping inheritance with no demand for behaviour conformity. They use the same set of place and transition fusions for message passing and for net composition. This means net composition results exclusively from message passing. Interestingly, this somehow mimics the problem, in the UML specification, where “the attempt to abstract with the same construct both the static structure of the system and the structure of interactions between objects is not free of problems.” [26].

The already mentioned Petri net tools implement Hierarchical Coloured Petri nets (HCPNs) [18, 17]. As demonstrated, in theory [8] and practice, HCPNs are

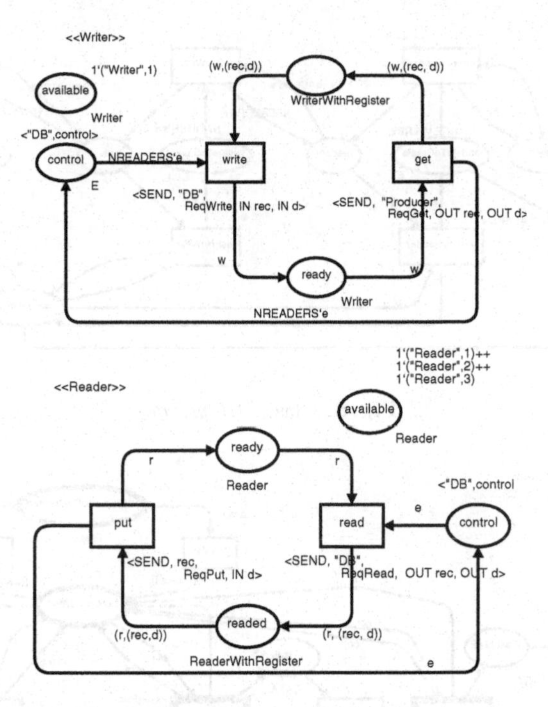

Fig. 6. Class *Writer* and class *Reader*.

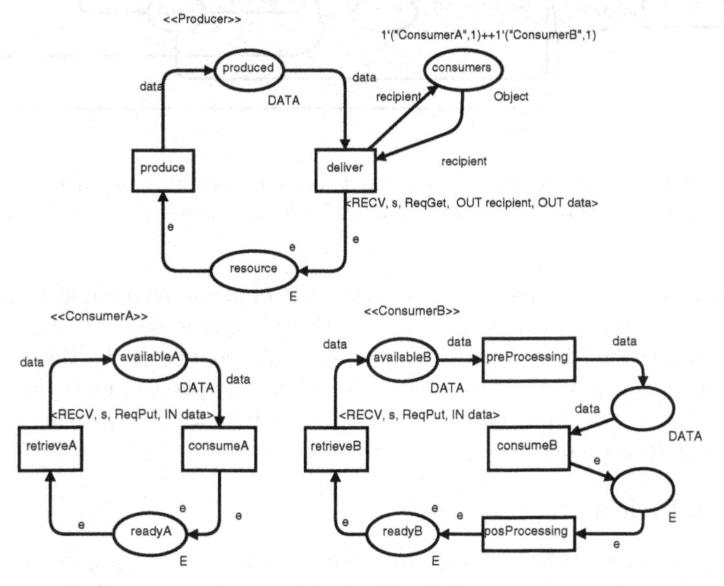

Fig. 7. *Producer*, *ConsumerA* and *ConsumerB* classes.

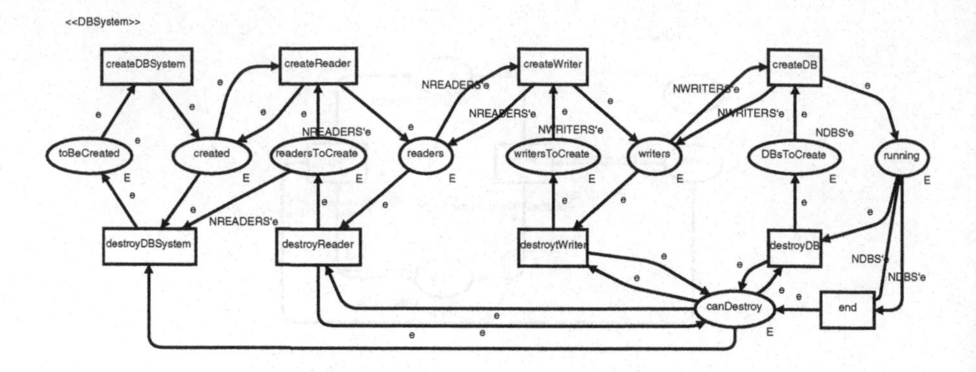

Fig. 8. Class *DBSystem*.

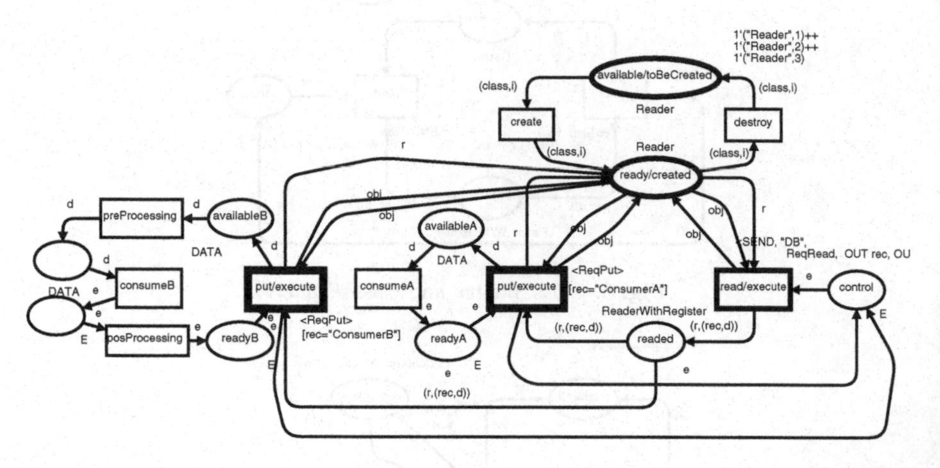

Fig. 9. Result after adding the *Reader* and *Creator* classes plus the invocation of the *ConsumerA* and *ConsumerB* request *ReqPut*. Node fusions are shown in bold.

translatable to CPNs. A HCPN can be seen as a set of connected non-hierarchical coloured Petri nets called pages. Each of these pages is seen by the others as a macro transition. These are a kind of sub-nets connected to the page they are in (super-page) by the fusion of places with port places inside the sub-net. As such, they can be readily integrated with CCPNs as they provide a natural complement to class decomposition and net addition.

8 Conclusion

Composable Coloured Petri Nets offer a way to minimise the gap between Petri nets and object-oriented design. To that end they rely on a well-known class of Coloured Petri nets, which are very well studied from a theoretical point of view and provide excellent tool support.

CCPNs provide an additional layer on top of CPNs, which allows the modelling of synchronous and asynchronous requests closer to the syntax usually found in mainstream object-oriented programming languages. Furthermore, the structural composition operator named net addition, supports the modelling of more "class-oriented" constructs, most notably inheritance. More dynamic compositions are supported by parameterised transition fusion. This allows further net customisation and composition while avoiding additional complexity in the net syntax and semantics.

The place fusion across CPNs is fundamentally the same as the one in Hierarchical Coloured Petri nets. Future work will research this similarity and, especially, the possibility of using or extending the CPN-Tools application to support the presented SEND/RECV transition fusion, while strengthening the connection between CCPNs and UML class diagrams.

Finally, we believe CCPNs offer an interesting alternative to statecharts (or state diagrams) for the modelling of distributed objects behaviour. This is specially true in the cases where intra-object concurrency, and resource modelling are more relevant, as Petri nets offer better scalability and better synchronisation visualisation. Moreover their syntactical and semantical similarity to CPNs and the exclusive use of well-known Petri net concepts as the underlying support, provide a stable and precise ground to object-oriented design.

Acknowledgements

The authors thank the anonymous reviewers whose comments helped to improve the paper.

References

1. Harel, D.: Statecharts: A visual formalism for complex systems. Science of Computer Programming **8** (1987) 231–274
2. Huber, P., Jensen, K., Shapiro, R.M.: Hierarchies in coloured Petri nets. In: Proceedings of the 10th International Conference on Application and Theory of Petri Nets, 1989, Bonn, Germany. (1989) 192–209
3. Brauer, W., Gold, R., Vogler, W.: A survey of behaviour and equivalence preserving refinements of Petri nets. Lecture Notes in Computer Science; Advances in Petri Nets 1990 **483** (1991) 1–46
4. Fehling, R.: A concept of hierarchical Petri nets with building blocks. In: Proceedings of the 12th International Conference on Application and Theory of Petri Nets, 1991, Gjern, Denmark. (1991) 370–389
5. He, X., Lee, J.A.N.: A methodology for constructing predicate transition net specifications. Software–Practice and Experience **21** (1991) 845–875
6. Christensen, S., Hansen, N.D.: Coloured Petri nets extended with channels for synchronous communication. Daimi PB-390 (1992)
7. Christensen, S., Petrucci, L.: Towards a modular analysis of coloured Petri nets. In Jensen, K., ed.: Lecture Notes in Computer Science; 13th International Conference on Application and Theory of Petri Nets 1992, Sheffield, UK. Volume 616., Springer-Verlag (1992) 113–133

8. Jensen, K.: Coloured Petri Nets. Basic Concepts, Analysis Methods and Practical Use - Volume 1–3. Monographs in Theoretical Computer Science. An EATCS Series. Springer-Verlag, Berlin, Germany (1992–1997)
9. Lakos, C.A.: On the abstraction of coloured Petri nets. In Azéma, P., Balbo, G., eds.: Lecture Notes in Computer Science: 18th International Conference on Application and Theory of Petri Nets, Toulouse, France, June 1997. Volume 1248., Berlin, Germany, Springer-Verlag (1997) 42–61
10. Gomes, L., Barros, J.: On structuring mechanisms for Petri nets based system design. In: Proceedings of the 2003 IEEE Conference on Emerging Technologies and Factory Automation (ETFA 2003), IEEE Catalog Number: 03TH8696 (2003) 431–438
11. Agha, G., de Cindio, F., Rozenberg, G., eds.: Concurrent Object-Oriented Programming and Petri Nets, Advances in Petri Nets. Volume 2001 of Lecture Notes in Computer Science. Springer (2001)
12. OMG: Unified modeling language specification, version 1.5. `http://www.omg.org/cgi-bin/doc?formal/03-03-01` (2003) Unified Modeling Language, v1.5, Object Management Group.
13. Reisig, W.: Petri nets: an Introduction. Springer-Verlag New York, Inc. (1985)
14. Desmond Francis D'Souza, Alan Cameron Wills: Objects, Components, and Frameworks With UML: The Catalysis Approach. Addison Wesley Longman (1998)
15. Maier, C., Moldt, D.: Object coloured Petri nets – a formal technique for object oriented modelling. Lecture Notes in Computer Science: Concurrent Object-Oriented Programming and Petri Nets, Advances in Petri Nets **2001** (2001) 406–427
16. Barros, J., Gomes, L.: Modifying Petri net models by means of crosscutting operations. In: Proceedings of the 3rd International Conference on Application of Concurrency to System Design, IEEE Computer Society (2003)
17. CPN Tools. `http://wiki.daimi.au.dk/cpntools` (2004)
18. Design/CPN. `http://www.daimi.au.dk/designCPN/` (2004)
19. Gamma, E., Helm, R., Johnson, R., Vlissides, J.: Design Patterns: Elements of Reusable Object-Oriented Software. Addison-Wesley (1995)
20. Fischer, C., Wehrheim, H.: Behavioural subtyping relations for object-oriented formalisms. In: AMAST 2000: Algebraic Methodology and Software Technology, LNCS 1816, Springer (2000)
21. Harel, D., Kupferman, O.: On object systems and behavioral inheritance. IEEE Transactions on Software Engineering **28** (2002) 889–903
22. van der Aalst, W., Basten, T.: Life-cycle inheritance - a Petri-net-based approach. In: Proceeding of the 18th International Conference on Application and Theory of Petri Nets, Toulouse, France, June 1997., Springer-Verlag (1997) 62–81
23. Balzarotti, C., de Cindio, F., Pomello, L.: Observation equivalences for the semantics of inheritance. In: Formal Methods for Open Object-based Distributed Systems (FMOODS'99), Kluwer Academic Publishers (1999) 67–82
24. Steimann, F., Gößner, J., Mück, T.: On the key role of compositioning object-oriented modelling. In: UML 2003 - The Unified Modeling Language, Modeling Languages and Applications, 6th International Conference San Francisco, CA, USA, October 2003 Proceedings. LNCS 2863, Springer (2003)
25. Meyer, B.: Object-oriented Software Construction. 2nd edn. Prentice-Hall (1997)
26. Génova, G., Llorens, J., Palacios, V.: Sending messages in UML. Journal of Object Tecnology **2** (2003) 99–115

Qualitative Modelling of Genetic Networks: From Logical Regulatory Graphs to Standard Petri Nets

Claudine Chaouiya[1], Elisabeth Remy[2], Paul Ruet[2], and Denis Thieffry[1]

[1] Laboratoire de Génétique et Physiologie du Développement
IBDM, CNRS – INSERM – Université de la Méditerranée
Campus de Luminy, 13288 Marseille Cedex 9, France
{chaouiya,thieffry}@ibdm.univ-mrs.fr

[2] Institut de Mathématiques de Luminy
CNRS – Université de la Méditerranée
Campus de Luminy, 13288 Marseille Cedex 9, France
{remy,ruet}@iml.univ-mrs.fr

Abstract. In this paper, a systematic rewriting of logical genetic regulatory graphs in terms of standard Petri net models is proposed. We show that, in the Boolean case, the combination of the logical approach with the standard Petri net framework enables the analysis of isolated regulatory circuits, confirming their most fundamental dynamical properties. Furthermore, two more realistic applications are also presented, the first dealing with the control of the early cell cycles in the developing fly, the second dealing with flower morphogenesis.
The combination of logical and Petri net formalisms open new prospects for the delineation of specific relationships between the feedback structure and the dynamical properties of complex regulatory systems. Moreover, this approach should ease the definition of integrated models of networks encompassing various kinds of interactions: genetic or metabolic regulations, signal transduction cascades...

Keywords: regulatory graphs, gene regulation, discrete dynamics, qualitative analysis.

1 Introduction

Regulatory networks are found at the core of all biological functions, from biochemical pathways, to gene regulation, and cell communication processes. Their complexity often defies the intuition of the biologist and calls for the development of proper mathematical methods to model their structure and simulate their dynamical behaviour (for a recent review, see [3]). A large variety of formal approaches has already been applied to biological regulatory networks, from ordinary or partial differential systems, to sets of stochastic equations. However, the lack of precise, quantitative information about the shape of regulatory functions or about the values of involved parameters pleads for the development of qualitative approaches.

J. Cortadella and W. Reisig (Eds.): ICATPN 2004, LNCS 3099, pp. 137–156, 2004.
© Springer-Verlag Berlin Heidelberg 2004

One qualitative approach consists in modelling regulatory networks in terms of logical equations (using either Boolean or multi-level discretisation) [4, 15]. The development of logical models for various biological networks has already led to interesting insight in network structures (in particular, regulatory feedback circuits) and the corresponding dynamical properties [16]. Relying on the **generalised logical approach** of R. Thomas, we have recently developed a software tool, GIN-sim, which enables the biologist to specify a regulatory model and check the qualitative temporal evolution of the system for given initial states [1]. However, as the number of qualitative (*i.e.* logical or discrete) states grows exponentially with the number of elements involved in the regulatory network, there is a pressing need for proper analytical approaches to cope with the rapid delineation of larger regulatory networks. In this respect, in a recent study, we have derived a series of analytical results concerning properties of the dynamics in the case of isolated regulatory circuits with arbitrary numbers of elements [13].

The Petri net (PN) formalism offers another, complementary framework to deal with the analysis of the dynamical properties of large systems, either from a qualitative or a quantitative point of view. Indeed, Petri nets have already been applied to various types of biological networks. In particular, metabolic networks, which are endowed with conservation laws, can be relatively easily represented into the PN framework [12, 6, 7]. Several applications of PN to genetic regulatory networks can also be found in the literature [5, 9]. However, these applications largely rely on simulations and consequently provide limited insights in the general properties of biological regulatory networks. Moreover, previous PN models of genetic networks largely rely on sophisticated (application-driven) representations rather than on the definition of a systematic method to represent genetic regulatory networks into standard PN.

At this stage, it should be interesting to articulate the logical generalised approach with the PN formalism, in order to combine the delineation of the dynamical roles of specific feedback structures, with the algebraic tools underlying the PN framework for the study of fundamental dynamical properties. In this paper, we propose a rigourous and systematic rewriting of logical regulatory models into specific PN, focusing on the Boolean case.

In the following section, the notion of logical regulatory graph is introduced. Next, we define the corresponding Petri net and prove some basic general properties. The case of isolated regulatory circuits is then analysed in more details, in order to illustrate the fact that their PN counterparts also enable the delineation of the fundamental properties of such circuits. Afterwards, referring to published logical models, we derive the Petri nets corresponding to two biological regulatory networks, the first involved in the control of the cell cycle during the early stages of *Drosophila melanogaster* development, the second involved in the control of flower morphogenesis in *Arabidopsis thaliana*. Finally, conclusions and prospects are proposed.

2 Logical Regulatory Graphs

In this section, we briefly describe regulatory graphs in the **Boolean** case, *i.e.* when the expression state of each gene takes its value in $\{0, 1\}$ (0 when the level of the regulatory product is negligible, 1 when the regulatory product is present at a "sufficient" level). For more details, see [1] where the formalism is described in the general multi-valued case.

A **regulatory graph** is a labelled directed graph which represents interactions between genes. Each interaction is oriented, and involves two genes: the source and the target. An interaction is said to be **operating** whenever the level of expression (or activity) of the source is sufficient (*i.e.* at level 1). An interaction is called an **activation** when its effect on the targeted gene tends to be positive, *i.e* to an increase of the level of expression of the target. It is called a **repression** (or inhibition) when its effect on the targeted gene tends to be negative *i.e* to a decrease of the level of expression. Note, however, that effective activatory or inhibitory effects generally depend on the presence or the absence of cofactors. Indeed, one gene can be the target of several interactions. For each gene g_j, we define the subset $\mathcal{I}(j)$, called *input of* g_j and containing the source genes of all incoming interactions on g_j. Note that as we consider here the Boolean case, there are no multi-arcs between genes and therefore the interactions are fully defined by their sources and targets. When the expression levels of the genes are given, we know which interactions are operating, and we represent their global effects through **logical parameters** defined as follows.

For each gene g_j, the application K_j, called **logical function**, associates a **parameter** $K_j(X)$ to each subset X of $\mathcal{I}(j)$. The value of this parameter defines the level to which g_j tends when X is the set of operating incoming interactions. As we consider here the Boolean case, these functions take their values in $\{0, 1\}$. Thus, for each gene, the corresponding logical function allows the qualitative specification of the effects of any combination of incoming interactions.

Consequently, we can describe a regulatory graph by three components:

- a set of nodes $\mathcal{G} = \{g_1, \ldots, g_n\}$,
- a set of arcs defined by the sets $\mathcal{I}(j), j = 1, \ldots, n$,
- a set of parameters $\mathcal{K} = \{K_j(X), j = 1, \ldots, n, X \subseteq \mathcal{I}(j)\}$.

In order to represent the (discrete) dynamics of the system, we define a second type of graphs, called **dynamical graphs**, where vertices represent states of the system (*i.e.* n-tuples giving the expression levels of the n genes), and edges represent transitions between states. In most applications, dynamical graphs are generated either on the basis of a fully synchronous assumption [19] or on the basis of a fully asynchronous approach [16].

3 Regulatory Petri Nets

Consider a Boolean regulatory graph $\mathcal{R} = (\mathcal{G}, \mathcal{I}, \mathcal{K})$. In this section, we define the Petri net corresponding to $\mathcal{R}$, *i.e.* whose dynamics simulates the dynamical behaviour of the genetic regulatory network.

3.1 Preliminary Rewriting

The most natural way is to define a finite capacity Petri net, with inhibitor arcs (Figure 1):

– to each gene corresponds a place g_j, $j = 1, \ldots, n$,
– to each parameter $K_j(X)$, $j \in \{1, \ldots, n\}, X \subseteq \mathcal{I}(j)$ corresponds a transition $t_{g_j,X}$.

Transition $t_{g_j,X}$ is enabled as soon as all the places of the set X are marked AND all the places of the complementary set of X in $\mathcal{I}(j)$ (denoted $\mathcal{I}(j) \setminus X$) are empty. So the places of $\mathcal{I}(j)$ are the *input* places of the transition $t_{g_j,X}$, with the places of X connected to $t_{g_j,X}$ by standard arcs, and the places of $(\mathcal{I}(j) \setminus X)$ connected to $t_{g_j,X}$ by inhibitor arcs.

Place g_j is an output (*resp.* input) of $t_{g_j,X}$ if $K_j(X) = 1$ (*resp.* if $K_j(X) = 0$). We now need to ensure two constraints:

1. there is no "consumption" of the tokens in the places of X; this constraint is satisfied by adding *read arcs* for these places (self-loops); indeed, the present regulatory products of the input genes activate the transcription of the regulatees, but are not consumed;
2. the number of tokens in each place should be limited to 1 (Boolean case); this constraint can be satisfied by adding a capacity restriction for each place.

Figure 1 illustrates the PN representation of activation *versus* inhibition.

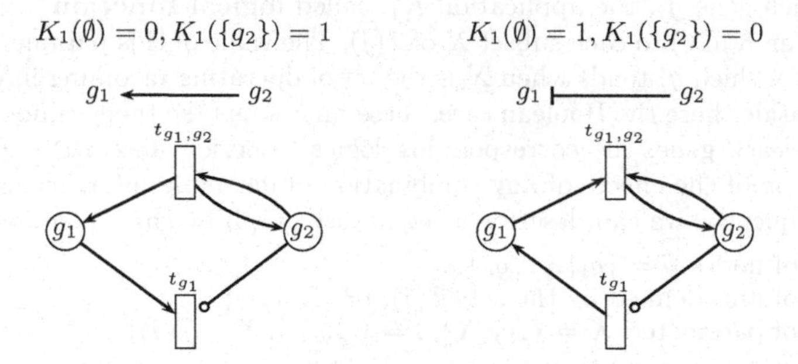

Fig. 1. Petri nets corresponding to the activation of g_1 by g_2 (left) and to the inhibition of g_1 by g_2 (right).

Remark 1. Note that the case of a self-regulation is slightly different. Indeed, if g_j is a self-regulator, there is one transition $t_{g_j,X}$ for each parameter $K_j(X)$, $j \in \{1, \ldots, n\}, X \subseteq \mathcal{I}(j)$ verifying: $K_j(X) = 0$ if $g_j \in X$ and $K_j(X) = 1$ if $g_j \notin X$ ($g_j \notin X$ means that g_j is absent, and $K_j(X) = 0$ does not lead to any change on g_j, and $g_j \in X$ means that g_j is present, and $K_j(X) = 1$ does not lead to any change on g_j). Consequently,

– when $g_j \in X$ ($K_j(X) = 0$), g_j is only an input of transition $t_{g_j,X}$.
– when $g_j \notin X$ ($K_j(X) = 1$), g_j is an input connected by an inhibitor arc, and a "standard" output of transition $t_{g_j,X}$.

Property 1. This Petri net is equivalent to the Boolean regulatory graph $\mathcal{R}$, *i.e.* $\mathcal{R}$ can be reconstructed from it.

3.2 Boolean Regulatory Petri Nets

Let us now consider the net obtained by a complementary-place transformation (cf [11], p. 543): a complementary place $\overline{g_i}$ is defined for each place g_i, such that the sum of tokens in places g_i and $\overline{g_i}$ equals 1 for each $i = 1, \ldots, n$ (Figure 2). The set of places P contains now $2\,n$ elements: $P = \mathcal{G} \cup \overline{\mathcal{G}}$, with $\overline{\mathcal{G}} = \{\overline{g_1}, \ldots, \overline{g_n}\}$. With this transformation, the capacity restriction is automatically satisfied, and inhibitory arcs are not further needed.

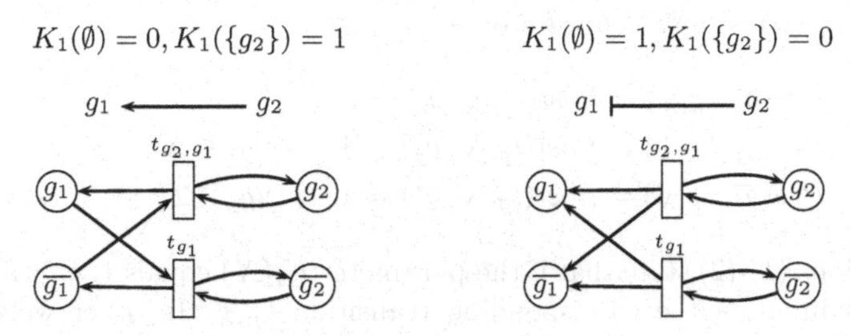

Fig. 2. Regulatory Petri net corresponding to the activation of g_1 by g_2 (left) and to the inhibition of g_1 by g_2 (right), respectively.

Definition 1. *Given a Boolean regulatory graph,* $\mathcal{R} = (\mathcal{G}, \mathcal{I}, \mathcal{K})$*, the associated* ***Boolean regulatory Petri net (BRPN)*** $\mathbf{N}(\mathcal{R}) = (P, T, Pre, Post)$ *is defined as follows:*

- $P = \mathcal{G} \cup \overline{\mathcal{G}} = \{g_1, \overline{g_1}, \ldots, g_n, \overline{g_n}\}$ *is the set of places,*
- $T = \{t_{g_i,X},\ i = 1, \ldots n,\ X \subseteq \mathcal{I}(i)\}$ *is the set of transitions,*
- $Pre : P \times T \to \{0,1\}$ *is the mapping defining arcs between places and transitions (*Pre-conditions*),*
- $Post : T \times P \to \{0,1\}$ *is the mapping defining arcs between transitions and places (*Post-conditions*).*

The functions Pre and Post are defined as follows:

1. *Case* $g_i \notin \mathcal{I}(i)$ *(*g_i *is not a self-regulator). For a given transition* $t_{g_i,X}$*, only the following terms have to be defined (all the other terms being equal to zero):*

$$Pre(g_i, t_{g_i,X}) = Post(t_{g_i,X}, \overline{g_i}) = 1 - K_i(X)\,, \tag{1}$$
$$Pre(\overline{g_i}, t_{g_i,X}) = Post(t_{g_i,X}, g_i) = K_i(X)\,, \tag{2}$$
$$Pre(g_j, t_{g_i,X}) = Post(t_{g_i,X}, g_j) = 1 \qquad \forall g_j \in X\,, \tag{3}$$
$$Pre(\overline{g_j}, t_{g_i,X}) = Post(t_{g_i,X}, \overline{g_j}) = 1 \qquad \forall g_j \in \mathcal{I}(i) \setminus X\,. \tag{4}$$

2. *Case $g_i \in \mathcal{I}(i)$ (g_i is a self-regulator). Consider a given transition $t_{g_i,X}$.*
   - *if $g_i \in X$, the only case to be considered is $K_i(X) = 0$ (cf. Remark 1). Therefore, the only terms to be defined are:*

$$Pre(g_i, t_{g_i,X}) = Post(t_{g_i,X}, \overline{g_i}) = 1\,, \tag{5}$$
$$Pre(g_j, t_{g_i,X}) = Post(t_{g_i,X}, g_j) = 1 \qquad \forall g_j \in X\,, g_j \neq g_i\,, \tag{6}$$
$$Pre(\overline{g_j}, t_{g_i,X}) = Post(t_{g_i,X}, \overline{g_j}) = 1 \qquad \forall g_j \in \mathcal{I}(i) \setminus X\,. \tag{7}$$

   - *if $g_i \notin X$, the only case to be considered is $K_i(X) = 1$. Therefore, the only terms to be defined are:*

$$Pre(\overline{g_i}, t_{g_i,X}) = Post(t_{g_i,X}, g_i) = 1\,, \tag{8}$$
$$Pre(g_j, t_{g_i,X}) = Post(t_{g_i,X}, g_j) = 1 \qquad \forall g_j \in X\,, \tag{9}$$
$$Pre(\overline{g_j}, t_{g_i,X}) = Post(t_{g_i,X}, \overline{g_j}) = 1 \qquad \forall g_j \in \mathcal{I}(i) \setminus X\,, g_j \neq g_i\,. \tag{10}$$

Equations (1)-(2) state that if the parameter $K_i(X)$ equals 1, g_i is an output and $\overline{g_i}$ an input of the corresponding transition $t_{g_i,X}$. In other words, there can be an increase of the level of the product of g_i if it is not already present. Symmetrically, if $K_i(X) = 0$, then $\overline{g_i}$ is an output and g_i is an input of the corresponding transition $t_{g_i,X}$.

In the case of a self-regulation, two situations are considered, both leading to a change of the value of g_i. Equation (5) states that if $g_i \in X$ (*i.e.* the product of g_i is present), then g_i is an input and $\overline{g_i}$ an output of the corresponding transition $t_{g_i,X}$. Conversely, equation (8) states that if $g_i \notin X$ (*i.e.* the product of g_i is absent) then $\overline{g_i}$ is an input and g_i an output of the corresponding transition $t_{g_i,X}$.

Equations (3)-(4), (6)-(7) and (9)-(10) state that the regulatory products contributing to the combination of interactions involved in $K_i(X)$ (*i.e.* which are in X) are the inputs of the corresponding transitions and are not consumed by these transitions.

Remark 2. Consider a BRPN $\mathbf{N}(\mathcal{R})$. Definition 1 gives a unique rewriting $\mathbf{N}(\mathcal{R})$. But if we consider two complementary places, p and $\overline{p}$, we have no way to determine which one corresponds to the presence of the regulatory product. Consequently, Property 1 does not hold anymore.

We shall use the following notation:

- for $i = 1, \ldots, n$: $\widehat{K}_i(X) = 2\,K_i(X) - 1$;
- d_i denotes the number of parameters for g_i. If $g_i \notin \mathcal{I}(i)$, then $d_i = 2^{\#\mathcal{I}(i)}$. Otherwise, if g_i is auto-regulated, then d_i is smaller and depends on the combined effect of incoming interactions together with the self-regulation (cf. Case 2 in Definition 1);
- $X_i^1, \ldots, X_i^{d_i}$ denote the subsets of $\mathcal{I}(i)$ which characterise the parameters for g_i. (The order between these subsets is a meaningless convention.)

Property 2. The **incidence matrix** $C = Post^T - Pre$ is a $2n \times (\sum_{i=1 \ldots n} d_i)$ matrix. Its components take their values in $\{-1, 0, 1\}$, and C has the following structure:

$$C = \begin{pmatrix} \boxed{C_1} & 0 & \cdots & 0 \\ 0 & \boxed{C_2} & \cdots & 0 \\ 0 & 0 & \ddots & 0 \\ 0 & 0 & \cdots & \boxed{C_n} \end{pmatrix}, \quad \text{where} \quad C_i = \begin{pmatrix} \widehat{K}_{i,X_i^1} & \cdots & \widehat{K}_{i,X_i^{d_i}} \\ -\widehat{K}_{i,X_i^1} & \cdots & -\widehat{K}_{i,X_i^{d_i}} \end{pmatrix}.$$

It is important to point out that the incidence matrix of a **BRPN** does not correspond to the structure of the underlying graph, because of the *read arcs.*

Definition 2. *Given a regulatory Petri net $(P, T, Pre, Post)$, a* ***valid marking*** *$M : P \rightarrow \{0, 1\}$ corresponds to a state of the Boolean regulatory graph $(\mathcal{G}, \mathcal{K})$ and verifies:*

$$\forall g_i \in \mathcal{G}, M(g_i) = 1 - M(\overline{g_i}). \tag{11}$$

The following property is straightforward and ensures the conservation of the validity of the markings as the system evolves.

Property 3. Given a regulatory Petri net $(P, T, Pre, Post)$ with a valid initial marking, any reachable marking is still valid and therefore the BRPN is *1-safe* (*i.e.* the marking of any place is at most 1).

Property 4 shows the equivalence between the asynchronous dynamical graph associated with the Boolean regulatory graph and the reachability graph of the corresponding BRPN. A state in the dynamical graph is described by a valid marking, *i.e.* the current values associated to each element (1 if the regulatory product g_i is present at this state, 0 otherwise). Note that we only consider the fully asynchronous assumption, as it is generally the case in the Petri net framework, but we could also consider the fully synchronous assumption, modifying the firing rule (in this case, all enabled transitions should fire simultaneously).

Property 4. There exists a transition between two states S_1 and S_2 in the asynchronous dynamical graph related to a Boolean regulatory graph $\mathcal{R} = (\mathcal{G}, \mathcal{I}, \mathcal{K})$, iff there exists an enabled transition t in the associated BRPN such that M_1 verifies $M_1[t\rangle M_2$ (t is enabled by M_1 and its firing leads to the marking M_2) with, for all $i = 1, \ldots n$,

$$\begin{aligned} M_1(g_i) &= S_1(i) \quad M_1(\overline{g_i}) = 1 - S_1(i)\,, \\ M_2(g_i) &= S_2(i) \quad M_2(\overline{g_i}) = 1 - S_2(i)\,. \end{aligned}$$

4 Regulatory Circuits

As mentioned in the introduction, in the context of the logical approach, several authors have emphasised the crucial dynamical roles of regulatory circuits. More specifically, **positive** circuits (*i.e.* involving an even number of inhibitions) have been associated to multi-stationary behaviour, whereas **negative** circuits (involving an odd number of inhibitions) can generate sustained periodic behaviour [16]. A circuit is said functional when it generates the corresponding dynamical property. In this section, we derive a general PN formulation for these two classes of regulatory circuits and check their dynamical properties.

In the case of isolated regulatory circuits, each gene g_i is the target of a unique interaction exerted by g_{i-1}, and is the source of a unique interaction towards g_{i+1} (here and in the sequel, indices are considered *modulo n*, *i.e.* $i + n = i$). Note that sets $\mathcal{I}$ are singleton sets, uniquely defined and therefore they will be omitted.

In this section, to simplify the notations, we will denote

$$\begin{array}{llll} K_i(\{g_{i-1}\}) & \text{by } K_i(g_{i-1}), & K_i(\emptyset) & \text{by } K_i, \\ t_{g_i,\{g_{i-1}\}} & \text{by } t_{i,i-1}, & t_{g_i,\emptyset} & \text{by } t_i. \end{array}$$

Let $\mathcal{C} = (\mathcal{G}, \mathcal{K})$ be a regulatory circuit, with $\mathcal{G} = \{g_1, \ldots, g_n\}$ and $\mathcal{K} = \{K_i, K_i(g_{i-1})\}_{i=1,\ldots,n}$ ($\mathcal{I}(i) = \{g_{i-1}\}$; cf. [13] for a more detailed description). In this simpler context, the notion of activation *versus* inhibition (cf. Section 2) can be simply expressed depending on the values of the parameters. Consider the interaction from g_i to g_{i+1},

- if $K_{i+1} = 0$ and $K_{i+1}(g_i) = 1$, *i.e.* the presence of the regulatory product of g_i increases the expression level of g_{i+1}, then we say that this interaction is an activation; it is labelled with the sign $\varepsilon_i = +1$,
- if $K_{i+1} = 1$ and $K_{i+1}(g_i) = 0$, *i.e.* the presence of the regulatory product of g_i decreases the expression level of g_{i+1}, then we say that this interaction is an inhibition; it is labelled with the sign $\varepsilon_i = -1$.

The structure of the regulatory Petri net $\mathbf{N}(\mathcal{C})$ corresponding to a Boolean circuit is described by the following property (see also Figure 3):

Property 5. Consider a Boolean regulatory circuit $\mathcal{C} = (\mathcal{G}, \mathcal{K})$. The corresponding regulatory Petri net $\mathbf{N}(\mathcal{C})$ is given by the following components:

- $P = \mathcal{G} \cup \overline{\mathcal{G}} = \{g_1, \overline{g_1}, \ldots, g_n, \overline{g_n}\}$ is the set of places, with cardinal $2n$.
- $T = \{t_i, t_{i,i-1}, i = 1, \ldots n\}$ is the set of transitions, with cardinal $2n$.
- $Pre : P \times T \to \{0, 1\}$ is a $2n \times 2n$ matrix. The only terms which may be non-zero are:

$$\begin{array}{llll} Pre(g_i, t_i) & = 1 - K_i & Pre(\overline{g_i}, t_i) & = K_i \\ Pre(g_i, t_{i,i-1}) & = 1 - K_i(g_{i-1}) & Pre(\overline{g_i}, t_{i,i-1}) & = K_i(g_{i-1}) \\ Pre(g_{i-1}, t_{i,i-1}) & = 1 & Pre(\overline{g_{i-1}}, t_i) & = 1. \end{array}$$

- $Post : T \times P \to \{0, 1\}$ is a $2n \times 2n$ matrix. The only terms which may be non-zero are:

$$\begin{array}{llll} Post(t_i, \overline{g_i}) & = 1 - K_i & Post(t_i, g_i) & = K_i \\ Post(t_{i,i-1}, \overline{g_i}) & = 1 - K_i(g_{i-1}) & Post(t_{i,i-1}, g_i) & = K_i(g_{i-1}) \\ Post(t_{i,i-1}, g_{i-1}) & = 1 & Post(t_i, \overline{g_{i-1}}) & = 1. \end{array}$$

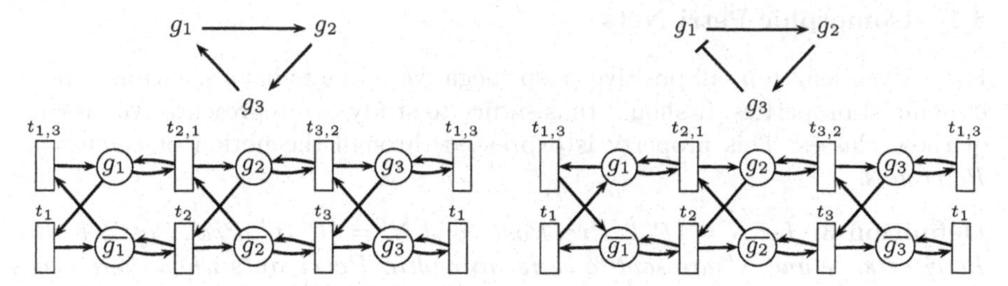

Fig. 3. Three-element regulatory circuits and the corresponding Petri nets. Left: positive circuit. Right: negative circuit. Note that transitions between g_1 and g_3 are repeated, illustrating the general structure of the net.

We use the following notation already introduced in 3.2, for $i = 1, \ldots, n$:

$$\widehat{K}_i = 2\,K_i - 1$$
$$\widehat{K}_{i,i-1} = 2\,K_i(g_{i-1}) - 1.$$

Using the Property 5, we can further characterize the regulatory Petri nets corresponding to Boolean circuits as follows:

Property 6. Let $\mathbf{N}(\mathcal{C}) = (\mathcal{G} \cup \overline{\mathcal{G}}, T, Pre, Post)$ be the regulatory Petri net corresponding to the Boolean regulatory circuit $\mathcal{C}$. The incidence matrix C has a block-diagonal $2n \times 2n$ form:

$$C = \begin{pmatrix} \boxed{C_1} & 0 & \cdots & 0 \\ 0 & \boxed{C_2} & \cdots & 0 \\ 0 & 0 & \ddots & 0 \\ 0 & 0 & \cdots & \boxed{C_n} \end{pmatrix},$$

where

$$C_i = \begin{pmatrix} -1 & 1 \\ 1 & -1 \end{pmatrix} \quad \text{if } g_{i-1} \text{ is an activator for } g_i \text{, and}$$

$$C_i = \begin{pmatrix} 1 & -1 \\ -1 & 1 \end{pmatrix} \quad \text{if } g_{i-1} \text{ is an inhibitor for } g_i.$$

Proof. It follows from Property 2 that the incidence matrix C is block-diagonal in the case of a circuit. The blocks are:

$$C_i = \begin{pmatrix} \widehat{K}_i & \widehat{K}_{i,i-1} \\ -\widehat{K}_i & -\widehat{K}_{i,i-1} \end{pmatrix};$$

using Property 5, we further have $K_i = 0$, $K_i(g_{i-1}) = 1$ if g_{i-1} activates g_i, and $K_i = 1$, $K_i(g_{i-1}) = 0$ if g_{i-1} inhibits g_i. ■

4.1 Isomorphic Petri Nets

For a given length n, all positive (resp. negative) circuits have the same type of dynamical properties. It should thus suffice to study one representative for each of these classes. This property is expressed through the notion of *isomorphic Petri nets.*

Definition 3. *Let $N = (P, T, Pre, Post)$ and $N' = (P', T', Pre', Post')$ be two Petri nets. N and N' are said to be* **isomorphic Petri nets** *when there exists a graph isomorphism $\phi : (P \cup T, Pre \cup Post) \to (P' \cup T', Pre' \cup Post')$ such that $\phi(p) \in P'$ for each $p \in P$ and $\phi(t) \in T'$ for each $t \in T$ (note that adjacency is preserved).*

For $n \in \mathbb{N}, n \geqslant 1$, let $\mathcal{C}_n^+$ be the regulatory circuit $g_1 \stackrel{+1}{\to} g_2 \stackrel{+1}{\to} \cdots \stackrel{+1}{\to} g_n \stackrel{+1}{\to} g_1$ with n genes and all interactions positive and let $\mathcal{C}_n^-$ be the regulatory circuit $g_1 \stackrel{-1}{\to} g_2 \stackrel{+1}{\to} \cdots \stackrel{+1}{\to} g_n \stackrel{+1}{\to} g_1$ with n genes and all interactions positive except the first one.

Property 7. Let $\mathcal{R}$ be a regulatory circuit $g_1 \stackrel{\varepsilon_1}{\to} g_2 \stackrel{\varepsilon_2}{\to} \cdots \stackrel{\varepsilon_{n-1}}{\to} g_n \stackrel{\varepsilon_n}{\to} g_1$, with $\varepsilon_i = +1$ or -1. If $\mathcal{R}$ is positive, then $\mathbf{N}(\mathcal{R})$ is isomorphic to $\mathbf{N}(\mathcal{C}_n^+)$. If $\mathcal{R}$ is negative, then $\mathbf{N}(\mathcal{R})$ is isomorphic to $\mathbf{N}(\mathcal{C}_n^-)$.

Proof. Let us assume $\mathcal{R}$ has at least 2 negative interactions.

Let $g_i \stackrel{-1}{\to} g_{i+1}$ and $g_j \stackrel{-1}{\to} g_{j+1}$ be two negative interactions, $i < j$. By exchanging in $\mathbf{N}(\mathcal{R})$ the transitions t_{k+1} and $t_{k+1,k}$, and the places g_k and $\overline{g_k}$ for each k, $i+1 \leqslant k \leqslant j$, one obtains an isomorphic Petri net N'. Now, N' is clearly the Petri net corresponding to the circuit obtained from $\mathcal{R}$ by replacing the sequence

$$g_i \stackrel{-1}{\to} g_{i+1} \stackrel{\varepsilon_{i+1}}{\longrightarrow} \cdots \stackrel{\varepsilon_{j-1}}{\longrightarrow} g_j \stackrel{-1}{\to} g_{j+1}$$

by

$$g_i \stackrel{+1}{\to} \overline{g_{i+1}} \stackrel{\varepsilon_{i+1}}{\longrightarrow} \cdots \stackrel{\varepsilon_{j-1}}{\longrightarrow} \overline{g_j} \stackrel{+1}{\to} g_{j+1}.$$

Hence, if the indices i and j correspond to two consecutive negative interactions (*i.e.* such that $\varepsilon_k = +1$ for each k, $i < k < j$), by iterating the above process, one obtains either $\mathcal{C}_n^+$ when $\mathcal{R}$ is positive, or a circuit with one negative interaction when $\mathcal{R}$ is negative. In the latter case, if $g_i \stackrel{-1}{\to} g_{i+1}$ is the negative interaction, then the "rotation" $\phi : g_k \mapsto g_{k+i-1}$, $\overline{g_k} \mapsto \overline{g_{k+i-1}}$, $t_{k+1} \mapsto t_{k+i}$, $t_{k+1,k} \mapsto t_{k+i,k+i-1}$ is an isomorphism between N' and $\mathbf{N}(\mathcal{C}_n^-)$. ■

Consequently, in the sequel, we restrict our study to $\mathcal{C}_n^+$ and $\mathcal{C}_n^-$.

4.2 Positive Circuits

In [13], we have proved that an isolated *functional* positive circuit (*i.e.* for proper logical parameter values) generates two stable states, which are *mirroring* each other (a component is "on" in one state iff it is "off" in the other state).

Property 8. Let $\mathbf{N}(\mathcal{R}) = (P, T, Pre, Post)$ be a regulatory Petri net corresponding to an isolated functional positive regulatory circuit, then there are exactly two dead valid markings M_d^1 and M_d^2 which are mirroring each other. Each of these two markings is reachable from any valid marking.

Proof. Using Property 7, we can restrict the proof for $\mathbf{N}(\mathcal{C}_n^+)$. We will first prove that there exists two mirroring dead markings M_d^1 and M_d^2.

For any $i = 1, \dots n$, two transitions have to be considered. Transition t_i (corresponding to parameter $K_i = 0$) has two input places which are g_i and $\overline{g_{i-1}}$. It is not enabled under a marking M satisfying: $M(g_i) = 0$ or $M(\overline{g_{i-1}}) = 0$. And transition $t_{i,i-1}$ (corresponding to parameter $K_i(g_{i-1}) = 1$) has two input places which are $\overline{g_i}$ and g_{i-1}. It is not enabled under a marking M satisfying: $M(\overline{g_i}) = 0$ or $M(g_{i-1}) = 0$. The conjunction of these two constraints leads to $M(g_i) = M(g_{i+1}), \forall i = 1, \dots n$. The only two markings satisfying these constraints are M_d^1 such that $M_d^1(g_i) = 0, M_d^1(\overline{g_i}) = 1$, $\forall i$, and M_d^2 such that $M_d^2(g_i) = 1, M_d^1(\overline{g_i}) = 0$, $\forall i$.

Now, let M be a valid marking different from M_d^1 and M_d^2. Then for M, there exists an index i such that $M(g_i) \neq M(g_{i+1})$. Two cases have to be considered:

1. $M(g_i) = 1$ and $M(g_{i+1}) = 0$: $t_{i+1,i}$ is enabled and its firing preserves all the values of M except for $i+1$ (the token in $\overline{g_{i+1}}$ has been transferred to g_{i+1}). Thus the firing of $t_{i+1,i}$ leads to a marking with a smaller number of 0 values for places g (consequently a smaller number of places $\overline{g}$ marked).
2. $M(g_i) = 0$ and $M(g_{i+1}) = 1$: t_i is enabled and its firing preserves all the values of M except for $i+1$ (there is now no token in g_{i+1}). Thus the firing of t_i leads to a marking with a smaller number of 1 values for places g (consequently a smaller number of places $\overline{g}$ unmarked).

More generally, for M different from M_d^1 and M_d^2, there obviously exists (because it is a circuit) an index i for which $M(g_i) = 1, M(g_{i+1}) = 0$ and an index k for which $M(g_k) = 0, M(g_{k+1}) = 1$. Therefore, using the above argument, there exists an enabled sequence σ^1 such that $M[\sigma^1\rangle M_d^1$ (iteratively decreasing the number of places g marked), and a sequence σ^2 such that $M[\sigma^2\rangle M_d^2$ (decreasing the number of places $\overline{g}$ marked). This proves that M_d^1 and M_d^2 are reachable from any valid marking. ■

4.3 Negative Circuits

As mentioned above, negative regulatory circuits typically generate periodic dynamical behaviour. In [13], we have shown that an isolated functional negative circuit leads to a dynamical graph where all states feed a specific dynamical circuit of length twice the number of elements in the circuit.

Property 9. Let $\mathbf{N}(\mathcal{R}) = (P, T, Pre, Post)$ be a regulatory Petri net corresponding to a negative regulatory circuit and E be the set of all the valid markings which enable exactly one transition.

1. No dead marking is reachable from any initial valid marking.
2. E has $2n$ elements and is organised as a cycle.
3. Each marking in E is reachable from any valid marking.
4. $\mathbf{N}(\mathcal{R})$ is live for any initial valid marking.

Proof. By Property 7, we can restrict the proof to the case of $\mathbf{N}(\mathcal{C}_n^-)$.

1. The argument is similar to that used in the positive case. If M is a dead marking reached from an initial valid marking, then it is valid by Property 3, and for any $i \neq 1$, the tuple $(M(g_i), M(\overline{g_i}), M(g_{i+1}), M(\overline{g_{i+1}}))$ is either $(0,1,0,1)$ or $(1,0,1,0)$, and the tuple $(M(1), M(2), M(3), M(4))$ is either C$(0,1,1,0)$ or $(1,0,0,1)$ because the first interaction is negative. These constraints are clearly incompatible: for instance, if the marking M starts with $(0,1,1,0)$, then M has to be $(0,1,1,0,1,0,\ldots,1,0)$, but then the tuple $(M(g_n), M(\overline{g_n}), M(g_1), M(\overline{g_1}))$ is $(1,0,0,1)$, whereas it should be either $(0,1,0,1)$ or $(1,0,1,0)$.
2. Let M be a valid marking. If $i \neq 1$, the interaction between g_i and g_{i+1} is positive, so
$$\begin{aligned} M[t_{i+1}\rangle &\Leftrightarrow M(\overline{g_i}) = M(g_{i+1}) = 1 && \text{(see Fig. 2)}\\ &\Leftrightarrow (M(g_i), M(\overline{g_i}), M(g_{i+1}), M(\overline{g_{i+1}})) = (0,1,1,0) && (M \text{ valid}). \end{aligned}$$
Similarly, $M[t_{i+1,i}\rangle \Leftrightarrow (M(g_i), M(\overline{g_i}), M(g_{i+1}), M(\overline{g_{i+1}})) = (1,0,0,1)$.
Otherwise, the interaction between g_1 and g_2 is negative, and thus
$$\begin{aligned} M[t_2\rangle &\Leftrightarrow (M(g_1), M(\overline{g_1}), M(g_2), M(\overline{g_2})) = (0,1,0,1)\\ M[t_{2,1}\rangle &\Leftrightarrow (M(g_1), M(\overline{g_1}), M(g_2), M(\overline{g_2})) = (1,0,1,0)\,. \end{aligned}$$
Let us define:
$$\begin{aligned} M_{i+1} &= (1,0,0,1,\ldots,0,1,\underbrace{\overset{g_i}{0},\overset{\overline{g_i}}{1},1,0}_{t_{i+1}},\ldots,1,0) \quad \text{for } i \neq 1,\\ M_{i+1,i} &= (0,1,1,0,\ldots,1,0,\underbrace{\overset{g_i}{1},\overset{\overline{g_i}}{0},0,1}_{t_{i+1,i}},\ldots,0,1) \quad \text{for } i \neq 1,\\ M_2 &= (\underbrace{0,1,0,1}_{t_2},0,1,\ldots,0,1),\\ M_{2,1} &= (\underbrace{1,0,1,0}_{t_{2,1}},1,0,\ldots,1,0). \end{aligned}$$
Then, we have, for all $j = 1,\ldots n$:
$$\begin{aligned} M \in E \text{ and } M[t_j\rangle &\Longleftrightarrow M = M_j\,,\\ M \in E \text{ and } M[t_{j,j-1}\rangle &\Longleftrightarrow M = M_{j,j-1}\,. \end{aligned}$$
Thus, $E = \{M_j, M_{j,j-1},\ j = 1,\ldots n\}$ and has exactly $2\,n$ elements.
Clearly E is organised as a cycle since $M_i[t_i\rangle M_{i+1,i}[t_{i+1,i}\rangle M_{i+1}$ for any i.

3. Given a valid marking M, we denote $\#M$ the number of indices i such that $M(g_i) = 1$. Note that M_2 is the unique marking s.t. $\#M_2 = 0$ and $M_{2,1}$ the only one s.t. $\#M_{2,1} = n$.
   Let $M \notin E$ be a valid marking. Obviously, $0 < \#M < n$. Consider $J(M) = \{j \mid 1 \leqslant j \leqslant n, M(g_j) = 0 \text{ and } M(g_{j+1}) = 1\}$. In particular, an index $j \neq 1$ belongs to $J(M)$ iff t_{j+1} is enabled.
   Follow a marking path from M to E along which $\#M$ decreases: while $J(M) \neq \emptyset$ and $J(M) \neq \{1\}$, for all $j \in J(M)$ and $j \neq 1$, we have $M[t_{j+1}\rangle M'$, with $\#M' = \#M - 1$. Finally,
   - if $J(M) = \emptyset$ then $M = M_2 \in E$,
   - if $J(M) = \{1\}$ then $M = M_{3,2} \in E$.
4. The liveness is a straighforward consequence of 2 and 3. ■

For regulatory Petri nets corresponding to negative circuits, it can also be proved that the number of enabled transitions at each valid marking is odd, and that firing one transition decreases this number either by 0 or 2.

5 Applications

In what follows, we present two biological illustrations of our rewriting of Boolean regulatory networks into Petri nets. The first application consists in a simplified model of the protein network controlling the cell cycle during the early stages of the development of the fly *Drosophila melanogaster*. The second application consists in a genetic regulatory network involved in the control of flower morphogenesis in the plant *Arabidopsis thaliana*. In both cases, the generalised logical approach (though still Boolean) led to interesting insight about the dynamical and biological roles of specific feedback circuits, as well as to the prediction of the behaviour of the system in new situations (mutations, perturbations) [18, 10].

5.1 Drosophila Cell Cycle

Extensive genetic and molecular data has been collected on the various components and individual interactions at the basis of the cell cycle and its properties. To integrate these diverse pieces of data, several authors have been working on the development of dynamical models [17]. Here, for the sake of simplicity, we consider a simple Boolean model involving the minimal number of components susceptible to generate the observed oscillatory behaviour during the early stages of Drosophila embryonic development [18]. At that time, the core of the cell cycle control network involves four main active regulatory compounds: the MPF complex (*i.e.*, the Mitosis Promoting Factor, made of the association of proteins Cyclin B and Cdc2), and the proteins Fizzy, Weel, and String. Cross-interactions between these four compounds can be summarised in the form of the regulatory graph shown in Figure 4. In the context of the logical approach, it proved possible to derive the parameter values enabling the generation of a unique periodic attractor, matching the periodic properties observed in the real

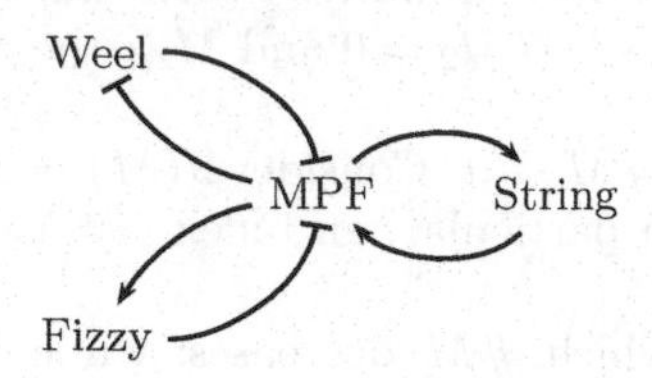

target	parameter
MPF	K_M
	$K_{M,WS}$
	$K_{M,S}$
	$K_{M,W}$
Fizzy	$K_{F,M}$
Weel	K_W
String	$K_{S,M}$

Fig. 4. A simple logical model of the gene network controlling the first cell cycles during Drosophila embryogenesis. Left: Regulatory graph. Right: non-zero parameters for the four core regulators.

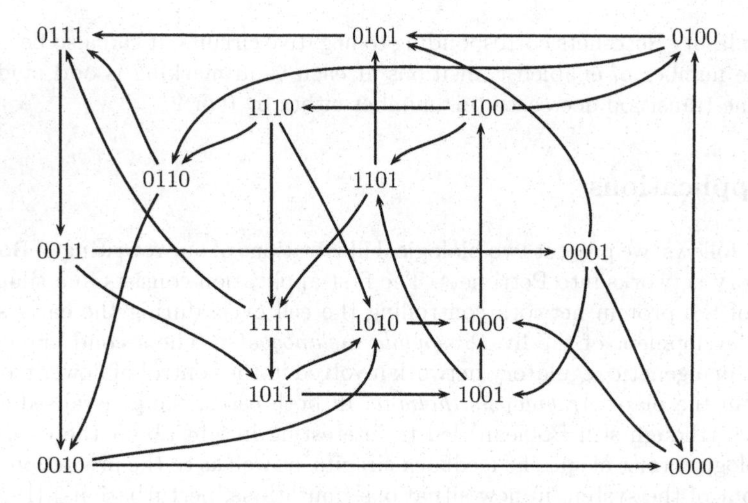

Fig. 5. Asynchronous dynamical graph encompassing all possible state transitions for the model defined in Figure 4.

system (see Figure 4 for the parameter values, and Figure 5 for the corresponding asynchronous dynamical graph). In order to illustrate the translation rules described above, we derive hereafter the BRPN of this parameterised model and check its properties in reference to the Petri net analytical framework.

In the sequel, M will stand for MPF, F for $Fizzy$, W for $Wee1$ and S for $String$. The resulting BRPN can be defined by the following matrices:

$$Pre = \begin{pmatrix} & t_M & t_{M,FWS} & t_{M,F} & t_{M,WS} & t_{M,FW} & t_{M,S} & t_{M,FS} & t_{M,W} & t_F & t_{F,M} & t_W & t_{W,M} & t_S & t_{S,M} \\ M & 0 & 1 & 1 & 0 & 1 & 0 & 0 & 1 & 0 & 1 & 0 & 1 & 0 & 1 \\ \overline{M} & 1 & 0 & 0 & 1 & 0 & 1 & 1 & 0 & 1 & 0 & 1 & 0 & 1 & 0 \\ F & 0 & 1 & 1 & 0 & 1 & 0 & 0 & 1 & 1 & 0 & 0 & 0 & 0 & 0 \\ \overline{F} & 1 & 0 & 0 & 1 & 0 & 1 & 1 & 0 & 0 & 1 & 0 & 0 & 0 & 0 \\ W & 0 & 1 & 0 & 1 & 1 & 0 & 1 & 0 & 0 & 0 & 0 & 1 & 0 & 0 \\ \overline{W} & 1 & 0 & 1 & 0 & 0 & 1 & 0 & 1 & 0 & 0 & 1 & 0 & 0 & 0 \\ S & 0 & 1 & 0 & 1 & 0 & 1 & 0 & 1 & 0 & 0 & 0 & 0 & 1 & 0 \\ \overline{S} & 1 & 0 & 1 & 0 & 1 & 0 & 1 & 0 & 0 & 0 & 0 & 0 & 0 & 1 \end{pmatrix}$$

$$Post = \begin{pmatrix} & M & \overline{M} & F & \overline{F} & W & \overline{W} & S & \overline{S} \\ t_M & 1 & 0 & 0 & 1 & 0 & 1 & 0 & 1 \\ t_{M,FWS} & 0 & 1 & 1 & 0 & 1 & 0 & 1 & 0 \\ t_{M,F} & 0 & 1 & 1 & 0 & 0 & 1 & 0 & 1 \\ t_{M,WS} & 1 & 0 & 0 & 1 & 1 & 0 & 1 & 0 \\ t_{M,FW} & 0 & 1 & 1 & 0 & 1 & 0 & 0 & 1 \\ t_{M,S} & 1 & 0 & 0 & 1 & 0 & 1 & 1 & 0 \\ t_{M,FS} & 0 & 1 & 1 & 0 & 0 & 1 & 1 & 0 \\ t_{M,W} & 1 & 0 & 0 & 1 & 1 & 0 & 0 & 1 \\ t_F & 0 & 1 & 0 & 1 & 0 & 0 & 0 & 0 \\ t_{F,M} & 1 & 0 & 1 & 0 & 0 & 0 & 0 & 0 \\ t_W & 0 & 1 & 0 & 0 & 0 & 1 & 0 & 0 \\ t_{W,M} & 1 & 0 & 0 & 0 & 1 & 0 & 0 & 0 \\ t_S & 0 & 1 & 0 & 0 & 0 & 0 & 0 & 1 \\ t_{S,M} & 1 & 0 & 0 & 0 & 0 & 0 & 1 & 0 \end{pmatrix}$$

$$Post^T - Pre = \begin{pmatrix} & t_M & t_{M,FWS} & t_{M,F} & t_{M,WS} & t_{M,FW} & t_{M,S} & t_{M,FS} & t_{M,W} & t_F & t_{F,M} & t_W & t_{W,M} & t_S & t_{S,M} \\ M & 1 & -1 & -1 & 1 & -1 & 1 & -1 & 1 & 0 & 0 & 0 & 0 & 0 & 0 \\ \overline{M} & -1 & 1 & 1 & -1 & 1 & -1 & 1 & -1 & 0 & 0 & 0 & 0 & 0 & 0 \\ F & 0 & 0 & 0 & 0 & 0 & 0 & 0 & 0 & -1 & 1 & 0 & 0 & 0 & 0 \\ \overline{F} & 0 & 0 & 0 & 0 & 0 & 0 & 0 & 0 & 1 & -1 & 0 & 0 & 0 & 0 \\ W & 0 & 0 & 0 & 0 & 0 & 0 & 0 & 0 & 0 & 0 & 1 & -1 & 0 & 0 \\ \overline{W} & 0 & 0 & 0 & 0 & 0 & 0 & 0 & 0 & 0 & 0 & -1 & 1 & 0 & 0 \\ S & 0 & 0 & 0 & 0 & 0 & 0 & 0 & 0 & 0 & 0 & 0 & 0 & -1 & 1 \\ \overline{S} & 0 & 0 & 0 & 0 & 0 & 0 & 0 & 0 & 0 & 0 & 0 & 0 & 1 & -1 \end{pmatrix}$$

In order to demonstrate that this system has a cyclical attractor, we can refer to the absence of *deadlock* property.

Property 10. The system is deadlock-free.

Proof. We will prove that there is no marking such that no transition is enabled in the BRPN $\mathcal{C}$. In fact, such a marking should satisfy the following set of constraints; each constraint defines the condition under which the corresponding transition is not enabled, and is given by the relevant column of the *Pre* matrix:

$$\begin{aligned} t_M &: M_d(M) = 1 \lor M_d(F) = 1 \lor M_d(W) = 1 \lor M_d(S) = 1 \\ t_{M,FWS} &: M_d(M) = 0 \lor M_d(F) = 0 \lor M_d(W) = 0 \lor M_d(S) = 0 \\ t_{M,F} &: M_d(M) = 0 \lor M_d(F) = 0 \lor M_d(W) = 1 \lor M_d(S) = 1 \\ t_{M,WS} &: M_d(M) = 1 \lor M_d(F) = 1 \lor M_d(W) = 0 \lor M_d(S) = 0 \\ t_{M,FW} &: M_d(M) = 0 \lor M_d(F) = 0 \lor M_d(W) = 0 \lor M_d(S) = 1 \\ t_{M,S} &: M_d(M) = 1 \lor M_d(F) = 1 \lor M_d(W) = 1 \lor M_d(S) = 0 \\ t_{M,FS} &: M_d(M) = 0 \lor M_d(F) = 0 \lor M_d(W) = 1 \lor M_d(S) = 0 \\ t_{M,W} &: M_d(M) = 1 \lor M_d(F) = 1 \lor M_d(W) = 0 \lor M_d(S) = 1 \\ t_F &: M_d(F) = 0 \lor M_d(M) = 1 \\ t_{F,M} &: M_d(F) = 1 \lor M_d(M) = 0 \\ t_W &: M_d(W) = 1 \lor M_d(M) = 1 \\ t_{W,M} &: M_d(W) = 0 \lor M_d(M) = 0 \\ t_S &: M_d(S) = 0 \lor M_d(M) = 1 \\ t_{S,M} &: M_d(S) = 1 \lor M_d(M) = 0 \end{aligned}$$

Using the constraints on t_F and $t_{F,M}$, we obtain $M_d(F) = M_d(M)$. Using the constraints on t_W and $t_{W,M}$, we obtain $M_d(W) \neq M_d(M)$. Using the constraints on t_S and $t_{S,M}$, we obtain $M_d(S) = M_d(M)$. The only two markings that verify these constraints are (omitting the values for the complementary places):

$$\begin{bmatrix} M_d^1(M) = 0 \\ M_d^1(F) = 0 \\ M_d^1(W) = 1 \\ M_d^1(S) = 0 \end{bmatrix} \quad \text{and} \quad \begin{bmatrix} M_d^2(M) = 1 \\ M_d^2(F) = 1 \\ M_d^2(W) = 0 \\ M_d^2(S) = 1 \end{bmatrix}$$

The marking M_d^1 enables $t_{M,W}$ and M_d^2 enables $t_{M,FS}$. Therefore, there is no marking such that no transition is enabled. ■

We can further generate the complete reachability graph (for all the valid initial markings), which exactly matches the graph of Figure 5 (taking into account that we omit the marking of the complementary places).

Note that if the values of parameters K_F and $K_{F,M}$ are exchanged (changing the effect of gene *MPF* upon *Fizzy*, *i.e.* specifying an inhibition rather than an activation), then Property 10 does not hold anymore. One can easily check that markings $[1, 0, 0, 1]$ and $[0, 1, 1, 0]$ are indeed dead. This exchange of values corresponds to the transformation of the unique negative circuit into a third positive circuit in the regulatory graph (see Figure 4), consequently leading to the loss of the oscillatory behaviour associated to the negative circuit (cf Property 9).

5.2 Flowering in Arabidopsis

Let us now turn to a larger network involved in the control of flower morphogenesis in the model plant *Arabidopsis thaliana*. On the basis of a mutant analysis, plant geneticists have proposed an abstract combinatory model ('ABC model') to account for the differentiation of the flower meristem into the four typical flower organs: carpels, stamens, sepals and petals [2]. More recently, on the basis of new molecular genetic data, Mendoza *et al.* have proposed a Boolean regulatory model involving 10 genes cross-regulating each other [10]. The corresponding regulatory graph is shown in Figure 6. For proper parameter value sets, this model encompasses 6 stable states, four of them matching the qualitative gene expression patterns observed in the different flower organs, while the two last stable states correspond to non-flowering situations.

For sake of simplicity, we focus here on a subset of six genes which play a crucial role in the selection of specific flowering differentiative pathways, leaving aside the genes which can be treated as simple inputs (EMF1, UFO, LUG and SUP). A selection of parameter values is presented in Table 1 (Right). For this parameter set, the system has four stable states, each corresponding to a gene expression pattern associated with a specific flower organ.

Contrasting with the model of Drosophila cell cycle, we are now facing a network, whose biological role consists essentially in selecting specific differentiation pathways. Consequently, we do not expect to observe oscillatory behaviour here, but rather a specific set of dead markings, each representing a logical stable state, *i.e.* a specific gene expression state.

In the associated BRPN, we have 12 places (corresponding to the 6 genes and the complementary places) and 22 transitions. Note that in the case of P and I, which are auto-regulated, we consider only two parameters for each, as

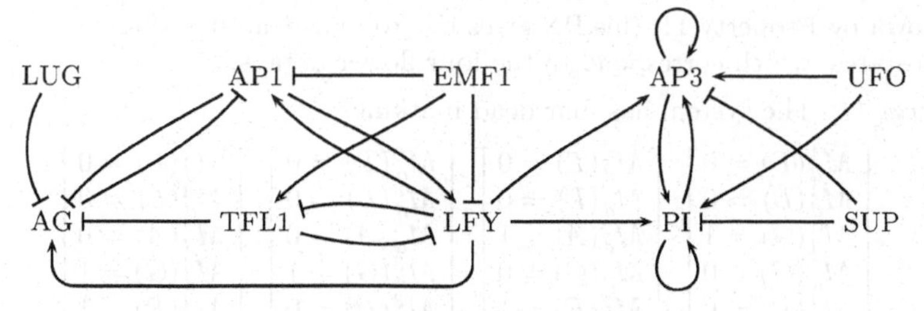

Fig. 6. Regulatory graph for the gene network controlling early flower morphogenesis in the plant *Arabidopsis thaliana*.

Table 1. Left: Names, symbols and variables for the six core genes selected in Figure 6. Right: The transitions and associated parameter values for each of these six genes.

Gene name	Symbol	Variable
Terminal Flower 1	TFL1	T
LEAFY	LFY	L
APETALA1	AP1	A
AGAMOUS	AG	G
APETALA3	AP3	P
PISTILLATA	PI	I

T		L		A		G		P		I	
t_T	0	t_L	0	t_A	1	t_G	1	$t_{P,P}$	0	$t_{I,I}$	0
$t_{T,L}$	0	$t_{L,A}$	0	$t_{A,L}$	1	$t_{G,L}$	1	$t_{P,LP}$	0	$t_{I,LI}$	0
		$t_{L,T}$	0	$t_{A,G}$	0	$t_{G,A}$	0				
		$t_{L,AT}$	0	$t_{A,LG}$	0	$t_{G,T}$	0				
						$t_{G,LA}$	0				
						$t_{G,TA}$	0				
						$t_{G,TL}$	0				
						$t_{G,TLA}$	0				

only $K_{P,P} = 0$ and $K_{P,LP} = 0$ (resp. $K_{I,I} = 0$ and $K_{I,LI} = 0$) are calling for a change of the value of P (resp. I).

The *Pre* and *Post* matrices of the BRPN associated (we give *Pre* transposed because of its size) are:

$Pre^T =$

	T	$\overline{T}$	L	$\overline{L}$	A	$\overline{A}$	G	$\overline{G}$	P	$\overline{P}$	I	$\overline{I}$
t_T	1	0	0	0	0	1	0	0	0	0	0	0
$t_{T,L}$	1	0	0	0	1	0	0	0	0	0	0	0
t_L	0	1	0	0	1	0	0	1	0	0	0	0
$t_{L,A}$	0	1	0	0	1	0	1	0	0	0	0	0
$t_{L,T}$	1	0	0	0	0	0	0	1	0	0	0	0
$t_{L,AT}$	1	0	0	0	1	0	1	0	0	0	0	0
t_A	0	0	0	0	1	0	0	1	0	1	0	0
$t_{A,L}$	0	0	0	0	0	1	1	0	1	0	0	0
$t_{A,G}$	0	0	0	0	0	1	1	0	1	0	0	0
$t_{A,LG}$	0	0	0	0	1	0	1	0	1	0	0	0
t_G	0	1	0	0	0	1	0	1	0	1	0	0
$t_{G,L}$	0	1	0	0	0	1	0	1	0	1	0	0
$t_{G,A}$	0	1	0	0	0	1	1	0	1	0	0	0
$t_{G,T}$	1	0	0	0	0	1	0	1	1	0	0	0
$t_{G,LA}$	0	1	0	0	1	0	1	0	1	0	0	0
$t_{G,TL}$	1	0	0	0	1	0	0	1	1	0	0	0
$t_{G,TA}$	1	0	0	0	0	1	1	0	1	0	0	0
$t_{G,TLA}$	1	0	0	0	1	0	1	0	1	0	0	0
$t_{P,P}$	0	0	0	1	0	1	0	0	0	0	1	0
$t_{P,LP}$	0	0	0	1	1	0	0	0	0	0	1	0
$t_{I,I}$	0	0	1	0	0	1	0	0	0	0	0	1
$t_{I,LI}$	0	0	1	0	1	0	0	0	0	0	0	1

$Post =$

	T	$\overline{T}$	L	$\overline{L}$	A	$\overline{A}$	G	$\overline{G}$	P	$\overline{P}$	I	$\overline{I}$
t_T	0	1	0	0	0	1	0	0	0	0	0	0
$t_{T,L}$	0	1	0	0	1	0	0	0	0	0	0	0
t_L	0	1	0	0	0	1	0	1	0	0	0	0
$t_{L,A}$	0	1	0	0	0	1	1	0	0	0	0	0
$t_{L,T}$	1	0	0	0	0	0	0	1	0	0	0	0
$t_{L,AT}$	1	0	0	0	0	1	1	0	0	0	0	0
t_A	0	0	0	0	1	0	1	0	0	1	0	0
$t_{A,L}$	0	0	0	0	1	0	1	0	0	1	0	0
$t_{A,G}$	0	0	0	0	0	1	0	1	1	0	0	0
$t_{A,LG}$	0	0	0	0	1	0	0	1	1	0	0	0
t_G	0	1	0	0	0	1	0	1	1	0	0	0
$t_{G,L}$	0	1	0	0	0	1	0	1	1	0	0	0
$t_{G,A}$	0	1	0	0	0	1	1	0	0	1	0	0
$t_{G,T}$	1	0	0	0	0	1	0	1	0	1	0	0
$t_{G,LA}$	0	1	0	0	1	0	1	0	0	1	0	0
$t_{G,TL}$	1	0	0	0	1	0	0	1	0	1	0	0
$t_{G,TA}$	1	0	0	0	0	1	1	0	0	1	0	0
$t_{G,TLA}$	1	0	0	0	1	0	1	0	0	1	0	0
$t_{P,P}$	0	0	0	1	0	1	0	0	0	0	0	1
$t_{P,LP}$	0	0	0	1	1	0	0	0	0	0	0	1
$t_{I,I}$	0	0	0	1	0	1	0	0	0	0	0	1
$t_{I,LI}$	0	0	0	1	1	0	0	0	0	0	0	1

As shown by Property 11, this PN gives rise to four dead marking matching the stable states, which correspond to the four flower organs.

Property 11. The system has four dead markings:

$$\begin{bmatrix} M_d^1(T)=0 \\ M_d^1(L)=0 \\ M_d^1(A)=1 \\ M_d^1(G)=0 \\ M_d^1(P)=0 \\ M_d^1(I)\;=0 \end{bmatrix} \begin{bmatrix} M_d^2(T)=0 \\ M_d^2(L)=0 \\ M_d^2(A)=1 \\ M_d^2(G)=0 \\ M_d^2(P)=1 \\ M_d^2(I)\;=1 \end{bmatrix} \begin{bmatrix} M_d^3(T)=0 \\ M_d^3(L)=0 \\ M_d^3(A)=0 \\ M_d^3(G)=1 \\ M_d^3(P)=0 \\ M_d^3(I)\;=0 \end{bmatrix} \begin{bmatrix} M_d^4(T)=0 \\ M_d^4(L)=0 \\ M_d^4(A)=0 \\ M_d^4(G)=1 \\ M_d^4(P)=1 \\ M_d^4(I)\;=1 \end{bmatrix}$$

Sepals *Petals* *Carpels* *Stamens*

The method for proving that these four markings are dead is very similar to that used to prove Property 10 and amounts to show that a set of Boolean constraints involving the *Pre* matrix is satisfied.

The four stable states corresponding to the four dead markings of Property 11 are reachable from an initial state with genes LEAFY (L) and APETALA 1 (A) ON, all the others being OFF. The reachability can be easily verified through the analysis of the reachability graph, but this method is not tractable for large networks. More generally, the question of the reachability of stable states is related to the notion of circuit functionality. We have shown that dead markings are reachable from any initial valid marking in the case of isolated positive circuits. However, when a positive circuit is embedded in a more complex regulatory network, this behaviour may only occur in specific regions of the logical state space, depending on the functionality domains of the corresponding circuit (work in progress). The general question of finding all the dead markings of a given regulatory Petri net goes well beyond the scope of the present paper and requires further investigation.

6 Conclusions

The combination of a logical approach with the standard Petri net framework offers a powerful set of analytical tools enabling the delineation of specific relationships between the feedback structure and the dynamical properties of complex regulatory systems. Our combined modelling approach encompasses two main steps:

1. A logical processing covers the model specification in terms of a generic regulatory graph, followed by its parameterisation, taking advantage of the flexibility of the definition of the logical parameters; the analysis of circuit functionality can be used here to select specific parameter constraints and/or to derive dynamical insights.
2. The PN corresponding to the resulting parameterised regulatory graph can then be systematically generated, allowing the application of existing algebraic methods to analytically evaluate dynamical properties such as the absence or the presence of deadlocks, the occurrence of specific paths in the reachability graph, etc.

In the case of isolated circuits, we have recovered the main results inferred from the logical framework. More specifically, an isolated functional positive circuit gives rise to two dead markings in the corresponding PN. On the other hand, a functional negative circuit leads to no dead marking but to a cyclical attractor in the corresponding reachability graph. Our approach has been further evaluated through the PN translation of two Boolean regulatory graphs involved in the control of Drosophila cell cycle and in the differentiation of Arabidopsis flower organs, respectively. Here again, we recovered the salient properties found in the original logical model analyses.

The results presented here are encouraging for the analysis of more complex regulatory networks, eventually combining genetic and metabolic interactions. They also point towards a systematic characterisation of the structure of the regulatory Petri nets, in order to derive specific theorems on induced dynamical properties such as liveness, reversibility or reachability. This approach should ease the analysis of large and complex regulatory systems which are difficult to explore through systematic simulations. However, the representation of complex regulatory networks requires the extension of our PN rewriting rules to encompass multi-level logical systems. Finally, it is important to note that the fully asynchronous approach used to generate the logical dynamical graphs covers in fact various (and often incompatible) temporal behaviours. In principle, the distinction between alternative temporal pathways can be forced through the delineation of specific assumptions on transition delays or on priority rules. In this context, the Generalised Stochastic Petri net approach [8] offers a framework enabling the representation of such assumptions taking into account experimental noise.

Acknowledgement

This work has been supported by a Bioinformatics Inter-EPST grant from the French Ministry for Research and Industry. We also wish to thank L.M. Porto, E. Simão, B. Mossé and R. Lima for valuable discussions on the qualitative dynamical analysis of biological regulatory networks.

References

1. Chaouiya, C., Remy, E., Mossé, B., Thieffry, D.: Qualitative analysis of regulatory graphs: a computational tool based on a discrete formal framework. In: L. Benvenuti, A. De Santis, L. Farina (eds.), Positive Systems, POSTA 2003, Springer Lect. Notes Cont. and Info. Sci. **294** (2003) 119–126.
2. Coen, E.S., Meyerowitz, E.M.: The war of the whorls: genetic interactions controlling flower development. Nature **353** (1991) 31–37.
3. de Jong, H.: Modeling and Simulation of Genetic Regulatory Systems: A Literature Review. J. Comput. Biol. **9** (2002) 67–103.
4. Glass, L., Kauffman, S.A.: The logical analysis of continuous, non-linear biochemical control networks. J. theor. Biol. **39** (1973) 103–129.

5. Goss, P.J.E., Peccoud, J.: Quantitative modeling of stochastic systems in molecular biology by using stochastic Petri nets. Proc. Natl. Acad. Sci. USA. **95** (1998) 6750–6755.
6. Hofestädt, R., Thelen, S.: Quantitative Modeling of Biochemical Networks. In Silico Biol. **1** (1998) 39–53.
7. Küfner, R., Zimmer, R., Lengauer, T.: Pathway analysis in metabolic databases via differential metabolic display (DMD). Bioinformatics **16** (2000) 925–836.
8. Marsan, M.A., Balbo, G., Conte, G., Donatelli, S., Franceschinis, G.: Modelling with Generalized Stochastic Petri Nets. Wiley (1995).
9. Matsuno, H., Tanaka, Y., Aoshima, H., Doi, A., Matsui, M., Miyano, S.: Biopathways representation and simulation on hybrid functional Petri net. In Silico Biol. **3** (2003) 389–404.
10. Mendoza, L., Thieffry, D., Alvarez-Buylla, E.R.: Genetic control of flower morphogenesis in Arabidopsis thaliana: a logical analysis: Bioinformatics **15** (1999) 593–606.
11. Murata, T.: Petri Nets: Properties, Analysis and Applications. Proceedings of the IEEE **77** (1989) 541–580.
12. Reddy, V.N., Liebman, M.N., Mavrovouniotis, M.L.: Qualitative analysis of biochemical reaction systems. Comput. Biol. Med. **26** (1996) 9–24.
13. Remy, E., Mossé, B., Chaouiya, C., Thieffry, D.: A description of dynamical graphs associated to elementary regulatory circuits. Bioinformatics **19** (2003) ii172–ii178.
14. Reisig, W.: Petri Nets. Springer-Verlag (1985).
15. Thomas, R.: Boolean formalization of genetic control circuits. J. theor. Biol. **42** (1973) 563–585.
16. Thomas, R., Thieffry, D., Kaufman, M. Dynamical behaviour of biological regulatory networks–I. Biological role of feedback loops and practical use of the concept of the loop-characteristic state. Bull. Math. Biol. **57** (1995) 247–276.
17. Tyson, J.J., Chen, K., Novak, B.: Network dynamics and cell physiology. Nat. Rev. Mol. Cell. Biol. Vol 2 **12** (2001) 908–16.
18. Vallet, M.C., Novak, B., Thieffry, D.: Qualitative modeling of the cell cycle in the fly. Proc. of JOBIM 2002 Conf. (St Malo, France) 329–331.
19. Wuensche, A.: Genomic regulation modeled as a network with basins of attraction. Proc. Pac. Symp. Biocomput. 1998 (Hawai, USA) 89–102.

Finite Unfoldings of Unbounded Petri Nets

Jörg Desel, Gabriel Juhás, and Christian Neumair

Lehrstuhl für Angewandte Informatik
Katholische Universität Eichstätt–Ingolstadt
Ostenstr. 28, 85071 Eichstätt, Germany
{joerg.desel,gabriel.juhas,christian.neumair}@ku-eichstaett.de

Abstract. The aim of this paper is to introduce a concept for an efficient representation of the behavior of an unbounded Petri net. This concept combines a known method for the description of unbounded Petri nets, namely coverability trees, with an efficient, partial order based method developed for bounded Petri nets, namely Petri net unfoldings.

1 Introduction

A crucial problem in applying Petri net theory is the efficient description of the behavior of a Petri net. Already in Petri nets with finitely many different states, so-called bounded nets, the state space explosion ([13]) arises. There is a worst case exponential growth in the number of reachable states with the increase of the size of the Petri net. For example, a Petri net consisting of n different and independent components, each with two different states, has a state space of size 2^n. Consequently, the representation of the behavior of bounded Petri nets by the marking graph, representing all reachable states and their mutual transitions, is too large and inefficient.

K. L. McMillan ([10]) introduced in his doctoral thesis a new approach to a compact representation of the behavior of bounded Petri nets, which counters the state space explosion. Later, the approach was extended by J. Esparza et al. ([7]). The approach is based on so-called branching processes, which are basically constructed like acyclic process nets, generating a partial order. In addition, branching processes can include events which are in conflict, so that alternative process nets can be represented simultaneously.

As different branching processes of a Petri net can be included in each other, one can define a partial order on them. The set of branching processes of a bounded net has a maximal element with regard to this partial order, called the unfolding. This unfolding of a bounded net can be infinite. However, by McMillan's approach a finite prefix of the unfolding can always be constructed. This prefix is complete in the sense that all reachable states are represented, even though it is much smaller than the state space in many cases. In addition, actual causality between the events is explicitly represented, because branching processes include concurrent events. A further advantage of the finite and complete prefix is that local states can be analyzed without focusing on global states.

J. Cortadella and W. Reisig (Eds.): ICATPN 2004, LNCS 3099, pp. 157–176, 2004.
© Springer-Verlag Berlin Heidelberg 2004

Considering unbounded Petri nets, one needs other concepts to find a finite representation of the state space. For this purpose, symbolic representations have been invented, describing infinite sets of states. One such approach is the coverability tree ([4]), where sets of reachable markings are combined to so-called ω-markings. ω can be interpreted as "arbitrarily many" and refers to the number of tokens in a place. A place gets an ω-entry by a ω-marking if it can be pumped by a partial repetition of some occurrence sequence. So an infinite set of reachable states of the Petri net is covered by an ω-marking with an ω-entry. Similar to occurrence sequences, so-called ω-occurrence sequences can be defined. Their compact representation is called the coverability tree. In addition, there exist related concepts like coverability graphs ([8]). The state space explosion problem also exists for coverability trees resp. graphs representing the modified state space, because they are based on the same technique as the marking graph for representing the state space.

The aim of this paper is to introduce a new concept for unbounded nets similar to the idea of a complete and finite prefix of McMillan. For this purpose, the concepts of coverability trees and branching processes are combined to define so-called ω-branching processes. Similar to branching processes for a Petri net, there exists a maximal ω-branching process with respect to set inclusion, called the ω-unfolding. All reachable states of the Petri net are covered by the ω-unfolding, what makes it a coverability unfolding. In analogy to the method of McMillan, a finite prefix of the ω-unfolding can be constructed. This prefix includes the same information about the boundedness of places as the coverability tree, but is often much smaller than the coverability tree, especially if the underlying Petri net includes much concurrency. In addition, the advantages of the McMillan prefix remain: local states can be analyzed without focusing on global states and the causal dependencies of the events are displayed.

In the following section, the construction of the ω-unfolding is illustrated by an example. The complete definition of the ω-unfolding, its finite prefix, theorems and proofs are given in the third section.

2 Principles of Constructing the Finite Unfolding of Unbounded Nets

ω-branching processes result from the combination of the concepts of coverability tree and branching process. The main idea is to introduce a new marking for ω-branching processes which corresponds to the ω-marking of the coverability tree. For this purpose, ω-branching processes are as Petri nets with a labelling function, similar to branching processes. In addition, a new function is introduced, called ω-function, which assigns to each condition of the ω-branching process either the number 1 one or ω. ω will be interpreted as 'arbitrarily many', like in the coverability tree. Each minimal condition, representing an initial token, is mapped to 1 by the ω-marking. According to the occurrence rule, for a transition t which can fire in the underlying Petri net, the minimal ω-branching process can be extended by an event labelled by t.

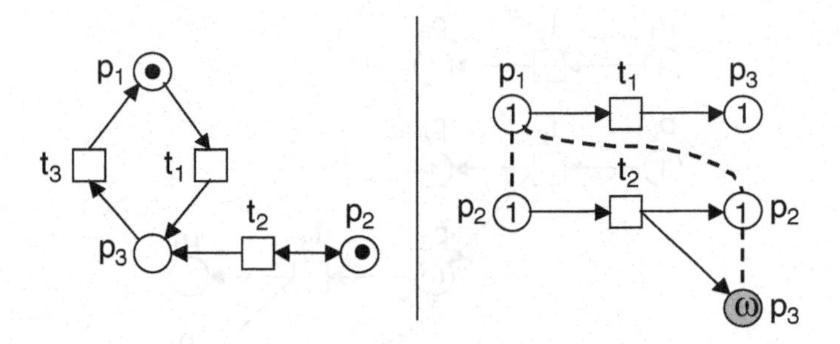

Fig. 1. On the left hand side an unbounded Petri net is shown. The right hand side shows the construction of an ω-branching process of this net. By comparing the cuts (indicated by dashed lines), p_3 is detected as unbounded place. The corresponding condition is colored grey. Firing t_2 causes the token growth in p_3.

As ω-branching processes have the same structure as branching processes, configurations and cuts can be defined in a similar way. In addition, a local configuration is introduced for each event e which includes all events smaller than e with regard to the partial order given by the acyclic structure. Using the labelling function and the ω-function, an ω-marking will be defined for each cut of an ω-branching process, called $Mark_\omega$. This marking is a mapping from the places into the set $\mathbb{N} \cup \{\omega\}$. The marking $Mark_\omega$ derived from an ω-branching process and the ω-marking of a coverability tree of the Petri net are very similar.

$Mark_\omega$ is defined by adding the values of the ω-function for all conditions of a cut with the same label, i.e. for all those conditions which represent tokens in a particular place. The result of the summation of ω with an arbitrary other value is ω again. According to the interpretation of ω, this is interpreted as 'arbitrarily many'.

The ω-marking assigns to a condition the value ω if the corresponding place of the Petri net (by which the condition is labelled) is unbounded. Similar to the coverability tree, this is detected by comparing the marking with earlier markings. Translated to ω-branching processes, this means to for each new condition the marking $Mark_\omega$ of the corresponding local configuration with $Mark_\omega$ of all earlier local configurations. An earlier configuration is a configuration included in the later one. If an ω-marking can be found so that the earlier marking is equal to or bigger than the later one on all places not mapped to ω, the conditions labelled by 'growing' places are mapped to ω by the ω-mapping (Figure 1). A condition having an ω-entry can be interpreted as the potential of arbitrarily many tokens in the corresponding place. So a further consequence for the construction of ω-branching processes is that conditions with the value ω are 'carried over'. This means that, for each ω-condition in the preset of an event, there exists one and only one ω-condition in the postset of the event with the same label. This happens according to the idea that an ω-condition represents 'arbitrarily many' tokens in a particular place, so if a token is removed from a particular place, there remain arbitrarily many tokens. So the occurrence rule

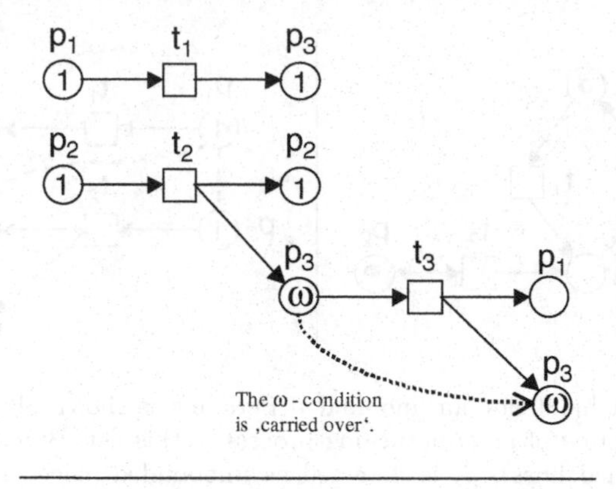

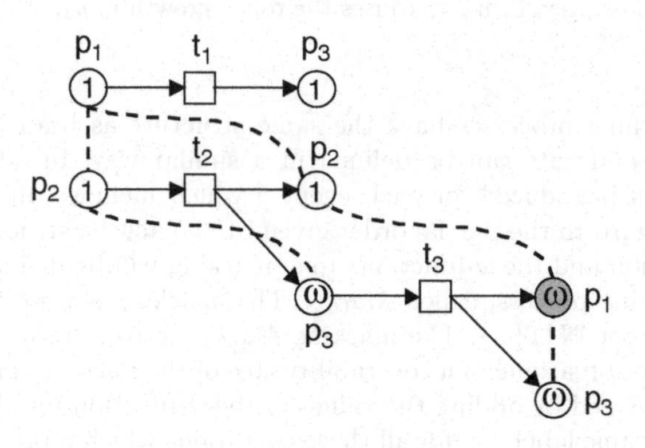

Fig. 2. Further construction of the ω-branching process of the Petri net of Figure 1. The upper part of the figure illustrates how an ω-condition is 'carried over'. In the lower part the unboundedness of place p_1 is detected.

in ω-branching processes is defined similar as in the coverability tree (Figure 2). Cut-off events of the ω-unfolding are defined very much like the cut-off events of McMillan ([10]). A prefix of the ω-unfolding is computed (called Pre_ω) by stopping the construction at cut-off events until no more extensions are possible (Figures 3 and 4).

3 Basic Definitions

In this section we give basic definitions necessary for defining ω-unfolding. Due to lack of space, we omit formal definitions of the coverability tree, unfoldings and related results. For more details about these concepts see e.g. [4] and [7].

Definition 1 (Net). *A net N is a triple (S, T, F), where S is a set of places, T is a set of transitions, satisfying $S \cap T = \emptyset$, and $F \subseteq (S \times T) \cup (T \times S)$ is a flow relation.*

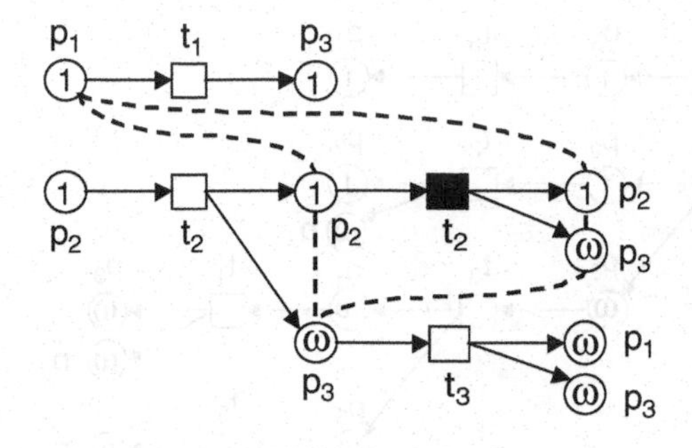

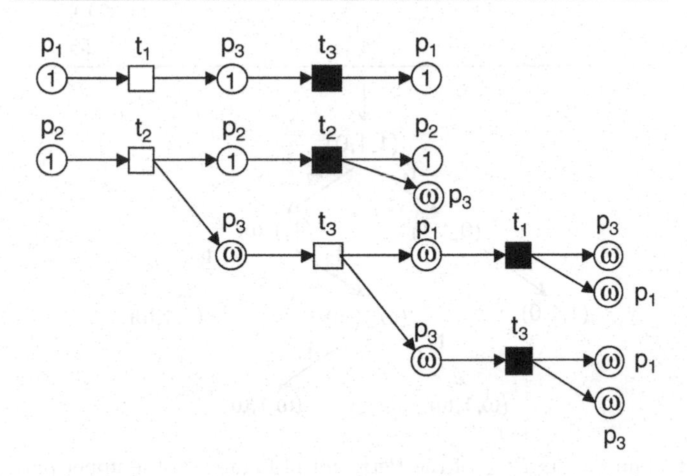

Fig. 3. Construction of the prefix Pre_ω. In the upper part the dashed lines indicate the two cuts with the same marking $Mark_\omega$. So the black colored cut-off event is detected. In the lower part of the figure all cut-off events are marked black.

Places and transitions of a net are also called net elements or elements for short.

Definition 2 (Subnet). *Let* $N = (S, T, F)$ *be a net. A net* $N' = (S', T', F')$ *is said to be a subnet of* N, *iff* $S' \subseteq S$, $T' \subseteq T$ *and* $F' = F \cap ((S' \times T') \cup (T' \times S'))$.

A net is finite if the sets of its elements are finite. To avoid confusion, sometimes we will write $N = (S_N, T_N, F_N)$ to denote $N = (S, T, F)$.

As usual, places are drawn as circles, transitions as boxes, and the flow relation is expressed using arcs connecting places and transitions.

Let $N = (S, T, F)$ be a net and $x \in S \cup T$ be an element of N. The *preset* $\bullet x$ is the set $\{y \in S \cup T \mid (y, x) \in F\}$, and the *postset* $x\bullet$ is the set $\{y \in S \cup T \mid (x, y) \in F\}$. Given a set $X \subseteq S \cup T$, this notation is extended to sets as follows:

$$\bullet X = \bigcup_{x \in X} \bullet x \quad \text{and} \quad X\bullet = \bigcup_{x \in X} x\bullet$$

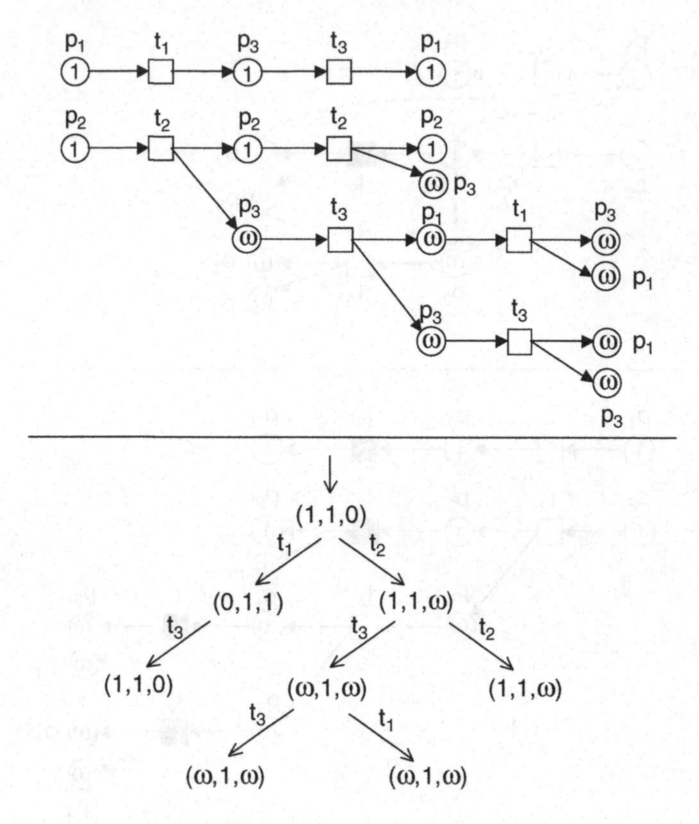

Fig. 4. The finite prefix Pre_ω of the Petri net of Figure 1in the upper part. The lower part shows a corresponding coverability tree.

For technical reasons, in the paper we consider only nets in which every transition has nonempty presets and postsets.

A marking of a net $N = (S, T, F)$ is a function $m : S \to \mathbb{N}$. Graphically, a marking is expressed using a respective number of black tokens in each place.

Definition 3 (Occurrence rule). *Let $N = (S, T, F)$ be a net. A transition $t \in T$ is enabled to occur in a marking m of N iff $m(s) > 0$ for every place $s \in \bullet t$. If a transition t is enabled to occur in a marking m, then its occurrence leads to a new marking m' defined by: $m'(s) = m(s) - |F \cap \{(s,t)\}| + |F \cap \{(t,s)\}|$ for every $s \in S$. We write $m \xrightarrow{t} m'$ to denote that t is enabled to occur in m and that its occurrence leads to m'.*

In words, m' is derived from m by removing one token from every place in the preset $\bullet t$ of the occurring transition t and adding one token to every place in its postset $t\bullet$.

Definition 4 (Occurrence sequence, reachability). *Let $N = (S, T, F)$ be a net and m be a marking of N.*

(i) A finite sequence of transitions $\sigma = t_1, \ldots, t_n$, $n \in \mathbb{N}$ *is called a finite occurrence sequence from* m *to* m_n *if there exists a sequence of markings* $m_1, \ldots, m_n$ *such that*

$$m \xrightarrow{t_1} m_1 \xrightarrow{t_2} \ldots \xrightarrow{t_n} m_n.$$

We say that the occurrence sequence σ *is enabled to occur in* m *and that its occurrence leads to* m_n. *The marking* m_n *is said to be reachable from the marking* m.

(ii) An infinite sequence of transitions $t_1, t_2, t_3 \ldots$ *is called an infinite occurrence sequence if there exists an infinite sequence of markings* $m_1, m_2 \ldots$ *such that*

$$m \xrightarrow{t_1} m_1 \xrightarrow{t_2} m_2 \xrightarrow{t_3} \ldots$$

Remark 1. Given two markings m, m' satisfying $m'(s) \geq m(s)$ for every place s, observe that if an occurrence sequence σ is enabled to occur in m' then it is also enabled to occur in m.

Definition 5 (Petri net). *A finite net with initial marking (Petri net) is a quadruple* $(S_\Sigma, T_\Sigma, F_\Sigma, m_0)$, *where* $(S_\Sigma, T_\Sigma, F_\Sigma)$ *is a finite net and* m_0 *is a marking of* $(S_\Sigma, T_\Sigma, F_\Sigma)$ *called initial marking.*

In a Petri net markings, reachable from the initial marking m_0, *are called reachable markings.*

Definition 6 (Boundedness). *Let* $\Sigma = (S_\Sigma, T_\Sigma, F_\Sigma, m_0)$ *be a Petri net.*

(i) A place $s \in S_\Sigma$ *is called* n*-safe* ($n \in \mathbb{N}$) *iff every reachable marking* m *satisfies* $m(s) \leq n$.

(ii) Σ *is called bounded if there exists an integer* n *such that all places of* Σ *are* n*-safe.*

Definition 7 (Net isomorphism).
Let $N = (S, T, F)$ *and* $N' = (S', T', F')$ *be two nets. An isomorphism from* N *to* N' *is a bijection* $h : S \cup T \longrightarrow S' \cup T'$ *satisfying:*

(i) $h(S) = S'$, $h(T) = T'$;

(ii) $\forall x, y \in S \cup T : (x, y) \in F \Leftrightarrow (h(x), h(y)) \in F'$.

Definition 8 (Conflict). *Let* x_1 *and* x_2 *be elements of a net* $N = (S, T, F)$. *Elements* x_1 *and* x_2 *are in conflict, denoted by* $(x_1, x_2) \in \sharp$ *or* $x_1 \sharp x_2$, *if there exist distinct transitions* $t_1, t_2 \in T$ *such that* $\bullet t_1 \cap \bullet t_2 \neq \emptyset$, *and* $(t_1, x_1), (t_2, x_2)$ *belong to the reflexive and transitive closure of* F. *In other words,* x_1 *and* x_2 *are in conflict if there exist two paths leading to* x_1 *and* x_2 *which start at the same place and are immediately branched (although later on they can be joined again).*

Definition 9 (Occurrence net). *An occurrence net is a net* $O = (B_O, E_O, F_O)$ *satisfying:*

(i) $|\bullet b| \leq 1$ *for every* $b \in B_O$.
(ii) $|\bullet e| \geq 1$ *for every* $e \in E_O$.
(iii) *O is acyclic, i.e. the (irreflexive) transitive closure of F_O is a (strict) partial order.*
The transitive closure of F_O is called the causal relation of O. It is denoted by $<$. The reflexive and transitive closure of F_O is denoted by $\leq$.
(iv) *The set* $\{y \in B_O \cup E_O \mid y < x\}$ *is finite for every* $x \in B_O \cup E_O$;
(Every element x of O has finitely many elements which are smaller than x according to the partial order $<$.)
(v) $(x, x) \notin \sharp$ *for every* $x \in B_O \cup E_O$;
(No element of O is in a self-conflict. Therefore, there can be branching places in O, but no joining places.)

Places of an occurrence net are called conditions and transitions of an occurrence net are called events.

Definition 10 (Concurrency). *Let x_1 and x_2 be elements of an occurrence net. Then x_1 and x_2 are concurrent, denoted by $(x_1, x_2) \in \text{co}$ or $x_1 \text{ co } x_2$, if neither $x_1 < x_2$ nor $x_2 < x_1$ nor $x_1 \sharp x_2$.*

Remark 2. The set of conditions of an occurrence net which are minimal according to $<$ are denoted by $Min(O)$. Obviously, every two elements of $Min(O)$ are concurrent.

Moreover, if x and y are in conflict then, for every $x' \in x\bullet$ and every $y' \in y\bullet$, x' and y' are in conflict, too.

Definition 11 (Configuration). *Let $O = (B_O, E_O, F_O)$ be an occurrence net. A configuration C of O is a set of events $C \subseteq E_O$, satisfying:*

(i) $\forall\, e, e' \in E_O : e \in C \wedge e' < e \Rightarrow e' \in C$ *(C is causally closed.)*
(ii) $\forall\, e, e' \in C : (e, e') \notin \sharp$ *(C is conflict-free.)*

Definition 12 (co–set, cut). *Let O be an occurrence net.*

(i) *A set B' of conditions of O is called a co–set if all conditions b, b' of B' satisfy $b \text{ co } b'$.*
(ii) *A cut is a maximal co–set (w.r.t. set inclusion).*

Remark 3. Every configuration C of an occurrence net defines the cut $Cut(C)$ given by

$$Cut(C) = (Min(O) \cup C\bullet) \setminus \bullet C.$$

Definition 13 (Local configuration). *Let $O = (B_O, E_O, F_O)$ be an occurrence net and $e \in E_O$ an event. Then*

$$[e] := \{e' \in E_O | e' \leq e\}$$

is called the local configuration of e.

It can be easily observed, that $[e]$ is a configuration.

4 Formal Results

4.1 ω-Branching Processes

A basis for the definition of ω-branching processes are (X, e)-extensions of occurrence nets, which are defined in the following way:

Definition 14 ((X, e)-extension of an occurrence net).
Let $O' = (B_{O'}, E_{O'}, F_{O'})$ be an occurrence net and $X \subseteq B_{O'}$ a co–set. Then $O = (B_O, E_O, F_O)$ is an (X, e)-extension of an occurrence net, if:

(i) $B_{O'} \subseteq B_O$ *and* $B_O \setminus B_{O'}$ *is a finite set;*
(ii) $E_{O'} \subseteq E_O$ *and* $E_O \setminus E_{O'} = \{e\}$*;*
(iii) $F_O = F_{O'} \cup \{(x, e)\} \cup \{(e, b)\}$ *for every* $x \in X$ *and every* $b \in B_O \setminus B_{O'}$*.*

So an (X, e)-extension of an occurrence net is generated by adding an event e such that $\bullet e = X$ is a set of pairwise concurrent conditions and $e\bullet$ is a finite set of new conditions.

Lemma 1. *An (X, e)-extension of an occurrence net O' is an occurrence net again.*

Proof. The definition implies immediately that every condition of the extended occurrence net has at most one event in its preset. As the postset of the new event includes only new conditions, the extended net remains acyclic. The number of new elements and arcs is finite. Therefore the property of occurrence nets that each element is preceded by finitely many elements still holds true. In addition, no element of the extended occurrence net is in self conflict, as the preset of the new event consists only of pairwise concurrent conditions. □

Remark 4. Let $O' = (B_{O'}, E_{O'}, F_{O'})$ be an occurrence net and $O = (B_O, E_O, F_O)$ the (X, e)-extension of O'. Then, for every $x \in (B_O \setminus B_{O'}) \cup \{e\}$ and every $y \in B_{O'} \cup E_{O'}$, we have $x \nless y$.

Definition 15. *Let be $\mathbb{N}_\omega := \mathbb{N} \cup \{\omega\}$. Then, for every $n \in \mathbb{N}$, we define*

(i) $n < \omega$*;*
(ii) $n + \omega = \omega + n = \omega + \omega = \omega$*.*

ω-branching processes are inductively defined, starting from the minimal ω-branching process representing the initial marking. Then new events and their modified postsets are added stepwise, generating new ω-branching processes.

Definition 16 (ω-branching process). *Let $\Sigma = (S_\Sigma, T_\Sigma, F_\Sigma, m_0)$ be a Petri net. Then ω-branching processes of Σ are inductively defined in the following way:*

1. *$\beta_\omega = (O, \varphi, \pi)$ is an ω-branching process of Σ, where:*
   (i) $O = (B_O, \emptyset, \emptyset)$*;*
   (O is an occurrence net without events, B_O is a set of conditions.)

(ii) $\varphi : B_O \to S_\Sigma$;
(φ is the labelling function.)

(iii) *for every place* $s \in S_\Sigma$, $|\{b \in B_O \mid \varphi(b) = s\}| = m_0(s)$;
(B_O represents the initial marking of Σ.)

(iv) $\pi : B_O \to \{1, \omega\}$ *is a mapping satisfying* $\pi(b) = 1$ *for every condition in* $b \in B_O$.
(The minimal ω-branching process includes no ω-conditions, i.e. no unbounded place of Σ is detected.)

2. *Let* $\beta'_\omega = (O', \varphi', \pi') = (B_{O'}, E_{O'}, F_{O'}, \varphi', \pi')$ *be an ω-branching process. Let* $X \subseteq B_{O'}$ *be a co-set (X is a set of concurrent conditions of $B_{O'}$) and* $X_\omega = \{b \in X \mid \pi'(b) = \omega\}$ *(X_ω is the set of ω-conditions of X). Assume a transition* $t \in T_\Sigma$ *satisfying* $\varphi'(X)$ [1] $= \bullet t$ *and* $|X| = |\bullet t|$ *(t is enabled in $\varphi(X)$) such that $E_{O'}$ includes no event e' satisfying* $\varphi'(e') = t$ *and* $\bullet e' = X$ *(Every occurrence of a transition should be represented only once).*
Then $\beta_\omega = (O, \varphi, \pi) = (B_O, E_O, F_O, \varphi, \pi)$ *is an ω-branching process of Σ, where*

(i) *O is an (X, e)-extension of the occurrence net O';*
(O is generated from O' by adding a new event e with postset $B_O \setminus B_{O'}$.)

(ii) $\varphi(e) = t$, $\varphi(e') = \varphi'(e')$ *for all* $e' \in E_{O'}$;
(The existing labelling function is extended by e.)

(iii) $\varphi(B_O \setminus B_{O'}) = \varphi(X_\omega) \cup t\bullet$, $|B_O \setminus B_{O'}| = |\varphi(X_\omega) \cup t\bullet|$, $\varphi(b') = \varphi'(b')$ *for all* $b' \in B_{O'}$;
(The existing labelling function is extended; $B_O \setminus B_{O'}$ includes a condition with the label of the corresponding place for the union of the places in $t\bullet$ and those places in $\bullet t$ which have an ω-condition.

(iv) *Define* $VS_e := \{Cut([e']) \mid e' \in E_{O'} \wedge [e'] \subsetneq [e]\} \cup Min(O')$;
(This is the set of the cuts corresponding to those local configurations of O' which are subsets of the local configuration of $[e]$. Recognize that $Min(O')$ representing the initial marking is included, too.)
Define $UE := \{b \in B_O \setminus B_{O'} \mid \varphi(b) \in \varphi(X_\omega)\}$ *and* $NUE = B_O \setminus (B_{O'} \cup UE)$;
(UE is the set of conditions which are 'carried over', as there exists an ω-condition with the same label in the preset of e; NUE is its complement.)
Define $VMark(s) := \omega$ *for every* $s \in \varphi(UE)$ *and*

$$VMark(s) := \sum_{\{b \in Cut([e]) \cap B_{O'} \mid \varphi(b) = s\}} \pi'(b) + |\{b \in NUE \mid \varphi(b) = s\}|$$

for all other $s \in S_\Sigma$. *(VMark is the comparison marking to detect ω-conditions.)*
Then $\pi(b_0) = \omega$ *for* $b_0 \in B_O \setminus B_{O'}$, *if one of the following conditions is satisfied:*

---

[1] φ is extended to a set of conditions X by $\varphi(X) = \{\varphi(x) \mid x \in X\}$.

(a) $b_0 \in \varphi(UE)$.
(ω-conditions are 'carried over' and remain ω-conditions.)

(b) *There exists an event* $e' \in E_{O'}$ *with* $Cut([e']) \in VS_e$, *so that every place* $s \in S_\Sigma \setminus \{\varphi(b_0)\}$ *satisfies*

$$\sum_{\{b \in Cut([e']) | \varphi(b)=s\}} \pi(b) \leq VMark(s)$$

and the place $\varphi(b_0)$ *satisfies*

$$\sum_{\{b \in Cut([e']) | \varphi(b)=\varphi(b_0)\}} \pi(b) < VMark(\varphi(b_0)).$$

(conditions with the label of unbounded places will be ω-conditions.)

(c) *There exists a* $b \in Cut([e])$ *satisfying* $\varphi(b) = \varphi(b_0)$ *and* $\pi(b) = \omega$.
(If the corresponding cut of the local configuration of e includes an ω-condition with the label of a certain place, so all new conditions with the label of this place are ω-conditions.)

For every condition $b \in B_O \setminus B_{O'}$ *satisfying* $\pi(b) \neq \omega$ *it holds* $\pi(b) = 1$. *In addition,* $\pi(b') = \pi'(b')$ *for all* $b' \in B_{O'}$.

As the basis of ω-branching processes are occurrence nets, the notions 'configuration' and 'cut' are also defined for ω-branching processes. Remember that in contrary to usual branching processes, where any cut represents exactly one marking, any cut in an ω-branching process containing an ω-condition represents a whole set of markings, which differ in the number of tokens in the place corresponding to the ω-condition. The question is, which of these markings should be relevant to deal with conflict situation in such a cut. In our definition of ω-branching processes two events are in conflict, if they share a condition in their presets labelled with ω. This corresponds to the marking, when in the place is exactly one token. Considering a marking with more than one token in the place, one could formulate an alternative definition in which these two events would be concurrent. This would correspond to the situation that sometime later when the corresponding place contains enough tokens, these transitions can occur concurrently.

A new marking $Mark_\omega$ can be defined by the combination of labelling function and ω-function. This marking is similar to the ω-marking of the coverability tree.

Definition 17 ($Mark_\omega$). *Let* $C \subseteq E_O$ *be a configuration of an ω-branching process* $\beta_\omega = (B_O, E_O, F_O, \varphi, \pi)$ *of a Petri net* $\Sigma = (S_\Sigma, T_\Sigma, F_\Sigma, m_0)$. *Then* $Mark_\omega(C) : S_\Sigma \longrightarrow \mathbb{N} \cup \{\omega\}$ *is a marking of* Σ *defined for every place* $s \in S_\Sigma$ *by*

$$Mark_\omega(C)(s) = \sum_{\{b \in Cut(C) | \varphi(b)=s\}} \pi(b).$$

Remark 5. The way of computation of $Mark_\omega$ derived ω-branching processes is similar to the marking (*Mark* for short) derived from branching processes. In

Mark, the number of conditions with the same label in the cut of a configuration is computed. For the computation of $Mark_\omega$, the values 1 and ω of the ω-function are added up for such conditions. So, if a cut of a configuration includes no ω-conditions, *Mark* and $Mark_\omega$ coincide.

4.2 ω-Unfoldings

By the subset relation, a partial order is given on the set of ω-branching processes of a Petri net. The aim of the following section is to prove that there exists a unique maximal ω-branching process with respect to this partial order, called the ω-unfolding.

Remark 6. Let β_ω be an ω-branching process of a Petri net. Let β'_ω be a new ω-branching process which is generated of β_ω by adding a new event e with label t to a *co*-set X according to the above definition. Then we write $\beta'_\omega = \beta_\omega \cup \overset{X}{\Uparrow} (e,t)$.

In the definition of ω-branching processes a new ω-branching process is generated from a given one by adding a new 'possible' event and the corresponding modified postset. This way, a partial order on ω-branching processes can be defined.

There is a unique maximal element (fix point) only if the order of the extensions does not matter.

In the following Lemma we prove that the extension of a given ω-branching process by two concurrent events and their modified postsets does not depend on the order. The generated ω-branching process which might only have different names for the new events and conditions. Such ω-branching processes are called isomorphic.

Definition 18. *Two ω-branching processes $\beta_\omega = (O, \varphi, \pi)$ and $\beta'_\omega = (O', \varphi', \pi')$ of a Petri net are called isomorphic if there exists a net isomorphism $i : O \to O'$ satisfying $\varphi' \circ i = \varphi$ and $\pi' \circ i = \pi$.*

Lemma 2. *Let $\beta_\omega = (B_O, E_O, F_O, \varphi, \pi)$ be an ω-branching process of a Petri net $\Sigma = (S_\Sigma, T_\Sigma, F_\Sigma, m_0)$.*

Let $C_1, C_2 \subseteq E_O$ be two configurations of β_ω and $Cut(C_1), Cut(C_2) \subseteq B_O$ be the corresponding cuts.

Let be $X_1 \subseteq Cut(C_1)$ and $X_2 \subseteq Cut(C_2)$.

(X_1 and X_2 are sets of concurrent conditions because they are subsets of cuts.)

Let $t_1, t_2 \in T$ be two transitions of Σ satisfying $\bullet t_1 = \varphi(X_1), |\bullet t_1| = |X_1|$ and $\bullet t_2 = \varphi(X_2), |\bullet t_2| = |X_2|$. Assume that β_ω includes no event e satisfying $\bullet e = X_1$ and $\varphi(e) = t_1$ resp. $\bullet e = X_2$ and $\varphi(e) = t_2$. (β_ω can be extended by two events e_1, e_2.)

Define $\beta^1_\omega = (B^1_O, E^1_O, F^1_O, \varphi^1, \pi^1) = \beta_\omega \cup \overset{X_1}{\Uparrow} (e_1, t_1)$ and

$\beta^2_\omega = (B^2_O, E^2_O, F^2_O, \varphi^2, \pi^2) = \beta_\omega \cup \overset{X_2}{\Uparrow} (e_2, t_2)$.

Then $\beta^1_\omega \cup \overset{X_2}{\Uparrow} (e_2, t_2)$ and $\beta^2_\omega \cup \overset{X_1}{\Uparrow} (e_1, t_2)$ are isomorphic.

Proof. It can be seen directly from the definition of ω-branching processes that $\beta^1_\omega \cup \overset{X_2}{\Uparrow} (e_2, t_1)$ and $\beta^2_\omega \cup \overset{X_1}{\Uparrow} (e_1, t_1)$ exist.

First it will be shown that the events e_1 and e_2 are either concurrent or in conflict in $\beta^1_\omega \cup \overset{X_2}{\Uparrow} (e_2, t_2)$ resp. $\beta^2_\omega \cup \overset{X_1}{\Uparrow} (e_1, t_1)$. For this purpose, we show that the assumption $e_1 < e_2$ in $\beta^1_\omega \cup \overset{X_2}{\Uparrow} (e_2, t_2)$ yields a contradiction. The corresponding conclusion for $\beta^2_\omega \cup \overset{X_1}{\Uparrow} (e_1, t_1)$ results then from the symmetry of e_1 and e_2.

By $\bullet e_2 \subseteq B_O$, $e_1 \bullet \cap B_O = \emptyset$ and Remark 4 we get $b_1 \not< b_2$ for every $b_1 \in e_1\bullet$ and every $b_2 \in \bullet e_2$ in $\beta^1_\omega \cup \overset{X_2}{\Uparrow} (e_2, t_2)$. Assuming $e_1 < e_2$, there would exist conditions $b_1 \in e_1\bullet$ and $b_2 \in \bullet e_2$ satisfying $b_1 < b_2$, or $\bullet e_2 \subseteq e_1\bullet$ would hold true, both leading to a contradiction. Obviously, $e_2 < e_1$ is not fulfilled in $\beta^1_\omega \cup \overset{X_2}{\Uparrow} (e_2, t_2)$ by Remark 4, because e_2 is added after e_1 and so it cannot be causally before e_1.

Let now be $\beta^1_\omega \cup \overset{X_2}{\Uparrow} (e_2, t_2) = (O_1, \varphi_1, \pi_1)$ and $\beta^2_\omega \cup \overset{X_1}{\Uparrow} (e_1, t_1) = (O_2, \varphi_2, \pi_2)$. As e_1 and e_2 are in no order in $\beta^1_\omega \cup \overset{X_2}{\Uparrow} (e_2, t_2)$ resp. $\beta^2_\omega \cup \overset{X_1}{\Uparrow} (e_1, t_1)$, it follows immediately from the definition of ω-branching processes that there exists a net isomorphism $i : O_1 \to O_2$ (choose the identity for all elements except of the postsets of e_1 and e_2) so that $\varphi_2 \circ i = \varphi_1$. In addition, for the computation of the ω-function of the conditions in the postset of e_1 resp. e_2 one only needs the preset of e_1 resp. e_2 and those events, which are smaller than e_1 resp. e_2 according to the partial order defined by the acyclic structure of the underlying occurrence net. These are those events which happened before e_1 resp. e_2. They are the same in $\beta^1_\omega \cup \overset{X_2}{\Uparrow} (e_2, t_2)$ and $\beta^2_\omega \cup \overset{X_1}{\Uparrow} (e_1, t_1)$ because e_1 and e_2 are not ordered. So $\pi_2 \circ i = \pi_1$. □

For technical reasons we now define special ω-branching processes of a Petri net, so-called ω_k-branching processes ($k \in \mathbb{N}$). The ω_0-branching process is the minimal one which represents the initial marking. An ω_k-branching process is created in an inductive way by adding all concurrent events resp. events in conflict and the corresponding postsets to an ω_{k-1}-branching process.

Definition 19 (ω_k-branching process). *Let $\Sigma = (S_\Sigma, T_\Sigma, F_\Sigma, m_0)$ be a Petri net. Then ω_k-branching processes β_{ω_k} ($k \in \mathbb{N}$) are inductively defined in the following way:*

(i) *β_{ω_0} is the minimal ω-branching process.*
(ii) *Let $\beta_{\omega_{k-1}} = (B_O, E_O, F_O, \varphi, \pi)$ be a ω_{k-1}-branching process.*
Let $\{X_1, \ldots, X_n\}$, $n \in \mathbb{N}$ be the set of all co-sets of B_O such that there are transitions $t_1, \ldots, t_n \in T_\Sigma$ satisfying $\bullet t_i = \varphi(X_i), |\bullet t_i| = |X_i|$ and such that $\beta_{\omega_{k-1}}$ includes no event e satisfying $\bullet e = X_i$ and $\varphi(e) = t_i$, $i = 1, \ldots, n$.
Then $\beta_{\omega_k} := \beta_{\omega_{k-1}} \cup \overset{X_1}{\Uparrow} (e_1, t_1) \cup \ldots \cup \overset{X_n}{\Uparrow} (e_n, t_n)$.

Remark 7. It follows immediately from Lemma 2 that an ω_k-branching process of a Petri net is unique up to isomorphism.

The ω_k-branching process for $k \to \infty$ will in the following be called *ω-unfolding.* This ω-unfolding is unique up to isomorphism because of the deterministic definition of ω_k-branching processes.

4.3 Constructing a Finite Prefix of the ω-Unfolding

The ω-unfolding of a Petri net is a unique structure, but it can be infinite. We now define cut-off events of the ω-unfolding similar to cut-off events used to generate a finite prefix of the unfolding in the case of bounded nets.

Definition 20. *Let $\beta_\omega = (B_O, E_O, F_O, \varphi, \pi)$ be the ω-unfolding of Petri net and let $e \in E_O$ be an event of β_ω.*

Then e is a cut-off event if β_ω contains an event $e' \in E_O$ satisfying:

(i) $Mark_\omega([e]) = Mark_\omega([e'])$;
(ii) $[e'] \subsetneq [e]$.

In [7] it is shown that each adequate order can be used to define cut-off events in order to generate a finite and complete prefix of the unfolding for bounded Petri nets. For simplicity, we use the partial order on configurations defined by the subset relation, which is obviously a partial order.

To construct smaller prefixes, the cutoffs are allowed to be made on the basis of a so-called 'additional imaginary starting event', whose final state is the initial marking.

A prefix of the ω-unfolding is generated by stopping the construction when a cut-off event is reached until no more extensions are possible. Stopping the construction at a cut-off event e means to add no events which are causally after e. In the following, the so constructed ω-branching process of a Petri net will be called Pre_ω. To show that Pre_ω of a Petri net is finite, the following lemma, known as *Dickson's Lemma*, is needed. The proof of the lemma can be found in [4].

Lemma 3 (Dickson). *Let S be a finite set and let $\varphi_1, \varphi_2, \varphi_3, \ldots$ be an infinite sequence of mappings from S to $\mathbb{N}_\omega$. There exists an infinite sequence of indices $(i_k)_{k \in \mathbb{N}}$ satisfying:*

(i) $i_l < i_m$ *for all* $l < m$, $l, m \in \mathbb{N}$;
(ii) $\varphi_{i_l}(s) \leq \varphi_{i_m}(s)$ *for all* $l < m$, $l, m \in \mathbb{N}$ *and for every* $s \in S$.

Now we prove the following theorem which shows that in an ω-branching process the same markings $Mark_\omega$ can always be found.

Theorem 1. *Let β_ω be the ω-unfolding of a Petri net $\Sigma = (S_\Sigma, T_\Sigma, F_\Sigma, m_0)$. Let $LK_\omega := \{[e_i] \mid i \in \mathbb{N}\}$ be an infinite sequence of local configurations of β_ω such that $[e_j], [e_k] \in LK_\omega$ and $j < k$ imply $[e_j] \subsetneq [e_k]$. Then there exist indices i_1 and i_2 satisfying $Mark_\omega([e_{i_1}]) = Mark_\omega([e_{i_2}])$.*

Proof. By contraposition, assume an infinite sequence of local configurations satisfying the property above so that all corresponding markings $Mark_\omega$ are different.

By Dickson's Lemma, there exists an infinite strongly monotonic sequence of indices $i_1, i_2, i_3, \ldots$, such that, for each place $s \in S_\Sigma$,

$$Mark_\omega([e_{i_1}])(s) \leq Mark_\omega([e_{i_2}])(s) \leq Mark_\omega([e_{i_3}])(s) \leq \ldots$$

Let i and j be two subsequent indices of the sequence $i_1, i_2, i_3, \ldots$. By assumption, $Mark_\omega([e_i])(s_0) \neq Mark_\omega([e_j])(s_0)$ for at least one place $s_0 \in S_\Sigma$. By the definition of ω-branching processes it follows $Mark_\omega([e_i])(s_0) \neq \omega$ and $Mark_\omega([e_j])(s_0) = \omega$. In addition, $Mark_\omega([e_i])(s) = \omega$ implies $Mark_\omega([e_j])(s) = \omega$ for every place $s \in S_\Sigma$. So the marking $Mark_\omega([e_j])$ has more ω-entries than $Mark_\omega([e_i])$. This implies, that the number of ω-entries grows to infinity, contradicting the finite number of places of Σ. □

With the last result we can now prove that Pre_ω of a Petri net is finite. We will show that the number of included conditions and events is finite.

Theorem 2. *Pre_ω of a Petri net is finite.*

Proof. By contradiction, assume a Petri net with infinite corresponding Pre_ω. Similarly to branching processes of bounded Petri nets ω-branching processes of unbounded Petri nets have finitely branching events. Therefore an extension of "König's-Lemma" can be applied (see. Prop. 1 of [9]), which says that a branching process is infinite if and only if it contains a path including an infinite number of events. So an infinite path can be found, such that all markings $Mark_\omega$ of the local configurations of the events of the path are different, because otherwise the criteria of a cut-off event would eventually be fulfilled. But the existence of such an infinite path contradicts Theorem 1. □

For technical reasons we define extensions of configurations of ω-branching processes:

Definition 21 (Extension of a Configuration). *Let C be a configuration of an occurrence net $O = (B_O, E_O, F_O)$ and $E \subseteq E_O$ satisfying $C \cap E = \emptyset$. Then $C \oplus E := C \cup E$ is an extension of the configuration C, if $C \cup E$ is a configuration again.*

Obviously, for two configurations C and C' satisfying $C \subsetneq C'$ a nonempty set of events E can be found satisfying $C \oplus E = C'$.

Definition 22. *Let $\beta_\omega = (B_{O'}, E_{O'}, F_{O'}, \varphi', \pi')$ be an ω-branching process and C be a configuration of β_ω. Then $\Uparrow C = (B_O, E_O, F_O, \varphi, \pi)$ is the subnet of β_ω satisfying $B_O \cup E_O = \{x \in B_{O'} \cup E_{O'} \mid x \notin C \cup \bullet C \wedge \forall y \in C : (x, y) \notin \sharp\}$ (φ resp. π are the restrictions of φ' resp. π' to the elements of $\Uparrow C$).*

($\Uparrow C$ is the part of β_ω 'after' C.)

Remark 8. Let C_1 and C_2 be two finite configurations of an ω-branching process satisfying $Mark_\omega(C_1) = Mark_\omega(C_2)$. Then there exists for each finite extension $C_1 \oplus E_1$ a finite extension $C_2 \oplus E_2$, such that E_1 and E_2 are isomorphic. We will then write $E_2 = I_1^2(E_1)$.

Theorem 3. *Let m be a reachable marking of a Petri net.*
Then there exists a configuration C_ω in Pre_ω satisfying $Mark_\omega(C_\omega)(s) = m(s)$ for every place s satisfying $Mark_\omega(C_\omega)(s) \neq \omega$.

Proof. The (potentially infinite) unfolding (cf. [6]) of the Petri net includes a configuration C satisfying $Mark(C) = m$.

The modified postset of an event e of an ω-branching process always includes at least the represented postset of the transition corresponding to e. So the ω-unfolding includes for every configuration of the unfolding an isomorphic configuration.

If Pre_ω includes a configuration C_ω which is isomorphic[2] to the configuration C, the occurrence rule in ω-branching processes implies that $Mark_\omega(C_\omega)$ and m are equal except for the places marked by ω.

If a configuration C' isomorphic to C cannot be found in Pre_ω, it can nevertheless be found in the corresponding ω-unfolding of Σ. As C' is not included in $Pre0_\omega$, there exists a cut-off event e_1 in C' and we have $C' = [e_1] \oplus E_1$ for a set of events E_1. By the definition of cut-off events there is a local configuration $[e_2]$ in Pre_ω satisfying $Mark_\omega([e_2]) = Mark_\omega([e_1])$ and additionally $[e_2]$ is a proper subset of $[e_1]$. By Remark 8, the ω-unfolding includes a configuration $C_2 = [e_2] \oplus I_1^2(E_1)$. As $[e_2]$ is a proper subset of $[e_1]$, C_2 includes less events than C'. In addition, $Mark_\omega(C')$ and $Mark_\omega(C_2)$ are equal on all places not mapped to ω. Assume that $Mark_\omega(C')$ and $Mark_\omega(C_2)$ are not equal for all places. Then $Mark_\omega(C')$ can only have more places marked by ω than $Mark_\omega(C_2)$. This is because the number of possible comparison cuts for the detection of ω-conditions is larger for the conditions of C' than for the conditions of C_2. If C_2 is a configuration not included in Pre_ω, the above procedure can be repeated and so a configuration C_3 can be found satisfying $Mark_\omega(C')(s)$=$Mark_\omega(C_3)(s)$ for every place s not marked to ω. The configuration C_3 then includes less events than C_2. So with every step the number of events of new configuration decreases. Therefore a configuration C_ω in Pre_ω can be found satisfying $Mark_\omega(C_\omega)(s) = Mark_\omega(C')(s) = Mark(C)(s) = m(s)$ for all places s not mapped to ω. □

Theorem 4. *Let $\Sigma = (S_\Sigma, T_\Sigma, F_\Sigma, m_0)$ be a Petri net and C a configuration of Pre_ω of Σ. Then there exists a reachable marking m for every $a \in \mathbb{N}$, such that every place $s \in S_\Sigma$ satisfies:*

$$\begin{array}{ll} m(s) = Mark_\omega(C)(s) & \text{if } Mark_\omega(C)(s) \neq \omega \\ m(s) \geq a & \text{if } Mark_\omega(C)(s) = \omega. \end{array}$$

[2] A configuration in an ω-branching process is isomorphic to a configuration in another ω-branching process if there is a bijection between events preserving labelling by transitions and partial order on events.

Proof. Let $Mark_\omega(C)$ be a marking of Pre_ω and let $S_\omega \subseteq S_\Sigma$ be the set of places s satisfying $Mark_\omega(C)(s) = \omega$. Let $a \in \mathbb{N}$.

We prove that there exists a reachable marking m of Σ which assigns at least a tokens to each place of S_ω and coincides with $Mark_\omega(C)$ on all places not in S_ω. To this end, we construct a linearization $e_1, \ldots, e_k$ of C satisfying either $e_i < e_j$ or e_i *co* e_j for all $i < j$, $i, j \in \{1, \ldots, k\}$. So every subsequence of the form $e_1, \ldots, e_l$, $1 \leq e_l \leq e_k$ is, as set, a configuration of Pre_ω. This determines the following linearization of markings:

$$m_0 = Mark_\omega(\emptyset) \xrightarrow{e_1} Mark_\omega(\{e_1\}) \xrightarrow{e_2} Mark_\omega(\{e_1, e_2\}) \xrightarrow{e_3} \ldots$$

$$\xrightarrow{e_k} Mark_\omega(\{e_1, \ldots, e_k\}) = Mark_\omega(C)$$

Now we construct an occurrence sequence with the corresponding reachable markings for Σ of the form

$$m_0 \xrightarrow{\varphi(e_1)} m_1 \xrightarrow{\sigma_1} m_1' \xrightarrow{\varphi(e_2)} m_2 \xrightarrow{\sigma_2} m_2' \xrightarrow{\varphi(e_3)} m_3 \ldots \xrightarrow{\sigma_{k-1}} m_{k-1}' \xrightarrow{\varphi(e_k)} m_k \xrightarrow{\sigma_k} m_k'.$$

The σ_i are defined in such a way that m_k' satisfies the properties of m.

Let be $s_0 \in S_\omega$. By the definition of ω-branching processes there is a unique index $i = i(s_0)$, such that:

$$\begin{array}{ll} Mark_\omega(\{e_1, \ldots, e_j\})(s_0) \neq \omega & \text{for every } e_j \in C \text{ satisfying } 1 \leq e_j < e_i \\ Mark_\omega(\{e_1, \ldots, e_j\})(s_0) = \omega & \text{for every } e_j \in C \text{ satisfying } e_i \leq e_j \leq e_k. \end{array}$$

In addition, there is an event $e_j \in C$ such that there exists a subsequence of $e_j, \ldots, e_i$ (we will write $e_j \Rightarrow e_i(s_0)$), satisfying the property, that the occurrence sequence $\varphi(e_j \Rightarrow e_i(s_0))$ increases the number of tokens in s_0 and does not change the token count in all places s of S satisfying $Mark_\omega(\{e_1, \ldots, e_i\})(s) \neq \omega$.

However, it is possible that the token count in places marked by ω before the event e_1 decreases. The token count in such a place will decrease at most for the number of events of the sequence $e_j \Rightarrow e_i(s_0)$ (we will write $|e_j \Rightarrow e_i(s_0)|$). These are those events which are necessary to pump the place s_0. Every marking which coincides with $Mark_\omega(\{e_1, \ldots, e_i\})$ in all places not marked by ω and which assigns at least $|e_j \Rightarrow e_i(s_0)|$ tokens to the other places enables the occurrence sequence $\varphi(e_j \Rightarrow e_i(s_0))$.

In the sequel, $(\varphi(e_j \Rightarrow e_i(s_0)))^x$ stands for firing the occurrence sequence $\varphi(e_j \Rightarrow e_i(s_0))$ x-times. Let be $S_n = \{s \in S_\omega | i(s) = n\}$ for $n = 1, \ldots, k$; we will write $S_n = \{s_n^1, \ldots, s_n^{|S_n|}\}$. We define the occurrence sequence

$$\sigma_k = (\varphi(e_j \Rightarrow e_i(s_k^1)), \varphi(e_j \Rightarrow e_i(s_k^2)), \ldots, \varphi(e_j \Rightarrow e_i(s_k^{|S_k|})))^a$$

to ensure that every place s in S_k has at least a tokens. As the number of tokens in S_{k-1} can be decreased by $\varphi(e_k)$ and σ_k, we define accordingly

$$\sigma_{k-1} = (\varphi(e_j \Rightarrow e_i(s_{k-1}^1)), \varphi(e_j \Rightarrow e_i(s_{k-1}^2)), \ldots, \varphi(e_j \Rightarrow e_i(s_{k-1}^{|S_{k-1}|})))^{(a+1+|\sigma_k|)}$$

to ensure, that the occurrence sequence $\varphi(e_k), \sigma_k$ is enabled after firing σ_{k-1} and assigns at least a tokens to each place of S_{k-1}.

In general, for $i = k, k-1, \ldots, 1$ the sequence σ_i is defined by

$$\sigma_i = (\varphi(e_j \Rightarrow e_i(s_i^1)), \varphi(e_j \Rightarrow e_i(s_i^2)), \ldots$$
$$\ldots, \varphi(e_j \Rightarrow e_i(s_i^{|S_i|})))^{(a+(k-i)+|\sigma_k|+|\sigma_{k-1}|+\ldots+|\sigma_{i+1}|)}$$

As the marking $Mark_\omega(\emptyset)$ has no ω-entries, the occurrence sequence σ_1 decreases the token count in no place and is therefore enabled in $Mark_\omega(\{e_1\})$.

As the token count of places not in S_ω is not changed by any occurrence sequence $\varphi(e_j \Rightarrow e_i)$, also σ_i does not change it. So $Mark_\omega(C)$ and m'_k are equal for those places. □

Remark 9. If a marking $Mark_\omega(C)$ has exactly one ω-entry for a place s_0, a minimal local configuration $[e]$ can be directly found in Pre_ω which increases the token count of s_0 and corresponds to a local configuration of the unfolding. e is the event which is in the preset of an ω-condition included in $Cut(C)$ and labelled by s_0.

If a marking $Mark_\omega$ has ω-entries for more than one place, a minimal local configuration which increases the token count of these places and corresponds to a local configuration of the unfolding cannot be found in general. But such a configuration can be constructed according to the method of the proof of Theorem 4.

Theorem 5. *A place s_0 of a Petri net is unbounded iff there exists a marking in Pre_ω satisfying $Mark_\omega(C)(s_0) = \omega$.*

Proof. ($\Leftarrow$) Follows immediately from Theorem 4.
($\Rightarrow$) Since Pre_ω is finite, there exists only a finite number of markings $Mark_\omega$ in Pre_ω. So there is a bound $a \in \mathbb{N}$ such that all markings satisfy $Mark_\omega(s_0) < a$ or $Mark_\omega(s_0) = \omega$. As the place s_0 is unbounded, there exists a reachable marking m in Σ satisfying $m(s_0) \geq a$. This marking does not coincide with a marking $Mark_\omega$ of Pre_ω, so by Theorem 3 it must be covered by a marking $Mark_\omega$. This implies, that there exists at least one marking in Pre_ω satisfying $Mark_\omega(s_0) = \omega$. □

By Theorem 5 the next Corollary follows immediately.

Corollary 1. *A Petri net is bounded iff no marking $Mark_\omega$ of Pre_ω has an ω-entry.*

The cut-off events used for the construction of Pre_ω are defined on the simplest adequate order. If we denote by Pre a complete and finite prefix of the unfolding using the same adequate order, we can formulate the following result:

Theorem 6. *If a Petri net is bounded, then Pre_ω and Pre are isomorphic, i.e there is a bijection between Pre_ω and Pre which preserves labelling and structure.*

Proof. By Corollary 1 bounded nets with initial marking have no ω-conditions. The result follows immediately by the definition of Pre and Pre_ω. □

5 Conclusion

We have presented a finite description of the infinite concurrent semantics for unbounded Petri nets that identifies boundedness properties and approximates the set of reachable markings employing coverability properties. The ω-unfolding is a coverability unfolding in the sense that a finite prefix Pre_ω can be constructed in which all reachable markings of the corresponding Petri net are covered.

As the formal definition is based on induction, an algorithm can easily be developed to compute Pre_ω. So the next step is the implementation and the analysis of such an algorithm. In this paper we did not focus on the construction of an algorithm and computing its complexity, but rather developed a formal basis for what we call Pre_ω. Our prefix of the ω-unfolding closes a gap in theory, namely a concept for unbounded nets which corresponds to the coverability tree resp. graph for a partial order based semantics.

A further goal that deserves theoretical investigations is to find a special adequate order to detect cut-off events. In [7] a special order for elementary net systems is presented providing an improvement of the algorithm which computes the complete finite prefix of the unfolding. A corresponding improvement might exist for unbounded Petri nets as well.

References

1. P. A. Abdulla, S. P. Iyer and A. Nylen. Unfoldings of Unbounded Petri nets. In *Proceedings of the 12th International Conference on Computer Aided Verification (CAV 2000)*, volume 1855 of *Lecture Notes in Computer Science*, pages 495–507. Springer–Verlag, Berlin, Germany.
2. E. Best, C. Fernande. Notations and Terminology on Petri Net Theory. In Arbeitspapiere der GMD, 1987.
3. E. Best, C. Fernandez. Nonsequential processes. In *EATCS Monographs*, volume 13. Springer–Verlag, Berlin, Germany, 1988.
4. J. Desel, W. Reisig. Place/Transition Petri Nets. In *W. Reisig, G. Rozenberg (Eds.): Lectures on Petri Nets I: Basic Models*, volume 1491 of *Lecture Notes in Computer Science*, pages 122–173. Springer–Verlag, Berlin, Germany, 1998.
5. *Lecture Notes in Computer Science* J. Desel, G. Juhas. What is a Petri Net? Informal Answers for the Informed Reader. In *H. Ehrig, G. Juhas, J. Padberg, G. Rozenberg (Eds.): Unifying Petri Nets*, volume 2128 of *Lecture Notes in Computer Science*, pages 1–25. Springer, Berlin, Germany, 2001.
6. J. Engelfriet. Branching Processes of Petri Nets. In *Acta Informatica 28*, pages. 575–591, 1991.
7. J. Esparza, S. Römer and W. Vogler. An Improvement of McMillan's Unfolding Algorithm. In *Formal Methods in System Design, 20* pages 285–310, 2002.
8. A. Finkel. The minimal coverability graph for Petri Nets. In *Advances in Petri Nets*, volume 674 of *Lecture Notes in Computer Science*, pages 210–243. Springer–Verlag, Berin, Germany, 1993.
9. V. Khomenko, M. Koutny and W. Vogler. Canonical Prefixes of Petri Net Unfoldings. In *Proceedings of the 14th International Conference on Computer Aided Verification (CAV 2002)*, volume 2404 of *Lecture Notes in Computer Science*, pages 582–595. Springer, Berlin, Germany, 2002.

10. K. L. McMillan. A Technique of State Space Search Based on Unfolding. In *Formal Methods in System Design, 6(1)*, pages 45–65, 1995.
11. K. L. McMillan. Using Unfoldings to Avoid the State Explosion Problem in the Verification of Asynchronous Circuits. In *4th Workshop on Computer Aided Verification*, volume 663 of *Lecture Notes in Computer Science*, pages 164–174. Springer, Berlin, Germany, 1992.
12. James L. Peterson. *Petri Net Theory and the Modelling of Systems.* Prentice-Hall, Englewood Cliffs, New Jersey, 1981.
13. A. Valmari. The State Explosion Problem. In *W. Reisig, G. Rozenberg (Eds.): Lectures on Petri Nets I: Basic Models*, volume 1491 of *Lecture Notes in Computer Science*, pages 429–528. Springer, Berlin, Germany, 1998.

Compositional Modeling of Complex Systems: Contact Center Scenarios in OsMoSys

Giuliana Franceschinis[1], Marco Gribaudo[2], Mauro Iacono[3], Stefano Marrone[3], Nicola Mazzocca[3], and Valeria Vittorini[4]

[1] Dipartimento di Informatica, Università del Piemonte Orientale, Piazza Ambrosoli 5, 15100, Alessandria, Italy
giuliana@mfn.unipmn.it

[2] Dipartimento di Informatica, Università di Torino, Corso Svizzera 185/B, 10149 Torino, Italy
marcog@di.unito.it

[3] Dipartimento di Ingegneria dell'Informazione, Seconda Università di Napoli, Via Roma 29, 81031 Aversa, Italy
{mauro.iacono,stefano.marrone,nicola.mazzocca}@unina2.it

[4] Dipartimento di Informatica e Sistemistica, Università degli Studi di Napoli "Federico II", Via Claudio 21, 80125 Napoli, Italy
vittorin@unina.it

Abstract. In this paper we present the application of a compositional modeling methodology to the re-engineering of Stochastic Well Formed net (SWN) models of a contact center. The modeling methodology is based on the definition of proper operators to connect submodels and it is supported by the OsMoSys modeling framework. The paper describes the implementation of a library of reusable SWN submodels of the contact center components and the definition of proper SWN connectors to easily develop models of different configurations of the system. We also describe the solving process of the composed models and its integration in the OsMoSys framework. Moreover, we discuss the advantages that this approach, based on the definition of classes and instances of submodels, can provide to the application of SWN to complex case studies.

1 Introduction

Compositionality is an effective mean to cope with the complexity of real system modeling and analysis. It can also be exploited to set up a framework where designers, who may not be expert in formal modeling languages, compose models of complex scenarios by just choosing building blocks from a library and connecting them. The building blocks could hide complex models built by experts.

In the context of Petri nets (PN) a number of proposals have appeared in the literature to add model (de)composition features to a formalism that is not compositional in its basic definition (e.g. [1, 3, 10, 11]). In particular, in [13, 14] PNs are used to describe the behavior of object classes and communication mechanisms are defined to allow the object models to inter-operate. The approach proposed in [7] adapts this model composition paradigm to Stochastic

J. Cortadella and W. Reisig (Eds.): ICATPN 2004, LNCS 3099, pp. 177–196, 2004.
© Springer-Verlag Berlin Heidelberg 2004

Well Formed nets (SWN) by extending it with a client-server model composition schema, inspired by the Cooperative Nets approach [12]. Such composition schema is supported by OsMoSys, a multi-formalism modeling framework providing the mechanisms to define and apply composition operators and rules on the top of formalisms that are not compositional in their original definition.

In this paper we present the application of a compositional modeling methodology to the re-engineering of the SWN models of a real contact center, and the work done to *implement* a library of reusable submodels, a feature that was considered as highly desirable in an earlier study on this application domain [6]. Both the contact center components modeling methodology and the implementation of a library of reusable building blocks, are based on the OsMoSys framework. The approach to modeling of contact center scenarios that we apply in this paper differs from the one described in [6] from two points of view: in [6] the composition was performed on low level operators, namely transition superposition, moreover the choice of the number of subnets to be composed, the labels assignment and the exact sequence of calls to the composer tool (*Algebra*) were all done manually, without any automatic support; in this paper instead higher level composition operators have been defined, that have the effect of better isolating the model of the components from the model of their interaction, as a consequence the correct configuration of the connection subnets and the needed sequence of calls to the composer tool can be automatically generated. This highly simplifies the task of generating different scenarios by a user with limited knowledge on the underlying formalism. In this paper we do not focus on the analysis phase and on the results that can be obtained, since they would be comparable to those already presented in [6], but rather on the advantages of the proposed methodology in promoting model reuse and in facilitating the composition task. An interesting next step would be to investigate the possibility of exploiting the model structure to improve the efficiency of the analysis phase through decomposition methods.

With respect to the work presented in [7], message sending is used as communication paradigm (instead of client-server) and more complex connection operators are introduced to represent the communication sessions between the client and the site components of the contact center.

The paper is organized as follows. Sect. 2 describes the contact center case study, Sect. 3 introduces the modeling approach and provides a brief introduction to the OsMoSys framework, Sect. 4 defines the set of operators used to build the compositional models of contact center scenarios and describes the SWN models of the contact center components. In Sect. 5 the automatic process that generates the final SWN model of the contact center scenarios is briefly presented. Finally, Sect. 6 discusses the advantages of the proposed compositional approach and highlights possible directions for future work.

2 Contact Center Scenarios

Contact centers are today an effective way for companies to manage client relationships and achieve some economical advantages, since they may lower post-sell assistance costs or allow companies to bypass usual distribution chains for selling products or services.

Effectiveness and inexpensiveness of contact centers can only be exploited through a correct design and proper management policies. Sophisticated policies can require a very complex structure. Performances of a contact center can be evaluated by simulation of a proper model of their architecture. A correct model of a contact center architecture should be fed with incoming calls. Calls are generated by a client pool. A client pool is influenced by the kind of product or service the contact center offers and is a collection of customers, grouped in customer classes. Customers can be grouped by the level of quality of service they are entitled to, by behavior, by the service they access. It is thus necessary to capture a good client pool model in order to build simulation scenarios able to properly represent the real behavior of the contact center under evaluation. In our modeling technique, site classes are developed, as well as client pool classes, which are "glued" by connectors representing the communication infrastructure.

In this work we present some models of client pools and different site architectures.

Client pool models describe the behavior of customer classes and are characterized by several parameters; from another point of view they can be seen as workload generators for sites. Customers in a pool can be partitioned into classes with different privileges and/or different propensity for hang up. The classes can have different call generation rates, to describe different service needs, and different hang up rates. Sites are more complex. Customers always use a single policy: they need a service, they call the contact center, they give up or wait, they access the service or give up again, they leave the contact center. On the other hand, sites can implement different policies in order to obtain different QoS (Quality Of Service) classes for customer, they can have a variable number of queues, a variable number of operators per queue, service times that depend on the queue or the operator. Some queues could be specialized in handling some type of requests, or different maximum queuing times before discarding a waiting call.

While client pools and sites are the basic components of our scenarios, connectors define how they can interact, and allow to define different contact center topologies. The communications between pools and sites follow routing and session handling rules, in other words they support the communication protocols used by clients and sites to interact. In this work message sending is used as the basis for implementing different protocols in a flexible way. Two protocols have been defined and modeled: the first, C2SP (Client to Server Protocol) is used in communications between a client pool (called in the following client for sake of simplicity) and a site (in the following server); the other, S2SP (Server to Server Protocol) in communications between sites. The C2SP is logically described in Fig. 1.

Two kinds of messages can be sent from the client to the server: the first (REQ) is used to initiate a communication session and request a service, the second (HUP) is used to drop the communication instantaneously (regardless

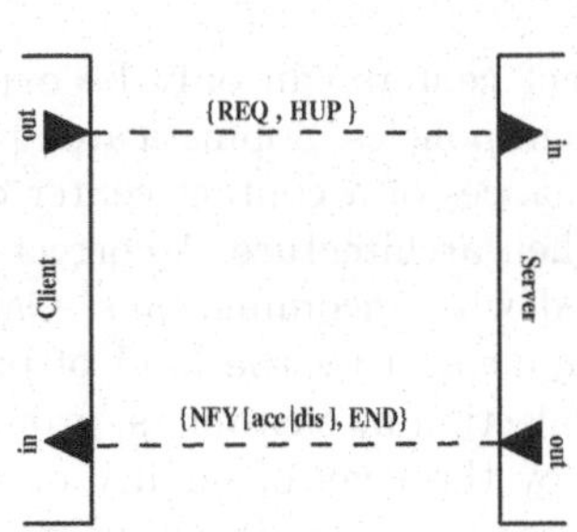

Fig. 1. C2SP protocol

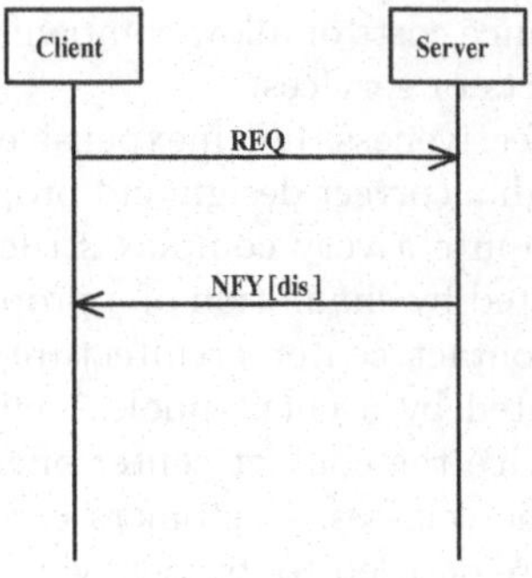

Fig. 2. C2SP – Service denied

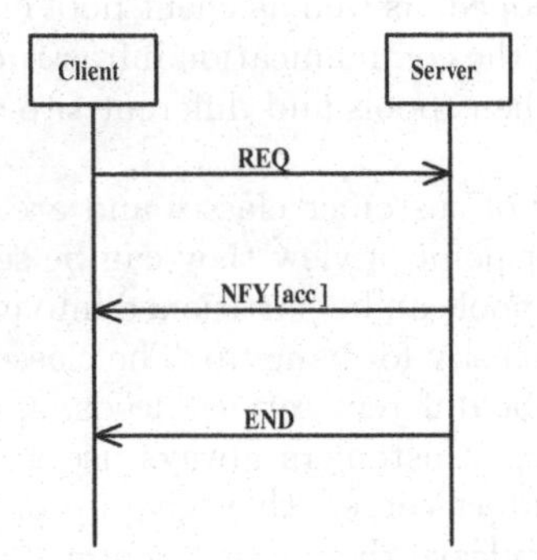

Fig. 3. C2SP – Service accepted

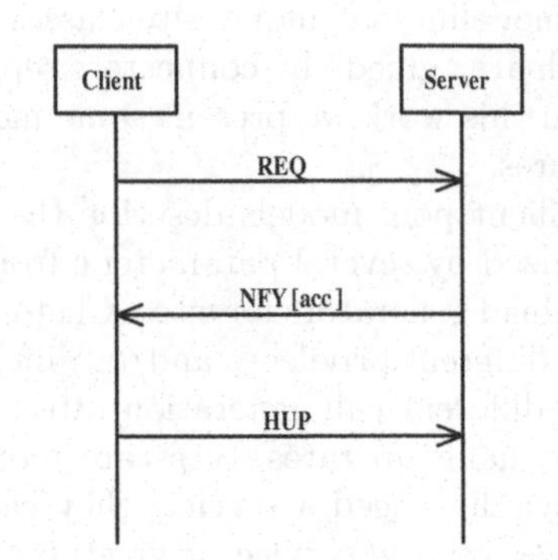

Fig. 4. C2SP – Hang Up

of whether the service has been furnished or not). Two other messages can be sent from the server to the client: the first (NFY) signals the acceptance or the refusal of a request (REQ), the second (END) signals the correct end of a service. Services are considered as transactions in data bases: they are either completely supplied or not at all. A server receiving a request (REQ) can either accept it for service or not, depending on its instantaneous service capability. After the server accepts the request, the client can either wait for service completion or hang up the telephone without obtaining the service.

Figures from 2 to 4 depict some possible sessions. In Fig. 2, a simple connection refusal is showed. In Fig. 3, a correct service session is showed. In Fig. 4, the service request is accepted by the server, but the client interrupts the service by issuing a hang up signal (it asynchronously closes the communication).

The example situations depicted in Fig. 3 and 4 show the importance of having message sending as a communication paradigm: in fact the protocol has been designed in order to be interrupted asynchronously in any phase with different behaviors.

The S2SP is logically described in Fig. 5. It is used whenever a site cannot serve an incoming call for lack of resources as well in servicing units as in queues.

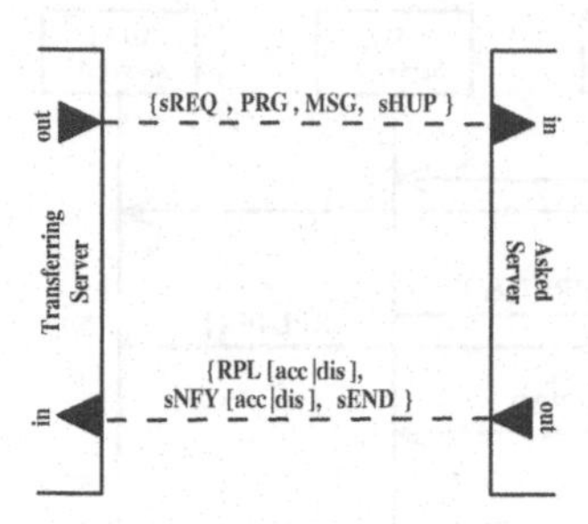

Fig. 5. S2SP protocol

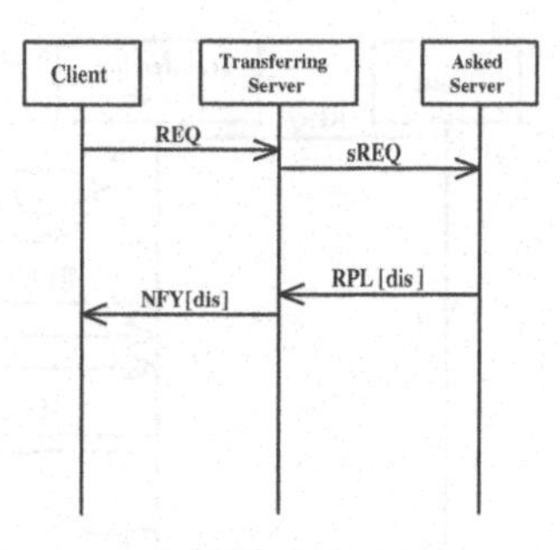

Fig. 6. S2SP – Transfer Failed

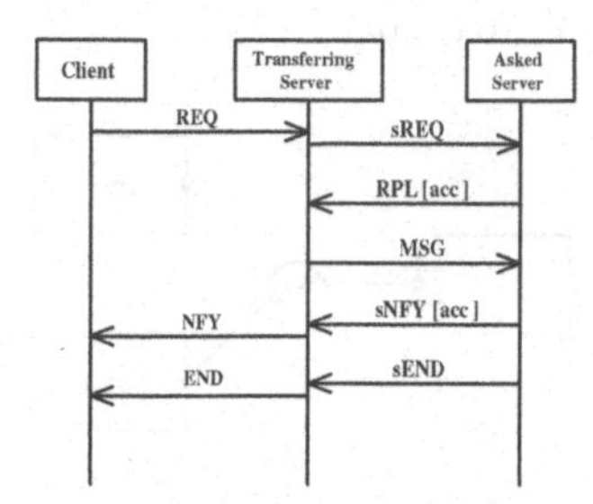

Fig. 7. S2SP – Transfer OK

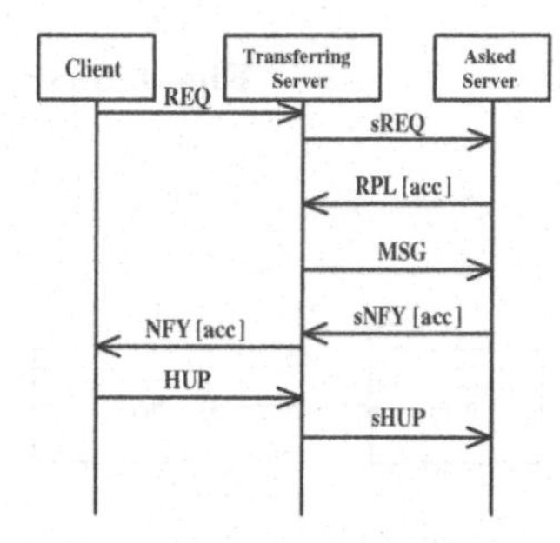

Fig. 8. S2SP – Hang Up

When a server is in such a situation, it asks all other servers, using the request (sREQ) signal, if any of them is able to service the request: all servers reply with a RPL signal either accepting or refusing the request. If at least one answering server is accepting, a place for the incoming call is reserved in the queues of the first accepting server which notifies the acceptance using the RPL signal, while all other servers receive a purge (PRG) message. If no answering server is accepting, the call is lost. When a server receives a PRG, it drops the reservations it holds. The acceptor server receives a MSG signal carrying the original client request and answers the requesting server with a sNFY signal, analogous to the NFY signal in C2SP. The rest of the protocol uses the sEND and sHUP signals, like END and HUP in C2SP. Figures from Fig. 6 to Fig. 8 depict some possible sessions. Fig. 6 describes the situation in which a call is transferred to another server which does not accept it (for lack of resources). Fig. 7 describes an accepted call transfer. A situation where the client hangs up is illustrated in Fig. 8.

An example of a simple complete session involving both protocols is given in Fig. 9. In this situation we have one client and four servers. The server connected with the client has no resources to handle an incoming call: it asks the other servers for a transfer. Two servers are available and the fourth refuses the request. One of the available servers receives the call for service while the other receives

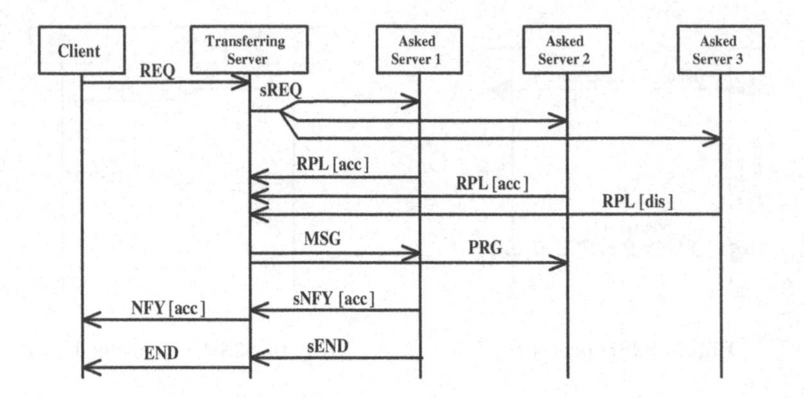

Fig. 9. S2SP – Call Transfer Accepted

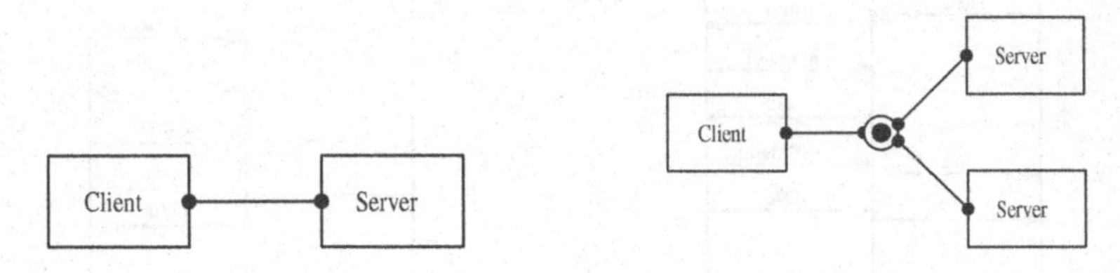

Fig. 10. Scenario 1

Fig. 11. Scenario 2

a PRG message in order to free booked resources. The service is completed by the "winner" server regularly.

Several scenarios have been selected in order to model and study the most interesting situations. The first scenario, depicted in in Fig. 10, is the simplest. It includes a single client and a single server, thus only the C2SP is involved, and a simple connector which is in charge of maintaining essential information about a single communication session. In Fig. 11, a more complex situation is showed: a single client can request a service to one of two servers, non-deterministically chosen. As in the previous scenario, only the C2SP is involved, because the service is to be supplied either by one server or the other, with no possibility of call transfer. Servers are not connected to each other but both are connected to the client. In this case, non-determinism drives the creation of a connection, while the rest of the protocol is fully deterministic (after the choice of a server, the whole communication continues between it and the client). In this case, the connector is in charge to manage both routing and maintaining session information.

The last scenario we considered is in Fig. 12. In this case, three clients are connected to three interconnected servers, so both protocols are involved. In this situation the modeler can evaluate the impact on global QoS of transferred calls. The scenario also allows for the evaluation of the effects of unbalanced clients (clients with different call generation rates) or asymmetric servers (servers with different call management policies or capabilities).

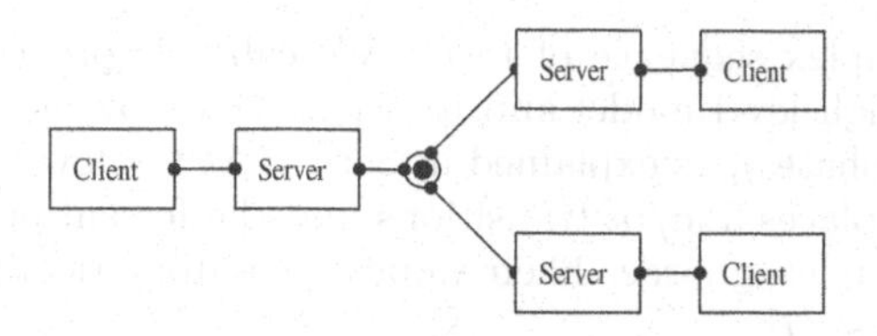

Fig. 12. Scenario 3

Several performance indexes can be evaluated on these scenarios, in order to evaluate the system load, effective QoS or customer satisfaction (in terms of number of successful calls), effects of a certain configuration on the client behavior, effects of call management policies on the overall performances. Relevant measures that can be obtained are: average number of busy operators, average number of waiting calls, number of refused calls, number of hang-ups, number of hang-ups while servicing, average waiting time, average service time (see [6]).

3 Methodological Issues

The compositional modeling methodology we apply in this paper to the call center scenarios described in the previous Section is aimed at aiding the user in the task of building complex SWN models. The goal is to provide predefined SWN submodels and proper SWN connectors to compose them. The submodels must represent the components of the system to be analyzed, in this case the components of the call center system. On the contrary, the connection operators do not depend on the specific application domain. They model general patterns of communication and can be used to realize networks of submodels according to different topologies, for multiple classes of applications. The operatord introduced this paper are general enough to be used in other client-server contexts.

From the user point of view both submodels and connection operators are "black boxes" that define input and output ports. The interface of a submodel (of a connection operator) is the set of its input and output ports. An input port may only be connected to an output port, and some information (parameters) must be specified when solving the model (see Section 5). Indeed, an interface can be understood as it was a function prototype: a list of formal parameters is associated to each interface that declares types and variable names. These formal parameters represent data that must be used in the communication or to provide a required service.

In the following we say *high level model* the model built by the user and consisting of a network of black boxes, and *low level model* the SWN "implementation" of each submodel or connection operator. The SWN implementations of submodels and operators differ since the SWN net of an operator must be generated on the fly: its net structure depends on the actual numbers of submodels that it connects in the system configuration. Indeed, once the end-user has built an high level model, the final SWN model is **automatically** built by performing a complex sequence of tasks. We call this procedure the "solution process" of the high level model and it is performed by means of proper tools (post-processing phases), as explained in Sect.5. At the low level, the input and output ports are places and/or transitions and the formal parameters are color classes, variable tuples, guards. Their values are automatically evaluated during the solution process [7].

This approach to system modeling is partially supported by the OsMoSys modeling framework [16] and the DrawNET tool [8, 9] that have been developed to implement and study multi-formalism and compositional approaches for the analysis of complex systems.

In OsMoSys family of models (model classes) can be defined and instantiated to obtain model objects. Model objects are used as building blocks to obtain composed models by means of proper interface elements and composition operators. The interface elements correspond to the input and output ports and the composition operators correspond to the connection operators. The instantiation of an object model may require that the user specify some information (e.g. a service rate, the number of clients in the pool, etc) needed to fully define the object.

The DrawNET tool is a configurable GUI for graph-based formalisms, to create and edit models. The basis for the tool configurability is a meta-formalism used to define the elements of any formalism a user may want to use. DrawNET allows to implement all the features of the OsMoSys modeling methodology and provides both a graphical representation and an XML (eXtended Markup Language) based description of the model. In this paper we use DrawNET to built both the high level model and the SWN implementations of the call center submodels.

Fig. 13 shows the high level model of the contact center Scenario 2 described in Sect. 2. The *clientPool* node is an instance of the *Client* model class whose SWN implementation, shown in Fig. 14, is provided in the OsMoSys library of submodels. In the OsMoSys terminology, the *clientPool* node is a model object. The same holds for the nodes *siteA* and *siteB* that are two different instances of the *Site* model class (in Fig. 15). The three model objects are connected by means of an instance of the Session Route operator (the *memoryRouter* node in the figure), also available in the OsMoSys library of operators. The Session Route operator is in charge of modeling the C2SP protocol and the communication session (see Section 4). The arcs between the nodes are used to link the interface elements that correspond in the model composition.

4 Modeling the Contact Center Scenarios

In this Section we describe the model objects used to build the contact center scenarios and introduce the connection operators defined to compose the client and the site model objects. The final SWN models are automatically generated by the OsMoSys framework according to the input format accepted by GreatSPN [5] which is the tool used for computing the performance indices. For sake of space

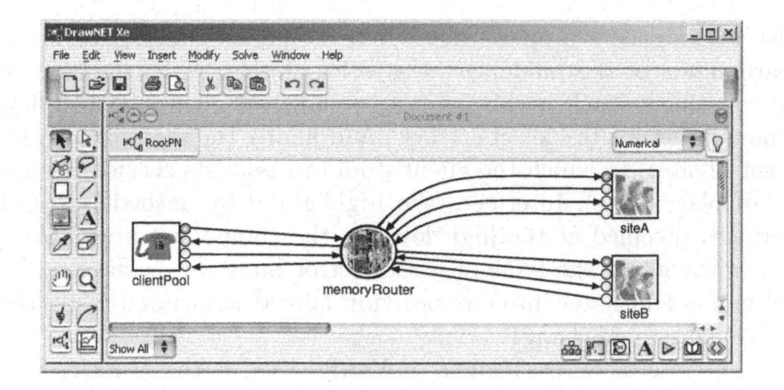

Fig. 13. The model of a contact center scenario in OsMoSys

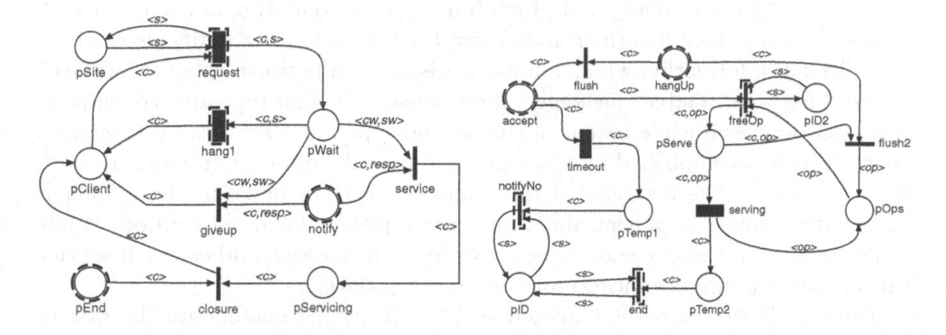

Fig. 14. The "Client pool" model class **Fig. 15.** The "Simple site" model class

the SWN formalism [4] is not described here; however, models are presented in an intuitive style without going into the technical details of the formalism.

4.1 Modeling the Client Pool and Site Components

In this section the low level models of the three components are described, namely the SWN models of the client pool, simple site and enterprise site. Each component includes a behavior (represented by the SWN net structure), a number of parameters (to be set upon instantiation) and a set of input/output interfaces: the interface information of the low level models is the only visible part at the high level model.

The client pool model is depicted in Fig. 14: the figure is a screenshot of the SWN model within the DrawNET GUI: some details are lacking, like the guards associated with transitions, but we shall highlight in the textual description the important information needed in the context of the complete model composition. The bold transitions in the model are timed, while the others are immediate: timed transitions (`request`, `hang1`) have an associated rate, which is parametric and must be assigned a proper value when the model class is instantiated. The other parameters of the model are the color classes, which include three sets of client identifiers (each corresponding to a set of clients with different QoS requirements) and a site identifier class, and finally the identifier of the specific site (or set of sites) to which the client pool can issue its service request (initial marking of place pSite). Interfaces are highlighted by dashed boxes. The output interfaces specified at the high level in the component are *outRequest* and *outHang*, representing the issue of a request or hang up message, which in the low level model translates into composition labels[1] associated respectively with transitions request and hangl.

The input interfaces are instead *inNotify* (used both to notify the start of a service or the impossibility of issuing the service) and *inEndServ* (used to notify the end of service): in the low level model these are connection labels associated with places notify and pEnd. The tokens circulating in the net are all "colored" with a client identifier; moreover, the tokens in place notify also include an additional information which allows to distinguish between "start of service" and "service not available" messages: the tokens of the first type are "consumed" by transition giveup while tokens of the second type are consumed by transition service (this is accomplished by associating a "guard", omitted in figure, to both the transitions). The behavior of the client is easily readable in the model: it starts with a token in pClient and waits (state pWait) for a notification. While in this condition it may decide to issue an hang up message; otherwise if service starts it waits (state pServicing) for the end of service.

The simple site model is depicted in Fig. 15: its parameters are the rate of transition serving, representing the service rate of operators at the site (which may depend on the type of operator and the type of client), and the rate of transition timeout (defining how long a client request should wait before sending a "service not available" notification). The client, site and operator identifier classes are also parametric, as well as the unique identifier associated with the specific site represented by the model (this model could be easily extended to represent a set of homogeneously behaving sites). The input interfaces of this model are *inAccept* and *inHangup*, which are the composition labels of places accept and hangUp respectively. The output interfaces are *outEnd* and *outNotify*, the first is a composition label associated with transition end, the second is associated with both transitions freeOp and notifyNo. Tokens circulating in the net carry the identifier of the clients being served, and an operator identifier is used for accounting of resources use. Interface transitions freeOp and notifyNo have an associated "guard" (omitted) indicating the content of the notification message ("start of service" and "service not available" respectively) they must originate. The behavior of the site upon issuing a request can be easily read on the net structure: notice that when an hang up message is received from the client, its identifier must be removed by the site net: this is done through the immediate transitions flush and flush2.

[1] Composition labels are used in the automatic construction of the complete low level model, to "glue" the components and connection operator SWN models.

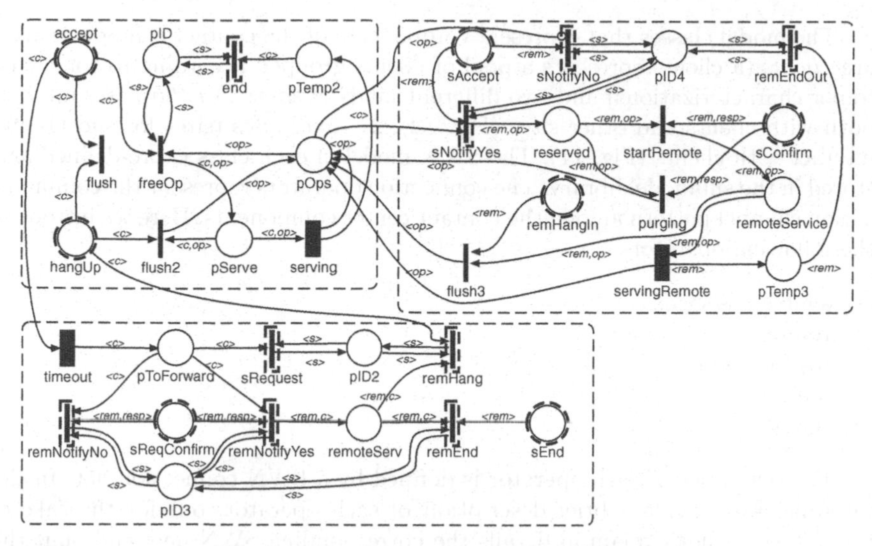

Fig. 16. The "Enterprise site" model class

Finally the more complex enterprise site model is depicted in Fig. 16. In the figure the submodel corresponding to the behavior of the simple site, still possible in the enterprise site, is highlighted (upper left corner), the only difference corresponds to the behavior after the firing of the timeout transition which activates the lower left part of the net, that models the attempt to forward the call to another site, instead of just sending back a "service not available" message to the client. Vice versa, the additional subnet in the right part of the net represents the reaction of the site to an incoming forwarding request by another analogous site. The only additional parameter in the enterprise site model is the service rate for remote clients (which might or might not be the same for local clients); the additional interfaces correspond to the messages of the S2S protocol: the output ports *outSrequest* (labeling transition sRequest), *outSnotify* (labeling transitions sNotifyNo and sNotifyYes), *outShangup* (labeling transition remHang), and *outSend* (labeling transition remEndOut); the input ports *inSconfirm* (labeling place sConfirm), *inShangup* (labeling place remHangIn and finally *inSend* (labeling place sEnd). Some new transitions in the model are labeled with input and output ports name that already existed in the simple site model: for example transition remEnd is labeled *outEnd*, while transitions remNotifyNo and remNotifyYes are labeled *outNotify* (and have an associated guard to send back the correct notification message). Indeed these transitions are used to forward the reply messages coming from the site that is actually serving its request to a client whose call has been redirected.

4.2 Interfaces and Connection Operators

In Sect. 3 we explained that model objects and connection operator nodes are composed by means of proper interface elements.

The model classes that represent components of the contact center scenarios are *clients* (a client represents a pool of clients grouped according to some customer characterizations) and two different kinds of sites: 1) *enterprises* interact both with clients and other sites (Fig. 12), 2) *small sites* (sites for short) only interact with clients (Fig. 11). The SWN model of each class is pre-defined and stored in the submodel library. The connection operators represent the communication channel pattern among the contact center components. Here we introduce the following operators:

- simple message;
- route;
- session route;
- anycast;
- reroute.

The semantics of each operator is defined by a SWN connection net. In the following we provide a brief description of each operator, but for the sake of brevity we do not explain in details the corresponding SWN nets and omit the formal parameter lists associated to their interface elements.

Simple message and route connection operators. The simple message is represented by an arc in the high level model and its corresponding low level SWN net is shown in Fig. 17, in which the interface elements of the two model objects Sender and Receiver (the transition labeled *portOut* and the place labeled *portIn*, respectively) are connected by means of the simple message operator.

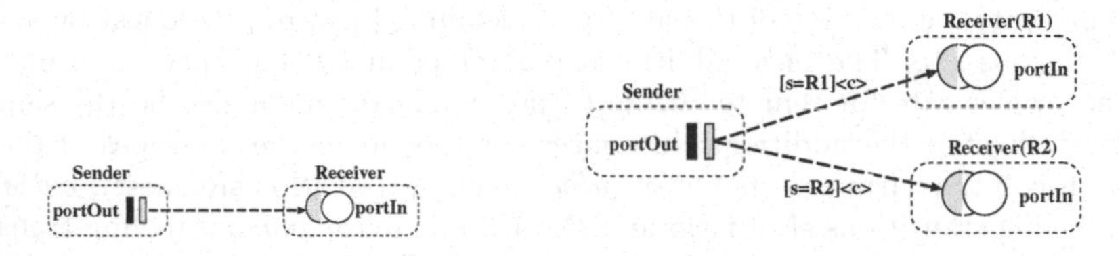

Fig. 17. The simple message connection operator

Fig. 18. The route connection operator

The simple message operator, in turn, consists of an arc connecting a transition with a place. The transition and the place (shadowed in the figure) are the interface elements of this operator. The end-user is not aware of the net implementations of these three blocks. He/she only knows that Sender has an output interface and Receiver has an input interface and uses them to connect the blocks by the message sending operator. When the low level models are composed during the solution process, the interface elements (in this case the two transitions and the two places) are superposed according to their labels.

The message sending paradigm adopted in developing the contact center models allows us to model communications at a lower level, with respect to the client-server paradigm used in [7]. The immediate advantage we obtained in this way is the possibility of capturing the characteristics of the model with a thinner grain, e.g. it is possible to avoid leaving a client blocked in a wait state during the service, with respect to the communication, and is easier to model communication faults situations.

The low level SWN net of the route connection operator is shown in Fig. 18. The interface elements are the transitions whose label name is *sendOut* and the places whose label name is *acceptIn*. During the composition phase the two transitions and the couple of places are superposed. When the receiver R1 (R2) is selected as a target for a request by the sender (by binding variable s to R1 (R2)), the arc between the transition *sendOut* and the place *acceptIn* of R2 (R1) vanishes, and a message is sent only to receiver R1 (R2).

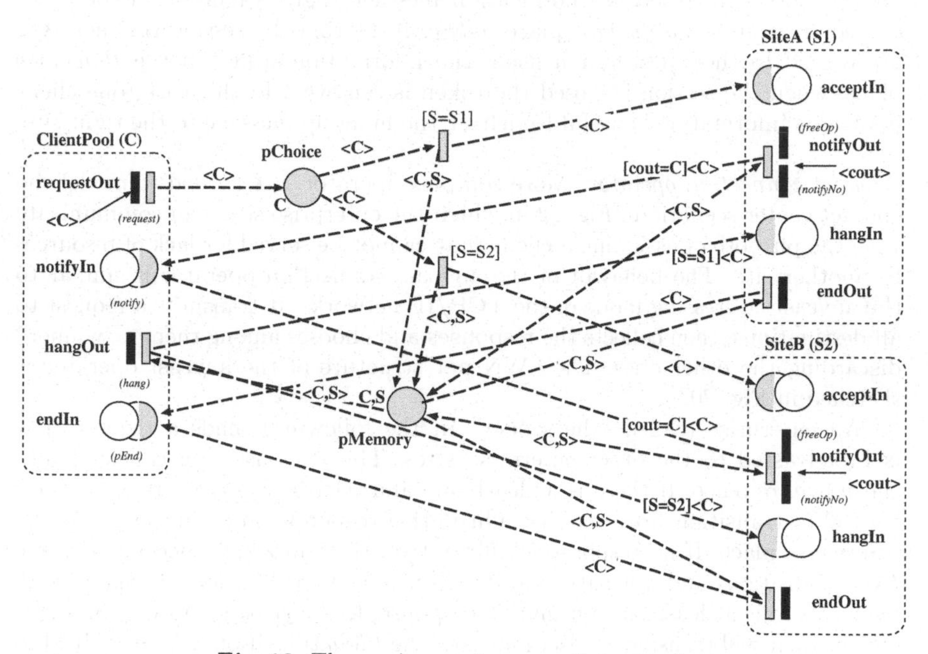

Fig. 19. The session route connection operator

Session route connection operator. In Fig. 19 the structure of the connection net corresponding to the session route operator is sketched, in the case of one client and two sites. It is used to implement the example in Fig. 13 and consists of 8 places and 7 transitions (shadowed in the figure). On the left side, the figure depicts the composition of the connection net with the client in Fig. 14. The transition labels are in bold, the names in brackets are the names of places and transitions in the low level model of the client (site) and have been added to make it clear which elements of the net correspond to the class model interface. In the composition phase the *RequestOut* transition and the respective grey transition of the operator are merged, the same for the two transitions *HangOut*. Analogously, the place *notifyIn* and the place *endIn* of the client are merged with the respective grey places of the operator interface. The same occurs on the sites side (on the right).

The session route operator is used to route messages for connection oriented communication sessions (for which it is required keeping the association between the identities of the participants along all the session). A session is opened when a request arrives from a client (transition *requestOut*). The destination site is randomly chosen between SiteA and SiteB (place *Pchoice*), but once the site is selected (if siteA (siteB) is selected variable s is assigned the value S1 (S2)), the couple client-site is determined for that session and this information is stored by a token in place *Pmemory*. According to the C2SP protocol described in Sect. 2, the client receives a notification message that its request has been accepted (transition *notifyOut* - place *notifyIn*). During the session the messages are routed to their destination using the information stored in the *Pmemory* place, when the session is closed the token is removed. In this case (one client pool) this information is useful to deliver the hung up message to the right site.

Anycast connection operator. More complex operators are needed to model the contact center scenario in Fig. 12 in which the enterprise sites can communicate with the purpose of assigning a client that cannot be served for lack of resources to another site. The behavior of the anycast connection operator is similar to the anycast addressing mode of the TCP/IP networks. It transmits a request to all destinations, then collects the responses and chooses among them a "winner" discarding the other ones. The SWN net structure of the anycast operator is sketched in Fig. 20.

An enterprise (called "the sender" in the following) sends a request that is broadcasted to the other enterprise sites. The responses are collected into the *pAcquire* place. If there is at least one affirmative response ([resp = true]) the *tOK* transition fires and the affirmative responses are transferred to the *pResponse* place. If no response is affirmative, they all are flushed out (by the *tNO* transition) and a negative notification is sent to the client. In the case in which there is at least one affirmative response, the anycast operator chooses to which site it will transfer the user request. The *tWin* transition will be enabled to fire only once due to the inhibitor arc from *pWinner* which will only contain the identity of the winner server. The *tConfirmNo* transition is enabled by *pWinner* to completely flush out other affirmative responses in the *pResponse* place and to provide "purge" messages to the enterprise sites that sent them. When all "purge" messages have been sent, the winner is notified by firing the *tConfirmYes* transition; concurrently the winner identity is notified to the sender. Notice that possible conflicts among transitions are solved by the priority levels $\pi 1$ and $\pi 2$ ($\pi 2 > \pi 1$) in order to avoid uncorrect negative answers propagation to the sender.

Reroute connection operator. The reroute connection operator is used to implement the S2SP protocol and it is based on the anycast and route operators, as

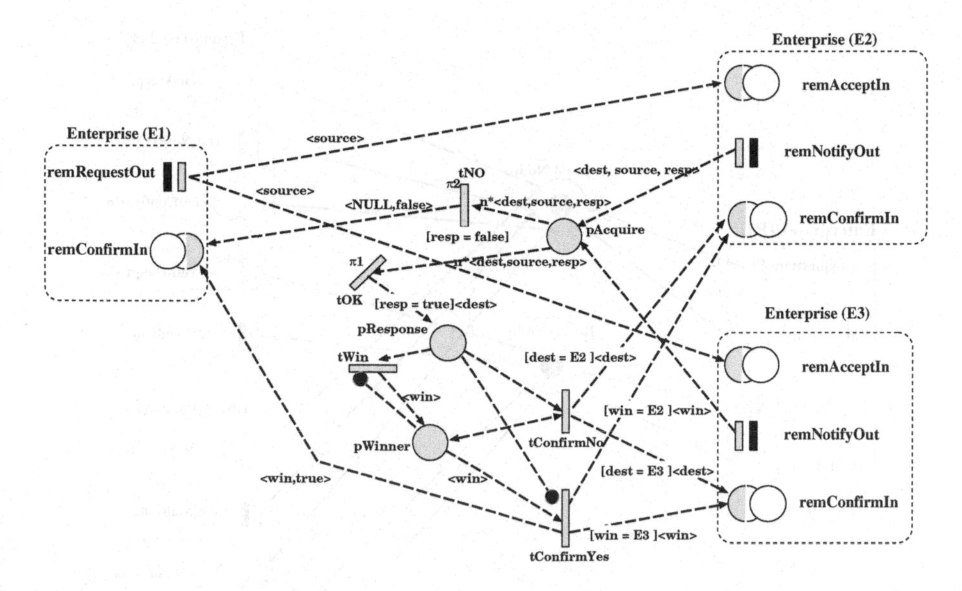

Fig. 20. The anycast connection operator

shown in Fig. 21. Indeed, the reroute operator could be interpreted as a strong aggregation of the anycast and route operators intended as an high level representation of the corresponding SWN model (class). *Strong aggregation* is used in OsMoSys to encapsulate model classes into new ones so creating new sub-models, whose components are not visible to the external environment. The interface of an aggregation may be the union of the interface elements of its sub-components, as for the reroute operator, but it is also possible to hide some of them (for example because they have been used to connect the components). The reroute operator is used to establish a communication session between two enterprises i and j in order to export a client from i to j. To this aim the anycast is used to assign the client to another site (if possible), then the route operator is used to route the messages needed to export the client.

Composition of scenarios The composition of scenarios is easy thanks to the simple interface of client and sites model classes and the very expressive connectors. Actually it is questionable whether the connectors should not be defined as model classes themselves: the only problem is due to the fact that the actual underlying low level model may change depending on the number of client pools/sites they link, this would require the possibility of defining model classes that are parametric also in their internal structure, a feature currently not allowed in OsMoSys.

Compared to the models and composition schemes proposed in [6] the client pool and site interface is simpler due to the separation of the interaction protocol (which has been moved into the connectors) from the component behavior. On

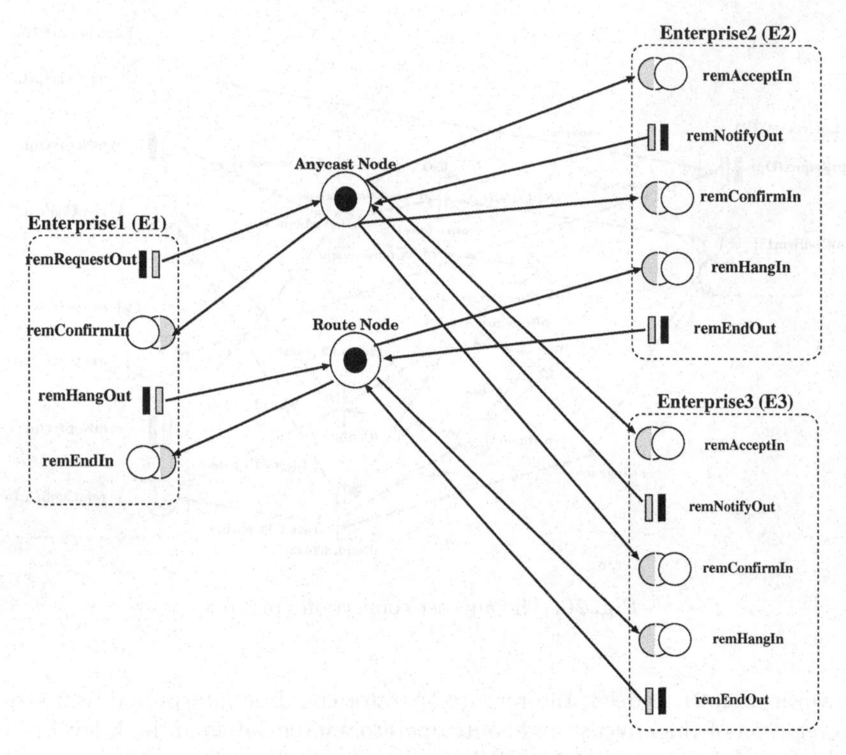

Fig. 21. The reroute connection operator

the other hand the components presented in this paper are not designed to represent several client pools (requesting service to different sites) or several sites folded together in a unique model: this can be done in a rather simple way as far as components are concerned, but poses some subtle problem in the way colors are propagated by connectors that we decided not to discuss in this paper for simplicity of presentation. Moreover from the high level modeler point of view it is simpler to see each black box as a separate object in his scenario, rather than as several objects.

5 The Solution Process

The solving process of contact center scenarios is performed in three main steps pursued by proper software tools of the OsMoSys framework called post-processors. The post-processing steps are: 1) High level model analysis; 2) Components instantiation; 3) Final model generation.

The input of the first phase is the high level model and its output is a graph representing the connections between the nodes (model objects and operators). These connections are semantically checked for preserve coherence and consistency. The second step is in charge of instantiating low level object models from classes by using the instantiation values provided by the end-user. The output is the low level models graph in which the model objects have been associated to their low level models. The core of the solution process is the last post-processing step. In this step the following tasks are performed:

- the SWN nets of the connection operators are generated;
- all the SWN components of the final model (expressed by means of the DrawNET XML format) are automatically translated into GreatSPN files;
- a composition script is created. For SWN this script contains the proper set of Algebra tool commands [2] in order to merge the SWN nets and obtain the final SWN global model.

The complete solution process is depicted in Fig. 22. For sake of brevity we do not provide a detailed description of the entire solution process, but we describe the automatic generation of the SWN nets of the connection operators that have been introduced in this paper.

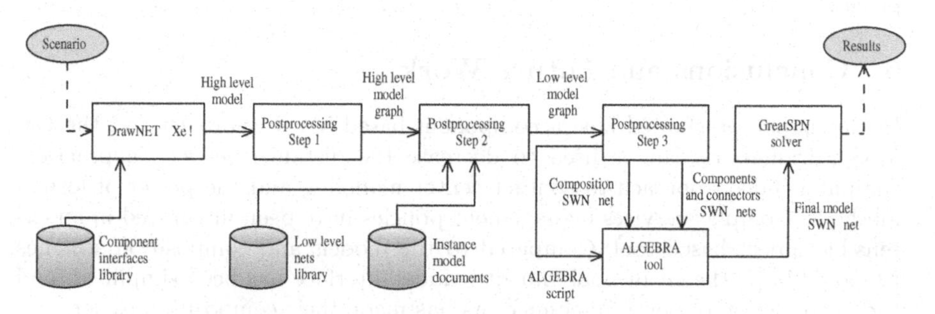

Fig. 22. Solution process of contact center models

The SWN net of a connection operator is created in three phases:

- internal structure generation;
- interface elements generation;
- net coloring.

The internal SWN net of a connection operator is customized for each operator (it describes its behavior) and its generation is performed by the post-processor according to a static pre-defined description.

On the contrary, the interface is dynamically defined according to the low level models graph and the proper set of interface transitions and places must be generated. Each SWN interface element of a connection operator must have a label identical to the label associated to the SWN interface element that implements the interface of the low level model to be composed with the operator. These labels are later used by the composition tool (Algebra for the SWN) in order to merge transitions and places. Then the right arcs must be added to connect the new transitions and places to the internal SWN net.

The last step colors the whole net starting from the interface transitions to the interface places through the internal elements of the net. This process is made with information taken by: the connected output interface, the internal structure of the operator and the SWN low level models of the objects.

We use the anycast operator in Fig. 20 as example. In the first step the internal structure of the net is incrementally built up; so the *tNo*, *tOk*, *tConfirmNo*, *tConfirmYes*, *tWin*, *tSourceConfirm* transitions and the *pAcquire*, *pResponse*, *pWinner* places are added to the compositional net independently from the involved objects interfaces; then arcs among these elements are generated. In the second step, the interface transitions (places) of the low level model are created and labeled with the value taken from the high level model interfaces linked by the anycast node; these labels are also propagated during the generation of the low level topological graph. In the last step we start coloring the net; consider for example the transition labeled *remRequestOut* on the client side. The list of the formal parameters associated to the *remRequestOut* transition has also been propagated from the high level interface description. In this (simple) case a list of variables are written as functions on the arcs towards the two *remAcceptIn* places.

6 Conclusions and Future Work

In this paper we showed how a component-based SWN approach and the OsMoSys framework can be used to decrease the distance between a practical computer aided approach to contact center modeling and the power of formal methods. Complex services management policies have been integrated in an extensible object based tool. Compared to the models and composition schemes proposed in [6] the client pool and site model interface has been simplified, and the introduction of powerful connectors has made their composition easier.

Our work on methodology will evolve toward a better comprehension of modeling power and opportunities offered by connectors. We want investigate if they are to be intended in an abstract way or they should model something that is strictly related to the application field, moreover we want to study and define the rules needed to perform the automatic net coloring of the connector operators that link folded models objects representing different instances of the same classes. We also will explore the chances of introducing composition of connectors in our methodology as another means to further exploit modularity. Another interesting methodological extension concerns the possibility of relating model classes through a behavior inheritance concept (e.g. the enterprise site model class inherits and extends the behavior of the simple site). Consequently the application modeling framework will be enriched by introducing several advanced aspects from the real contact center world.

Acknowledgements

The work of G.Franceschinis and M.Gribaudo has been partially funded by the MIUR FIRB project "Perf: RBNE019N8N"; the work of M. Iacono has been partially funded by Centro Regionale Di Competenza per l'ICT of Regione Campania local government.

References

1. P. Ballarini, S. Donatelli, and G. Franceschinis. Parametric stochastic well-formed nets and compositional modelling. In *Proc. 21st International Conference on Application and Theory of Petri Nets*, Aarhus, Denmark, June 2000.
2. S. Bernardi, S. Donatelli, and A. Horváth. Compositionality in the greatspn tool and its use to the modelling of industrial applications. *Software Tools for Technology Transfer*, 2001.
3. E. Best, H. Flrishhacl ANF W. Fraczak, R. Hopkins, H. Klaudel, and E. Pelz. A class of composable high level Petri nets with an application to the semantics of $B(PN)^2$. In *Proc. of the 16th international conference on Application and Theory of Petri nets 1995*, Torino, Italy, 1995. Springer Verlag. Volume LNCS 935.
4. G. Chiola, C. Dutheillet, G. Franceschinis, , and S. Haddad. Stochastic well-formed coloured nets for symmetric modelling applications. *IEEE Transactions on Computers*, 42(11):1343–1360, November 1993.
5. G. Chiola, G. Franceschinis, R. Gaeta, and M. Ribaudo. GreatSPN 1.7: Graphical Editor and Analyzer for Timed and Stochastic Petri Nets. *Performance Evaluation, special issue on Performance Modeling Tools*, 24(1&2):47–68, November 1995.
6. G. Franceschinis, C. Bertoncello, G. Bruno, G. Lungo Vaschetti, and A. Pigozzi. SWN models of a contact center: a case study. In *Proc. of the 9th Int. Workshop on Petri Nets and Performance Models*, Aachen, Germany, September 2001. IEEE C.S. Press.
7. G. Franceschinis, S. Marrone, N. Mazzocca and V. Vittorini. SWN Client-server composition operators in the OsMoSys framework. In *Proc. of the 10th Int. Workshop on Petri Nets and Performance Models*, Urbana, Illinois, USA, September 2003.
8. G. Franceschinis, M. Gribaudo, M. Iacono, V. Vittorini, C. Bertoncello. DrawNet++: a flexible framework for building dependability models. In *Proc. of the Int. Conf. on Dependable Systems and Networks*, Washington DC, USA, June, 2002.
9. M. Gribaudo, M. Iacono, N. Mazzocca, V. Vittorini. The OsMoSys/DrawNET Xe! Languages System: A Novel Infrastructure for Multi-Formalism Object-Oriented Modelling. In *Proc. of the 15th European Simulation Symposium and Exhibition*, Delft, The Netherlands, October 2003.
10. K. Jensen. *Coloured Petri Nets, Basic Concepts, Analysis Methods and Practical Use. Volume 1.* Springer Verlag, 1992.
11. I. Rojas. *Compositional construction and analysis of Petri net systems.* PhD thesis, University of Edinburgh, 1997.
12. C. Sibertin-Blanc. A client-server protocol for the composition of petri nets. In *Proc. of the 14th Int. Conf. on Application and Theory of Petri Nets*, Chicago, IL, USA, June 1993. LNCS 691, Springer Verlag.
13. C. Sibertin-Blanc. Comunicative and cooperative nets. In *Proc. of the 15th Int. Conf. on Application and Theory of Petri Nets*, Zaragoza, Spain, June 1994. LNCS 815, Springer Verlag.
14. C. Sibertin-Blanc. CoOperative Objects: Principles, use and implementation. In G. Agha, F. De Cindio, and Grzegorz Rozenberg, editors, *Concurrent Object-Oriented Programming and Petri Nets, Advances in Petri Nets*, volume 2001 of *Lecture Notes in Computer Science*. Springer Verlag, 2001.

15. V. Vittorini, G. Franceschinis, M. Gribaudo, M. Iacono, and N. Mazzocca. Drawnet++: Model objects to support performance analysis and simulation of complex systems. In *Proc. of the 12th Int. Conference on Modelling Tools and Techniques for Computer and Communication System Performance Evaluation (TOOLS 2002)*, London, UK, April 2002.
16. V. Vittorini, M. Iacono, N. Mazzocca, and G. Franceschinis. OsMoSys: a new approach to multi-formalism modeling of systems. *Journal of Software and System Modeling*, vol. 3(1):68–81, March 2004.

Generalised Soundness of Workflow Nets Is Decidable*

Kees van Hee, Natalia Sidorova, and Marc Voorhoeve

Department of Mathematics and Computer Science
Eindhoven University of Technology
P.O. Box 513, 5600 MB Eindhoven, The Netherlands
{k.m.v.hee,n.sidorova,m.voorhoeve}@tue.nl

Abstract. We investigate the decidability of the problem of generalised soundness for Workflow nets: "Every marking reachable from an initial marking with k tokens on the initial place terminates properly, i.e. it can reach a marking with k tokens on the final place, for an arbitrary natural number k". We start with considering simple correctness criteria for Workflow nets and reduce them to the check of structural properties formulated in terms of traps and siphons, which can be easily checked. We call the nets that possess those properties Batch Workflow nets (BWF-nets). We show that every sound WF-net can be transformed to a BWF-net with the same behaviour. Then we use algebraic methods to prove that generalized soundness is decidable for BWF-nets and give a decision procedure.

Keywords: Petri nets; workflows; verification; soundness, decidability.

1 Introduction

Petri nets are widely used for the modelling and verification of workflows. In [1], the class of Workflow (Petri) nets (WF-nets) was defined. A Petri net is a WF-net iff it satisfies certain structural properties, namely it possesses one source place (initial place) and one sink place (final place) and all other nodes lie on paths from the source to the sink place. The main correctness criterion introduced there was *soundness.* A WF-net is sound iff (1) from any marking reachable from the initial marking it is possible to reach the final marking, (2) whenever the final place becomes marked from the initial marking, the final marking is reached, (3) no transitions are dead in the initial marking. (The initial (final) marking of WF-nets consist of a single token in the initial (final) place.) Soundness for WF-nets is decidable, and the decision procedure has been implemented e.g. in the WOFLAN tool [8].

In [4] we showed that the notion of soundness from [1] is *not compositional*, and moreover, it does not allow for handling of multiple cases in the WF-net. We introduced there a generalized soundness notion that amounts to *proper*

* This research has been supported by the MoveBP NWO project (project number 612.000.315).

J. Cortadella and W. Reisig (Eds.): ICATPN 2004, LNCS 3099, pp. 197–215, 2004.
© Springer-Verlag Berlin Heidelberg 2004

termination of all markings obtained from markings with multiple tokens on the initial place, which corresponds to the processing of batches of cases in the WF-net. With proper termination for marking m obtained from a marking with k tokens on the initial place, we mean that there exists a firing sequence leading from m to the marking with k tokens on the final place. We proved that generalised soundness is *compositional*. The original soundness notion from [1] corresponds to 1-soundness in our case. Deciding generalised soundness is harder than deciding 1-soundness, since the straightforward approach involves an infinite number of checks of proper termination. We did not solve the problem of soundness in [4] but defined a class of nets (ST-nets) that are sound by construction. In this paper we prove that the problem of (generalised) soundness is *decidable* for arbitrary WF-nets and describe a decision procedure for it.

We start with considering simple *behavioural* correctness criteria for WF-nets: *non-redundancy* and *non-persistency*. Non-redundancy means that every place can be marked and every transition can fire, provided that the initial place contains enough tokens, while non-persistency means that all places (except for the final one) can become empty again, lest some garbage would be left after the processing of the case is finished. We show that the WF-nets meet non-redundancy and non-persistency requirements iff they satisfy a simple *structural* characterisation: all proper siphons of these nets contain the initial place and all proper traps contain the final place. We call this class of WF-nets *Batch Workflow nets (BWF-nets)*. We show that every WF-net is either not sound (in case it contains persistent places) or it can be transformed to a BWF-net with the same behaviour (by removing redundant places and transitions).

In the second half of the paper we consider the problem of (generalised) soundness for BWF-nets and prove that this problem is *decidable*. The decidability proof is based on two ideas. First, we extend the set $\mathcal{R}$ of markings reachable from the initial markings upto a set $\mathcal{G}$ that has a 'regular' structure and show that the notion of soundness can be equally defined by requiring proper termination of all markings of this extended set $\mathcal{G}$. And second, we use the regularity of $\mathcal{G}$ to show that it is enough to check proper termination for the markings of a finite subset Γ of $\mathcal{G}$ to prove the proper termination of all markings from $\mathcal{G}$. Thus, the infinite set of markings in the initial formulation of the problem of generalised soundness can be reduced to a finite set due to the clean algebraic model of Petri nets.

The rest of the paper is organised as follows. In Section 2, we sketch the basic definitions related to Petri nets and Workflow nets. In Section 3 we introduce the notion of Batch Workflow Nets. In Section 4 we prove that the problem of generalised soundness is decidable. In Section 5 we illustrate the decision procedure for soundness on a concrete example. We conclude in Section 6 with discussion of the obtained results and directions for future work.

2 Preliminaries

$\mathbb{N}$ denotes the set of natural numbers, $\mathbb{Z}$ the set of integers and $\mathbb{Q}$ the set of rational numbers. $\mathbb{Q}^+$ stands for the set of non-negative rational numbers.

Let P be a set. A *bag (multiset)* m over P is a mapping $m : P \rightarrow \mathbb{N}$. The set of all bags over P is $\mathbb{N}^P$. We use $+$ and $-$ for the sum and the difference of two bags and $=, <, >, \leq, \geq$ for comparisons of bags, which are defined in a standard way. We overload the set notation, writing $\emptyset$ for the empty bag and $\in$ for the element inclusion. We write $m = 2[p] + [q]$ for a bag m with $m(p) = 2$, $m(q) = 1$, and $m(x) = 0$ for all $x \notin \{p, q\}$. For the sum over the elements of a bag m we write $\sum_{p \in m} f(p)$ (assuming that every p appears in the sum $m(p)$ times) rather than $\sum_{p \in m} m(p) \cdot f(p)$.

Transition Systems. A *transition system* is a tuple $E = \langle S, Act, T \rangle$ where S is a set of *states*, Act is a finite set of *action names* and $T \subseteq S \times Act \times S$ is a *transition relation*. A *process* is a pair $\langle E, s_0 \rangle$ where E is a transition system and $s_0 \in S$ an initial state. We denote (s_1, a, s_2) from T as $s_1 \xrightarrow{a} s_2$, and we say that a leads from s_1 to s_2. For a sequence of transitions $\sigma = \langle t_1, \ldots, t_n \rangle$ we write $s_1 \xrightarrow{\sigma} s_2$ when $s_1 = s^0 \xrightarrow{t_1} s^1 \xrightarrow{t_2} \ldots \xrightarrow{t_n} s^n = s_2$, and $s_1 \xrightarrow{\sigma}$ when $s_1 \xrightarrow{\sigma} s_2$ for some s_2. In this case we say that σ is a trace of $\langle E, s_0 \rangle$. Finally, $s_1 \xrightarrow{*} s_2$ means that there exists a sequence of transitions $\sigma \in T^*$ such that $s_1 \xrightarrow{\sigma} s_2$.

Petri Nets. A *Petri net* is a tuple $N = \langle P, T, F^+, F^- \rangle$, where:

- P and T are two disjoint non-empty finite sets of *places* and *transitions* respectively, the set $P \cup T$ are the *nodes* of N;
- F^+ and F^- are mappings $(P \times T) \rightarrow \mathbb{N}$ that are *flow functions* from transitions to places and from places to transitions respectively.

$F = F^+ - F^-$ is the *incidence matrix* of net N.
We present nets with the usual graphical notation.

Markings are states (configurations) of a net. Depending on the context, we interpret a *marking* m of N either as a bag over P (in Section 3) or as a vector from $P \rightarrow \mathbb{N}$ (in Sections 4 and 5)).

Given a transition $t \in T$, the *preset* $^\bullet t$ and the *postset* $t^\bullet$ of t are the *bags* of places where every $p \in P$ occurs in $^\bullet t$ $F^-(p, t)$ times and in $t^\bullet$ $F^+(p, t)$ times. Analogously we write $^\bullet p, p^\bullet$ for pre- and postsets of places. We overload this notation further and apply preset and postset operations to a set B of places: $^\bullet B = \{t \mid \exists p \in B : t \in {}^\bullet p\}$ and $B^\bullet = \{t \mid \exists p \in B : t \in p^\bullet\}$. Note that $^\bullet B$ and $B^\bullet$ are not bags but sets. We will say that node n is a *source* node iff $^\bullet n = \emptyset$ and n is a *sink* node iff $n^\bullet = \emptyset$. A *path* of a net is a sequence $\langle x_0, \ldots, x_n \rangle$ of nodes such that $\forall\, i : 1 \leq i \leq n : x_{i-1} \in {}^\bullet x_i$.

A transition $t \in T$ is *enabled* in marking m iff $^\bullet t \leq m$. An enabled transition t may fire. This results in a new marking m' defined by $m' \stackrel{\text{def}}{=} m - {}^\bullet t + t^\bullet$. We interpret a Petri net N as a transition system/process where markings play the role of states, firings of the enabled transitions define the transition relation and the initial marking corresponds to the initial state. The notion of reachability for Petri nets is inherited from the transition systems. For a firing sequence σ in a net N, we define $^\bullet \sigma$ and $\sigma^\bullet$ respectively as $\sum_{t \in \sigma} {}^\bullet t$ and $\sum_{t \in \sigma} t^\bullet$, which are the sums of all tokens consumed/produced during the firings of σ. So $m \xrightarrow{\sigma} (m + \sigma^\bullet - {}^\bullet \sigma)$.

We denote the set of all markings reachable in net N from marking m as $\mathcal{R}(m)$. The set of markings from which marking m is reachable is denoted as $\mathcal{S}(m)$.

We will use the well-known *Marking Equation Lemma*:

Lemma 1 (Marking Equation). *Given a finite firing sequence σ of a net N: $m \xrightarrow{\sigma} m'$, the following equation holds: $m' = m + F^{+} \cdot \overrightarrow{\sigma} - F^{-} \cdot \overrightarrow{\sigma}$, or in other words, $m' = m + F \cdot \overrightarrow{\sigma}$.*

Note that the reverse is not true: not every marking m' that is representable as a sum $m + F \cdot v$ for some $v \in \mathbb{N}^T$ is reachable from the marking m.

Traps and Siphons. (see [2]) A set R of places is a *trap* if $R^{\bullet} \subseteq {}^{\bullet}R$. The trap is a *proper trap* iff it is not empty. A set R of places is a *siphon* if ${}^{\bullet}R \subseteq R^{\bullet}$. The siphon is a *proper siphon* iff it is not empty. Important properties of traps and siphons are that *marked traps remain marked* and *unmarked siphons remain unmarked* whatever transition firings would happen. As follows from the definition, traps and siphons are dual by their nature.

Place Invariants. (see [5]) A *place invariant* is a row vector $I : P \rightarrow \mathbb{Q}$ such that $I \cdot F = 0$. When talking about invariants, we consider markings as *vectors*. We will say that markings m_1 and m_2 *agree on a place invariant* I if $I \cdot m_1 = I \cdot m_2$ (see [3]).

Lemma 2. *Two markings m_1, m_2 agree on all place invariants iff the equation $m_1 + F \cdot x = m_2$ has some rational-valued solution for x.*

The main property of place invariants is thus that any two markings m_1, m_2 such that $m_1 \xrightarrow{*} m_2$ agree on all place invariants. The check whether the two markings m_1, m_2 agree on all place invariants can be done by a simple check whether $\mathcal{I} \cdot m_1 = \mathcal{I} \cdot m_2$, where $\mathcal{I}$ is a matrix that consists of all basis place invariants as rows.

Workflow Petri Nets. In this paper we primarily focus upon the *Workflow Petri nets (WF-nets)* [1]. As the name suggests, WF-nets are used to model the ordering of tasks in workflow processes. The initial and final nodes indicate respectively the initial and final states of processed cases.

Definition 3. *A Petri net N is a* Workflow net (WF-net) *iff:*

1. *N has two special places: i and f. The initial place i is a source place, i.e. ${}^{\bullet}i = \emptyset$, and the final place f is a sink place, i.e. $f^{\bullet} = \emptyset$.*
2. *For any node $n \in (P \cup T)$ there exists a path from i to n and a path from n to f. (We call this property the* path property *of WF-nets.)*

In this paper, we study the processing of *batches* of tasks in Workflow nets, meaning that the initial place of a Workflow net may contain an arbitrary number of tokens. Our goal is to provide correctness criteria for the design of these nets. One natural correctness requirement is *proper termination*, which is called *soundness* in the WF-net theory. We will use the generalised notion of soundness for WF-nets introduced in [4]:

Definition 4. *Let N be a WF-net. We say that marking $m \in \mathcal{R}(k[i])$* terminates properly *in N iff $m \xrightarrow{*} k[f]$.*
N is k-sound *for some $k \in \mathbb{N}$ iff all $m \in \mathcal{R}(k[i])$ terminate properly.*
N is sound *iff it is k-sound for all $k \in \mathbb{N}$.*

We will use terms *initial* and *final* markings for markings $k[i]$ and $k[f]$ respectively ($k \in \mathbb{N}$). We will write $\mathbf{0}$ for the vector representation of marking $\emptyset$, $\mathbf{i}$ for the vector representation of marking $[i]$ and $\mathbf{f}$ for the vector representation of $[f]$. For every marking m reachable from an initial marking $k[i]$ holds: $\mathcal{I}\cdot m = \mathcal{I}\cdot(k\cdot\mathbf{i})$, and if marking m terminates properly in N, then $\mathcal{I}\cdot m = \mathcal{I}\cdot(k\cdot\mathbf{f})$.

3 Batch Workflow Nets

We are interested in the correct and optimal design of WF-nets. Ideally, correctness requirements should be formulated as requirements on the structure of the net (thus, they can be easily checked) and they should guarantee the correctness of the net behaviour. In this section, we consider behavioural criteria of the correct design and reduce them to structural ones.

3.1 Structural Non-redundancy for Workflow Nets

Besides soundness, a logical requirement for the correct design of a WF-net is *non-redundancy,* namely: every transition of the net can potentially fire and every place of the net can potentially obtain tokens, provided that there are enough tokens in the initial place. WF-net N_1 in Fig. 1 does not satisfy this requirement because transition d can never fire and place s can never get tokens. So d and s are *redundant.* At the same time, it should be possible for every place (except for f) to become unmarked again – otherwise the net is guaranteed to be not sound, as e.g. net N_2 in Fig. 1 – place s can obtain tokens but it can never become unmarked after that, i.e. this place is *persistent.* In formal terms:

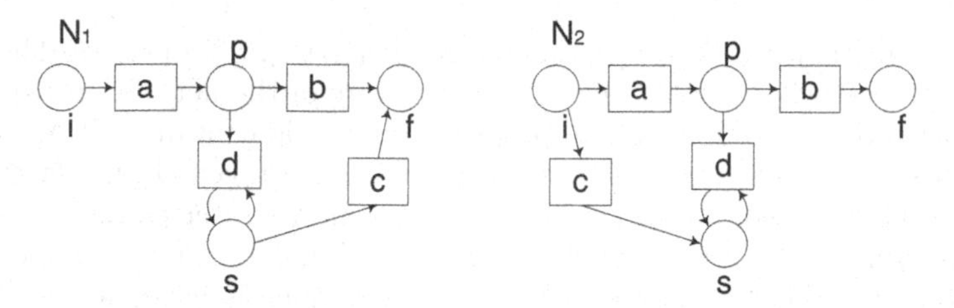

Fig. 1. Redundant and persistent places

Definition 5. *Let $N = \langle P, T, F^+, F^- \rangle$ be a WF-net.*
A place $p \in P$ is non-redundant *iff there exist $k \in \mathbb{N}$ and $m \in \mathbb{N}^P$ such that $k[i] \xrightarrow{*} m \wedge p \in m$.*

A place $p \in P$ is non-persistent *iff there exist $k \in \mathbb{N}$ and $m \in \mathbb{N}^P$ such that $p \in m \wedge m \xrightarrow{*} k[f]$.*
A transition t is non-redundant *iff there exist $k \in \mathbb{N}$ and $m \in \mathbb{N}^P$ such that $k[i] \xrightarrow{*} m \xrightarrow{t}$.*

The following lemma presents these desirable behavioural properties in more general terms:

Lemma 6. (1) *A WF-net N has no redundant places iff every marking is majorated by a marking reachable from some initial marking $k[i]$, i.e.*
$\forall\, m \in \mathbb{N}^P : \exists\, k \in \mathbb{N}, m' \in \mathcal{R}(k[i]) : m' \geq m.$
(2) *A WF-net N has no persistent places iff every marking is majorated by a marking from which some final marking $k[f]$ is reachable, i.e.*
$\forall\, m \in \mathbb{N}^P : \exists\, k \in \mathbb{N}, m' \in \mathcal{S}(k[f]) : m' \geq m.$

Proof. (1) If every marking can be majorated by a marking reachable from some $k[i]$, then every marking $[p]$, $p \in P$ can be majorated and p is non-redundant. In the opposite direction: for every p there exist k_p, m_p, such that $k_p[i] \xrightarrow{*} m_p$ where $m_p \geq [p]$. Then we can majorate a given marking m by a marking $m' = \sum_{p \in m} m_p$ reachable from $\left(\sum_{p \in m} k_p\right)[i]$.
(2) can be proved similarly. □

As an immediate consequence we obtain the following property:

Lemma 7. *A WF-net N has no redundant places iff it has no redundant transitions.*

Proof. Let N has no redundant places. Consider an arbitrary transition $t \in T$. By applying property (1) of Lemma 6 to $^\bullet t$ we obtain that t can get enabled, and hence it is non-redundant.

Now assume that N has no redundant transitions. Consider an arbitrary place $p \in P \setminus [i]$. Since N is a WF-net, $^\bullet p \neq \emptyset$, and since all transitions are non-redundant, transitions from $^\bullet p$ can fire and so p can get marked. Thus p is non-redundant. □

Non-redundancy and non-persistency are behavioural properties. They imply though the following restrictions on the structure of the net: all proper siphons of the net should contain i and all proper traps should contain f. If N contained a proper siphon without i, the transitions consuming tokens from places of that siphon would be dead, no matter how many tokens are inserted into i. Similarly, if N contained a trap without f, the net could not terminate properly. It is not surprising that the absence of traps and siphons is a necessary condition for the correctness of the design. What is more interesting is that the absence of such siphons and traps is a *sufficient* condition for the absence of redundant and persistent places respectively: if a net has a redundant place, there exists a proper siphon without i, and if a net has a persistent place, there exists a proper trap without f, i.e. these behavioural and structural characteristics are equivalent for WF-nets:

```
input : A Petri net N = (P, T, F+, F−) and S ⊆ P;
output: X ⊆ S;

X = S;
while  there exists p ∈ X and t ∈ •p such that t ∉ X•  do  X = X \ {p};
return(X);
```

Fig. 2. Algorithm for finding the maximal siphon in a set of places S

Theorem 8. *Let $N = \langle P, T, F^+, F^- \rangle$ be a WF-net. Then the following holds:*
(1) N has no redundant places iff $P \setminus \{i\}$ contains no proper siphon.
(2) N has no persistent places iff $P \setminus \{f\}$ contains no proper trap.

Proof. (1) Let $X \subseteq P \setminus \{i\}$ be a proper siphon. Note that $i \notin X$ and, hence, X is unmarked in any initial marking $k[i]$. Since an unmarked siphon stays unmarked, places from X are redundant.

In the opposite direction: Let $X \subseteq P \setminus \{i\}$ be the set of all redundant places of N. We will prove that X is a siphon. Consider some $t \notin X^\bullet$; ${}^\bullet t$ contains no places from X and hence all places from ${}^\bullet t$ are non-redundant. Then for every place p in ${}^\bullet t$ there exists a marking $m_p \geq [p]$ reachable from some $k_p[i]$, $k_p \in \mathbb{N}$. Taking a sum of corresponding initial markings we obtain an initial marking from which a marking $m \geq {}^\bullet t$ can be reached. Thus t can fire and all places from $t^\bullet$ can obtain tokens, i.e. they are non-redundant. Therefore, $t^\bullet \cap X = \emptyset$ and so $t \notin {}^\bullet X$. Hence $(T \setminus X^\bullet) \subseteq (T \setminus {}^\bullet X)$, and so X is a siphon. Thus every WF-net with redundant places contains a proper siphon in $P \setminus \{i\}$.
(2) can be proved analogously. □

To check that $P \setminus \{i\}$ contains no proper siphon it is enough to compute the largest siphon X in $P \setminus \{i\}$ in a standard manner [7] (see Fig.2): initialize X with $P \setminus \{i\}$ and remove places that belong to $t^\bullet$ for some t such that $t \notin X^\bullet$ until the fixed point is reached. The largest trap not containing f can be computed with a similar algorithm.

As a spin-off of the check for absence of traps and siphons, we get a check of the path property of a WF-net:

Lemma 9. *Let $N = \langle P, T, F^+, F^- \rangle$ be a Petri net with a single source place i and a single sink place f, and every transition of N has at least one input and one output place. Moreover, $P \setminus \{i\}$ contains no proper siphon and $P \setminus \{f\}$ contains no trap. Then N is a WF-net.*

Proof. Consider an arbitrary node n and the set X of all places such that for every place $p \in X$ there is a path from p to n. We will show that X is a proper siphon, which implies that $i \in X$ and so there exists a path from i to n.

The set X is nonempty: If n is a place then $n \in X$ (since there is a path from n to n), and if n is a transition then its input places are in X. X it is a proper siphon: For every transition t, if $t \in {}^\bullet X$ then there is a path from t to

n, and hence there is a path from places from ${}^{\bullet}t$ to n and so every input place of t is in X. Since ${}^{\bullet}t \neq \emptyset$, we have $t \in X^{\bullet}$. Thus, X is a proper siphon indeed.

Similarly, the set of places to which there is a path from n is a trap, and so there is a path from n to f. Thus N is indeed a WF-net. □

Thus we obtained a characterization that guarantees non-redundancy and non-persistency for WF-nets, and moreover it serves as a check of the path property.

3.2 Batch Workflow Nets

Since we are interested in the class of WF-nets that have no redundant or persistent places, we introduce the notion of *Batch Workflow nets* by imposing requirements on the structure of the net:

Definition 10. *A Batch Workflow net (BWF-net) N is a Petri net that has the following properties:*

(1) N has a single source place i and a single sink place f;
(2) every transition of N has at least one input and one output place;
(3) every siphon of N contains i;
(4) every trap of N contains f.

The purpose of imposing structural requirements in the BWF-net definition resembles the purpose of one of the requirements on sound WF-nets from [1]. Sound WF-nets are defined there as nets where
(1) $\forall\, m \in \mathcal{R}([i]) : m \xrightarrow{*} [f]$,
(2) $\forall\, m \in \mathcal{R}([i]) : m \geq [f] \Rightarrow m = [f]$,
(3) $\forall\, t \in T : \exists\, m, m' : [i] \xrightarrow{*} m \xrightarrow{t} m'$.
Our definition of soundness is stronger than requirement (1) from the above definition: it corresponds to 1-soundness in our definition. We do not use requirement (2) since it follows immediately from (1) (we prove the implication in a generalized form that is applicable to nets with multiple tokens on i):

Lemma 11. *Let N be a WF-net such that $\forall\, m \in \mathcal{R}(k[i]) : m \xrightarrow{*} k[f]$. Then $\forall\, m \in \mathcal{R}(k[i]) : m \geq k[f] \Rightarrow m = k[f]$.*

Proof. Consider a marking $m \in \mathcal{R}(k[i])$ such that $m \geq k[f]$, i.e. $m = m' + k[f]$ for some $m' \geq \emptyset$. Since $m \in \mathcal{R}(k[i])$, $m \xrightarrow{*} k[f]$, i.e. $m' + k[f] \xrightarrow{*} k[f]$. Since $f^{\bullet} = \emptyset$, we have $m' \xrightarrow{*} \emptyset$. However, every transition of a WF-net has at least one output place. Thus $m' = \emptyset$ and so $m = k[f]$. □

We do not include requirement (3) in the definition of soundness. In fact we do not require all the transitions to be live in $(N, [i])$, since we allow batches of tasks to be processed in the net. The definition of BWF-nets implies that the net has no redundant transitions, which corresponds to (3).

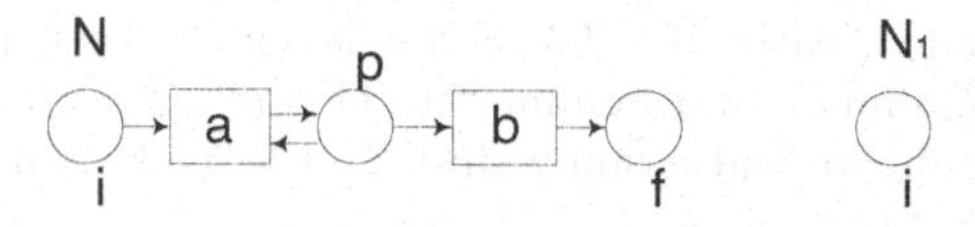

Fig. 3. Removing redundant places and transitions from WF-net N

In the rest of the paper we focus on the problem of soundness for BWF-nets. Working with BWF-nets instead of WF-nets does not limit the applicability of our approach:
Let a WF-net N be given. First, we find a maximal siphon X in $P \setminus \{i\}$. All places from X are redundant (see Theorem 8). Thus, transitions from $X^\bullet$ never get enabled and are redundant as well. Hence, by removing places from X and transitions from $X^\bullet$ together with the corresponding ingoing and outgoing arcs we obtain a net N_1 with the same behavior as the one of net N, whatever initial marking $k[i]$ is chosen for these nets. N_1 is either not a WF-net any more (the path condition gets violated, or the place f is removed – see Fig. 3) and so we can make the conclusion that N was ill-designed, or N_1 is a WF-net with the same behavior as the original net N but without redundant places. So N_1 is an improved version of N. Further, we check whether net N_1 has persistent places. If yes, we may conclude that N_1 is not a sound WF-net. Otherwise, N_1 is a BWF-net.

4 Soundness Is Decidable

In this section, we show that the problem of soundness is decidable for BWF-nets. As indicated above, this implies decidability of (general) soundness for WF-nets. We first discuss the necessary theoretical issues and then move to the actual decision procedure.

4.1 Decidability of Soundness

Proper termination for a given marking can be easily checked by using standard reachability algorithms. Deciding soundness is more intricate: A straightforward approach would require an *infinite* number of checks of proper termination (by checking proper termination for every marking reachable from some *arbitrary initial marking*). We shall try and reduce the check of soundness to the check of proper termination for a *finite* set of markings.

Consider the set of all markings reachable from some initial marking in a BWF-net N: $\mathcal{R} = \bigcup_{k \in \mathbb{N}} \mathcal{R}(k \cdot \mathbf{i})$. Every $\mathcal{R}(k \cdot \mathbf{i})$ is a subset of the set

$$\mathcal{G}_k = \{k \cdot \mathbf{i} + F \cdot v \mid v \in \mathbb{Z}^T\} \cap \mathbb{N}^P$$

(see Lemma 1). Note that $\mathcal{R}(k \cdot \mathbf{i}) \subseteq \mathcal{G}_k$ but in general $\mathcal{G}_k \not\subseteq \mathcal{R}(k \cdot \mathbf{i})$. For every marking m from $\mathcal{G}_k$ holds $\mathcal{I} \cdot m = \mathcal{I} \cdot (k \cdot \mathbf{i})$ ($\mathcal{I}$ is a matrix with basis place invariants as rows). Clearly, if a BWF-net is sound, then $\mathbf{i} \xrightarrow{*} \mathbf{f}$, and hence $\mathcal{I} \cdot \mathbf{i} = \mathcal{I} \cdot \mathbf{f}$ (see Lemma 1 and Lemma 2). This is the *first soundness check* we perform. Further on, we will assume that $\mathcal{I} \cdot \mathbf{i} = \mathcal{I} \cdot \mathbf{f}$ holds for the net under consideration.

We now formulate the lemma that is fundamental for deciding soundness (in the proof we use the fact that N is a *BWF-net*, and so it has no redundant places).

Lemma 12. *Let N be a sound BWF-net and let $m \in \mathcal{G}_k$ for some $k \in \mathbb{N}$. Then there exists $\ell \in \mathbb{N}$ such that $(k + \ell) \cdot \mathbf{i} \xrightarrow{*} m + \ell \cdot \mathbf{f}$.*

Proof. Let m be a marking from $\mathcal{G}_k$, i.e. $m = k \cdot \mathbf{i} + F \cdot v$ for some $v \in \mathbb{Z}^T$. Then there exist $v_1, v_2 \in \mathbb{N}^T$ such that $v = v_1 - v_2$. Note that $F = F^+ - F^-$. So

$$m = k \cdot \mathbf{i} + F^+ \cdot v_1 + F^- \cdot v_2 - F^- \cdot v_1 - F^+ \cdot v_2.$$

By Lemma 6, we can majorate markings $F^+ \cdot v_1$, $F^- \cdot v_2$: There exist $a, b \in \mathbb{N}$ and markings A, B such that $a \cdot \mathbf{i} \xrightarrow{*} A + F^+ \cdot v_1$ and $b \cdot \mathbf{i} \xrightarrow{*} B + F^- \cdot v_2$. Then $(k + a + b) \cdot \mathbf{i} \xrightarrow{*} k \cdot \mathbf{i} + A + F^+ \cdot v_1 + B + F^- \cdot v_2 = m + A + F^- \cdot v_1 + B + F^+ \cdot v_2$.

Let γ_2 be an arbitrary firing sequence with $\overrightarrow{\gamma_2} = v_2$. Then $b \cdot \mathbf{i} \xrightarrow{*} B + F^- \cdot v_2 \xrightarrow{\gamma_2} B + F^+ \cdot v_2$, and since N is sound, $B + F^+ \cdot v_2 \xrightarrow{*} b \cdot \mathbf{f}$. Now consider a marking $A + F^- \cdot v_1$. For an arbitrary firing sequence γ_1 with $\overrightarrow{\gamma_1} = v_1$, $A + F^- \cdot v_1 \xrightarrow{\gamma_1} A + F^+ \cdot v_1$. Moreover, we have $a \cdot \mathbf{i} \xrightarrow{*} A + F^+ \cdot v_1$, and since N is sound, $A + F^- \cdot v_1 \xrightarrow{*} A + F^+ \cdot v_1 \xrightarrow{*} a \cdot \mathbf{f}$. Thus we obtain $m + A + F^- \cdot v_1 + B + F^+ \cdot v_2 \xrightarrow{*} m + (a + b) \cdot \mathbf{f}$. So with $\ell = a + b$ the lemma holds. □

This immediately leads us to the conclusion that every marking from $\mathcal{G}_k$ should enable some firing sequence leading to $k \cdot \mathbf{f}$, lest the soundness condition gets violated.

Lemma 13. *Let N be a sound BWF-net and $m \in \mathcal{G}_k$. Then $m \xrightarrow{*} k \cdot \mathbf{f}$.*

Proof. By Lemma 12, there exists $\ell \in \mathbb{N}$ such that $(k+\ell) \cdot \mathbf{i} \xrightarrow{*} m + \ell \cdot \mathbf{f}$. Since N is sound, $m + \ell \cdot \mathbf{f} \xrightarrow{*} (k + \ell) \cdot \mathbf{f}$. Since f is a sink place, we deduce $m \xrightarrow{*} k \cdot \mathbf{f}$. □

One more conclusion we can draw now is the following:

Corollary 14. *Let N be a sound BWF-net. Then $\mathcal{I} \cdot x = \mathbf{0}$ for $x \in (\mathbb{Q}^+)^P$ iff $x = \mathbf{0}$* [1].

Proof. Let $\mathcal{I} \cdot x = \mathbf{0}$ for some $x > \mathbf{0}$. Then we can find $\ell \in \mathbb{N}$ such that $y = \ell \cdot x$ is in $\mathbb{N}^P$. Note that $\mathcal{I} \cdot y = 0$ as well. So y agrees with marking $\mathbf{0}$ on all place invariants. By Lemma 2, $y = F \cdot v$ for some $v \in \mathbb{Q}^T$. Then there exists $n > 0$ such that $w = n \cdot v \in \mathbb{Z}^T$ and thus $z = n \cdot y = F \cdot w$ is also the solution for the equation $\mathcal{I} \cdot x = \mathbf{0}$. Note that z is a non-empty marking from $\mathcal{G}_0$. By Lemma 13, we have $z \xrightarrow{*} \mathbf{0}$. This is in contradiction with the fact that for any transition of N, $t^\bullet \neq \emptyset$. □

[1] We overload the notation and use $\mathbf{0}$ for a zero-vector of an arbitrary dimension.

Thus, the *second soundness check* we perform is the check whether the equation $\mathcal{I} \cdot x = \mathbf{0}$ has only the trivial solution on $\mathbb{N}^P$. If so, we can conclude that $\mathcal{G}_k$'s are *disjoint* sets:

Corollary 15. *Let N be a BWF-net such that the equation $\mathcal{I} \cdot x = \mathbf{0}$ has only the trivial solution in $\mathbb{N}^P$. Then $\mathcal{G}_k \cap \mathcal{G}_\ell \neq \emptyset$ implies $k = \ell$.*

Proof. Since $m \in \mathcal{G}_k$ implies $\mathcal{I} \cdot m = \mathcal{I} \cdot (k \cdot \mathbf{i})$ and $m \in \mathcal{G}_\ell$ implies $\mathcal{I} \cdot m = \mathcal{I} \cdot (\ell \cdot \mathbf{i})$, $m \in (\mathcal{G}_k \cap \mathcal{G}_\ell)$ implies $\mathcal{I} \cdot (k \cdot \mathbf{i}) = \mathcal{I} \cdot (\ell \cdot \mathbf{i})$. By Corollary 14, $k = \ell$. □

Further on, we assume that $\mathcal{I} \cdot x = \mathbf{0}$ has only the trivial solution on $\mathbb{N}^P$ and thus $\mathcal{G}_k$'s are disjoint sets. We define the *i-weight function* $w(m)$ of a marking m as a natural number k such that $\mathcal{I} \cdot m = \mathcal{I} \cdot (k \cdot \mathbf{i})$ ($w(m)$ is undefined if such a value k does not exist). All markings in $\mathcal{G}_k$ have i-weight k.

Now we introduce the set $\mathcal{G} = \bigcup_{k \in \mathbb{N}} \mathcal{G}_k$, i.e.

$$\mathcal{G} = \{k \cdot \mathbf{i} + F \cdot v \mid k \in \mathbb{N} \wedge v \in \mathbb{Z}^T\} \cap \mathbb{N}^P,$$

and extend the notion of proper termination for all markings from $\mathcal{G}$: We say that a marking $m \in \mathcal{G}$ *terminates properly* in N iff $m \xrightarrow{*} w(m) \cdot \mathbf{f}$.

Lemma 16. *Let $m_1, m_2 \in \mathcal{G}$ be markings that terminate properly and $m = \lambda_1 m_1 + \lambda_2 m_2$ for some $\lambda_1, \lambda_2 \in \mathbb{N}$. Then $m \in \mathcal{G}$ and it terminates properly.*

Proof. By the definition of $\mathcal{G}_k$, $(\lambda_1 m_1 + \lambda_2 m_2) \in \mathcal{G}_{\lambda_1 k_1 + \lambda_2 k_2}$ and $\mathcal{I} \cdot (\lambda_1 m_1 + \lambda_2 m_2) = \mathcal{I} \cdot ((\lambda_1 k_1 + \lambda_2 k_2) \cdot \mathbf{i})$. Thus $w(\lambda_1 m_1 + \lambda_2 m_2) = \lambda_1 w(m_1) + \lambda_2 w(m_2)$. Since $m_1 \xrightarrow{*} w(m_1) \cdot \mathbf{f}$ and $m_2 \xrightarrow{*} w(m_2) \cdot \mathbf{f}$, we have $\lambda_1 m_1 + \lambda_2 m_2 \xrightarrow{*} (\lambda_1 w(m_1) + \lambda_2 w(m_2)) \cdot \mathbf{f}$. □

We now formulate a necessary and sufficient condition for soundness.

Theorem 17. *Let N be a BWF-net. Then N is sound iff all markings in $\mathcal{G}$ terminate properly.*

Proof. ($\Rightarrow$): Suppose N is sound. Consider any marking $m \in \mathcal{G}_k$ for an arbitrary $k \in \mathbb{N}$. By Lemma 13, $m \xrightarrow{*} k \cdot \mathbf{f}$.
($\Leftarrow$): We must prove that N is sound, i.e. that all markings from $\bigcup_{k \in \mathbb{N}} \mathcal{R}(k \cdot \mathbf{i})$ terminate properly. Since $\bigcup_{k \in \mathbb{N}} \mathcal{R}(k \cdot \mathbf{i}) \subseteq \mathcal{G}$ and all markings in $\mathcal{G}$ terminate properly, N is sound. □

We thus obtained a characterization of soundness involving the set $\mathcal{G}$ rather than reachable markings. We shall use the regularity of the structure of $\mathcal{G}$ to reduce the problem of proper termination of markings of $\mathcal{G}$ to the problem of proper termination of some *finite subset* Γ of $\mathcal{G}$.

In order to construct Γ, we extend the set $\mathcal{G}$ even further by making a step from integers to rational numbers and considering the set

$$\mathcal{H} = \{a \cdot \mathbf{i} + F \cdot v \mid a \in \mathbb{Q}^+ \wedge v \in \mathbb{Q}^T\} \cap (\mathbb{Q}^+)^P.$$

(We refer to the appendix for the definitions of algebraic notions that we use in the rest of this section.)

Lemma 18. *The set $\mathcal{H}$ is a convex polyhedral cone. Moreover, there exists a finite set $e_1, \ldots, e_n$ of generators such that $e_1, \ldots, e_n \in \mathcal{G}$.*

Proof. $\mathcal{H}$ is a convex polyhedral cone (the proof is given in Appendix B). By Theorem 26, we can find generators $E_1, \ldots, E_n$ of $\mathcal{H}$. Each E_i is a linear combination of $\mathbf{i}$ and the column vectors of F with *rational* coefficients. The lcm of the denominators divided by the gcd of the numerators gives for a given E_i the smallest rational number γ_i such that $\gamma_i E_i$ can be written as $k_i \cdot \mathbf{i} + F \cdot v_i$ with $k_i \in \mathbb{N}$, $v_i \in \mathbb{Z}^T$. Set $e_i = \gamma_i E_i$ for $i = 1, \ldots, n$. Then the $e_1, \ldots, e_n$ are generators of $\mathcal{H}$ and $e_1, \ldots, e_n \in \mathcal{G}$. □

We define our *finite* set Γ as

$$\Gamma = \{\sum_i \alpha_i \cdot e_i \mid 0 \leq \alpha_i \leq 1\} \cap \mathcal{G}$$

and show that the proper termination of any marking from $\mathcal{G}$ can be reduced to the proper termination of markings from Γ. Note that $e_1, \ldots, e_n$ are in Γ.

Theorem 19. *Let N be a BWF-net such that $\mathcal{I} \cdot \mathbf{i} = \mathcal{I} \cdot \mathbf{f}$ and $\mathcal{I} \cdot x = \mathbf{0}$ has only the trivial solution in $(\mathbb{Q}^+)^P$. Further, let $\mathcal{G} = \{k \cdot \mathbf{i} + F \cdot v \mid k \in \mathbb{N} \wedge v \in \mathbb{Z}^T\} \cap \mathbb{N}^P$, $\mathcal{H} = \{a \cdot \mathbf{i} + F \cdot v \mid a \in \mathbb{Q}^+ \wedge v \in \mathbb{Q}^T\} \cap (\mathbb{Q}^+)^P$, $e_1, \ldots, e_n \in \mathcal{G}$ be the generators of the cone $\mathcal{H}$ and $\Gamma = \{\sum_i \alpha_i \cdot e_i \mid 0 \leq \alpha_i \leq 1\} \cap \mathcal{G}$. Then N is sound iff all markings from Γ terminate properly.*

Proof. ($\Rightarrow$): Let N be a sound WF-net. By Theorem 17 all markings of $\mathcal{G}$ terminate properly. Since $\Gamma \subseteq \mathcal{G}$, all markings of Γ terminate properly.

($\Leftarrow$): Let all markings from Γ terminate properly. Consider an arbitrary marking $m \in \mathcal{G}$. Since $m \in \mathcal{G}$, $m = n \cdot \mathbf{i} + F \cdot w$ for some $n \in \mathbb{N}$, $w \in \mathbb{Z}^T$. Since $\mathcal{G} \subseteq \mathcal{H}$ and so $m \in \mathcal{H}$, $m = \sum_i \lambda_i \cdot e_i$ with the $\lambda_i \in \mathbb{Q}^+$. We can represent m as $\sum_i \ell_i \cdot e_i + \sum_i \mu_i \cdot e_i$ where $\ell_i = \lfloor \lambda_i \rfloor$ (the integer part of λ_i) and $\mu_j = \lambda_j - \lfloor \lambda_j \rfloor$ (the fractional part of λ_i), i.e. $0 \leq \mu_i < 1$. We will prove that $m' = m - \sum_i \ell_i \cdot e_i = \sum_i \mu_i \cdot e_i$ is a marking from $\mathcal{G}$. First note that $m \in \mathbb{N}^P$, $e_i \in \mathbb{N}^P$, $\ell_i \in \mathbb{N}$ for all i. Thus $m' \in \mathbb{Z}^P$. Moreover, $\mu_i \geq 0$, which implies that $m' \geq \mathbf{0}$. Thus $m' \in \mathbb{N}^P$. Since $e_i \in \mathcal{G}$, they can be represented as $k_i \cdot \mathbf{i} + F \cdot v_i$ where $k_i \in \mathbb{N}$, $v_i \in \mathbb{Z}^T$. Since $m' = m - \sum_i \ell_i \cdot e_i$, we have $m' = k \cdot \mathbf{i} + F \cdot v$ with $k = n - \sum_i k_i \cdot \ell_i$ and $v = w - \sum_i \ell_i \cdot v_i$. Therefore, we can conclude that $k \in \mathbb{Z}$ and $v \in \mathbb{Z}^T$. Now we only have to show that $k \in \mathbb{N}$.

Note that since columns of F are not necessarily linearly independent, the representation of m' as $k \cdot \mathbf{i} + F \cdot v$ is not necessarily unique. However, k has the same value in every such a representation: Let $a = \mathcal{I} \cdot m'$ and $k \cdot \mathbf{i} + F \cdot v$ is some representation of m'. Then $\mathcal{I} \cdot m' = \mathcal{I} \cdot (k \cdot \mathbf{i}) + \mathcal{I} \cdot (F \cdot v)$. By the definition of place invariants, $\mathcal{I} \cdot (F \cdot v) = 0$. Thus $a = \mathcal{I} \cdot (k \cdot \mathbf{i})$. Due to Corollary 14, k is uniquely defined. Since $k = \sum_i \mu_i \cdot k_i$, $k \geq 0$ and so $k \in \mathbb{N}$. Therefore $m' \in \mathcal{G}$. Since $0 \leq \mu_i < 1$, $m' \in \Gamma$.

Since $e_1, \ldots, e_n$ and m' are markings from Γ and thus terminate properly, m terminates properly as well (Lemma 16). Thus, all markings in $\mathcal{G}$ terminate properly and, due to Theorem 17, N is sound. □

Thus we have reduced the problem of soundness to the problem of proper termination of a finite number of markings and hereby proved its decidability.

4.2 Decision Procedure

In this subsection, we describe the decision procedure for soundness in a systematic way. We do not claim algorithm status: we focused on clarity rather than efficiency here.

We start with trivial checks that can lead to the negative answer. First we find a set of basis place invariants and check that $\mathcal{I} \cdot \mathbf{i} = \mathcal{I} \cdot \mathbf{f}$. If not, the net is not sound. Then we check that the only solution of the equation $\mathcal{I} \cdot x = \mathbf{0}$ on $(\mathbb{Q}^+)^P$ is the trivial solution $x = \mathbf{0}$; otherwise the net is not sound (see Corollary 14). This second condition also guarantees the boundedness of the net:

Lemma 20. *Let N be a BWF-net such that the only solution of $\mathcal{I} \cdot x$ in $(\mathbb{Q}^+)^P$ is $x = \mathbf{0}$. Then $\mathcal{R}(k \cdot \mathbf{i})$ and $\mathcal{S}(k \cdot \mathbf{f})$ are finite sets for any $k \in \mathbb{N}$.*

Proof. Assume some $\mathcal{R}(k \cdot \mathbf{i})$ is an infinite set. Then there exist $m_1, m_2 \in \mathcal{R}(k \cdot \mathbf{i})$ such that $m_1 < m_2$ and $\mathcal{I} \cdot m_1 = \mathcal{I} \cdot m_2 = \mathcal{I} \cdot (k \cdot \mathbf{i})$. Then $\mathcal{I} \cdot (m_2 - m_1) = \mathbf{0}$ and thus $m_2 - m_1 = \mathbf{0}$. This is a contradiction with $m_1 < m_2$. Thus $\mathcal{R}(k \cdot \mathbf{i})$ is finite. Analogously, we prove that $\mathcal{S}(k \cdot \mathbf{f})$ is a finite set. □

At the next step, we compute generators $E_1, \ldots, E_n$ of $\mathcal{H}$ and rescale them to obtain generators $e_1, \ldots, e_n \in \mathcal{G}$ of $\mathcal{H}$. Now our goal is to enumerate the markings of Γ.

The generators of the cone $\mathcal{H}$ are not necessarily linearly independent (e.g. the set of generators can include vectors e, f, g and $f + g - e$). This implies that the representation of elements of the cone as nonnegative linear combinations of the cone generators are not necessarily unique. However, by Carathéodory's theorem (Theorem 27 in Appendix A), we can represent $\mathcal{H}$ as $\mathcal{H} = \bigcup_j \mathcal{H}_j$, where $\mathcal{H}_j$'s are cones generated by vectors from some maximal subset $\mathcal{E}_j \subseteq \{e_1, \ldots, e_n\}$ of *linearly independent vectors.* We define $\mathcal{G}_j \subseteq \mathcal{G}$ as $\mathcal{G}_j = (\mathcal{G} \cap \mathcal{H}_j)$ and $\Gamma_j \subseteq \Gamma$ as

$$\Gamma_j = \{ \sum_{e_i \in \mathcal{E}_j} \lambda_i \, e_i \mid 0 < \lambda_i \leq 1\} \cap \mathcal{G}_j .$$

Note that $\mathcal{G} = \bigcup_j \mathcal{G}_j$ but in general $\Gamma \neq \bigcup_j \Gamma_j$ though $\Gamma \supseteq \bigcup_j \Gamma_j$. Still, the proper termination of markings from $\bigcup_j \Gamma_j$ guarantees the proper termination of markings in $\mathcal{G}$. We do not give a complete proof but sketch the main idea. Every $x \in \mathcal{G}$ is also an element of $\mathcal{H}$ and so an element of some $\mathcal{H}_j$. Thus it can be represented as $\sum_{e_i \in \mathcal{E}_j} \lambda_i \cdot e_i$. Now we can use the same construction with the integer and the fractional part of λ_i as in the proof of Theorem 19 to prove that x is a nonnegative integer linear combination of markings that terminate properly, x terminates properly as well.

Now we will construct an algorithm for the enumeration of the elements of Γ_j. Since vectors in $\mathcal{E}_j$ are linearly independent, any $x \in \mathcal{G}_j$ has a unique representation as a of vectors from $\mathcal{E}_j$. Note that the dimension m of any set $\mathcal{E}_j$

---

input : a BWF-net N, a set of linearly independent vectors $\mathcal{E}_j$, translation function θ_j;
output : the set X of markings;

$X = \{\mathbf{0}\}$;
repeat
  $X = X \cup \{\theta_j(x + \mathbf{i}) \mid x \in X\} \cup \{\theta_j(x \pm F_t) \mid x \in X \wedge t \in T\}$
until the fixed point is reached;
return(X);

Fig. 4. Algorithm for enumeration of Γ

equals the dimension of the vector space $\{\lambda \cdot \mathbf{i} + F \cdot v \mid \lambda \in \mathbb{Q} \wedge v \in \mathbb{Q}^T\}$, thus all vectors from this vector space have unique representations as $\sum_{e_i \in \mathcal{E}_j} \lambda_i \cdot e_i$ where $\lambda_i \in \mathbb{Q}$. Now consider the set $\mathcal{U} \supseteq \mathcal{G}_j$ defined as $\mathcal{U} = \{k \cdot \mathbf{i} + F \cdot v \mid k \in \mathbb{N} \wedge v \in \mathbb{Z}^T\}$ (note that we do not require elements of $\mathcal{U}$ to be nonnegative vectors). Every element of $\mathcal{U}$ has a unique representation as a linear combination of vectors from $\mathcal{E}_j$. We define the translation functions $\theta_j : \mathcal{U} \to \Gamma_j$ as follows:

$$\theta_j\Big(\sum_{e_i \in \mathcal{E}_j} \lambda_i \cdot e_i\Big) = \sum_{e_i \in \mathcal{E}_j} \mu_i \cdot e_i \text{ where } \mu_i = \lambda_i - \lfloor \lambda_i \rfloor \text{ (the fractional part of } \lambda_i).$$

We will use a simple property of the translation function:

Lemma 21. *For any markings $x, y \in \mathcal{H}_j$, $\theta_j(x + y) = \theta_j(x + \theta_j(y))$.*

Lemma 22. *$\Gamma_j = \{\theta_j(x) \mid x \in \mathcal{U}\}$.*

Proof. $\Gamma_j \subseteq \mathcal{G}_j \subseteq \mathcal{U}$ and for any $x \in \Gamma_j$, $\theta_j(x) = x$ (by the definition of θ), so $\Gamma_j \subseteq \{\theta_j(x) \mid x \in \mathcal{U}\}$.

For every $x \in \mathcal{G}_j$, $\theta_j(x) \in \Gamma_j$ (cf. the proof of Theorem 19).

For any $x \in (\mathcal{U} \setminus \mathcal{G}_j)$, there exist representations of x as $k \cdot \mathbf{i} + F \cdot v$ where $k \in \mathbb{N}, v \in \mathbb{Z}^T$ and as $\sum_i \lambda_i e_i$ with $e_i \in \mathcal{E}_i$ and $\lambda_i \in \mathbb{Q}$. We choose $y \in \mathcal{G}_j$, $y = \ell \cdot \mathbf{i} + F \cdot w = \sum_i \alpha_i e_i$ with $\ell \in \mathbb{N}, w \in \mathbb{Z}^T, \alpha \in \mathbb{N}, e_i \in \mathcal{E}_j$, such that $k + \ell \geq 0$ and $\alpha_i + \lambda_i \geq 0$ for all i. Now consider the marking $z = x + y$. By the choice of y, $z \geq \mathbf{0}$ and $z = (k + \ell) \cdot \mathbf{i} + F \cdot (v + w)$ with $(k + l) \in \mathbb{N}$ and $(v + w) \in \mathbb{Z}^T$, i.e. $z \in \mathcal{G}$. Moreover, $z = \sum_i (\alpha_i + \lambda_i) \cdot e_i$, i.e. it is a nonnegative linear combination of vectors from $\mathcal{E}_i$ and thus $z \in \mathcal{G}_j$. Note that $\theta_j(y) = 0$, which means $\theta_j(z) = \theta_j(x)$. Since $z \in \mathcal{G}_j$, $\theta_j(z) \in \Gamma_j$ and so $\theta_j(x) \in \Gamma_j$. Thus $\Gamma_j \supseteq \{\theta_j(x) \mid x \in \mathcal{U}\}$. □

Fig. 4 gives the algorithm that enumerates the elements of Γ_j. F_t stands there for the column of F that corresponds to transition t, i.e. that is a vector corresponding to $t^\bullet - {}^\bullet t$.

Theorem 23. *The algorithm in Fig. 4 terminates and its output equals Γ_j.*

Proof. First we prove by induction on the iteration step number that $X \subseteq \Gamma_j$ at every iteration step. Let $X_{\ell-1}$ be the value of X at the beginning of iteration l. $X_0 = \{\mathbf{0}\}$ and by the definition of Γ_j, $\mathbf{0} \in \Gamma_j$. Now let $X_\ell \subseteq \Gamma_j$ at some step. Consider some $x \in X_\ell$ (and hence $x \in \Gamma_j$ and $x \in \mathcal{U}$). The elements added to X at iteration $(\ell + 1)$ are $\theta(x + \mathbf{i})$ and $\theta(x \pm F_t)$ for all $x \in X$ and all $t \in T$. By the definition of $\mathcal{U}$, $(x + \mathbf{i}) \in \mathcal{U}$ and thus $\theta_j(x + \mathbf{i}) \in \Gamma_j$ by the definition of θ_j. Now consider a marking $(x \pm F_t)$ for any $t \in T$. By the definition of $\mathcal{U}$, $(x \pm F_t) \in \mathcal{U}$ and so $\theta_j(x \pm F_t) \in \Gamma_j$. Thus, $X \subseteq \Gamma_j$ at every iteration step. Since X grows monotonously and Γ_j is bounded, the algorithm terminates and it returns a subset X of Γ_j.

Now we will prove that $\Gamma_j \subseteq X$. Any marking $x \in \Gamma_j$ is a marking from $\mathcal{U}$ and hence it can be represented as $m = k \cdot \mathbf{i} + F \cdot v$ for some $k \in \mathbb{N}$ and $v \in \mathbb{Z}^T$ with components $v_1, \ldots, v_n \in \mathbb{Z}$. We define $\|x\| = k + \sum_j |v_j|$. Then for any x such that $\|x\| \leq l$, $\theta_j(x) \in X_l$. We prove it by induction on l. For $l = 0$ it holds trivially. Assume that it holds for some l. Consider $x = k \cdot \mathbf{i} + F \cdot v$ such that $\|x\| = l + 1$. If $k > 0$, $x = \mathbf{i} + ((k-1) \cdot \mathbf{i} + F \cdot v)$. Note that for $y = (k-1) \cdot \mathbf{i} + F \cdot v$, $\|y\| = l$ and thus $\theta_j(y) \in X_l$. By Lemma 21, $\theta_j(x) = \theta_j(\mathbf{i} + y) = \theta_j(\mathbf{i} + \theta_j(y))$. Since $\theta_j(y) \in X_l$, $\theta_j(\mathbf{i} + \theta_j(y))$ is in X_{l+1} by the definition of the algorithm, and thus $\theta_j(x) \in X_{l+1}$.

If $k = 0$ then $v \neq \mathbf{0}$, i.e. $v_j \neq 0$ for some j. We represent x as $F_{t_j} + (k \cdot \mathbf{i} + F \cdot v')$ where all components of v' equal the corresponding components of v, except for v'_j: $|v'_j| = |v_j| - 1$ and $v_j \cdot v'_j \geq 0$. Then we use a similar induction scheme to prove the statement.

Thus at iteration $l = \max_{x \in \Gamma_j} \|x\|$, X_l contains all elements of Γ_j. □

When we found all markings of Γ_j's, we check proper termination for every marking. To check whether a marking m terminates properly, it is enough to check whether $m \in \mathcal{S}(w(m) \cdot \mathbf{f})$. Since $\mathcal{S}(w(m) \cdot \mathbf{f})$ is a finite set (see Lemma 20), we can construct it by a backward reachability analysis, starting with $X = \{w(m) \cdot \mathbf{f}\}$ and consequently augmenting this set by adding markings $\{x - F_t \mid x \in X \wedge t \in T \wedge (x - F_t) \geq \mathbf{0}\}$.

5 Example

We illustrate the decision procedure for soundness with an example. Consider net N depicted in Figure 5. (The net has weighted arcs as indicated.) We have $P = \{i, a, b, f\}$, $T = \{t, u, v, w\}$ and the incidence matrix

$$F = \begin{pmatrix} -1 & 0 & 0 & 0 \\ 3 & -2 & 8 & -1 \\ 1 & 2 & -8 & -3 \\ 0 & 0 & 0 & 1 \end{pmatrix}.$$

First, we find the place invariants of the net. Solutions of the equation $I \cdot F = 0$ are vectors $I = k \cdot (4, 1, 1, 4)$, $k \in \mathbb{Q}$. Thus the matrix of basis invariants $\mathcal{I}$ is $(4, 1, 1, 4)$. Now we check that $\mathcal{I} \cdot \mathbf{i} = \mathcal{I} \cdot \mathbf{f}$ indeed (the first and the last columns of $\mathcal{I}$ coincide).

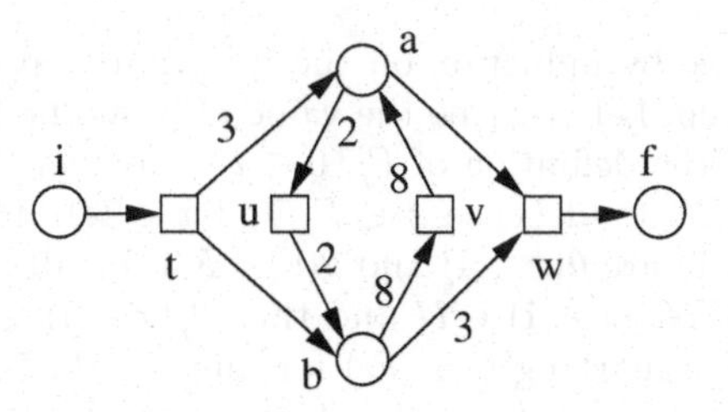

Fig. 5. Example net N

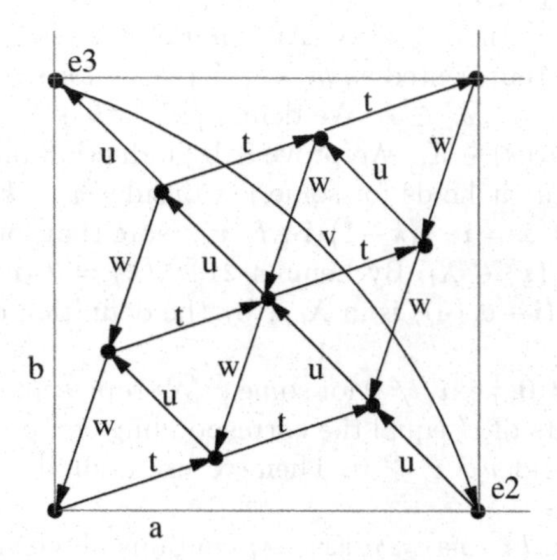

Fig. 6. Grid within Γ of net N

The cone $\mathcal{H}$ for this example is the set $(\mathbb{Q}^+)^P$, i.e. $\mathcal{H}$ is the cone generated by vectors $\mathbf{i}, \mathbf{a}, \mathbf{b}, \mathbf{f}$ corresponding to markings $[i], [a], [b], [f]$. These are vectors from $\mathcal{H}$, so they are representable as $\alpha \cdot \mathbf{i} + F \cdot x$ for some $\alpha \in \mathbb{Q}^+, x \in \mathbb{Q}^T$. By solving linear equations, we obtain the following representations of $\mathbf{a}$ and $\mathbf{b}$.

$$\mathbf{a} = \frac{1}{4} \cdot \mathbf{i} + F \cdot \begin{pmatrix} 1/4 \\ -1/8 \\ 0 \\ 0 \end{pmatrix} \text{ and } \mathbf{b} = \frac{1}{4} \cdot \mathbf{i} + F \cdot \begin{pmatrix} 1/4 \\ 3/8 \\ 0 \\ 0 \end{pmatrix}.$$

To find generators from $\mathcal{G}$ we need to rescale these vectors. Thus, we obtain $e_1 = \mathbf{i}$, $e_2 = 8 \cdot \mathbf{a}$, $e_3 = 8 \cdot \mathbf{b}$, $e_4 = \mathbf{f}$. So we have defined our set Γ and we can find all the points inside it. In Figure 6, the points of Γ are depicted, projected on the $(\mathbf{a}, \mathbf{b})$-plane. We have also depicted the transitions between them, from which it can be inferred that all grid points terminate. The net N is thus sound.

If the transition v were removed from N, the resulting net would not be sound, since the grid point e_3 no longer terminates. The net then is 1-sound but not 2-sound.

6 Conclusion

We have introduced a subclass of Workflow nets: *Batch Workflow nets (BWF-nets)* that have a simple structural characterisation based on traps and siphons. Batch Workflow nets are Workflow nets without redundant places and transitions and without persistent places. We have shown that Workflow nets with redundant places/transitions or with persistent places are ill-designed. Therefore, we moved from the study of Workflow nets to the study of Batch Workflow nets.

Since we are interested in the processing of batches of tasks in the net, we investigated the generalised notion of soundness introduced in [4] for Batch Workflow nets. We have proved that the generalised soundness is decidable and have described the decision procedure.

The decidability of soundness implies trivially the decidability of weak separability [4]. It remains still unclear whether strong separability is decidable.

Future Work

For the soundness decision procedure, we focused on the clarity rather than efficiency. It is still to be investigated how to solve the problem of soundness in an efficient manner and what complexity this algorithm would have. Besides, soundness preserving Petri net reduction techniques can be employed prior to the use of the soundness decision procedure to speed up the check.

Acknowledgements

We want to thank our colleague Aart Blokhuis for useful discussions on algebraic issues.

References

1. W. van der Aalst. Verification of workflow nets. In P. Azéma and G. Balbo, editors, *Application and Theory of Petri Nets 1997, ICATPN'1997*, volume 1248 of *Lecture Notes in Computer Science*. Springer-Verlag, 1997.
2. F. Commoner. *Deadlocks in Petri Nets*. Applied Data Research, Inc., Wakefield, Massachusetts, Report CA-7206-2311, 1972.
3. J. Desel and J. Esparza. *Free Choice Petri nets.*, volume 40 of *Cambridge Tracts in Theoretical Computer Science*. Cambridge University Press, 1995.
4. K. van Hee, N. Sidorova, and M. Voorhoeve. Soundness and separability of workflow nets in the stepwise refinement approach. In W. van der Aalst and E. Best, editors, *Application and Theory of Petri Nets 2003, ICATPN'2003*, volume 2679 of *Lecture Notes in Computer Science*. Springer-Verlag, 2003.
5. K. Lautenbach. *Liveness in Petri Nets*. Internal Report of the Gesellschaft für Mathematik und Datenverarbeitung, Bonn, Germany, ISF/75-02-1, 1975.
6. A. Schrijver. *Theory of Linear and Integer Programming*. Wiley-Interscience series in discrete mathematics. John Wiley & Sons, 1986.
7. P. Starke. *Analyse von Petri-Netz-Modellen*. Teubner, 1990.
8. H. Verbeek, T. Basten, and W. van der Aalst. Diagnosing workflow processes using Woflan. *The Computer Journal*, 44(4):246–279, 2001.

Appendix A

Here we give the mathematical definitions and results we need (see e.g. [6] for more detail).

Let E_n stand for an n-dimensional vector space over some number field (in our case the rational numbers). Let u be a vector. We write $u(x)$ for the value of the vector component corresponding to coordinate x.

A *convex combination* of the vectors $u_1, u_2, \ldots, u_n$ from E_n is a vector $u = \alpha_1 \cdot u_1 + \alpha_2 \cdot u_2 + \ldots + \alpha_n \cdot u_n$ where the α_i are nonnegative scalars such that $\sum_i \alpha_i = 1$. A subset S of E_n is *convex* iff for all pairs of vectors u_1, u_2 any convex combination of them is also in S.

Lemma 24. *The intersection of two convex sets is convex.*

A set of vectors C is called a *cone* if, for every vector $u \in C$, $\lambda\, u \in C$ for every nonnegative λ. A convex cone C is *polyhedral* if $C = \{x \mid A \cdot x \leq \mathbf{0}\}$ for some matrix A, i.e. C is the intersection of finitely many linear half-spaces.

Lemma 25. *Let X, Y be convex polyhedral cones. Then $Z = X \cap Y$ is a convex polyhedral cone as well.*

Proof. Since X, Y are convex polyhedral cones, $X = \{x \mid A \cdot x \leq \mathbf{0}\}$ and $Y = \{y \mid B \cdot y \leq \mathbf{0}\}$ for some matrix A, B. Then $X \cap Y$ is defined as $\{z \mid C \cdot z \leq \mathbf{0}\}$ where C is the matrix composed of matrices A and B, namely a matrix whose rows are the ones of A and B. Thus $X \cap Y$ is a convex polyhedral cone. □

A cone C is *finitely generated* if there exist vectors $x_1, \ldots, x_n$ such that $C = \{\sum_i \lambda_i \cdot x_i \mid \lambda_i \geq 0\}$. These vectors are called *generators* of C.

Theorem 26 (Farkas-Minkowski-Weyl). *A convex cone is polyhedral iff it is finitely generated.*

A set of generators of a convex polyhedral cone can be obtained by solving linear equations defining the cone.

Theorem 27 (Carathéodory's theorem). *Let C be a cone generated by vectors of the (finite) set X. Then for any vector $x \in C$ there exists a set $Y \subseteq X$ of* linearly independent vectors *such that x is a vector of the cone generated by the vectors of Y.*

We define the + operation on the sets X, Y of vectors in the following way: $X + Y = \{x + y \mid x \in X \wedge y \in Y\}$. Then the following lemma holds:

Lemma 28. *Let X, Y be convex polyhedral cones. Then $Z = X + Y$ is a convex polyhedral cone as well.*

Proof. Let $z_1, z_2 \in Z$ and $z = \alpha \cdot z_1 + \beta \cdot z_2$ be some convex combination of them. Since $z_1 = x_1 + y_1$ and $z_2 = x_2 + y_2$ for some $x_1, x_2 \in X, y_1, y_2 \in Y$,

$z = (\alpha \cdot x_1 + \beta \cdot x_2) + (\alpha \cdot y_1 + \beta \cdot y_2)$. Due to convexity, $(\alpha \cdot x_1 + \beta \cdot x_2) \in X$ and $(\alpha \cdot y_1 + \beta \cdot y_2) \in Y$, and thus $z \in Z$. So Z is convex.

Now consider some vector $\alpha \cdot z$ with $\alpha \geq \mathbf{0}$ and $z \in Z$. Since $z = x + y$ for some $x \in X, y \in Y$, $\alpha z = \alpha \cdot x + \alpha \cdot y$. Since X and Y are cones, $(\alpha \cdot x) \in X$ and $(\alpha \cdot y) \in Y$. Thus, $z \in Z$.

Finally, since X, Y are convex polyhedral cones, there exist generators $x_1, \ldots, x_m$ of X and $y_1, \ldots, y_n$ of Y (Theorem 26). We will prove that $x_1, \ldots, x_m, y_1, \ldots, y_n$ are generators of Z. Let $z \in Z$. Then $z = x + y$ for some $x \in X, y \in Y$, $x = \sum_i \mu_i x_i$, $y = \sum_i \lambda_i y_i$. Thus any vector $z \in Z$ can be represented as a linear combination of $x_1, \ldots, x_m, y_1, \ldots, y_n$. Conversely, if $z = \sum_i x_i + \sum_i y_i$, z is the sum of some $x \in X$, $y \in Y$. So Z is finitely generated. By Theorem 26, Z is a convex polyhedral cone. □

Appendix B

Lemma 29. *The set* $\mathcal{H} = \{a \cdot \mathbf{i} + F \cdot v \mid a \in \mathbb{Q}^+ \wedge v \in \mathbb{Q}^T\} \cap (\mathbb{Q}^+)^P$ *is a convex polyhedral cone.*

Proof. Since the set $A = \{a \cdot \mathbf{i} \mid a \in \mathbb{Q}^+\}$ is a convex cone generated by vector $\mathbf{i}$, A is a convex polyhedral cone. The set $B = \{F \cdot v \mid v \in \mathbb{Q}^T\}$ is a convex cone generated by the column vectors of matrices F and $-F$, and thus B is a convex polyhedral cone as well. The set $\{a \cdot \mathbf{i} + F \cdot v \mid a \in \mathbb{Q}^+ \wedge v \in \mathbb{Q}^T\}$ is $A + B$ and due to Lemma 28 it is a convex polyhedral cone. Finally, $C = (\mathbb{Q}^+)^P$ is clearly a convex polyhedral cone and the intersection $\mathcal{H} = (A + B) \cap C$ is a convex polyhedral cone (Lemma 25). □

Petri Net Based Model Validation in Systems Biology

Monika Heiner[1] and Ina Koch[2]

[1] Brandenburg University of Technology Cottbus, Department of Computer Science, Postbox 10 13 44, 03013 Cottbus, Germany
monika.heiner@informatik.tu-cottbus.de
http://www.informatik.tu-cottbus.de/~wwwdssz

[2] Technical University of Applied Sciences Berlin, Department of Bioinformatics, Seestrasse 64, 13347 Berlin, Germany
ina.koch@tfh-berlin.de
http://www.tfh-berlin.de/bi/

Abstract. This paper describes the thriving application of Petri net theory for model validation of different types of molecular biological systems. After a short introduction into systems biology we demonstrate how to develop and validate qualitative models of biological pathways in a systematic manner using the well-established Petri net analysis technique of place and transition invariants. We discuss special properties, which are characteristic ones for biological pathways, and give three representative case studies, which we model and analyse in more detail. The examples used in this paper cover signal transduction pathways as well as metabolic pathways.

1 Introduction

The interest in biological systems is as old as mankind. It was always of great interest to understand the complex interactions and thus the behaviour in an organism, in a cell, or in a special metabolism. These questions are still a major challenge in systems biology. *Escherichia coli* – a famous bacterium often used as typical model organism (see e.g. [19]), because it is rather simple in comparison to higher organisms – contains already more than thousands of biochemical components. A human is composed of some five octillion atoms [17]. The human brain is estimated to be built of about 100 billion neural components with hundreds of trillions interconnections [13]. Just the description, not to speak of analysis or simulation, of such huge systems requires computational techniques. Biological molecular systems are characterized by a large number of components and interactions between components, which are manifold, strongly coupled, and associative rather than additive. There is almost no real understanding of the design principles that govern intact biological systems [30].

A recent goal in systems biology is to understand the processes in a living cell. Similar to computer scientists, biologists use "divide and conquer"-techniques to investigate subsystems experimentally. Processes in the cell are divided into

J. Cortadella and W. Reisig (Eds.): ICATPN 2004, LNCS 3099, pp. 216–237, 2004.
© Springer-Verlag Berlin Heidelberg 2004

special metabolisms or pathways, which often correspond to special main functions of the cell, e.g. carbon metabolism, energy metabolism, purine metabolism, glycolytic pathway, and many others. Due to newly developed techniques in experimental biology, e.g. microarray analysis, a lot of data of biological processes have been produced over the last years. With the huge and rapidly increasing amount of data of biological systems, the following difficulties arise:

1. Data Collection Data of biochemical pathways are basically stored in the literature. But typically, special subsystems in special organisms under very special aspects are investigated, such that the entire data publication is widespread over various scientific fields. E. g., results of pharmaceutical research on special human pathways with respect to a special disease are published in other journals than data resulting from fundamental research on the same human pathways or data of the same pathway in other organisms. Cross references are the exception. Intelligent data mining techniques might be of help to search for known data of special biological processes. Up to now, much effort must be spent for gathering and evaluating relevant biological literature to get the appropriate data which one is interested in.

2. Data Representation To handle the arising amount of data it is necessary to represent and store them computationally using a unique description of biochemical pathways. There are several special databases of pathway or interaction data, e.g. *EC Enzyme Database* [1], *EMP Enzymology Database* [33], *MPW Metabolic Pathway Database* [34]. All of them use different description techniques. The most commonly used database, containing many pathways of different species, is the *KEGG Database* [11], but its representation of concurrently behaving pathways by monochromatic graphs is not free of ambiguities, as discussed e. g. in [7].

Biochemical pathways are modelled at different abstraction levels. It must be distinguished between quantitative (kinetic) models and qualitative (stoichiometric or even purely causal) models. The first ones represent the actual objective and real purpose in the long-term. They are used as soon as kinetic parameters, such as substance concentration, equilibrium constants, and reaction rates of a pathway are known. The aim of these models is to predict the system's dynamics. Related evaluation methods are typically based on solutions of systems of differential equations [8], [10], [28], [29]. Related tools for simulation were developed, such as GEPASI [20], and E-CELL [36]. Contrary, qualitative models are commonly used only, if kinetic parameters are not available or incomplete. All these qualitative models are based on more or less graphtheoretical descriptions of the system topology, which are defined in case of stoichiometric models by the known stoichiometric equations.

3. Data Validation Models are widely used for biotechnological questions, e.g. for the optimization of pathways in order to maximize the yield of special products. With the increasing number of known metabolites and known interactions between them, a validation of the interaction network becomes more and more important. The net behaviour is not understandable and predictable anymore by using merely human skills.

But, available evaluation packages for quantitative models are not able to check the model for validity. There is a strong demand for mathematical methods to validate a model for consistency and to answer questions on general structural and dynamic properties like liveness, dead states, traps, structural deadlocks, and invariant properties.

Moreover, existing methods are dedicated to a certain system type or a certain pathway represented by special graphs, see [15], [42], and many others.

Hence, a crucial point seems to be the concise and unambiguous representation of biological networks to handle computationally these highly integrated networks in an efficient manner. It is necessary to get a consistent view of the entire current state of knowledge about a particular pathway. For that purpose, a readable language with a formal, and hence unambiguous semantics would be obviously of great help as a common intermediate language in order to avoid the production of just larger patchwork, exposed to even more interpretation choices. Independently of the given description level and the particular view extension, all pathways exhibit inherently very complex structures, exploiting all the patterns well-known in software engineering, like sequence, branching, repetition, and concurrency, in any combination. But, opposite to technical networks, natural networks tend to be much more complex and apparently unstructured, making the understandability of the full network of interactions extremely difficult and therefore error-prone.

Petri nets could play an integrating role by serving as a common intermediate representation. They are able to provide a mathematically unique representation of biochemical pathways, whereby different biochemical processes may be depicted hierarchically at different abstraction levels. Moreover, established Petri net analysis techniques may be used for the validation of qualitative biochemical models, before they are extended to quantitative ones. Altogether, Petri nets enjoy the following features which might be of great help for systems biology: (1) readability – to support understanding, and therefore enable fault avoidance in the model construction process, (2) executability (animation techniques) – to experience a model in order to get really familiar with it, (3) validation techniques – for consistency checks to ensure the model integrity and correspondence to reality, and (4) analysis techniques – for qualitative as well as quantitative behaviour prediction.

In this paper, we focus on model validation by means of qualitative models, because it is obviously necessary to check at first a model for consistency and correctness of its biological interpretation before starting further analyses, aiming in the long-term at behaviour prediction by means of quantitative models. The expected results, justifying the additional expense of a preliminary model validation, consist in concise, formal and therefore unambiguous models, which are provably self-consistent and correspond to the modelled reality.

This paper is organized as follows. In the next section we describe shortly the main modelling principles for building Petri net models of biological systems. In section three we discuss and motivate the basic principles of model validation we used. Three representative case studies are presented in section four, the apoptosis in mammalian cells, the sucrose breakdown pathway in the potato tuber, and the combined glycolysis and pentosephosphate pathway in erythrocytes. To model and validate them, we use (low-level) place/transition nets as well as (high-level) coloured nets. Section five gives an overview on related work, and the final section provides a summary with an outlook on future work.

2 Modelling

There are three main types of molecular biological networks – metabolic pathways, signal transduction pathways, and gene expression networks.

Living organisms require a continuous influx of free energy to carry out their various functions. The term *metabolism* refers to the overall process, through which living systems acquire and utilize the free energy they need. During this process many chemical reactions take place, by which chemical compounds are converted into other chemical compounds, often catalysed by special enzymes. Despite of the complexity of their internal processes, living systems maintain – under normal conditions – a *steady state*, which means that the concentrations of the inner compounds are constant. This steady state is maintained by a sophisticated mesh of metabolic controls. Metabolic pathways are series of consecutive enzymatic reactions producing specific products. Their reactants, intermediates, and products are called *metabolites*. In metabolic pathways the chemical reactions of metabolites, given by their stoichiometric equations, the metabolite concentration, and the enzyme concentrations are usually known. We have here a flux of chemical substances. Two of the presented case studies are metabolic pathways – case study two, the sucrose breakdown pathway, and case study three, the combined glycolysis and pentosephosphate pathway.

Signal transduction pathways, also called *information metabolism*, have molecular on/off switches that transmit information when "on"; i. e. there is a signal, which will be passed on to the next substance. They describe how the cell receives, processes, and responds to information from the environment. These information processing circuits are widespread and diverse, e.g. half of the 25 largest protein families, encoded by the human genome, deal primarily with information processing [2]. Often signal-transduction cascades mediate the sensing and processing of stimuli. They detect, amplify, and integrate diverse external signals to generate responses such as changes in enzyme activity, gene expression, or ion-channel activity. Apoptosis, described in the first case study, is a typical signal transduction pathway.

Gene expression is the combined process of transcription of a gene into mRNA and its translation into protein. The level of gene expression can be measured by microarray experiments, giving information about expressed genes and protein-protein interactions. Based on these data, biological networks have been constructed to analyse the underlying processes. Special databases have been built for these data, e.g. [40], [41], and many others are emerging. Corresponding case studies may be found in [18], [19], [43].

Popular models used for all three types of biopathways are schematic and informal ones, usually they need additional verbose explanations how to read them. To get a unifying as well as unambiguous knowledge representation, allowing at the same time some consistency checks to get the unification approved, we have to apply representation techniques enjoying a formal and therefore unambiguous semantics. To model the different biopathway types as a Petri net[1], possibly consisting of several components, each biochemical compound (metabolite) is assigned to a place. The relations between some biochemical compounds, established by chemical reactions, are represented by transitions, modelling a biochemical atomic event. The corresponding arcs reflect the given stoichiometric relations. However, while the modelling of the biochemical compounds is quite straightforward, the elaboration of the transition structures tends to be rather time-consuming, requiring a lot of reading and interpretation of various verbose or graphical statements. Figure 1 shows a simple Petri net, modelling just one chemical reaction, given by its stoichiometric equation.

This easy-to-use modelling principle has been applied successfully to a variety of biological pathways, see [39] for a bibliography of related papers. The Petri net structure then truly reflects the biochemical topology, and the incidence matrix of the net is identical to the stoichiometric matrix of the modelled metabolic system. The Petri net behaviour gives the set of all partial order sequences of chemical reactions from the input to the output compounds respecting the given stoichiometric relations.

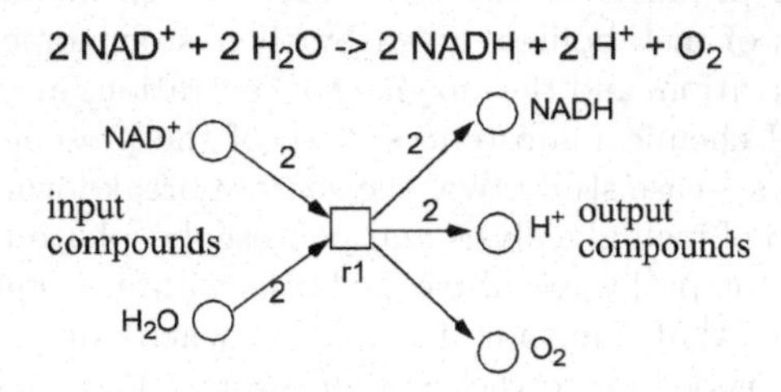

Fig. 1. Petri net model of a single chemical reaction (light-induced phosphorylation), given by its stoichiometric equation.

Moreover, the same modelling idea may be applied on a more abstract level, where stoichiometric details are not known or do not matter, resulting into a partial order description of binary causal relations of the basic (re-) actions involved.

To get readable representations we utilize three widely used short-hand notations: (1) Test arcs, represented as bidirectional arcs, stand shortly for two unidirectional arcs. Usually, signal transduction does not involve the immediate resetting of the triggering signal(s). Therefore, such events are modelled by test arcs. Similarly, enzyme reactions are catalytic reactions, i.e. there is no consumption of the biochemical compound, so they are modelled by test arcs, too. (2) Fusion nodes, given in grey, serve as connectors to glue together distributed net components. They are often used to highlight special molecules like ADP, ATP,

[1] We consider the reader to be familiar with the basic Petri net terminology, otherwise please take a look in appropriate literature, e.g. [27].

NAD^+, Pi etc., which play a slightly particular role in metabolic networks. They are called ubiquitous because they are found in sufficiently large amounts in the cell. For ease of distinction, the remaining substances are named primary. (3) When appropriate, we exploit hierarchical structuring techniques. Transition-bordered subnets are abstracted by so-called macro transitions, represented by two centrically nested squares. We are going to apply this notation to abstract from the two directions of reversible reactions and to abstract from purely linear reaction sequences.

Implementing these principles, we usually get place-bordered models, where the input compounds appear as source nodes (no predecessors) and the output compounds as sink nodes (no successors). To animate and analyse such a model, we need an environment to produce the input compounds and to remove the output compounds. There are basically two styles, how such an environment behaviour can be described.

(1) The tokens for all input compounds are generated by auxiliary input transitions (source nodes, having no predecessor), while the tokens of all output compounds are consumed by auxiliary output transitions (sink nodes, having no successor). To high-light the special meaning of these transitions for the whole model and in order to distinguish them from the ordinary ones, they are drawn as flat hollow bars. Doing so, no assumptions about the quantitative relations of input/output compounds are made. Because there are transitions without pre-places, we get automatically unbounded Petri nets (at least the postplaces of the input transitions are unbounded).

(2) The tokens for all input compounds are generated by one auxiliary transition *generate*, while the tokens of all output compounds are consumed by one auxiliary transition *remove*. Both transitions are connected by an auxiliary place *cycle*, compare Figure 2. The arc weights represent the stoichiometric relations of the sum equation of the whole network. This kind of environment behaviour reflects explicit assumptions about the quantitative relations of input/output compounds. Because now there are no transitions without preplaces, we have a chance to get bounded models.

3 Model Validation

The transformation from an informal to a formal model involves the resolution of any ambiguities, which must not happen necessarily in the right way. Therefore, the next step in a sound model-based technology for behaviour prediction should be devoted to model validation.

Model validation aims basically at increasing our confidence in the constructed model. There is no doubt that this should be a prerequisite before raising more sophisticated questions, where the answers are supposed to be found by help of the model and where we are usually ready to trust the answers we get. So, before thinking about model analysis, we are concerned with model validation.

To accomplish model validation, we need validation criteria, establishing consistency checks for the model. Looking for such criteria, we should take into

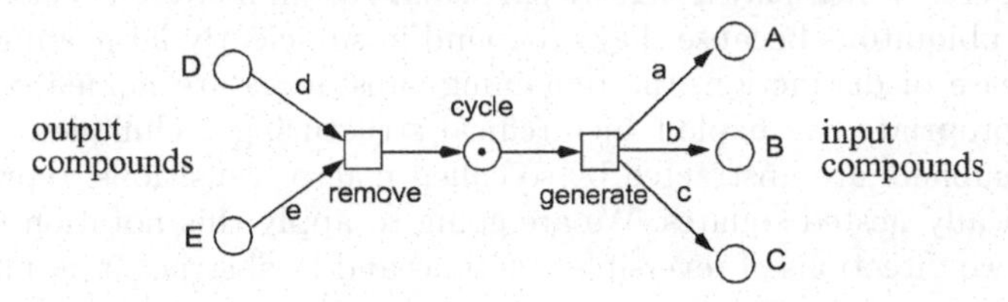

Fig. 2. Petri net environment component, style 2.

account that our models are usually the outcome of a quite heuristic procedure assembling together separate pieces, perhaps with different possible interpretations, into a larger picture.

Thus, a first and very evident question concerning a model of increasing size is the one, whether all the former basic behaviour of the smaller pieces, i. e. model components, are still maintained in the larger model, and that there are no unwanted additional ones. Due to the model's size and inherent complexity, such a property can hardly be decided without computational support.

For that purpose, we exploit two of the fundamental behavioural properties, a Petri net may exhibit – the transition invariants (T-invariants for short), and the place invariants (P-invariants for short), which have been introduced 1973 in [16].

T-invariants are multi-sets of transitions, reproducing a given marking, i.e. in the context of metabolic Petri nets – sets of chemical reactions, reproducing a given distribution of chemical compounds, or more generally spoken in the context of arbitrary biological Petri nets – sets of actions, reproducing a given system state. Due to the fact of state reproduction, a behaviour, establishing a T-invariant, may happen infinitely often, resulting into cyclic system behaviour.

To describe all possible behaviour in a given cyclic system, it would be obviously of great help to have all system's basic (cyclic) behaviour. In [31], [32] they are called elementary modes of pathways, where compounds have reached a dynamic concentration equilibrium, i. e. steady state. Then, any system behaviour may be decomposed into a positive linear combination of basic behaviour. Having this, model validation means to compare the calculated basic behaviour with the expected one.

To implement these considerations, we use – opposite to [31], [32] – standard Petri net analysis techniques and tools. To make life easy, we take the empty Petri net (no tokens at all), whereby all input and output nodes are transitions (environment style 1, compare chapter 2). An input transition may fire forever, each time generating tokens on all its postplaces. Consequently, such a net structure represents an unbounded net (there is no finite upper bound for the total token number in the net), which are generally harder to handle than bounded ones. Contrary, an output transition consumes by each firing the tokens of its preplaces, therefore decreasing the total number of tokens.

So, if we now take the empty net, we are able to look for all T-invariants, i.e. for all multi-sets of transitions reproducing the empty marking. That seems to be – at least currently – the best way to handle inherently unbounded systems, without assuming anything about the system environment. But, to give the net a real chance to reproduce the empty marking, any read arcs in the model under discussion have to be transformed into unidirectional ones, reflecting the main flow direction.

As it is well-known, we get all minimal semi-positive T-invariants by solving the following integer linear programming problem by determining the generating system:

$\mathcal{C} \cdot x = 0,$	whereby	$\mathcal{C}$	–	$(P \times T)$-incidence matrix
$x \neq 0, x \geq 0$		x	–	transition vector

Similarly, we can calculate all minimal semi-positive P-invariants by solving the following integer linear programming problem by determining the generating system:

$y \cdot \mathcal{C},$	whereby	$\mathcal{C}$	–	$(P \times T)$-incidence matrix
$y \neq 0, y \geq 0$		y	–	place vector

and interpret them as substance preservation rules in the given biological system. To be able to apply this kind of validation rules to the whole network, we need the second style of environment behaviour introduced in the former modelling section.

The calculation of T-/P-invariants requires only structural reasoning, the state space need not to be generated. Therefore, the danger of the famous state space explosion problem does not apply here. However, solving integer linear programming problems is known to be NP-complete.

Because of the given application, we are interested only in the minimal semi-positive T-/P-invariants. Therefore, they are called T-/P-invariants for short in the following.

4 Case Studies

In the following we sketch three case studies, demonstrating the general principles established above. They are ordered according the modelling convenience of the used Petri net class. See the appendix for all the acronyms used throughout this section.

4.1 Apoptosis

4.1.1 Biological Background. To demonstrate that even incomplete and uncertain knowledge may be subject of our technology, we start with apoptosis. This term refers to the regulated cell suicide program, which is of central importance in the cells' life cycle. It allows the organism to control cell numbers and tissue sizes and to protect itself from morbid cells. Neurodegenerative diseases, e.g. Alzheimer's, Huntington's, and Parkinson's disease, and other diseases as AIDS and cancer, exhibit often disturbances in apoptosis or its regulation.

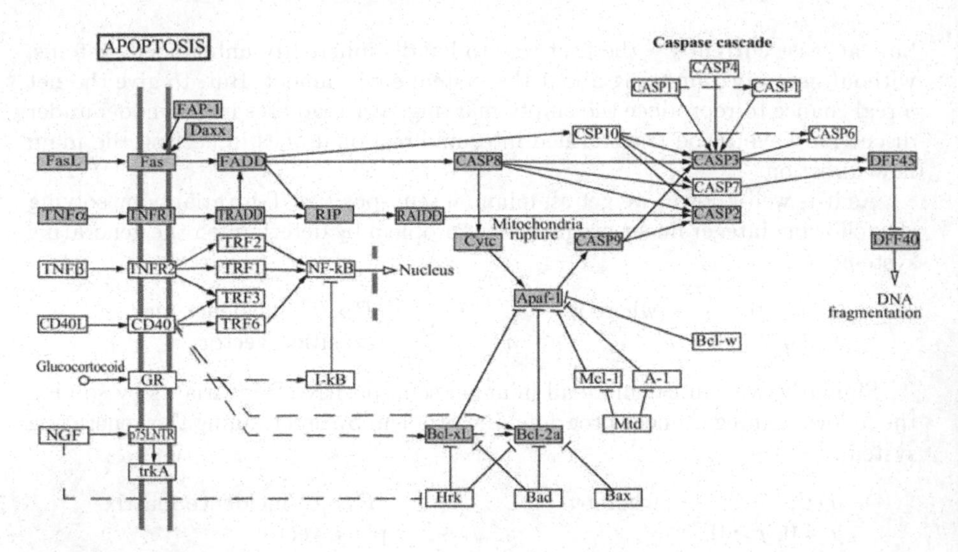

Fig. 3. The KEGG [11] representation of apoptosis. Crossbar arrowheads indicate inhibition. Branching arcs go to alternative as well as to concurrent successors. For further ambiguities see [7]. The fragment considered here is highlighted in grey.

A variety of different cellular signals initiate activation of apoptosis in distinctive ways, depending on the various cell types and their biological states, compare Figure 3. Caspases (cysteinyl-aspartate-specific proteinases) play a crucial role in the apoptotic signal transduction pathways. In living cells, caspases exist as inactive zymogens, which are activated by proteolytic cleavage. The caspases convey the apoptotic signal in a proteolytic cascade, whereby caspases cleave and activate other caspases which then degrade other cellular targets [9].

Getting a survey on the current state of the art, comprising numerous assumptions, requires a lot of reading, including the creative interpretation of various graphical representations. Figure 3 gives one of them, which is commonly used.

4.1.2 Petri Net Model. Due to the accuracy of available knowledge, which lacks particularly any stoichiometric relations of the chemical reactions involved, ordinary place/transition nets (no arc weights) are sufficient here, see Figure 4. Two further modelling aspects are worth mentioning. Usually, signal transduction does not involve the resetting of triggering signal(s). Hence, such circumstances are modelled by test arcs. Biological systems are typically full of inhibitors, i.e. compounds, preventing by their presence a certain reaction. To reflect these situations in a qualitative model adequately, we need inhibitor arcs.

4.1.3 Model Validation. Before analysing the model, two adaptations are done: (1) Test arcs are replaced by normal ones, corresponding to the main flow.

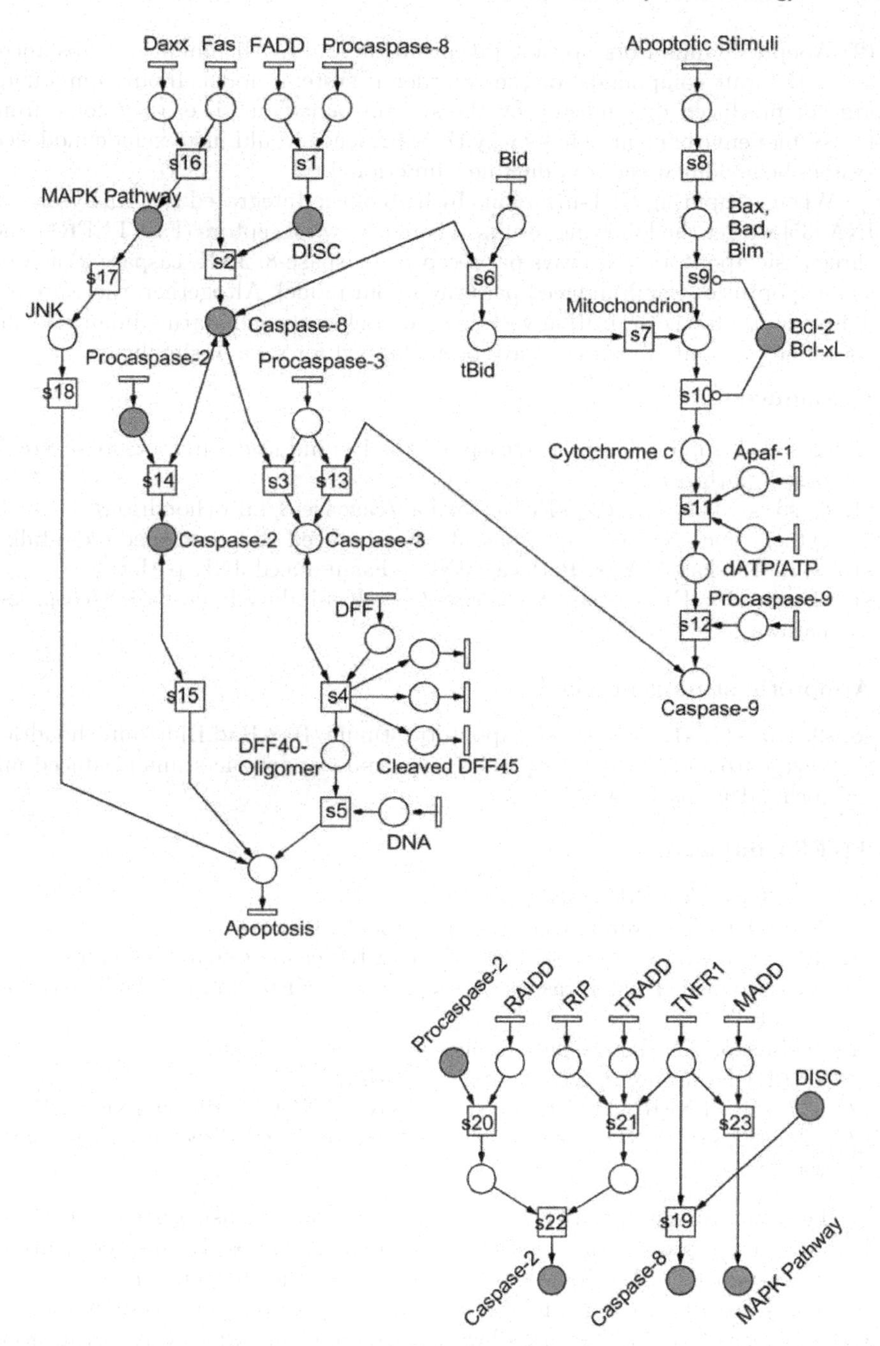

Fig. 4. Petri net of apoptosis in two net components. Please note, input and output transitions are drawn as flat bars and have the same name as the place, they are related to. Grey nodes represent fusion nodes and realizes the connection between different net components. Read arcs reflect the signal transduction principle.

(2) Apoptosis inhibitors are not taken into account. All inhibiting substances are only input compounds for the considered system model. Input compounds are not produced dynamically by the system behaviour, i. e. they come from the system environment. That's why their presence would just exclude modelled system behaviour without adding new functionality.

When computing all T-invariants by help of the Integrated Net Analyser tool INA [35] we get the following results. There are two receptors (Fas, TNFR1) and three basic apoptotic pathways per receptor (caspase-8, JNK, caspase-2) as well as an apoptotic stimuli-induced pathway in our model. Altogether, there are ten T-invariants. In the transition vectors given below the generating input and the consuming output transitions have been skipped for sake of simplicity.

Fas-induced:

s1, s2, s3, s4, s5: Fas/caspase-8/caspase-3 – Fas-induced direct caspase-8/caspase-3 pathway
s1, s2, s6, s7, s10, s11, s12, s13, s4, s5: Fas/caspase-8/mitochondrion/cytochrome c/caspase-9/caspase-3 – Fas-induced Bid-controlled cross-talk
s16, s17, s18: Fas/MAPK-Pathway/JNK – Fas-induced JNK pathway
s1, s2, s14, s15: Fas/caspase-8/caspase-2 – Fas-induced caspase-8/caspase-2 pathway

Apoptotic stimuli-induced:

s8, s9, s10, s11, s12, s13, s4, s5: apoptotic stimuli/Bax,Bad,Bim/mitochondrion /cytochrome c/Apaf-1/caspase-9/caspase-3 - apoptotic stimuli-induced mitochondrial pathway

TNFR1-induced:

s1, s19, s3, s4, s5: TNFR1/caspase-8/caspase-3 – TNFR1-induced direct caspase-8/caspase-3 pathway
s1, s19, s6, s7, s10, s11, s12, s13, s4, s5: TNFR1/caspase-8/mitochondrion/cytochrome c/ Apaf-1/caspase-9/caspase-3 - TNFR1-induced Bid-controlled cross-talk
s1,s19,s14,s15: TNFR1/caspase-8/caspase – TNFR1-induced caspase-8/caspase-2 pathway
s23, s17, s18: TNFR1/MAPK-Pathway/JNK – TNFR1-induced JNK pathway
s20, s21, s22, s15: TNFR1/caspase-2 – TNFR1-induced direct caspase-2 pathway

The minimal semi-positive T-invariants describe the basic system behaviour, because they represent the linearly independent semi-positive integer solutions of the system of linear equations resulting from the incidence matrix C. All possible system behaviour can be described by a positive linear combinations of these T-invariants. All known pathways in the modelled apoptosis fragment are reflected in a corresponding T-invariant, and there is no computed T-invariant without an apoptosis-related interpretation. Due to the given environment style, there are no P-invariants here.

4.2 Carbon Metabolism in Potato Tuber

4.2.1 Biological Background. The accumulation of starch in the Solanum tuberosum (potato) tuber is a crucial point in biotechnology. The conversion of sucrose through hexose phosphates is the major flux in the potato tuber carbon metabolism. Nearly all genes, believed to be directly involved in the sucrose breakdown transition, have been cloned by transgenic approaches. However, some fundamental questions are still open.

Figure 5 gives an overview of the sucrose breakdown pathway in potato tuber, for details see [5]. Sucrose delivered to the tuber can be cleaved in the cytosol by *invertase* to yield glucose and fructose, or by *sucrose synthase* to yield fructose and UDP-glucose. By *hexokinase*, *fructokinase*, and *UDPglucose pyrophosphorylase* hexosephosphates are produced, which are equilibrated by the action of *phosphoglucose isomerase* and *phosphoglucomutase*, and could lead either to starch synthesis, to glycolysis, or to sucrose synthesis through *sucrose phospahate synthase* and *sucrose phosphate phosphatase*. The following 16 chemical stoichiometric equations characterise the pathway.

R1. SuSy:	*sucrose synthase*	Suc + UDP <---> UDPglc + Frc
R2. UGPase:	*UDPglucose pyrophosphorylase*	UDPglc + PP <---> G1P + UTP
R3. PGM:	*phosphoglucomutase*	G6P <---> G1P
R4. FK:	*fructokinase*	Frc + ATP ---> F6P + ADP
R5. PGI:	*phosphoglucose isomerase*	G6P <---> F6P
R6. HK:	*hexokinase*	Glc + ATP ---> G6P +ADP
R7. Inv:	*invertase*	Suc ---> Glc + Frc
R8. Glyc(b):	*glycolysis*	F6P + 29 ADP + 28 P_i ---> 29 ATP
R9. SPS:	*sucrose phospahate synthase*	F6P + UDPglc <---> S6P + UDP
R10. SPP:	*sucrose phosphate phosphatase*	S6P ---> Suc + P_i
R11. NDPkin:	*NDP kinase*	UDP + ATP <---> UTP +ADP
R12. SucTrans:	*sucrose transporter*	eSuc ---> Suc
R13. ATPcons(b):	*ATP consumption*	ATP ---> ADP + P_i
R14. StaSy(b):	*starch synthesis*	G6P + ATP ---> starch + ADP + PP
R15. AdK:	*adenylate kinase*	ATP + AMP <---> 2 ADP
R16: PPase:	*pyrophosphatase*	PP ---> 2 P_i

4.2.2 Petri Net Model. We have validated the pathway using the Petri net given in Figure 6, reflecting the stoichiometric equations by a non-ordinary place/transition net. Transitions are named by the enzyme catalysing the chemical reaction or by summarized processes (R8, R13, R14). Reversible reactions are modelled by macro transitions, i.e. by hierarchical nodes, hiding the forward and backward reactions. The places for external sucrose (eSuc) and starch represent the source and sink, respectively. Correspondingly, the interface to the environment consists of one transition generating tokens for eSuc and of one transition consuming the tokens from starch, both drawn as flat bars.

4.2.3 Model Validation. The following set of 19 minimal T-invariants has been calculated using INA [35], all enjoy a sensible biological interpretation. The reversible reactions yield seven T-invariants, consisting of only two transitions.

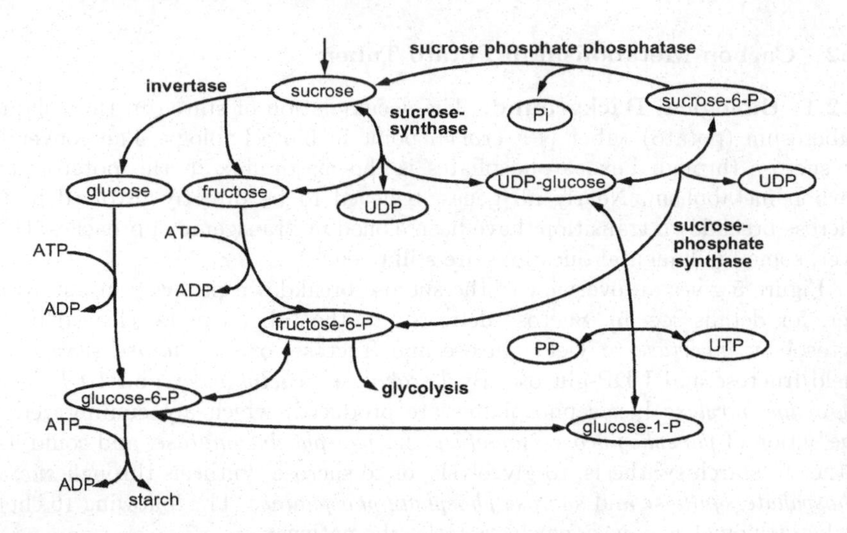

Fig. 5. Schematic overview of the central carbon metabolism in potato tuber. All equations, except R11, R13, R15 and R16, are represented.

The remaining 12 invariants are given below by the transition names, followed by their amount in brackets (if unequal to one). The net is covered by T-invariants, consequently all reactions contribute to the pathway.

T-Invariants with sucrose cleavage by sucrose synthase:

TI-8: geSuc, SucTrans, SuSy(29), UGPase, PGM_rev, FK(29), Glyc(b), StaSy(b), rStarch, SPS(28), SPP(28), NDPkin_rev.

TI-9: geSuc, SucTrans, SuSy, UGPase, PGM_rev, FK, Glyc(b), StaSy(b), rStarch, ATPcons(28), NDPkin_rev.

TI-10: geSuc(15), SucTrans(15), SuSy(15), PGI_rev(14), UGPase(15), PGM_rev(15), FK(15), Glyc(b), StaSy(b)(29), rStarch(29), NDPkin_rev(15), PPase(14).

T-Invariants with sucrose cleavage by invertase

TI-11: geSuc, SucTrans, Inv(14), UGPase_rev(13), PGM(13), HK(14), FK, Glyc(b), StaSy(b), rStarch, SuSy_rev(13), NDPkin(13), PPase(14).

TI-12: geSuc(3), SucTrans(3), Inv(29), UGPase_rev(26), PGM(26), HK(29), FK(29), Glyc(b)(3), StaSy(b)(3), rStarch(3), SPS(26), SPP(26), NDPkin(26), PPase(29).

TI-13: geSuc, SucTrans, Inv, HK, FK(27), Glyc(b), StaSy(b), rStarch, SuSy(26), SPS(26), SPP(26), PPase.

TI-14: geSuc, SucTrans, Inv, HK, FK, Glyc(b), StaSy(b), rStarch, ATPcons(26), PPase.

TI-15: geSuc(15), SucTrans(15), Inv(15), HK(15), FK(15), PGI_rev(13), Glyc(b)(2), StaSy(b)(28), rStarch(28), PPase(28).

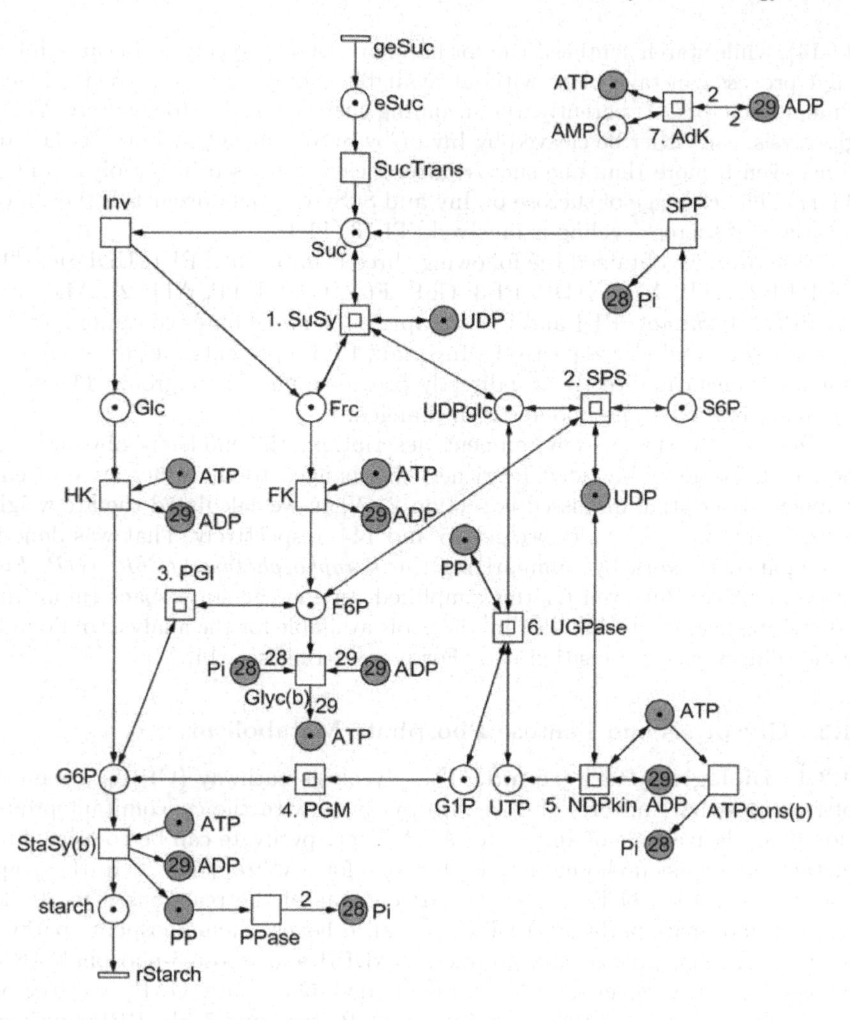

Fig. 6. Petri net of the central carbon metabolism in potato tuber, see Figure 5 and the stoichiometric reaction equations given in subsection 4.2.1.

TI-16: geSuc, SucTrans, Inv(29), HK(29), FK, PGI, UGPase_rev(28), PGM(28), Glyc(b)(2), SuSy_rev(28), NDPkin(28), PPase(28).

TI-17: geSuc(3), SucTrans(3), Inv(59), HK(59), FK(59), UGPase_rev(56), PGM(56), PGI(3), Glyc(b)(6), SPS(56), SPP(56), NDPkin(56), PPase(56).

TI-18: geSuc, SucTrans, Inv, HK, FK(57), PGI, Glyc(b)(2), SuSy(56), SPS(56), SPP(56).

TI-19: geSuc, SucTrans, Inv, HK, FK, PGI, Glyc(b)(2), ATPcons(56).

There are two possibilities for sucrose cleavage, two for hexose phosphate utilisation, and four for ATP utilisation. Both possibilities for hexose utilisation (starch synthesis and glycolysis) can occur together in one invariant (TI-11 –

TI-14), while starch synthesis cannot happen without glycolysis, because no cellular process can take place without available energy in form of ATP or other similar cofactors. Apparently, the incoming sucrose, needed to generate ATP in glycolysis, can either be cleaved by Inv or by SuSy, but not by both at the same time, even if more than one sucrose after cleavage is used in glycolysis (TI-12, TI-17). The cleavage of sucrose by Inv and SuSy can only occur together in one invariant, if sucrose cycling is involved (TI-13, TI-18).

Moreover, we obtained the following three P-invariants: PI-1: UDPglc, UTP, UDP.PI-2: ATP, AMP, ADP. PI-3: G6P, F6P, G1P, UTP, ATP(2), ADP, S6P, Pi, PP(2). Invariants PI-1 and PI-2 comprise the metabolites containing uridine or adenosine residues, respectively. Invariant PI-3 represents the sum of all compounds, which can directly or indirectly provide a phosphate group. The sum of the phosphorylated metabolites is unchanged.

Due to the kind of environment description, the model is obviously unbounded. To get a bounded Petri net, we changed to the other style of environment description discussed in section 2. When we calculated the arc weights for the environment model, we get 75 and 144, respectively. That was done for a simplified network by summarizing the *hexophosphatases (G6P, G1P, F6P)* into one place. But even for this simplified version the state space (more than 1010 states) cannot be handled by the tools available for the analysis of bounded non-ordinary place/transition nets. For more details see [14].

4.3 Glycolysis and Pentose Phosphate Metabolism

4.3.1 Biological Background. The glycolysis pathway (GP) is a sequence of reactions that converts glucose into pyruvate with the concomitant production of a relatively small amount of ATP. Then, pyruvate can be converted into lactate. The version chosen here is that one for erythrocytes [2]. In the graphical representation of Figure 7, the GP consists of the reactions 9 to 20. The pentose phosphate pathway (PPP), also called hexose monophosphate pathway, again starts with glucose and produces NADPH and ribose-5-phosphate (R5P) which then is transformed into glyceraldehyde-3-phosphate (GAP) and fructose-6-phosphate (F6P) and thus flows into the GP. In Figure 7, the PPP consists of the reactions 9, 1 to 8, and 15 to 20.

Whereas the GP generates primarily ATP with glucose as a fuel, the PPP generates NADPH, which serves as electron donor for biosyntheses in cells. The interplay of the glycolytic and pentose phosphate pathways enables the levels of NADPH, ATP, and building blocks for biosyntheses, such as R5P and Pyr, to be continuously adjusted to meet cellular needs. This interplay is quite complex, even in its somewhat simplified version discussed here.

4.3.2 Petri Net Model. The application of Petri nets to this field began in the nineties with the publications of Reddy et al. [25], [26], presenting a place/transition net to model the structure of the combined GP/PPP of erythrocytes. They compute also S- and T-invariants, but provide neither a full P-invariant nor a non-trivial T-invariant. A thorough analysis of an extended

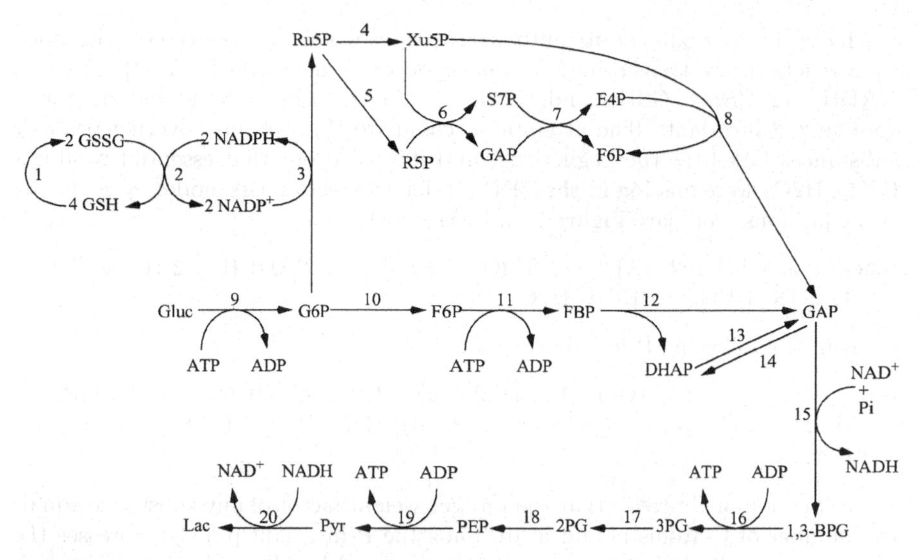

Fig. 7. Typical representation of the glycolysis and pentose phosphate metabolism according to [26]. The arc numbers identify the chemical reactions. Prefix numbers in node names stand for multiplicities. It's open for interpretation what the meaning of equally named nodes might be.

form of this pathway was performed in [12], which is presented and analysed in [38] as a coloured Petri net. In the Petri net of Figure 8, the GP consists of the reactions l1 to l8, while the PPP consists of the reactions l1, m1 to m3, r1 to r5, and l3 to l8.

In this section, we use Design/CPN [4] to edit and execute models. The calculations reuse an experimental software package SY, written by H. Genrich in Standard ML, for the symbolic analysis of coloured Petri nets, for details see [38].

4.3.3 Model Validation. The crucial point of using coloured instead of low-level Petri nets is the possibility to discriminate between different molecules of the same metabolite via their identifiers (colours). This allows to separate different branches of a composite pathway and to distinguish among molecules on the same place according to their origin and destination reaction. Alternative metabolic paths (splitting at conflict places, here G6P, GAP, Ru5P) most often result in different metabolic overall reactions. Therefore, they have to be discriminated, in order to combine them afterwards deliberately. This discrimination is performed by attributing different identifiers to the molecules and by additionally blocking certain transitions for particular molecules (using guards).

Using the software package SY, we were able to calculate an environment net component (not shown in this paper) to get a bounded model, which already had been found (but just by guessing) in [12] for a low-level model version of this metabolic pathway.

There are four pairs of ubiquitous substrates which, if produced or consumed by a reaction, are transformed into each other, namely (ADP, ATP), (NAD^+, NADH), (2 GSSG, GSH), and (NAD^+, NADH). This is validated by corresponding P-invariants. The attempt to calculate P-invariants covering primary substances failed (at the beginning), and it turned out that essential products (CO_2, H2O) were missing in the PPP model. Correcting the model by replacing two equations (compare Figure 7 and Figure 8)

reaction 3: G6P + 2 $NADP^+$ + H2O → Ru5P + 2 NADPH + 2 H^+ + CO_2,
reaction 18: DPG → PEP + H2O

leads to the following P-invariant:

p-inv_C = [(Gluc, 6), (G6P, 6), (F6P, 6), (FBP, 6), (CO_2, 1), (Ru5P, 5), (R5P, 5), (Xu5P, 5), (S7P, 7), (E4P, 4), (DHAP, 3), (GAP, 3), (BPS, 3), (Lac, 3)].

An inspection of C reveals that the integer weight factor of any substance equals the number of C-atoms bound in it. Thus the P-invariant p-inv_C expresses the conservation rule that the sum of C-atoms bound by all involved substances is constant. This represents obviously a sensible biochemical interpretation. Similarly, the conservation of Oand P-atoms could be shown in the improved model by help of P-invariants [38].

There are three T-invariants, enjoying sensible interpretation and covering the net, given in short-hand notation:

glycolysis pathway (l1, C), (l2, D), (l3, C), (l4, D), (l5, D), (l7, 2*D), (l8, 2*D), (s1, D), (s2, D)
pentose phosphate pathway (l1, G + 2*H), (l3, 2*H), (l4, 2*H), (l5, 2*H), (l7, 5*H), (l8, 5*H), (m1, G + 2*H), (m2, 6*D), (m3, 6*D), (r1, G), (r2, 2*H), (r3, (G,H)), (r4, (G,H)), (5, H), (s1, D), (s2, D)
pathway including the reverse reaction l2′ (l1, G'), (l2', 2*H'), (l7, H'), (l8, H'), (m1, G' + 2*H'), (m2, 6*D), (m3, 6*D), (r1, G'), (r2, 2*H'), (r3, (G',H')), (r4, (G',H')), (5, H'), (s1, D), (s2, D)

5 Related Work

The idea to represent chemical systems, consisting of chemical compounds and chemical reactions, by net models has already been mentioned 1976 by C. A. Petri in his paper on interpretations of net theory [23]. The first paper really demonstrating the modelling of metabolic processes by Petri nets appeared 1993 [25]. In the meantime, several research groups followed this line, and Petri nets have been applied to all three types of molecular biological networks [18], [43]. But a closer look on the literature (see [39] for a bibliography) reveals that the majority of papers, applying Petri nets for modelling and analysis of biological systems, concentrate on quantitative aspects – despite the severe restriction, often encountered in the construction of these models, the imperfect knowledge of the kinetic parameters. Typical examples of used Petri net extensions

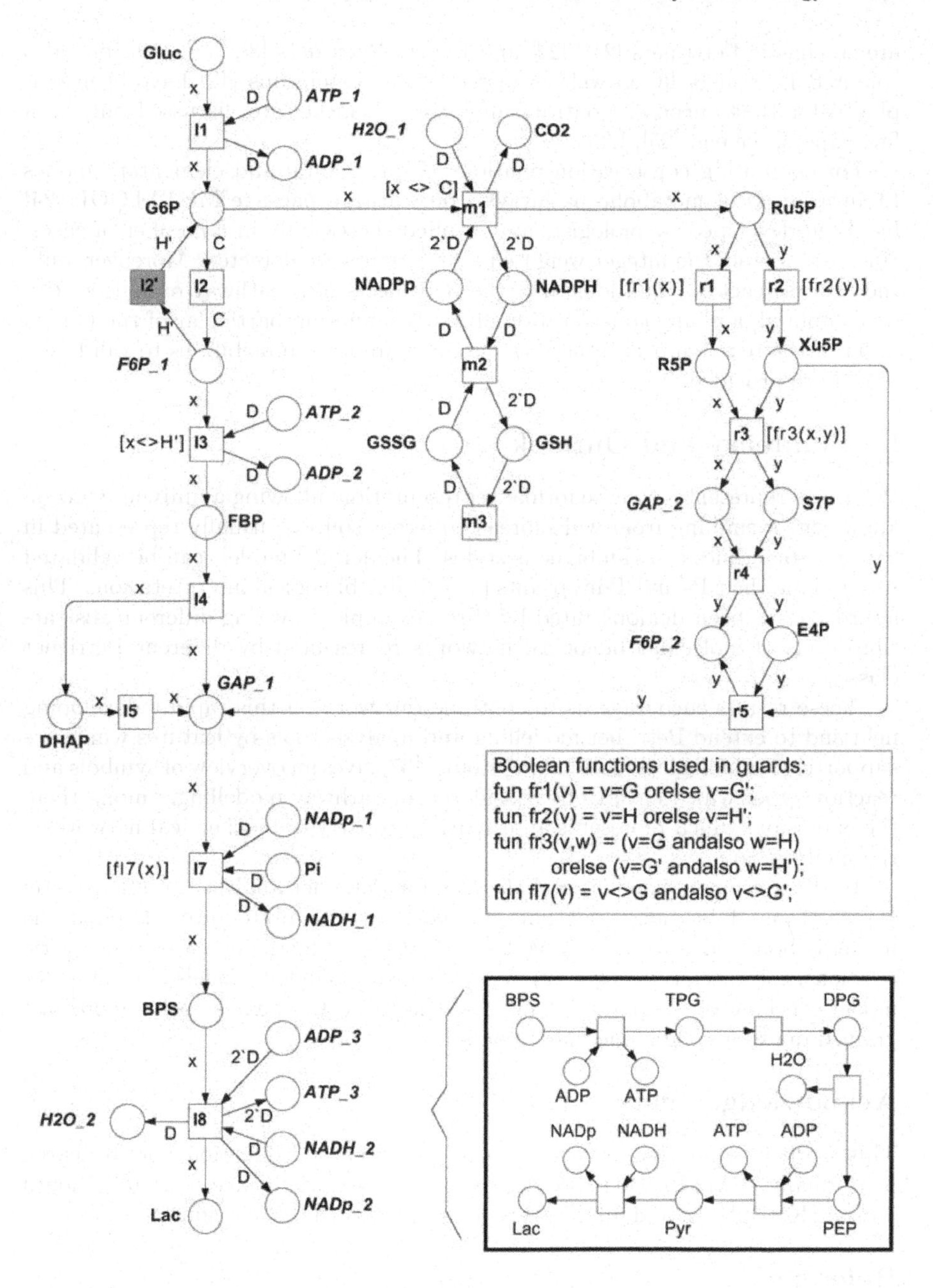

Fig. 8. The metabolism of Figure 7 as Coloured Petri net. All places have the colour set CS = {C, D, F, G, H, G′, H′}. A term in brackets [] is a transition guard. A transition t′, highlighted in grey, denotes the reverse counterpart of the reversible reaction t. The (substitution) transition l8 summarizes the reaction sequence going from BPS to Lac. Names of fusion nodes are given in italic, numbered consecutively.

are stochastic Petri nets [21], [22] and hybrid Petri nets [3], [18], [19], but also coloured Petri nets [6] as well as discrete time extensions [12] have been employed for that purpose. Contrary, qualitative aspects are discussed only in a few papers, see e.g. [25], [26].

For computing conservation relations (P-invariants) and elementary modes (T-invariants) of metabolic pathways, the software package METATOOL [24] has been developed by biologists and applied successfully in a number of cases. However, merely the integer weighted P-invariants are detected. Moreover, only the overall reaction equations, i. e. the net effects of a pathway execution, can be computed, and any consideration of its dynamics, in particular of the partial order of the reaction occurrences, is missing. No other possibilities to validate a model are provided.

6 Summary and Outlook

Petri nets represent a concise formal representation, allowing a unifying view on knowledge stemming from wide-spread different sources, usually represented in various, sometimes even ambiguous styles. The derived models can be validated by checking their P- and T-invariants for suitable biological interpretations. This approach has been demonstrated by three examples covering different abstraction levels of molecular biological networks represented by different Petri net classes.

These results encourage us to continue our work on this rapidly developing field and to extend Petri net modelling and analysis tools by features which are supportive for biological molecular systems. [37] gives an overview of symbols and reaction types which have to be considered in pathway modelling, among them the extension, which represents an essential property of biochemical networks – the inhibitor arc.

In the given context, major challenges for Petri net analysis techniques are: (1) Analysis of bounded, but not safe nets, with inhibitor arcs. Even if the model is bounded, the state space tends to be very huge due to stoichiometric relations, see case study two. (2) Analysis of unbounded models. In any case, model checking of temporal formulae seems to be the next step for more fine grained questions the model has to pass.

Acknowledgements

This work is supported by the Federal Ministry of Education and Research of Germany (BMBF), BCB project 0312705D. We would like to thank Bjoern Junker, Klaus Voss, and Jürgen Will for many fruitful discussions.

References

1. Bairoch, A.: The ENZYME Data Bank. NucleicAcids Res. 21 (3) (1993) pp. 3155-3156
2. Berg, J. M.; Tymoczko, J. L.; Stryer, L.: Biochemistry. 5th ed., Freeman, New York 2002

3. Chen, M.; Hofestaedt, R.: Quantitative Petri Net Model of Gene Regulated Metabolic Networks in the Cell. Silico Biol. 3, 0030 (2003)
4. Design/CPN. http://www.daimi.au.dk/designCPN/
5. Fernie, A.R.; Willmitzer, L.; Trethewey, R.N.: Sucrose-to-starch: a Transition in Molecular Plant Physiology. Trends in Plant Sci. 7(1) (2002) pp. 35-41
6. Genrich, H.; Küffner, R.; Voss, K.: Executable Petri Net Models for the Analysis of Metabolic Pathways. 21th International Conference on Application and Theory of Petri Nets, *Workshop Proc. "Practical Use of High-level Petri Nets"*, Aarhus, June 2000, pp. 1-14
7. Heiner, M.; Koch, I.; Will, J.: Model Validation of Biological Pathways Using Petri Nets – Demonstrated for Apoptosis. Journal on BioSystems, in press (2003)
8. Heinrich, R.; Rapoport, T.A.: A Linear Steady-state Treatment of Enzymatic Chains: General Properties, Control and Effector Strength. Eur. J. Biochem. 42 (1974) pp. 89-95
9. Hengartner, M. O.: The Biochemistry of Apoptosis. Nature **407** (2000) Oct., pp. 770-776
10. Kacser, H; Burns, J.A.: The Control of Flux. Symp. Soc. Exp. Bio. 27 (1973) pp. 65-104
11. KEGG Kyoto Encyclopedia of Genes and Genomes. http://www.genome.ad.jp/kegg/pathway/hsa/hsa04210.html
12. Koch, I.; Schuster, S.; Heiner, M.: Simulation and Analysis of Metabolic Networks Using Time-Dependent Petri Nets. Proc. of the German Conference on Bioinformatics (GCB'99), Hannover, October 1999, pp. 208-209
13. Koch, C.; Laurent, G.: Complexity and the Nervous System. Science 284 (1999) pp. 96-98
14. Koch, I.; Junker, B.; Heiner, M.: Application Petri Net Theory to Model Validation of the Sucrose-to-starch Pathway in Potato Tuber. submitted to *Bioinformatics* (2003)
15. Kohn, M. C.; Letzkus, W. J.: A Graph-theoretical Analysis of Metabolic Regulation. *J. Theoret. Biol.* **100** (1983) pp. 293-304
16. Lautenbach, K.: Exakte Bedingungen der Lebendigkeit für eine Klasse von Petrinetzen. Berichte der GMD 82, Bonn 1973 (in German)
17. Lazlo, E.: The Systems View of the World. George Braziller (1972) p. 5
18. Matsuno, H.; Fujita, S.; Doi, A.; Nagasaki, M.; Miyano, S.: Towards Biopathway Modeling and Simulation. Proc. ICATPN 2003, LNCS 2679, pp.3-22
19. Matsuno, H.; Tanaka, Y.; Aoshima, H.; Doi, A.; Matsui, M.; Miyano, S.: Biopathways Representation and Simulation on Hybrid Functional Petri Net. *Silico Biol.* **3**, 0032 (2003)
20. Mendes, P.: Biochemistry by Numbers: Simulation of Biochemical Pathway with Gepasi 3. *Trends Biochem. Sci.* **22** (1999) pp. 361-363.
21. Narahari, Y.; Suryanarayanan, K.; Reddy, N. V. S.: Discrete Event Simulation of Distributed Systems Using Stochastic Petri Nets. Energy, Electronics, Computers, Communications (1989) pp. 622-625
22. Peccoud, J.: Stoch. PN for Genetic Networks. MS-Medicine *Sciences* 14(1998) 991-993
23. Petri, C. A.: Interpretations of Net Theory. GMD, Interner Bericht 75-07, 2nd improved edition December 1976, 26 p.
24. Pfeiffer, T.; Sánchez-Valdenebro, I.; Nuño, J. C.; Montero, F.; Schuster, S.: METATOOL: For Studying Metabolic Networks. Bioinformatics 15(1999) pp. 251-257

25. Reddy, V. N.; Mavrovouniotis, M. L.; Liebman, M. N: Petri Net Representation in Metabolic Pathways. Proc. First International Conference on Intelligent Systems for Molecular Biology, AAAI Press, Menlo Park, 1993, pp. 328-336
26. Reddy, V. N.; Liebman, M. N.; Mavrovouniotis, M. L.: Qualitative Analysis of Biochemical Reaction Systems. Comput. Biol. Med. 26(1) (1996) pp. 9-24
27. Reisig, W.: Petri Nets; An Introduction; Springer 1982
28. Savageau, M.A.: Biochemical Systems Analysis, I. Some Mathematical Properties of the Rate Law for the Component Ezymatic Reactions. J. Theoret. Biol. 25(1969) pp. 365-369
29. Savageau, M.A.: Biochemical Systems Analysis, II. The Steady-state Solutions for an n- Pool System Using a Power-law Approximation. J. Theoret. Biol. 25(1969) pp. 370-379
30. Savageau, M.A.: Reconstructionist Molecular Biology. The New Biologist 3(1991) pp. 190-197
31. Schuster, S.; Hilgetag, C.; Schuster, R.: Determining Elementary Modes of Functioning in Biochemical Reaction Networks at Steady State. Proc.Second Gauss Symposium (1993) pp. 101-114
32. Schuster, S.; Pfeiffer, T.; Moldenhauer, F.; Koch, I.; Dandekar, T.: Structural Analysis of Metabolic Networks: Elementary Flux Modes, Analogy to Petri Nets, and Application to Mycoplasma pneumoniae. Proc. of the German Conference on Bioinformatics 2000. Logos Verlag, Berlin, 2000, pp. 115-120
33. Selkov, E.; Basmanova, S.; Gaasterland, T.; Goryanin, I.; Gretchkin, Y.; Maltsev, N.; Nenashev, V.; Overbeek, R.; Panyushkina, E.; Pronevitch, L.; Selkov Jr., E.; Yunus, I.: The Metabolic Pathway Collection from ERM: the Enzyme and Metabolic Pathway Database. Nucleic Acids Res. 24(1) (1996) pp. 26-29
34. Selkov Jr., E.; Gretchkin, Y.; Mikhailova, N.; Selkov, E.: MPW: The Metabolic Pathways Database. Nucleic Acids Res. 26(1) (1998) pp. 43-45
35. Starke, P. H.: INA - Integrated Net Analyzer. Manual, Berlin 1998
36. Tomita, M.; Hashimoto, K.; Takahashi, K.; Shimuzu, T.S., Matsuzaki, Y.; Miyoshi, F.; Saito, K.; Tanida, S.; Yugi, K.; Venter, J.C.; HUtchinson, 3rd, C. A.: E-CELL: Software Enironment for Whole-cell Simulation. *Bioinformatics* **15** (199) pp. 72-84.
37. Voit, E. O.: Computational Analysis of Biochemical Systems. Cambridge University Press (2000)
38. Voss, K.; Heiner, M.; Koch, I.: Steady State Analysis of Metabolic Pathways Using Petri Nets; *In Silico Biol.* **3**, 0031 (2003)
39. Will, J.; Heiner, M.: Petri Nets in Biology, Chemistry, and Medicine - Bibliography; Computer Science Reports 04/02, BTU Cottbus, November 2002, 36 p.
40. Wingender, E.; Chen, X.; Fricke, E.; Geffers, R.; Hehl, R.; Liebich, I.; Krull, M.; Matys, V.; Michael, H.; Ohnhauser, R.; Pruss, M.; Schacherer, F.; Thiele, S.; Urbach, S.: The Transfac System on Gene Expression Regulation. *Nucl. Acids Res.* **29** (1)(2001)pp.281-283
41. Xenarios, I.; Rice, D.W.; Salwinski, L.; Baron, M.K.; Marcotte, E.M.; Eisenberg, D.: Dip: the Database of Interacting Proteins. *Nucl. Acids Res.* **28** (1) (2000) pp. 289-291.
42. Zeigarnik, A.V.: A Graph-Theoretical Model of Complex Reaction Mechanisms: Special Graphs for Characterization of the Linkage between the Routes in Complex Reactions Having Linear Mechanisms. Kinetics and Catalysis 35(5)(1994) pp. 656-658
43. Zimmer, R.: Petri Net Based Pathway Models of Biochemical Networks for the Interpretation of Expression Data; Supplement to the Proc. of the 2nd Workshop on Computation of Biochemical Pathways and Genetic Networks. EML Heidelberg 2001, 10 p.

Appendix: Abbreviations

Case Study 1: Apoptosis, Biochemical Compounds

Apaf-1:	apoplectic protease activating factor 1
Ask1:	apoptosis signal-regulating kinase-1
ATP:	adenosine triphosphate
Bad:	Bcl-x_L/Bcl-2 associated death promoter
Bax:	Bcl-2 associated X protein
Bcl-2:	B-Cell lymphoma 2
Bid:	Bcl-2 interacting domain
CASP:	caspase
Caspase:	cysteinyl aspartate-specific protease
CrmA:	cytokine response modifier A
Cyt c:	cytochrome c
dATP:	desoxyadenosine triphosphate
Daxx:	Fas death domain associated protein xx
DFF:	DNA fragmentation factor
DFF40:	40 kDa unit of DFF
DFF45:	45 kDa unit of DFF
DISC:	death inducing signaling complex
FADD:	Fas associating protein with death domain
FAP-1:	Fas associated phosphatase-1
Fas:	Fas receptor
Fas-L, FasL:	Fas ligand
FLIP:	FLICE inhibitory protein
JNK:	c-Jun amino-terminal kinase
MADD:	mitogen activated kinase activating death domain
MAPK:	mitogen activated protein kinase
RAIDD:	RIP associated Ich-1/CED homologous protein with death domain
RIP:	receptor interacting protein
tBid:	truncated Bid
TNF, TNFα:	tumor necrosis factor
TNFα-R, TNFR1:	tumor necrosis factor receptor
TRADD:	TNF receptor 1 associated death domain

Case Study 2: Potato Tuber, Metabolites

ADP	adenosine diphosphate
AMP	adenosine monophosphate
ATP	adenosine triphosphate
eSuc	extern sucrose
F6P	fructose-6-phosphate
Frc	fructose
G1P	glucose-1-phosphate
G6P	glucose-6-phosphate
Glc	glucose
Pi	phosphate ionized
PP	pyrophosphate
S6P	sucrose-6-phosphate
Suc	sucrose
UDP	uridine diphosphate
UDPglc	uridine diphosphate glucose
UTP	uridine triphosphate

Case Study 3: Glycolysis and Pentose Phosphate Metabolism

Biochemical Compounds

ADP	Adenosine diphosphate
ATP	Adenosine triphosphate
BPS	1,3-Biphosphoglycerate
DHAP	Dihydroxyacetone phosphate
DPG	2-Phosphoglycerate
E4P	Erythose-4-phosphate
FBP	Fructose biphosphate
F6P	Fructose-6-phosphate
GAP	Glyceraldehyde-3-phosphate
Gluc	Glucose
GSH	Glutathione
GSSG	Glutathionedisulfide
G6P	Glucose-6-phosphate
Lac	Lactate
NADH	Nicotinamide adenine dinucleotide, reduced form
NAD+/NADp	Nicotinamide adenine dinucleotide, oxidized form
NADPH	Nicotinamide adenine dinucleotide phosphate, reduced form
NADP+/NADPp,	Nicotinamide adenine dinucleotide phosphate, oxidized form
PEP	Phosphoenolpyruvate
Pi	Orthophosphate, ionic form
Pyr	Pyruvate
Ru5P	Ribulose-5-phosphate
R5P	Ribose-5-phosphate
S7P	Sedoheptulose-5-phosphate
TPG	3-Phosphoglycerate
Xu5P	Xylulose-5-phosphate

Correspondence between Petri net transitions/reaction numbers and enzymatic reactions

l1/9	Hexokinase
l2/10	Phosphoglucose isomerase
l3/11	Phosphofructokinase
l4/12	Aldolase
l5/13	Triosephosphate isomerase (forw.)
l6/-	Triosephosphate isomerase (backw.)
l7/15	GAP dehydrogenase
l8/16-20	Reaction path consisting of: phosphoglycerate kinase, phosphoglycerate mutase, enolase, pyruvate kinase, and lactate dehydrogenase
m1/3	G6P oxidation reactions
m2/2	Glutathione reductase
m3/1	Glutathione oxidation reaction
r1/5	Ribulose-5-phosphate isomerase
r2/4	Ribulose-5-phosphate epimerase
r3/6	Transketolase
r4/7	Transaldolase
r5/8	Transketolase

Synthesis of Controlled Behavior with Modules of Signal Nets

Gabriel Juhás, Robert Lorenz, and Christian Neumair*

Lehrstuhl für Angewandte Informatik
Katholische Universität Eichstätt, 85071 Eichstätt, Germany
{gabriel.juhas,robert.lorenz,christian.neumair}@ku-eichstaett.de

Abstract. In this paper we present a methodology for synthesis of controlled behavior for systems modelled by modules of signal sets. Modules of signal nets are modules, which are based on Petri nets enriched by two kinds of signals and an signal input/output structure. They are also known as net condition/event systems (or modules) [5, 4, 6]. Given an uncontrolled system (a plant) modelled by a module of a signal net, and a control specification given as a regular language representing the desired signal output behavior of this system, we show how to synthesize the maximal permissive and nonblocking behavior of the plant respecting the control specification. Such a behavior serves as an input for an algorithm (presented in [11]), which computes a controller realized as a module of a signal net which in combination with the plant module ensures this behavior.

1 Introduction

In classical control theory it is given a system which can interfere with environment via inputs and outputs. The aim of its control is to ensure desired behavior by giving the system right inputs in order to get the right outputs. The central idea in control theory is, that system and control build a so called *closed loop* (or *feedback loop*), which means, roughly speaking, that the control gives inputs to the system based on the system outputs which are observed by the control. In this paper, we are interested in control of discrete event systems, where the dynamic behavior of a system is described by occurrence of discrete events changing the states of the system. The crucial question to be answered when choosing a formalism for modelling control systems is how to formalize "giving inputs and observing outputs".

An event of a system can have two kinds of inputs: Actuators, which try to force the event, or sensors, which can prohibit the event. Events associated to inputs are called *controllable*. Of course there can be uncontrollable events in the system. Regarding for example a printer, a "paper jam" event can occur without any influence from the control. The following two kinds of outputs can be observed: Either the occurrence of an event (via actuators) or the fact that a state is reached (via sensors). Events resp. states associated to outputs are called *observable*. Of course there can be unobservable events resp. states. It would be natural to model control of a discrete event system by

* Supported by DFG: Project "SPECIMEN".

J. Cortadella and W. Reisig (Eds.): ICATPN 2004, LNCS 3099, pp. 238–257, 2004.
© Springer-Verlag Berlin Heidelberg 2004

influencing its behavior by actuators and sensors in order to observe desired outputs as described above.

However, the solution in the discrete event control community, which is now quite accepted, is to use only the sensors. More exactly, in supervisory control [2, 14] the events of the system to be controlled are divided as above into controllable and uncontrollable. But the controllable events can only be enabled/prohibited by a supervisor. Thus, in supervisory control actuators can only be modelled indirectly using the "sensor principle" by prohibiting all controllable events, except the event which is actuated ([1], pp. 185 - 202). There arises the natural question, why not directly model actuators? In our paper, we adapt the framework of supervisory control providing a methodology for control of discrete event systems using **both** concepts, namely actuators and sensors. Such a methodology with the slogan "forcing and prohibiting instead of only prohibiting" would be more appropriate for the class of discrete event systems, where actuators and commands are used in practice.

As a modelling formalism, we use modules communicating by means of the above described signals. This formalism was developed in the series of papers [5, 4, 6] under the name net condition/event systems. In this paper we are using the name signal nets. One reason is that the name condition/event nets is used in the Petri net context for a well known basic net class. A signal net is a Petri net enriched by *event signals*, which force the occurrence of (enabled) events (typically switches), and *condition signals* which enable/prohibit the occurring of events (typically sensors). Adding input and output signals to a signal net, one gets a *module of a signal net*. Modules of signal nets can be composed by connecting their respective input and output signals. There are several related works employing modules of signal nets in control of discrete event systems. In [5, 4, 6] effective solutions for particular classes of specifications, such as forbidden states, or simple desired and undesired sequences of events, are described. Recently, an approach for control specifications given by cycles of observable events was presented in [13]. However, in [13] the actuators are used only to observe events of the controlled system, but surprisingly, for control actions only condition signals for prohibiting events are taken. In our paper, we consider a general class of control specification in form of a language over steps of event outputs (steps of observable events). We have steps (i.e. sets) of outputs, rather then simple outputs, because some outputs can be simultaneously synchronized by an event of the system. We allow also steps containing an input with some outputs. Such a situation describes that an input signal is trying to synchronize a controllable event of the system, which is also observable. So the controller can immediately (i.e. in the same step) observe whether the input signal has forced the event to occur or not. However, since the control is assumed to send inputs based on observed outputs (as stated in the beginning), we do not allow the symmetric situation: observable events can not synchronize inputs in the same step.

In our framework we identify which input signals have to be sent to the module of the plant in order to observe only such sequences of (steps of) output signals, which are prefixes of the control specification, and every sequence of (steps of) output signals can be completed to a sequence of output signals belonging to the control specification (i.e. the behavior is nonblocking). We construct a language over steps of input and output signals of the module of the plant, which represents the maximally permissive nonblocking behavior and fulfills the control specification.

In [11], moreover the construction of a control module (of a signal net), which will in composition with the plant module realize this maximally permissive nonblocking behavior, is shown. It is proven that such a control module always exists, if there is such a behavior of the plant fulfilling the control specification.

The paper is organized as follows: After introducing some preliminary mathematical notations in Section 2, in Section 3 we present *modules of signal nets* with definition of step semantics, composition rules and input/output behavior. In Section 4 we outline our control framework implementing the "forcing and prohibiting"-paradigm by means of modules of signal nets. It is compared in detail to classical supervisory control. The main result is the synthesis of the maximally permissive nonblocking behavior of the module of a signal net (representing the plant) respecting a given regular specification language. The Section splits into two parts. In Subsection 4.1 the special case of a prefix closed specification language is considered. In Subsection 4.2 the general situation is addressed.

2 Mathematical Preliminaries

We need the following language theoretic notations ([8]). For a finite set A we denote $2^A = \{B \mid B \subseteq A\}$ the set of all subsets of A and $A^* = \{a_1 \dots a_n \mid n \in \mathbb{N}_0, a_1, \dots, a_n \in A\}$ the set of all finite words over the alphabet A. Let $L \subseteq A^*$ be a language over a finite alphabet A. The empty word is denoted by ϵ, $\overline{L} = \{v \in A^* \mid \exists x \in A^* : \ vx \in L\}$ is the prefix closure of L, and $post(L) = \{v \in A^* \mid \exists w \in L, \exists x \in A^* : \ v = wx\}$ is the language of all possible extensions of words of L. Observe that for a regular language L also $post(L)$ is regular. We will use another operation on languages preserving regularity: For two languages $L_1, L_2 \subseteq A^*$: $L_1/L_2 = \{w \in A^* \mid \exists v \in L_2 : \ wv \in L_1\}$ is the *quotient of* L_1 *and* L_2. We will consider languages over alphabets $A = 2^X$ for finite sets X. We need an extension of the projection operator to such languages for subsets $Y \subseteq X$. Define the *hiding operator* λ_Y *w.r.t.* Y by:
For a character $\xi \in A$: $\lambda_Y(\xi) = \xi \setminus Y$ if $\xi \setminus Y \neq \emptyset$, and $\lambda_Y(\xi) = \epsilon$ otherwise.
For a word $w \in A^*$: $\lambda_Y(w) = \lambda_Y(\xi_1) \dots \lambda_Y(\xi_n)$ if $w = \xi_1 \dots \xi_n$, and $\lambda_Y(w) = \epsilon$ if $w = \epsilon$. For a language $L \subseteq A^*$: $\lambda_Y(L) = \{\lambda_Y(w) \mid w \in L\}$.
The hiding operator defines equivalence classes over A^* in the following way. For a $w \in A^*$ denote $[w]_Y = \{v \in A^* \mid \lambda_Y(w) = \lambda_Y(v)\}$. Regular languages are represented by regular expressions or finite automata. Remember that states of a deterministic finite automata DFA can be denoted as equivalence classes over A^*:
$[w]_{DFA} = \{v \in A^* \mid \text{Execution of } v \text{ and } w \text{ lead to the same state}\}$.

3 Modules of Signal Nets

We use an extension of elementary Petri nets (1-safe Petri nets) equipped with the so called *first consume, then produce* semantics (since we want to allow loops, e.g. [10]). The first step in the extension is to add two kinds of signals, namely active signals, which force the occurrence of (enabled) events (typically switches or actuators), and passive signals which enable/prohibit the occurrence of events (typically sensors).

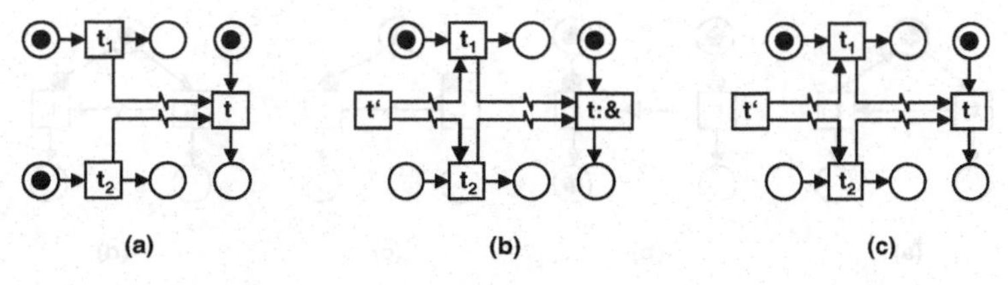

Fig. 1. In Figure (a) the enabled steps are $\{t_1, t\}$ and $\{t_2, t\}$. Figure (b) shows a signal net in AND-semantics: here the only enabled step is $\{t', t_1\}$, i.e. t is not synchronized. In Figure (c) the same net is shown in OR-semantics: here we have the enabled step $\{t', t_1, t\}$, i.e. t is synchronized.

These signals are expressed using two kind of arcs. A Petri net extended with such signals is simply called a *signal net*. Active signals are represented using arcs connecting transitions and can be interpreted in the following way: an active signal arc, also called *event arc*, leading from a transition t_1 to a transition t_2 specifies that if transition t_1 occurs and transition t_2 is enabled to occur then the occurrence of t_2 is forced (synchronized) by the occurrence of t_1, i.e. transitions t_1 and t_2 occur in one (synchronized) step. If t_2 is not enabled, t_1 occurs without t_2, while an occurrence of t_2 without t_1 is not allowed. Event arcs are not allowed to build cycles. In general (synchronized) steps of transitions are build inductively in the above way. Every step starts at a unique transition, which is not synchronized itself. Consider a transition t which is synchronized by several transitions $t_1, \ldots, t_n$, $n \geqslant 2$. Then two situations can be distinguished. For simplicity consider the case $n = 2$. If the transitions t_1, t_2 do not build a synchronized step themselves, either t_1 or t_2 can synchronize transition t in the above sense, but never transitions t_1, t_2 can occur in one synchronized step. As an example you can think of several switches to turn a light on (see Figure 1, part (a)). If the transitions t_1, t_2 build a synchronized step themselves, then there are two dialects in literature to interpret such a situation: In the first one ([5, 4, 6]) both transitions t_1, t_2 have to agree to synchronize t. Thus the only possible step of transitions involving t has to include transitions t_1, t_2, too. We call this dialect AND-semantics (see Figure 1, part (b)). In the second one ([3]) the occurrence of at least one of the transitions t_1 and t_2 synchronizes transition t, if t is enabled. It is also possible, that t_1, t_2 and t occur in one synchronized step. We call this dialect OR-semantics (see Figure 1, part (c)). In general the relation given by event arcs builds a forest of arbitrary depth. In this paper we introduce the most general interpretation, where both semantics are possible and are interpreted locally backward. That means we distinguish between OR- and AND-synchronized transitions. An OR-synchronized transition demands to be synchronized by at least one of its synchronizing transitions, whereas an AND-synchronized transition demands to be synchronized by all of its synchronizing transitions.

Since we allow loops w.r.t. single transitions, we also allow loops w.r.t. steps of transitions (see Figure 2, part (a)).

Passive signals are expressed by so called *condition arcs* (also called read arcs or test arcs in the literature) connecting places and transitions. A condition arc leading from a place to a transition models the situation that the transition can only occur if

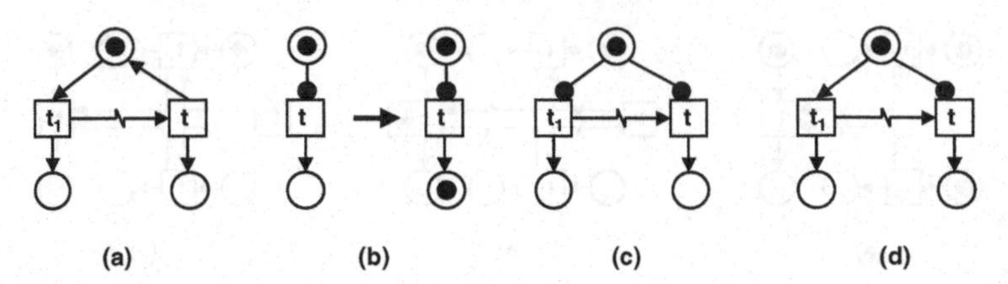

Fig. 2. Figure (a) shows an enabled step $\{t_1, t\}$. The left part of Figure (b) shows an enabled transition t, which tests a place to be marked. The occurrence of t leads to the marking shown in the right part of Figure (b). Figures (c) and (d) again present situations of an enabled step $\{t_1, t\}$.

the place is in a certain state but this state remains unchanged by the transition's occurrence (read operation) (see Figure 2, part (b)). Of course several transitions belonging to a synchronized step can test a place to be in a certain state via passive signals simultaneously, since the state of this place is not changed by their occurrence (see Figure 2, part (c)). We also allow that a transition belongs to a synchronized step of transitions testing a place to be in a certain state via a passive signal, whereas the state of this place is changed by the occurrence of another transition in this step. That means we use the so called *a priori* semantics ([9]) for the occurrence of steps of transitions, where testing of states precedes changing of states by occurrence of steps of transitions (see Figure 2, part (d)).

Definition 1 (Signal nets). *A signal net is a six-tuple* $N = (P, T, F, CN, EN, m_0)$ *where* P *denotes the finite set of* places, $T = T_{AND} \dot\cup T_{OR}$ *the distinct union of the finite sets of* AND-synchronized transitions T_{AND} *and* OR-synchronized transitions T_{OR} *(*$P \cap T = \emptyset$*),* $F \subseteq (P \times T) \cap (T \times P)$ *the* flow relation, $CN \subseteq (P \times T)$ *the set of* condition arcs *(*$CN \cap (F \cup F^{-1}) = \emptyset$*),* $EN \subseteq (T \times T)$ *the acyclic set of* event arcs *(*$EN^+ \cap id_T = \emptyset$*), and* $m_0 \subseteq P$ *the* initial marking.

Places, transitions and the flow relation are drawn as usual using circles, boxes and arrows. To distinguish between AND- and OR-synchronized transitions, AND-synchronized transitions are additionally labelled by the symbol "&". Event arcs and condition arcs are visualized using arcs of a special form given in Figure 1 and Figure 2.

For a place or a transition x we denote ${}^\bullet x = \{y \mid (y, x) \in F\}$ the *preset* of x and $x^\bullet = \{y \mid (x, y) \in F\}$ the *postset* of x. For a transition t we denote ${}^+t = \{p \mid (p, t) \in CN\}$ the *positive context* of t, ${}^{\rightsquigarrow}t = \{t' \mid (t', t) \in EN\}$ the *synchronization set* of t, $t^{\rightsquigarrow} = \{t' \mid (t, t') \in EN\}$ the *synchronized set* of t. Given a set $\xi \subseteq T$ of transitions, we extend the above notations to: ${}^\bullet\xi = \bigcup_{t \in \xi} {}^\bullet t$ and $\xi^\bullet = \bigcup_{t \in \xi} t^\bullet$, ${}^{\rightsquigarrow}\xi = \bigcup_{t \in \xi} {}^{\rightsquigarrow}t$, $\xi^{\rightsquigarrow} = \bigcup_{t \in \xi} t^{\rightsquigarrow}$.

Definition 2 (Enabling of transitions). *A transition* $t \in T$ *is* enabled *at a marking* $m \subseteq P$*, if* ${}^\bullet t \cup {}^+t \subseteq m$ *and* $(t^\bullet \setminus {}^\bullet t) \cap m = \emptyset$.

The following definition introduces a notion of steps of transitions which is different to the usual one used in Petri nets. A step denotes a set of transitions connected by event arcs, which occur synchronously. A transition, which is not synchronized by another transition, is a step. Such transitions are called *spontanuous*. In general, steps are sets of transitions such that for every non-spontaneous OR-synchronized transition in this step at least one of it's synchronizing transitions belongs also to this step, and for every AND-synchronized transition in this step all of it's synchronizing transitions belong also to this step.

Definition 3 (Steps). *Given a signal net N, steps are sets of transitions ξ defined inductively by*
– If $t \in T$ with $\rightsquigarrow t = \emptyset$ (t is spontaneous), then $\xi = \{t\}$ is a step.
– If ξ is a step, and $t \in T \setminus \xi$ is a transition, then $\xi \cup \{t\}$ is a step, if either $t \in T_{OR}$ and $\rightsquigarrow t \cap \xi \neq \emptyset$, or $t \in T_{AND}$ and $\emptyset \neq \rightsquigarrow t \subseteq \xi$.
The set of all steps of N is denoted by Σ_N.

Now we introduce how a step is enabled to occur. A step ξ is said to be potentially enabled at a marking m if every transition $t \in \xi$ is enabled at m and no transitions $t_1, t_2 \in \xi$ are in conflict, except for possible loops $p \in {}^\bullet\xi \cap \xi^\bullet$ w.r.t. ξ, where $p \in m$ is required. From all steps potentially enabled at a marking only those are enabled which are maximal with this property.

Definition 4 (Potential enabling/enabling of steps). *A step ξ is* potentially enabled *in a marking m if*
– $\forall t \in \xi$: ${}^\bullet t \cup {}^+t \subseteq m$ and $(t^\bullet \setminus {}^\bullet\xi) \cap m = \emptyset$ and
– $\forall t, t' \in \xi, t \neq t'$: ${}^\bullet t \cap {}^\bullet t' = t^\bullet \cap (t')^\bullet = \emptyset$ (t, t' are not in conflict).
The step ξ is enabled, *if ξ is potentially enabled, and there is not a potentially enabled step $\eta \supsetneq \xi$ (ξ is maximal).*

Definition 5 (Occurrence of steps and follower markings). *The occurrence of an enabled step ξ yields the follower marking $m' = (m \setminus {}^\bullet\xi) \cup \xi^\bullet$. In this case we write $m[\xi\rangle m'$.*

Definition 6 (Reachable markings, occurrence sequences). *A marking m is called reachable from the initial marking m_0 if there is a sequence of markings $m_1, \ldots, m_k = m$ and a sequence of steps $\xi_1, \ldots, \xi_k$, such that $m_0[\xi_1\rangle m_1, \ldots, m_{k-1}[\xi_k\rangle m_k$. Such a sequence of steps is called an occurrence sequence. The set of all reachable markings is denoted by $[m_0\rangle$.*

Adding some inputs and outputs to signal nets, i.e. adding condition and event arcs coming from or going to an environment, we get modules of signal nets with input and output structure.

Definition 7 (Modules of signal nets). *A module of a signal net is a triple $M = (N, \Psi, c_0)$, where $N = (P, T, F, CN, EN, m_0)$ is a signal net, and $\Psi = (\Psi^{sig}, \Psi^{arc})$ is the input/output structure, where $\Psi^{sig} = C^{in} \cup E^{in} \cup C^{out} \cup E^{out}$ is a set of input/output signals, and $\Psi^{arc} = CI^{arc} \cup EI^{arc} \cup CO^{arc} \cup EO^{arc}$ is a set of arcs connecting input/output signals with the elements of the net N. Namely, C^{in} denotes a finite set of condition inputs, E^{in} a finite set of event inputs, C^{out} a finite set of condition outputs, E^{out} a finite set of event outputs (all these sets are pair-wise disjoint), $CI^{arc} \subseteq C^{in} \times T$ a set of condition input arcs, $EI^{arc} \subseteq E^{in} \times T$ a set of event input arcs, $CO^{arc} \subseteq P \times C^{out}$ a set condition output arcs and $EO^{arc} \subseteq T \times E^{out}$ a set of event output arcs. $c_0 \subseteq C^{in}$ is the initial state of the condition inputs.*

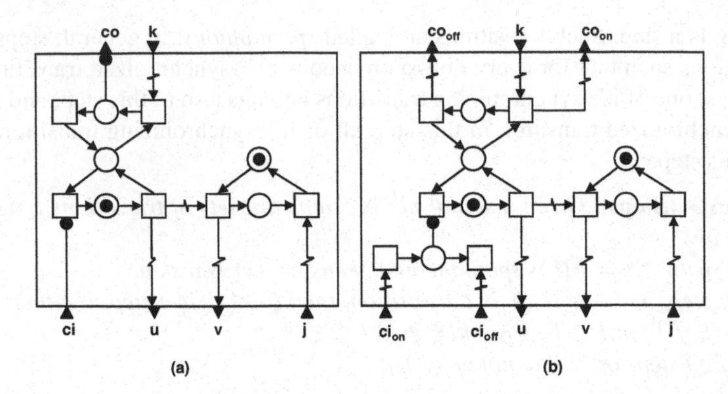

Fig. 3. Figure (a) shows a module of a signal net. Figure (b) shows the same module, where condition inputs and outputs are replaced by equivalent structures involving event inputs and outputs.

We extend the notions of preset, postset, positive context, synchronization set and synchronized set to the elements of Ψ^{sig} in the obvious way. An example of a module of a signal net, with $C^{in} = \{ci\}$, $E^{in} = \{j, k\}$, $C^{out} = \{co\}$ and $E^{out} = \{u, v\}$ is shown in the Figure 3, part (a).

Two modules can be composed by identifying some inputs of the one module M_1 with appropriate outputs of the other module M_2 and replacing the connections of the nets to the involved identified inputs and outputs by direct signal arcs respecting the identification (see Figure 4).

Definition 8 (Composition of modules of signal nets). *Let* $M_1 = (N_1, \Psi_1, c_{01})$, $M_2 = (N_2, \Psi_2, c_{02})$ *be modules of signal nets with input/output structures* $\Psi_i = (\Psi_i^{sig}, \Psi_i^{arc})$, $i = 1, 2$ *and initial markings* m_{01}, m_{02}. *Let further* $Q \subseteq \Psi_1^{sig}$ *and* $\Omega : Q \to \Psi_2^{sig}$ *be an injective mapping, such that the initial markings are compatible with the initial states of the condition inputs in the sense:* $(p, co) \in CO_1^{arc} \wedge \Omega(co) \in c_{02} \Rightarrow p \in m_{01}$ *and* $(p, co) \in CO_2^{arc} \wedge \Omega^{-1}(co) \in c_{01} \Rightarrow p \in m_{02}$. *Moreover* Ω *has to satisfy* $\Omega(E_1^{in} \cap Q) \subseteq E_2^{out}$, $\Omega(E_1^{out} \cap Q) \subseteq E_2^{in}$, $\Omega(C_1^{in} \cap Q) \subseteq C_2^{out}$, *and* $\Omega(C_1^{out} \cap Q) \subseteq C_2^{in}$, *such that no cycles of event arcs are generated.*

Then the composition $M = M_1 *_\Omega M_2$ of M_1 and M_2 w.r.t. Ω *is the module* $M = (N, \Psi, c_0)$ *with* $N = (P_1 \cup P_2, T_1 \cup T_2, F_1 \cup F_2, CN, EN, m_{01} \cup m_{02})$ *and* $\Psi = (\Psi^{sig}, \Psi^{arc})$, *where involved inputs, outputs and corresponding signal arcs are deleted, i.e.*

$$\begin{aligned}
\Psi^{sig} &= (\Psi_1^{sig} \setminus Q) \cup (\Psi_2^{sig} \setminus \Omega(Q)), \\
\Psi^{arc} &= (\Psi_1^{arc} \setminus ((^\bullet Q \times Q) \cup (Q \times Q^\bullet))) \cup \\
&\quad (\Psi_2^{sig} \setminus ((^\bullet \Omega(Q) \times \Omega(Q)) \cup (\Omega(Q) \times \Omega(Q)^\bullet))), \\
c_0 &= (c_{01} \setminus Q) \cup (c_{02} \setminus \Omega(Q)),
\end{aligned}$$

and new signal arcs are added according to Ω *in the following way:*

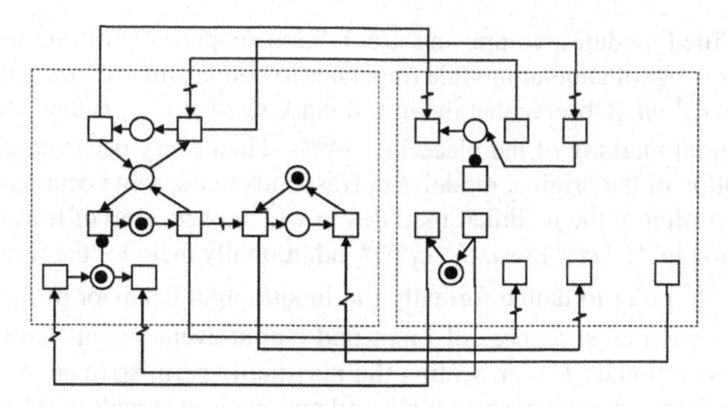

Fig. 4. The standalone of the module of a signal net in Figure 3 (a). The composed modified modules are indicated by dashed boxes. The input/output behavior of the module is given by the set of all occurrence sequences of this standalone, where the transitions in the left box are hidden. The control module with respect to a given specification can be synthesized by adding appropriate net structure to the maximally environment module represented by the right box.

$$
\begin{aligned}
CN = {} & CN_1 \cup CN_2 \cup \\
& \{(p,t) \mid \exists co \in C_1^{out} : (p, co) \in CO_1^{arc} \wedge (\Omega(co), t) \in CI_2^{arc}\} \cup \\
& \{(p,t) \mid \exists ci \in C_1^{in} : (ci, t) \in CI_1^{arc} \wedge (p, \Omega(ci)) \in CO_2^{arc}\}, \\
EN = {} & EN_1 \cup EN_2 \cup \\
& \{(t,t') \mid \exists eo \in E_1^{out} : (t, eo) \in EO_1^{arc} \wedge (\Omega(eo), t') \in EI_2^{arc}\} \cup \\
& \{(t,t') \mid \exists ei \in E_1^{in} : (ei, t') \in EI_1^{arc} \wedge (t, \Omega(ei)) \in CO_2^{arc}\}
\end{aligned}
$$

In the following, we define the input/output behavior of Petri modules as the set of all possible sequences of input signals sent to the module and output signals sent from the module. Since condition signals are signals with duration, we replace them for this purpose by equivalent structures of event signals in the following way (see Figure 3):

Definition 9. *Let $M = (N, \Psi, c_0)$ be a module with $\Psi = (\Psi^{sig}, \Psi^{arc})$. Define $M_m = (N_m, \Psi_m, c_{0m})$, $\Psi_m = (\Psi_m^{sig}, \Psi_m^{arc})$ by deleting condition inputs and outputs via*

$$C_m^{in} = C_m^{out} = CI_m^{arc} = CO_m^{arc} = \emptyset,$$

and adding new structures involving event inputs and outputs in the following way:

$$
\begin{aligned}
\text{For } c \in C^{out} : \; & E_m^{out} = \{c.on,\ c.off \mid c \in C^{out}\} \\
& EO_m^{arc} = \{(t, c.on),\ (t', c.off) \mid c \in C^{out}, t \in {}^\bullet({}^+c), t' \in ({}^+c)^\bullet\} \\
\text{For } c \in C^{in} : \; & P_m = P \cup \{p_{c.on} \mid c \in C^{in}\} \\
& T_m = T \cup \{t_{c.on}^{net},\ t_{c.off}^{net} \mid c \in C^{in}\} \\
& F_m = F \cup \{(t_{c.on}^{net}, p_{c.on}),\ (p_{c.on}, t_{c.off}^{net}) \mid c \in C^{in}\} \\
& E_m^{in} = \{c.on,\ c.off \mid c \in C^{in}\} \\
& EI_m^{arc} = \{(c.on, t_{c.on}^{net}),\ (c.off, t_{c.off}^{net}) \mid c \in C^{in}\}
\end{aligned}
$$

Finally every $p_{c.on}$ is marked if $c \in c_0$. M_m is called modified module (of M).

For modified modules, composing a condition output c_1^{out} of one module with a condition input c_2^{in} of another module translates to composing $c_1^{out}.on$ with $c_2^{in}.on$ and $c_1^{out}.off$ with $c_2^{in}.off$ (Observe that the initial marking of $p_{c_2^{in}.on}$ is then chosen according to the initial marking of the place in $^+c_1^{out}$). Then every occurrence sequence of the composition of the original modules corresponds to the same occurrence sequence of the composition of the modified modules in which every step of transitions involving a transition in $^\bullet(^+c_1^{out})$ resp. $(^+c_1^{out})^\bullet$ additionally includes the transition $t^{net}_{c_2^{in}.on}$ resp. $t^{net}_{c_2^{in}.off}$. In order to define formally the input/output behavior of a module M as the set of all possible sequences of input and output event signals we first compose M with another module E representing the maximally permissive environment of M. Then we can represent sequences of input and output event signals of M by occurrence sequences of the composed module restricted to the transition set of E (see Figure 4). As the mentioned maximally permissive environment we recognize a module E, which

- at any moment can send event inputs to M: so each event signal of M is modelled in E by a corresponding always enabled transition;
- at any moment can enable and disable condition inputs of M: so each condition input of M is modelled in E by a corresponding place, which can be marked and unmarked by associated transitions;
- can observe outputs of M: every output of M is modelled in E by a corresponding transition, which is synchronized in the case of an event output, and enabled in the case of an condition output;
- does not allow synchronization between its transitions: in particular, inputs should not be sent in steps from E to M, and outputs M should only be observed by E and not synchronize inputs of M via E.

Definition 10 (Maximally permissive environment). *Let $M = (N, \Psi, c_0)$ be a module with $\Psi = (\Psi^{sig}, \Psi^{arc})$. Define the maximally permissive environment module $E_M = (N_E, \Psi_E, c_{0E})$, $\Psi_E = (\Psi_E^{sig}, \Psi_E^{arc})$, w.r.t. M by $EN_E = CN_E = \emptyset$ and*

$$
\begin{aligned}
\textit{For } c \in C^{out} :\ & T_E = \{t_c \mid c \in C^{out}\}, \\
& C_E^{in} = \{ci_c \mid c \in C^{out}\}, CI_E^{arc} = \{(ci_c, t_c) \mid c \in C^{out}\}, \\
\textit{For } c \in C^{in} :\ & T_E = T_E \cup \{t_{ci.on}, t_{ci.off} \mid ci \in C^{in}\},\ P_E = \{p_{ci.on} \mid ci \in C^{in}\} \\
& F_E = \{(t_{ci.on}, p_{ci.on}), (p_{ci.on}, t_{ci.off}) \mid ci \in C^{in}\} \\
& m_{0E} = \{p_{ci.on} \mid ci \in C^{in} \cup c_0\} \\
& C_E^{out} = \{co_c \mid c \in C^{in}\},\ CO_E^{arc} = \{(p_c, co_c) \mid c \in C^{in}\}, \\
\textit{For } e \in E^{out} :\ & T_E = T_E \cup \{t_e \mid e \in E^{out}\}, \\
& E_E^{in} = \{ei_e \mid e \in E^{out}\}, EI_E^{arc} = \{(ei_e, t_e) \mid e \in E^{out}\} \\
\textit{For } e \in E^{in} :\ & T_E = T_E \cup \{t_e \mid e \in E^{in}\}, \\
& E_E^{out} = \{eo_e \mid e \in E^{in}\}, EO_E^{arc} = \{(t_e, eo_e) \mid e \in E^{in}\}.
\end{aligned}
$$

We call the composition of the modified module of M with the modified module of its maximally permissive environment the *standalone of M* (observe that this composition has empty input/output structure) (as an example see Figure 4). The restriction of the occurrence sequences of the standalone to the transition set of the environment is then formalized by using the hiding operator defined in section 2, which is used to make the inner transitions of a module invisible.

Definition 11. (Standalones) *Let M be a module of a signal net, E be the maximally permissive environment module of M and $M_m = (N, \Psi)$, $E_m = (N_E, \Psi_E)$ be their modified modules. The* standalone *of M is the composition module $M_S = (N_S, \Psi_S) = N_m *_\Omega E_m$ w.r.t. the following composition mapping $\Omega : \Psi^{sig} \to \Psi_E^{sig}$:*

$$\Omega(e) = ei_e \text{ for } e \in E_E^{out}$$
$$\Omega(e) = eo_e \text{ for } e \in E_E^{in}$$

The set $I = \{t \in T_S \cap T_E \mid t^{\leadsto} \neq \emptyset\}$ is called the set of input transitions of M_S, the set $O = \{t \in T_S \cap T_E \mid {}^{\leadsto}t \neq \emptyset\}$ is called the set of output transitions of M_S.

Definition 12. (Input/output behavior of modules of signal nets) *Let M be a module of a signal net with the set of transitions T. Let L_M be the set of all finite occurrence sequences of the standalone M_S of M. Then the language $\lambda_T(L_M)$ is called input/output behavior of the module M.*

L_M represents the set of all possible sequences of steps of input and output signals under the assumptions: Output signals of M can not synchronize input signals of M via the maximally permissive environment module. Input signals of M are not sent in steps from the maximally permissive environment module.

4 Controller Synthesis

Throughout this section we consider a module of a signal net $\mathcal{P}$ as a model of an uncontrolled plant. As in the previous section T denotes the set of transitions of $\mathcal{P}$, and I resp. O denote the sets of input resp. output transitions of the standalone of $\mathcal{P}$.

In our modelling formalism we consider the inside of $\mathcal{P}$ as a black box: We only can send input signals to $\mathcal{P}$ and meanwhile observe sequences of output signals. In particular, the behavior of the DES (represented by $\mathcal{P}$) is forced, not only restricted from outside. Of course this approach leads to formal and technical differences to the classical supervisory control approach:

Mainly, all events of $\mathcal{P}$ are assumed to be uncontrollable and unobservable. Controllable are only the input signals, modelled by the set of transitions I of the maximally permissive environment of $\mathcal{P}$, and observable are, beside the input signals, exactly the output signals, modelled by the set of transitions O of the maximally permissive environment of $\mathcal{P}$.

We specify a desired behavior of $\mathcal{P}$ by a set of desired sequences only of output signals. Observe moreover that, since the event arc relation produces a step semantics, we observe sequences of steps of output signals. Therefore we suppose a specification to be given as a language $L_c \subseteq (2^O)^*$. Since sets of occurrence sequences of signal nets are regular languages[1] over an alphabet of steps, we assume L_c also to be regular.

[1] Observe that we use elementary nets, which have a finite reachability graph.

The question is, whether it is possible to force the behavior of $\mathcal{P}$ via input signals to respect the given specification of output signals in a *maximally permissive way*. This can be formalized by asking, whether there is a module of a signal net $\mathcal{C}$ modelling the control and a composition $\mathcal{P} *_\Omega \mathcal{C}$, such that the set of occurrence sequences of $\mathcal{P} *_\Omega \mathcal{C}$ respects $L_\mathcal{C}$.

The answer of the question above splits into two parts. One part is to decide whether a control exists for the given specification and in the positive case to compute the behavior of the desired control module resp. the resulting composed module, also called the *behavior of the controlled plant*. The other part is, to implement the given controlled behavior via synthesizing the control module $\mathcal{C}$ from the behavior of the controlled plant and composing this module with the plant module $\mathcal{P}$, such that the resulting composition has exactly this controlled behavior. The scope of this paper is the first part. We will finally give the behavior of the controlled plant by a deterministic finite automaton, which is intended to represent the marking graph of the composed module, where only input and output transitions are visible.

The second part is presented in [11]. The main idea is to synthesize $\mathcal{C}$ from the automaton representing the behavior of the controlled plant, by adding new net structure to the modified maximally permissive environment module $\mathcal{E}$ of $\mathcal{P}$ (see Figure 4). That means the control module $\mathcal{C}$ is constructed from $\mathcal{E}$, and composed with $\mathcal{P}$ via the given connections between $\mathcal{P}$ and $\mathcal{E}$.

For a detailed running example of the computation of the behavior of the controlled plant and the implementation of this behavior via synthesizing a control module we recommend the paper [12].

We will formulate our approach language theoretically similarly as it is done in classical supervisory control. We will see, that despite the mentioned differences, some algorithms of classical supervisory control can at least be adapted to our framework. While omitting therefore most details of these algorithms, out paper remains self contained, i.e. can be understood without previous knowledge of supervisory control.

Without knowing any control module $\mathcal{C}$ we search for a sublanguage K (of occurrence sequences) of the language $L_\mathcal{P}$ (of all occurrence sequences of the standalone of $\mathcal{P}$) which represents the behavior of a composition of the plant module $\mathcal{P}$ and a control module $\mathcal{C}$ (i.e. the behavior of the controlled plant):

- If an occurrence sequence in K can be extended by a step of output transitions or invisible transitions to an occurrence sequence in $L_\mathcal{P}$, then also this extended occurrence sequence should be in K. This follows the paradigm: "*what cannot be prevented, should be legal*".
- According to unobservability of some events, some occurrence sequences in $L_\mathcal{P}$ cannot be distinguished by the control. As a consequence, following the paradigm "*what cannot be distinguished, cannot call for different control actions*", if an input is sent to the plant after a sequence w of steps has occurred, then the same input has to be sent after occurrence of any other sequence, which is undistinguishable to w.

Observe that the first condition corresponds to the classical one supervisory control. The second one is due to our step semantics, where an input can synchronize different invisible and output transitions depending on the state of $\mathcal{P}$, in combination with the notion of *observability* in supervisory control. Such a sublanguage K is called controllable w.r.t. $L_{\mathcal{P}}$, I and O:

Definition 13 (Controllable Language). *Given three finite, disjoint sets T, I and O and a prefix closed regular language L over the alphabet $2^{T \cup I \cup O}$, then a prefix closed sublanguage K of L is said to be* controllable w.r.t. L, I and O *(or simply controllable, if the sets are clear), if*

- $\forall w \in K, \forall o \in 2^{OUT} : wo \in L \Rightarrow wo \in K$,
- $\forall vj \in K, j \in 2^{I \cup OUT}, j \cap I \neq \emptyset, \forall j' \in 2^{I \cup OUT}, \lambda_{OUT}(j) = \lambda_{OUT}(j'), \forall v' \in K, \lambda_T(v) = \lambda_T(v') : v'j' \in L \Rightarrow v'j' \in K$.

Of course we are searching for such a K, which additionally respects L_c and is maximal with this property.

Definition 14 (Maximally Permissive Controllable Language). *Let T, I and O be finite, disjoint sets, L be a prefix closed regular language over the alphabet $2^{T \cup I \cup O}$ and L_c be a regular language over the alphabet 2^O.*

Let $K \subseteq L$ be controllable w.r.t L, I and O satisfying

$$\lambda_{T \cup I}(K) \subseteq \overline{L_c}.$$

We say that K is maximally permissive controllable *w.r.t. L_c, L, I and O (or simply maximally permissive controllable, if the sets are clear), if there exists no language K' satisfying $K \subsetneq K' \subseteq L$, which is controllable w.r.t. L, I and O and fulfills $\lambda_{T \cup I}(K') \subseteq \overline{L_c}$.*

It is possible to get the result $K = \{\epsilon\}$ as maximally permissive controllable language, what means that the maximal behavior respecting the specification is empty, but there happens nothing wrong without inputs from outside. If even without any input the specification can be violated, we call L_c unsatisfiable w.r.t. L, I and O.

Definition 15. *L_c is said to be* unsatisfiable w.r.t. L, I and O *(or simply unsatisfiable, if the sets are clear), if*

$$\exists w \in (2^{OUT})^* : w \in L \wedge \lambda_T(w) \notin \overline{L_c}. \tag{1}$$

We call this condition unsatisfiability condition.

Consider a maximally permissive controllable language K: by definition every occurrence sequence in K is a prefix of an occurrence sequence respecting L_c. But it can happen there are such occurrence sequences that cannot be extended within K to an occurrence sequence respecting L_c, i.e. the desired behavior is blocked. We require additionally K to be nonblocking:

Definition 16 (Nonblocking Language). *Let T, I and O be finite , disjoint sets, L be a prefix closed regular language over the alphabet $2^{T \cup I \cup O}$ and L_c be a regular language over the alphabet 2^O.*

Let $K \subseteq L$ be maximally permissive controllable w.r.t L_c, L, I and O.

Let $M \subseteq K$ be controllable w.r.t. L, I and O satisfying

$$\forall r \in M : \exists x \in (2^{O \cup I \cup T})^* \text{ with } rx \in M, \lambda_{I \cap T}(rx) \in L_c. \quad (2)$$

We say that M is nonblocking controllable *w.r.t. L_c, L, I and O (or simply nonblocking controllable, if the sets are clear). If it is maximal with this property, M is called* maximally permissive nonblocking controllable language.

In the next two paragraphs we synthesize the maximally permissive nonblocking controllable language M, if it exists. First we examine the case, when L_c is prefix closed. In this case the maximally permissive controllable sublanguage of $L_{\mathcal{P}}$ is already nonblocking. In particular safety properties can be formalized via a prefix closed specification L_c.

4.1 Safety Properties

Safety properties specify undesired behavior, that should not happen (for example forbidden states of the system). If some undesired behavior is realized by an occurrence sequence, the whole possible future of this occurrence sequence is undesired too.

The searched (control) language M as defined in the last paragraph is computed in several steps. First we define the (potentially safe) language L_{psafe} as the set of all occurrence sequences of $L_{\mathcal{P}}$ respecting L_c.

Definition 17. *We define* $L_{psafe} = \{w \in L_{\mathcal{P}} \mid \lambda_{I \cup T}(w) \in L_c\}$ *and* $L_{unsafe} = L_{\mathcal{P}} \setminus L_{psafe}$.

Observe that L_{psafe} is only a first approximation to M, since it is in general not controllable. In particular it may contain occurrence sequences which are not closed under extensions by outputs (first condition in definition 13). Such occurrence sequences must be cut at the last possible input (the last possibility of control), if there is one. The prefixes ending with these inputs are collected in the language L_{danger}. Deleting the futures of occurrence sequences in L_{danger} from L_{psafe} gives the language L_{safe}, which we will prove below to be the searched language M.

Definition 18. *We define*

$$\begin{aligned} L_{danger} = \{vj \in L_{psafe} \mid\ & j \in 2^{T \cup I \cup O}, j \cap I \neq \emptyset, (\exists v' \in [v]_T) \\ & \wedge (\exists j' \in 2^{T \cup I \cup O} : \lambda_{T \cup O}(j) = \lambda_{T \cup O}(j')) \\ & \wedge (\exists y \in (2^{T \cup O})^* : v'j'y \in L_{unsafe})\}. \end{aligned}$$

$$L_{safe} = L_{psafe} \setminus post(L_{danger}).$$

It is obvious from the definitions of L_{psafe}, L_{unsafe}, L_{danger} and L_{safe} that every equivalence class $[w]_T \cap L_{\mathcal{P}}$ is either subset of or disjoint to these languages.

The main result of this subsection is the following theorem:

Theorem 1. *L_{safe} is maximally permissive nonblocking controllable, if L_c is not unsatisfiable.*

Before proving this theorem we give an algorithm to compute L_{safe}: It is essentially shown, that L_{safe} can be constructed by appropriate operations on regular languages. We want to remark here that for computing the maximally permissive controllable language also the more involved framework presented in [2] could be adapted (since our different notion of controllability is still compatible with the union operation $\cup$). There can also be found some hints to the complexity of the computation.

Let us first see that the language L_{psafe} is regular. We use the following general construction to pump regular languages by new characters:

Definition 19. *Let A, A' be two finite, disjoint sets and α be a regular expression over the alphabet 2^A. We construct a regular expression $\alpha_{ext(A')}$ over the alphabet $2^{A\cup A'}$ by replacing every $x \in 2^A$ in α by*

$$ext_{A'}(x) = \left(\sum_{y\in 2^{A'}} y\right)^* \left(\sum_{y\in 2^{A'}} x\cup y\right) \left(\sum_{y\in 2^{A'}} y\right)^* .$$

By this construction we want to generate a regular expression whose corresponding language $L(\alpha_{ext})$ contains exactly all those words, which belong to $L(\alpha)$ when all characters of the alphabet A' are hidden. Later on, this construction will be used to pump L_c by $I \cup T$. Then L_{psafe} can be computed as the intersection of such pumped L_c and $L_{\mathcal{P}}$. Pumping a regular language by A' in the above way computes the preimage $\lambda_{A'}^{-1}()$ of this language w.r.t. the appropriate hiding operator:

Lemma 1. *Let A, A' be finite disjoint sets and α be a regular expression over the alphabet 2^A. Each $w \in \left(2^{A\cup A'}\right)^*$ satisfies*

$$\lambda_{A'}(w) \in L(\alpha) \Leftrightarrow w \in L(\alpha_{ext(A')}).$$

Proof. We will show both directions of the above Lemma.
'$\Rightarrow$': we show by structural induction over the construction rules of regular expressions that the above property holds true:

Let $\alpha = x \in 2^A \cup \{\epsilon\}$ (these are the constants of a regular expression over 2^A), and $w \in (2^{A\cup A'})^*$ satisfy $\lambda_{A'}(w) = x$. Then w is of the form $w = e_1 \ldots e_n$, $e_i \in 2^{A\cup A'}$, such that there exists an index i_0 satisfying $\lambda_{A'}(e_{i_0}) = x$ and $\lambda_{A'}(e_j) = \epsilon$ for $j \neq i_0$. It follows immediately from the construction above, that $w \in L(ext_{A'}(x)) = L(\alpha_{ext(A')})$.
Let α^1 and α^2 be two regular expressions over the alphabet 2^A satisfying the induction hypothesis, α^1_{ext} and α^2_{ext} be the corresponding extensions according to the above construction, and $w \in (2^{A\cup A'})^*$.

(i) Let $\alpha = \alpha^1 + \alpha^2$ and $\lambda_{A'}(w) \in L(\alpha)$. Then $\lambda_{A'}(w) \in L(\alpha^i)$, $i = 1$ or $i = 2$. By induction hypothesis $w = w_i$, $w_i \in L(\alpha^i_{ext})$, $i = 1$ or $i = 2$. So $w \in L(\alpha^1_{ext} + \alpha^2_{ext}) = L(\alpha_{ext})$, what follows immediately by the above construction.
(ii) Let $\alpha = \alpha^1\alpha^2$ and $\lambda_{A'}(w) \in L(\alpha)$. Then $w = w_1w_2$, such that $\lambda_{A'}(w_i) \in L(\alpha^i)$, $i = 1, 2$. By induction hypothesis $w = w_1w_2 \in L(\alpha^1_{ext})L(\alpha^2_{ext})$. So $w \in L(\alpha^1_{ext}\alpha^2_{ext}) = L(\alpha_{ext})$.

(iii) Let $\alpha = (\alpha^1)*$ and $\lambda_{A'}(w) \in L(\alpha)$. Then $w = w_1 \ldots w_n, \lambda_{A'}(w_i) \in L(\alpha^1), n \in \mathbb{N}$. By induction hypothesis $w = w_1 \ldots w_n \in L(\alpha^1_{ext})^*$. So $w \in L((\alpha^1_{ext})^*) = L(\alpha_{ext})$.

Observe that obviously $\alpha^1_{ext} + \alpha^2_{ext} = (\alpha^1 + \alpha^2)_{ext}$, $\alpha^1_{ext}\alpha^2_{ext} = (\alpha^1\alpha^2)_{ext}$ and $(\alpha^1_{ext})^* = ((\alpha^1)^*)_{ext}$.
'$\Leftarrow$': let be $w \in L(\alpha_{ext})$. It follows immediately by the construction of α_{ext} that $\lambda_{A'}(w) \in L(\alpha)$, because each constant $x \in 2^A$ of α is pumped up only by characters of $2^{A'}$. □

Corollary 1. *L_{psafe} is regular.*

Proof. Let α be a regular expression with $L(\alpha) = L_c$. Then according to Lemma 1

$$L_{psafe} = \{w \in L_{\mathcal{P}} \mid \lambda_{I \cup T}(w) \in L_c\} = L_{\mathcal{P}} \cap L(\alpha_{ext(I \cup T)}),$$

i.e. L_{psafe}, as the intersection of regular languages, is regular. □

Since $post(L)$ is obviously a regular language for regular languages L, it remains to show L_{danger} to be regular in order to show L_{safe} to be regular. Since with L_{psafe} also the language L_{unsafe} is regular, we can give L_{danger} as a simple formula on the regular languages $(2^{OUT})^*$ and L_{unsafe}. First observe that the regular language

$$L^{real}_{danger} = (L_{unsafe}/(2^{OUT})^*) \cap L_{\mathcal{P}},$$

where the symbol "/" denotes the quotient operation on languages, is the set of those words $vj \in L_{danger}$, which themselves can be extended by an $y \in (2^{OUT})^*$ to a word in L_{unsafe}. The remaining words in L_{danger} are of the form $v'j'$ with $v' \in [v]_T$ and $j' \cap I = j \cap I$ for a word $vj \in L^{real}_{danger}$. We get these words by means of a special defined hiding operator $\overline{\lambda} : (2^{OUIUT})^* \to (2^{OUIUT})^*$, defined by

$$\text{For } w = vx,\ v \in (2^{OUIUT})^*,\ x \in 2^{OUIUT} : \overline{\lambda}(w) = \lambda_T(v)\lambda_{T \cup O}(x),$$

and extended in the obvious way to languages. Obviously the operators $\overline{\lambda}$ and $\overline{\lambda}^{-1}$ preserve regularity of languages, since this is the case for the hiding operator λ as argued in lemma 1. We get

Lemma 2. $L_{danger} = \overline{\lambda}^{-1}[\overline{\lambda}(L^{real}_{unsafe})] \cap L_{psafe}$.

From now on we assume L_c to be satisfiable. The main theorem 1 now is shown in two steps by the following lemmata.

Lemma 3. *L_{safe} is controllable.*

Proof. First we have to show, that

$$\forall w \in L_{safe}, \forall o \in 2^{OUT} : wo \in L_{\mathcal{P}} \Rightarrow wo \in L_{safe}.$$

Assume $w \in L_{safe}$ and $o \in 2^{OUT}$ satisfying $wo \in L_{\mathcal{P}}$, but $wo \notin L_{safe}$. There are two cases:

- $wo \in L_{psafe}$: Then $wo \in post(L_{danger})$. This implies obviously $w \in post(L_{danger})$, what contradicts $w \in L_{safe}$.
- $wo \notin L_{psafe}$: Then by definition $wo \in L_{unsafe}$. Since L_c is satisfiable, w has a prefix in L_{danger}. This again contradicts $w \in L_{safe}$.

It remains to show, that

$$\forall vj \in L_{safe}, j \in 2^{I \cup OUT}, j \cap I \neq \emptyset, \forall j' \in 2^{I \cup OUT}, \lambda_{OUT}(j) = \lambda_{OUT}(j'),$$
$$\forall v' \in L_{safe}, v' \in [v]_T : \qquad v'j' \in L_{\mathcal{P}} \Rightarrow v'j' \in L_{safe}.$$

For vj and $v'j'$ as above we have according to the definition of L_{danger}:
$vj \in post(L_{danger}) \Leftrightarrow v'j' \in post(L_{danger})$. □

Lemma 4. *There is no language $K \subseteq L_{\mathcal{P}}$ satisfying $L_{safe} \subsetneq K$, which is controllable, and which fulfills $\lambda_{IUT}(K) \subseteq \overline{L_c}$.*

Proof. We choose a $w \in K \setminus L_{safe}$ and construct from w a $w' \in K$ satisfying $\lambda_{IUT}(w') \notin \overline{L_c}$. As $w \in L_{\mathcal{P}}$, there are two cases:

- $w \notin L_{psafe}$: Then $w \in L_{unsafe}$ and thus $\lambda_{IUT}(w) \notin L_c$.
- $w \in L_{psafe}$: Then $w \in post(L_{danger})$, i.e. w has a prefix $vj \in L_{danger}$. That means, there are $v' \in [v]_T$, $j' \in 2^{I \cup OUT}$ with $j \cap I = j' \cap I$ and $y \in 2^{OUT}$, such that $v'j'y \in L_{unsafe}$, i.e. $\lambda_{IUT}(v'j'y) \notin L_c$. Since K is controllable, $v'j'$ also belongs to K (second property), and consequently $v'j'y \in K$ (first property). □

It follows immediately from the above proof, that L_{safe} is the unique maximally permissive (nonblocking) language, analogously to related results in supervisory control.

4.2 Nonblocking Control

More general properties as for example the full execution of certain tasks cannot be formalized by a regular language L_c which is prefix closed. Of course a maximally permissive controllable language K w.r.t. a not prefix closed L_c, $L_{\mathcal{P}}$, I and O should contain occurrence sequences of the standalone of $L_{\mathcal{P}}$ which represent prefixes of words in L_c, but only such ones, which can be extended to a word in L_c within K, i.e. which are nonblocking. In other words the set of blocking words of K

$$K_{blocking} = \{r \in K \mid \nexists x \in (2^{O \cup IUT})^* : rx \in K \wedge \lambda_{IUT}(rx) \in L_c\}.$$

should be empty.

For this purpose replace L_c by $\overline{L_c}$ in the definitions of L_{psafe} and L_{unsafe} (definition 17). We now search for a sublanguage L_{nbsafe} of L_{safe}, which is controllable, nonblocking and maximal with these two properties according to definition 16. Since in our framework the special property, namely that every controllable event is also observable, is fulfilled, we are able to adapt a result in supervisory control ([2], subsection 3.7.5), which states under the assumption of this property: If there is at least one controllable language respecting L_c which is nonblocking, then there is a unique maximal one.

We compute L_{nbsafe}, if it exists, in two steps:
First we represent L_{safe} by a finite automaton A which
– separates by its states words respecting L_c from words not respecting L_c:
$\forall w \in (2^{I \cup OUT})^*, \forall v \in [w]_A : \lambda_{I \cup T}(v) \in L_c \Leftrightarrow \lambda_{I \cup T}(w) \in L_c$, and
– separates by its states words, for which undistinguishable words with different futures exist.
Second we recursively delete input edges in A, which can lead to blocking words.

Such an automaton always exists. We omit the construction of A due to space limitations. For a detailed investigation see [11].

According to the first property of A we can define the set of states $S_c = \{[w]_A \mid \lambda_{I \cup T}(w) \in L_c\}$ representing L_c. So finding blocking words translates to finding blocking states, from which there is no continuation in A to a state in S_c. The following lemma directly follows:

Lemma 5. *Let $K = L_{safe}$ and let $r \in K$.*

- *$r \in K_{blocking}$, iff every $x \in (2^{I \cup OUT})^*$ with $rx \in L_{safe}$ fulfills $[rx]_A \notin S_c$.*
- *If $r \in K_{blocking}$ and $[r]_A = [r']_A$ for a word $r' \in K$, then $r' \in K_{blocking}$.*
- *If $r \in K_{blocking}$ and $r \in M \subseteq K$, then $r \in M_{blocking}$.*

That means we can construct L_{nbsafe} from L_{safe} by deleting edges in A representing inputs, which lead to blocking states in the sense of the above lemma in a maximally permissive way. Of course such edges must be deleted in all undistinguishable paths in A (see step 4 of the algorithm later on). The second property of A ensures that two different words ending with the same such edge in A, are both undistinguishable from words with a blocking future (i.e. by deleting such edges no futures of words, which have no blocking future, are cut). Since the deletion of an edge can produce new blocking states, the procedure is iterative. As for controllability (condition (1)), there is a condition saying when we cannot find a controllable nonblocking sublanguage: Every controllable sublanguage of L_{safe} contains all words in L_{safe} of the form $w \in (2^{OUT})^*$. Therefore:

Lemma 6. *Let $M \subseteq L_{safe}$. If there is a $w \in M_{blocking} \cap (2^{OUT})^*$, then there is no controllable sublanguage of M, which is nonblocking w.r.t. L_c.*

We call this condition *blocking condition*. It is only sufficient for nonexistence of a controllable nonblocking sublanguage, but not necessary. We are now prepared to state the algorithm:

Input: Automaton $A^0 = A$, Integer $k = 0$
Output: Automaton A_{nbsafe}, if L_{nbsafe} exists
Step 1: If $L(A^k)$ fulfills the blocking condition, **then return** "L_{nbsafe} *does not exist*"
Step 2: If $(L(A^k))_{blocking} = \emptyset$, **then return** A^k
Else Choose $w \in (L(A^k))_{blocking}$
Step 3: Compute a prefix vj of w with $j \cap I \neq \emptyset$, such that $w = vjy$ for $y \in (2^{OUT})^*$
Step 4: Compute the set of states $S^k_{delarc} = \{[u]_{A^k} \mid u \in [v]_T\}$
Step 5: Delete every edge starting in any state $[u]_{A^k} \in S^k_{delarc}$ with a label i fulfilling $i \cap I = j \cap I$

Step 6: Set $k = k + 1$
Set A^k to be the new constructed automaton
Goto Step 1

Let us state the main theorem of this subsection:

Theorem 2. *There exists a maximally permissive nonblocking controllable sublanguage of L_{safe}, if and only if the previous algorithm outputs an automaton A_{nbsafe}. In this case $L(A_{nbsafe})$ is this searched sublanguage.*

Proof. Let $A_{safe} = A^0, \ldots, A^{N_0}$ be the sequence of automata the algorithm has computed until it has stopped.

We first show the "only if"-part:
Assume the previous algorithm outputs "L_{nbsafe} *does not exist*". We have to show, that there is no maximally permissive nonblocking controllable sublanguage of L_{safe}. For this it is enough to prove, that every controllable sublanguage K of L_{safe} must contain a blocking word w.r.t. K.

According to lemma 5 it is enough to find a word in K which is blocking w.r.t. L_{safe}. By **Step 1** $L(A^{N_0})$ fulfills the blocking condition, i.e.

$$\exists v_0 \in (2^{OUT}) : \; v_0 \in L^{N_0}_{blocking}.$$

Since K is assumed to be controllable, K contains all words in $L_{safe} \cap [v_0]_T$. If one of these words is blocking w.r.t. L_{safe}, we are done.

So assume all $u \in L_{safe} \cap [v_0]_T$ not to be blocking w.r.t. L_{safe}. That means every such u can be extended by an $y \neq \epsilon$ (remark that $\lambda_{T \cup I}(v_0) \notin L_c$!) such that $[uy]_{A^0} \in S_c$, in particular v_0. Observe that

- If none of all such possible extensions y of v_0 is in K, we are done (this would imply v_0 to be blocking w.r.t. K). So assume that there is at least one such extension $v_0 y_0 \in K$ with $[v_0 y_0]_{A^0} \in S_c$.
- We have $v_0 y_0 \in L(A^0) = L_{safe}$ but $v_0 y_0 \notin L(A^{N_0})$ (since v_0 blocking w.r.t. L^{N_0}).

Therefore there must be an index $N_1 < N_0$ and a prefix xi of y_0 (with $x \in (2^{I \cup OUT})^*$, $i \in 2^{I \cup OUT}$, $i \cap I \neq \emptyset$), such that $[v_0 x]_{A^{N_1}} \in S^{N_1}_{delarc}$ and the edge starting in state $[v_0 x]_{A^{N_1}}$ with label i was deleted from A^{N_1} in **Step 5**. This deletion was caused by the existence of a $v_1 \in L^{N_1}_{blocking}$ of the form $v_1 = w_1 j w_2$, where $j \in 2^{I \cup OUT}$, $w_1 \in (2^{I \cup OUT})^*$, and

$$(a)\; w_1 \in [v_0 x]_T \cap L_{safe}, \quad (b)\; j \cap I = i \cap I \quad \text{and} \quad (c)\; w_2 \in (2^{OUT})^*.$$

Remember now that all prefixes of $v_0 y_0$, in particular $v_0 x$ and $v_0 xi$, belong to K. Since K is assumed to be controllable, K contains all words in $L_{safe} \cap [v_0 x]_T$, in particular w_1 (property (a)). From the second condition of controllability and (b) we get further $w_1 j \in K$, and therefore $v_1 = w_1 j w_2 \in K$ (first condition of controllability and (c)).

By repeating this construction we get an strictly decreasing sequence of natural numbers $N_0 > N_1 > \ldots$ and associated words $v_0, v_1, \ldots \in K$, such that $v_i \in L_{blocking}^{N_i}$, $i = 0, 1, \ldots$. Finally $N_k = 0$ for some k, which implies $v_k \in K$ blocking w.r.t. K.

Next we consider the "if"-part:
By construction $L(A^{N_0}) = L(A_{nbsafe})$ is controllable and nonblocking. It remains to show that it is maximally permissive with these two properties. Assume another language K to be maximally permissive nonblocking controllable w.r.t. L_c, $L_{\mathcal{P}}$, I and O satisfying $L(A_{nbsafe}) \subsetneq K \subseteq L_{safe}$. We will construct inductively a blocking word in K.

There is a $x \in K \setminus L(A_{nbsafe})$. As $L(A_{nbsafe})$ and K are prefix closed we can assume (without loss of generality) $x = wj$, $w \in L(A_{nbsafe})$, $j \in 2^{I \cup OUT}$, $j \cap I \neq \emptyset$. Because K is controllable, we have

$$\forall y \in (2^{OUT})^*, \forall w' \in [w]_T, \forall j' \in 2^{I \cup OUT}, j' \cap I = j \cap I : \\ w'j'y \in L_{safe} \Rightarrow w'j'y \in K.$$

Since in some step $N_1 < N_0$ the edge starting from the state $[w]_{A^0}$ with label j was deleted, one of these words $w'j'y$, call it v_0, must be blocking w.r.t. $L(A^{N_1})$. Moreover, analogously to the "only if"-part, either one of these words is blocking w.r.t. L_{safe} (and we are done), or all of them must have an extension within K to a word respecting L_c, in particular v_0. Let $v_0 y_0$ be this extension of v_0. Proceed now as in the "only if"-part. □

5 Conclusion

In this paper we have presented a methodology for synthesis of the controlled behavior of discrete event systems employing actuators which try to force events and sensors which can prohibit event occurrences. As a modelling formalism, we have used modules of signal nets. The signal nets offer a direct way to model typical actuators behavior. Another advantage of such modules consists in supporting input/output structuring, modularity and compositionality in an intuitive graphical way.

In the paper we were not focusing on complexity issues. It is known that the complexity of the supervisory control problem is in general PSPACE-hard, and sometimes even undecidable ([16], pp. 15 - 36). To get efficient algorithms one has to restrict the setting in some way, for example by considering only special kinds of specifications.

References

1. B. Caillaud, P. Darondeau, L. Lavagno and X. Xie (Eds.). Synthesis and Control of Discrete Event Systems Kluwer Academic Press, 2002
2. C. G. Cassandras and S. Lafortune. Introduction to Discrete Event Systems. Kluwer, 1999.
3. J. Desel, G. Juhás and R. Lorenz. Input/Output Equivalence of Petri Modules. In *Proc. of IDPT 2002*, Pasadena, USA, 2002.

4. H-M. Hanisch and A. Lüder. A Signal Extension for Petri nets and its Use in Controller Design. *Fundamenta Informaticae*, 41(4) 2000, 415–431.
5. H.-M. Hanisch, A. Lüder, M. Rausch: Controller Synthesis for Net Condition/Event Systems with Incomplete State Observation, European Journal of Control, Nr. 3, 1997, S. 292-303.
6. H.-M. Hanisch, J. Thieme and A. Lüder. Towards a Synthesis Method for Distributed Safety controllers Based on Net Condition/Event Systems. *Journal of Intelligent Manufacturing*, 5 ,1997, 8, 357-368.
7. L.E. Holloway, B.H. Krogh and A. Giua. A Survey of Petri Net Methods for Controlled Discrete Event Systems. *Discrete Event Dynamic Systems: Theory and Applications*, 7, 1997), 151–190.
8. J.E. Hopcroft, R. Motwani and J.D.Ullman. Introduction to Automata Theory, Languages, and Computation. Addison Wesley, 2001.
9. R. Janicki and M. Koutny. Semantics of Inhibitor Nets. *Information and Computations*, 123, pp. 1–16, 1995.
10. G. Juhás. On semantics of Petri nets over partial algebra. In J. Pavelka, G. Tel and M. Bartosek (Eds.) *Proc. of 26th Seminar on Current Trends in Theory and Practice of Informatics SOFSEM'99*, Springer, LNCS 1725, pp. 408-415, 1999.
11. G. Juhás, R. Lorenz and C. Neumair. Modelling and Control with Modules of Signal Nets. To appear in Advanced in Petri Nets, LNCS, Springer 2004.
12. J. Desel, H. -M- Hanisch, G. Juhás, R. Lorenz and C. Neumair. A Guide to Modelling and Control with Modules of Signal Nets. To appear in LNCS, Springer 2004.
13. L.E. Pinzon, M.A. Jafari, H.-M. Hanisch and P. Zhao Modelling admissible behavior using event signals submitted
14. P.J. Ramadge, W.M. Wonham: The Control of Discrete Event Systems. Proceedings of the IEEE, 77 (1989) 1, S. 81-98.
15. G. Rozenberg, and J. Engelfriet. Elementary Net Systems. In W. Reisig and G. Rozenberg (Eds.) *Lectures on Petri Nets I: Basic Models*, Springer, LNCS 1491, pp. 12-121, 1998.
16. P. Darondeau and S. Kumagai (Eds.). Proceedings of the *Workshop on Discrete Event System Control*. Satellite Workshop of ATPN 2003.
17. M.C. Zhou und F. DiCesare. *Petri Net Synthesis for Discrete Event Control of Manufacturing Systems.* Kluwer Adacemic Publishers, Boston, MA, 1993.

New Canonical Representative Marking Algorithms for Place/Transition-Nets

Tommi A. Junttila*

Helsinki University of Technology, Lab. for Theoretical Computer Science
P.O. Box 5400, FIN-02015 HUT, Finland
Tommi.Junttila@hut.fi
Currently at ITC-IRST, Via Sommarive 18, 38050 Povo, Trento, Italy

Abstract. Symmetries of a Place/Transition-net can be exploited during the state space analysis by considering only one representative marking in each orbit induced by the symmetries. This paper describes two new algorithms for the core problem of transforming a marking into an equivalent, canonical representative marking. The algorithms are based on a backtrack search through all the symmetries of the net. The first algorithm prunes the search with the marking in question and its stabilizers that are found during the search. The second algorithm first applies a standard preprocessing technique in graph isomorphism algorithms to obtain an ordered partition of the net elements in a symmetry-respecting way. The partition is then used to further prune the search through the symmetries. The efficiency of the proposed algorithms is experimentally evaluated. The results show that the new algorithms usually outperform the previous ones implemented in the LoLA tool.

1 Introduction

Symmetries of a Place/Transition-net produce symmetries into its state space [1]. These state space symmetries can be exploited during the state space analysis by considering only one (or few) marking(s) in each set of markings equivalent under the symmetries. This may result in exponential savings in both memory and time requirements of the state space analysis. The core problem during the generation of the reduced reachability graph is to decide whether a marking equivalent to the newly generated one is already visited during the search. This can be accomplished by either (i) comparing the new marking for equivalence pairwise with each already visited one, or (ii) transforming each generated marking into an equivalent *canonical representative* and storing only these into the reduced reachability graph. Some algorithms for these tasks are presented in [2, 3], while the computational complexity of the tasks is analyzed in [4, 5].

This paper describes and experimentally evaluates the efficiency of two new algorithms for producing canonical representative markings. The algorithms presented require that the symmetry group of the net is known. This is in contrast

* The financial support of the Academy of Finland (project no. 53695) is gratefully acknowledged. This work has also been sponsored by the CALCULEMUS! IHP-RTN EC project, contract code HPRN-CT-2000-00102.

J. Cortadella and W. Reisig (Eds.): ICATPN 2004, LNCS 3099, pp. 258–277, 2004.
© Springer-Verlag Berlin Heidelberg 2004

to some algorithms described in [2, 3]. The fact that the algorithms depend on precalculation of a Schreier-Sims representation for the symmetry group of the net is not a serious drawback. This is because it is beneficial to first compute the symmetry group of the net in order to see if there are any non-trivial symmetries, i.e., to see whether the symmetry reduction method can help at all. In addition, the performance of symmetry reduction algorithms may depend on the size of the symmetry group, see [3] and Sect. 5.2, and thus knowing it may help in selecting an appropriate algorithm.

The first proposed algorithm in Sect. 3 is a backtrack search in the Schreier-Sims representation of the symmetry group. The algorithm returns the smallest marking produced by symmetries that are "compatible" with the marking as the canonical representative. The search is pruned (i) by considering only symmetries that are "compatible" with the marking, (ii) by using the smallest already found equivalent marking, and (iii) by exploiting the stabilizers of the marking (that are found during the search). The algorithm is a variant of the backtrack search algorithms developed in the computational group theory, see e.g. [6]. However, the compatibility definition between symmetries and markings is, to author's knowledge, novel.

The second algorithm in Sect. 4 exploits a standard preprocessing technique applied in graph isomorphism algorithms such as [7, 8]. Given a marking, this technique produces an ordered partition of the net elements in a symmetry-respecting way. The partition is then used to prune the backtrack search in the Schreier-Sims representation by considering only symmetries that are compatible with the partition.

The proposed algorithms have been implemented in the LoLA reachability analyzer [9]. Some experimental results are provided, too. They show that the new algorithms are competitive againts the previous ones implemented in LoLA.

Due to space limitations, proofs are omitted but can be found in [5].

1.1 Related Work

Some algorithms for the marking equivalence problems under Place/Transition-nets are described in [2, 3]. The first algorithm, "iterating the symmetries", applies all the symmetries to the new marking and checks whether the resulting marking has already been visited during the reduced reachability graph construction. The facts that (i) the symmetries are stored in a Schreier-Sims representation and (ii) the set of already visited markings is stored as a prefix sharing decision tree, are exploited to prune the set of symmetries considered. The second algorithm, "iterating the states", checks the newly generated marking pairwise with each already visited marking for equivalence by using an algorithm described in [2]. The set of necessary equivalence checks is reduced by using symmetry-respecting hash functions. The third algorithm, "canonical representatives", computes a (non-canonical) representative for the newly generated marking. This is done by a bounded search with greedy heuristics in the Schreier-Sims representation of symmetries, trying to find the lexicographically smallest equivalent marking. The first algorithm in this paper can be seen as a complete, canonical version of this algorithm augmented with some non-trivial pruning techniques.

In addition to Place/Transition-nets, symmetries can be exploited in the state space analysis of other formalisms, too. For the symmetry reduction method in general and temporal logic model checking under symmetry, see e.g. [10–15].

In the Murφ system [13] and in high-level Petri nets [10, 16, 17], the symmetries are produced by permuting the values of data types. Usually the symmetries in these formalisms don't have to be represented explicitly but are defined by declaring some data types to be permutable (e.g., the scalar set data type in Murφ). Despite these differences to Place/Transition-nets, the algorithms in this paper have some similarities to some proposed for these formalisms.

1. The obvious state canonization algorithm enumerating through all the states that are equivalent to the given state is not usually very effective for the above mentioned formalisms, see e.g. [18]. The first algorithm in this paper uses pruning techniques that make this approach work reasonably well for Place/Transition-nets.
2. Partitions are used to prune the set of symmetries that have to be considered not only in the second algorithm in this paper but also in many other algorithms, e.g. [18–21]. However, the second algorithm in this paper is to author's knowledge the first one that can work under general symmetries (i.e., symmetries that are not direct products of symmetric and/or cyclic groups like those induced by scalar sets). This is achieved by using a Schreier-Sims representation for the symmetry group and the novel compatibility definition between permutations and partitions.

Some other canonical representative algorithms for high-level Petri nets are presented in [16, 22].

2 Place/Transition-Nets and Their Symmetries

The following standard definitions of Place/Transition-nets are based on [1–3]. A *Place/Transition-net* (or simply a *net*) is a tuple $N = \langle P, T, F, W, M_0 \rangle$, where

1. P is a finite, nonempty set of *places*,
2. T is a finite, nonempty set of *transitions* such that $P \cap T = \emptyset$,
3. $F \subseteq (P \times T) \cup (T \times P)$ is the set of *arcs*,
4. $W : F \to \mathbb{N} \setminus \{0\}$ associates each arc a positive *weight*, and
5. $M_0 : P \to \mathbb{N}$ is the *initial marking* of N. A *marking* of N is a multiset on P, i.e., a function $M : P \to \mathbb{N}$ [1]. The *set of all markings* of N is denoted by $\mathbb{M}$.

The weight function W is implicitly extended to $(P \times T) \cup (T \times P) \to \mathbb{N}$ by $W(\langle x, y \rangle) = 0$ if $\langle x, y \rangle \notin F$. A transition $t \in T$ is *enabled* in a marking M if

[1] A multiset $m : A \to \mathbb{N}$ over a set A can be denoted by the formal sum $\sum_{a \in A} m(a)'a$. Elements with multiplicity 0 can be dropped and unit multiplicities omitted, e.g., the multi-set $M = \{p_1 \mapsto 1, p_2 \mapsto 3, p_3 \mapsto 0\}$ can be denoted by $p_1 + 3'p_2$.

$W(\langle p,t\rangle) \le M(p)$ for each $p \in P$. If t is enabled in M, it may *fire* and transform M into M', denoted by $M\ [t\rangle\ M'$ and defined by $M'(p) = M(p) - W(\langle p,t\rangle) + W(\langle t,p\rangle)$ for each $p \in P$. The *state space* of N is the labeled transition system $\langle \mathbb{M}, T, [\rangle, M_0\rangle$, where $[\rangle = \{\langle M,t,M'\rangle \in \mathbb{M} \times T \times \mathbb{M} \mid M\ [t\rangle\ M'\}$. The *reachability graph* of N is the part of the state space reachable from the initial marking.

Example 1. Consider the variant of a railroad system net [23] shown in Fig. 1. All the arc weights in the net equal to 1 and are not drawn here or in any subsequent figures. The initial marking of the net is $U_{a0} + U_{b3} + V_1 + V_4$ and its reachability graph is shown in the left hand side of Fig. 2.

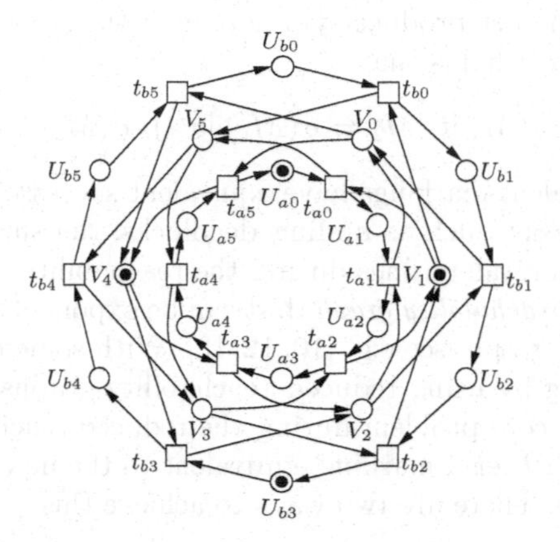

Fig. 1. A railroad net.

A *symmetry* (an *automorphism*) of the net N is a permutation σ of $P \cup T$ that respects

- the node types: $\sigma(P) = P$ and $\sigma(T) = T$,
- the arcs: $\langle x,y\rangle \in F \Leftrightarrow \langle \sigma(x), \sigma(y)\rangle \in F$, and
- the arc weights: $W(\langle x,y\rangle) = W(\langle \sigma(x), \sigma(y)\rangle)$ for each $\langle x,y\rangle \in F$.

The set of *all symmetries of* N (the *automorphism group* of N) is denoted by $\mathrm{Aut}(N)$ and is a group under the function composition operator $\circ$. A symmetry $\sigma \in \mathrm{Aut}(N)$ *acts* on the markings of N by mapping a marking M to the marking $\sigma(M)$ defined by $\sigma(M) : \sigma(p) \mapsto M(p)$ for each place $p \in P$. Two markings, M_1 and M_2, are *equivalent under a subgroup G of* $\mathrm{Aut}(N)$, denoted by $M_1 \equiv_G M_2$, if there is a $\sigma \in G$ such that $\sigma(M_1) = M_2$. By the group properties of G, $\equiv_G$ is an equivalence relation on the set $\mathbb{M}$. The equivalence class of a marking M, called the *G-orbit of M*, is denoted by $[M]_G$. In the case G is understood, one may simply speak of equivalent markings and orbits, and omit the subscript G. Furthermore, the *stabilizer subgroup of a marking M in G* is $\mathrm{Stab}(G,M) = \{\sigma \in G \mid \sigma(M) = M\}$, i.e., the set of all symmetries in G that fix M.

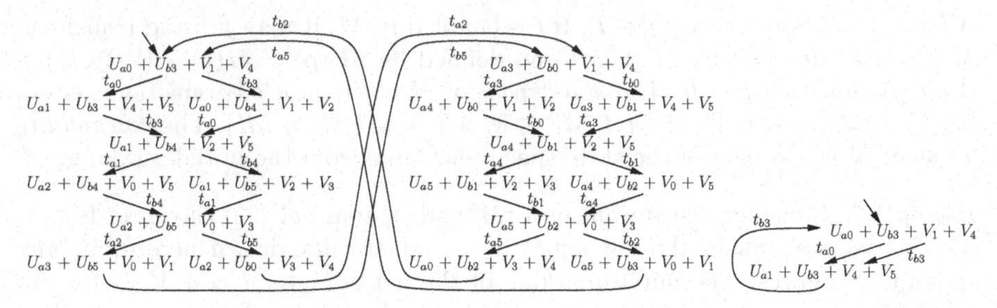

Fig. 2. The reachability graph and a reduced reachability graph of the net in Fig. 1.

Symmetries of the net produce symmetries to the state space of the net [1]: for each symmetry σ it holds that

$$M_1\ [t\rangle\ M_2 \Leftrightarrow \sigma(M_1)\ [\sigma(t)\rangle\ \sigma(M_2)$$

meaning that equivalent markings have equivalent successor markings. Thus for many verification tasks, such as finding deadlocks, the successor markings can be "redirected" to equivalent ones during the reachability graph generation, resulting in a *reduced reachability graph* that can be exponentially smaller than the original reachability graph, see e.g. [10, 12, 13]. With some extensions, temporal logic model checking by using reduced reachability graphs is also possible, see e.g. [11, 14, 15]. The core problem during the reduced reachability graph generation is to decide whether a marking equivalent to the newly generated one has already been visited. There are two ways to achieve this.

1. The newly generated marking is compared for equivalence pairwise with all the markings in the set of already visited markings. Symmetry-respecting hash functions can be used to prune the set of markings that have to be tested [2].
2. The new marking is transformed into an equivalent *canonical representative marking.* Only the canonical representatives are stored in the reachability graph. Formally, a function $repr : \mathbb{M} \to \mathbb{M}$ is a *representative marking function* if $repr(M) \equiv M$ for all markings $M \in \mathbb{M}$. A representative marking function *repr* is *canonical* if $M_1 \equiv M_2$ implies $repr(M_1) = repr(M_2)$. In this case, the marking $repr(M)$ is the *canonical representative* of M (under *repr*).

This paper contributes to the second approach, describing new algorithms for computing canonical representative functions.

Example 2. The symmetries of the net N in Fig. 1 are those that are generated by the compositions of (i) the rotation of the railroad sections 0–5:

$$\sigma_{\text{rot}} = \begin{pmatrix} U_{a0} & U_{a1} & U_{a2} & U_{a3} & U_{a4} & U_{a5} & U_{b0} & \cdots & U_{b5} & V_0 & \cdots & V_5 & t_{a0} & \cdots & t_{a5} & t_{b0} & \cdots & t_{b5} \\ U_{a1} & U_{a2} & U_{a3} & U_{a4} & U_{a5} & U_{a0} & U_{b1} & \cdots & U_{b0} & V_1 & \cdots & V_0 & t_{a1} & \cdots & t_{a0} & t_{b1} & \cdots & t_{b0} \end{pmatrix}$$

and (ii) the swapping of train identities a and b:

$$\sigma_{\text{swap}} = \begin{pmatrix} U_{a0} & \cdots & U_{a5} & U_{b0} & \cdots & U_{b5} & V_0 & \cdots & V_5 & t_{a0} & \cdots & t_{a5} & t_{b0} & \cdots & t_{b5} \\ U_{b0} & \cdots & U_{b5} & U_{a0} & \cdots & U_{a5} & V_0 & \cdots & V_5 & t_{b0} & \cdots & t_{b5} & t_{a0} & \cdots & t_{a5} \end{pmatrix}.$$

The group $\mathrm{Aut}(N)$ has 12 elements. The initial marking $M_0 = U_{a0} + U_{b3} + V_1 + V_4$ is equivalent (under $\mathrm{Aut}(N)$) to the marking $M = U_{a4} + U_{b1} + V_2 + V_5$ as $(\sigma_{\mathrm{swap}} \circ \sigma_{\mathrm{rot}})(M_0) = \sigma_{\mathrm{swap}}(\sigma_{\mathrm{rot}}(M_0)) = \sigma_{\mathrm{swap}}(U_{a1} + U_{b4} + V_2 + V_5) = M$. The $\mathrm{Aut}(N)$-orbit of M_0 consists of the markings M_0, $U_{a1} + U_{b4} + V_2 + V_5$, $U_{a2} + U_{b5} + V_0 + V_3$, $U_{a3} + U_{b0} + V_1 + V_4$, $U_{a4} + U_{b1} + V_2 + V_5$, and $U_{a5} + U_{b2} + V_0 + V_3$. A reduced reachability graph for the net is shown in the right hand side of Fig. 2.

2.1 The Schreier-Sims Representation

Although a permutation group on a set of n elements may have up to $n!$ permutations, there are polynomial size representations for permutation groups. The following describes one such representation having some useful properties. For more on permutation groups, see [6]. The presentation here is based on [8].

Assume a finite set X of size n and a permutation group G on X. Order the elements in X in any order $\beta = [x_1, x_2, \ldots, x_n]$ called the *base*. Let $G_0 = G$ and $G_i = \{g \in G_{i-1} \mid g(x_i) = x_i\}$ for $1 \leq i \leq n$. The groups $G_0, G_1, \ldots, G_n$ are subgroups of G such that $G = G_0 \geq G_1 \geq \cdots \geq G_n = \{\mathbf{I}\}$, where $\mathbf{I}$ denotes the identity permutation. A permutation $g \in G_i$, $0 \leq i \leq n$, fixes each element $x_1, \ldots, x_i$. For each $1 \leq i \leq n$, let $[x_i]_{G_{i-1}} = \{g(x_i) \mid g \in G_{i-1}\}$ denote the orbit of x_i under G_{i-1}. Assume that $[x_i]_{G_{i-1}} = \{x_{i,1}, x_{i,2}, \ldots, x_{i,n_i}\}$ for some $1 \leq n_i \leq n$. For each $1 \leq j \leq n_i$, choose a $h_{i,j} \in G_{i-1}$ such that $h_{i,j}(x_i) = x_{i,j}$ and let $U_i = \{h_{i,1}, h_{i,2}, \ldots, h_{i,n_i}\}$. Now U_i is a left transversal of G_i in G_{i-1}, i.e., $h_{i,j} \circ G_i \neq h_{i,k} \circ G_i$ for $j \neq k$ and $G_{i-1} = h_{i,1} \circ G_i \cup \cdots \cup h_{i,n_i} \circ G_i$, where $h \circ G_i$ denotes the left coset $\{h \circ g \mid g \in G_i\}$. The structure $\boldsymbol{G} = [U_1, U_2, \ldots, U_n]$ is a *Schreier-Sims representation* of the group G. Each element in $g \in G$, and only those, can be uniquely written as a composition $g = h_1 \circ h_2 \circ \cdots \circ h_n$, where $h_i \in U_i$, and thus $|G| = \prod_{i=1}^{n} |U_i|$. It can be and is assumed from now on that each U_i contains the identity permutation $\mathbf{I}$. There are at most $n(n+1)/2$ permutations in the Schreier-Sims representation $\boldsymbol{G}$ [2]. Many operations, such as testing whether a permutation belongs to the group, can be performed in polynomial time by using Schreier-Sims representations. The ground sets in [2, 3] are actually Schreier-Sims representations. Thus the algorithm presented in [2] for computing the symmetries of a net produces Schreier-Sims representations.

Example 3. Consider the net in Fig. 3. Its automorphism group, call it G, under the base $\beta = [p_1, p_2, p_3, p_4, t_{1,2}, t_{2,1}, t_{2,3}, t_{3,2}, t_{3,4}, t_{4,3}, t_{4,1}, t_{1,4}]$ has a Schreier-Sims representation $\boldsymbol{G} = [U_1, U_2, \ldots, U_{|P|+|T|}]$, where

$$U_1 = \left\{ \begin{array}{l} h_{1,1} = \mathbf{I} \\ h_{1,2} = \begin{pmatrix} p_1 & p_2 & p_3 & p_4 & t_{1,2} & t_{2,1} & t_{2,3} & t_{3,2} & t_{3,4} & t_{4,3} & t_{4,1} & t_{1,4} \\ p_2 & p_3 & p_4 & p_1 & t_{2,3} & t_{3,2} & t_{3,4} & t_{4,3} & t_{4,1} & t_{1,4} & t_{1,2} & t_{2,1} \end{pmatrix} \\ h_{1,3} = \begin{pmatrix} p_1 & p_2 & p_3 & p_4 & t_{1,2} & t_{2,1} & t_{2,3} & t_{3,2} & t_{3,4} & t_{4,3} & t_{4,1} & t_{1,4} \\ p_3 & p_4 & p_1 & p_2 & t_{3,4} & t_{4,3} & t_{4,1} & t_{1,4} & t_{1,2} & t_{2,1} & t_{2,3} & t_{3,2} \end{pmatrix} \\ h_{1,4} = \begin{pmatrix} p_1 & p_2 & p_3 & p_4 & t_{1,2} & t_{2,1} & t_{2,3} & t_{3,2} & t_{3,4} & t_{4,3} & t_{4,1} & t_{1,4} \\ p_4 & p_1 & p_2 & p_3 & t_{4,1} & t_{1,4} & t_{1,2} & t_{2,1} & t_{2,3} & t_{3,2} & t_{3,4} & t_{4,3} \end{pmatrix} \end{array} \right\},$$

[2] A more compact representation consisting of at most $n-1$ permutations could also be used instead [24].

$$U_2 = \left\{ \begin{array}{l} h_{2,1} = \mathbf{I} \\ h_{2,2} = \begin{pmatrix} p_1 & p_2 & p_3 & p_4 & t_{1,2} & t_{2,1} & t_{2,3} & t_{3,2} & t_{3,4} & t_{4,3} & t_{4,1} & t_{1,4} \\ p_1 & p_4 & p_3 & p_2 & t_{1,4} & t_{4,1} & t_{4,3} & t_{4,2} & t_{3,2} & t_{2,3} & t_{2,1} & t_{1,2} \end{pmatrix} \end{array} \right\},$$

and $U_i = \{\mathbf{I}\}$ for $3 \leq i \leq |P| + |T|$. Therefore, $|G| = 8$.

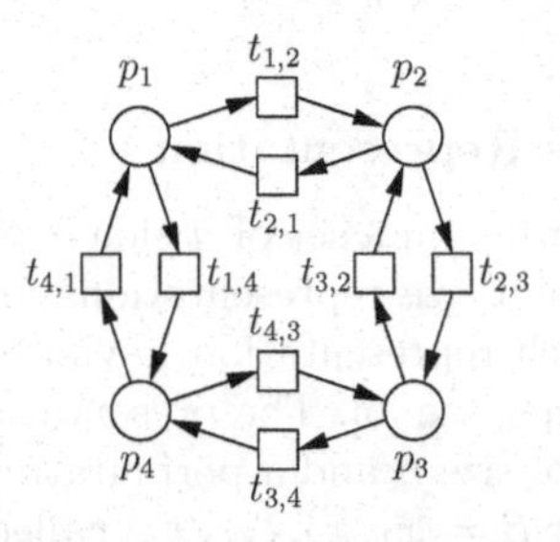

Fig. 3. A toy example net.

2.2 Place Valuations and Compatible Permutations

In addition to the standard Schreier-Sims representation definitions above, some new concepts are needed. Assume a net $N = \langle P, T, F, W, M_0 \rangle$, a subgroup G of $\mathrm{Aut}(N)$, and a Schreier-Sims representation $\boldsymbol{G} = [U_1, \ldots, U_{|P|+|T|}]$ of G under a base $\beta = [p_{\beta,1}, \ldots, p_{\beta,|P|}, t_{\beta,1}, \ldots, t_{\beta,|T|}]$ in which the places are enumerated before the transitions[3]. The Schreier-Sims representation $\boldsymbol{G}$ can be seen as a tree. The levels of the tree correspond to the base of the representation and each node at a level i has $|U_i|$ children at the level $i+1$, the edges to the children being labeled with the permutations in U_i. For instance, Fig. 4 shows a prefix of the tree corresponding to the Schreier-Sims representation in Ex. 3 (the permutation labels are of form $h_{i,j}$). Consider a path in the tree starting from the root and ending in a node v at a level i. Composing the labels of the edges in the path defines the corresponding permutation in $g \in U_1 \circ \cdots \circ U_{i-1}$. Thus the full paths ending in leaf nodes of the tree define exactly the permutations in the group. The node v has $|U_i|$ child nodes, and extending the path to any of them defines an extension permutation of g which is in $g \circ U_i$. The set $\{g(h(x_i)) \mid h \in U_i\}$ is now the set of $|U_i|$ possible images of the ith base element x_i under all the permutations corresponding to the paths going through the node v. Figure 4 also shows the base element images (that are of form p_i).

Now consider a *place valuation* function $pval : P \to \mathbb{N}$ assigning each place a natural number[4]. Observe that the definition is exactly the same as for markings, a different name is only used in order to avoid confusions later. The action of

[3] For representation reasons, it is assumed that the subgroup of G fixing all the places is the trivial group so that $U_j = \{\mathbf{I}\}$ for each $|P| < j \leq |P| + |T|$. That is, the net does not contain identical transitions.

[4] *pval* is implicitly extended to $P \cup T$ by defining that $pval(t) = 0$ for each $t \in T$.

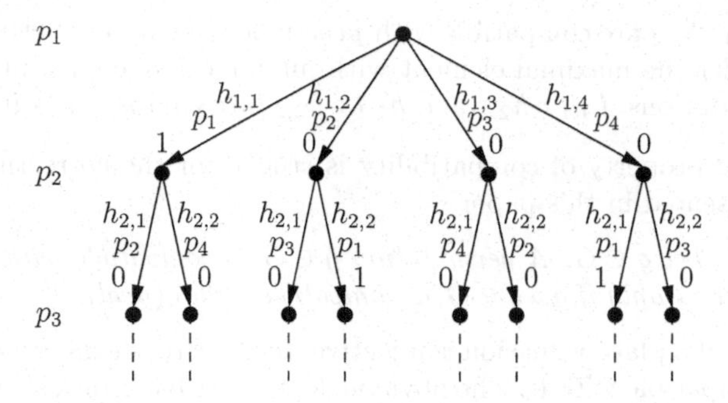

Fig. 4. Schreier-Sims representation seen as a tree.

permutations in $\mathrm{Aut}(N)$ on place valuations is defined similarly to that on markings: $g(\mathit{pval}) : g(p) \mapsto \mathit{pval}(p)$ for each $p \in P$. The set $\{g(h(x_i)) \mid h \in U_i\}$ of base element images in a node v now defines the multiset $\sum_{h \in U_i} 1'\mathit{pval}((g \circ h)(x_i))$ over natural numbers, see Fig. 4 for an example when the place valuation is $\mathit{pval} = \{p_1 \mapsto 1, p_2 \mapsto 0, p_3 \mapsto 0, p_4 \mapsto 0\}$.

The idea now is to select only a subset of children of the node v based on the multiset $\sum_{h \in U_i} 1'\mathit{pval}((g \circ h)(p_{\beta,i}))$. This is done by applying a *multiset selector* to the multiset. Formally, a multiset selector is a function from nonempty multisets over natural numbers to natural numbers such that the image has a non-zero multiplicity in the argument multiset. That is, if *select* is a multiset selector and $\mathit{select}(m) = n$, then $m(n) \geq 1$. For instance, the *minimal element* multiset selector returns the smallest number that has non-zero multiplicity in the argument multiset: $\mathit{select}_{\mathrm{min}}(3'2 + 2'4 + 2'5 + 4'7) = 2$. The *maximal element with minimal frequency* multiset selector returns the largest number among those that have the smallest non-zero multiplicity: $\mathit{select}_{\mathrm{maxminfreq}}(3'2 + 2'4 + 2'5 + 4'7) = 5$.

In the following definition, the above discussed pruning procedure is formulated by defining which permutations corresponding to the full paths in the tree survive the pruning. Such permutations are called compatible. As each node has at least one child that is not pruned, there always is at least one permutation compatible with the place valuation. Assume a fixed multiset selector *select*.

Definition 1. *A permutation $g_1 \circ \cdots \circ g_{|P|} \circ g_{|P|+1} \cdots \circ g_{|P|+|T|} \in G$, where $g_i \in U_i$, is* compatible *with a place valuation pval (under select) if*

$$\mathit{pval}((g_1 \circ \cdots \circ g_{i-1} \circ g_i)(p_{\beta,i})) = \mathit{select}\left(\sum_{h \in U_i} 1'\mathit{pval}((g_1 \circ \cdots \circ g_{i-1} \circ h)(p_{\beta,i}))\right)$$

holds for each $1 \leq i \leq |P|$ (when $i = 1$, $g_1 \circ \cdots \circ g_{i-1} = \mathbf{I}$).

Example 4. Recall the net in Fig. 3 and the Schreier-Sims representation of its automorphism group G described in Ex. 3. Assume a place valuation $\mathit{pval} = \{p_1 \mapsto 1, p_2 \mapsto 0, p_3 \mapsto 0, p_4 \mapsto 0\}$. The permutations $h_{1,2} \circ h_{2,1}$, $h_{1,3} \circ h_{2,1}$, $h_{1,3} \circ$

$h_{2,2}$, and $h_{1,4} \circ h_{2,2}$ are compatible with *pval* under the minimal element multiset selector. Under the maximal element with minimal frequency multiset selector, only 2 permutations, $h_{1,1} \circ h_{2,1}$ and $h_{1,1} \circ h_{2,2}$, are compatible with *pval*.

The following property of compatibility is crucial for the correctness of the algorithms presented in this paper.

Theorem 1. *Let $g \in G$. A permutation $\hat{g} \in G$ is compatible with a place valuation pval if and only if $g \circ \hat{g} \in G$ is compatible with $g(pval)$.*

Furthermore, if a place valuation is injective, then there is exactly one element in G that is compatible with it. The obvious depth-first backtrack search algorithm enumerating all permutations compatible with a place valuation is given in Fig. 5.

1. procedure *compatible_permutations*(*pval*)
2. call *backtrack*$(1, \mathbf{I})$

3. procedure *backtrack*(*level*, $\hat{g}$)
4. if $level = |P| + 1$
5. report $\hat{g} \circ g'$ for each $g' \in U_{|P|+1} \circ \cdots \circ U_{|P|+|T|}$
6. return
7. evaluate $s = select(\Sigma_{h \in U_{level}} 1'pval(\hat{g}(h(p_{\beta,level}))))$
8. for all $h \in U_{level}$ such that $pval(\hat{g}(h(p_{\beta,level}))) = s$ do
9. call *backtrack*$(level + 1, \hat{g} \circ h)$

Fig. 5. Enumerating all compatible permutations.

3 Marking Guided Schreier-Sims Search

The algorithm presented in this section is based on selecting a permutation that is compatible with the marking in question. That is, *the marking itself is interpreted as a place valuation.* A canonical representative function is obtained by performing a backtracking search in the Schreier-Sims representation for the lexicographically smallest marking produced by a compatible permutation. Some pruning techniques for the search are discussed, too.

First, assume a base $\beta = [p_{\beta,1}, \ldots, p_{\beta,|P|}, t_{\beta,1}, \ldots, t_{\beta,|T|}]$ where the places are enumerated before the transitions and a Schreier-Sims representation $\boldsymbol{G} = [U_1, \ldots, U_{|P|+|T|}]$ of any subgroup G of $\mathrm{Aut}(N)$ under the base. In addition, a fixed multiset selector is implicitly assumed in this and following sections. Let

$$posreps(M) = \{\hat{g}^{-1}(M) \mid \hat{g} \in G \text{ and } \hat{g} \text{ is compatible with } M\}$$

denote the set of *possible representative markings* for M. Obviously, $M' \in posreps(M)$ implies $M' \equiv_G M$. For equivalent markings, the sets of possible representative markings are the same:

Theorem 2. *For each $M \in \mathbb{M}$ and each $g \in G$, $posreps(M) = posreps(g(M))$.*

Example 5. Recall the net in Fig. 3 and the Schreier-Sims representation of its automorphism group G described in Ex. 3. In Ex. 4, it was shown that the symmetries $h_{1,2} \circ h_{2,1} = \begin{pmatrix} p_1 & p_2 & p_3 & p_4 & t_{1,2} & \dots \\ p_2 & p_3 & p_4 & p_1 & t_{2,3} & \dots \end{pmatrix}$, $h_{1,3} \circ h_{2,1} = \begin{pmatrix} p_1 & p_2 & p_3 & p_4 & t_{1,2} & \dots \\ p_3 & p_4 & p_1 & p_2 & t_{3,4} & \dots \end{pmatrix}$, $h_{1,3} \circ h_{2,2} = \begin{pmatrix} p_1 & p_2 & p_3 & p_4 & t_{1,2} & \dots \\ p_3 & p_2 & p_1 & p_4 & t_{3,2} & \dots \end{pmatrix}$, and $h_{1,4} \circ h_{2,2} = \begin{pmatrix} p_1 & p_2 & p_3 & p_4 & t_{1,2} & \dots \\ p_4 & p_3 & p_2 & p_1 & t_{4,3} & \dots \end{pmatrix}$ are compatible with the marking $M = 1'p_1$ (under the minimal element multiset selector). Thus $posreps(M) = \{1'p_3, 1'p_4\}$.

A very simple non-canonical representative marking algorithm would be to simply take an arbitrary permutation $\hat{g}$ that is compatible with the marking M in question and then return $\hat{g}^{-1}(M)$ as the representative marking. By Thm. 2 it is possible, although not guaranteed, that the same representative marking is selected for equivalent markings.

Assuming a fixed total order between all the markings, a *canonical* representative marking algorithm can be obtained by selecting the smallest marking in the set of possible representative markings to be the representative. A natural choice for the ordering is the lexicographical ordering. Formally, a marking M_1 is *lexicographically smaller than* a marking M_2 under the base β, denoted by $M_1 <_\beta M_2$, if there is a number $1 \leq k \leq |P|$ such that $M_1(p_{\beta,j}) = M_2(p_{\beta,j})$ for all $1 \leq j < k$ and $M_1(p_{\beta,k}) < M_2(p_{\beta,k})$. Now define $canrepr(M)$ to be the $<_\beta$-smallest marking in $posreps(M)$. As $posreps(M) = posreps(g(M))$, $canrepr(M) = canrepr(g(M))$ for all markings M and for all $g \in G$. Furthermore, $canrepr(M) \equiv_G M$. The function *canrepr* can be computed by the backtracking depth-first search algorithm shown in Fig. 6, derived from the one in Fig. 5.

1. procedure $canrepr(M)$
2.   set $BestMarking = p \mapsto \infty$ for each $p \in P$
3.   set $pval(p) = M(p)$ for each place p
4.   call $backtrack(1, \mathbf{I})$
5.   return $BestMarking$

6. procedure $backtrack(level, \hat{g})$
7.   if $level = |P| + 1$
8.   if $\hat{g}^{-1}(M) <_\beta BestMarking$
9.   set $BestMarking = \hat{g}^{-1}(M)$
10.   return
11.   evaluate $s = select(\Sigma_{h \in U_{level}} 1'pval(\hat{g}(h(p_{\beta,level}))))$
12.   for all $h \in U_{level}$ such that $pval(\hat{g}(h(p_{\beta,level}))) = s$ do
13.   call $backtrack(level + 1, \hat{g} \circ h)$

Fig. 6. Finding the lexicographically smallest marking in the set $posreps(M)$.

The "hardness" of a marking M can be classified by defining that M is

1. *trivial* if there is exactly one permutation in G compatible with the marking,
2. *easy* if it is not trivial but the set $posreps(M)$ contains only one marking,
3. *hard* if it is neither trivial nor easy.

Note that this classification depends on the applied Schreier-Sims representation and multiset selector. Furthermore, the classes are closed under G. For both trivial and easy markings, the set *posreps*(M) contains only one marking. The difference is that easy markings have several permutations in G that are compatible with the marking.

Pruning with the already fixed prefix. Consider a permutation $g = g_1 \circ \cdots \circ g_i$ in G, where $1 \leq i \leq |P|$ and $g_j \in U_j$ for each $1 \leq j \leq i$. Now each "extended" permutation $\tilde{g} = g_1 \circ \cdots \circ g_i \circ g_{i+1} \circ \cdots \circ g_{|P|+|T|}$ in G maps $p_{\beta,1}$ to $g(p_{\beta,1})$, $p_{\beta,2}$ to $g(p_{\beta,2})$, and so on up to and including $p_{\beta,i}$ that is mapped to $g(p_{\beta,i})$. Thus the values of the first i places in $\tilde{g}^{-1}(M)$ are known: $(\tilde{g}^{-1}(M))(p_{\beta,1}) = M(\tilde{g}(p_{\beta,1})) = M(g(p_{\beta,1})), \ldots$, and $(\tilde{g}^{-1}(M))(p_{\beta,i}) = M(\tilde{g}(p_{\beta,i})) = M(g(p_{\beta,i}))$. If a marking $M' \in$ *posreps*(M) such that (i) $M'(p_{\beta,j}) = M(g(p_{\beta,j}))$ for each $1 \leq j < k$ and (ii) $M'(p_{\beta,k}) < M(g(p_{\beta,k}))$ for a $1 \leq k \leq i$ has already been found during the search, one knows that $M' <_\beta \tilde{g}^{-1}(M)$ for all extensions $\tilde{g}$ of g and can therefore skip all such $\tilde{g}$.

To improve the possibilities of this pruning technique to work efficiently, the Schreier-Sims representation can be optimized to have the fixed elements as early as possible in the base. Let $p_{\beta,i}$ be the last element in the base where a place $p_{\beta,j}$, $j \geq i$, may be permuted i.e. $h(p_{\beta,j}) \neq p_{\beta,j}$ for an $h \in U_i$. Now the base can be changed so that $p_{\beta,j}$ is after $p_{\beta,i}$ but before any $p_{\beta,k}$ for which $U_k \supset \{\mathbf{I}\}$.

Finding and pruning with stabilizers. Take any permutation $\tilde{g} = g_1 \circ \cdots \circ g_{i-1} \in U_1 \circ \cdots \circ U_{i-1}$ for an $1 \leq i \leq |P|$. Consider two left cosets (i.e. subtrees of the Schreier-Sims search tree), $(\tilde{g} \circ g_i) \circ G_{i+1}$ and $(\tilde{g} \circ g_i') \circ G_{i+1}$, where $g_i, g_i' \in U_i$. Let α be a stabilizer of a marking M that (i) fixes each place $\tilde{g}(p_{\beta,1}), \ldots, \tilde{g}(p_{\beta,i-1})$, and (ii) maps the place $(\tilde{g} \circ g_i)(p_{\beta,i})$ to $(\tilde{g} \circ g_i')(p_{\beta,i})$. Now, if a permutation g' belongs to the left coset $(\tilde{g} \circ g_i') \circ G_{i+1}$, then $\alpha^{-1} \circ g'$ belongs to the left coset $(\tilde{g} \circ g_i) \circ G_{i+1}$. Furthermore, for each marking M, $(\alpha^{-1} \circ g')^{-1}(M) = (g'^{-1} \circ \alpha)(M) = g'^{-1}(M)$. Therefore, the left cosets $(\tilde{g} \circ g_i') \circ G_{i+1}$ and $(\tilde{g} \circ g_i) \circ G_{i+1}$ produce the same markings. In addition, if g is compatible with M, then $\alpha^{-1} \circ g$ is compatible with $\alpha^{-1}(M) = M$ and therefore the sets of possible representative markings in the left cosets are the same. To sum up, if all the permutations in a left coset $(\tilde{g} \circ g_i) \circ G_{i+1}$ have already been searched and there is a stabilizer α with the above mentioned properties, one can ignore the left coset $(\tilde{g} \circ g_i') \circ G_{i+1}$.

Stabilizers of markings can be found during the backtrack search in the Schreier-Sims representation. Consider that M' is a marking that has been found earlier during the search by traversing a path $g = g_1 \circ \cdots \circ g_{i-1} \circ g_i \circ g_{i+1} \circ \cdots g_{|P|}$ meaning that $g^{-1}(M) = M'$. For instance, M' could be the lexicographically smallest marking found so far. Assume that the currently traversed path is $g' = g_1 \circ \cdots \circ g_{i-1} \circ g_i' \circ g_{i+1}' \circ \cdots g_{|P|}'$, where $g_i' \neq g_i$. If it holds that $g'^{-1}(M) = M' = g^{-1}(M)$, then (i) $g' \circ g^{-1}$ is a stabilizer of M and (ii) $g' \circ g^{-1}$ fixes each $(g_1 \circ \cdots \circ g_j)(p_{\beta,j})$, $1 \leq j < i$, and (iii) $g' \circ g^{-1}$ maps $(g_1 \circ \cdots \circ g_{i-1} \circ g_i)(p_{\beta,i}) = g(p_{\beta,i})$ to $g'(p_{\beta,i}) = (g_1 \circ \cdots \circ g_{i-1} \circ g_i')(p_{\beta,i})$. Thus $g' \circ g^{-1}$ is a stabilizer of M fulfilling the properties discussed above (the prefix $\tilde{g}$ being $g_1 \circ \cdots \circ g_{i-1}$), and the search can be "back-jumped" to the level $i - 1$. This is the most trivial (and easiest to implement) way to prune with the found stabilizers. There are many ways to achieve even larger degree of pruning by composing the found stabilizers, see [6–8].

4 Partition Guided Schreier-Sims Search

It is possible to combine the backtracking search in the Schreier-Sims representation described in the previous section with a standard preprocessing technique applied in graph isomorphism algorithms. Assuming a fixed subgroup G of $\mathrm{Aut}(N)$ and given a marking M, an ordered partition of $P \cup T$ is first computed in a way that respects the symmetries in G. The procedure computing the partition for M is based on the use of invariants and is a variant of a standard technique used in graph isomorphism algorithms, see e.g. [7, 8]. The place valuation corresponding to the partition is then used to prune the search in the Schreier-Sims representation of G. That is, instead of searching through the permutations that are compatible with the marking as is done in Sect. 3, the permutations compatible with the constructed place valuation are searched. The hope is that the place valuation is closer to being injective than the original marking, i.e., that it can distinguish the places from each other better. The experimental results in Sect. 5 show that this is indeed the case for the nets considered there.

Some notation and preliminaries of ordered partitions is defined first. An *ordered partition* of a nonempty set A is a list $\pi = [C_1, \ldots, C_n]$ such that the set $\{C_1, \ldots, C_n\}$ is a partition of A. An ordered partition is *discrete* if all its *cells* C_i are singleton sets and *unit* if it contains only one cell (the set A). The function *incell* from the ordered partitions of A and the elements of A to natural numbers is defined by $\mathit{incell}([C_1, \ldots, C_n], x) = i \Leftrightarrow x \in C_i$. An ordered partition π_1 of A is a *refinement* of an ordered partition $\pi_2 = [C_{2,1}, \ldots, C_{2,n}]$, denoted by $\pi_1 \preceq \pi_2$, if π_1 is of form $[C_{1,1,1}, \ldots, C_{1,1,d_1}, \ldots C_{1,n,1}, \ldots, C_{1,n,d_n}]$ such that $\bigcup_{1 \le j \le d_i} C_{1,i,j} = C_{2,i}$ for all $1 \le i \le n$. For instance, $[\{a\}, \{c\}, \{b\}] \preceq [\{a\}, \{b, c\}]$. but *not* $[\{b\}, \{a\}, \{c\}] \preceq [\{a\}, \{b, c\}]$. A permutation γ of A acts on ordered partitions of A by $\gamma([C_1, \ldots, C_n]) = [\gamma(C_1), \ldots, \gamma(C_n)]$.

4.1 Partition Generators and the Algorithm

Assume a net N, a subgroup G of $\mathrm{Aut}(N)$, and a Schreier-Sims representation $\boldsymbol{G} = [U_1, \ldots, U_{|P|+|T|}]$ of G under a base $\beta = [p_1, \ldots, p_{|P|}, t_1, \ldots, t_{|T|}]$ in which the places are enumerated before the transitions. Denote the set of all ordered partitions of $P \cup T$ by Π.

First, the marking M in question is assigned an ordered partition of $P \cup T$. The idea is to try to distinguish between the elements in $P \cup T$ in a way that respects the symmetries in G so that distinguishable elements are put in different cells. Formally, define the following.

Definition 2. *A function* $f : \mathbb{M} \to \Pi$ *is a* G-partition generator *if* $f(g(M)) = g(f(M))$ *holds for each marking* $M \in \mathbb{M}$ *and for each* $g \in G$.

That is, for permuted markings, similarly permuted ordered partitions are assigned. A technique for obtaining G-partition generators is described in Sect. 4.2. For now, assume a fixed G-partition generator f.

An ordered partition can be interpreted as a place valuation by simply assigning each place its cell number in the partition. Formally, the place valuation $pval_\pi$ corresponding to an ordered partition $\pi \in \Pi$ is defined by

$$pval_\pi : p \mapsto incell(\pi, p)$$

for each place $p \in P$. After building the ordered partition $f(M)$ for the marking M and the corresponding place valuation $pval_{f(M)}$, let

$$posreps(M) = \{\hat{g}^{-1}(M) \mid \hat{g} \in G \text{ and } \hat{g} \text{ is compatible with } pval_{f(M)}\}$$

denote the set of *possible representative markings* for M (recall Sect. 3). Again, equivalent markings have the same possible representatives (cf. Thm. 2).

Theorem 3. *For each $M \in \mathbb{M}$ and each $g \in G$, $posreps(M) = posreps(g(M))$.*

Example 6. Consider the net in Fig. 3 and the Schreier-Sims representation $\mathbf{G}$ of its automorphism group G described in Ex. 3. Assume a marking $M = 1'p_1$ and a G-partition generator f mapping M to

$$f(M) = [\{p_3\}, \{p_2, p_4\}, \{p_1\}, \{t_{1,2}, t_{1,4}\}, \{t_{2,1}, t_{4,1}\}, \{t_{3,2}, t_{3,4}\}, \{t_{2,3}, t_{4,3}\}].$$

This is one of the finest partitions that any G-partition generator can produce as $g = \left(\begin{smallmatrix} p_1 & p_2 & p_3 & p_4 & t_{1,2} & t_{2,1} & t_{2,3} & t_{3,2} & t_{3,4} & t_{4,3} & t_{4,1} & t_{1,4} \\ p_1 & p_4 & p_3 & p_2 & t_{1,4} & t_{4,1} & t_{4,3} & t_{3,4} & t_{3,2} & t_{2,3} & t_{2,1} & t_{1,2} \end{smallmatrix}\right) \in \mathrm{Stab}(G, M)$ and thus $g(f(M)) = f(g(M)) = f(M)$ must hold. The corresponding place valuation is $pval_{f(M)} = \{p_1 \mapsto 3, p_2 \mapsto 2, p_3 \mapsto 1, p_4 \mapsto 2\}$ and the symmetries in G compatible with $pval_{f(M)}$ under the minimal element multiset selector are $h_{1,3} \circ h_{2,1}$ and $h_{1,3} \circ h_{2,2}$. Now $(h_{1,3} \circ h_{2,1})^{-1}(M) = (h_{1,3} \circ h_{2,2})^{-1}(M) = 1'p_3$. Thus $posreps(M) = \{1'p_3\}$ (cf. the set $posreps(M)$ in Ex. 5).

The lexicographically smallest marking in $posreps(M)$, i.e. a canonical representative for M, can be searched by using the backtrack search algorithm in Fig. 6 when line 3 is changed to refer to the valuation $pval_{f(M)}$ instead of M. Obviously, the pruning technique based on the fixed prefix is sound. The soundness of the stabilizer pruning technique can be established by noticing that the place valuations assigned to equivalent markings are equivalent, too: for each $g \in G$ and each marking M, $pval_{f(g(M))} = g(pval_{f(M)})$. Consequently, stabilizers of M are stabilizers of $pval_{f(M)}$:

Corollary 1. *For each stabilizer $g \in \mathrm{Stab}(G, M)$, $g(pval_{f(M)}) = pval_{f(M)}$.*

Similarly to that in Sect. 3, a hardness measure can be defined by declaring that a marking M is

1. *trivial* if the partition $f(M)$ is discrete,
2. *easy* if it is not trivial but the set $posreps(M)$ contains only one marking,
3. *hard* if it is neither trivial nor easy.

Again, this classification depends on the applied (i) Schreier-Sims representation, (ii) G-partition generator, and (iii) multiset selector. Note that if a marking M is trivial, then the set *posreps*(M) contains only one marking. The definition of triviality defined here is stronger than that in Sect. 3: there may be markings M for which there is only one permutation compatible with $f(M)$ although $f(M)$ is not discrete. The definition here is chosen because it reveals the efficiency of the applied G-partition generator better (the more trivial markings, the better).

4.2 Partition Refiners and Invariants

The G-partition generators discussed above can be obtained by using G-partition refiners.

Definition 3. *A* G-partition refiner *is a function* $\mathcal{R} : \mathbb{M} \times \Pi \to \Pi$ *such that both (i)* $\mathcal{R}(M, \pi) \preceq \pi$ *and (ii)* $\mathcal{R}(g(M), g(\pi)) = g(\mathcal{R}(M, \pi))$ *hold for each* $g \in G$.

That is, the resulting partition must be a refinement of the argument partition and for permuted arguments the result has to be similarly permuted. The composition $\mathcal{R}_2 \star \mathcal{R}_1$ of two G-partition refiners, $\mathcal{R}_1$ and $\mathcal{R}_2$, defined by $(\mathcal{R}_2 \star \mathcal{R}_1)(M, \pi) = \mathcal{R}_2(M, \mathcal{R}_1(M, \pi))$, is also a G-partition refiner. When a G-partition refiner is applied to the unit partition, the result is a G-partition generator.

Lemma 1. *For each G-partition refiner $\mathcal{R}$, the function $f_{\mathcal{R}} : \mathbb{M} \to \Pi$ defined by $f_{\mathcal{R}}(M) = \mathcal{R}(M, [P \cup T])$ is a G-partition generator.*

A way to obtain G-partition refiners is based on the use of G-invariants.

Definition 4. *A function* $\mathcal{I} : \mathbb{M} \times \Pi \times \{P \cup T\} \to \mathbb{Z}$ *is a* G-invariant *if it holds that* $\mathcal{I}(M, \pi, x) = \mathcal{I}(g(M), g(\pi), g(x))$ *for each* $g \in G$.

The following are G-invariants for each subgroup G of $\mathrm{Aut}(N)$.

- Assume a fixed order between the orbits of G, e.g. the one obtained by considering the smallest element in each orbit under the base. The *G-orbit invariant* is: $\mathcal{I}_{G\text{-orbit}}(M, \pi, x) = i$ if and only if x belongs to the ith orbit.
- The *marking invariant* is defined by $\mathcal{I}_{\text{marking}}(M, \pi, x) = M(x)$ if $x \in P$ and -1 if $x \in T$.
- The *partition dependent weighted in- and out-degree invariants* are $\mathcal{I}_{\text{in } w,c}(M, \pi, x) = |\{x' \mid \langle x', x\rangle \in F \wedge W(\langle x', x\rangle) = w \wedge \mathit{incell}(\pi, x') = c\}|$ and $\mathcal{I}_{\text{out } w,c}(M, \pi, x) = |\{x' \mid \langle x, x'\rangle \in F \wedge W(\langle x, x'\rangle) = w \wedge \mathit{incell}(\pi, x') = c\}|$.

A partition is *refined according to an invariant* by splitting the cells according to the values assigned to nodes by the invariant. Formally, for a G-invariant $\mathcal{I}$, define the function $\mathcal{R}_{\mathcal{I}} : \mathbb{M} \times \Pi \to \Pi$ by $\mathcal{R}_{\mathcal{I}}(M, \pi) = \pi_r$ such that

1. $\mathit{incell}(\pi_r, x) = \mathit{incell}(\pi_r, x')$ if and only if $\mathit{incell}(\pi, x) = \mathit{incell}(\pi, x')$ and $\mathcal{I}(M, \pi, x) = \mathcal{I}(M, \pi, x')$, and

2. $incell(\pi_r, x) < incell(\pi_r, x')$ if and only if either
   (a) $incell(\pi, x) < incell(\pi, x')$, or
   (b) $incell(\pi, x) = incell(\pi, x')$ and $\mathcal{I}(M, \pi, x) < \mathcal{I}(M, \pi, x')$.

Lemma 2. *The function $\mathcal{R}_\mathcal{I}$ is a G-partition refiner.*

Partition refiners with respect to some invariants can also be defined procedurally so that in the resulting partition two nodes are in the same cell if and only if their invariant values in that partition are the same. This is especially the case for the partition dependent weighted in- and out-degree invariants, where the procedure corresponds to the method of computing the so-called equitable partition [7, 8].

Example 7. Consider again the net in Fig. 3 and the Schreier-Sims representation $\boldsymbol{G}$ of its automorphism group G given in Ex. 3. Assume a marking $M = 1'p_1$. Initially, the partition is $[\{p_1, p_2, p_3, p_4, t_{1,2}, \ldots\}]$. Refining this according to the G-orbit invariant yields $[\{p_1, p_2, p_3, p_4\}, \{t_{1,2}, \ldots\}]$. Further refinement with the marking invariant gives $[\{p_2, p_3, p_4\}, \{p_1\}, \{t_{1,2}, \ldots\}]$. When this partition is refined according to the in-degree of weight 1 from cell 1 invariant $\mathcal{I}_{\text{in } 1,1}$, the result is $[\{p_2, p_3, p_4\}, \{p_1\}, \{t_{1,2}, t_{1,4}\}, \{t_{2,1}, t_{2,3}, t_{3,2}, t_{4,3}, t_{3,4}, t_{4,1}\}]$. A further systematic refinement with the partition dependent weighted in- and out-degree invariants results in $[\{p_3\}, \{p_2, p_4\}, \{p_1\}, \{t_{1,2}, t_{1,4}\}, \{t_{2,1}, t_{4,1}\}, \{t_{3,2}, t_{3,4}\}, \{t_{2,3}, t_{4,3}\}]$, equaling to the partition in Ex. 6.

5 Experimental Results

The experimental results in this section are obtained by extending the LoLA reachability analyzer, version 1.0 beta [9]. The extended version is available at `http://www.tcs.hut.fi/~tjunttil`.

5.1 Net Classes

The following net classes are used in the experiments.

Mutual exclusion in grid-like networks, based on the nets in [3]. A net "grid d n" models a d-dimensional hypercube with n agents in each dimension. Each agent has two states, critical and non-critical, and can move from the non-critical state into the critical one only if none of its neighbors is in the critical state. The automorphism group of a d-dimensional grid net is isomorphic to the automorphism group of a d-dimensional hypercube.

Dining philosophers. A version of the classic dining philosophers net. A net "ph n" has n philosophers and its automorphism group is isomorphic to the cyclic group of order n.

Database managers. An unfolding of the Colored Petri net presented in [25]. "db n" denotes the net with n managers, having the automorphism group isomorphic to the symmetric group of degree n.

Graph enumeration nets. These nets resemble the one appearing in [4, Lemma 11], inspired by the system in the proof of Thm. 3.4 in [18]. A net "digraphs n"

enumerates all the directed, unlabeled graphs having no self-loops over a vertex set $V = \{1, \ldots, n\}$ in its reachable markings. See Fig. 7 for an example when $n = 3$. The net has the place p_v for each vertex $v \in V$, and the place p_{v_1,v_2} for each possible edge $\langle v_1, v_2 \rangle \in V \times V$, $v_1 \neq v_2$. A place p_{v_1,v_2} has one token in a marking if and only if the graph corresponding to the marking has an edge $\langle v_1, v_2 \rangle$. The transitions of the net are as described in Fig. 7. The transitions guarantee that the automorphism group of the net is isomorphic to the permutation group consisting of all permutations of V (i.e., to the symmetric group of degree n). The action of a permutation γ of V on the places corresponds to the usual action on the adjacency matrix of a graph: each p_v is permuted to $p_{\gamma(v)}$ and each p_{v_1,v_2} is permuted to $p_{\gamma(v_1),\gamma(v_2)}$. Thus two reachable markings are equivalent if and only if their corresponding graphs are isomorphic. Consequently, a minimal reduced reachability graph consisting only of one marking of each orbit has exactly one marking for each class of mutually isomorphic graphs. A similar net, call it "graphs n", enumerating all undirected, unlabeled graphs over n vertices having no self-loops can be constructed by similar principles, see [5].

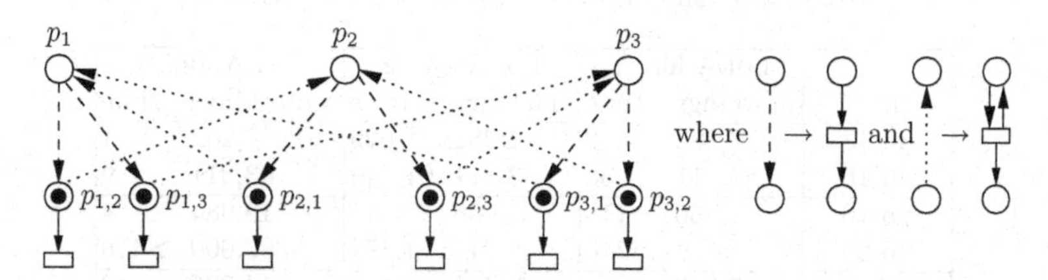

Fig. 7. A net enumerating all directed graphs without self-loops over three vertices.

Properties. Table 1 lists the properties of the nets used in the experiments. The columns $|P|$ and $|T|$ describe the number of places and transitions in the net, respectively, and $|G|$ gives the size of the symmetry group stabilizing the initial marking (the group that is used in the experiments). The number of reachable markings as well as the run time of LoLA in seconds without the symmetry reduction method is given in the last two columns, respectively. For some nets the number of reachable markings is too large and running LoLA would result in running out of memory. In such cases, the run time of LoLA is not given but the number of reachable markings is given analytically.

5.2 Results

The experimental results were obtained in a Debian Linux PC machine with 1GHz AMD Athlon processor and 1Gb of memory. The extended LoLA was compiled with the GNU g++ compiler with the -O3 optimization flag switched on. All run-times were obtained by the Unix `time` command and are user times rounded up to full seconds unless otherwise stated. The available memory was limited to 900Mb and time to 24 hours by the Unix `ulimit` command.

Table 1. Properties of the nets.

net	$\|P\|$	$\|T\|$	$\|G\|$	reachable markings	LoLA time
ph 13	52	39	13	94,642	4
ph 16	64	48	16	1,331,714	90
db 10	301	200	3,628,800	196,831	15
db 20	1201	800	20!	23×10^9	
grid 2 5	50	50	8	55,447	3
grid 3 3	54	54	48	70,633	4
grid 5 2	64	64	3840	254,475	20
graphs 7	28	63	5,040	2,097,152	86
graphs 8	36	84	40,320	$2^{\binom{8}{2}} = 2^{28}$	
graphs 9	45	108	362,880	$2^{\binom{9}{2}} = 2^{36}$	
digraphs 4	16	36	24	4,096	1
digraphs 5	25	60	120	1,048,576	39
digraphs 6	36	90	720	$2^{6\times(6-1)} = 2^{30}$	

Table 2. Results for the original LoLA algorithms.

	LoLA alg. 1		LoLA alg. 2		LoLA alg. 3	
net	markings	time	markings	time	markings	time
ph 13	7,282	2	7,282	629	7,282	1
ph 16	83,311	33	83,311	81,440	83,311	9
db 10	56	118	56	11	19,684	4
db 20		>24h	211	1,477	>399,000	>418
grid 2 5	7,567	1	7,471	183	14,236	2
grid 3 3	2,154	2	2,103	62	10,847	2
grid 5 2	296	7	287	14	3,020	1
graphs 7	1,044	17	1,022	27	37,195	3
graphs 8	12,346	2,358	12,095	3,662	1,536,698	246
graphs 9	>47,683	>24h	>55,400	>24h	>5,128,600	>801
digraphs 4	218	1	215	1	347	1
digraphs 5	9,735	3	9,567	1,197	40,078	3
digraphs 6	1,598,555	1,810	>85,469	>24h	>4,581,000	>512

The original symmetry reduction algorithms in LoLA described in [3] are numbered as follows: 1 refers to the "iterating the symmetries" algorithm, 2 is the "iterating the states" algorithm, and 3 is the "canonical representative" algorithm[5]. The results of these algorithms are shown in Table 2. The LoLA implementation seems to contain some bugs because the algorithms 1 and 2 should both produce minimal reduced reachability graphs but the numbers of the markings in the generated reduced reachability graphs are not the same.

Table 3 shows the results for the marking guided Schreier-Sims search algorithm described in Sect. 3. The maximal element with minimal frequency mul-

[5] Not a canonical representative marking function by the terms used in this paper.

Table 3. Results of the marking and partition guided Schreier-Sims searches.

		marking guided					partition guided				
net	markings	time	trivial %	easy %	max dead	av. dead	time	trivial %	easy %	max dead	av. dead
ph 13	7,282	1	2.85	0.00	12	2.09	4	99.997	0.003	-	-
ph 16	83,311	15	1.04	0.01	15	2.98	66	99.91	0.09	-	-
db 10	56	1	0	3.81	259	60.36	1	0	100	-	-
db 20	211	172	0	0.86	40,152	1,844.12	27	0	100	-	-
grid 2 5	7,471	1	0	3.88	7	2.07	9	90.86	8.71	2	1.19
grid 3 3	2,103	1	0	1.59	46	12.14	4	60.82	33.22	16	1.86
grid 5 2	288	1	0	1.03	278	126.94	1	2.26	72.00	15	2.30
graphs 7	1,044	3	0	0.01	1,413	272.48	1	24.70	50.58	196	3.18
graphs 8	12,346	82	0	0.00	8,770	580.28	15	40.52	42.95	535	3.71
graphs 9	274,668	5,036	0	0.00	70,017	226.80	586	57.46	33.92	2,045	3.02
digraphs 4	218	1	0	4.05	7	3.15	1	78.29	21.18	3	2.29
digraphs 5	9,608	2	0	0.09	27	7.97	7	89.15	10.22	10	1.52
digraphs 6	1,540,944	929	0	0.00	93	17.19	2,404	95.68	4.05	34	1.10

tiset selector is used because it seems to usually give the best results. Pruning with the fixed prefix, the trivial pruning with the found stabilizers, and the base optimization described in Sect. 3 are applied, too. The "trivial %" and "easy %" columns show the percentage of trivial and easy canonized markings, respectively, as defined in Sect. 3. The "max dead" and "av. dead" columns show the maximum and average number of dead nodes, respectively, in the search trees for hard markings. As can be seen, practically all markings are usually hard and the number of bad nodes in a search tree can grow quite large. One reason for this behavior is that all the nets are 1-safe. Thus the multiset selector cannot usually prune the search tree efficiently.

Table 3 also shows the results for the partition guided Schreier-Sims search algorithm described in Sect. 4. The applied partition generator first refines the unit partition according to the orbit and marking invariants, and then refines the resulting partition with the partition dependent weighted in- and out-degree invariants until no improvement is achieved. As can be seen from the results, the amounts of trivial and easy markings are now much higher, compared to the marking guided Schreier-Sims search algorithm discussed above. Furthermore, hard markings are also easier, and although the number of dead nodes can be still in thousands, on the average it is very low. For nets with small symmetry groups, the overhead of computing the ordered partition sometimes makes the algorithm slower than the marking guided Schreier-Sims search (e.g., the dining philosophers nets and the nets "grid 2 5", "grid 3 3", and "digraphs 6").

6 Conclusions

Symmetries of a Place/Transition-net can be exploited during the state space analysis to alleviate the combinatorial state explosion problem. This paper describes two new algorithms for the core problem of transforming markings into symmetry equivalent canonical representatives. The algorithms are based on a backtrack search in the Schreier-Sims representation of the symmetry group. They return the smallest symmetry equivalent marking they find as the canonical representative. In the first algorithm, the search tree is pruned by (i) a novel compatibility definition between symmetries and markings, (ii) the smallest already found equivalent marking and (iii) the stabilizers of the marking found during the search. The second algorithm improves the first one by combining it with a standard preprocessing technique applied in graph isomorphism algorithms. That is, an ordered partition for the marking is first built by applying a set of invariants. The partition is then used to further prune the search in the Schreier-Sims representation by considering only the symmetries that are compatible with the partition.

The experimental results show that the proposed algorithms are very competitive against the previous ones in [2, 3]. The second proposed algorithm is more robust, working well with many kinds of symmetry groups, even with very large ones. This is because the constructed partition is usually able to prune the search tree in the Schreier-Sims representation better than the marking alone.

Finally, notice that both of the new algorithms could be approximated (i.e., made non-canonical) by performing only a limited search in the Schreier-Sims representation. For instance, an upper bound for the traversed nodes could be set. Doing so avoids the exponential worst case running time of the algorithms with the risk that a non-minimal reduced reachability graph is produced.

References

1. Starke, P.H.: Reachability analysis of Petri nets using symmetries. Systems Analysis Modelling Simulation **8** (1991) 293–303
2. Schmidt, K.: How to calculate symmetries of Petri nets. Acta Informatica **36** (2000) 545–590
3. Schmidt, K.: Integrating low level symmetries into reachability analysis. In Graf, S., Schwartzbach, M., eds.: Tools and Algorithms for the Construction and Analysis of Systems, TACAS 2000. Volume 1785 of LNCS., Springer (2000) 315–330
4. Junttila, T.A.: Computational complexity of the Place/Transition-net symmetry reduction method. Journal of Universal Computer Science **7** (2001) 307–326
5. Junttila, T.: On the symmetry reduction method for Petri nets and similar formalisms. Research Report A80, Helsinki Univ. of Technology, Lab. for Theoretical Computer Science (2003) Doctoral dissertation.
6. Butler, G.: Fundamental Algorithms for Permutation Groups. Volume 559 of LNCS. Springer (1991)
7. McKay, B.D.: Practical graph isomorphism. Congressus Numerantium **30** (1981) 45–87
8. Kreher, D.L., Stinson, D.R.: Combinatorial Algorithms: Generation, Enumeration and Search. CRC Press, Boca Raton, Florida, USA (1999)
9. Schmidt, K.: LoLA: A low level analyser. In Nielsen, M., Simpson, D., eds.: Application and Theory of Petri Nets 2000, ICATPN 2000. Volume 1825 of LNCS., Springer (2000) 465–474

10. Jensen, K.: Coloured Petri Nets: Basic Concepts, Analysis Methods and Practical Use: Volume 2, Analysis Methods. Monographs in Theoretical Computer Science. Springer (1995)
11. Emerson, E.A., Sistla, A.P.: Symmetry and model checking. Formal Methods in System Design **9** (1996) 105–131
12. Clarke, E.M., Enders, R., Filkorn, T., Jha, S.: Exploiting symmetry in temporal logic model checking. Formal Methods in System Design **9** (1996) 77–104
13. Ip, C.N., Dill, D.L.: Better verification through symmetry. Formal Methods in System Design **9** (1996) 41–76
14. Emerson, E.A., Sistla, A.P.: Utilizing symmetry when model checking under fairness assumptions: An automata-theoretic approach. ACM Transactions on Programming Languages and Systems **19** (1997) 617–638
15. Gyuris, V., Sistla, A.P.: On-the-fly model checking under fairness that exploits symmetry. Formal Methods in System Design **15** (1999) 217–238
16. Chiola, G., Dutheillet, C., Franceschinis, G., Haddad, S.: On well-formed coloured nets and their symbolic reachability graph. [26] 373–396
17. Junttila, T.: Detecting and exploiting data type symmetries of algebraic system nets during reachability analysis. Research Report A57, Helsinki Univ. of Technology, Lab. for Theoretical Computer Science (1999)
18. Ip, C.N.: State Reduction Methods for Automatic Formal Verification. PhD thesis, Department of Computer Science, Stanford University (1996)
19. Huber, P., Jensen, A.M., Jepsen, L.O., Jensen, K.: Towards reachability trees for high-level Petri nets. Technical Report DAIMI PB 174, Datalogisk Afdeling, Matematisk Institut, Aarhus Universitet (1985)
20. Sistla, A.P., Gyuris, V., Emerson, E.A.: SMC: A symmetry-based model checker for verification of safety and liveness properties. ACM Transactions on Software Engineering and Methodology **9** (2000) 133–166
21. Junttila, T.: Symmetry reduction algorithms for data symmetries. Research Report A72, Helsinki Univ. of Technology, Lab. for Theoretical Computer Science (2002)
22. Lorentsen, L., Kristensen, L.M.: Exploiting stabilizers and parallelism in state space generation with the symmetry method. In: Proceedings of the Second International Conference on Application of Concurrency to System Design (ACSD 2001), IEEE Computer Society (2001) 211–220
23. Genrich, H.J.: Predicate/transition nets. [26] 3–43
24. Jerrum, M.: A compact representation for permutation groups. Journal of Algorithms **7** (1986) 60–78
25. Jensen, K.: Coloured Petri Nets: Basic Concepts, Analysis Methods and Practical Use: Volume 1, Basic Concepts. Second edn. Monographs in Theoretical Computer Science. Springer (1992)
26. Jensen, K., Rozenberg, G., eds.: High-level Petri Nets; Theory and Application, Springer (1991)

Properties of Object Petri Nets

Michael Köhler and Heiko Rölke

University of Hamburg, Department of Computer Science
Vogt-Kölln-Str. 30, D-22527 Hamburg
{koehler,roelke}@informatik.uni-hamburg.de

Abstract. In this presentation the structure of formalisms are studied that allow Petri nets as tokens. The relationship towards common Petri net models and decidability issues are studied. Especially for "elementary object-net systems" defined by Valk [20] the decidability of the reachability and the boundedness problem is considered. It is shown that reachability becomes undecidable while boundedness remains decidable for elementary object-net systems. Furthermore it is shown that even for minimal extensions the formalism obtains the power of Turing machines.

Keywords: decidability, nets within nets, object Petri nets

1 Introduction

In this presentation decidability issues for different kinds of Object Petri Nets are studied. Object Petri Net formalisms use complex objects (defined in some object-oriented specification language) as tokens. The *"nets within nets"* approach of Valk assumes that these objects are Petri nets again: In [20] "unary elementary object systems" (UEOS) are introduced – a basic model restricted to a two-level hierarchy. In the following the term "object-net" is used for the "nets within nets" interpretation. Related models of the "nets within nets" approach are Nested Petri Nets [13], Linear Logic Object Petri Nets [5], Reference Nets [12] and Mobile Object-Net Systems [10]. Furthermore, there is a close connection to mobility calculi, like the ambient calculus [2].

The main question is whether results for ordinary Petri net formalisms carry over for object-nets. The analysis of decidable properties of "nets within nets" formalisms gives a deeper insight in the question whether they are just a more convenient representation of another – possibly larger – Petri net model, or whether they are a real extension with more computational power – similar to Self-Modifying Nets [19] which are a real extension.

The paper is structured as follows. Section 2 gives an informal introduction into the *nets within nets* approach. Section 3 recalls the formal definition of "unary elementary object systems" together with two variants of the firing rule, namely reference and value semantics. Some basic properties concerning the relationship of the two semantics or the existence of linear invariants are proven. We analyse the decidability of the reachability and boundedness problem for UEOS. Section 4 defines the generalised model of object- nets which allow an arbitrary deep nesting structure. Furthermore, it is proven, that the generalised formalism has the power of Turing machines. The work closes with a conclusion.

J. Cortadella and W. Reisig (Eds.): ICATPN 2004, LNCS 3099, pp. 278–297, 2004.
© Springer-Verlag Berlin Heidelberg 2004

2 Object-Oriented Petri Nets and Nets within Nets

The paradigm of nets within nets due to Valk [20] formalises the aspect that tokens of a Petri net can be nets again. Taking this point of view it is possible to model e.g. mobility very naturally: A mobile entity is described by a Petri net which is a token of another Petri net describing the whole surrounding system (for a detailed discussion of mobility in the context of Petri nets cf. [9]).

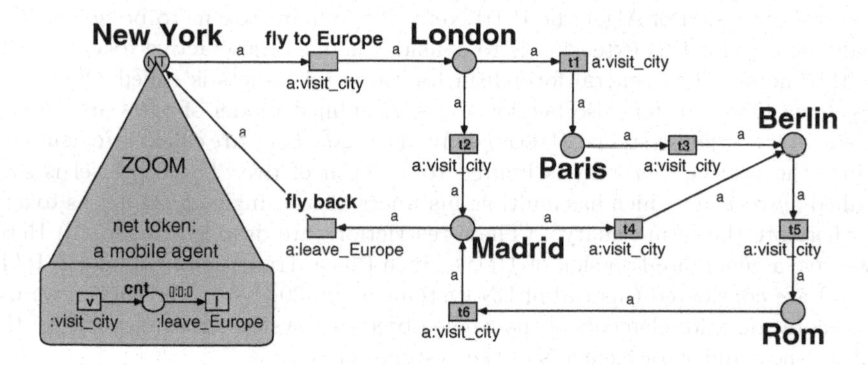

Fig. 1. A Mobile Agent as a Net-Token

To give an example we consider a situation in Figure 1 where we have a two-level hierarchy[1]. The net-token is then called the "object-net", the surrounding net is called the "system-net". An intuitive interpretation of this model is a scenario, where each object-net models a mobile agent and the system-net models the agent system. In this example the agent a wants to travel from New York to Europe and he does not come back until he has visited at least three European cities. Initially the agent is a net-token on the place New York – indicated by the ZOOM. The places in the system-net describes the cities of the scenario, the transitions movements between them.

Object and system-nets synchronise via *channels*. The channels are denoted as transition inscriptions of the form a:visit_city in the system-net and :visit_city in the object-net. The asymmetry is due to the fact, that the object-net (the agent a) is known in the system-net but not vice versa. Transitions with corresponding channels must fire synchronously. Transitions without an inscription can fire autonomously and concurrently to other enabled transitions.

Let us look at an example process of the system in Figure 1: The first firing step is a synchronous firing of the transitions fly to Europe and the transition v ("visit") wrt. the channel visit_city. As a result one black token is generated on the place cnt (the counter) inside the object-net and the whole object-net is located in London. Then the agent moves to Madrid, generating a second token on the place cnt. This time the agent cannot fly back to New York, since he has

[1] The example uses the syntax of the RENEW-tool (cf. www.renew.de).

only two tokens on the place cnt, while three are needed to activate the channel leave_Europe. So, the transitions fly back and l ("leave") are not activated. The sole possibility is to travel to Berlin, to Rom, and afterwards to Madrid again. Now, the agent can fly back, since he has five tokens on the place cnt.

3 Unary Elementary Object Systems

As Stehr, Meseguer and Ölveczky [18] mentioned in their outlook, it is a quite natural extension of Algebraic Petri Nets [17] to allow tokens to be *active*. The canonic way for this extension is to consider nets as these active tokens ("nets within nets"). The general formalism for these token-nets is called *Object-Net Systems* ONS [10] (cf. also Section 4). A simplified model of ONS are "unary elementary object systems" (UEOS) defined by [20]. They are called "elementary" since the nesting hierarchy is limited to a depth of two. The top level is a so called *system-net* which has multiple instances of one single *object-net* as tokens – therefore the term "unary". (These restrictions are dropped in Sec. 4.) Here, we give a generalised version of UEOS, since Place/Transition nets (short: P/T-nets) are considered (instead of EN systems as in [20])[2]. By convention, we use the notation $\mathbf{x}$ for elements of the whole object-net system, $\hat{x}$ for elements of the object-net, and x for elements of the system-net.

Definition 1. *An* unary elementary object system *is a tuple* $OS = (N, \hat{N}, \rho)$*, such that:*

- *The* system-net $N = (P, T, pre, post, M_0)$ *is a P/T-net with* $|M_0| = 1$ *and* $|^\bullet t|, |t^\bullet| > 0$ *for all* $t \in T$.
- *The* object-net $\hat{N} = (\hat{P}, \hat{T}, \hat{pre}, \hat{post}, \hat{M}_0)$ *is a P/T-net disjoint from the system-net:* $(P \cup T) \cap (\hat{P} \cup \hat{T}) = \emptyset$.
- $\rho \subseteq T \times \hat{T}$ *is the* interaction relation.

The set of transitions in the system-net that have to fire synchronously is $T_\rho := (\cdot\rho\hat{T}) := \{t \mid \exists\hat{t} : (t, \hat{t}) \in \rho\}$. The set of autonomous (i.e. synchronisation free) transitions is $T_{\bar{\rho}} := T \setminus T_\rho$. Analogously, the set of transitions in the object-net that have to fire synchronously is $\hat{T}_\rho := (T\rho\cdot) := \{\hat{t} \mid \exists t : (t, \hat{t}) \in \rho\}$ and $\hat{T}_{\bar{\rho}} := \hat{T} \setminus \hat{T}_\rho$. The set of transitions of the object system is $\mathbf{T} := (T_{\bar{\rho}} \cup \hat{T}_{\bar{\rho}} \cup \rho)$.

3.1 Locality and Distribution

Object-nets can be investigated wrt. the so called *reference* and *value semantics* – similarly to "call by reference" and "call by value" in programming languages. The main difference between value and reference semantic is due to the handling of distribution. Reference semantics assumes a global name space.

[2] In the following the standard notations for Petri nets are used: $MS(P)$ denotes the set of multisets over P with $+$ as multiset addition and 0 as the empty multiset. A P/T-net is a tuple $N = (P, T, \text{pre}, \text{post}, M_0)$, such that $\text{pre}, \text{post} : T \to MS(P)$ are the pre- and post-condition functions, resp., mapping transitions to multisets of places; $M_0 \in MS(P)$ is the initial marking.

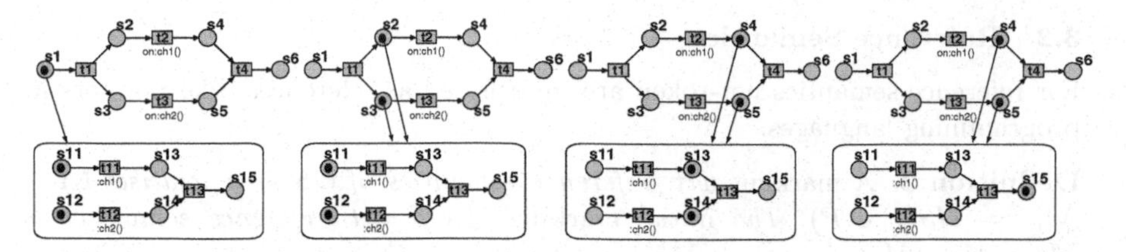

Fig. 2. Firing Sequence wrt. Reference Semantics

In Figure 2 a firing sequence wrt. reference semantics is illustrated. First, the firing of transition t_1 creates two references on the places s_2 and s_3. The concurrent firing of (t_2, t_{11}) and (t_3, t_{12}) moves the references to s_2 and s_3 resp. and creates the marking $s_{13} + s_{14}$ in the object-net. So, the transition t_{13} of the object-net is activated and fires $s_{13} + s_{14}$ to s_{15}.

It is easy to see, that this sequence contradicts our intuition to deal with two *independent* net-tokens. Value semantics provides the intended behaviour (cf. Fig. 3): The firing of t_1 created two net-tokens, which are copies of the original one. The marking of the net-token on p_1 is distributed two the copies (to preserve the number of tokens in total). The distribution chosen in this sequence allows the two synchronisations (t_2, t_{11}) and (t_3, t_{12}). The effect of firing (t_2, t_{11}) modifies the marking of the net-token on place s_2, but not the copy on s_3 – similarly for (t_3, t_{12}). Since the tokens on s_{13} und s_{14} are in different net-tokens, transitions t_{13} is *not* enabled. The two net copies have to be recombined by the system-net transition t_4, resulting in a net-token with the marking $s_{13} + s_{14}$.

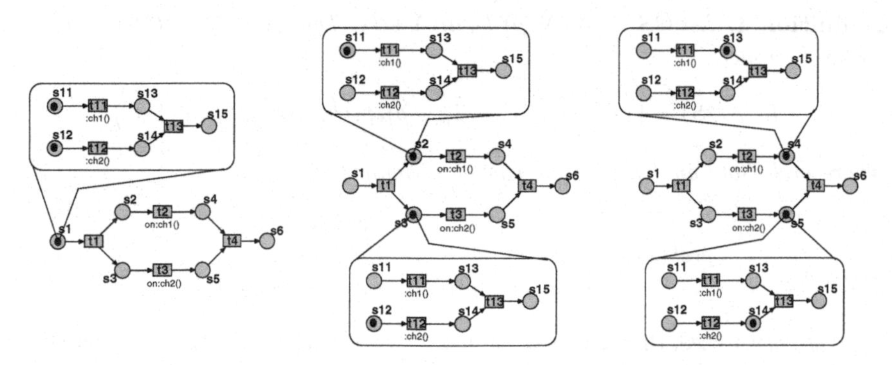

Fig. 3. Firing Sequence wrt. Value Semantics

It can be concluded, that value semantics is morc intuitivc when dealing with mobile objects in a distributed system, since for value semantics each place p denotes a location for the net-tokens independent from all other locations.

3.2 Reference Semantics

For reference semantics net-token are interpreted as references, like pointers in programming languages.

Definition 2. *A* marking *wrt. reference semantics of OS is a multiset* $\mathbf{M} \in \mathcal{M}_r := MS(P \cup \hat{P})$. *The* initial marking *of OS wrt. reference semantics is* $\mathbf{M}_0 = M_0 + \hat{M}_0$.

The firing rule is defined for three cases: system-autonomous firing, object-autonomous firing, and synchronised firing. Again, a transition can only occur autonomously if there exists no synchronisation partner in ρ, otherwise only synchronous firing is possible.

Definition 3. *Let $OS = (N, \hat{N}, \rho)$ be an unary elementary object system. Let $M \in \mathcal{M}$ be a marking wrt. reference semantics of OS. Firing can take place:*

1. *System-autonomous: A transition $t \in T_{\bar{\rho}}$ is activated in $\mathbf{M} = M + \hat{M}$, iff $M \geq pre(t)$. The successor marking is $\mathbf{M}' = M' + \hat{M}$ where $M' = M - pre(t) + post(t)$.*
2. *Synchronous: A pair $(t, \hat{t}) \in \rho$ is activated in $\mathbf{M} = M + \hat{M}$, iff $M \geq pre(t)$ and $\hat{M} \geq \hat{pre}(\hat{t})$. The successor marking is $\mathbf{M}' = M' + \hat{M}'$ where $M' = M - pre(t) + post(t)$ and $\hat{M}' = \hat{M} - \hat{pre}(\hat{t}) + \hat{post}(\hat{t})$.*
3. *Object-autonomous: A transition $\hat{t} \in \hat{T}_{\bar{\rho}}$ is activated in $\mathbf{M} = M + \hat{M}$, iff $\hat{M} \geq \hat{pre}(\hat{t})$. The successor marking is $\mathbf{M}' = M' + \hat{M}$ where $\hat{M}' = \hat{M} - \hat{pre}(\hat{t}) + \hat{post}(\hat{t})$.*

Since there is exactly one system-net and one object-net the whole UEOS can easily be described by the union of system and object-net, where synchronising transitions $(t, \hat{t}) \in \rho$ are fused.

Definition 4. *Let $OS = (N, \hat{N}, \rho)$ be an* UEOS. *The* fusion $fuse(OS)$ *is defined as the P/T-net:*

$$fuse(OS) = \Big((P \cup \hat{P}), (T_{\bar{\rho}} \cup \hat{T}_{\bar{\rho}} \cup \rho), pre^f, post^f, (M_0 + \hat{M}_0)\Big)$$

where pre^f (and analogously for $post^f$) is defined by:

$$pre^f(x) = \begin{cases} pre(x), & \text{if } x \in T_{\bar{\rho}} \\ \hat{pre}(x), & \text{if } x \in \hat{T}_{\bar{\rho}} \\ pre(t) + \hat{pre}(\hat{t}), & \text{if } x \in \rho \wedge x = (t, \hat{t}) \end{cases}$$

It is easy to see, that a transition from $\tau \in \mathbf{T} = (T_{\bar{\rho}} \cup \hat{T}_{\bar{\rho}} \cup \rho)$ is activated for an UEOS OS if it is activated in its fusion $fuse(OS)$. So, the P/T-net $fuse(OS)$ can be used to define the semantics of an UEOS[3].

[3] However, $fuse(OS)$ has in general a much more greater size (counting the number of places and transitions) than sizes of system and object-net together – in the worst case, where all transitions synchronise with all possible partners, i.e. $\rho = T \times \hat{T}$, the fusion $fuse(OS)$ has $|T| \cdot |\hat{T}|$ transitions - compared to $|T| + |\hat{T}|$ for the components in isolation. So, the UEOS is usually easier to understand compared to its fusion.

Proposition 1. *Let OS be an* UEOS. *A transition $\tau \in \mathbf{T}$ is activated in OS wrt. reference semantics iff it is activated in fuse(OS):*

$$\mathbf{M} \xrightarrow[OS]{\tau} \mathbf{M}' \iff \mathbf{M} \xrightarrow[fuse(OS)]{\tau} \mathbf{M}'$$

Proof. Immediately from Definition 3. □

The definition is a conservative extension of P/T-nets, since an UEOS with an object-net without any transitions behaves like the system-net, i.e. like a P/T-net.

Proposition 2. *Let $OS_\emptyset = (N_\emptyset, \hat{N}_\emptyset, \rho_\emptyset)$ be an* UEOS *with empty sets of places and transitions for the object-net: $\hat{P} = \hat{T} = \emptyset$. A transition $\tau \in (T_{\bar{\rho}} \cup \hat{T}_{\bar{\rho}} \cup \rho)$ is enabled in the object system $OS_\emptyset$ iff it is enabled in its system-net N.*

$$\mathbf{M} \xrightarrow[OS_\emptyset]{\tau} \mathbf{M}' \iff \mathbf{M} \xrightarrow[N]{\tau} \mathbf{M}'$$

Proof. In $OS_\emptyset$, neither synchronous firing nor object-autonomous firing are possible, because $\hat{T} = \emptyset$. Due to Def. 4 we have $fuse(OS_\emptyset) = N_\emptyset$. Since $MS(\emptyset) = \{0\}$, we only have to deal with empty object-net markings $\hat{M} = 0$ and $\mathbf{M} = M$. Due to Prop. 1 a transition in $OS_\emptyset$ is activated iff it is in $fuse(OS_\emptyset) = N_\emptyset$. □

3.3 Value Semantics

For value semantics markings are described by nested multisets, since independent copies are considered. For UEOS the nesting level is fixed with one level of nesting.

Definition 5. *A* marking *wrt. value semantics of OS is a multiset $\mathbf{M} \in \mathcal{M}_v := MS(P \times MS(\hat{P}))$. The* initial marking *of an* UEOS *OS wrt. value semantics is $\mathbf{M}_0 = (p, \hat{M}_0)$ for $M_0 = p$.*

By using projections on the first or last component of a UEOS marking $\mathbf{M}$, it is possible to compare object system markings. The projection $\Pi^1(\mathbf{M})$ on the first component abstracts away the substructure of a net-token, while the projection $\Pi^2(\mathbf{M})$ on the second component can be used as the abstract marking of the net-tokens without considering their local distribution within the system-net.

Definition 6. *Let $\mathbf{M} = \sum_{i=1}^{n}(p_i, \hat{M}_i)$ be a marking of a* UEOS *$OS = (N, \hat{N}, \rho)$.*

1. *Abstraction from the token substructure: $\Pi^1(\sum_{i=1}^{n}(p_i, \hat{M}_i)) = \sum_{i=1}^{n} p_i$*
2. *The distributed object-net copy's markings: $\Pi^2(\sum_{i=1}^{n}(p_i, \hat{M}_i)) = \sum_{i=1}^{n} \hat{M}_i$*

The UEOS firing rule is defined for three cases: system-autonomous firing, object-autonomous firing, and synchronised firing. A transition can only occur autonomously if there exists no synchronisation partner in ρ, i.e. if $\tau \in (T_{\bar{\rho}} \cup \hat{T}_{\bar{\rho}})$. Otherwise if $\tau \in \rho$, only synchronous firing is possible.

The autonomous firing of a system-net transition t removes net-tokens in the pre-conditions together with their individual internal markings. Since the markings of UEOS are higher-order multisets, we have to consider terms $\mathbf{PRE} \in \mathcal{M}_v$ that correspond to the pre-set of t in their first component: $\Pi^1(\mathbf{PRE}) = \text{pre}(t)$. In turn, a multiset $\mathbf{POST} \in \mathcal{M}_v$ is produced, that corresponds with the post-set of t in its first component. Thus, the successor marking is $\mathbf{M}' = \mathbf{M} - \mathbf{PRE} + \mathbf{POST}$, in analogy to the successor marking $M' = M - \text{pre}(t) + \text{post}(t)$ of P/T-nets. The firing of t must also obey the *object marking distribution condition* $\Pi^2(\mathbf{PRE}) = \Pi^2(\mathbf{POST})$, ensuring that the sum of markings in the copies of a net-token is preserved[4]. An object-net transition $\hat{t}$ is enabled autonomously if in $\mathbf{M}$ there is an addend $(p, \hat{M})$ in the sum and $\hat{M}$ enables $\hat{t}$. For synchronous firing, a combination of both is required[5].

Definition 7. *Let $OS = (N, \hat{N}, \rho)$ be an* UEOS. *Let $\mathbf{M}, \mathbf{M}' \in \mathcal{M}_v$ be markings wrt. value semantics of OS. Then $\mathbf{M} \xrightarrow[OS]{\tau} \mathbf{M}'$ iff exists $\mathbf{PRE}, \mathbf{POST} \in \mathcal{M}_v$ such that $\mathbf{PRE} \leq \mathbf{M}$, $\mathbf{M}' = \mathbf{M} - \mathbf{PRE} + \mathbf{POST}$, and the following holds:*

$$\begin{aligned}
\tau = t \in T_{\bar{\rho}} \Rightarrow{} & \Pi^1(\mathbf{PRE}) = pre(t) \wedge \Pi^1(\mathbf{POST}) = post(t) \wedge \\
& \Pi^2(\mathbf{POST}) = \Pi^2(\mathbf{PRE}) \wedge \\
\tau = (t, \hat{t}) \in \rho \Rightarrow{} & \Pi^1(\mathbf{PRE}) = pre(t) \wedge \Pi^1(\mathbf{POST}) = post(t) \wedge \\
& \Pi^2(\mathbf{PRE}) \geq \hat{pre}(\hat{t}) \wedge \\
& \Pi^2(\mathbf{POST}) = \Pi^2(\mathbf{PRE}) - \hat{pre}(\hat{t}) + \hat{post}(\hat{t}) \wedge \\
\tau = \hat{t} \in \hat{T}_{\bar{\rho}} \Rightarrow{} & \exists p : \Pi^1(\mathbf{PRE}) = \Pi^1(\mathbf{POST}) = p \wedge \\
& \Pi^2(\mathbf{PRE}) \geq \hat{pre}(\hat{t}) \wedge \\
& \Pi^2(\mathbf{POST}) = \Pi^2(\mathbf{PRE}) - \hat{pre}(\hat{t}) + \hat{post}(\hat{t})
\end{aligned}$$

The following properties characterise the symmetries of UEOS.

Proposition 3. *Let OS be an* UEOS *as in Def. 1.*

1. *The abstract behaviour – determined by the projection Π^1 – on the system level of an* UEOS *corresponds to the behaviour of the system-net viewed as a P/T-net: $\mathbf{M} \xrightarrow[OS]{t} \mathbf{M}'$ implies $\Pi^1(\mathbf{M}) \xrightarrow[N]{t} \Pi^1(\mathbf{M}')$, and $\Pi^1(\mathbf{M}')$ is uniquely determined.*
2. *The distributed object-net marking is invariant under autonomous actions of the system-net: $\mathbf{M} \xrightarrow[OS]{t} \mathbf{M}'$ implies $\Pi^2(\mathbf{M}) = \Pi^2(\mathbf{M}')$.*

---

[4] This represents the main difference of UEOS compared with Valk's object systems [20], which require that each net-token in $\mathbf{POST}$ hold all tokens $\Pi^2(\mathbf{PRE})$ of the net-tokens from the pre-set. This kind of multiplication of system resources is inhibited in UEOS. The structural symmetry of our definition is needed to obtain an invariant calculus and to establish a formal connection between the reference and the value semantics. These results rely on Prop. 3 (2) and (5) which are invalid for the semantics presented in [20].

[5] In contrast to P/T-nets, the successor marking is not uniquely defined for UEOS, due to their vertical substructure. The firing rule can be made deterministic by using the pair $(\mathbf{PRE}, \mathbf{POST})$ as an index for transitions in the system-net.

3. *The behaviour of the distributed object-net is determined by* Π^2 *and is a sub-behaviour of the object-net viewed as a P/T-net:* $\mathbf{M} \xrightarrow[OS]{\hat{t}} \mathbf{M}'$ *implies* $\Pi^2(\mathbf{M}) \xrightarrow[\hat{N}]{\hat{t}} \Pi^2(\mathbf{M}')$*, and* $\Pi^2(\mathbf{M}')$ *is uniquely determined.*
4. *The abstract state of the system-net is invariant under object-autonomous actions:* $\mathbf{M} \xrightarrow[OS]{\hat{t}} \mathbf{M}'$ *implies* $\Pi^1(\mathbf{M}) = \Pi^1(\mathbf{M}')$*.*
5. *Synchronisation is an action composed of two sub-actions:* $\mathbf{M} \xrightarrow[OS]{(t,\hat{t})} \mathbf{M}'$ *implies* $\Pi^1(\mathbf{M}) \xrightarrow[N]{t} \Pi^1(\mathbf{M}')$ *as well as* $\Pi^2(\mathbf{M}) \xrightarrow[\hat{N}]{\hat{t}} \Pi^2(\mathbf{M}')$*. Furthermore,* $\Pi^1(\mathbf{M}')$ *and* $\Pi^2(\mathbf{M}')$ *are uniquely determined.*

Proof. Immediate from Definition 7. □

By using a degenerated UEOS, which has neither places nor transitions, it can easily be shown that UEOS are a canonical extension of P/T-nets with respect to interleaving semantics (i.e. firing sequences). We thereby define a UEOS whose tokens have no inner structure. Therefore, this UEOS is equivalent to a P/T-net.

Proposition 4. *Let* $OS_\emptyset$ *be a* UEOS *with empty sets of places and transitions for the object-net:* $\hat{P} = \hat{T} = \emptyset$*. A transition* t *is enabled in the object system* $OS_\emptyset$ *iff it is enabled in its system-net* N*.*

$$\mathbf{M} \xrightarrow[OS_\emptyset]{t} \mathbf{M}' \iff \mathbf{M} \xrightarrow[N]{t} \mathbf{M}'$$

Proof. In $OS_\emptyset$, neither synchronous firing nor object-autonomous firing are possible, because $\hat{T} = \emptyset$. Since $MS(\emptyset) = \{0\}$, we only have to deal with empty object-net markings $\hat{M} = 0$, the object marking distribution condition is always met. The equivalence follows from Prop. 3 (1), stating that the projection $\Pi^1(\mathbf{M})$ is isomorphic to the marking M of the system-net N. □

3.4 Relation of Value and Reference Semantics

The translation of the notion of "activation" from value to reference semantics are based on the translation from value-markings to reference-markings. Nested multisets can be mapped directly to unnested ones. Define $\phi : \mathcal{M}_v \to \mathcal{M}_r$ by the linear extension of $\phi((p, \hat{M})) = p + \hat{M}$.

Proposition 5. *Let* OS *be an* UEOS *as in Def. 1 and let* $\mathbf{M}, \mathbf{M}' \in \mathcal{M}_v$ *be markings wrt. value semantics. If* $\mathbf{M} \xrightarrow[OS]{w} \mathbf{M}'$*,* $w \in \mathbf{T}$ *is a possible firing wrt. value semantics then* $\phi(\mathbf{M}) \xrightarrow[OS]{w} \phi(\mathbf{M}')$ *is a possible firing sequence wrt. reference semantics.*

Proof. From the definitions of the firing rules in Def. 7 (value semantics) and Def. 3 (reference semantics) it can be seen directly, that activation wrt. value semantics implies activation wrt. reference semantics. □

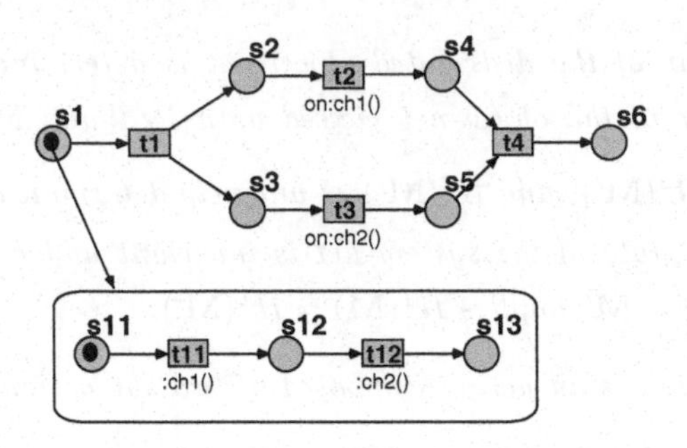

Fig. 4. An example UEOS

The converse direction does not hold, which can be demonstrated using the UEOS in Fig. 4. For reference semantics the place s_1 contains initially a reference to the object-net: $\mathbf{M} = s_1 + \hat{s}_{11}$. Firing of t_1 duplicates this reference onto s_2 and s_3 resulting in $\mathbf{M}_1 = s_2 + s_3 + \hat{s}_{11}$. This marking activates the transition pair $(t_2, \hat{t}_{11})$ while $(t_3, \hat{t}_{12})$ is not. The resulting marking is $\mathbf{M}_2 = s_4 + s_3 + \hat{s}_{12}$. Since the effect in the object-net is visible in the whole system, the pair $(t_3, \hat{t}_{12})$ is now activated. Firing leads to $\mathbf{M}_3 = s_4 + s_5 + \hat{s}_{13}$.

For value semantics we have $\mathbf{M} = (s_1, \hat{s}_{11})$. Here $\phi(\mathbf{M}) = s_1 + \hat{s}_{11}$ is the corresponding marking wrt. reference semantics. Firing of t_1 can result in in the marking $\mathbf{M}_1 = (s_2, \hat{s}_{11}) + (s_3, 0)$ – corresponding to $\phi(\mathbf{M}_1) = s_2 + s_3 + \hat{s}_{11}$ for reference semantics. This marking activates the transition pair $(t_2, \hat{t}_{11})$ while $(t_3, \hat{t}_{12})$ is not. The resulting marking is $\mathbf{M}_2 = (s_4, \hat{s}_{12}) + (s_3, 0)$. Since the effect in the object net is only local the pair $(t_3, \hat{t}_{12})$ is not activated. So $w = t_1(t_2, \hat{t}_{11})(t_3, \hat{t}_{12}) \in \mathbf{T}^*$ is a possible firing sequence for reference but not for value semantics.

The difference of reference and value semantics is the concept of "location" for net-tokens which is explicit for value but not for reference semantics, since it is unclear which reference can be considered as the location of a net-token. The localisation of net-tokens is expressed by a mapping from reference semantics markings to value semantics ones.

Definition 8. *A mapping* $\sigma : \mathcal{M}_r \to \mathcal{M}_v$ *is called a* localisation *iff for each* $\mathbf{M} \in \mathcal{M}_r$ *with* $\mathbf{M} = M_1 + \hat{M}_2$, $M_1 \in MS(P)$, $\hat{M}_2 \in MS(\hat{P})$ *the following holds:*

$$\Pi^1(\sigma(\mathbf{M}_r)) = M_1 \qquad \text{and} \qquad \Pi^2(\sigma(\mathbf{M}_r)) = \hat{M}_2$$

A localisation σ *is* compatible *wrt. the firing rule iff for each* $\mathbf{M}, \mathbf{M}' \in \mathcal{M}_r$ *and* $\tau \in \mathbf{T}$ *we have:*

$$\mathbf{M}_r \xrightarrow[OS]{\tau} \mathbf{M}_r' \Rightarrow \sigma(\mathbf{M}_r) \xrightarrow[OS]{\tau} \sigma(\mathbf{M}_r')$$

For the special case that the system-net is a state machine the ambiguity of locations cannot appear, since there always is exactly one reference.

Definition 9. *An* UEOS OS *is called* simple *iff the system-net N is a state machine, i.e. $|M_0| = 1$ and for all $t \in T$ we have $|pre(t)| \leq 1$ and $|post(t)| \leq 1$.*

For a simple UEOS there is exactly one localisation mapping $\sigma : \mathcal{M}_r \to \mathcal{M}_v$: Since N is a state machine for each $\mathbf{M} \in \mathcal{M}_r$ with $\mathbf{M} = M + \hat{M}$ we have exactly one net-token reference: $|M| = 1$. So, with $M = p$ the only map with $\Pi^1(\sigma(\mathbf{M}_r)) = p$ and $\Pi^2(\sigma(\mathbf{M}_r)) = \hat{M}$ is $\sigma_{sm}(p + \hat{M}) := (p, \hat{M})$. For a simple UEOS reference and value semantics coincide.

Proposition 6. *Let OS be a simple* UEOS. *Then σ_{sm} is a compatible localisation: $\mathbf{M}_r \xrightarrow[OS]{\tau} \mathbf{M}'_r \Rightarrow \sigma_{sm}(\mathbf{M}_r) \xrightarrow[OS]{\tau} \sigma_{sm}(\mathbf{M}'_r)$ for all $\tau \in \mathbf{T}$.*

Proof. Follows from the property of a localisation map σ (Def. 8) and the definition of the value semantics firing rule (Def. 7) and of the reference semantics firing rule (Def. 3). □

The general condition for a firing sequence wrt. reference semantics being also possible for value semantics is given in [8] where the concept of locality that is present for value semantics but not for reference semantics is combined with Petri net processes.

3.5 Invariants in Ueos

The compositionality of invariants is interesting for UEOS. In the following we investigate how the invariant calculus of P/T-nets extends for UEOS. Let $N = (P, T, \mathrm{pre}, \mathrm{post}, M_0)$ be a P/T-net. The incidence matrix Δ is defined by $\Delta(p,t) := \mathrm{post}(t)(p) - pre(t)(p)$. A *P*-invariant $\boldsymbol{i} \in \mathbb{Z}^{|P|}, \boldsymbol{i} \neq \mathbf{0}$ is a vector that fulfils $\boldsymbol{i} \cdot \Delta = \mathbf{0}$. Then every reachable marking $\boldsymbol{M}$ fulfils the linear equation $\boldsymbol{i} \cdot \boldsymbol{M} = \boldsymbol{i} \cdot \boldsymbol{M}_0$.

The following theorem states that invariants for a UEOS can easily be composed from the invariants of the components.

Proposition 7. *Let $OS = (N, \hat{N}, \rho)$ be an* UEOS *as in Def. 1 and let $\boldsymbol{i} \in \mathbb{Z}^{|P|}$ be an invariant of the system-net and $\hat{\boldsymbol{i}} \in \mathbb{Z}^{|\hat{P}|}$ one of the object-net.*

Let $\mathbf{M} \in \mathcal{M}_r$ be a reference semantics marking. Then $\boldsymbol{i} \cdot \mathbf{M}_{|P} = \boldsymbol{i} \cdot \mathbf{M}_{0|P}$ and $\hat{\boldsymbol{i}} \cdot \mathbf{M}_{|\hat{P}} = \hat{\boldsymbol{i}} \cdot \mathbf{M}_{0|\hat{P}}$ holds for every reachable marking $\mathbf{M} \in \mathcal{M}_r$ wrt. reference semantics. Additionally, the vector $(\boldsymbol{i}, \hat{\boldsymbol{i}}) \in \mathbb{Z}^{|P|+|\hat{P}|}$ is an invariant of the fusion net fuse(OS).

Let $\mathbf{M} \in \mathcal{M}_v$ be a value semantics marking. Then $\boldsymbol{i} \cdot \Pi^1(\mathbf{M}) = \boldsymbol{i} \cdot \Pi^1(\mathbf{M}_0)$ and $\hat{\boldsymbol{i}} \cdot \Pi^2(\mathbf{M}) = \hat{\boldsymbol{i}} \cdot \Pi^2(\mathbf{M}_0)$ holds for every reachable marking $\mathbf{M} \in \mathcal{M}_v$ wrt. value semantics.

Proof. For reference semantics the propositions follows from Prop. 1.

For value semantics: For a system-autonomous step $\mathbf{M} \xrightarrow{\tau} \mathbf{M}'$ we have $\Pi^1(\mathbf{M}) \xrightarrow[N]{\tau} \Pi^1(\mathbf{M}')$ by Prop. 3(1). Since $\boldsymbol{i}$ is an invariant of N we have $\boldsymbol{i} \cdot \Pi^1(\mathbf{M}) = \boldsymbol{i} \cdot \Pi^1(\mathbf{M}')$. Also, we have $\hat{\boldsymbol{i}} \cdot \Pi^2(\mathbf{M}) = \hat{\boldsymbol{i}} \cdot \Pi^2(\mathbf{M}_0)$, since $\Pi^2(\mathbf{M}) = \Pi^2(\mathbf{M}')$ by Prop. 3(2). Analogously with Prop. 3(3) and (4) for object-autonomous and steps and with Prop. 3(5) for synchronisations. □

In the following some decidability question related to Petri nets are studied for "nets within nets" formalisms, especially the reachability and the boundedness problem is considered.

3.6 Reachability

It is a well known fact that Petri nets (including EN systems or P/T-nets) are not Turing powerful, since the reachability problem is decidable [14], whereas the halting problem for Turing machines is not.

Reachability is decidable for 1-safe unary elementary object-nets, since the set of possible markings $\mathcal{M}_v = MS(P \times MS(\hat{P}))$ reduces to $\mathcal{M}_v = 2^{P \times 2^{\hat{P}}}$, which is a finite set. So, $\mathcal{M}_v$ can be used as the set of places, i.e. each place has the form $(p, \hat{M})$. It can be seen directly, that the resulting Petri net has isomorphic marking and firing sequences. The result can be generalised directly for UEOS where system and object-net are n-safe.

For the unbounded case the reachability problem for UEOS wrt. reference semantics is decidable.

Proposition 8. *Reachability is decidable for* UEOS *wrt. reference semantics.*

Proof. The fusion of an UEOS OS results in the P/T-net $fuse(OS)$ which has the same markings and firing sequences (cf. Prop. 1). So, the reachability problem for UEOS reduces to that for P/T-nets. □

On the other hand the reachability problem for UEOS wrt. value semantics is undecidable, since they can simulate nets with transfer arcs. Transfer arcs (cf. [3]) are used to transfer all tokens of one place to another place. So, transfer arcs are related to reset arcs which remove all tokens from one place.

Proposition 9. *Reachability is undecidable for* UEOS *wrt. value semantics.*

Proof. This can be seen by giving a simulation of Petri nets with transfer arcs using UEOS and the fact that the reachability problem for Petri nets with at least two transfer arcs is undecidable (cf. Theorem 11 in [4]). In the simulation the system-net describes the original Petri net N and the object-net acts as a container for the tokens. Whenever a token is added to a place p in N the system-net removes the object-net on p, adds one token inside using the synchronisation mechanism, and puts the object-net back. Whenever the whole marking of the place p is transferred in N the object-net is removed as a whole together with its marking $\hat{M}$. □

A related result holds for Linear Logic Petri Nets [5], which are Petri nets with Linear Logic formulas as tokens. It is assumed that the token-formulas can be rewritten using the calculus of Linear Logic independently of the firing of transitions. By using undecidability results for deduction in Linear Logic, Farwer [5] shows that the reachability problem is undecidable for Linear Logic Petri Nets.

3.7 Boundedness

A natural question is whether UEOS are as expressive as Nested Petri Nets [13]. The answer to this question is "no" since it can be shown that boundedness is a decidable property of UEOS, while boundedness is undecidable for Nested Petri Nets.

Proposition 10. *Boundedness is undecidable for Nested Petri Nets.*

Proof. Nested Petri Nets can simulate Reset Nets (cf. [13]) and boundedness for Reset Nets is undecidable (cf. [4, Theorem 8]). □

In general, the boundedness problem can be decided using the coverability graph construction of Karp and Miller [7]: To construct the coverability graph we are looking for a marking sequence $m_0 \xrightarrow{*} m \xrightarrow{+} m'$ where $m < m'$ holds, i.e. m' covers m on a non-empty subset of places $A \subseteq P$: $m(p) < m'(p)$ for all $p \in A$ and $m(p) = m'(p)$ for all $p \in P \setminus A$. Since Petri nets enjoy the property of monotonicity (i.e. if $m_1 < m_2$ and $m_1 \xrightarrow{*} m_1'$ then there exists a sequence $m_2 \xrightarrow{*} m_2'$ with $m_1' < m_2'$) this sequence can be repeated infinitely often showing that any bound for a place $p \in A$ can be covered, i.e. the place p is unbounded.

To apply this technique for UEOS the partial order $\leq$ on multisets can be extended to a partial order on nested multisets.

Definition 10. *Let* $\mathbf{M}, \mathbf{M}' \in \mathcal{M}_v$ *two nested multisets. Define* $\mathbf{M} \preceq \mathbf{M}'$ *iff there exists a total and injective mapping* f *from* $\mathbf{M} = \sum_{i=1}^{n}(p_i, \hat{M}_i)$ *to* $\mathbf{M}' = \sum_{j=1}^{n}(p_j', \hat{M}_j')$ *with* $f((p_i, \hat{M}_i)) = (p_j', \hat{M}_j')$ *implying* $p_i = p_j'$ *and* $\hat{M}_i \leq \hat{M}_i'$.

Note, that with this definition $\mathbf{M} \preceq \mathbf{M}'$ implies $\Pi^1(\mathbf{M}) \leq \Pi^1(\mathbf{M}')$.

Proposition 11. *Boundedness is decidable for* UEOS *wrt. value semantics.*

Proof. A partial order $\leq$ has the property of strict transitive compatibility (monotonicity), i.e. if $m_1 < m_2$ and $m_1 \rightarrow m_1'$ then there exists a sequence $m_2 \xrightarrow{*} m_2'$ with $m_1' < m_2'$. Generalising the result of [7] it is shown in [6], that the boundedness problem is decidable iff $\leq$ is a decidable, strict partial order and the set of successor markings is decidable. Obviously $\preceq$ is decidable and strict transitive compatibility and set the of successors is effective constructible. □

4 Anonymous Object-Net Systems

In the following a generalised model of object-net systems is defined, which drops the restriction to exactly two levels of nesting: Anonymous Object-Net Systems (ONS)[6] are defined to give a precise definition of nets within nets using rewriting logic [15]. Let $\mathcal{N}$ be a finite set of net sorts in the following[7]. The black token net N_* is defined as the object-net with no places and no transitions: $P(N_*) = T(N_*) = \emptyset$.

[6] These Object-Net Systems are called "anonymous" since the net-tokens carry no identity (see also Prop. 12 below).

[7] Therefore the notation conventions for elements in UEOS (like $\mathbf{x}$, $\hat{x}$, and x) are not applicable anymore.

Channels. Object-nets synchronise via a set of channels $V = \bigcup_{N \in \mathcal{N}} V_N$ with disjoint V_N. Channels are either directed downwards (called *downlinks* – the "synchronising" side, where the net-token is known) or upwards (*uplinks* – the "synchronised" side)[8]. The labelling function $\lambda^\uparrow : T \to MS(V)$ maps each transition to a multiset of uplinks while $\lambda^\downarrow : T \to MS(V)$ maps a transition to a multiset of downlinks. Using multisets of labels allows to describe multiple synchronisation. The neutral element 0 means that no synchronisation is needed. There is at most one uplink, i.e. $|\lambda^\uparrow(t)| \leq 1$ for each transition $t \in T$, which is required to obtain a tree-like synchronisation structure.

Nets as Tokens. Markings are multisets of net-tokens (following the "Petri nets are monoids" paradigm [16]). Define the sorts of tokens and markings with the usual multiset addition (being associative and commutative) and the neutral element 0_N:

sorts $Token_N$ MS_N .
subsort $Token_N < MS_N$.
op $0_N : \to MS_N$.
op _ + _ : $MS_N \times MS_N \to MS_N$ [assoc comm id: 0_N] .

Object-nets describe the structure of net-tokens. Net-tokens are obtained if a marking is considered as one single entity. This is done by the operator $\text{net}_N : MS_N \to N$.

Tokens are described by their value and the place that is marked. It is natural to use the place name as operators. Note that these operators are needed to formulate the non-zero multisets. Each place $p \in P$ is mapped by $d : P \to \mathcal{N}$ to a place sort. Define for all $p \in P(N)$ the place operators: $p : d(p) \to Token_N$. In the following the token term $p(\text{net}_{d(p)}(M))$ is abbreviated as $p[M]$. Let "$\bullet$" be the abbreviation of $[0]$ for N_*. We assume, that the initial marking of all nets contains only black-tokens: $M_0(N)(p) > 0 \Rightarrow d(p) = N_*$ for all $N \in \mathcal{N}$. The *initialisation* of a net-token given the net sort N is defined as:

$$init_N := \text{net}_N\left(\sum\nolimits_{p \in P(N)} M_0(N)(p) \cdot p(\bullet)\right)$$

Define for all $v \in V_N$ the constants $v^{?!}, v^{!?} : \to Token_N$. Informally the place $v^{?!}$ stands for the start of the atomically executed sub-synchronisation and $v^{!?}$ stands for its end.

Object-Net Systems. An anonymous Object-Net System (ONS) consists of a set of Petri Nets. For simplicity we denote this set also by $\mathcal{N}$, so N denotes a sort as well as a net. This re-use is harmless, since it can be resolved by introducing a set $\{ON_N : N \in \mathcal{N}\}$ isomorphic to $\mathcal{N}$. One net $N_{sn} \in \mathcal{N}$ is the system-net, building the top level of the system. It is assumed that places and transitions are disjoint for all nets and all nets use a common set of up- and downlinks V.

---

[8] Note, that a horizontal synchronisation of two net-tokens occupying the same place can be simulated by attaching a transition to that place which synchronises twice vertically. Therefore synchronisation is restricted to the vertical case.

Definition 11. *An* anonymous Object-Net System *is the tuple*

$$ONS = (\mathcal{N}, d, V, \lambda)$$

1. $\mathcal{N} = \{N_1, \ldots N_n\}$ *is a set of pairwise disjoint P/T-nets.* $\mathcal{N}$ *includes the black token net* $N_* \in \mathcal{N}$ *and the system-net* $N_{sn} \in \mathcal{N}$*. Let* P *be the union of all components:* $P := \bigcup_{N \in \mathcal{N}} P(N)$*. Analogously for* T*, pre, and post.*
2. $d : P \to \mathcal{N}$ *is the place typing.*
3. $V = \bigcup_{N \in \mathcal{N}} V_N$ *is the set of channels.*
4. $\lambda = (\lambda^\uparrow : T \to MS(V), \lambda^\downarrow : T \to MS(V))$ *are mappings of transitions to multisets of up- resp. downlinks. There is at most one uplink, i.e.* $|\lambda^\uparrow(t)| \leq 1$ *for all transition* $t \in T$.

The initial marking is $M_0 = init_{N_{sn}}$.

An Object Petri Net is called *pairwise synchronised* iff every transition t has either uplinks or downlinks but not both (similar to process calculi):

$$\forall t \in T : (|\lambda^\uparrow(t)| > 0 \Rightarrow |\lambda^\downarrow(t)| = 0) \wedge (|\lambda^\downarrow(t)| > 0 \Rightarrow |\lambda^\uparrow(t)| = 0)$$

This condition is needed to obtain a finite set of possible synchronisations.

Firing Rule. The presentation of Petri nets as tokens leads to the problem, how synchronised actions spanning over several net levels should be formalised. Due to the tree-like structure of synchronisation there are infinitely many possible synchronisations in general, so we cannot formalise each synchronisation as a transition. Instead we formalise sub-synchronisations as conditional rewrites. Define for all $t \in T$ the following conditional rule:

$$\begin{aligned} t : \lambda^\uparrow(t)^{?!} &+ \sum_{p \in {}^\bullet t} \sum_{i=1}^{W(p,t)} p[A_{p,t,i} + B_{p,t,i}] \\ \to \lambda^\uparrow(t)^{!?} &+ \sum_{p \in t^\bullet} \sum_{j=1}^{W(t,p)} p[A'_{p,t,j} + B'_{p,t,j}] \quad \text{if } \psi_1(t) \wedge \psi_2(t) \wedge \psi_3(t) \end{aligned}$$

1. The uplink (if any) $v \in V$ is denoted as $\lambda^\uparrow(t)^{?!}$ in the pre-set and as $\lambda^\uparrow(t)^{!?}$ in the post-set.
2. The markings of net-tokens are distributed into two groups of multisets: $A_{p,t,i}, A'_{p,t,j}$ for the tokens that are transported and $B_{p,t,i}, B'_{p,t,j}$ for the tokens that are used for synchronisation. Unprimed variables are used for variables of the pre-set and single primed ones for in the post-set.
   - For all $p \in {}^\bullet t$ and $i \in \{1, \ldots, W(p,t)\}$: $A_{p,t,i}, B_{p,t,i} : MS_{d(p)}$.
   - For all $p \in t^\bullet$ and $j \in \{1, \ldots, W(t,p)\}$: $A'_{p,t,j}, B'_{p,t,j} : MS_{d(p)}$.
   - For all $v \in V_N$ and $l \in \{1, \ldots \lambda^\downarrow(t)(v)\}$: $M_{v,l}, M'_{v,l} : MS_N$.
3. The markings of the net-tokens are related by the requirement, that firing preserves the overall sum of all transported net-tokens. The sum is preserved for each net sort N:

$$\psi_1(t) := \forall N \in \mathcal{N} : \sum_{p \in ({}^\bullet t \cap d^{-1}(N))} \sum_{i=1}^{W(p,t)} A_{p,t,i} = \sum_{p \in (t^\bullet \cap d^{-1}(N))} \sum_{j=1}^{W(t,p)} A'_{p,t,j}$$

4. The sub-synchronisations fire $M_{v,l}$ to $M'_{v,l}$. These markings are distributed on the synchronisation markings $B_{p,t,i}$ for $M_{v,l}$. Analogously, the $B'_{p,t,j}$ must sum up to $M'_{v,l}$:

$$\psi_2(t) := \forall N \in \mathcal{N} : \sum_{v \in V_N} \sum_{l=1}^{\lambda^{\downarrow}(t)(v)} M_{v,l} = \sum_{p \in ({}^{\bullet}t \cap d^{-1}(N))} \sum_{i=1}^{W(p,t)} B_{p,t,i} \wedge$$
$$\sum_{v \in V_N} \sum_{l=1}^{\lambda^{\downarrow}(t)(v)} M'_{v,l} = \sum_{p \in (t^{\bullet} \cap d^{-1}(N))} \sum_{j=1}^{W(t,p)} B'_{p,t,j}$$

5. Each downlink $v \in V$ with $\lambda^{\downarrow}(t)(v) > 0$ creates one sub-synchronisation on this channel in the conditional:

$$\psi_3(t) := \forall N \in \mathcal{N} : \forall v \in V_N : \forall l \in \{1, \ldots, \lambda^{\downarrow}(t)(v)\} : v^{?!} + M_{v,l} \to v^{!?} + M'_{v,l}$$

By this construction every outgoing channel results in a sub-synchronisation which can be executed by the corresponding incoming channel side.

Example 1. Have a look at a simple example, pictured in Fig. 5: The transition t fires synchronously via the channel v with the object-net transition $\hat{t}$. The transition in the system-net is:

$$p[M] \to p'[M'] \quad \text{if} \quad v^{?!} + M \to v^{!?} + M'$$

The transition in the object-net is $\hat{t} : v^{?!} + \hat{p}(\bullet) \to v^{!?} + \hat{p}'(\bullet)$. For the binding $M = M'' + \hat{p}(\bullet)$ and $M' = M'' + \hat{p}'(\bullet)$ the conditional is fulfilled. So the synchronous firing of t and $\hat{t}$ is the step $p[M'' + \hat{p}(\bullet)] \to p'[M'' + \hat{p}'(\bullet)]$.

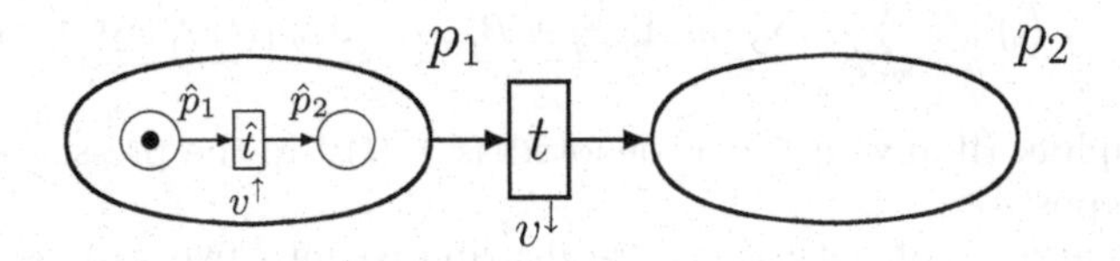

Fig. 5. An Object-Net System

An even stronger result than the undecidability of reachability is that some of the above mentioned formalisms also have the power of Turing machines. This can be shown by reduction to other extended Petri net formalisms which have Turing power or by simulating Turing machines or two-counter-machines directly.

A generalisation of UEOS are minimal OO-nets [11] which have objects as tokens. Objects are characterised by their identity (and maybe several other attributes which are not considered). The only assumption of OO-nets is a countable set of object *identities* which can be tested for equality and that fresh identities can be created at run-time. This minimal assumption should be met by every existing object-oriented extension of Petri nets. Under this assumptions OO-nets can simulate two-counter-machines. This implies that reachability is also undecidable for OO-nets and therefore for almost every object-oriented extension of Petri nets.

Proposition 12 ([11]). *Minimal OO-nets are Turing powerful.*

It is easy to see, that the fusion construction of Def. 4 can be extended for an arbitrary but fixed number of nets, i.e. instances with the restriction that there are only finitely many synchronisations. From Prop. 1 we know, that the fusion net can always be simulated by a P/T-net which is not Turing powerful. So, we are only considering value semantics.

In the following it it shown that it is the recursive structure that creates the expressiveness, since the "nets within nets" model is Turing powerful even if *no identities* are used. It is sufficient to allow the operation of vertical nesting and synchronisation to simulate Turing machines.

Definition 12. *A* 2-counter automaton *is represented by the tuple*

$$A = (Q, \delta_{0,0}, \delta_{0,1}, \delta_{1,0}, \delta_{1,1}, q_0, Q_f),$$

where Q is a finite set of states, $q_0 \in Q$ is the initial state, and $Q_f \subseteq Q$ is the set of final states. The transitions function $\delta_{0,0}$ to $\delta_{1,1}$ are defined depending on the values of the counters:

- $\delta_{0,0} : Q \to Q \times \{1\} \times \{1\}$ *(if both counters are equal zero),*
- $\delta_{0,1} : Q \to Q \times \{1\} \times \{-1, 1\}$ *(if the first counter equals zero),*
- $\delta_{1,0} : Q \to Q \times \{-1, 1\} \times \{1\}$ *(if the second counter equals zero), and*
- $\delta_{1,1} : Q \to Q \times \{-1, 1\} \times \{-1, 1\}$ *(if both counters are positive).*

A configuration *is a triple $(q, n_1, n_2) \in Q \times \mathbb{N} \times \mathbb{N}$. The successor configuration of (q, n_1, n_2) is $(q', n_1 + d_1, n_2 + d_2)$ where $k = \min(n_1, 1)$, $l = \min(n_2, 1)$, and $(q', d_1, d_2) = \delta_{k,l}(q)$. Note, that $\delta_{k,l}$ is actually a test for zero which guarantees $n_1 + d_1 \geq 0$ and $n_2 + d_2 \geq 0$.*

The following proof of Prop. 13 encoded natural numbers by exploiting the nesting structure of the net-tokens, like $\mathsf{suc}[\mathsf{suc}[\ldots\mathsf{suc}[\mathsf{zero}]\ldots]]$. Incrementation is implemented by synchronising downwards the marking structure and modifying token zero into $\mathsf{suc}[\mathsf{zero}]$. Similarly for decrementation.

Proposition 13. *Anonymous Object Net Systems are Turing powerful.*

Proof. Let A be a 2-counter machine, then define the object net $N(A) = (P_1 \cup P_2, T_1 \cup T_2, \mathrm{pre}, \mathrm{post}, M_0)$ as follows: Let $P_1 = \{q_1, q_2 \mid q \in Q\}$, where the place q_1 holds the first and q_2 the second counter (encoded as a net-token) of A when being in state q. The places $\{q_1, q_2 \mid q \in Q_f\}$ describe the final states. Let $T_1 = \{\mathsf{f}_{(q,q',k,l)}, \mathsf{j}_{(q,q',k,l)} \mid \delta_{k,l}(q) = (q', d_1, d_2)\}$, where $k = 0$ ($l = 0$) iff the first (second) counter equals zero.

Let $P_2 = \{p_{(q,q',k,l),i}, p'_{(q,q',k,l),i}, p''_{(q,q',k,l),i} \mid i \in \{1,2\} \wedge \delta_{k,l}(q) = (q', d_1, d_2)\}$ and $T_2 = \{t_{(q,q',k,l),i}, t'_{(q,q',k,l),i} \mid i \in \{1,2\} \wedge \delta_{k,l}(q) = (q', d_1, d_2)\}$, used to modify the counters. Let $V = \{\mathsf{inc}_i, \mathsf{set}_i, \mathsf{dec}_i, \mathsf{test}_{k,i} \mid i \in \{1,2\} \wedge k \in \{0,1\}\}$ be the set of channels. The initial marking is $M_0 = q_{0,1} + q_{0,2}$, i.e. the initial state of A. We use the abbreviation $c := (q, q', k, l)$ for subscripts.

Define the type $d(q_1) = d(p_{c,1}) = d(p'_{c,1}) = d(p''_{c,1}) = N_{suc,1}$ and $d(q_2) = d(p_{c,2} = d(p'_{c,2}) = d(p''_{c,2}) = N_{suc,2}$. Pre- and post-condition are given as:

- $\text{pre}(\mathsf{f}_c) = q_1 + q_2$, $\text{post}(\mathsf{f}_c) = p_{c,1} + p_{c,2}$, $\lambda^\uparrow(\mathsf{f}_c) = 0$, and $\lambda^\downarrow(\mathsf{f}_c) = \mathsf{test}_{k,1} + \mathsf{test}_{l,2}$.
- $\text{pre}(\mathsf{j}_c) = p''_{c,1} + p''_{c,2}$, $\text{post}(\mathsf{j}_c) = q'_1 + q'_2$, and $\lambda^\uparrow(\mathsf{j}_c) = \lambda^\downarrow(\mathsf{j}_c) = 0$
- $\text{pre}(t_{c,i}) = p_{c,i}$, $\text{post}(t_{c,i}) = p'_{c,i}$, $\lambda^\uparrow(t_{c,i}) = 0$ and $\lambda^\downarrow(t_{c,i}) = \begin{cases} \mathsf{inc}_i, & \text{if } d_i = +1 \text{ where } \delta_{k,l}(q) = (q', d_1, d_2) \\ \mathsf{dec}_i & \text{if } d_i = -1 \text{ where } \delta_{k,l}(q) = (q', d_1, d_2) \end{cases}$.
- $\text{pre}(t'_{c,i}) = p'_{c,i}$, $\text{post}(t'_{c,i}) = p''_{c,i}$, $\lambda^\uparrow(t'_{c,i}) = 0$, and $\lambda^\downarrow(t'_{c,i}) = \begin{cases} \mathsf{set}_i, & \text{if } d_i = +1 \text{ where } \delta_{k,l}(q) = (q', d_1, d_2) \\ 0, & \text{if } d_i = -1 \text{ where } \delta_{k,l}(q) = (q', d_1, d_2) \end{cases}$.

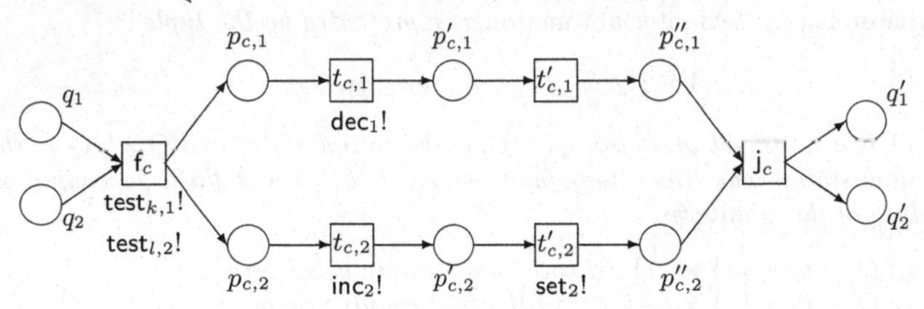

Fig. 6. Simulation of the state change $\delta_{k,l}(q) = (q', -1, 1)$

Figure 6 shows the state change $\delta_{k,l}(q) = (q', -1, 1)$ incrementing the first and decrementing the second counter (where uplinks are denoted with "?" and downlinks with "!").

Figure 7 shows the structure of the successor net $N_{suc,1}$ and $N_{suc,2}$ which is the same (the index i of nodes are omitted in the inscriptions). The net $N_{suc,i}$ implements a successor cell. The type mapping is defined by $d(\mathsf{zero}_i) = d(\mathsf{one}_i) = d(\mathsf{pos}_i) = N_\bullet$ and $d(\mathsf{suc}_i) = N_{suc,i}$.

Define the encoding of the counters using $cod_i(n+1) = \mathsf{pos}_i + \mathsf{suc}_i[cod_i(n)]$ and $cod_i(0) = \mathsf{zero}_i$ (where $i \in \{1,2\}$ denotes the first or the second counter, resp.). So, the counter n is represented by the nested marking structure:

$$cod_i(n) := \underbrace{\mathsf{suc}_i[\mathsf{pos}_i + \mathsf{suc}_i[\ldots \mathsf{pos}_i + \mathsf{suc}_i[}_{n-\text{times}} \mathsf{zero}_i \underbrace{]\ldots]]}_{n-\text{times}}$$

The object system $ONS(A) = (\mathcal{N}, d, V)$ with $\mathcal{N} = \{N(A), N_{suc,1}, N_{suc,2}, N_\bullet\}$ simulates A. Its initial marking is

$$\mathbf{M}_0 = q_{0,1}[\mathsf{zero}_1] + q_{0,2}[\mathsf{zero}_2] = q_{0,1}[cod_1(0)] + q_{0,2}[cod_2(0)]$$

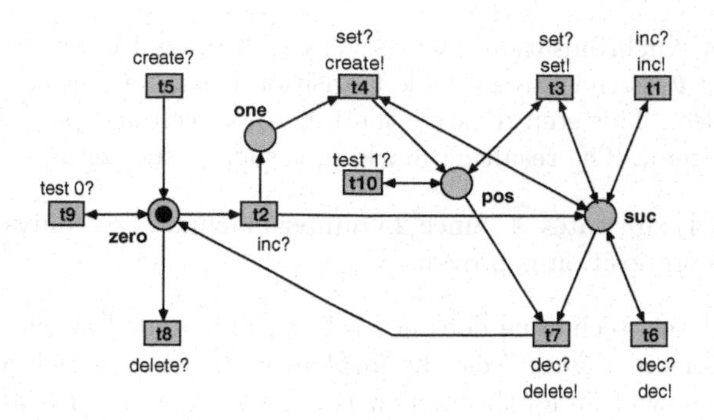

Fig. 7. The "Successor" Net N_{suc}

We prove, that whenever the automaton A changes its configuration (q, n_1, n_2) to (q', n_1+d_1, n_2+d_2), then the object system $ONS(A)$ changes from $q_1[cod_1(n_1)] + q_2[cod_2(n_2)]$ to $q_1'[cod_1(n_1 + d_2)] + q_2'[cod_2(n_2 + d_2)]$:

1. Initially, A makes two tests for zero by choosing the appropriate $\delta_{k,l}$. This is simulated by the synchronisation on the channels $\mathsf{test}_{k,1}$ und $\mathsf{test}_{l,2}$ of the transition $\mathsf{f}_{(q,q',k,l)}$.
2. Incrementation (i.e. $d_i = +1$) of $cod(n_i) = \mathsf{suc}_i[\cdots[\mathsf{pos}_i + \mathsf{suc}_i[\mathsf{zero}_i]]\cdots]$ involves two steps: The first one fires
$$\theta_1 = t_{(q,q',k,l),i} /\!/_{\mathsf{inc}_i} \underbrace{t_{1,i} /\!/_{\mathsf{inc}_i} \cdots /\!/_{\mathsf{inc}_i} t_{1,i} /\!/_{\mathsf{inc}_1}}_{n\text{-times}} t_{2,i}$$
which is a synchronisation along the channel inc_1. Note, that θ_1 is the only possible transition when starting to synchronise on channel inc_1. This step removes the token from the place zero_i and creates a new cell on the place suc_i with the empty marking (due to the preservation of the marking sum). So, we obtain $\mathsf{suc}_i[\cdots[\mathsf{pos}_i + \mathsf{suc}_i[\mathsf{one}_i + \mathsf{suc}_i[0]]]\cdots]$. The token on the place one_i indicates the net that contains the cell, that still has to be initialised. By firing n-times via the channel set_i and once via create_i:
$$\theta_2 = t'_{(q,q',k,l),i} /\!/_{\mathsf{set}_1} \underbrace{t_{3,i} /\!/_{\mathsf{set}_1} \cdots /\!/_{\mathsf{set}_i} t_{3,i} /\!/_{\mathsf{set}_i}}_{n\text{-times}} t_{4,i} /\!/_{\mathsf{create}} t_{5,i}$$
The resulting marking is $\mathsf{suc}_i[...[\mathsf{pos}_i + \mathsf{suc}_i[\mathsf{pos}_i + \mathsf{suc}_i[\mathsf{zero}_i]]]...] = cod(n_i+1)$. Note, that θ_2 is the only possible sequence when starting to synchronise on channel set_i. The token on one_i is used to switch the synchronisation sequence from set_i to create_i. The token on pos_i indicates cells describing a counter greater than zero.
3. Decrementation (i.e. $d_i = -1$) of $cod(n_i+1) = \mathsf{suc}_i[...\mathsf{suc}_i[\mathsf{pos}_i + \mathsf{suc}_i[\mathsf{zero}_i]]...]$ is implemented by firing
$$\theta_3 = t_{(q,q',k,l),i} /\!/_{\mathsf{dec}_i} \underbrace{t_{6,i} /\!/_{\mathsf{dec}_i} \cdots /\!/_{\mathsf{dec}_i} t_{6,i} /\!/_{\mathsf{dec}_i}}_{n\text{-times}} t_{7,i} /\!/_{\mathsf{delete}_i} t_{8,i}$$

which is a synchronisation n-times via the channel dec_i and once via delete_i. Note, that θ_3 is the only possible transition when starting to synchronise on channel dec_i. This step removes the innermost cell and puts one token onto the place zero_i. The resulting marking is $\mathsf{suc}_i[\cdots \mathsf{suc}_i[\mathsf{zero}_i] \cdots] = cod(n_i)$.

So, $ONS(A)$ simulates A. Since 2-counter machines are equivalent to Turing machines the proposition is proven. □

Note, that this technique does not rely on unbounded synchronisation: The synchronisations in Figure 7 can be implemented using paired synchronisation only. So, it is shown that for the "nets within nets" approach Turing completeness can be established even if the operation of vertical synchronisation is limited to span over one nesting level, i.e. with paired synchronisation only.

5 Conclusion

In this contribution we presented decidability results for "nets within nets" formalisms, i.e. formalisms that allow for Petri nets as tokens. The analysis of decidable properties has showed that UEOS are more than just a convenient representation of another – possibly larger – Petri net model – since the reachability problem is undecidable for UEOS. So, UEOS are a real extension with greater computational power. Nevertheless, interesting questions – like boundedness – remain decidable, making UEOS weaker than Petri nets with reset arcs.

As an extension the formalism of anonymous ONS without object identities has been investigated, which allows nets as tokens in a recursive way. It has been shown that for this minimal extension of UEOS the formalism allowing arbitrary deep nesting Turing power is obtained. This generalises the result of Kummer who showed that Petri net formalism having identities are Turing powerful. So, the investigation of formalisms that are less expressive is worthwhile. An interesting starting point are ONS with bounded nesting depth.

References

1. A. Bouhoula, J.-P. Jouannaud, and J. Meseguer. Specification and proof in membership equational logic. *Theoretical Computer Science*, 236:35–132, 2000.
2. L. Cardelli, A. D. Gordon, and G. Ghelli. Mobility types for mobile ambients. In *Automata, Languages, and Programming (ICALP'99)*, volume 1644 of LNCS, pages 230–239. Springer-Verlag, 1999.
3. G. Ciardo. Petri nets with marking-dependent arc cardinality: properties and analysis. In R. Valette, editor, *Application and Theory of Petri Nets*, volume 815 of LNCS, pages 179–199. Springer-Verlag, 1994.
4. C. Dufourd, A. Finkel, and P. Schnoebelen. Reset nets between decidability and undecidability. In K. Larsen, editor, *Automata, Languages, and Programming (ICALP'98)*, volume 1443 of LNCS, pages 103–115. Springer-Verlag, 1998.
5. B. Farwer. A linear logic view of object Petri nets. *Fundamenta Informaticae*, 37(3):225–246, 1999.

6. A. Finkel and P. Schnoebelen. Well-structured transition systems everywhere! *Theoretical Computer Science*, 256(1-2):63–92, 2001.
7. R. M. Karp and R. E. Miller. Parallel program schemata. *Journal of Computer and System Sciences*, 3(2):147–195, May 1969.
8. M. Köhler and B. Farwer. Mobile object-net systems and their processes. *Fundamenta Informaticae*, 59:1-17, 2004.
9. M. Köhler, D. Moldt, and H. Rölke. Modelling mobility and mobile agents using nets within nets. In W. v. d. Aalst and E. Best, editors, *Application and Theory of Petri Nets 2003*, volume 2679 of LNCS, pages 121–140. Springer-Verlag, 2003.
10. M. Köhler and H. Rölke. Concurrency for mobile object-net systems. *Fundamenta Informaticae*, 54(2-3), 2003.
11. O. Kummer. Undecidability in object-oriented Petri nets. *Petri Net Newsletter*, 59:18–23, 2000.
12. O. Kummer. *Referenznetze.* Logos Verlag, 2002.
13. I. A. Lomazova and P. Schnoebelen. Some decidability results for nested Petri nets. In *Perspectives of System Informatics (PSI'99)*, volume 1755 of LNCS, pages 208–220. Springer-Verlag, 2000.
14. E. W. Mayr. An algorithm for the general Petri net reachability problem. In *13th Annual ACM Symposium on Theory of Computing*, pages 238–246, 1981.
15. J. Meseguer. Conditional rewriting logic as a unified model of concurrency. *Theoretical Computer Science*, 96:73–155, 1992.
16. J. Meseguer, U. Montanari. Petri nets are monoids, *Information and Computation*, 88(2), 105–155, 1990.
17. W. Reisig. Petri nets and algebraic specifications. *Theoretical Computer Science*, 80:1–34, 1991.
18. M.-O. Stehr, J. Meseguer, and P. C. Ölveczky. Rewriting logic as a unifying framework for Petri nets. In H. Ehrig, G. Juhas, J. Padberg, and G. Rozenberg, editors, *Unifying Petri Nets*, LNCS. Springer-Verlag, December 2001.
19. R. Valk. Self-modifying nets, a natural extension of Petri nets. In Ausiello, G. and Böhm, C., editors, *Automata, Languages and Programming*, volume 62 of LNCS, pages 464–476. Springer-Verlag, 1978.
20. R. Valk. Petri nets as token objects: An introduction to elementary object nets. In J. Desel and M. Silva, editors, *Application and Theory of Petri Nets*, volume 1420 of LNCS, pages 1–25, 1998.

LTL Model Checking for Modular Petri Nets

Timo Latvala* and Marko Mäkelä

Laboratory for Theoretical Computer Science
Helsinki University of Technology
P.O. Box 5400, FIN-02015 HUT, Finland
{Timo.Latvala, Marko.Makela}@hut.fi

Abstract. We consider the problem of model checking modular Petri nets for the linear time logic LTL-X. An algorithm is presented which can use the synchronisation graph from modular analysis as presented by Christensen and Petrucci and perform LTL-X model checking. We have implemented our method in the reachability analyser Maria and performed experiments. As is the case for modular analysis in general, in some cases the gains can be considerable while in other cases the gain is negligible.

Keywords: Modular Petri nets, LTL-X, model checking, Maria.

1 Introduction

Modelling using Petri nets can be made easier in many ways. Examples of extensions of simple Place/Transition nets which ease modelling include, adding types as in Coloured Petri Nets [13], or allowing modular specifications as in [1]. If, however, we are to reap the full benefits of easier modelling, analysis methods must also scale up to take advantage of the new features.

One of the most powerful methods of analysing the behaviour of a Petri net is *reachability analysis*. By constructing the set of reachable markings we can decide important properties such as if the net is live, does a certain invariant hold for all states, etc. Perhaps the most flexible method of analysis we can use is *model checking* [2]. Model checking allows us to check if the behaviour of the net corresponds to a specification given in a temporal logic such as the linear-time temporal logic (LTL).

Modular Petri nets as presented by Christensen and Petrucci [1] allow designers to specify a system as communicating modules. Modules communicate using shared transitions or fusion places. Although one of the chief motivations for using modular specifications is to ease the design of complex systems, another reason is facilitating *compositional reasoning.* All analysis methods suffer from the so called state explosion problem (see, e.g., [22]). Compositional analysis tries to alleviate the problem by considering modules in isolation and then reason about the system as a whole.

* The financial support of Helsinki Graduate School in Computer Science and Engineering and the Academy of Finland (project 53695) is gratefully acknowledged.

J. Cortadella and W. Reisig (Eds.): ICATPN 2004, LNCS 3099, pp. 298–311, 2004.
© Springer-Verlag Berlin Heidelberg 2004

Christensen and Petrucci [1] presented a way to perform invariant analysis, reachability analysis and how to prove several important properties for modular nets in a modular way. Modular reachability analysis works by hiding internal moves of the modules. Mäkelä [17] extended their approach for reachability analysis and implemented modular reachability analysis for hierarchical modular High-level nets. The work also covers model checking of safety properties using the translation from LTL safety properties to finite automata implemented in [15]. Full LTL model checking is something which has been missing.

In this work we show how model checking for the temporal logic LTL-X of the synchronisation graph produced by modular analysis can be achieved. We have implemented our method in the reachability analyser Maria [18] and present experimental results. Our results indicate that the overhead of the method is fairly small, which means that the method works well when modular analysis is efficient. Our work is inspired by a somewhat similar LTL model checking method for unfoldings of Petri nets [8]. There are also similarities with the work done on testers [24, 12].

There are many methods which try to use some form of compositional analysis. As summarised in the survey article [7], using Kronecker algebra, especially with stochastic Petri nets, can allow analysis of the system as whole using the components. Compositional reasoning as described by Valmari [23] advocates the use of process algebraic equivalences for minimisation and composition of components, to form an equivalent smaller system where properties can be proved easily. Another method of compositional reasoning is to use assume guarantee reasoning (see, e.g., [3]). The basic idea is to prove properties of the modules separately and then conclude that the wanted property holds for the global system.

2 Definitions

Definition 1. *A Place/Transition net (PT-net) is a tuple* $N = (P, T, W, M_0)$ *where,*

- P *is a finite set of places,*
- T *is a finite set of transitions such that* $P \cap T = \emptyset$*,*
- $W : (P \times T) \cup (T \times P) \to \mathbb{N}$ *is the arc weight function,*
- $M_0 : P \to \mathbb{N}$ *is the is initial marking.*

A marking is a multiset over P. For a transition $t \in T$ we identify $t^\bullet$ ($^\bullet t$) with the multiset given by $t^\bullet(p) = W(t, p)$ ($^\bullet t(p) = W(p, t)$) for any $p \in P$.

A transition $t \in T$ is *enabled* in a marking M iff $^\bullet t \subseteq M$. A transition t enabled in a marking M can *occur* resulting in the marking $M' = M - {}^\bullet t + t^\bullet$. This is denoted $M \xrightarrow{t} M'$. A marking in which no transition is enabled is called a *deadlocking* marking.

An *execution* of a net is an infinite sequence of markings $\xi = M_0 M_1 M_2 \ldots$ such that $M_0 \xrightarrow{t_0} M_1 \xrightarrow{t_1} \ldots$. The corresponding *trace* is the infinite sequence of

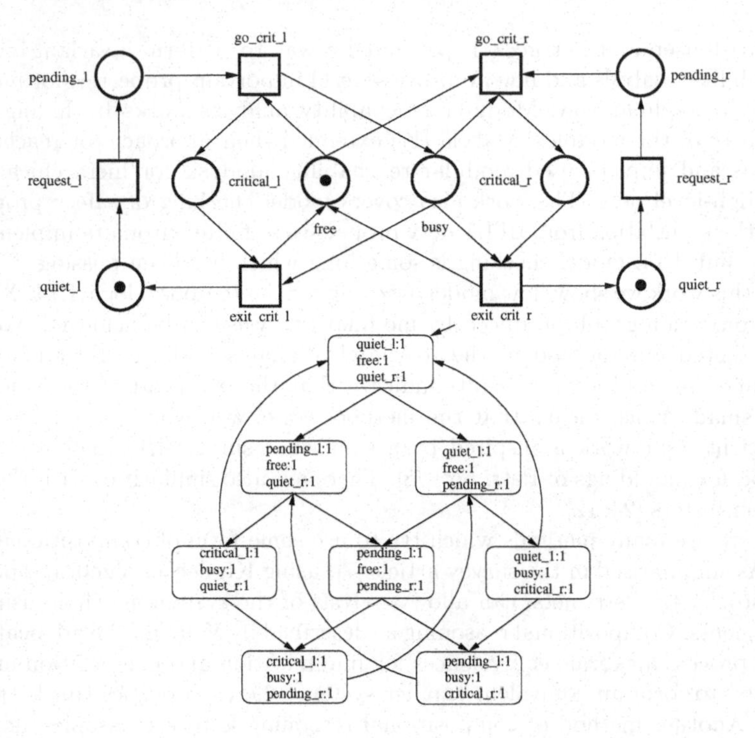

Fig. 1. A simple model of a mutual exclusion algorithm and its reachability graph.

transitions $\sigma = t_0t_1t_2\ldots$. A finite sequence $M_0 \xrightarrow{t_0} M_1 \xrightarrow{t_1} \ldots \xrightarrow{t_{n-1}} M_n$, where M_n is a deadlocking marking, can be seen as an execution by repeating the last marking forever, i.e., $M_0M_1\ldots M_nM_nM_n\ldots$.

Definition 2. *The* reachability graph $G = (V, E, v_0)$ *of a net* $N = (P, T, W, M_0)$ *is defined inductively as follows:*

- $v_0 = M_0 \in V$.
- *If* $M \in V$ *and* $M \xrightarrow{t} M'$ *then* $M' \in V$ *and* $(M, t, M') \in E$.
- V *and* E *contain no other elements.*

The reachability graph of a net describes the dynamic behaviour of the net. In general, the reachability graph can be infinite but in this work we will assume it is finite unless explicitly stated otherwise. Many properties of a net, such as home markings and deadlocks, can be decided in linear time w.r.t. the size of the reachability graph. A simple PT-net modelling a mutual exclusion algorithm and its reachability graph can be found in Figure 1. The model shows two processes (l and r) competing for a critical section which is guarded by a lock. A quick inspection of the reachability graph confirms that under no circumstances are both processes in the critical section at the same time.

Although PT-nets have great modelling power, the lack of structure can sometimes be a problem. It is conceptually easier to deal with large systems as modules, because it allows the designer to consider different parts of the system in relative isolation. Furthermore, the structural information can in some cases be utilised to reduce the complexity of analysis.

Modular PT-nets introduce structure to PT-nets by letting modules be specified separately. The modules communicate either by using shared transitions or place fusion. We restrict ourselves to nets which communicate using shared transitions. Christensen and Petrucci [1] have shown that modular nets with place fusion can be transformed to nets using only shared transitions.

Definition 3. *A modular PT-net is tuple $\Sigma = (S, TF)$ where:*

- *S is a finite set of* modules*:*
  - *each module $s \in S$ is a PT-net $s = (P_s, T_s, W_s, M_{0_s})$,*
  - *the sets of nodes corresponding to different modules are pairwise disjoint, i.e., for all $s_1, s_2 \in S : s_1 \neq s_2 \Rightarrow (P_{s_1} \cup T_{s_1}) \cap (P_{s_2} \cup T_{s_2}) = \emptyset$.*
- *Let $T = \bigcup_{s \in S} T_s$ be the set of all transitions. $TF \subseteq 2^T$ is a finite set of* transition fusion sets *such that for all $tf \in TF$ we have that if $t_i, t_j \in tf$ and $i \neq j$ then $t_i \in T_s \Rightarrow t_j \notin T_s$. In other words, a module may contribute only one transition to a fusion transition.*

Transition fusion sets model synchronising actions. Because the nodes of the modules are pairwise disjoint, the global marking of a modular net is simply the union of the markings of the modules. In this work, we mostly use the global marking of the net and simply denote it M as before. We denote by $ET \subseteq T$ the set of transitions which belong to a transition fusion set and by $IT = T \setminus ET$ the set of internal transitions.

Since there are both fusion transitions and internal transitions in a modular net, we need a uniform way to refer to them. Here, we call them transition groups and they are essential equivalent to the transition concept in a standard PT-net.

Definition 4. *A* transition group *$tg \subseteq T$ is a set of transitions such that it consists of a single internal transition $t \in IT$ or is equal to a transition fusion set $tf \in TF$. The set of all transition groups is denoted TG.*

We extend the preset and postset notation to transition groups. Let $^\bullet tg$ denote the multiset given by

$$^\bullet tg(p) = \sum_{t \in tg} W(t,p) \qquad \text{where } W = \bigcup_{s \in S} W_s$$

The notation for the postset of a transition group is generalised in a similar manner. With this notation, enabledness for a transition group generalises in a natural way.

Definition 5. *A transition group $tg \in TG$ is* enabled *in a marking M if*

$$^\bullet tg \subseteq M$$

The result of occurrence of a fusion transition generalises as expected. An enabled transition group tg can occur in marking M resulting in a marking M' given by

$$M' = M - {}^{\bullet}tg + tg^{\bullet}$$

It is useful to differentiate between the firing of an internal or an external transition. We therefore introduce the following notation.

- $M[[t\rangle M'$ denotes that M' is reachable from M by firing an internal transition.
- $M[tf\rangle\rangle M'$ denotes that M' is reachable from M by firing a fusion transition tf.
- $M[[\sigma\rangle\rangle M'$, where $\sigma = t_0t_1t_2 \ldots t_n tf$, denotes that M' is reachable from M by a sequence of internal transitions followed by a fused transition.

In Figure 2 we show the same mutual exclusion algorithm as in Figure 1 modelled as a modular net. We have split the net into three modules. One module each for the competing processes and one module for the lock. The fusion sets are indicated by the labels on the transitions. Note that the transitions 'lock' and 'unlock' belong to several transition fusion sets.

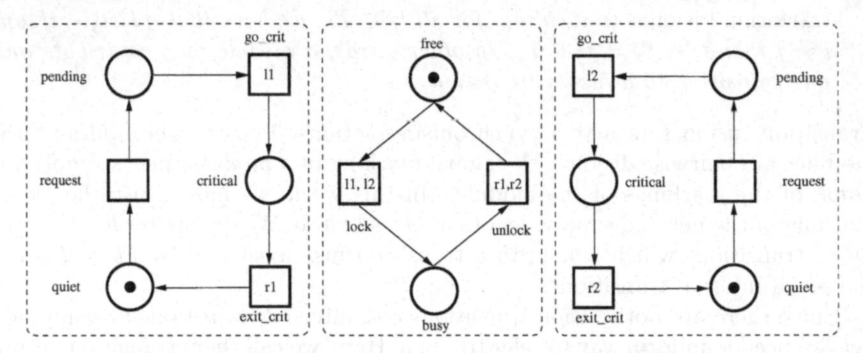

Fig. 2. A mutual exclusion algorithm as a modular net.

3 Modular Reachability Analysis

One of the chief motivations for using modular Petri nets is the possibility of using modular analysis [1]. With modular analysis we can analyse the behaviour of the net without explicitly constructing the full reachability graph. Instead we only construct the so called synchronisation graph between the modules. The synchronisation graph only includes the external moves, i.e., moves where several modules participate – all internal moves are hidden. For loosely coupled systems the synchronisation graph can be significantly smaller than the full reachability graph.

The key idea behind modular analysis is having a single marking represent all markings which can be reached from that marking with internal transitions. We define $\Pi(M)$ to be the set of markings reachable from M using a possibly empty sequence of internal transitions. Two markings M, M' are considered equal iff $\Pi(M) = \Pi(M')$.

Definition 6. *Let $\Sigma = (S, TF, M_0)$ be a modular net. The* synchronisation graph $\mathbf{G} = (\mathbf{V}, \mathbf{E}, \mathbf{v}_0)$ *of the net is defined inductively as follows:*

- $\mathbf{v}_0 = M_0 \in \mathbf{V}$.
- *If* $M \in \mathbf{V}$ *and* $\exists M' \in \Pi(M) : M'[tf\rangle\rangle M''$ *then* $(M, tf, M'') \in \mathbf{E}$ *and* $M'' \in \mathbf{V}$.
- $\mathbf{V}$ *and* $\mathbf{E}$ *contain no other elements.*

The second item of the definition describes which markings are stored in the graph. The successor markings of a global marking M can be computed, e.g., by exploring in each module the set of local markings reachable from M via internal transitions and recording where there are enabled external transitions, and then composing global markings and firing the corresponding transition fusion sets [17, Section 4.1].

Christensen and Petrucci [1] describe how the synchronisation graph can be computed for modular Petri nets. Their approach was extended to hierarchical high-level nets by Mäkelä [17], who also considered verification of simple safety properties.

In Figure 3 we show the synchronisation graph of the modular net given in Figure 2. The synchronisation graph has three states and four arcs compared to original reachability graph which has eight states and fourteen arcs. The arcs are labelled by the fusion sets given in the net description. From the graph it is easy see that the mutual exclusion property holds. What the graph does not show is the interleaving between the processes when they change from quiet to pending.

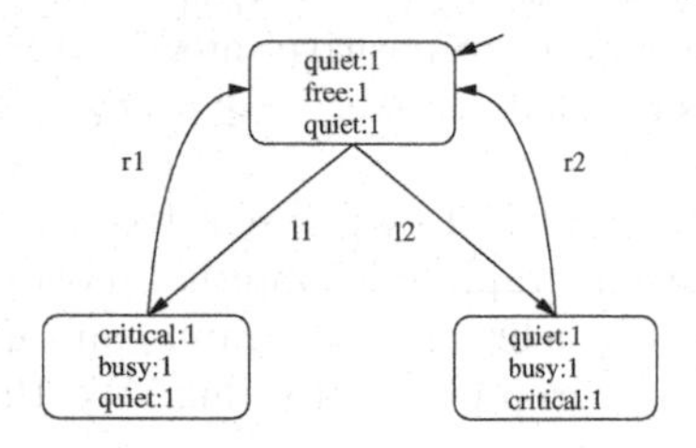

Fig. 3. The synchronisation graph for the mutual exclusion algorithm.

4 Model Checking LTL-X

Model checking the synchronisation graph would be interesting as it can be significantly smaller than the reachability graph. However, it is not immediately clear how the synchronisation graph should be used in model checking. The traditional automata-theoretic solution [14, 25, 26], to synchronise the graph with a Büchi automaton representing the given LTL formula is not directly applicable, as the synchronisation graph hides information. However, by considering what information is preserved by the synchronisation graph an automata-theoretic model checking method can be devised.

An LTL formula φ is defined over a set of atomic propositions AP. The models of the formula are infinite words over 2^{AP}. An LTL formula has the following syntax:

1. $\psi \in AP$ is an LTL formula.
2. If ψ and φ are LTL formulae then so are $\neg\psi$, $\mathbf{X}\psi$, $\psi\,\mathbf{U}\,\varphi$ and $\psi \vee \varphi$.

We denote the suffix of a model $\pi = \sigma_0\sigma_1\sigma_2 \ldots \in (2^{AP})^\omega$ by $\pi^i = \sigma_i\sigma_{i+1}\sigma_{i+2}\ldots$. The semantics of LTL are inductively defined using the 'models' relation $\models$:

- $\pi^i \models \psi$ iff $\psi \in \sigma_i$ for $\psi \in AP$.
- $\pi^i \models \neg\psi$ iff $\pi \not\models \psi$.
- $\pi^i \models \psi \vee \varphi$ iff $\pi \models \psi$ or $\pi \models \varphi$.
- $\pi^i \models \mathbf{X}\psi$ iff $\pi^{i+1} \models \psi$.
- $\pi^i \models \psi\,\mathbf{U}\,\varphi$ iff $\exists k \geq i$ such that $\pi^k \models \varphi$ and $\pi^j \models \psi$ for all $i \leq j < k$.

If $\pi^0 \models \psi$ we simply write $\pi \models \psi$. Common abbreviations used are $\top = p \vee \neg p$ for some arbitrary $p \in AP$, the usual abbreviations for the Boolean operators $\wedge, \Rightarrow$ and $\Leftrightarrow$, and the temporal operators 'finally' $\mathbf{F}\psi \equiv \top\,\mathbf{U}\,\psi$ and 'globally' $\mathbf{G}\psi \equiv \neg\mathbf{F}\neg\psi$.

The subset of LTL where the 'next' operator $\mathbf{X}$ is not allowed is denoted LTL-X. The 'next' operator allows LTL to specify properties which can differentiate between sequences which only have different internal moves, i.e., moves which do not affect the truth of relevant atomic propositions, or so called *stuttering*. Formally, two models $\pi = \sigma_0\sigma_1\ldots, \pi' = \sigma'_0\sigma'_1\ldots$ are stuttering equivalent if there are two infinite sequences of positive integers $0 = i_0 < i_1 < i_2 < \ldots$ and $0 = j_0 < j_1 < j_2 < \ldots$ such that for every $k \geq 0$: $\sigma_{i_k} = \sigma_{i_k+1} = \cdots = \sigma_{i_{k+1}-1} = \sigma'_{j_k} = \sigma'_{j_k+1} = \cdots = \sigma'_{j_{k+1}-1}$.

It is a well-known fact that LTL-X is insensitive to stuttering (see, e.g., [4]). Because the synchronisation graph hides internal moves of the modules, the sequences generated by the synchronisation graph can differ from sequences generated by the reachability graph by stuttering. We therefore focus on model checking for LTL-X.

A formula defines a language $\mathcal{L}(\varphi) = \{w \in (2^{AP})^\omega \mid w \models \varphi\}$. The language of the LTL formula can be captured by a *Büchi automaton* (see, e.g., [21]). In the recent years several papers have dealt with the problem of translating an LTL formula to a Büchi automaton [20, 10, 11].

Definition 7. *A Büchi automaton is a tuple $\mathcal{A} = (Q, A, \rho, q_0, Q_F)$, where Q is a finite set of states, A is a finite alphabet, $\rho \subseteq Q \times A \times Q$ is the transition relation, $q_0 \in Q$ is the initial state, and $Q_F \subseteq Q$ is a set of accepting states.*

An infinite word $w \in A^\omega$ generates a run $r = q_0q_1q_2\ldots$ of the automaton such that q_0 is the initial state of the automaton and for all $i \geq 0$ we have $(q_i, w(i), q_{i+1}) \in \rho$. If the automaton is non-deterministic, one word can generate several runs. A word w is accepted iff it has run $r : \mathbb{N} \to Q$ such that

$r(i) \in Q_F$ for infinitely many i. We use $\mathcal{A}^q$ to denote the automaton $\mathcal{A}$ with q set as the initial state. The language of the automaton, denoted $\mathcal{L}(\mathcal{A})$, is defined as the set of strings accepted by the automaton. Here we assume that when Büchi automata are used to represent LTL formulae the alphabet is $A = 2^{2^{AP}}$.

Definition 8. *Given a marking M, the function $eval(M)$ returns the set of atomic propositions which hold in M. The notation is extended to sequences of markings in the normal way.*

For a Petri net N, we write $N \models \psi$ iff for all executions ξ of the net, we have that $eval(\xi) \models \psi$. The executions of the Petri net define a language when projected with the *eval* function. A Petri net has a given temporal property if the language of the net, as defined above, is a subset of the language of the property automaton.

The traditional way of model checking a Petri net using the automata theoretic approach, is to check if intersection of the language of the Petri net with the language of the *negation* of the property is empty. This uses the fact that for any two languages L_1, L_2 the equivalence $L_1 \subseteq L_2 \Leftrightarrow L_1 \cap \overline{L_2} = \emptyset$ holds, which is why it is referred to making an emptiness check. The intersection of the languages is computed by synchronising the reachability graph of the net with a Büchi automaton representing the negation of the property. If the synchronisation has no accepting run, the property holds.

The usual way of computing the intersection with automaton and the reachability graph requires that the Büchi automaton synchronises with every move of the net. Because modular analysis relies on hiding the internal moves of the modules this is not a good approach, because it would make all moves external and forfeit the potential benefit of using modular analysis.

It is, however, possible to do model checking by only synchronising with the visible transitions. The price we pay is a more complex model checking algorithm. The approach we present is similar to the work on model checking using unfoldings of Petri nets [8] and has common elements with the tester approach [24] but as we are synchronising with different constructs there are technical differences.

Let φ be a formula over a set AP of atomic propositions (Boolean expressions on markings of the net). We call a place $p \in P$ *visible* if the truth of an atomic proposition can be changed by altering the marking of the place. Let $P_V \subseteq P = \bigcup_{s \in S} P_s$ be the set of visible places. The set of visible transitions is defined as $T_V = {}^\bullet P_V \cup P_V^\bullet$. Because all changes in atomic proposition must be visible in the synchronisation graph we require that $T_V \subseteq ET$ (in an implementation we could automatically detect which transition need to be treated as external transitions).

Our goal is to find the illegal executions of the modular net by synchronising the Büchi automaton with the *visible* transitions. Formally, we define the synchronisation in the following way. Let $\mathcal{A}_{\neg\varphi}$ be a Büchi automaton accepting the language of the negation of the property φ and Σ a modular net. If the net is in the marking M, the automaton is in the state q, and $(q, a, q') \subset \rho$ such that $eval(M) \in a$, then the net and the automaton will synchronise in the following way:

- All visible transitions must *always* be synchronised with the Büchi automaton. If a visible transition $t \in T_V$ is enabled in $\Pi(M)$, it is synchronised with the automaton, and the product moves to the state (M', q') where $M[tf\rangle\rangle M'$.
- Invisible transitions can occur in the system without synchronising with the automaton.

A state in the synchronisation $s = (M, q)$ is accepting if $q \in Q_F$. The synchronisation state (M, q) belongs to a livelock set I if $\mathcal{A}^q_{\neg\varphi}$ accepts $eval(M)^\omega$. The livelock set I can be very large and should thus be computed on demand when model checking.

We say that the net has an illegal ω-execution if there is an execution of the synchronisation where the corresponding trace has infinitely many visible transitions and there are infinitely many occurrences of an accepting state in the execution. The net has an illegal livelock if there is an execution of the synchronisation $s_0 s_1 s_2 \ldots s_n s_{n+1} \ldots$ such that $s_n \in I$ and the corresponding trace from s_n onward $\sigma^n = t_n t_{n+1} t_{n+2} \ldots$ does not contain any visible transition.

We can detect all illegal ω-executions and illegal livelocks by computing a synchronised product of the Büchi automaton and the synchronisation graph of the modular net.

Definition 9. *Let $\mathcal{A}_{\neg\psi} = (Q, A, \rho, q_0, Q_F)$ be a Büchi automaton and $\mathbf{G} = (\mathbf{V}, \mathbf{E}, \mathbf{v}_0)$ be a synchronisation graph. Their* product (V_p, E_p, p_0, F, I) *is defined in the following way:*

- $p_0 = (\mathbf{v}_0, q_0) \in V_p$.
- *Given $(M, q) \in V_p$, $(M, t, M') \in \mathbf{E}$, and $(q, a, q') \in \rho$ there are two possibilities:*
  - *If $t \in T_V$ and $eval(M) \in a$, the net and the automaton synchronise and we have $((M, q), \{t, a\}, (M', q')) \in E_p$ and $(M', q')' \in V_p$.*
  - *If $t \notin T_V$, the system moves: $((M, q), t, (M', q)) \in E_p$ and $(M', q) \in V_p$.*
- *V_p and E_p contain no other elements.*
- $F = \{(M, q) \in V_p \| \ q \in Q_F\}$.
- $I = \{(M, q) \in V_p \| \ \mathcal{A}^q_{\neg\psi}$ *accepts* $eval(M)^\omega\}$.

We claim that any execution of the net which breaks the given LTL specification will induce either an illegal ω-execution or an illegal livelock, which will also show up in the product of the synchronisation graph and the Büchi automaton.

Theorem 1. *Given a modular net Σ and a Büchi automaton $\mathcal{A}_{\neg\psi}$, $\Sigma \not\models \psi$ iff the product of the automaton and the synchronisation graph of the net has an illegal ω-trace or an illegal livelock.*

The proof is fairly similar to the proof of Theorem 2 given in [9] but as there are some technical differences we present it here.

Proof. Let $\xi = M_0 M_1 M_2 \ldots$ be an execution of the net Σ such that $eval(\xi) \not\models \psi$. By construction, we know that $\mathcal{A}_{\neg\psi}$ accepts $w = eval(\xi)$. Let $w' = eval'(\xi)$ be the same sequence with all stuttering removed. There are now two possible cases: w' can either be an infinite (a) or a finite sequence (b).

(a) Because all properties specified by LTL-X are immune to stuttering, $\mathcal{A}_{\neg\psi}$ accepts the infinite w'. Let $r' = q_0q_1q_2\ldots$ be one of the runs accepting w'. The product has the following illegal ω-execution. Set (M_0, q_0) as the initial state. Set $j = 0$. For each $i \geq 0$ do: (i) fire the transition t_i which leads to M_{i+1}, (ii) if $t_i \in IT$ then goto (i), otherwise if $t_i \in T_V$ set $j = j + 1$ (iv) (M_{i+1}, q_j) is the next state in the run. This sequence will exist as the synchronisation graph contains all possible visible sequences. Step (ii) deals with invisible internal transitions which are not present in the synchronisation graph while (iii) makes sure that the Büchi automaton advances only when we have a visible external transition. Because w' is an accepted sequence, the run r' has a final state occurring infinitely often and thus the product has an illegal ω-trace.
(b) In the same manner as in (a) we can argue that the finite w' will induce a finite run of the product. Let (M_i, q_i) be the final state of this run. Because $\mathcal{A}_{\neg\psi}$ accepts the full word w, we know that w' can be extended by stuttering to an accepting word. Thus, by the definition of I we know that $(M_i, q_i) \in I$. Since ξ was infinite we know it is possible to fire an infinite sequence of invisible transitions from M_i onward and consequently the product has an illegal livelock.

Let ξ_p be an illegal ω-execution of the product. By projecting the Büchi component of the run onto $\mathcal{A}_{\neg\psi}$ it is clear that this is an accepting run for the automaton. Similarly we can easily build a run of Σ from the net component of the execution. All it requires is finding the fired internal transition which occur between the external transitions. This is be possible due to the properties of the synchronisation graph. Essentially, we only need to compute how to enable the next fusion transition by firing internal transitions. By the properties of $\mathcal{A}_{\neg\psi}$ we can then conclude that $\Sigma \not\models \psi$.

Let ξ_p be an illegal livelock of the product and (M_i, q_i) be the state after which only invisible transitions are executed. We can project the Büchi component onto $\mathcal{A}_{\neg\psi}$ such that the trace ends in q_i. We know that (M_i, q_i) is in I and thus by implication that $\mathcal{A}^q_{\neg\psi}$ will accept $eval(M_i)^\omega$. The loop of invisible transitions corresponds to this infinite stuttering. Building an infinite execution of Σ from ξ_p is again easy. Thus we can again conclude that $\Sigma \not\models \psi$. □

Finding an illegal livelock or an illegal infinite trace from the product is equivalent to finding an error specified by a tester as described by Valmari [24]. As suggested in [12], we then have a solution which traverses the product three times in the worst case by first using Valmari's one-pass algorithm [24] to find any illegal livelocks. If no illegal livelocks are found we can use the standard nested depth-first algorithm [5] to find any illegal infinite traces. Some small modifications are required for the algorithms to function correctly. For the one-pass algorithm we need an efficient way of deciding if a state belongs the set I. This corresponds to model checking a reachability graph consisting of a single marking with a self loop. If a state M belongs to the set I, not only must we check if there is a loop of invisible transitions in the product starting from M, but also if internal transitions can loop in any of the modules. This can be implemented by depth-first traversal of the states reachable from the local states of the modules corresponding to M. Figure 4 describes the model checking algorithm at an abstract level.

```
Input: The product (V_p, E_p, p_0, F, I)
proc model check(V_p, E_p, p_0, F, I) begin
    for all states (M, q) ∈ V_p do
        if (M, q) ∈ I then
            if a loop of invisible transitions starts from M then
                return "illegal livelock found"
        fi
        if (M, q) ∈ F then
            if (M, q) is reachable from (M, q) in the product then
                return "illegal ω-execution found"
        fi
    od
end
```

Fig. 4. Abstract algorithm for model checking.

5 Experiments

We have implemented our method in the reachability analyser Maria [18]. In order to evaluate our implementation we have conducted some experiments. We compared modular model checking against the basic Maria model checker [16]. Additionally, we also ran a few benchmarks against PROD [27], a tool with advanced partial order reduction methods. This benchmark gives us some indication on how modular analysis compares with another method producing stuttering equivalent structures.

We used three different models of which two were parametric. The first model (AGV) describes a system of automated guided vehicles, first modelled by Petrucci [19]. The second model (SW) is a variant of a sliding window protocol and the third one (LE) models the leader election protocol in a unidirectional ring [6]. The results of the experiments can be found in Table 1. The statistics we recorded were number of states and arcs in the reachability graph, number of states in the product, time used for state space construction and model checking, size of the synchronisation graph, number of states in the product, the time used, and the type of formula. All tests were run on a machine with 1 GB of RAM with an AMD Athlon XP 2000 processor.

It would appear that in some cases the modular algorithm is faster while in some cases it is slower. For very loosely coupled models as AGV, the modular algorithm does well. Analysis of a model with a fair amount of synchronisation, such as LE, also shows gains using our modular algorithm. When the gains of using modular analysis are questionable as for the SW models, the overhead of using modular analysis and the modular algorithm for model checking is significant but not prohibitively expensive.

Table 1. Experimental results.

System	Flat state space $G = (V, E, v_0)$ $\|V\|$	$\|E\|$	product	time/s	Modular $\mathbf{G} = (\mathbf{V}, \mathbf{E}, \mathbf{v_0})$ $\|\mathbf{V}\|$	$\|\mathbf{E}\|$	product	time/s	ψ
AGV	30,965,760	216,489,984	N/A	N/A	87,480	464,616	87,492	27.3	$\mathbf{GF}\varphi$
SW_4	6,360	16,608	14,857	1.3	4,456	16,016	8,889	2.6	$\mathbf{GF}\varphi$
SW_5	24,270	68,760	52,891	5.8	16,930	72,660	31,991	13.1	$\mathbf{GF}\varphi$
SW_6	82,884	248,400	169,645	20.6	57,564	286,488	103,477	118	$\mathbf{GF}\varphi$
LE_3	159	303	314	0.0	35	65	68	0.0	$\mathbf{FG}\varphi$
LE_4	716	1,851	1,428	0.2	92	229	182	0.1	$\mathbf{FG}\varphi$
LE_5	3,432	11,198	6,860	1.3	253	802	504	0.2	$\mathbf{FG}\varphi$
LE_6	16,792	66,043	33,580	8.0	715	2,748	1,428	0.8	$\mathbf{FG}\varphi$
LE_7	82,667	380,267	165,330	49.3	2,043	9,212	4,084	2.9	$\mathbf{FG}\varphi$
LE_8	407,699	2,146,965	815,394	295	5,865	30,308	11,728	10.2	$\mathbf{FG}\varphi$

Our benchmarks against PROD were conducted with a special version of the sliding window protocol model with more realistic timeout conditions. The model was separately optimised for PROD and Maria in order to make the comparison as fair as possible. Results can be found from Table 2.

Table 2. Benchmarks against PROD.

System	PROD $G = (V, E, v_0)$ $\|V\|$	$\|E\|$	product	time/s	Modular (Maria) $\mathbf{G} = (\mathbf{V}, \mathbf{E}, \mathbf{v_0})$ $\|\mathbf{V}\|$	$\|\mathbf{E}\|$	product	time/s	ψ
$SW_{2,2}$	8,384	13,388	17,622	8.0	7,376	48,860	9,709	8.0	$\mathbf{GF}\varphi$
$SW_{3,2}$	131,555	198,466	270,142	245	86,995	802,650	101,551	148	$\mathbf{GF}\varphi$
$SW_{3,3}$	422,484	590,298	859,724	969	267,192	2,885,022	302,551	757	$\mathbf{GF}\varphi$
$SW_{4,2}$	1,434,750	2,056,176	2,914,484	7556	762,870	9,379,788	836,275	3031	$\mathbf{GF}\varphi$

The results seem to indicate that the number of states produced is fairly similar while partial order reductions eliminate arcs much more successfully. Modular analysis produces results faster but the difference is not large. The more relaxed synchronisation for computing the product in modular analysis can obviously lead to smaller products. A combination of modular analysis and partial order reductions could produce good results and would be very interesting.

6 Conclusions

In this paper we have shown how LTL-X model checking can be done on the synchronisation graph resulting from modular analysis of modular Petri nets as presented in [1]. Our method requires a different form of synchronisation compared to the traditional automata theoretic model checking and a somewhat more complicated emptiness checking algorithm. The time complexity overhead is, however, only linear compared to conventional emptiness checking algorithms for Büchi automata.

As the model checking algorithm has a reasonable overhead (not much worse than the traditional algorithm), the performance of model checking for modular nets is heavily dependent on how well modular analysis performs. This means that for many models where modular analysis does not reduce the state explosion, compared to reachability analysis of flat models, using our method will not result in excessive waiting. Our implementation is available as a patch to the standard Maria distribution from http://www.tcs.hut.fi/Software/maria/.

As future work we are considering refining the concept of visibility. A more relaxed view of visibility could potentially improve the performance of the model checking algorithm. In general, we believe that the efficiency of modular analysis could be improved by developing partial order methods which are compatible with modular analysis. The two problems are related as visibility is also an issue in partial order reductions.

Acknowledgements

We thank Kimmo Varpaaniemi for the PROD benchmarks.

References

1. S. Christensen and L. Petrucci. Modular analysis of Petri Nets. *The Computer Journal*, 43(3):224–242, 2000.
2. E. Clarke, O. Grumberg, and D. Peled. *Model Checking.* MIT Press, Cambridge, Massachusetts, 1999.
3. E.M. Clarke, D.E. Long, and K.L. McMillan. Compositional model checking. In R. Parikh, editor, *Proc. 4th IEEE Symposium on Logic in Computer Science*, pages 353–362. IEEE Computer Society Press, 1989.
4. E.M Clarke and B-H. Schlingloff. Model checking. In A. Robinson and A. Voronkov, editors, *Handbook of Automated Reasoning*, pages 1637–1790. Elsevier, 2001.
5. C. Courcoubetis, M.Y. Vardi, P. Wolper, and M. Yannakakis. Memory efficient algorithms for the verification of temporal properties. *Formal Methods in System Design*, 1:275–288, 1992.
6. Danny Dolev, Maria Klawe, and Michael Rodeh. An $O(n \log n)$ unidirectional distributed algorithm for extrema finding in a circle. *Journal of Algorithms*, 3(3):245–260, September 1982.
7. S. Donatelli. Kronecker algebra and (stochastic) Petri nets: Is it worth the effort. In J-M. Colom and M. Koutny, editors, *Application and Theory of Petri Nets 2001*, volume 2075 of *LNCS*, pages 1–18. Springer, 2001.
8. J. Esparza and K. Heljanko. Implementing LTL model checking with net unfoldings. In *SPIN 2001*, volume 2057 of *LNCS*, pages 37–56. Springer, 2001.
9. Javier Esparza and Keijo Heljanko. A new unfolding approach to LTL model checking. Technical Report HUT-TCS-A60, Helsinki University of Technology, April 2000. Available from http://www.tcs.hut.fi/Publications.
10. P. Gastin and D. Oddoux. Fast LTL to Büchi automata translation. In *Computer Aided Verification (CAV'2001)*, volume 2102 of *LNCS*, pages 53–65. Springer, 2001.

11. D. Giannakopoulou and F. Lerda. From states to transitions: Improving translation of ltl formulae to büchi automata. In D.A. Peled and M.Y. Vardi, editors, *Formal Techniques for Networked and Distributed Systems - FORTE 2002*, pages 308 – 326. Springer, 2002.
12. Henri Hansen, Wojchech Penczek, and Antti Valmari. Stuttering-insensitive automata for on-the-fly detection livelock properties. In *Formal Methods for Industrial Critical Systems*, volume 66(2) of *Electronic Notes in Theoretical Computer Science*. Elsevier, 2002.
13. K. Jensen. *Coloured Petri Nets*, volume 1. Springer, Berlin, 1997.
14. R.P. Kurshan. *Computer-Aided Verification of Coordinating Processes: The Automata-Theoretic Approach*. Princeton University Press, 1994.
15. T. Latvala. Efficient model checking of safety properties. In T. Ball and S.K. Rajamani, editors, *Model Checking Software. 10th International SPIN Workshop*, volume 2648 of *LNCS*, pages 74–88. Springer, 2003.
16. Timo Latvala. Model checking LTL properties of high-level Petri nets with fairness constraints. In *Applications and Theory of Petri Nets (ICAPTN'2001)*, volume 2075 of *LNCS*, pages 242–262. Springer, 2001.
17. Marko Mäkelä. Model checking safety properties in modular high-level nets. In *Application and Theory of Petri Nets (ICATPN'2003)*, volume 2679 of *LNCS*, pages 201–220. Springer, 2003.
18. Marko Mäkelä. Maria: modular reachability analyser for algebraic system nets. In Javier Esparza and Charles Lakos, editors, *Application and Theory of Petri Nets 2002*, volume 2360 of *LNCS*, pages 434–444. Springer, 2002.
19. Laure Petrucci. Design and validation of a controller. In *Proceedings of the 4th World Multiconference on Systemics, Cybernetics and Informatics*, volume VIII, pages 684–688, Orlando, FL, USA, July 2000. International Institute of Informatics and Systemics.
20. F. Somenzio and R. Bloem. Efficient Büchi automata from LTL formulae. In *Proceedings of the International Conference on Computer Aided Verification (CAV2000)*, volume 1855 of *LNCS*, pages 248–263. Springer, 2000.
21. W. Thomas. Automata on infinite objects. In J. Leeuwen, editor, *Handbook of Theoretical Computer Science*, volume B, pages 133–191. Elsevier, 1990.
22. A. Valmari. The state explosion problem. In *Lectures on Petri Nets I: Basic Models*, volume 1491 of *LNCS*, pages 429–528. Springer, 1998.
23. A. Valmari. Composition and abstraction. In F. Cassez, C. Jard, B. Rozoy, and M. Ryan, editors, *Modelling and Verification of Parallel Processes*, volume 2067 of *LNCS*, pages 58–99. Springer, 2001.
24. Antti Valmari. On-the-fly verification with stubborn sets. In *Computer Aided Verification (CAV'93)*, volume 697 of *LNCS*, pages 397–408. Springer, 1993.
25. M. Y. Vardi and P. Wolper. Automata-theoretic techniques for modal logic of programs. *Journal of Computer and System Sciences*, 32:183–221, 1986.
26. M.Y. Vardi. An automata-theoretic approach to linear temporal logic. In *Logics for Concurrency: Structure versus Automata*, volume 1043 of *LNCS*, pages 238–266. Springer, 1996.
27. K. Varpaaniemi, K. Heljanko, and J. Lilius. PROD 3.2 – an advanced tool for efficient reachability analysis. In *Computer Aided Verification: 9th International Conference, CAV'97, Haifa, Israel, June 22–25, 1997, Proceedings*, volume 1254 of *LNCS*, pages 472–475. Springer-Verlag, 1997.

Covering Fairness against Conspiracies

Edward Ochmański

Faculty of Mathematics and Computer Science, Nicolaus Copernicus University, Toruń, Poland
and Institute of Computer Science, Polish Academy of Sciences, Warszawa, Poland
edoch@mat.uni.torun.pl

Abstract. The paper takes advantage and develops the fundamental Best's ideas about fairness hierarchy and conspiracies in concurrent systems. Near to the start we characterize liveness with the notion of ∞-fairness. Next we show that the conspiracy-freeness problem is decidable for elementary nets and undecidable for place/transition nets. The main aim of the paper was to put places into work against conspiracies. We show, how marking fairness effectively fights with conspiracies in elementary nets. Next, we introduce the notion of covering fairness, as a tool against conspiracies in place/transition systems. Results of the paper say that one can ensure global fairness on a local level of executions of the systems.

1 Introduction

Twenty years ago Eike Best [1] proposed a precise formal tool to fight against conspiracies. The tool, the infinite hierarchy of fairness levels, were investigated and applied by many authors. Most of them were concentrated on the transition-oriented fairness. The notion of marking-oriented fairness was introduced, in 1987, by Agathe Merceron. This notion itself is certainly less interesting than the original one, but it seems that it should be more precisely watched. This is the main aim of this paper.

First two sections recall basic definitions concerning Petri nets and fairness. Section 3 presents an easy, but nice characterization of liveness with the notion of ∞-fairness. It is known that a similar characterization cannot be done by a weaker fairness notion. In section 4 we show that the Conspiracy-Freeness Problem (if any 0-fair computation is ∞-fair) is decidable for elementary nets and undecidable for place/transition nets.

As it was mentioned, the aim of this paper was to put places into work against conspiracies. They successfully played this role in the field of elementary nets. We prove, in section 5, that 0-fair behaviour of transitions and markings ensures ∞-fairness for transitions.

Main ideas and results are presented in the last two sections of the paper. We introduce the notion of covering fairness, and the results show, how useful for fairness and how harmful for conspiracies is this notion. Contrary to the traditional marking fairness, covering fairness perfectly fights with conspiracies in place/transition systems.

J. Cortadella and W. Reisig (Eds.): ICATPN 2004, LNCS 3099, pp. 312–330, 2004.
© Springer-Verlag Berlin Heidelberg 2004

2 Basic Notions

2.1 Formal Denotations

An *alphabet* A is a finite set of symbols. Strings of symbols of A are *words* over A. Then A^* and A^ω are sets of all finite and infinite (respectively) words over A, and $A^\infty = A^* \cup A^\omega$ is the set of all (finite and infinite) words over A. The empty word is denoted by ε. The *length* of $w \in A^\infty$ is denoted by $|w|$, with $|w|=\omega$ for infinite words. A word $u \in A^*$ is a *(finite) prefix* of a word $w \in A^\infty$ iff there is a word $v \in A^\infty$ such that $w=uv$. The number of occurrences of the symbol $a \in A$ in the word $w \in A^\infty$ is denoted by $|w|_a$. The infinite repetition of a finite word $w \in A^*$ is denoted by w^ω.

Subsets of A^* are called *languages*. Basic operations on languages: set-theoretical operations of union, intersection and difference, *concatenation*: $XY=\{xy \mid x \in X \ \& \ y \in Y\}$, *power*: $X^0=\{\varepsilon\}$, $X^{n+1}=XX^n$, *iteration* (or the *star operation*): $X^*=\bigcup\{X^n \mid n \geq 0\}$. A language $L \subseteq A^*$ is said to be *prefix-closed* iff it contains all prefixes of its members: $(\forall u,v \in A^*)$ if $uv \in L$ then $u \in L$.

The set of non-negative integers is denoted by $\mathbb{N}$. Given a set X, the cardinality (number of elements) of X is denoted by |X|, the powerset (set of all subsets) by 2^X, the cardinality of the powerset is $2^{|X|}$. Multisets over X are members of $\mathbb{N}^X$, i.e. functions from X into $\mathbb{N}$. For convenience, if the set X is finite, multisets of $\mathbb{N}^X$ will be treated as vectors of $\mathbb{N}^{|X|}$.

2.2 Petri Nets and Their Computations

The definitions concerning Petri nets are mostly based on [4] and [11].

Definition 1. *Nets*
Net is a triple $N=(P,T,F)$, where:

- P and T are finite disjoint sets, of *places* and *transitions*, respectively;
- $F \subseteq P \times T \cup T \times P$ is a relation, called the *flow relation.*

For all $a \in T$ we denote:

$^\bullet a = \{p \in P \mid (p,a) \in F\}$ – the set of *entries* to a
$a^\bullet = \{p \in P \mid (a,p) \in F\}$ – the set of *exits* from a

Petri nets admit a natural graphical representation. Nodes represent places and transitions, arcs represent the flow relation. Places are indicated by circles, and transitions by boxes.

Definition 2. Elementary Net Systems
Elementary net system (shortly, *EN-system*) is a quadruple $S=(P,T,F,M_0)$, where:

- (P,T,F) is a net, as defined above;
- $M_0 \subseteq P$ is a subset of places, named the *initial marking*; it is marked by *tokens* inside the circles.
- Moreover, we assume that entries and exits of any transition are nonempty and disjoint: $(\forall a \in T)\ ^\bullet a \neq \emptyset \ \& \ a^\bullet \neq \emptyset \ \& \ ^\bullet a \cap a^\bullet = \emptyset$.

Any subset of places is named a *marking*. A transition $a\in T$ is *enabled* in a marking $M\subseteq P$ iff ${}^\bullet a\subseteq M$ (all its entries are marked) and $a^\bullet\subseteq P-M$ (all its exits are empty). If $a\in T$ is enabled in M, then it can be executed, but the execution is not forced. The execution of a transition a changes the current marking M to the new marking $M'=(M-{}^\bullet a)\cup a^\bullet$ (tokens are removed from entries, then put to exits).

Definition 3. Place Transition Systems
Place transition system (shortly, *P/T-system*) is a quadruple $S=(P,T,F,M_0)$, where:

- (P,T,F) is a net, as defined above;
- $M_0\in \mathbb{N}^P$ is a multiset of places, named the *initial marking*; it is marked by *tokens* inside the circles, but in this case number of possible tokens is not bounded.

Like in EN-systems, multisets of places are named *markings*. In the context of P/T-systems, however, they are mostly regarded as nonnegative integer vectors of dimension |P|. The natural generalizations, for multisets, of arithmetic operations + and –, as well as the inclusion relation ≤, called here the relation of *covering*, are well known and their formal definitions are omitted. A transition $a\in T$ is *enabled* in a marking M iff ${}^\bullet a\leq M$ (all its entries are marked). If $a\in T$ is enabled in M, then it can be executed, but the execution is not forced. The execution of a transition a changes the current marking M to the new marking $M'=(M-{}^\bullet a)+a^\bullet$ (tokens are removed from entries, then put to exits).

The following settlements are common for both classes of Petri nets, considered here.

We shall denote: Ma for "a is enabled in M" and MaM' or $M'=Ma$ for "a is enabled in M and M' is the resulting marking". This denotation we extend, in a natural way, to strings: Mw means "w is enabled in M" and MwM' or $M'=Mw$ means "w is enabled in M and M' is the final marking"; then M' is said to be *reachable from M*. A marking M is said to be *reachable* in $S=(P,T,F,M_0)$ iff it is reachable from the initial marking M_0; the set of all reachable markings is denoted by $[M_0\rangle$. A marking M is *dead* iff there is no action enabled in M, and *alive* otherwise.

The sequence $w\in T^*$ is a *finite computation* of S iff it is enabled in M_0; the set of all finite computations of S (notice, it is always prefix-closed) is said to be a *language* of S and denoted by $L(S)$. An infinite sequence $\sigma\in T^\omega$ is an *infinite computation* of S iff any of its finite prefixes is a finite computation of S. The set of all infinite computations of S is denoted by $L^\omega(S)$, and the set of all (finite and infinite) computations of S, i.e. the union $L(S)\cup L^\omega(S)$, is denoted by $L^\infty(S)$.

Denotational convention. Notice that each computation $w=a_1a_2...a_i$ determines, in a unique way, the intermediate markings $M_1,M_2,\ldots,M_i$ and this way the complete form $M_0a_1M_1a_2M_2\ldots M_{i-1}a_iM_i$ of the computation w. We will write the markings or omit some of them, depending on current needs. For instance, we can write $w=MuM'vM''$ or $w=MuvM''$ or simply $w=uv$ for the same w.

Remark. The number of reachable markings in EN-systems is finite, as it is $\leq 2^{|P|}$, the number of all subsets of P. Clearly, it is not true, in general, for P/T-systems.

3 Fairness and Conspiracies

Classical fairness (strong fairness) definition [7] says: "A computation is fair iff any action infinitely often enabled is infinitely often executed." Best proposed, in [1], an extension of the notion to a (potentially infinite) hierarchy of fairness notions.

3.1 The Fairness Hierarchy: An Abstract Model

In this subsection we formulate, following Best [1], a general definition of fairness hierarchy, which abstracts from particular models of computing devices, and even from objects of computations.

Let A be an alphabet, and $L \subseteq A^*$ be a prefix-closed language. An infinite sequence $\sigma \in A^\omega$ belongs to L^ω iff all its finite prefixes belongs to L. Then L^∞ denotes the union $L \cup L^\omega$. A word $w \in L$ is *unextendable* (in L) iff $(\forall a \in A)\ wa \notin L$.

Definition 4. Infinite Fairness Hierarchy
Let L be a fixed prefix-closed language over A.
Let $w \in L$, $a \in A$ and $k \in \mathbb{N}$. Then a is said to be:

- k-*enabled* (in L) after w iff $\exists u \in A^*$ s.t. $|u| \leq k$ and $wua \in L$,
- ∞-*enabled* (in L) after w iff $\exists u \in A^*$ s.t. $wua \in L$.

Let $\sigma = a_1 a_2 \ldots \in L^\omega$, $a \in A$ and $k \in \mathbb{N} \cup \{\infty\}$. We say:

- $a \in A$ is *infinitely often* k-*enabled* in σ iff
  the set $\{i \in \mathbb{N} \mid a$ is k-enabled after $a_1 a_2 \ldots a_i\}$ is infinite;
- $a \in A$ *occurs infinitely often* in σ iff the set $\{i \in \mathbb{N} \mid a = a_i\}$ is infinite.

Fairness (w.r.t. L):

- σ is k-*fair* iff $(\forall a \in A)$ If a is infinitely often k-enabled in σ, then a occurs infinitely often in σ;
- σ is $\forall$k-*fair* iff $(\forall k \in \mathbb{N})$ σ is k-fair;
- σ is ∞-*fair* iff $(\forall a \in A)$ If a is infinitely often ∞-enabled in σ, then a occurs infinitely often in σ.

For finite words $w \in L$:

- w is k-*fair* iff w is (k+1)-*fair* iff w is $\forall$k-*fair* iff w is ∞-*fair* iff w is unextendable (in L). Hence, all the notions are equivalent for finite words.

This way we have defined the (potentially) infinite family of sets:

$$\text{k-Fair}(L) = \{\sigma \in L^\infty \mid \sigma \text{ is k-fair}\}, \text{ for } k \in \mathbb{N}$$
$$\forall\text{k-Fair}(L) = \{\sigma \in L^\infty \mid \sigma \text{ is } \forall\text{k-fair}\}$$
$$\infty\text{-Fair}(L) = \{\sigma \in L^\infty \mid \sigma \text{ is } \infty\text{-fair}\}$$

By the above definition, for any prefix-closed language $L \subseteq A^*$, the family of sets k-Fair(L) form an infinite decreasing hierarchy:

$$\text{0-Fair}(L) \supseteq \text{1-Fair}(L) \supseteq \ldots \quad \ldots \supseteq \forall\text{k-Fair}(L) \supseteq \infty\text{-Fair}(L)$$

The decreasing chain will be called *Best fairness hierarchy* (*bf-hierarchy*, for short).

The following proposition says that it is never too late for fairness. Such a property is called *feasibility*. It holds for any system with prefix-closed behaviour, so for any kind of Petri nets.

Proposition 5. Let $L \subseteq A^*$ be a prefix-closed language over a countable alphabet A.

For any $u \in L$ there is $v \in A^\infty$ such that $uv \in$ ∞-Fair(L).

i.e., any finite computation can be extended to an ∞-fair computation.

Proof. Easy and omitted. See [9], for instance. □

Remark. Since ∞-Fair is the least member of bf-hierarchy), any ∞-fair extension is thereby a k-fair extension for all $k \in \mathbb{N}$. Thus Prop. 5 gives an extension to a k-fair word, for all $k \in \mathbb{N}$, as well as to an $\forall$k-fair word.

3.2 Conspiracies and Conspiracy-Freeness

We will distinguish, following Best [1], a special kind of computations, called conspiracies. Informally speaking, they are "unfair" computations, pretending to be "fair", or computations which are fair locally (0-fair), but not globally (∞-fair).

Definition 6. *Conspiracies and Conspiracy-Freeness*

Let $L \subseteq A^*$ be a prefix-closed language.

- A computation $\sigma \in L^\infty$ is said to be a *conspiracy* iff it is 0-fair but not ∞-fair.
- The language L is *conspiracy-free* iff 0-Fair(S)=∞-Fair(S), i.e. any 0-fair computation of L^∞ is ∞-fair.

Example. *Conspiracy-free and conspiring EN-systems*

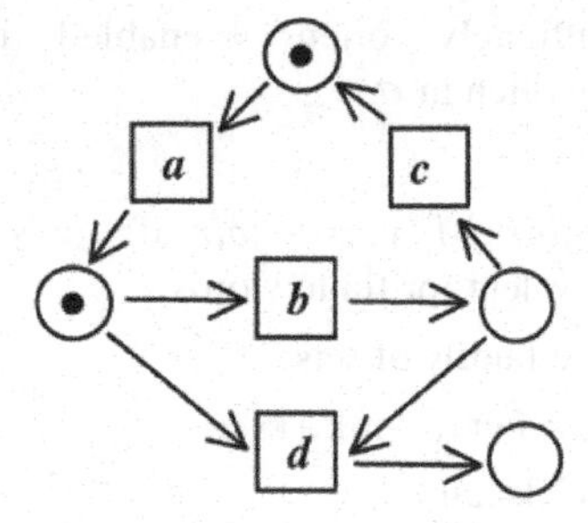

Fig. 1. A conspiracy-free EN-system. The only infinite computation $(bac)^\omega$ is not 0-fair, because d is infinitely often 0-enabled, and never executed

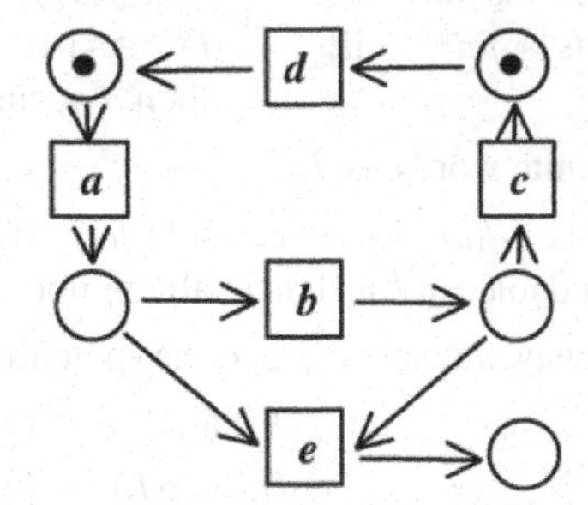

Fig. 2. An EN-system admitting conspiracy. The infinite computation $(abdc)^\omega$ is 0-fair (because e is never 0-enabled) and not ∞-fair (because e is still ∞-enabled)

3.3 Liveness, Fairness and Recurrence

The tittle of this subsection is a plagiarism of [6], where the authors investigate the mutual relations of the three notions. We want to show here how the notion of ∞-fairness can help to characterize liveness with fairness.

Let $L \subseteq T^*$ be a prefix-closed language. One can imagine that L is a set of finite computations of some concurrent system, say Petri net.

Definition 7. *Liveness and recurrence*

- A language $L \subseteq T^*$ is *live* iff $(\forall x \in L)$ $(\forall a \in T)$ $(\exists\ y \in T^*)$ s.t. $xya \in L$, i.e., any action is ∞-enabled in any state;
- A computation $\sigma \in L^\omega$ is *recurrent* iff $(\forall a \in T)$ $|\sigma|_a = \omega$, i.e., any action occurs in σ infinitely often;
- A language $L \subseteq T^*$ is *0-recurrent* iff $(\forall \sigma \in \text{0-Fair}(L))$ $(\forall a \in T)$ $|\sigma|_a = \omega$, i.e., any 0-fair computation is recurrent;
- A *system S* is said to be *live* iff its language $L(S)$ is live;
- A *system S* is said to be *0-recurrent* iff its language $L(S)$ is 0-recurrent.

Proposition 8. If L is 0-recurrent, then L is live.

Proof. Easily from Def. 7 and Prop. 5 (existence of 0-fair extensions). □

Example. *The converse of Prop. 8 is not true*

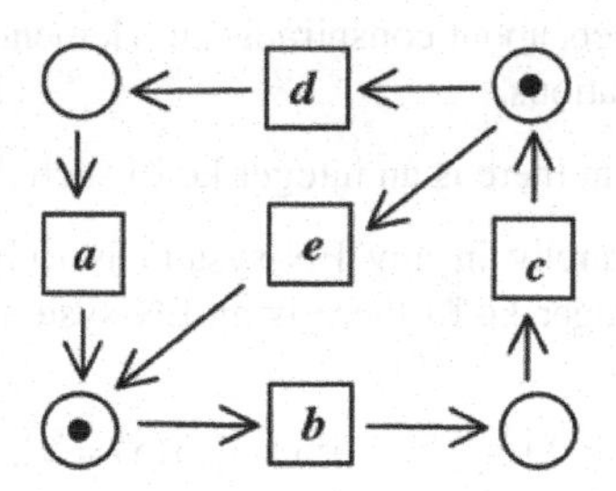

Fig. 3. A live EN-system which is not 0-recurrent. The 0-fair computation $(dbac)^\omega$ is 0-fair and not recurrent (the action e does not occur)

Let us reach for ∞-fairness. It can help us.

Definition 9. *∞-recurrence*

- A language $L \subseteq T^*$ is *∞-recurrent* iff $(\forall \sigma \in \infty\text{-Fair})$ $(\forall a \in T)$ $|\sigma|_a = \omega$, i.e., any ∞-fair computation is recurrent.

Remark. As ∞-Fair ⊆ 0-Fair, any 0-recurrent system is ∞-recurrent. But not conversely: the system of Fig. 3 shows a counterexample.

Now we are able to characterize liveness with ∞-recurrence:

Theorem 10. L is live iff L is ∞-recurrent.

Proof. (⇒) Directly from definitions: As L is live, any $a \in T$ is ∞-enabled in any $x \in L$, thus in any prefix of any $\sigma \in L^{\infty}$. If $\sigma \in$ ∞-Fair, then $|\sigma|_a = \omega$ for any $a \in T$. Hence, σ is ∞-recurrent.

(⇐) Let $x \in L$ and $a \in T$. Let $\sigma = xu$ be an ∞-fair extension of x (it exists, by Prop. 5). Since L is ∞-recurrent, $|\sigma|_a = \omega$, thus $|u|_a = \omega$ and there are $y, z \in T^*$ such that $u = yaz$. As L is prefix-closed, $xya \in L$. □

Notice that the EN-system of Fig. 3 is clearly ∞-recurrent.

Remark that Theorem 10 was proved without any particular assumption on the language (except it is prefix-closed). So it is true for all kinds of Petri nets, not only for EN-systems.

4 Transition Fairness

In this section we define the hierarchy of fairness in Petri nets, oriented on transitions. The definition is based on the general Def. 4. Let S be an EN-system or P/T-system. As we have remarked earlier, the set $L(S)$ of all finite computations of S is prefix-closed. Setting, in Def. 4, T for A and $L(S)$ for L, we get the present one. We will write k-Fair(S) instead of k-Fair($L(S)$).

4.1 Fairness and Conspiracies in Elementary Nets

We mention some basic facts about conspiracies in elementary nets. See [9] for proofs and more detailed investigations.

Fact 11. For any EN-system there is an integer $k \in \mathbb{N}$ such that k-Fair=∞-Fair.

Fact 11 states that bf-hierarchy in any EN-system is finite. Nevertheless, it can be arbitrary long. For any integer $k \in \mathbb{N}$ there is an EN-system S with strictly decreasing bf-hierarchy

$$\text{0-Fair}(S) \supset \text{1-Fair}(S) \supset \dots \quad \dots \supset \text{k-Fair}(S) = \forall\text{k-Fair}(S) = \infty\text{-Fair}(S)$$

Let us formulate the basic decision problem about conspiracies:

Conspiracy Problem: Instance: EN-system S Question: 0-Fair(S)=∞-Fair(S)?

Theorem 12. A problem "Is a given EN-system conspiracy-free?" is decidable.

4.2 Fairness and Conspiracies in Place/Transition Nets

The following facts are fairly obvious and well known:

Fact 13. Any infinite sequence of markings contains an infinite increasing subsequence.

Fact 14 (monotonicity property). Let S be a P/T-system.

If MwX and $M \le M'$, then $M'wX'$ and $X \le X'$.

Lemma 15. Let S be a P/T -system. For any $a \in T$ there is an integer $k_a \in \mathbb{N}$ such that in any marking M the transition a is k_a-enabled in M or dead in M.

Proof. Suppose it is not true for some $a \in T$. Hence, there are infinite sets of

markings: $M_1, M_2, \ldots$ and integers: $k_1 < k_2 < \ldots$

such that a is live in all M_i and it is not k_i-enabled in M_i for all i=1,2,...

Let us select (by Fact 13) an infinite increasing subsequence

$$M_{i1} \leq M_{i2} \leq \ldots$$

As a is live in M_{i1}, it is k-enabled in M_{i1}, for some $k \in \mathbb{N}$. As the strictly increasing sequence $k_1 < k_2 < \ldots$ is infinite, $k < k_{ij}$ for some j. By Fact 14 (monotonicity), a is k-enabled in M_{ij}, thus k_{ij}-enabled in M_{ij}. Contradiction. □

Theorem 16. The bf-hierarchy in any P/T-system is finite,

i.e. there is $K \in \mathbb{N}$ such that K-Fair=∞-Fair.

Proof. Take $K = \max\{k_a \mid a \in T\}$, where k_a are those of Lemma 15. □

Remark that the proof of Theorem 16 is purely existential, without any hint how to find it. For an effective algorithm computing K look at the final section.

4.3 Decision Problems in Place/Transition Nets

Decision problems on Petri nets are nicely presented in a survey of Esparza/Nielsen [5]. Let us mention some of them, related to the subject of the paper.

Theorem 17 (Carstensen [3]). 0-fair NonTermination Problem:

"If a given P/T-system has an infinite 0-fair computation?" is undecidable.

Theorem 18 (Howell/Rosier/Yen [14]). ∞-fair NonTermination Problem:

"If a given P/T-system has an infinite ∞-fair computation?" is decidable.

Let us reformulate:

0-Fair NonTermination Problem:	If 0-Fair⊆T*?	undecidable
∞-Fair NonTermination Problem:	If ∞-Fair⊆T*?	decidable

Conspiracy Problem: If 0-Fair = ∞-Fair?

And now undecidability of Conspiracy Problem for P/T-system: "If a given P/T-system is conspiracy-free?" is a simple corollary of both results.

Theorem 19. The Conspiracy Problem is undecidable.

Proof. Remark that: 0-Fair⊆T^* iff ∞-Fair⊆T^* and 0-Fair = ∞-Fair.

⇒: Since always ∞-Fair⊆0-Fair and 0-Fair∩T^*=∞-Fair∩T^*, we have ∞-Fair⊆ 0-Fair⊆T^* and 0-Fair=0-Fair∩T^*=∞-Fair∩T^*=∞-Fair; ⇐: obvious.

Now, because of decidability of "∞-Fair NonTermination Problem" (Th. 18), decidability of "Conspiracy Problem" implies decidability of "0-Fair NonTermination Problem". It contradicts Theorem 17. □

5 Marking Fairness

The definition of the fairness hierarchy of the previous section is based on occurrences of transitions. In this section a related notion, oriented on markings, will be defined. Originally, the notion was proposed in Merceron [8], in the context of processes. The notion itself will turn out to be rather weakly interesting. But together with the transition fairness they form a perfect couple against conspiracies.

Definition 20. k-*reachability*, ∞-*reachability*

Let $S=(P,T,F,M_0)$ be an EN-system or P/T-system.

- A marking M' is k-*reachable* from a marking M iff $(\exists w)$ MwM'' and $|w|\leq k+1$;
- A marking M' is ∞- *reachable* from a marking M iff $(\exists w)$ MwM'' and $M'\leq M''$.

Definition 21. *Marking Fairness*

Let $S=(P,T,F,M_0)$ be an EN-system or P/T-system. An infinite computation $\sigma\in L^{\omega}(S)$ is

- Mk-fair iff any marking infinitely often k-reachable from σ infinitely often occurs in σ
- M∀k-fair iff $(\forall k)$ σ is Mk-fair
- M∞-fair iff any marking infinitely often ∞-reachable from σ infinitely often occurs in σ

For finite computations $w\in L(S)$ all the notions are defined to be equivalent:

- w is Mk-fair iff w is M(k+1)-fair iff w is M∀k-fair iff w is M∞-fair iff w is unextendable in S.

Remark. As both fairness notions, marking fairness and transition fairness, will occur together in this section, the classical transition-oriented fairness of sections 3 and 4 will be preceded by T. For example, we will write T0-fair and T∞-fair instead of 0-fair and ∞-fair, respectively.

5.1 Marking Fairness in Elementary Nets

Like in the case of T-fairness, the M-fairness hierarchy in any EN-system is finite:

Fact 22. For any EN-system there is an integer $k\in\mathbb{N}$ such that Mk-Fair=M∞-Fair.

Proof. Again from finiteness of the set of reachable markings. □

But in the M-fairness case the whole marking oriented fairness hierarchy collapses.

Proposition 23. For any EN-system, M0-Fair = M∀k-Fair.

Proof. As bf-hierarchy is decreasing, we need to prove only the inclusion M0-Fair $\subseteq$ M∀k-Fair. Let $\sigma\in$ M0-Fair. If σ is finite, it belongs to M∀k-Fair, by Def. 3.1. Let σ be infinite. We show inductively that $\sigma\in$ M0-Fair implies $\sigma\in$ Mk-Fair, for any $k\in\mathbb{N}$. *Initial step*: M0-Fair$\subseteq$M0-Fair.

Induction hypothesis: $(\forall i \leq k)$ M0-Fair $\subseteq$ Mk-Fair.

Induction step: We have to prove that if σ is M0-fair then σ is M(k+1)-fair. Let us denote by X the set $X=\{M \mid M$ is infinitely often (k+1)-reachable from $\sigma\}$. We have to prove that any $M \in X$ occurs infinitely often in σ. If M is infinitely often k-reachable from σ, then, by the induction hypothesis, M occurs infinitely often in σ. Suppose M is only finitely often k-reachable from σ. Then there is a marking $M' \in \text{Prec}(M)$, where $\text{Prec}(M)= \{M' \mid (\exists a \in T)\ M'aM\}$ denotes the set of markings preceding M, which is infinitely often k-reachable from σ (since the set Prec(M) is finite). By the induction hypothesis, the marking M' occurs in σ infinitely often. Hence, M is infinitely often 0-reachable from σ. As σ is M0-fair, M occurs in σ infinitely often. So $\sigma \in$ M(k+1)-Fair. That way we have proved that M0-Fair = Mk-Fair, for all $k \in \mathbb{N}$, that is M0-Fair = M$\forall$k-Fair. □

Theorem 24. For any EN-system, M0-Fair = M∞-Fair.

Proof. Proposition 23 + Fact 22. □

The notions of T-fairness and M-fairness are related in some way, but the relationship is rather loose. One can show an EN-system with a T∞-fair computation that is not M0-fair, as well as an EN-system with a M∞-fair computation that is not T0-fair.

Example. *T-fairness vs M-fairness*

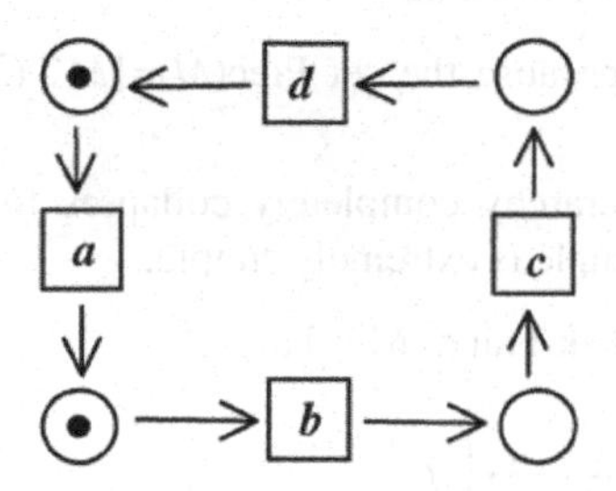

Fig. 4. The computation $(bcad)^\omega$ is T∞-fair but not M0-fair

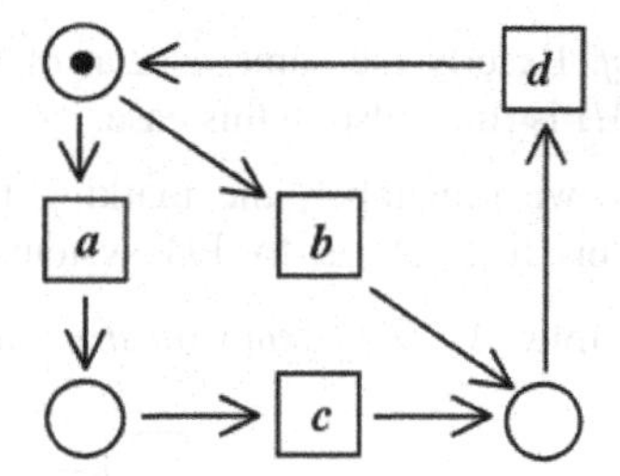

Fig. 5. The computation $(acd)^\omega$ is M∞-fair but not T0-fair

The next theorem shows the power of T-fairness and M-fairness, playing together:

Theorem 25. Let $S=(P,T,F,M_0)$ be an EN-system, and σ be a computation of S.

If σ is T0-fair and M0-fair, then it is T∞-fair.

Proof. If σ is finite T0-fair, then T∞-fair, by Def. 4. Assume σ is infinite:

$$\sigma = M_0 a_1 M_1 a_2 M_2 \ldots\ldots$$

We show inductively that if σ is T0-fair and M0-fair, then σ is Tk-fair, for any $k \in \mathbb{N}$.

The initial step, for k=0, is obvious.

Let us denote by $I_a(k)$, for $a \in T$ and $k \in \mathbb{N}$, the set $I_a(k)=\{i \in \mathbb{N} \mid a$ is k-enabled in $M_i\}$ of indices of such markings of σ that a is k-enabled in them.

Induction hypothesis: ($\forall$i$\leq$k) If σ is T0-fair and M0-fair, then σ is Ti-fair.

Induction step: We have to prove that if σ is T0-fair then σ is T(k+1)-fair. In other words, we have to prove $|\sigma|_a=\infty$, whenever $a\in T$ is infinitely often (k+1)-enabled in σ. Let $a\in T$ be such an action. Then the set I_a(k+1) is infinite. If also the set I_a(k) is infinite, then $|\sigma|_a=\infty$, by the induction hypothesis. Consider the case I_a(k) is finite. Then the set $J=I_a$(k+1)$-I_a$(k) is infinite. As a is (k+1)-enabled (and not k-enabled) in any M_j ($j\in J$), there is, for any $j\in J$, an extension $M_jx_jM_j'wa$ with $x_j\in T$ and $w\in T^*$ such that $|w|$=k, i.e. a is k-enabled in each M_j'. As number of markings in EN-systems is finite, at least one of the states M_j' is repeated infinitely often, so it is infinitely often 0-reachable from σ. By M0-fairness, it occurs infinitely often in σ. Hence, a is infinitely often k-enabled in σ, thus, by induction hypothesis, a occurs infinitely often in σ. That way we have proved: If σ is T0-fair and M0-fair, then it is T$\forall$k-fair. And Fact 11 completes the proof. □

5.2 Marking Fairness in Place/Transition Nets

It was remarked very early, by the author of this notion (Merceron [8]), that marking fairness, in its original formulation, is rather useless for investigations of unbounded P/T-systems. We will show some examples. But first a result.

Proposition 26. For any P/T-system, M0-Fair = M$\forall$k-Fair.

Proof. Exactly the same as that of Prop. 23, because the set Prec(M)=$\{M'|\ (\exists a\in T)\ M'aM\}$ is finite also in this case. □

As we remember, the marking fairness hierarchy completely collapses for EN-systems. It is not true for P/T-systems. The example is extremely simple.

Example. *A P/T-system with strict inclusion* M$\forall$k-Fair $\supset$ M∞-Fair

Fig. 6. The computation $(aab)^\omega$ is M$\forall$k-fair but not M∞-fair

And what if M-fairness will work with T-fairness? Is there a hope for something like Th. 25? No chances, says the next example.

Example. *Conspiracy against d*

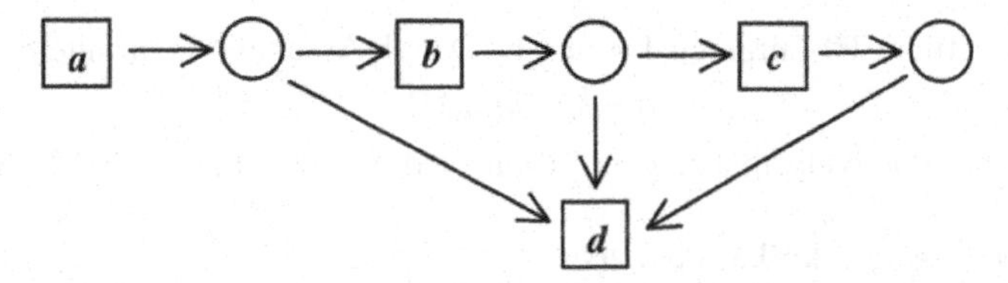

Fig. 7. The computation $(abc)^\omega$ is T0-fair and M0-fair, but not T∞-fair

6 Covering Fairness

Both fairness notions, considered up to now, transition and marking fairness, were based on an idea of enabling (reachability is, in fact, "enabling" for markings) and occurence. The aim of the paper is a fight with conspiracies. The enabling/occurrence method gave quite nice fruits for elementary nets (Theorem 25), but fell by P/T nets. In this section we will attack the problem with a slightly modified notion of marking fairness, concentrated on coverability rather than reachability. System, up to the end of the paper, means always P/T-system.

Definition 27. *Coverability and covering*

Let $S=(P,T,F,M_0)$ be a P/T-system.

- A marking M' is *covered* by a marking M'' (or M'' *covers* M') iff $M'\leq M''$ (i.e. $M'(p)\leq M''(p)$ for any $p\in P$);
- A marking M' is k-*coverable* from a marking M iff $(\exists w)$ MwM'' and $|w|\leq k+1$ and $M'\leq M''$, i.e. there is a marking M'' which is k-reachable from M and covers M';
- A marking M' is ∞-*coverable* from a marking M iff $(\exists w)$ MwM'' and $M'\leq M''$, i.e. there is a marking M'' which is ∞-reachable from M and covers M'.

Definition 28. *Covering Fairness*

Let $S=(P,T,F,M_0)$ be a P/T-system. An infinite computation $\sigma\in L^{\omega}(S)$ is

- Ck-fair iff any marking infinitely often k-coverable from σ is infinitely often covered in σ
- C$\forall$k-fair iff $(\forall k)$ σ is Ck-fair
- C∞-fair iff any marking infinitely often ∞-coverable from σ is infinitely often covered in σ

For finite computations $w\in L(S)$ all the notions are defined to be equivalent:

- w is Mk-fair iff w is M(k+1)-fair iff w is M$\forall$k-fair iff w is M∞-fair iff w is unextendable in S.

6.1 Covering Fairness Hierarchy Collapses

Let $Y\in\mathbb{N}^P$ be a marking and let $a\in T$ be a transition. Let (aY) be a marking, built from Y and a with the rules defined by Table 1, and illustrated by Fig. 8. It can be regarded as something like a "generalized preceding marking" of Y wrt a, because it is always defined, and equal to the "real" preceding marking, if the latter exists.

Table 1. Definition of a marking (aY) for given marking Y and transition a

If	$^{\bullet}a(p)$=	and	$a^{\bullet}(p)$=	then	$(aY)(p)$=
	0		0		$Y(p)$
	1		1		If $Y(p)$=0 then 1 else $Y(p)$
	1		0		$Y(p)$+1
	0		1		If $Y(p)$=0 then 0 else $Y(p)$-1

Example. *An example for the definition of (aY)*

The example illustrates all six cases of rules of Table 1 by the six places of the exemplary net.

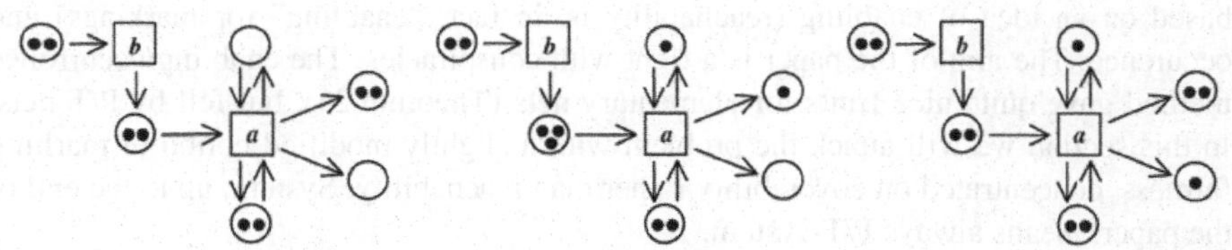

Fig. 8a. An example of a marking Y **Fig. 8b.** The marking $X=(aY)$ **Fig. 8c.** The marking $Z=Xa$ covering Y

The following lemma says that the marking (aY), defined above, is the smallest marking X enabling a (i.e. containing ${}^\bullet a$) s.t. Xa covers Y.

Lemma 29. Let M, M', Y be markings, let $a \in T$ be a transition, and let $X=(aY)$ be a marking given by Table 1. If MaM' and $0 \leq Y \leq M'$, then $0 \leq X \leq M$ and Y is 0-coverable from X, namely XaY' and $Y \leq Y' = X - {}^\bullet a + a^\bullet$.

Proof. By the table, $0 \leq X$. Now we prove that $X \leq Y - a^\bullet + {}^\bullet a \leq M' - a^\bullet + {}^\bullet a = M$, where only $X \leq Y - a^\bullet + {}^\bullet a$ needs a justification. If $X(p) \leq Y(p) - a^\bullet(p) + {}^\bullet a(p)$ then $X(p) \leq M(p)$, since $Y - a^\bullet + {}^\bullet a \leq M' - a^\bullet + {}^\bullet a = M$. The only different cases are:

${}^\bullet a(p)$	$a^\bullet(p)$	$Y(p)$	$X(p)$		
1	1	0	1	then	$M(p) \geq 1$, since a is 0-enabled in M
0	1	0	0	then	obviously $X(p)=0 \leq M(p)$

Now, directly from the table defining $X(p)$, we conclude that $(\forall p \in P)\ {}^\bullet a(p) \leq X(p)$, thus a is 0-enabled in X. Finally, we prove that $Y \leq Y'$, where $Y' = X - {}^\bullet a + a^\bullet$. In most cases $Y(p) = Y'(p)$. The only different cases are:

${}^\bullet a(p)$	$a^\bullet(p)$	$Y(p)$	$X(p)$	$Y'(p)$	
1	1	0	1	1	
0	1	0	0	1	In both cases $Y(p)=0<1=Y'(p)$.

This way the lemma is proved. □

Proposition 30. C0-Fair = C∀k-Fair

Proof. We prove inductively that $(\forall k \in \mathbb{N})$ C0-Fair $\subseteq$ Ck-Fair, i.e. any C0-fair computation is Ck-fair, for any $k \in \mathbb{N}$.

Initial step: C0-Fair $\subseteq$ C0-Fair

Induction hypothesis: $(\forall i \leq k)$ C0-Fair $\subseteq$ Ci-Fair

Induction step: We have to prove C0-Fair $\subseteq$ C(k+1)-Fair. Let σ be a C0-fair computation and let Y be a marking infinitely often (k+1)-coverable from σ.

There is a transition $a \in T$ which leads infinitely often from k-reachability level to (k+1)-reachability level. By Lemma 29, the marking X obtained from Y by the rules of

Table 1 is covered by all precedent markings (see Fig. 9). Thus X is infinitely often k-coverable from σ and, by induction hypothesis, X is infinitely often covered in σ. As Y is 0-coverable from X (by Lemma 29), so Y is infinitely often 0-coverable from σ, thus infinitely often covered in σ, since σ is C0-fair. This ends the induction step. We have proved this way that C0-Fair$\subseteq$C$\forall$k-Fair. Since the inverse inclusion C0-Fair$\supseteq$C$\forall$k-Fair always holds, the proof is finished. □

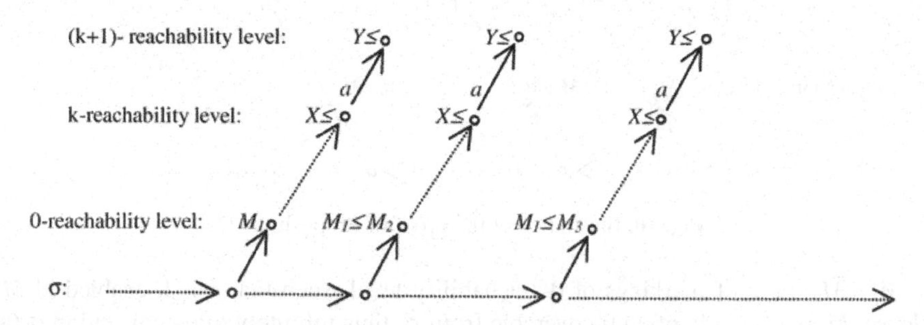

Fig. 9. Illustration of the proof of Proposition 30

Let us recall the monotonicity lemma: "If $M \leq M'$ and MwX, then $M'wX'$ and $X \leq X'$". It makes the next proof quite easy.

Proposition 31. C$\forall$k-Fair = C∞-Fair

Proof. Let $\sigma \in$ C$\forall$k-Fair. Let X be a marking infinitely often (thus constantly) ∞-coverable from σ. Let $\sigma = M_0 a_1 M_1 a_2 M_2 \ldots$ and let $M_{i1} \leq M_{i2} \leq \ldots$ be an increasing subsequence. There is an integer $k \in \mathbb{N}$ such that X is k-coverable from M_{i1}. It follows from the monotonicity that X is k-coverable from all M_{in}, hence X is infinitely often covered in σ, since σ is Ck-fair. □

As a corollary of Propositions 30 and 31 we conclude that the whole covering fairness hierarchy completely collapses, like marking fairness hierarchy in the case of elementary nets.

Theorem 32. C0-Fair = C∞-Fair

Proof. Prop.6.4+Prop.6.5. □

6.2 Covering Fairness against Conspiracies

Proposition 33. T0-Fair $\cap$ C0-Fair $\subseteq$ T$\forall$k-Fair
i.e. any computation which is T0-fair and C0-fair is then Tk-fair, for any $k \in \mathbb{N}$.

Proof. We prove inductively that $(\forall k \in \mathbb{N})$ T0-Fair $\cap$ C0-Fair $\subseteq$ Tk-Fair, i.e. any computation which is T0-fair and C0-fair is Tk-fair, for any $k \in \mathbb{N}$.

Initial step: T0-Fair $\cap$ C0-Fair $\subseteq$ T0-Fair

Induction hypothesis: $(\forall i \leq k)$ T0-Fair $\cap$ C0-Fair $\subseteq$ Ti-Fair

Induction step: We have to prove T0-Fair ∩ C0-Fair ⊆ T(k+1)-Fair. Let σ be a computation T0-fair and C0-fair, and let $a \in T$ be a transition infinitely often (k+1)-enabled in σ. If so, then there is (by Fact 13) an infinite monotonic sequence

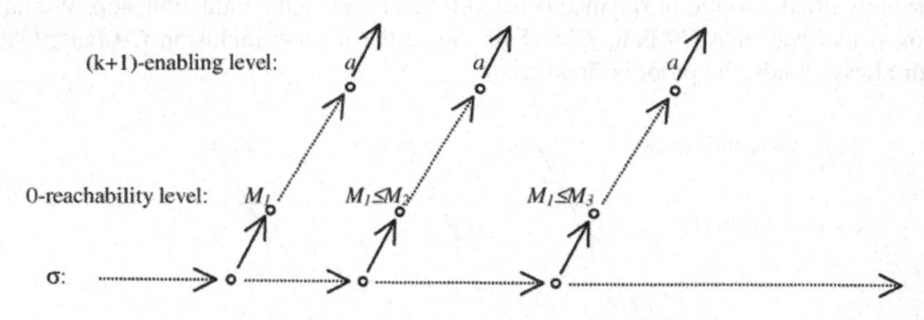

Fig. 10. Illustration of the proof of Proposition 33

$M_1 \leq M_2 \leq \ldots$ of markings of 0-reachability level, such that a is k-enabled id M_1. Hence, M_1 is infinitely often 0-coverable from σ, thus infinitely often covered in σ (as σ is C0-fair). Therefore, a is infinitely often k-enabled, so it occurs infinitely often in σ, by induction hypothesis. It means that σ is T(k+1)-fair. □

Theorem 34. T0-Fair ∩ C0-Fair ⊆ T∞-Fair
i.e. any T0-fair and simultaneously C0-fair computation is T∞-fair.

Proof. Directly from Proposition 33 and Theorem 16. □

7 Slim Covering Fairness

Now we know that covering fairness makes possible to force a conspiracy-free behaviour of P/T-systems. Setting the set of all markings as the control set for fair covering, we make conspiring computations impossible. All the control takes place locally, on the 0-level, and does not require to check k-enabling or k-coverability for k>0. The only, but crucial problem is the size of the set $\mathbb{N}^P$. The control based on Theorem 34 needs to check covering for infinite set of markings (in fact, in any step we have to check a finite number of markings, but the number may be quickly increasing, and is unbounded).

The aim of this section is to find a finite set of markings which could keep control on conspiracy-free executions of unbounded P/T-systems. (For bounded P/T-systems Theorem 25 can be easy adopted.)

7.1 Flat Covering Fairness Fails

Definition 35. *Flat markings*
Let $C_1 = \{0,1\}^P$ be the set of all markings with at most one token in any place. Markings of C_1 will be called *flat markings*.

Flat markings are enough for some systems. For instance, the net of Fig. 7 will behave ∞-fair under control of T0-fairness and $C_1$0-fairness. Unfortunately, flat markings are not always sufficient for this purpose, as the following example shows.

Example. *Flat covering fairness lets conspiracies through*

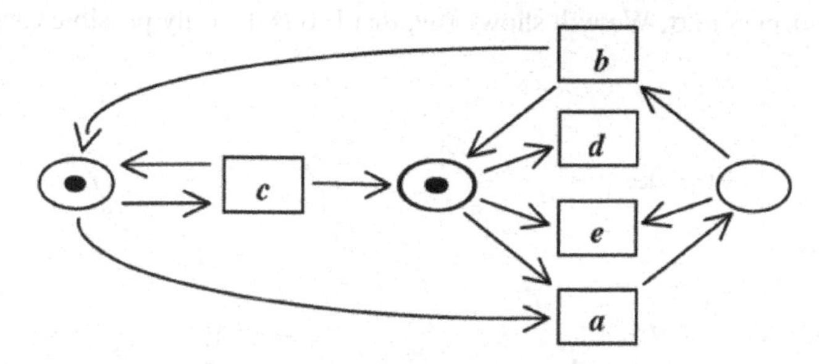

Fig. 11. The computation $(dcab)^\omega$ is T0-fair and $C_1$0-fair, but not T∞-fair

7.2 ...but Slim Covering Fairness Succeeds

At the start of this subsection we prove a lemma, which will be crucial in the proof of the main result of this section.

Definition 36. Let M be a marking over P. The *thickness* of M is the maximal number of tokens in a place: $\text{th}(M) = \max\{M(p) \mid p \in P\}$. The n-*layer* of M, for $n \in \mathbb{N}$, is a submarking $M_{\leq n}$ of M defined as follows: $(\forall p \in P)\ M_{\leq n}(p) = \min\{M(p), n\}$.

Lemma 37. Let $S=(P,T,F,M_0)$ be a P/T-system and $M \in \mathbb{N}^P$ be a marking. Any sequence $w \in T^*$ of a length $n=|w|$ executable from M is executable from the n-layer $M_{\leq n}$ of M.

Proof. Induction with respect to $n=|w|$. If $n=0$ then $w=\varepsilon$ and ε is executable from $\varnothing$. Assume the lemma holds for all $i \leq n$. Let w be executable from M and $|w|=n+1$, thus $w=ua$ for some u of the length $|u|=n$ and some $a \in T$. By the induction hypothesis, u is executable from $M_{\leq n}$, thus also from $M_{\leq n+1}$, as $M_{\leq n} \leq M_{\leq n+1}$. Suppose a is not enabled in $M_{\leq n+1}u$. It means there is a place $p \in {}^\bullet a$ which is empty in $M_{\leq n+1}u$. If so, then $M_{\leq n+1}(p) \leq n$ (because any single execution of a transition takes at most one token from any place), thus $M(p)=M_{\leq n+1}(p) \leq n$. But then the place p is empty also in Mu, so $ua=w$ is not executable from M. Contradiction. □

Definition 38. *Slim markings*
Let $K \in \mathbb{N}$ be an integer such that TK-Fair=T∞-Fair. We know that such K exists (Th. 16). Let $C_K=\{M \in \mathbb{N}^P \mid (\forall p \in P)\ M(p) \leq K\}$, i.e. C_K is the set of all markings (not necessarily reachable) with at most K tokens in any place. Such markings will be called *slim markings*.

Theorem 39. T0-Fair $\cap$ C_K0-Fair $\subseteq$ T∞-Fair

Proof. As TK-Fair=T∞-Fair, it is enough to prove that $T0\text{-Fair} \cap C_K0\text{-Fair} \subseteq TK\text{-Fair}$. Let σ be a computation T0-fair and C_K0-fair, and let $a \in T$ be a transition infinitely often K-enabled in σ. If so, then there is an I-enabling level (see Fig. 12), for some I≤K, such that the transition *a* is infinitely often I-enabled. Let us take the minimal I of this property. If I=0, then *a* is infinitely often 0-enabled, thus, by T0-fairness, infinitely often occurs in σ. We will show now, that I=0 is the only possible case. Suppose I>0.

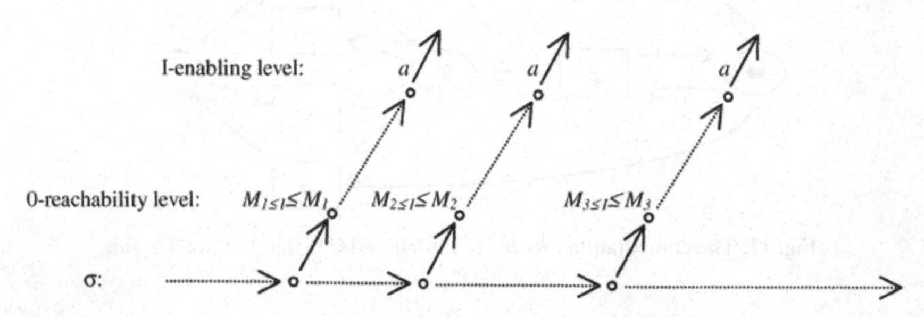

Fig. 2. Illustration of the proof of Theorem 39

Then, by Lemma 37, the transition *a* is I-enabled in any I-layer $M_{i \leq I}$ of any M_i of the 0-reachability level. But the number of all different I-layers is finite, hence one (at least) of them is infinitely often 0-coverable from σ. And belongs to C_K, since I≤K. By C_K0-fairness of σ it is covered infinitely often in σ. So *a* is infinitely often (I-1)-enabled in σ. But σ was a minimal integer of this property. Contradiction. □

7.3 ...and Can Be Effectively Applied

In [13] Valk informs about an unpublished(?) result of Carstensen. He proved, in [2], using the theory of residue sets [12], that the number K of Th. 16 can be effectively computed. That result makes Th. 39 practically applicable, as K for slim markings of Th. 39 is just that of Th. 16. This subsection contains a sketch of Carstensen's proof, reconstructed by the author.

Definition 40. A set of markings $\mathbb{K} \subseteq \mathbb{N}^P$ is said to be *right-closed* iff

$$(\forall X \in \mathbb{K})\ (\forall Y \in \mathbb{N}^P)\ X \leq Y \Rightarrow Y \in \mathbb{K},$$

i.e., along with any member, $\mathbb{K}$ contains all its covers.

Definition 41. Let $\mathbb{K} \subseteq \mathbb{N}^P$ be a set of markings. A marking $M \in \mathbb{K}$ is said to be a *residue* of $\mathbb{K}$ iff $\{X \in \mathbb{K} \mid X \leq M\} = \{M\}$, i.e. *M* is a minimal member of $\mathbb{K}$ wrt ≤. The set of all such elements, denoted by *res*($\mathbb{K}$), is called the *residue set* of $\mathbb{K}$.

Fact 42. For any set of markings $\mathbb{K} \subseteq \mathbb{N}^P$, the residue set *res*($\mathbb{K}$) is finite.

Proof. Directly by Fact 13.

Definition 43. A set of markings $\mathbb{K} \subseteq \mathbb{N}^P$ has *property RES* iff the predicate $P(M) = \textbf{if } (\exists X \in \mathbb{K})\ X \leq M \textbf{ then } 1 \textbf{ else } 0$ is computable for any marking $M \in (\mathbb{N} \cup \omega)^P$.

Theorem 44 (Valk/Jantzen [12]). Let $I\!K \subseteq I\!N^P$ be a right-closed set of markings. The residue set *res(IK)* is effectively constructable iff $I\!K$ has property RES.

Definition 45. Let $I\!K_a \subseteq I\!N^P$, for $a \in T$, be the set of all markings ∞-enabling the transition *a*:

$$I\!K_a = \{X \in I\!N^P \mid a \text{ is } \infty\text{-enabled in } X\}$$

Lemma 46. The set $I\!K_a$ has property RES.

Proof. By Def. 43, the set $I\!K_a$ has property RES iff the predicate

$$P_a(M) = \textbf{if } (\exists X \in I\!K_a)\ X \leq M \textbf{ then } 1 \textbf{ else } 0$$

is computable. Remark that $P_a(M)=1$ iff *a* is ∞-enabled in *M*. It is well known that it is decidable with a help of coverability tree rooted with *M* (see [4], for instance). □

Corollary 47. For any $a \in T$, the set res($I\!K_a$) is effectively constructable.

Proof. By Fact 14 (monotonicity property), the set $I\!K_a$ is right-closed. And Th. 44 + Lemma 46 yield the result. □

Lemma 48. For any $a \in T$, the integer k_a (of Lemma 15) is effectively computable.

Proof. If $M \in I\!K_a$, then $M \geq X$ for some $X \in$ res($I\!K_a$). Since "monotonicity property", it is enough to find, for all $X \in$ res($I\!K_a$), minimal integers x_a such that *a* is x_a-enabled in *X* (building initial parts of reachability tree rooted with *X*). Finally, set $k_a = \max\{x_a \mid X \in \text{res}(I\!K_a)\}$. □

Theorem 49 (Carstensen [2]). An integer K such that K-Fair=∞-Fair (that of Th. 16) is effectively computable.

Proof. Set $K = \max\{k_a \mid a \in T\}$, where k_a's are those computed by Lemma 48. □

Remark that the algorithm computes, in a unique way, the integer K such that K-Fair=∞-Fair, but K is not a minimal integer of this property. Even for the simplest producer/consumer net of Fig. 6, with single residue [0], we get K=1, whereas the minimal k is obviously 0. One can say more: there is no algorithm producing the minimal such integer. It follows directly from undecidability of the "Cospiracy Problem" – Theorem 19.

Acknowledgments

The author thanks four anonymous referees for their valuable remarks and comments.

References

1. E. Best: Fairness and Conspiracies. Information Processing Letters 18, pp. 215-220, 1984. Erratum: IPL 19, p.162, 1984.
2. H. Carstensen: Fairness bei nebenläufigen Systemen, eine Untersuchung am Modell der Petrinetze. Bericht des FB Informatik, Univ. Hamburg, 1986.

3. H. Carstensen: Decidability Questions for Fairness in Petri Nets. LNCS 247, pp. 396-407, Springer, 1987.
4. J. Desel, W. Reisig: Place/Transition Petri Nets. In [10], LNCS 1491, pp. 122-173. Springer, 1998.
5. J. Esparza, M. Nielsen: Decidability Issues for Petri Nets. EATCS Bulletin 52, pp. 245-262, 1994.
6. E. Kindler, W. van der Aalst: Liveness, Fairness and Recurrence in Petri Nets. Information Processing Letters 70, pp. 269-274, 1999.
7. D. Lehman, A. Pnueli, J. Stavi: Impartiality, Justice and Fairness: the Ethics of Concurrent Termination. LNCS 115, pp. 264-277. Springer, 1981.
8. A. Merceron: Fair Processes. Advances in Petri Nets, LNCS 266, pp. 181-195. Springer, 1987.
9. E. Ochmański: Best Fairness Hierarchy in Elementary Nets. Proc. of CS&P 2003 Workshop, pp. 382-396, Warsaw University, 2003.
10. W. Reisig, G. Rozenberg (eds.): Lectures on Petri Nets. LNCS 1491. Springer, 1998.
11. G. Rozenberg, J. Engelfriet: Elementary Net Systems. In [10], LNCS 1491, pp. 12-121. Springer, 1998.
12. R. Valk, M. Jantzen: The Residue of Vector Sets with Applications to Decidability Problems in Petri Nets. Acta Informatica 21, pp. 643-674, 1985.
13. R. Valk: Infinite Behaviour and Fairness. LNCS 254, pp. 377-396, Springer, 1986.
14. R. Howell, L. Rosier, H. Yen: A taxonomy of fairness and temporal logic problems for Petri nets. Theoretical Computer Science 82, pp. 341-372, 1991.

Modeling and Analysis of Margolus Quantum Cellular Automata Using Net-Theoretical Methods

Leo Ojala, Olli-Matti Penttinen, and Elina Parviainen

Helsinki University of Technology,
Laboratory for Theoretical Computer Science,
P.O. Box 5400, FIN-02015 TKK, Finland
{Leo.Ojala,Olli-Matti.Penttinen,Elina.Parviainen}@hut.fi

Abstract. Petri net methods have been very successful in modeling the operation of classical parallel systems. In this work, these methods are applied to designing semi-classical parallel quantum computers. The demonstration object of our study is a quantum Billiard Ball Model Cellular Automaton (BBMCA) suggested by Margolus. Firstly, a high-level Petri net model of a classical reversible version of this automaton is constructed. Subsequently, this Petri net model is used as a so-called kernel net of the quantum BBMCA. The time-independent Hamiltonian needed to generate the time-evolution of a quantum computer can be automatically generated from the reachability graph of a kernel net. Also, a new numerical method for solving the resulting Schrödinger differential equation system needed for time simulation of the quantum automaton is given. QUANTUM MARIA, a software package for modeling and numerical simulation of quantum computers, is introduced.

> Quantum mechanics, that mysterious, confusing discipline, which none of us really understands, but which we know how to use.
>
> Murray Gell-Mann

1 Introduction

Quantum computing, a cross-disciplinary merge of quantum mechanics and computing, is a continuous source of new research problems and challenges. The idea of a quantum mechanical computer was first published by Benioff in [1]. His ideas were further developed by the likes of Feynman [3, 4], Margolus [13, 14] and Deutsch [2]. Our research group has studied how to apply net-theoretical methods in modeling and simulation of quantum computers. The demonstration object of our study has been Feynman's quantum computer (FQC) [4]. We have modeled and simulated several incarnations of it such as the $\sqrt{\text{NOT}}$-computer [16] and the SWAP-computer [17]. However, FQC has a serial architecture; for the more

J. Cortadella and W. Reisig (Eds.): ICATPN 2004, LNCS 3099, pp. 331–350, 2004.
© Springer-Verlag Berlin Heidelberg 2004

challenging parallel computation, new configurations are needed. For our studies a proper platform has been the two-dimensional *Margolus Quantum Cellular Automaton* (MQCA) [13]. In our systematic design procedure, we demonstrate how to model, analyze and simulate MQCAs using Petri nets and related tools. Throughout the paper, we use a variant [10] of Reisig's *Algebraic System Nets* (ASN) [22] as our modeling language.

In Sect. 2, we describe the classical *Margolus Cellular Automaton* (MCA) [12], its block automaton structure and a method used for local synchronization. An application example is introduced as well: the Margolus *Billiard Ball Model Cell Automaton* (BBMCA) [12]. A Petri net model of this automaton is given in Sect. 3. In Sect. 4, we briefly introduce quantum computing concepts relevant to our work, which is based on so-called *semi-classical* computing; i.e., quantum mechanical realizations of reversible classical logic. We also give a method for constructing the *kernel net* of a quantum computer. In Sect. 5, we give the kernel net of a quantum BBMCA and analyze its time behavior. Appendix A gives an overview of QUANTUM MARIA, our analysis software.

2 Margolus Cellular Automata

2.1 Traditional Cellular Automata

Cellular automata (CA) were originally introduced by von Neumann for studying self-reproduction in biological systems [24]. Since then, they have been used for various purposes including modeling of physical systems. A new era in cellular automata research was started by Wolfram, who shifted the focus from applications of CA to the study of their mathematical properties, e.g. statistical parameters, entropy and topology [25]. An introduction to CA is found e.g. in [23].

A CA consists of several identical units, called *cells* that change their states according to some fixed set of rules. At each time step, the current state of some *neighborhood*; i.e, a set of cells that may but does not need to contain the cell itself, determines the new state of the cell. A cellular automaton may be of any dimension, one- and two-dimensional being the most common. In this paper, we only consider *cyclic* two-dimensional CA. The grid of a cyclic 2-D CA forms a finite, Cartesian *torus-like lattice*, whose left- and rightmost columns and whose top and bottom rows are connected to each other.

2.2 Partitioning Cellular Automata

In [12], Margolus introduces a slightly different kind of CA, the *partitioning cellular automaton*, whose grid is partitioned into disjoint *blocks*, and the evolution rules are given as functions from an entire block to itself, instead of functions from a neighborhood to a single cell. At each time step, the partitioning is changed, causing blocks at the previous time step to partly overlap the blocks at the current step. This creates a form of interaction between blocks and turns the automaton into a dynamic system. These CA are very useful for modeling physics-like microscopically reversible systems, since it has been shown in [12] that local reversibility of block rules implies global reversibility of a partitioning CA. Any rule defined by a bijection is reversible.

Margolus Cellular Automata (MCA) are a special case of partitioning CA. The grid of a MCA is divided into disjoint 2-by-2 cell blocks. Partitioning is changed so that at odd evolution steps the upper left-hand corner of a block has odd coordinates, and at even steps it has even coordinates. We call them *odd* and *even blocks*, respectively.

2.3 Local Synchronization

Traditional CA are *globally synchronized*; i.e., all blocks (or cells) are updated simultaneously. For simulations on a conventional computer, this is acceptable, but troublesome for the kinds of massively parallel implementations we have in mind: we want each block to execute its evolution on a separate miniature processor as independently from each other as possible. In a quantum mechanical setting, which is our ultimate goal, coupling each block to a fine-grained global clock that runs at a uniform pace not only introduces lots of unwanted long-range communications; it may actually cause the entire system to halt [19, 20].

It is possible, however, to perform an effectively synchronous computation using only local synchronizations [13]. This is achievable by augmenting each cell with an *age variable*. It is their rôle to record the *computational age* (i.e., the number of updates performed) of the cells. Each time a block is updated, the age variables of all cells in the block are incremented. A block is updatable if and only if all its cells have the same computational age. Consequently, parts of the grid lattice will have different computational ages, but the age difference of adjacent cells never exceeds one time unit. The operation of this *Asynchronous Cellular Automaton* (ACA) is analogous to a chain of people walking hand in hand: in order not to break the chain, a person can take a step forward only if both of his neighbors follow him "near enough". Otherwise, he must wait for them to catch up with him.

Since the age difference of adjacent cells cannot exceed one unit, a one bit age variable that holds the computational age of the cell modulo two is sufficient to enforce a correct temporal ordering of events. The age variables of all cells are initialized to one. The update rule for odd blocks requires all cells to have an odd age (initially true) and for even blocks to have an even age (initially false).

Figure 1 shows a high-level Petri net model of this updating scheme. Each square in the grid is represented by a tuple giving its Cartesian coordinates. If the age of cell (i, j) is even, there is a token $\langle i, j\rangle$ in place $\mathsf{Age}_{\mathsf{even}}$, otherwise there is a token in place $\mathsf{Age}_{\mathsf{odd}}$. Initially, all cells have an odd age. Correspondingly, the initial marking of place $\mathsf{Age}_{\mathsf{odd}}$ has a token for every cell in the grid and the initial marking of place $\mathsf{Age}_{\mathsf{even}}$ is empty. Note, that the number of columns (n_c) and the number of rows (n_r) must be even numbers. The symbol $\uplus$ denotes the *modular*[1] successor operator.

[1] modulo the arity of the operand

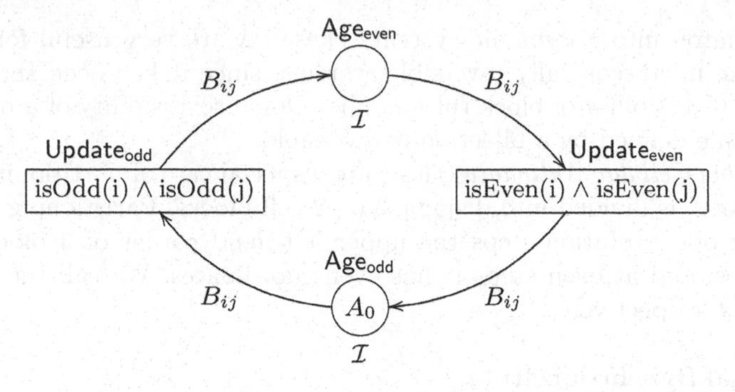

Declarations	
$n_c \in \mathbb{N}$	number of columns
$n_r \in \mathbb{N}$	number of rows
$\mathcal{X}$	$\{1, \ldots, n_c\}$
$\mathcal{Y}$	$\{1, \ldots, n_r\}$
$\mathcal{I}$	$\mathcal{X} \times \mathcal{Y}$
$\uplus$	modular successor operator
B_{ij}	$\langle i, j \rangle + \langle \uplus i, j \rangle + \langle i, \uplus j \rangle + \langle \uplus i, \uplus j \rangle$
A_0	$\sum_{i=1}^{n_c} \sum_{j=1}^{n_r} \langle i, j \rangle$

Fig. 1. Petri net model of updating an ACA

2.4 The Billiard Ball Model Cellular Automaton

The partitioning technique given in the previous section was developed [12] to construct an invertible CA to model Fredkin's Billiard Ball Model (BBM) [6]. The BBM is a continuous physical system that is turned into a digital one by constraining its initial conditions and the times at which we observe it. In BBM, the computations are constructed out of elastic collisions of incompressible spheres. Each ball starts at a grid point of a two-dimensional Cartesian lattice, moving diagonally along the grid in one of the four directions. All balls travel at a uniform velocity, advancing from one grid point to the next in exactly one time unit. The grid spacing is chosen in such a fashion that balls collide at grid points. All collisions take place at right-angles, which implies that one time step after the collision, both colliding balls are still at grid points.

To make a CA model of the BBM, we represent the finite diameter billiard balls in our CA by spatially separated pairs of particles one following the other – the leading edge of the ball followed by the trailing edge.

In Fig. 2, we have the BBMCA rule set given by Margolus. The rules are reversible and conserve the number of particles (black squares). Note, that only two rules bring a change in the positions of particles: rule 'B' is responsible for particle propagation and rule 'D' for changing directions upon a collision of two particles. This set of rules is applied alternatively to even and odd 2-by-2 blocks.

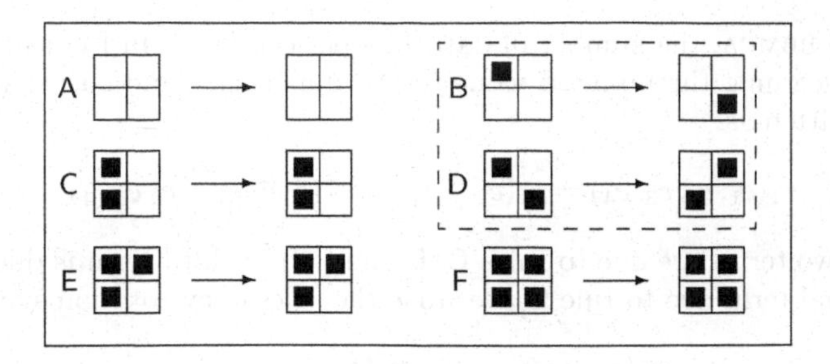

Fig. 2. BBMCA rules. All rules are rotation symmetric. Only rules B and D inside the dashed box (and their rotations) induce a change in the state of the CA

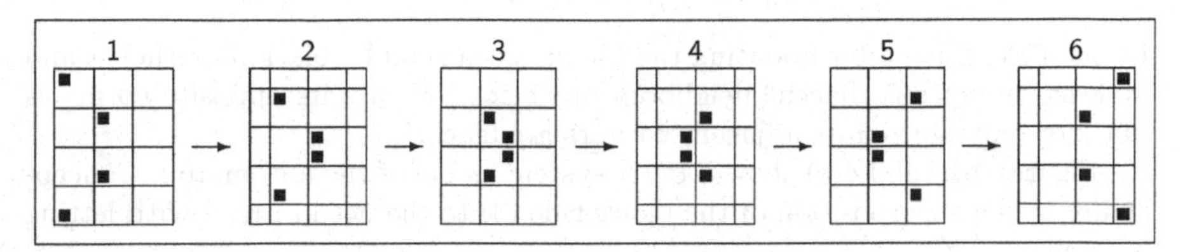

Fig. 3. BBMCA evolution of a collision of two balls

Figure 3 shows a BBMCA collision of two minimum-size balls. Until the balls get close to each other, the particles all propagate independently (step 1). When two leading-edge particles find themselves in the same block, the collision begins (step 2). The particles are stuck together for one step – there is no change in their positions (step 3). Meanwhile, the trailing-edge particles catch up and collide head-on with their corresponding leading-edge particles, causing the ball to change its orientation (step 4). The new leading-edge particles come out at right-angles to the original direction (step 5). They are followed by the new trailing-edge particles in such a fashion that each two-particle ball has been displaced from its original path (step 6).

3 Petri System of the Billiard Ball Model

Consider an m-by-n cell asynchronous MCA with l-bit cells. Its block update rule may be symbolically written as

$$\begin{array}{ll} c_1' = r_1(c1, c2, c3, c4) & c_2' = r_2(c1, c2, c3, c4) \\ c_3' = r_3(c1, c2, c3, c4) & c_4' = r_4(c1, c2, c3, c4) \end{array} \tag{1}$$

where cells are labeled according to Fig. 4(a) and

$$r_1, \ldots, r_4 : \{\bot, \top\}^l \times \{\bot, \top\}^l \times \{\bot, \top\}^l \times \{\bot, \top\}^l \to \{\bot, \top\}^l \tag{2}$$

give the new state of each cell as a function of the current state of the block.

For the BBMCA, the number of data bits per cell $l = 1$. In Boolean logic, the rule r_1 concerning the updated value of the upper left-hand corner (cf. Fig. 2) may be written as

$$r_1(c_1, c_2, c_3, c_4) = c_1(c_2 + c_3) + \overline{c_1}(\overline{c_2}\,\overline{c_3}\,c_4 + c_2\,c_3\,\overline{c_4}) \ . \tag{3a}$$

The first two terms are due to rules C, E and F. The third term is due to rule B and the final term due to rule D. Because the rules obey rotation symmetrics,

$$\begin{aligned} r_2(c_1, c_2, c_3, c_4) &= r_1(c_2, c_4, c_1, c_3) \ , \\ r_3(c_1, c_2, c_3, c_4) &= r_1(c_3, c_1, c_4, c_2) \ , \\ r_4(c_1, c_2, c_3, c_4) &= r_1(c_4, c_3, c_2, c_1) \ . \end{aligned} \tag{3b}$$

In plain English: "after updating the block, a cell shall be black, iff either it and at least one of its adjacent neighbors are black, or only its opposite corner is black or only both of its adjacent corners are black."

Figures 4(b) and 4(c) show a Petri system model of the automaton. Conceptually, it is a superposition of the block rule (3) to the net in Fig. 1 with letting $n_c = m, n_r = n$.

3.1 Places

Places $\mathsf{Age}_{\mathsf{even}}$ and $\mathsf{Age}_{\mathsf{odd}}$ model the cells whose computational ages are even and odd, respectively. Their markings consist of pairs $\langle i, j \rangle$ representing cell coordinates. The upper left hand cell has coordinates $(1, 1)$; coordinates of the lower right hand corner are (m, n).

Place Data models values of data bits of each cell. BBMCA uses two-color cells. In the Petri system, the markings of place Data are triplets $\langle i, j, b \rangle$, where (i, j) denotes the coordinates of the cell and b is a Boolean value denoting cell data. 'False' ($\perp$) models a white and 'true' ($\top$) models a black cell.

In any "legal" marking of the net, either $\mathsf{Age}_{\mathsf{even}}$ or $\mathsf{Age}_{\mathsf{odd}}$ shall hold a token $\langle i, j \rangle$ for all $i \in \mathcal{X}$ and all $j \in \mathcal{Y}$. Similarly, Data shall hold either a token $\langle i, j, \perp \rangle$ or a token $\langle i, j, \top \rangle$ for all i, j.

3.2 Transitions

Transitions $\mathsf{Update}_{\mathsf{odd}}$ and $\mathsf{Update}_{\mathsf{even}}$ model updating odd and even blocks, respectively. Both of them have six input variables $i, j, c_1, \ldots, c_4$, whose values define the *mode* of the transition. Transition $\mathsf{Update}_{\mathsf{odd}}$ takes an input arc $B_{ij} = \langle i, j \rangle + \langle \uplus i, j \rangle + \langle i, \uplus j \rangle + \langle \uplus i, \uplus j \rangle$ from place $\mathsf{Age}_{\mathsf{odd}}$ and has a gate expression isOdd(i) $\wedge$ isOdd(j) associated with it. Thus, it may be enabled only in modes in which all cells of block i, j have an odd age (due to the input arc B_{ij}) and whose coordinates are odd themselves (due to the gate expression). When $\mathsf{Update}_{\mathsf{odd}}$ fires in mode $i, j, \ldots$, the output arc to place $\mathsf{Age}_{\mathsf{even}}$ deposits the four tokens representing the updated age variables of all cells that belong to this block. For any mode satisfying the above-mentioned conditions, the input

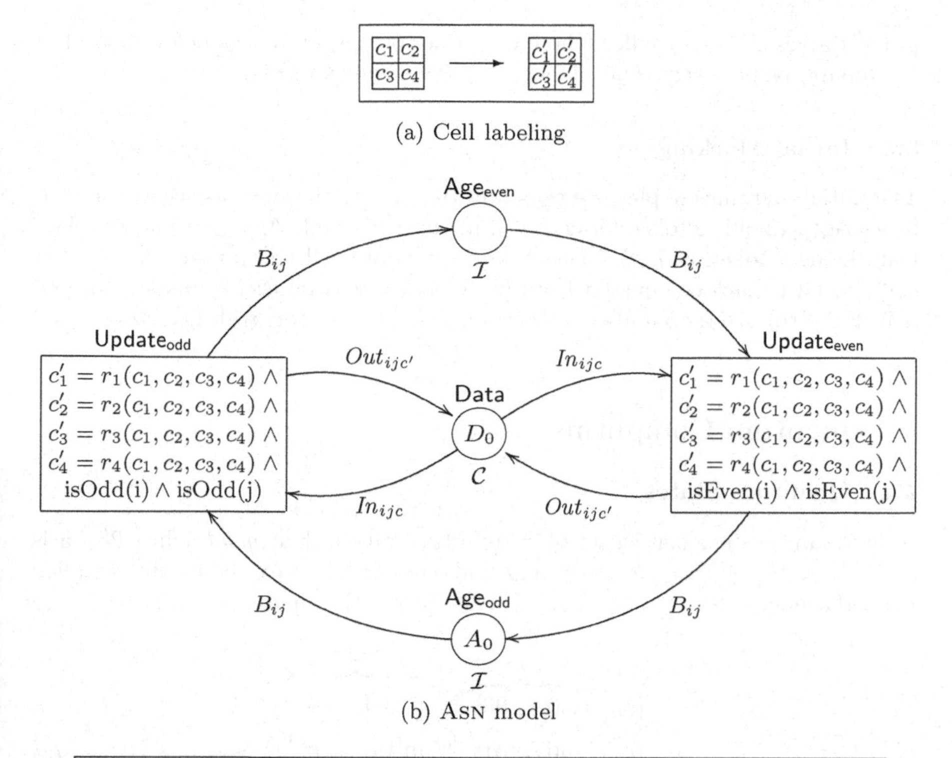

(b) ASN model

$\mathcal{X} = \{1, \ldots, m\}$ $\quad \mathcal{Y} = \{1, \ldots, n\}$ $\quad \mathbb{B} = \{\bot, \top\}$ $\mathcal{I} = \mathcal{X} \times \mathcal{Y}$ $\quad \mathcal{C} = \mathcal{X} \times \mathcal{Y} \times \mathbb{B}$	
r_1, r_2, r_3, r_4	see text
$\uplus$	modular successor operator
B_{ij}	$\langle i, j \rangle + \langle \uplus i, j \rangle + \langle i, \uplus j \rangle + \langle \uplus i, \uplus j \rangle$
In_{ijc}	$\langle i, j, c_1 \rangle + \langle \uplus i, j, c_2 \rangle + \langle i, \uplus j, c_3 \rangle + \langle \uplus i, \uplus j, c_4 \rangle$
$Out_{ijc'}$	$\langle i, j, c'_1 \rangle + \langle \uplus i, j, c'_2 \rangle + \langle i, \uplus j, c'_3 \rangle + \langle \uplus i, \uplus j, c'_4 \rangle$
D_0	$\left(\sum_{i=1}^{m} \sum_{j=1}^{n} \langle i, j, \bot \rangle\right) - \langle 1, 1, \bot \rangle + \langle 1, 1, \top \rangle - \langle 1, 8, \bot \rangle + \langle 1, 8, \top \rangle$ $-\langle 3, 3, \bot \rangle + \langle 3, 3, \top \rangle - \langle 3, 6, \bot \rangle + \langle 3, 6, \top \rangle$
A_0	$\sum_{i=1}^{m} \sum_{j=1}^{n} \langle i, j \rangle$

(c) Declarations

Fig. 4. Petri system of an asynchronous BBMCA

arc $In_{ijc} = \langle i, j, c_1 \rangle + \langle \uplus i, j, c_2 \rangle + \langle i, \uplus j, c_3 \rangle + \langle \uplus i, \uplus j, c_4 \rangle$ from place Data, the output arc $Out_{ijc'} = \langle i, j, c'_1 \rangle + \langle \uplus i, j, c'_2 \rangle + \langle i, \uplus j, c'_3 \rangle + \langle \uplus i, \uplus j, c'_4 \rangle$ to place Data and the rest of the gate expressions are always satisfiable; thus, they place no further restrictions on values i, j of modes in which the transition may fire. The enabling conditions for transition $\mathsf{Update}_{\mathsf{even}}$ are the same except that the block must have even coordinates and an even age. Input variables $c_1, \ldots, c_4$ and output variables $c'_1, \ldots, c'_4$ reflect the data values of the cell blocks before and after an update, respectively. Their labeling is the same as in Fig. 4(a).

3.3 Initial Marking

The initial markings of places $\mathsf{Age_{odd}}$ and $\mathsf{Age_{even}}$ are the same as those in Fig. 1. Place $\mathsf{Age_{odd}}$ holds a token for each cell in the grid, while $\mathsf{Age_{even}}$ is empty. Place Data holds a token $\langle i, j, \bot \rangle$ for each initially white cell and a token $\langle i, j, \top \rangle$ for each initially black cell in step 1 of Fig. 3. The actual values of constants m and n in Fig. 4(c) are the number of columns and rows of the grid; i.e., $m = 6$ and $n = 8$.

4 Quantum Computing

4.1 Quantum States

A quantum system consisting of k quantum bits (a.k.a. *qubits*) has 2^k basis states $|\psi_0\rangle, \ldots, |\psi_{2^k-1}\rangle$. Any orthonormal basis will do; we only use the so-called *computational basis* $\mathcal{B}_{2^k} = \{|\psi_0\rangle, |\psi_1\rangle, \ldots, |\psi_{2^k-1}\rangle\}$, where

$$\begin{aligned}
|\psi_0\rangle &= |\overbrace{0 \cdots 00}^{k}\rangle \stackrel{\text{def}}{=} \overbrace{[1\,0\,0 \cdots 0]}^{2^k}{}^{\mathsf{T}} , \\
|\psi_1\rangle &= |0 \cdots 01\rangle \stackrel{\text{def}}{=} [0\,1\,0 \cdots 0]^{\mathsf{T}} , \\
&\vdots \\
|\psi_{2^k-1}\rangle &= |1 \cdots 11\rangle \stackrel{\text{def}}{=} [0 \cdots 0\,0\,1]^{\mathsf{T}} .
\end{aligned} \tag{4}$$

Generally, the quantum state $|\psi\rangle$ of a k-qubit system is a unit length linear combination (*superposition*)

$$|\psi\rangle \stackrel{\text{def}}{=} \sum_{i=0}^{2^k-1} \omega_i |\psi_i\rangle = [\omega_0\,\omega_1 \cdots \omega_{2^k-1}]^{\mathsf{T}} , \qquad \sum_{i=0}^{2^k-1} |\omega_i|^2 = 1 , \tag{5}$$

of the basis states $|\psi_i\rangle \in \mathcal{B}_{2^k}$. Thus, $\mathcal{B}_{2^k}$ is a basis of the 2^k-dimensional *Hilbert space* $\mathcal{H}_{2^k}$. For an in-depth introduction, we refer the reader to e.g. textbook [8].

The coefficients $\omega_i \in \mathbb{C}$ are called *probability amplitudes.* Since representing the state of a k qubit system on a classical computer requires storage for 2^k complex numbers, a straight-forward simulation of large quantum systems is all but intractable (cf. Sect. 4.3).

4.2 Operators

Qubits in a quantum memory register are manipulated by *quantum gates.* An l-qubit quantum gate performs a linear, reversible, quantum logical operation simultaneously on l qubits. Mathematically, the operation performed by a quantum gate is usually represented as such an *operator matrix* M that when applied to a state $|\psi\rangle$, the resulting new state is $|\phi\rangle = \mathsf{M}\,|\psi\rangle$ [8]. In semi-classical computing, the only allowed gate operations are permutations. Because both $|\psi\rangle$ and $|\phi\rangle$ have 2^l components, the dimension of M has to be $2^l \times 2^l$. For M to be a permutation matrix, each of its rows and columns must have a single "1" in them, while the rest of the matrix entries have to be "0".

When a permutation matrix $\mathsf{M_P}$ is applied to a basis state $|\psi_i\rangle$, the resulting state $|\phi_i\rangle$ is another basis state. Applied to a superposition $|\psi\rangle = \omega_0\,|\psi_0\rangle + \cdots + \omega_{2^l-1}\,|\psi_{2^l-1}\rangle$, the operation yields

$$|\phi\rangle = \mathsf{M_P}\,|\psi\rangle = \mathsf{M_P}\left(\sum_{i=0}^{2^l-1} \omega_i\,|\psi_i\rangle\right) = \sum_{i=0}^{2^l-1} \omega_i\big(\mathsf{M_P}\,|\psi_i\rangle\big) = \sum_{i=0}^{2^l-1} \omega_i\,|\phi_i\rangle = |\phi\rangle \ . \quad (6)$$

In other words, $\mathsf{M_P} = \big[m_p(i,j)\big]$ may also be viewed as a *state transition matrix*:

$$m_p(i,j) = 1 \iff \mathsf{M_P}\,|\psi_i\rangle = |\psi_j\rangle \ . \quad (7)$$

An isolated quantum gate is of little practical use. In a quantum computer, there are a number of gates, each operating on only a small fraction of the k-qubit quantum memory. Nevertheless, the operator matrix of an l-qubit gate ($l < k$) has to consider all 2^k basis states. It is written as if the identity operation were performed on all unconnected qubits.

4.3 Control

To be able to implement a given algorithm with quantum gates and memory registers, we need a mechanism for ensuring a correct temporal ordering of gate applications; i.e., synchronization. For serial quantum processors, Feynman found a solution in [4]. Margolus extended the idea to his quantum cellular automata in [13, 14]. The approach works as follows: the quantum memory representing data registers is augmented with so-called *control qubits*[2] whose state determines which gate(s) are to operate next. Each gate is coupled to some of these control qubits.

For MQCA, we utilize the one-bit age variables (cf. Sect. 2.3). Thus, each cell consists of qubits representing the state of the cell plus one additional qubit representing the age. MQCA, whose grid size is m-by-n (both m and n must be even numbers) have $mn/4$ odd and $mn/4$ even blocks. Each block is operated by a quantum gate that works on $2 \cdot 2$ control qubits plus the $2 \cdot 2 \cdot l$ data qubits, assuming l-qubit cells. The state space of the MQCA has $2^{(l+1)mn}$ basis states.

When the control qubits are taken into account, the operator matrix $\mathsf{F}_{xy} = \big[f_{xy}(i,j)\big]$ that operates the block whose upper left hand coordinates are (x, y) is no longer a permutation matrix, because it has a non-zero entry only in columns that correspond to states in which the gate is active. We say that a gate is active

[2] Feynman called them *cursor qubits* and Margolus *guard qubits*.

if and only if the respective control qubits are in a state that allows the gate to operate.

The sum of all gate operators

$$F = \sum_{x=1}^{m} \sum_{\substack{y=1 \\ x,y \text{ Odd} \vee \\ x,y \text{ Even}}}^{n} F_{xy} \tag{8}$$

gives all possible state transitions of the MQCA. Generally, it is not a permutation matrix, because in a given state, any number of gates (including zero) might be active; and, consequently, the successor state is not necessarily uniquely determined by the current state of the MQCA. Thus, the interpretation of an entry $f_{xy}(i, j) = 1$ is: the gate located at (x, y) is active in state $|\psi_i\rangle$. If the corresponding block (x, y) is updated, the resulting state will be $|\psi_j\rangle$. Note that under any circumstances, $F|\psi_i\rangle \neq |\psi_i\rangle$ because a gate operation always produces a different state of the control qubits. This is also the reason, why identity operations have to be included in the system models. If none of the gates are active in $|\psi_i\rangle$, $F|\psi_i\rangle = \mathbf{0}$.

4.4 Orbits

Consider a quantum system Q with a computational base $\mathcal{B}$ and a transition matrix $F = [f(i, j)]$. Further, let $|\psi\rangle \in \mathcal{B}$ be the initial state of Q. The set of states

$$\mathcal{O}_{|\psi\rangle} \stackrel{\text{def}}{=} \{|\phi\rangle : |\psi\rangle \xrightarrow{*} |\phi\rangle\} \subset \mathcal{B} .$$

is called the *orbit* of $|\psi\rangle$. The symbol $\xrightarrow{*}$ denotes the closure of reachable states via zero or more gate applications. A single orbit of Q only consists of a tiny subset of $\mathcal{B}$. Furthermore, the orbits are orthogonal to each other. We denote with the symbol $F_{|\psi\rangle}$ the rows and columns of F that correspond to the states of $\mathcal{O}_{|\psi\rangle}$. Notice, that $F_{|\psi\rangle}$ is the transition matrix of the computation performed by Q with an initial state $|\psi\rangle$ in the subspace of $\mathcal{H}$ spanned by $\mathcal{O}_{|\psi\rangle}$.

A Petri net model N has been coined by us a *kernel net* of Q, iff for each $k \in \{0, \ldots, |\mathcal{B}| - 1\}$ there exists a distinct marking M_k such that M_k taken as the initial marking, the reachability graph of N is isomorphic to $F_{|\psi_k\rangle}$. In other words, there exists a set M of $|\mathcal{B}|$ distinct markings of N and a bijective mapping $\mathcal{L}$ from M to $\mathcal{B}$. When $\mathcal{L}$ is applied to each marking of a matrix representation of the reachability graph $\mathrm{RG}(N, M_k)$, the resulting matrix is isomorphic to $F_{\mathcal{L}(M_k)}$ up to state ordering.

4.5 Time Evolution

An isolated quantum system evolves according to the *Schrödinger differential equation*. For quantum computers with control qubits, it has the form

$$\mathrm{i}\hbar \frac{\mathrm{d}|\psi\rangle}{\mathrm{d}t} = H|\psi\rangle \ , \tag{9}$$

where the *Hamiltonian* H is a constant *Hermitian* matrix ($H = H^{\dagger}$; i.e., the matrix equals its conjugate transpose). We use a time scale in which $\hbar = 1$, so from now on, the Planck constant will be dropped.

The Hamiltonian describes all *forward* and *backward* state transitions of a quantum system. According to quantum theory, the evolution from state to state is a reversible process: every computation that a quantum computer does, it also undoes – unconditionally. Therefore H needs to be Hermitian. For our quantum computer models,

$$H = F + F^{\dagger} = \sum_{i} \left(F_i + F_i^{\dagger} \right) , \tag{10a}$$

where i is taken over all quantum gates in the system. Since all entries of F are real, $F^{\dagger} = F^{\mathsf{T}}$ and H is real and symmetric ($H = H^{\mathsf{T}}$). But we do not need the general solution of (9). We only want to solve the initial value problem with $|\psi(0)\rangle$. In the previous section we established that the computation actually takes place in the subspace spanned by the orbit of $|\psi(0)\rangle$. In that space, $F_{|\psi(0)\rangle}$ gives the forward state transitions. Consequently, we let

$$H = F_{|\psi(0)\rangle} + \left(F_{|\psi(0)\rangle}\right)^{\mathsf{T}} . \tag{10b}$$

Symbolically, the solution of (9) is

$$|\psi(t)\rangle = \mathrm{e}^{-\mathrm{i}Ht} \, |\psi(0)\rangle . \tag{11}$$

We recall that $|\psi(t)\rangle = \sum_i \omega_i(t) \, |\psi_i\rangle$. If the state of the system were measured at time t, it would be found to be $|\psi_i\rangle$ with probability $p_i(t) = |\omega_i(t)|^2$. Hence the term probability amplitude. Except for isolated points in time, all $p_i(t) < 1$; i.e., the system is usually in a superposition of its basis states. Among other things, this means that the *logical* ordering and the actual *time ordering* of events triggered by gate applications are *not* related [21].

5 Simulation Model of a Quantum Billiard Ball Model Cellular Automaton

5.1 Quantum Billiard Ball Model Cellular Automaton

Margolus gives a serial quantum BBMCA in [13] and a parallel asynchronous MQCA in [14]. In a quantum BBMCA, each cell is represented by a pair of qubits: one for cell data and one for the age variable. A white cell is represented by $|0\rangle$ and a black cell by $|1\rangle$. Similarly, even age is represented by $|0\rangle$ and odd age by $|1\rangle$. Each 2×2 block is governed by a quantum gate that is coupled to four data qubits and four control qubits. Even and odd gates are active if and only if all connected control qubits are in states $|0\rangle$ and $|1\rangle$, respectively. As a result of the gate application, the resulting values of control qubits are the complements of their original values; i.e., $|1\rangle$ and $|0\rangle$, respectively. Data qubits are manipulated according to the rules (3) with $|0\rangle$ denoting boolean $\perp$ and $|1\rangle$ denoting boolean $\top$.

5.2 Kernel Net

We model the collision of two billiard balls on a six columns by eight rows grid with periodic boundary conditions (i.e., a torus) – a situation we introduced in Sect. 2.4. We build our model by refining the classical construction given in Fig. 4. Since our grid has six columns and eight rows, we let $m = 6$ and $n = 8$ in the declarations of Fig. 4(c). We let place Data model data qubits and places $\mathsf{Age}_{\mathsf{even}}$ and $\mathsf{Age}_{\mathsf{odd}}$ model control qubits as outlined in Sects. 4.3 and 5.1.

For a given initial marking M_0, the matrix form of the reachability graph of the net in Fig. 4 is isomorphic to a submatrix $F_{|\psi_i\rangle}$, where M_0 models $|\psi_i\rangle$. Thus, it serves as a kernel net of the quantum BBMCA model without modifications.

As already noted, the major difference between the reachability graph and the full transition matrix F is that the reachability graph contains no spurious entries. As its name implies, it only describes transitions among precisely those states that are reachable from the initial marking, whereas the F matrix includes entries for all potential states of the system. In this sense, the generation of the reachability graph can be said to perform a *projection* from the potential to the actual state space of the system. This technique was first used by the third author in [18].

In practice, this projection is extremely useful, because unreachable states are not needed for anything during a simulation, and the savings in storage space and processing time are substantial. The dimension of the operator matrix F is $2^{(l+1)mn} \times 2^{(l+1)mn}$, which is huge even for modest m, n and l. For a given computation that starts at state $|\psi_{(0)}\rangle$, we may classify the non-zero entries of F into three disjoint sub-classes: the entries corresponding to the orbit $\mathcal{O}_{|\psi_{(0)}\rangle}$, the entries corresponding to the orbits of other potential initial states, and extraneous entries that correspond to illegal states. The last sub-class is actually the largest: the vast majority of potential states of control qubits are *ill-formed*: since each block update changes exactly four control qubits from $|0\rangle$ to $|1\rangle$ or vice versa, any state in which the number of non-zero control qubits is not divisible by four is not in the orbit any well-formed initial state.

5.3 Initial Marking

Margolus pointed out in [14] that the synchronization scheme for an ACA does not work satisfactorily in a quantum setting: with it, a MQCA computes forward only at a diffusive rate. (His argumentation is decisive but beyond the scope of this paper.) Fortunately, he found a partial solution. If instead of allowing full 2-D parallelism, parallel operation is constrained to a 1-D "wave", the computer operates satisfactorily. The computer itself, and thus our Petri net model of it need no changes, but, unfortunately, setting the initial state of both control and data qubits becomes a rather complicated task.

Recall, that the original local synchronization scheme allows different parts of the grid to have different computational ages, as long as the age difference of adjacent cells never exceeds one. With the new scheme, cells in different columns of the grid are *forced* to have different ages. Consider Fig. 5(a). It shows the

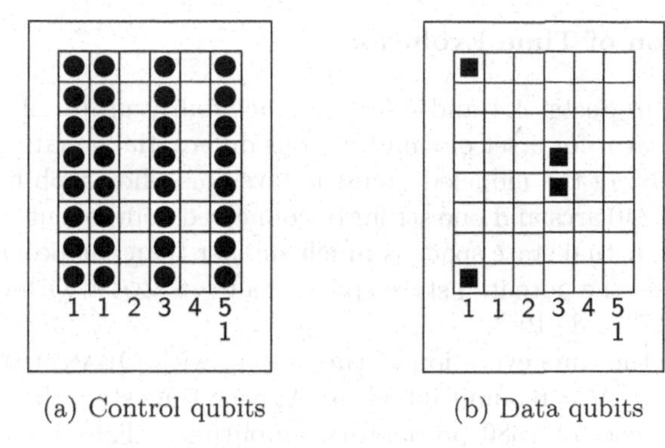

(a) Control qubits (b) Data qubits

Fig. 5. Initial state of a quantum BBMCA

initial states of control qubits according to the new scheme. Initially, only blocks in column one[3] are updatable. As soon as at least two adjacent blocks have been updated, blocks in column two become active and so on. Finally, blocks in column six go forward, whereafter the updating process begins anew from column one. All in all, updates occur in a wobbly wave from left to right.

The initial states of data qubits in Fig. 5(b) have to reflect this. Recall, that upon updating, all cells in a block must have exactly the same computational age. Thus, except for a three-column wide "flat" region (initially columns six, one and two), the computational ages of columns have to form a staircase-like pattern. The numbers under the columns of Fig. 5 are these initial ages of cells belonging to the column. The last column needs a special treatment. When a block in column five becomes active for the first time, its age is five (because originally it was four and it has been updated once by blocks in column four). Thus, the initial age of column six has to be five, as well. On the other hand, when a block in column six is active for the first time, its age has to be two, because that is the age of column one that has the right-hand side cells of the block! For this to work, the states of all cells in column six have to be the same at ages one and five. We pick the proper cell contents from appropriate columns of Fig. 3. In our Petri net model, the initial markings are

$$M_0(\mathsf{Age}_{\mathsf{even}}) = \sum_{j=1}^{8} \langle 3, j\rangle + \langle 5, j\rangle \ ,$$

$$M_0(\mathsf{Age}_{\mathsf{odd}}) = \Big(\sum_{i=1}^{6}\sum_{j=1}^{8} \langle i, j\rangle\Big) - M_0(\mathsf{Age}_{\mathsf{even}}) \ ,$$

$$M_0(\mathsf{Data}) = \Big(\sum_{i=1}^{6}\sum_{j=1}^{8} \langle i, j, \bot\rangle\Big) - \langle 1,1,\bot\rangle + \langle 1,1,\top\rangle - \langle 1,8,\bot\rangle + \langle 1,8,\top\rangle$$
$$- \langle 4,4,\bot\rangle + \langle 4,4,\top\rangle - \langle 4,5,\bot\rangle + \langle 4,5,\top\rangle \ .$$

[3] We say that a block is in the row and in the column in which its upper left hand corner cell lies.

5.4 Simulation of Time Evolution

As pointed out in Sects. 4.4 and 4.5, the reachability graph of the kernel net serves as a generator for a set of simultaneous differential equations that govern the time evolution of the modeled quantum system. The graph has 1680 reachable markings, 3840 arcs and one strongly connected component. In Sect. 4.4 we claimed that the actual state space is much smaller than the potential one. This indeed is the case: the potential state space of a 6×8 MQCA with one data qubit cells is $2^{2 \cdot 6 \cdot 8} = 2^{96} \approx 8 \cdot 10^{28}$.

We simulate the time evolution of the system with QUANTUM MARIA up to 1000 time units at $\Delta t = 0.5$ unit intervals. At each time step, the raw simulation output data consists of 1680 probability amplitudes. That is not the kind of data we would like. Instead, we are interested in the compound probability of all of the states we consider final; i.e., states that correspond to squares (6,1) and (6,8) both being black. With the help of the so-called *state property feature*, the reachability analyzer allows markings satisfying a Boolean condition to be grouped together. QUANTUM MARIA automatically computes the compound probabilities of each group.

For this simulation, only the group consisting of final states is defined. It has a total of 180 markings, approximately 10.7 % of the total of 1680. Figure 6 shows the simulation results. All three curves give the compound probability of finding the system in one of its final states if it were measured at time t and it had not been measured prior to time t.

Figure 6(a) shows that for $t \leq 3$ units, the probability is practically zero. Thereafter, it begins to rise rapidly towards initial local maxima, which are substantially higher than the eventual average value. This behavior is offset by a decline to a minimum at $t \approx 65$ visible in Fig. 6(b). Figure 6(c) reveals that eventually the probability will fluctuate around its expectation in a chaotic, non-periodic manner.

6 Conclusions

Through our design procedure, we have demonstrated in a systematic way the applicability of Petri nets to modeling and analysis of parallel quantum computers. Especially, constructing the Hamiltonian of a quantum system directly from the reachability graph of a corresponding kernel net leads to a huge reduction in the state space and makes a numerical time simulation of non-trivial quantum computers possible. Also, with our approach, quantum computers can be described in terms familiar to a far larger audience than those familiar with quantum mechanics.

In a forthcoming paper, we will formalize the notion of kernel nets and broaden our focus to encompass also fully quantum logical systems instead of the semi-classical ones discussed here. Towards this need, a new formal concept more general and powerful than the concept kernel net will be introduced: *Quantum nets*.

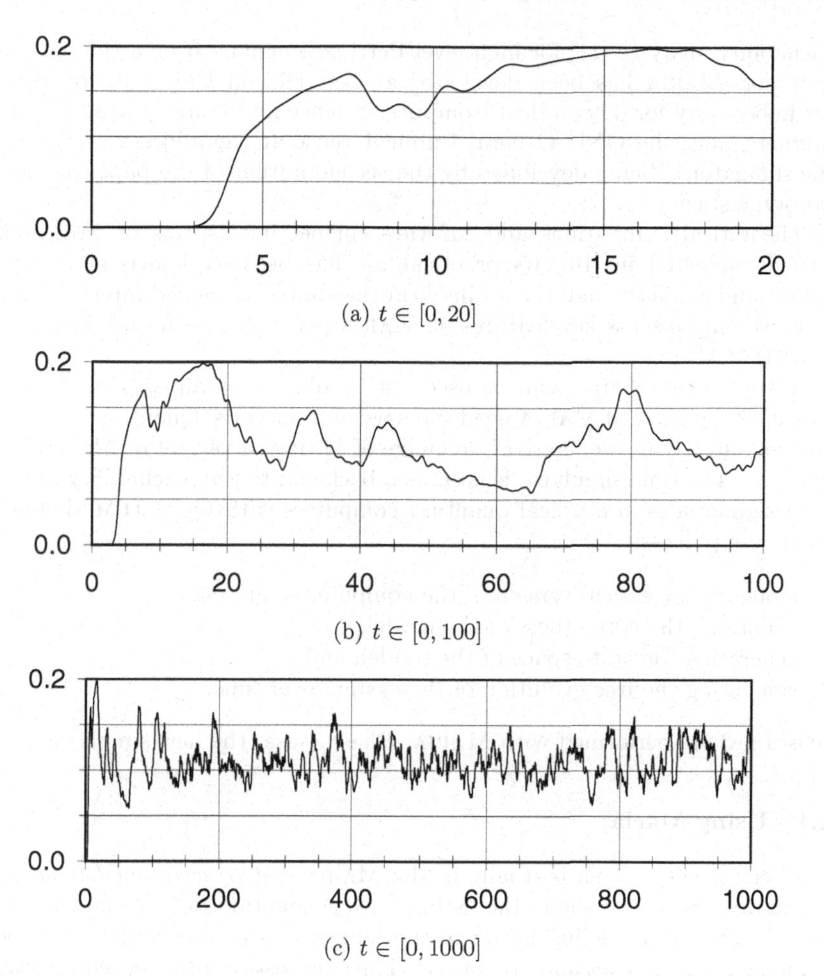

Fig. 6. Probability of finding the quantum BDMCA in a final state vs. time in three different time spans

We also continue to develop software tools and associated algorithms to aid in the analysis of quantum computers; specifically, tools for simulating multiple measurements.

In addition, we want to be able to utilize data symmetries present in quantum systems and construct automatically symbolic reachability graphs for quantum nets that only have one symbolic marking representing an entire equivalence class of ordinary markings [9].

A Software

This Appendix contains an overview of QUANTUM MARIA, a software package for modeling and simulating quantum computers. It consists of the MARIA reachability analyzer [11] for high-level Petri nets enhanced by a time-evolution simulator. MARIA has been developed at the Helsinki University of Technology Laboratory for Theoretical Computer Science and is freely available in the Internet under the GNU General Public License at the address given in [15]. The simulator is being developed by the second author of the paper and is in a prototype stage.

Theoretical foundations and software engineering aspects of MARIA have been documented in [10]. Its practical use has been explained in an on-line manual and auxiliary material available at the above-mentioned Internet address. Here, we only discuss key features of MARIA pertaining to its use as a part of QUANTUM MARIA.

MARIA is modular: it can be used with problem-domain-specific front- and back-ends. QUANTUM MARIA needs no special front-ends. Quantum systems are modeled directly in the "native" language of MARIA, a variant of Algebraic System Nets. The time simulator is used as a back-end to the reachability analyzer.

Designing a semi-classical quantum computers with QUANTUM MARIA is a four-phase process:

1. modeling a classical version of the computer as an ASN,
2. validating the correctness of the model,
3. generating the state space of the model, and
4. simulating the free evolution of the system over time.

Phases 1–3 are performed with MARIA. Phase 4 uses the new simulator.

A.1 Using Maria

The Petri net is written textually in the MARIA *Net Description Language.* As an example, Sect. A.3 shows the listing for our quantum BBMCA model given in Sect. 5. The full modeling power of the language is at the modeler's disposal, including all data types and operations. The only restriction imposed is a one-to-one correspondence between states of the modeled quantum system and markings of the Petri net.

MARIA has a rich set of features to aid model checking. Safety and liveness properties can be verified with *Linear Temporal Logic* (LTL) formulae. Quite a few useful safety properties can be defined in the Net Description Language as well: places may have capacity constraints, sets of places can be declared complementary, and markings that violate a Boolean property can be rejected. Forbidden markings can also be found with fact transitions. Experience has shown the usefulness of these features to catch modeling errors as early as possible; for instance, an ill-formed initial marking often leads to a constraint violation – the system will catch this immediately. Examples of use of these features can be found in the listing in Sect. A.3.

The reachability graph of the net is exported in the form of a *labeled state transition system* in machine readable form. In this language, reachable markings are represented as state numbers. Therefore, a mechanism is needed that maps these states back to markings of the Petri net. MARIA includes a state property feature that can be used for this. Sets of markings in which a given Boolean condition holds, are reported. Any number of sets can be defined, and they need not be disjoint.

A.2 Time-Evolution Simulator

The QUANTUM MARIA simulator has to numerically solve the linear Schrödinger differential equation (9), where the Hamiltonian H has the form (10b) with an initial condition $|\psi(0)\rangle$. Currently, the prototype simulator has implementations of two variants of our own experimental algorithm that is based on the so-called *randomization method* [7] and two variants of the Runge-Kutta algorithm used mainly for validation of the new algorithms, which we discuss next[4].

The randomization method is based on a clever computing of the matrix exponential in (11). By definition, e^{Mt} has the Maclaurin series expansion

$$e^{Mt} = \sum_{k=0}^{\infty} \frac{(Mt)^k}{k!} , \tag{A.1}$$

which converges for all square M and all $t \geq 0$. Let $N = M + I$. Now,

$$e^{Mt} = e^{Nt}e^{-t} = \sum_{k=0}^{\infty} \frac{(Nt)^k e^{-t}}{k!} = \sum_{k=0}^{\infty} N^k \frac{e^{-t}t^k}{k!} = \sum_{k=0}^{\infty} N^k \beta(k,t) , \tag{A.2}$$

where

$$\beta(k,t) = P[X = k], \quad X \sim \text{Poisson}(t) . \tag{A.3}$$

The Poisson distribution is useful here because it thins very rapidly before some $L(t)$ and after some $R(t)$. We use these as "cutoff" points for an approximation of the infinite series.

Recall, that $H = F + F^\dagger$. Since F is real, H is real, and $-iHt$ is pure imaginary. By a result given by Peres in [21],

$$H = \lambda F + (\lambda F)^\dagger \qquad \lambda \in \mathbb{C} : |\lambda| = 1 \tag{A.4}$$

give equivalent time behaviors for all λ. We let $\lambda = i$. Thus

$$-iHt = -i\big(iF + (iF)^\dagger\big)t = \big(F - F^\mathsf{T}\big)t . \tag{A.5}$$

All this put together, we compute with $A = F - F^\mathsf{T} + I$

$$|\psi(t)\rangle \approx \sum_{k=L(t)}^{R(t)} \alpha_k \beta_{k,t} , \tag{A.6a}$$

[4] We give the algorithm in the form that is applicable to the kind of semi-classical computing we have performed. For general quantum gates, a slightly modified version is used.

where

$$\begin{aligned} \alpha_0 &= |\psi(0)\rangle \ , \\ \alpha_m &= A\alpha_{m-1} \qquad (m \geq 1) \ , \end{aligned} \tag{A.6b}$$

and

$$\beta_{k,t} = P[X = k], \quad X \sim \text{Poisson}(t) \ . \tag{A.6c}$$

The Poisson probabilities are computed with a stable algorithm by Fox and Glynn [5]. $L(t)$ and $R(t)$ are parameters defined by the algorithm. For example, we use $L(.5) = 0, R(.5) = 13$.

With this algorithm, the time evolution simulation described in Sect. 5.4 takes less than two minutes of processing time on a modern workstation. The results have been verified to be correct to six decimal places.

A.3 Example of the Net Description Language

```
int m = 6; int n = 8;
typedef int (1..m) idx_t; typedef int (1..n) idy_t;
typedef struct { idx_t x; idy_t y; bool val } cell_t;
typedef struct { idx_t x; idy_t y; } ind_t;

bool isEven(int x) x % 2 == 0;
cell_t c(idx_t i, idy_t j, bool v) {i,j,v};
ind_t xy(idx_t i, idy_t j) {i,j};
cell_t balls(bool x)
  c(1,1,x) union c(4,4,x) union c(4,5,x) union c(1,8,x);
bool r(bool a, bool b, bool c, bool d)
  ((a&&(b || c )) || (!a&&((!b && !c && d) || (b && c && !d))));

place age_even(0..#ind_t) ind_t: (idy_t y: xy(3,y) union xy(5,y));
place age_odd(0..#ind_t) ind_t: ind_t ij: ij minus place age_even;
place data(#ind_t) cell_t: ((idy_t y: idx_t x: c(x,y,false))
  minus balls(false)) union balls(true);

trans :bbm {idx_t x; idy_t y; bool c1; bool c2; bool c3; bool c4;}
    in { data: {x,y,c1}, {+x,y,c2}, {x,+y,c3}, {+x,+y,c4}; }
    out { data: {x,y,r(c1,c2,c3,c4)}, {+x,y,r(c2,c4,c1,c3)},
      {x,+y,r(c3,c1,c4,c2)}, {+x,+y,r(c4,c3,c2,c1)}; } ;
trans odd:trans bbm
    in { place age_odd: {x,y}, {+x,y}, {x,+y}, {+x,+y}; }
    out { place age_even: {x,y}, {+x,y}, {x,+y}, {+x,+y}; }
    gate !(isEven(x) || isEven(y));
trans even:trans bbm
    in { place age_even: {x,y}, {+x,y}, {x,+y}, {+x,+y}; }
    out { place age_odd: {x,y}, {+x,y}, {x,+y}, {+x,+y}; }
    gate (isEven(x) && isEven(y));
```

```
trans fact_d in { data: {x,y,a}, {x,y,b}; } gate fatal;
reject
cardinality(idx_t x: idy_t y: c(x,y,false) minus place data) != 4;
reject (cardinality place age_even) % 4 != 0;

prop final : (c(6,1,true) union c(6,8,true)) subset place data;
```

References

1. Paul Benioff. "The Computer as a Physical System: A Microscopic Quantum Mechanical Hamiltonian Model of Computers as Represented by Turing Machines". *Journal of Statistical Physics*, **22(5)**:563–591, 1980.
2. David Deutsch. "Quantum Theory, the Church–Turing Principle and the Universal Quantum Computer". *Proceedings of Royal Society London A*, **400**:97–117, 1985.
3. Richard P. Feynman. "Simulating Physics with Computers". *International Journal of Theoretical Physics*, **21(6/7)**:467–488, 1982.
4. Richard P. Feynman. "Quantum Mechanical Computers". *Foundations of Physics*, **16(6)**:507–531, 1986.
5. Bennett L. Fox and Peter W. Glynn. "Computing Poisson Probabilities". *Communications of the ACM*, **31(4)**:440–445, April 1988.
6. Edward Fredkin and Tommaso Toffoli. "Conservative Logic". *International Journal of Theoretical Physics*, **21(3/4)**:219–253, 1982.
7. Donald Gross and Douglas R. Miller. "The Randomization Technique as a Modeling Tool and Solution Procedure for Transient Markov Processes". *Operations Research*, **32(2)**:345–361, March/April 1984.
8. Jozef Gruska. *Quantum Computing.* McGraw–Hill, UK, 1999.
9. Tommi Junttila. "On the Symmetry Reduction Method for Petri Nets and Similar Formalisms". Research Report HUT–TCS–A80, Helsinki University of Technology, Department of Computer Science and Engineering, Laboratory for Theoretical Computer Science, Espoo, Finland, September 2003. Dissertation for the degree of Doctor of Science in Technology.
10. Marko Mäkelä. "A Reachability Analyser for Algebraic System Nets". Research Report HUT–TCS–A69, Helsinki University of Technology, Department of Computer Science and Engineering, Laboratory for Theoretical Computer Science, Espoo, Finland, June 2001.
11. Marko Mäkelä. "Maria: Modular Reachability Analyser for Algebraic System Nets". In *Application and Theory of Petri Nets 2002: Proceedings of the 23rd International Conference (ICATPN 2002)*, number 2360 in Lecture Notes in Computer Science, pages 434–444, Adelaide, Australia, June 2002. Springer-Verlag, Berlin, Germany.
12. Norman Margolus. "Physics-Like Models of Computation". *Physica D*, **10**:81–95, 1984.
13. Norman Margolus. "Quantum Computation". *Annals of the New York Academy of Sciences*, **480**:487–497, 1986.
14. Norman Margolus. "Parallel Quantum Computation". In W.H. Zurek, editor, *Complexity, Entropy, and the Physics of Information*, volume VIII of *SFI Studies in the Sciences of Complexity*, pages 273–287. Addison–Wesley, 1990.
15. MARIA homepage. URL: `http://www.tcs.hut.fi/maria/`.

16. Leo Ojala and Olli-Matti Penttinen. "Simulating Quantum Interference in Feynman's $\sqrt{\text{NOT}}$-computer with Stochastic Petri Nets". In *Proceedings of the European Simulation and Modelling Conference (ESMc2003)*, pages 494–502, Naples, Italy, October 2003.
17. Leo Ojala, Olli-Matti Penttinen, and Heikki Rantanen. "A Novel Application of Stochastic Petri Nets: Simulation of Serial Quantum Computers – Feynman's Swap Computer". In Peter Kemper, editor, *On-site Proceedings of ICALP03 Satellite Workshop on Stochastic Petri Nets and Related Formalisms*, number 780/2003 in Forschungsberichte des Fachbereichs Informatik der Universität Dortmund, pages 103–122, Eindhoven, the Netherlands, June 2003.
18. Elina Parviainen. "Reducing Size of Quantum Gate Matrices Using Pr/T Nets". In *Proceedings of the 2002 IEEE International Conference on Systems, Man and Cybernetics*, volume 2, pages 634–639, Hammamet, Tunisia, October 2002. IEEE (Institute of Electrical and Electronics Engineers, Inc.).
19. Asher Peres. "Measurement of Time by Quantum Clocks". *American Journal of Physics*, **48**(**7**):552–557, July 1980.
20. Asher Peres. "Zeno Paradox in Quantum Theory". *American Journal of Physics*, **48**(**11**):931–932, November 1980.
21. Asher Peres. "Reversible Logic and Quantum Computers". *Physical Review A*, **32**(**6**):3266–3276, December 1985.
22. Wolfgang Reisig. "Petri Nets and Algebraic Specifications". *Theoretical Computer Science*, **80**:1–34, March 1991.
23. Tommaso Toffoli and Norman Margolus. *Cellular Automata Machines: A New Environment for Modeling.* MIT Press, Cambridge, MA, USA, 1987.
24. John von Neumann. *Theory of Self-reproducing Automata.* University of Illinois Press, Urbana, IL, USA, 1966. Completed and edited by Arthur W. Burks.
25. Stephen Wolfram. *Theory and Applications of Cellular Automata: Including Selected Papers 1983–1986.* World Scientific Publishing Co., Inc., River Edge, NJ, USA, 1986.

A Framework for the Modelling and Simulation of Distributed Transaction Processing Systems Using Coloured Petri Nets

M. José Polo Martín, Luis A. Miguel Quintales, and María N. Moreno García

Departament of Informatic and Automatic, University of Salamanca,
Plaza Merced s/n, 37008 Salamanca, Spain
{mjpolo,lamq,mmg}@usal.es

Abstract. In this paper, a flexible framework is presented that allows us to model and to simulate different types of Distributed Transaction Processing Systems which implement algorithms to solve certain specific problems related to security, concurrence, deadlock, etc. Based on the X/Open DTP model and using the Coloured Petri Nets as a tool, the basic model is broken down, maintaining its main components and presenting the new interfaces that appear. This allows us to simulate Distributed Transaction Processing Systems with different characteristics, to incorporate to the model any existing or future algorithm and to study the performance of the system both as a function of its characteristics and of the algorithms used.

1 Introduction

The characteristics of atomicity, consistency, isolation and durability (ACID) that transactions should have, both in centralized and in distributed environments, have given rise to the proliferation of numerous algorithms. These algorithms try to maintain the ACID properties of the transactions by solving the problems that appear with possible failures, concurrent access, deadlocks, etc. Most of these works, however, are centred on the study of a specific problem without treating the study of a general system where all the components are integrated: recovery, concurrence, deadlocks, consolidation, etc. The recovery and concurrence control protocols used in most of these works are based on the writing of a log file and the strict two-phase locking protocol [25]. There are many works that study the mechanisms for detecting deadlock, from the simplest, based on timeout, to other more complex ones based on wait-for-graph or send probes ([6], [7], [9], [8], [16], [19], [18], [20], [23], [24], [28], [29], [30]). The consolidation control of transactions has also been broadly studied ([1], [2], [3], [4], [5], [13], [22], [26], [27]). Many of these works also propose new algorithms, but they all treat the problem they undertake separately. The present work provides a framework that allows the modelling and simulation of the behaviour of these or other algorithms, integrating them in a complete Distributed Transaction Processing System model.

J. Cortadella and W. Reisig (Eds.): ICATPN 2004, LNCS 3099, pp. 351–370, 2004.
© Springer-Verlag Berlin Heidelberg 2004

To obtain a system with these characteristics we start from the X/Open Distributed Transaction Processing model (X/Open DTP), proposed by the International Organization of Open Systems, Open Group ([12], [11]). It is a standard supported by most of the commercial systems in the domain of transactions and database management. It presents an architecture that permits the design of multiple applications that can share resources by coordinating their work through global transactions. This architecture is based on five basic functional components in each node of the distributed system and six interfaces that allow the connection among these components, specifying the communication requirements among them. In this work the system is expanded with new components and interfaces that will allow the change of protocols without affecting the basic functional components of the model.

Coloured Petri Nets have been used to model, visualize and analyse this Distributed System ([14]). This tool makes it possible to build the general model through a set of separate submodels with well-defined interfaces. In [15] a series of reasons for the use of Coloured Petri Nets (CPN) in the design and analysis of systems are enumerated. They offer a simple form of expressing processes that require synchrony and allow the model to be taken to extreme conditions that would be difficult or very expensive to obtain in a real system. The CPN will allow the structure and behaviour of the system to be modelled by means of an abstraction of the same and represented through a composition of modules, which interact and can be considered a system in themselves. This characteristic will allow us to study its state and behaviour separately, bearing in mind their interaction with other modules. In [17] the theoretical justification of a simple mechanism is presented, external transitions, which makes it possible to incorporate distributed algorithms to distributed systems. This mechanism is based on distinguishing which actions of the system activate the execution of an algorithm. This work tries to distinguish what these external actions are and to define them through a set of input/output interfaces. These interfaces will be the actions that activate the algorithms that are implemented in each one of the components of the Distributed Transaction Processing System.

This paper is structured as follows. In section 2 the basic characteristics of the X/Open DTP model are introduced. The general design of the Distributed System through CPNs is in section 3. The versatility of the system built is approached in section 4, where we present, by way of example, how the system makes it possible to change from one commit protocol to another. We thus take as a basis the two-phase commit protocol but the system is prepared to use other different protocols without any additional obligation than to maintain the appropriate interfaces with the rest of the components of the system. Section 5 presents some results obtained for different protocols and section 6, some conclusions on the work developed.

2 Architecture of Transaction Processing Standard

The X/Open DTP model ([12]) defines a transaction as a unit of work characterized by four properties: atomicity, consistency, isolation and durability (ACID).

It also defines a reference model for Distributed Transaction Processing by means of a functional break-down with five main components (figure 1): Application Program (AP), Transaction Manager (TM), Resource Manager (RM), Communication Resource Manager (CRM) and a protocol that provides the communication services (OSI TP) used by the distributed applications and supported by the Communication Resource Managers.

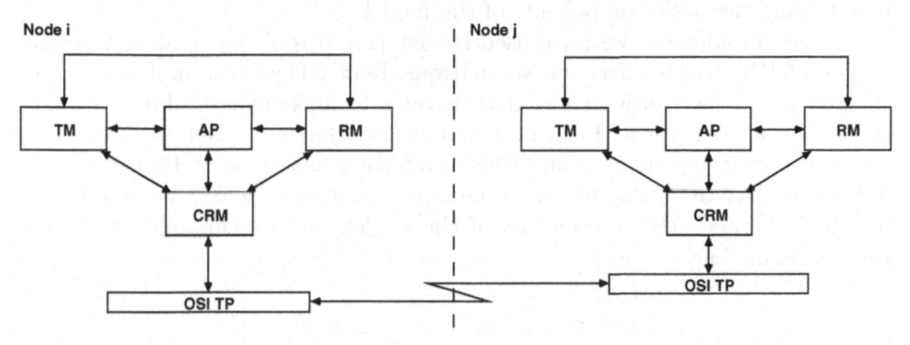

Fig. 1. Functional Components and interfaces in X/Open DTP model

A transaction is a complete unit of work that includes many computational tasks, such as data searches and communications. The AP specifies a sequence of operations that accesses the resources of a database and defines the beginning and end of a distributed transaction. TM controls all the resources that the AP accesses within the limits of a transaction. This component manages distributed transactions; it coordinates the decision to commit them or roll them back and the recovery when failure occurs. TM assigns an identifier to the global transaction and informs each RM of the global identifier on behalf of the local RM where the transaction was launched (branches of the transaction). The access to all the shared resources of the system should be made through the services that the RM of each node provides, but if the access is to a remote resource, the appropriate communication should be made through the CRM. The model defines as an AP, a TM and one or more RMs which cooperate in an environment of distributed transaction processing. These goals can be achieved via the specification of six interfaces among the five basic components: AP-RM, TM-RM, AP-TM, AP-CRM, TM-CRM and CRM-OSI TP interface.

3 Modeling X/Open DTP with CPNs

To develop the X/Open DTP model using Coloured Petri Nets, several nets were created. Some of them correspond to the basic functional components of the model while others simulate typical components of a Databases Management System whose blocks or pages will be the resources shared by the system. The designed model maintains a net for each one of the main components so that they implement the basic operations of each one of them. However, from the TM and RM components, some internal functionalities have been detailed (see figure 2), mainly those that guarantee the ACID properties of the transactions that access the database. These functionalities can be implemented by different protocols or algorithms and they are assigned to new components, so that each new component can be substituted by a net, the one that models the required algorithm. This characteristic will make it possible to change protocol easily, maintaining the basic components of the model.

In the development we have used Renew [21], a tool that makes it possible to build CPNs and to carry out simulations. Renew is written in Java, the nets themselves are Java objects and it is possible to make calls to Java code from the nets and vice versa. Each designed net or Reference Net becomes a net instance during the simulation. Other interesting aspects that Renew presents in its formalism are the synchronous channels, that will allow us to establish the interfaces between the components of the model, and its capability to include the notion of time.

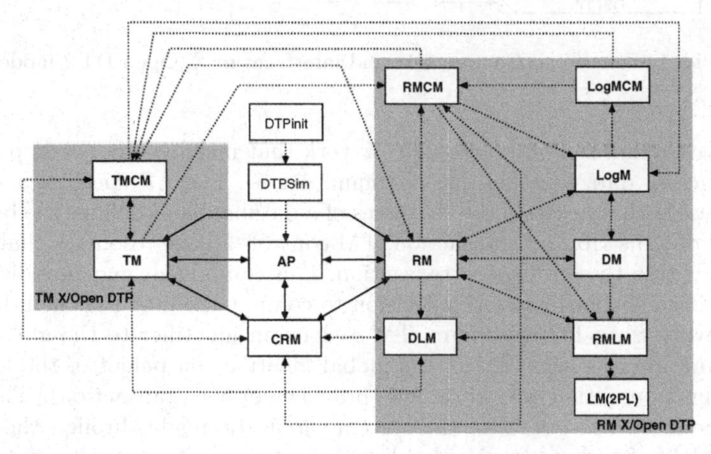

Fig. 2. Expansion of functional components in the designed model. The gray squares contain the new functional components included and show whether they are part of TM or RM in the general X/Open DTP model. The increase in components entails the appearance of new interfaces that are presented with dotted lines

3.1 Components of the Model

Our model includes a component for the Application Program (AP) and another for the Communication Resource Manager (CRM), which coincide with two of the basic components of the X/Open model.

The TM X/Open component breaks down into two: TM for the basic operations and TMCM for the operations of the commit protocol corresponding to the coordinator process.

The RM X/Open component breaks down into eight, maintaining an RM component for the basic operations and extracting from it all the operations corresponding to recovery, concurrence, deadlock detection and consolidation control. In the basic model we opted for the Two-Phases Lock (2PL) and the writing in the log file as mechanisms of concurrence and recovery control: LM (Lock Manager) and LogM (Log Manager) components. To be able to use protocols different to 2PL, the lock petitions that RM requests were extracted from the LM component, and included in the RMLM component. To be able to use different commit protocols, the operations that communicate the writing of a record of the protocol have been extracted from LogM, and included in the LogMCM component. The Disk Manager (DM) component represents the read/write operations in the database. Neither should we forget the possible deadlocks of the transactions when they access shared resources. Therefore a component to represent the detection deadlock protocol (DLM) is necessary.

Figure 2 shows the basic functional components of the model, each of which corresponds to a different net in the implementation. It can also be observed that in addition to a net for each basic functional component, two nets appear called DPTinit and DTPSim. These nets have been designed to begin the simulation and they will also allow its parameters to be modified as well as make it possible to decide which commit and deadlock detection protocols are used. The figure also shows how the TM and RM components of the X/Open model (shadowed rectangles) have been broken down into several modules, maintaining the basic component but extracting from it the functions that can be implemented through other algorithms. This characteristic will permit, simply by changing one net for another, the study of a different algorithm. The increase in components entails the appearance of new interfaces; in section 4 some of the new interfaces that arise from the break-down of the basic components of the X/Open model will be described.

3.2 Parameters

For the simulation, the implemented model is similar to the one used in [13] and [5]. The database is modelled like a collection of *DBSize* pages uniformly distributed in the *NumSites* nodes of the net. The level of multiprogramming of each node establishes the number of concurrent transactions that can be executed in it and is specified by the *MPL* parameter. The number of transactions to simulate in each case is indicated in the *NumTran* parameter. A distributed transaction consists of a master process that is executed in the node where the transaction begins, and a group of cohort or branch processes that are executed in several sites, nodes with database pages that the transaction accesses. A transaction is a sequence of access operations to a distributed database limited by the beginning and end operations. Each internal operation includes information on the node that contains the page and the access mode (reading or writing). The number of internal operations follows a normal distribution centred in *LongTran*. Of these, the percentage of local operations is given by *PorcLocal*. The rest of the operations are assigned to a number of remote nodes that follow a normal distribution

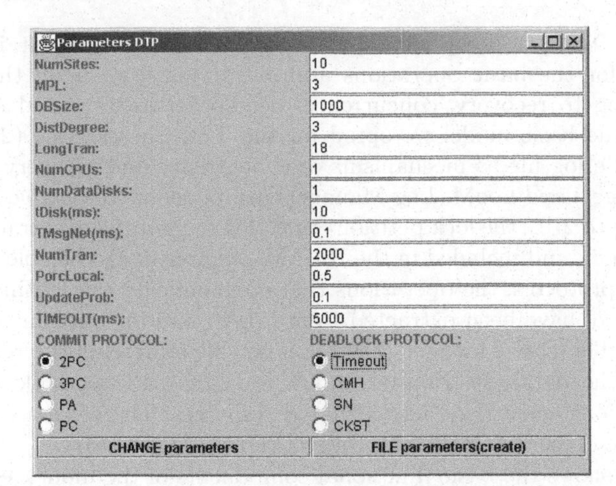

Fig. 3. Parameters of Distributed Transaction Processing System

with center in *DistDegree* on those *NumSites* nodes of the net, except the node that executes the local operations if it exists. The page to access each operation is chosen randomly form among those *DBSize* pages that each node contains. The number of upgrade operations of a transaction is specified by *UpdateProb*. When a transaction finishes its execution, a new one is launched in the node where it began. In each node a machine is simulated with *NumCPUs* processors and *NumDataDisk* data disks. The data disks store the database pages and there is another disk for storing the log files. The petitions to disk are processed in the order they arrive by means of a FIFO queue. The *tDisk* parameter captures the access time to disk for page of data. The net of communications among the nodes is simulated with a time of communication of messages *tMsgNet*. It is also necessary to establish a value of time limit, *timeout* parameter, to indicate that a deadlock can exist and the deadlock detection protocol should be activated.

The DPTinit net begins the simulation by taking the value of these parameters, reading them from a file or showing a dialogue panel that makes it possible to decide the appropriate parameters for simulation (see figure 3). In this figure it can be seen that not only is it possible to change the simulation parameters but also decide some commit and deadlock detection protocols. These protocols are implemented in different nets but in each simulation it decides which is the desired protocol and only the selected nets are activated. The possibility to change the nets that simulate different protocols will make it possible, for example, to evaluate the performance of several commit or deadlock detection protocols, just by changing the appropriate nets. The possibility of modifying the parameters will make it possible to compare the performances of Distributed Systems with different characteristics.

Once the simulation parameters have been decided on and the protocols chosen, the DTPinit net activates the DTPsim. In the latter the rest of the nets necessary for the chosen simulation are created and/or initialized dynamically and the net AP begins to launch transactions.

4 Modeling of Commit Protocols with CPNs

In a DTP system the users launch transactions from the different nodes of the distributed system. The transaction that is launched in a node can access any resource of the system. This means that transactions assigned in different nodes can compete for the same resources. It is therefore necessary to establish mechanisms that assure database consistency: protocols that allow the database to be reestablished to a consistent state after a failure as well as concurrence protocols. To illustrate the work form and its modularity we opted for considering the most common protocols in database systems: writing in the log as security protocol and 2PL for concurrence control. These protocols are part of RM as can be seen in figure 2.

When a distributed transaction finishes, the atomicity property of the transaction requires that all their branches complete the transaction or that they all rollback. This requires a cooperative procedure among all the nodes where the transaction executes some branch: the commit protocol. The TM in the node where the transaction began assumes the role of coordinator, beginning a protocol that assures the same result in all the nodes. The RMs in the nodes where the transaction executes a branch become participants and cooperate with the coordinator to assure that all the branches have consolidated their operations.

Among the different protocols of atomic consolidation, the Two Phase Commit Protocol (2PC) and their variants Presume Abort (PA), Presume Commit (PC) and Three Phase Commit Protocol (3PC) are the most used ([25], [10]). They are all characterized by the exchange of messages between the TM of the node where the transaction was launched (coordinator) and the RMs of nodes where the transaction has branches (participants). During the execution of the protocol these managers should register information in the log file for the recovery of a possible failure. Some of these records should be in stable memory before continuing with some steps of the protocol. The functional components of the model (figure 2) implied in the commit protocol are three:

- TMCM. This represents the coordinating process that manages the commit protocol in the node where the transaction was launched.
- RMCM. This represents the participant process that is executed in each node where the transaction has a branch.
- LogMCM.This communicates to the coordinator and the participants that the forced writings are in stable memory so that they continue with the protocol.

The nets that are activated in these components depend on the commit protocol that the user chooses in each simulation. Table 1 shows the nets that have already been designed for each of the protocols and the functional component of the model that they implement. The system is open so that other protocols can

Table 1. Nets designed for different commit protocols

		Protocols			
		2PC	PA	PC	3PC
Components	TMCM	TM2PC	TMPA	TMPC	TM3PC
	RMCM	RM2PC	RMPA	RMPC	RM3PC
	LogMCM	LogM2PC	LogMPA	LogMPC	LogM3PC

be integrated in it just by developing the appropriate nets and respecting the input/output interfaces to establish communication with the rest of the components. In this paper, for the sake of simplicity, only the nets corresponding to 2PC are shown, and in the following section the generic input and output interfaces are described. For the same reason, only the components implied in commit protocol are shown. To show all components and their interfaces would make this paper too extensive.

4.1 Component TMCM: COORDINATOR

TM begins 2PC protocol when it receives the *End Transaction* operation from the AP, activating the coordinator process. In 2PC protocol, the TMCM component corresponds to the TM2PC net that is shown in figure 4. The communication of this component with the rest is established by means of an interface that is defined through two input transitions (`I01` and `I1`) and several output transitions (`O1`, `O2`, `O3`, `E1`, `E2`, `E3`, `E4`, `E5` and `E6`). Two output transition types are identified: `Ox` are transitions that represent petitions to other components while the coordinator continues processing its work (no forced writing in log), while the exits through `Ex` indicate that the coordinator has processed a petition and is waiting for others that will arrive through the input transition `I1`. Figure 5 shows these input/output interfaces schematically.

Inputs. The input transition `I01(:INITCP(tag,N,T,ln))` represents the communication interface from TM to TMCM. It is always activated from TM and it implies the beginning of the protocol. This transition communicates to the coordinator the information that it needs to begin the protocol: the `tag`, which indicates where the input comes from; identification of the node where the transaction was launched (`N`); identifier of the transaction (`T`) and a list (`ln`) that includes the identifiers where the transaction has a branch, nodes that will execute the participant processes.

The rest of the petitions that activate the coordinating process will arrive through the input transition `I1(:IN(tag,N,T,op,nd))` that represents several communication interfaces (RMCM-TMCM, LogMCM-TMCM and CRM-TMCM). The `tag` indicates where the petition comes from and hence the interface that it represents; information also arrives on the identifier of the node that executes the coordinating process (`N`) for the transaction (`T`), the operation in course (`op`) that can be either the arrival of a protocol message, or the confirmation that a record is in stable memory in log and identifier of the node (`nd`) that

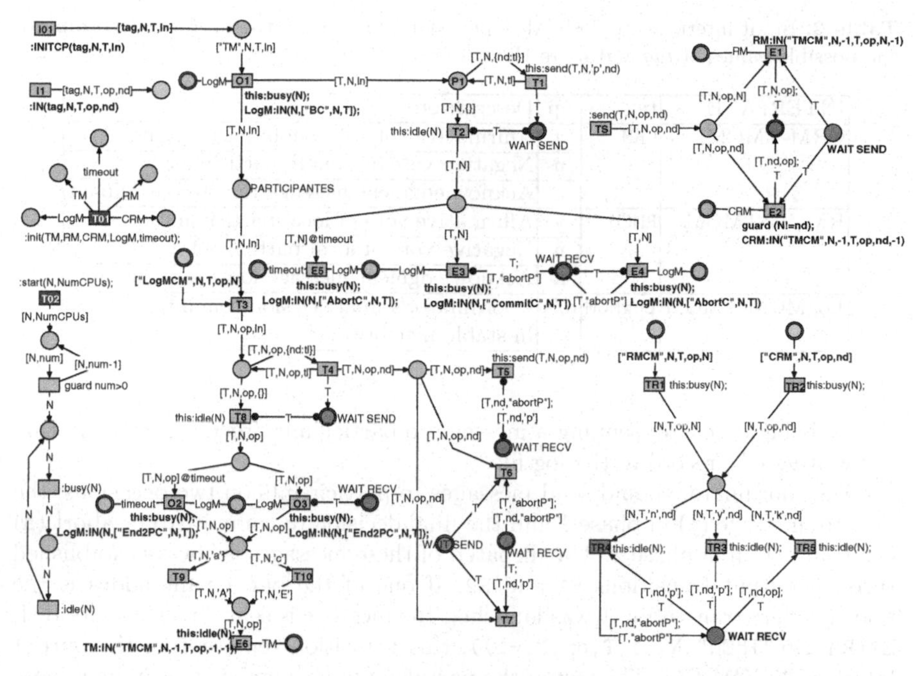

Fig. 4. Net TM2PC. This net is instanced in TMCM to implement 2PC. It represents the coordinator process in the Two Phase Commit Protocol

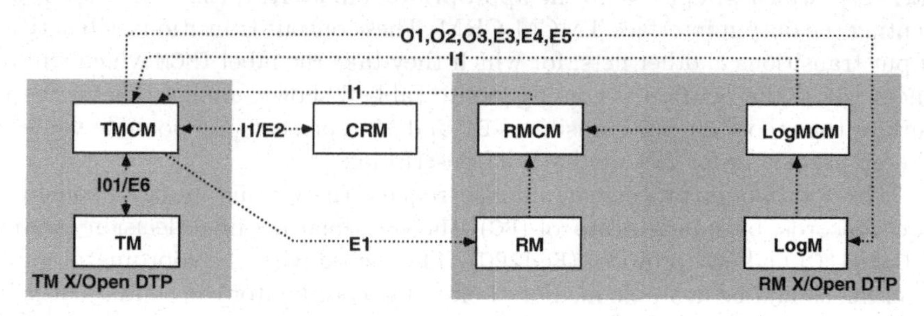

Fig. 5. Input/output interfaces for the TMCM component that represents the coordinator process in the Commit Protocol. The labels of the interfaces correspond to the input/output transactions of net TM2PC shown in figure 4

has executed the operation. Table 2 shows the interface represented by transition `I1` according to `tag` value, possible values of `op` for 2PC and a brief description of what the interface represents. The symbol that generates this input transition activates different internal transitions of the net which simulate the behaviour of the 2PC protocol.

Outputs. The exit transitions simulate the passage of messages from the coordinator process to appropriate processes depending on the part of the protocol

Table 2. Input interfaces to TMCM. Operation in course of the protocol according to the possible values of *tag* and *op* in the transition *I1*

INTERFACE	tag	op	Description
CRM-TMCM	CRM	y	Affirmative vote of remote participant
		n	Negative vote of remote participant
		k	Acknowledgment message remote participant
RMCM-TMCM	RMCM	y	Affirmative vote of local participant
		n	Negative vote of local participant
		k	Acknowledgment message local participant
LogMCM-TMCM	LogMCM	c/a	Coordinator's final decision commit/abort is in stable memory

that is being executed: sending a message to participant processes or requesting the writing of a record in the log file.

The coordinator should send messages to participants on two occasions: the prepare message (`p`) in phase 1 and the final decision to commit (`c`) or abort (`a`) the transaction in phase 2. The dispatch of these messages is always established through output transitions `E1` and `E2`. If one of the participant nodes is the node `N` where transaction `T` was launched the message is sent directly to its RM, `E1(RM:IN("TMCM",N,-1,T,op,N,-1))`, this transition representing the output interface TMCM-RM. The rest of the participants (or maybe all) will be remote and the message should be sent through CRM, `E2(CRM:IN("TMCM",N,-1,T,op, nd,-1))`, which will pass it to the appropriate remote RM, this transition representing the output interface TMCM-CRM. These output transitions will activate input transitions in other nets, for which they take the label `TMCM` which will tell them where the petition is coming from. Table 3 shows, among others, the interface that those output transitions `E1` and `E2` represent, the possible values of the `op` parameter for 2PC and a brief description.

The coordinator process should also request the writing in log of some protocol records: beginning protocol (`BC`), the coordinator's final decision (`AbortC` / `CommitC`) and end protocol (`End2PC`). The record with the coordinator's final decision should be in stable memory before the coordinator can send its decision to the participants while for the beginning and end records this is not required. The forced writing petition is made through the output transitions `E3`, `E4` and `E5(LogM:IN(N,[reg,N,T]))`, while non-forced writing petitions are made by means of `O1`, `O2` and `O3(LogM:IN(N,[reg,N,T]))`. They all represent the output interface TMCM-LogM and they pass to LogM of node `N` information in the way `[reg,N,T]` indicated by the coordinator petition of the writing in log of the corresponding record. These transitions activate the input transition `I1` in the general LogM net which, as has been mentioned previously, is broken down into two to be able to change protocol easily. The LogM net takes charge of simulating the writing in log and it will pass the control to LogMCM, which is the one that takes charge of communicating to the coordinator, in the cases of forced writing, that the record is in stable memory (see section 4.3). This has been extracted from the general process of writing in log, because the records with

Table 3. Output interfaces of TMCM. Possible values of *op/reg* and brief description of the interface they represent

INTERFACE	Transition	Parameter	Description
		op	Sending messages
TMCM-RM	`E1`	`p`	Prepare local participant
		`c`	Commit local participant
		`a`	Abort local participant
TMCM-CRM	`E2`	`p`	Prepare remote participant
		`c`	Commit remote participant
		`a`	Abort remote participant
		reg	Force write in log
TMCM-LogM	`E3`	`CommitC`	The coordinator reaches a global commit decision, because all the participants vote commit
	`E4`	`AbortC`	The coordinator reaches a global abort decision, even it just one participant votes to abort
	`E5`	`AbortC`	The coordinator decides to globally abort, because the timeout for the local decisions of the participants has run out
		reg	No force write in log
TMCM-LogM	`01`	`BC`	Init protocol
	`02`	`End2PC`	All the messages that the coordinator was waiting for have arrived
	`03`	`End2PC`	The coordinator receives no answer from a participant and decides to continue the protocol if the timeout has run out
		op	End protocol
TMCM-TM	`E6`	`C/A`	The coordinator returns the control to TM so that it can communicate to AP the end of the transaction

forced writing change from one protocol to another, and the net corresponding to the protocol is instanced in LogMCM. Table 3 shows the possible values for `reg` and a brief description of these transitions (`E3`, `E4`, `E5`, `01`, `02` and `03`).

The last output interface of component TMCM corresponds to output transition `E6(TM:IN("TMCM",N,-1,T,op,-1,-1))` which means the end of the protocol and the passing of the coordinator's final decision to TM, which will communicate to AP the final destination of transaction. This is also shown in Table 3.

4.2 Component RMCM: PARTICIPANT

Component RMCM represents, in each node in which the transaction executes a branch, the participant process of the commit protocol. Figure 6 shows net RM2PC which corresponds to this component for 2PC. The interfaces with other nets are established through an input transition and several output transitions;

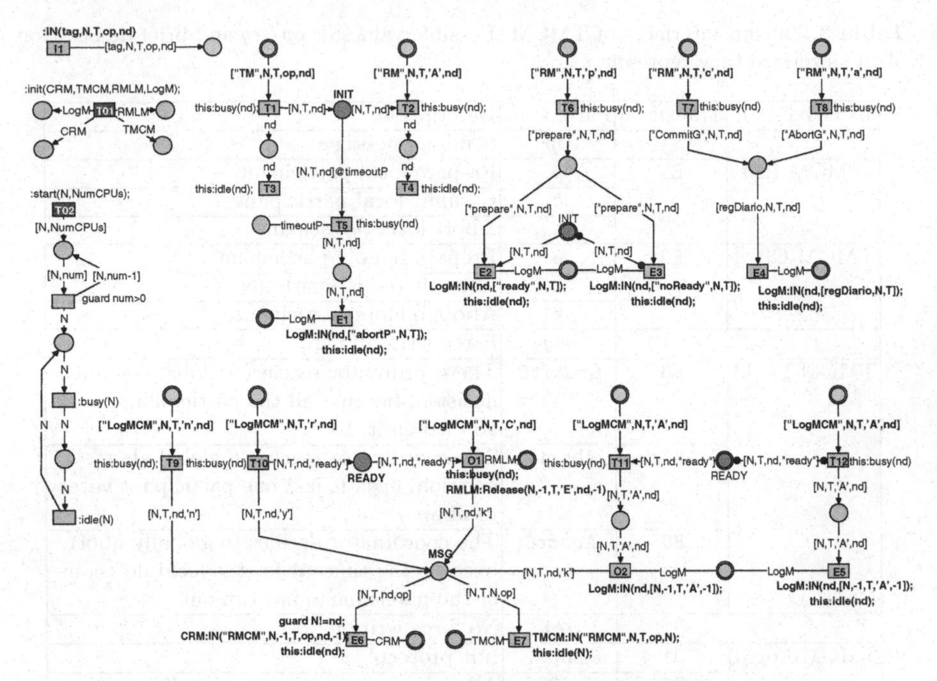

Fig. 6. RM2PC Net. This net is instanced in RMCM for 2PC. It represents the participant processes in each node where the transaction executes a branch

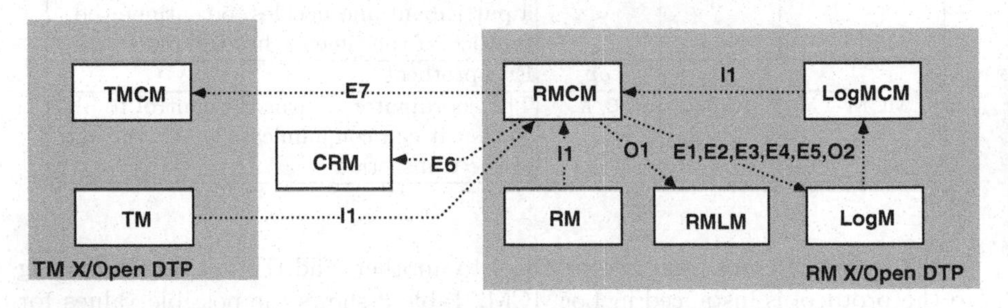

Fig. 7. Input/output interfaces for component RMCM which represents a participant process of the Commit Protocol. The tags of the interface correspond to input/output transitions in net RM2PC, which is shown in figure 6

figure 7 shows the correspondence between these transitions and the interfaces with other components of the model.

Inputs. The input transition `I1(:IN(tag,N,T,op,nd))` represents several input interfaces, depending on the `tag` value. When it is activated, it means the arrival of a message from the coordinator process (RM-RMCM), the confirmation of the writing in stable memory of a record in the log (LogMCM-RMCM),

Table 4. Input interfaces to RMCM according to *tag* value in the transition *I1*. Possible values for *op* in each case

INTERFACE	tag	op	Description
RM-RMCM	RM	p	Reception of prepare message from the coordinator
		c/a	Reception of coordinator's final decision (Commit/Abort)
		A	The transaction aborts before beginning the protocol
TM-RMCM	TM	S	TM informs to nd node its participation in a branch of T
LogMCM-RMCM	LogMCM	r/n	The log's record that indicates if participant is prepared to commit transaction is found in stable memory (ready/no ready)
		C/A	The log's record that indicates the destination of the transaction (commit/abort) it is in stable memory

the information from TM of the participation of the node in the execution of a branch of transaction (TM-RMCM) or the information from RM concerning the abnormal termination of the transaction (for example, that it has been the victim of a deadlock detection process). Table 4 shows the interfaces that this input transition represents according to tag values, with a brief description.

As was observed in section 4.1, the coordinator sends the messages to the remote participants through CRM; the latter will pass them by means of its exit interface to RM, which will detect that it is a message of the commit protocol and will send it to the participant process that has been extracted from RM in order to be able to change the protocol easily. This explains why RMCM does not have an input interface from CRM. These messages, the same as those of read/write page petitions, always arrive from CRM and they are passed to the RM which will send them to the corresponding component depending on the operation type.

Outputs. Figure 7 shows the output interfaces of the RMCM component tagged with the transitions corresponding to the RM2PC net. The interfaces RMCM-CRM and RMCM-TMCM correspond to shipment of messages of protocol from the participant (remote or local) to the coordinator; RMCM-LogM corresponds to the petition of writing log records and if the transaction aborts, to the roll-back petition of the effects of the branch of the transaction. If, on the contrary, the transaction commits, its modifications must be made permanent and the corresponding resources released; the interface RMCM-RMLM makes it possible to inform LM of this situation.

The participant should send messages to the coordinator on two occasions: its vote (y/n) to indicate whether or not it is prepared to commit transaction in phase 1 and an acknowledgment message (k) to confirm that it has executed its final decision in phase 2. Before sending these messages, the participant should store in log the corresponding record.

Table 5. Output interfaces of RMCM, possible values of *op/reg* and brief description of the interface they represent

INTERFACE	Transition	Parameter	Description
		op	Sending of messages to coordinator
RMCM-TMCM	E6	y	Local participant is prepared
		n	Local participant is not prepared
		k	The local participant's acknowledgement
RMCM-CRM	E7	y	Remote participant is prepared
		n	Remote participant is not prepared
		k	The remote participant's acknowledgement
		reg	Writing in log
RMCM-LogM	E2	ready	The participant is prepared to commit
	E3	noReady	The participant is not prepared to commit
	E4	CommitG/ AbortG	The participant accepts the coordinator's final decision (Commit/Abort)
	E1	AbortP	The participant aborts unilaterally
		op	Rollback participant's branch
RMCM-LogM	O2	A	Unilateral abort (timeoutP)
	E5	A	Coordinator's final decision
		op	Commit branch
RMCM-RMLM	O1	E	Relsease of resources for the coordinator's final decision of commit

The output transitions `E1`, `E2`, `E3` and `E4(LogM:IN(nd,[reg,N,T])` make these petitions to LogM so that the pertinent records can be stored in stable memory (`ready`, `noReady`, `CommitG`, `AbortG` or `AbortP`). The LogM component makes all these writings, and then passes the control to LogMCM, which will communicate to RMCM the cases of forced writing (all in 2PC for the participant process). If the coordinator's decision was to commit, the effects of the transaction should remain; for this, the exit transition `O1(RMLM:IN(N,-1,T,'E',nd,-1)` tells RMLM that it can release all the resources that the branch has. Transition `E1` represents the participant's unilateral abort in the face of the coordinator's possible failure; in this case the participant decides to abort its branch of `T` by writing the record `AbortP` in the log. Later, all their operations will be subjected to rollback: transition `E5(LogM:IN(nd,[N,-1,T,'A',-1])`. The same interface is represented by the output transition `O2`, but in this case the abort decision came from the coordinator and the participant must send it an acknowledgement message. In these cases LogM undoes the effects of the branch of the transaction and it releases all the resources that it has.

The protocol messages are always sent by the output transitions `E6(CRM:IN("RMCM",N,-1,T,op,nd,-1))` and `E7(TMCM:IN("RMCM",N,T,op,N))`, once the appropriate record is in the stable memory. If the participant is remote the message should be sent through CRM (`E6`), but if it is a local participant it is sent to the coordinating TMCM directly (`E7`).Table 5 shows the different interfaces according to the transition and the values possible for `op/reg`.

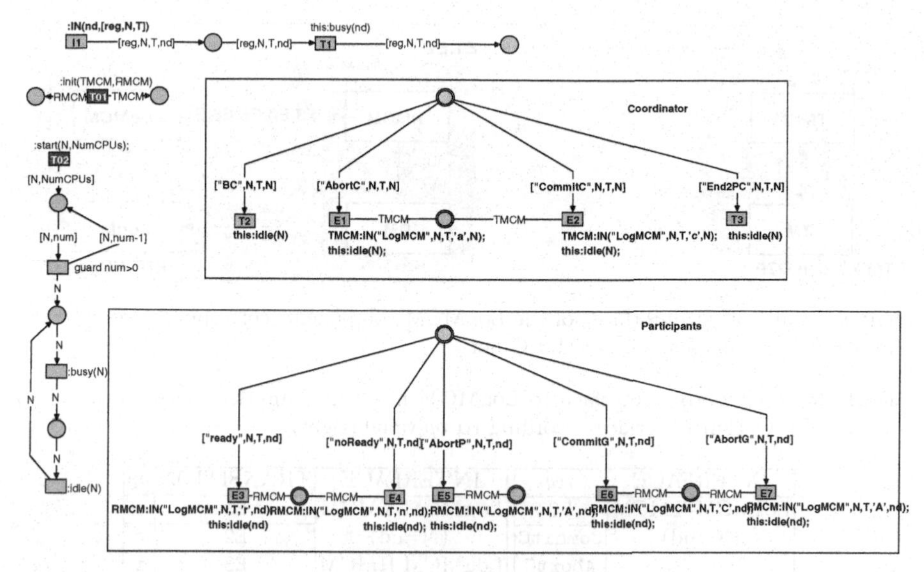

Fig. 8. LogM2PC Net instance in LogMCM. The upper part represents the interface with a coordinator process and the lower part with a participant process

4.3 Component LogMCM: Writings in log

The writing petitions (from both participants and the coordinator) in log are sent to LogM net that simulates the basic part. The interface between coordinator/participant processes and logM is established through LogMCM that instances the LogM2PC net (figure 8) for the case of 2PC. Figure 9 shows the correspondence between the input/output transitions and the interfaces and the rest of components.

Inputs. The net is activated from LogM by its only input transition `I1(:IN(nd, [reg,N,T]))` once the record is in stable memory. The record that has been stored in stable memory can correspond to a record of a coordinator process (upper part of figure 8) or of a participant process (lower part of figure 8). In this transition `nd` indicates the node that executes the coordinator/participant process for transaction `T`; `reg` is the record that has just been stored; and `N`, the node where the coordinator process is executed (it coincides with `nd` if it is the writing of a record of the coordinator process).

Table 6 shows the input interface represented by this transition according to values of `reg`. Description is not included because the meaning is the same in all the cases; a protocol record has been stored in log.

Outputs. The output transitions represent the LogMCM-TMCM interfaces (`E1` and `E2`) and LogMCM-RMCM (`E3`, `E4`, `E5`, `E6` and `E7`). The transitions `T2` and `T3` represent the non-forced writing of the coordinator's records `BC` and `End2PC`,

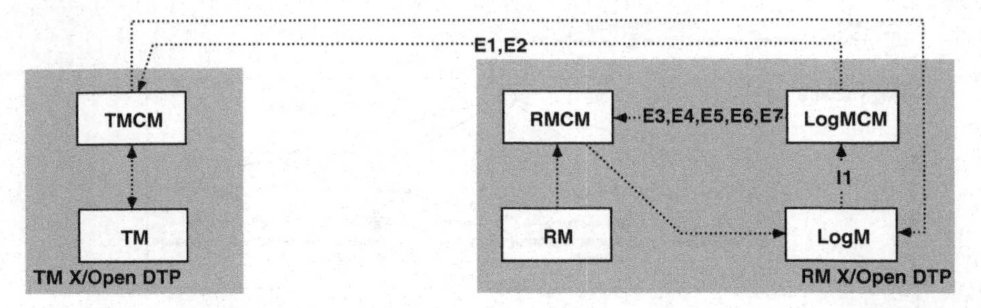

Fig. 9. Input/output interfaces of the LogMCM component; correspondence with the input/output transitions in LogM2PC net

Table 6. Input/output interface to LogMCM net. Input interfaces according to *reg* value (left). Output insterface according to *op* value (right)

INTERFACE	*reg*
LogMCM-TMCM (N=nd)	BC
	CommitC
	AbortC
	End2PC
LogMCM-RMCM	ready
	noReady
	AbortP
	CommitG
	AbortG

INTERFACE	TRANSITION	*op*
LogMCM-TMCM (N=nd)	E1	a
	E2	c
LogMCM-RMCM	E3	r
	E4	n
	E5	A
	E6	C
	E7	A

while transitions E1 and E2(TMCM:IN("LogMCM",N,T,op,N)) pass the control to the coordinator to tell it that the records that recall its final decision are in stable memory. The coordinator waits for this answer to send its final decision to the participants (see section 4.1).

The log's records corresponding to a participant process are all of mandatory writing. The output transitions E3, E4, E5, E6 and E7(RMCM:IN("LogMCM",N,T, op,nd)) pass the control to the corresponding participant process so that this sends the appropriate message to the coordinator (see section 4.2).

Table 6 shows the interfaces represented by these output transitions. Description is not included because the meaning is the same in all the cases. Some record of forced writing of the coordinator/participant is in stable memory.

5 Simulations Results

Using the Distributed Transaction Processing System model described in the previous sections, we present in this section the result of a small simulation experiment that makes it possible to show one of the possible applications of the general system. To do this we compared the number of transactions per second and the mean execution time of the transactions for the four commit protocols presented in section 4 (2PC, PA, PC and 3PC). In each node the multiprogramming level has been varied from 3 to 19 maintaining for all the cases a system with 10 nodes, each of which stores 1000 pages of the shared database. Timeout has been used as deadlock detection protocol. An ethernet (*tMsg* of 1ms), a data disk and a log disk with a mean time of access to disk of 10ms were considered. The Application Program has simulated homogeneous transactions of some 18 operations on average. The percentage of remote operations was 50% with a degree of distribution of 3 and 10% upgrade petitions.

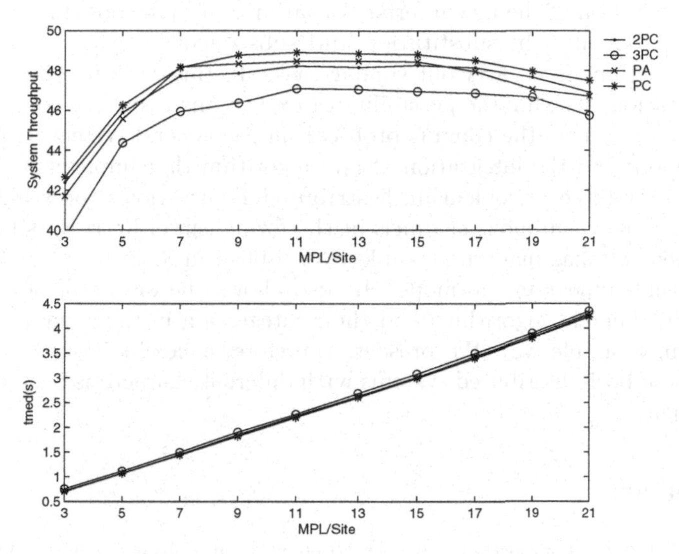

Fig. 10. System throughput (upper) and mean time (seconds) of transaction execution (lower) by the multiprogramming level per node

In the upper part of figure 9 it can be seen how the system performance decreases for 3PC, and increases slightly for PC, staying practically the same for 2PC and PA. These experimental results are in agreement with the behaviour of these algorithms in the absence of failures: 3PC decreases performance because it increases the number of messages exchanged in the protocol; 2PC and PA, in the absence of failures, behave in the same way, with the same performance and number of messages; PC decreases the number of messages for the commit transactions and for this reason the performance increases slightly. It can also be seen that between 9 and 17 concurrent transactions per node, the system throughput is maintained, but as shown in the lower part of figure 9 the time of execution increases progressively with MPL. This is because the transactions begin to compete for the resources of the system giving rise to timeout situations and deadlocks that can cause their rollback and subsequent rebooting.

6 Conclusions

The work developed here has a flexible framework that allows the study of different algorithms integrated inside a model of a Distributed Transaction Processing System. The properties of the algorithms are usually proved separately and it is not obvious how the algorithm is integrated into a large system [17]. The differentiation of the actions of the overall system that activate an algorithm has allowed the definition of a set of input/output interfaces that enables its integration.

The breakdown of the basic components defined in the X/Open DTP model and the definition of the new interfaces that appear make possible not only this integration but also the substitution and subsequent analysis and comparison of the algorithms that carry out similar tasks. In this work we present a possible application, showing the possibility of exchanging a specific component, the one that implements the commit protocol, in the general system. However, the system is open to the integration of any algorithm that implements a solution related to the typical problems in Distributed Transaction Processing.

The use of synchronous channels in the formalism of Petri Nets that Renew [21] implements has made it possible to establish in a simple way the new interfaces that appear in the model. It also allows the design of new nets that implement different algorithms and their integration in the general system developed in a simple way. We present, therefore, a flexible system that allows the study of both distributed systems with different characteristics and different algorithms.

References

1. M. Abdallah, R. Guerraoui, and P. Pucheral. One-phase commit: Does it make sense. In *Int. Conf. on Parallel and Distributed Systems (ICPADS)*, pages 182–192, 1998.
2. M. Abdallah and P. Pucheral. A single-phase non-blocking atomic commitment protocol. In *Proc. of the 9th International Conference on Database and Expert Systems Applications (DEXA)*, pages 584–595, Vienna, August 1998.
3. Y. Al-Houmaily and P. Chrysanthis. The implicit-yes vote commit protocol with delegation of commitment. In *Proceedings of 9th Intl. Conf. On Parallel and Distributed Computing Systems*, September 1996.
4. P. Ancilotti, B. Lazzerini, C. A. Prete, and M. Sacchi. A distributed commit protocol for a multicomputer system. *IEEE Transactions on Computers*, 39(5):718–724, May 1990.
5. M.J. Carey and M. Livny. Distributed concurrency control performance: A study of algorithms, distribution, and replication. In *14th International Conference on VLDB*, pages 13–25, Los Angeles, CA, August 1988.
6. K. M. Chandy, J. Misra, and L. M. Haas. Distributed deadlock detection. *ACM Transactions on Computer Systems*, 1(2):144–156, May 1983.
7. S. Chen, Y. Deng, W. Sun, and N. Rishe. Efficient algorithms for detection and resolution of distributed deadlocks. In *7th IEEE Symposium on Parallel and Distributeed Processing*, pages 10–16, San Antonio, Texas, October 25-28 1995.

8. A. N. Choudhary, W. H. Kohler, J. A. Stankovic, and D. Towsley. Correction to 'a modified priority based probe algorithm for distributed deadlock detection and resolution'. *IEEE Transaction on Software Engineering*, 15(12):1644, December 1989.
9. A. N. Choudhary, W. H. Kohler, J. A. Stankovic, and D. Towsley. A modified priority based probe algorithm for distributed deadlock detection and resolution. *IEEE Transaction on Software Engineering*, 15(1):10–17, January 1989.
10. Jim Gray and Andreas Reuter. *Transaction Processing: Concepts and Techniques.* Morgan Kaufmann, 1993.
11. Object Management Group. *Transaction Service Specification, Version 1.3.* 2002.
12. The Open Group. *Snapshot Distributed Transaction Processing: The XA+ Specification Version.* The Open Group, 1994.
13. Ramesh Gupta, Jayant Haritsa, and Krithi Ramamritham. Revisiting commit processing in distributed database systems. *Proc. of the ACM SIGMOD International Conference on Management of Data, Tucson, Arizona*, pages 486–497, June 1997.
14. Kurt Jensen. *Coloured Petri Nets. Basic Concepts, Analysis Methods and Practical Use. Vol. 1: Basic Concepts. Vol 2: Analysis Methods. Vol 3: Practical Use.* Monographs in Teorical Computer Science. Springer-Verlag, 1997.
15. Kurt Jensen. An introduction to the practical use of coloured petri nets. *Lectures on Petri Nets II: Applications, LNCS*, 1492:237–292, 1998.
16. Y. M. Kim, T. H. Lai, and N. Soundarajan. Efficient distributed deadlock detection and resolution using probes, tokens, and barriers. In *International Conference on Parallel and Distributed Systems (ICPADS '97)*, pages 584–591, Seoul, KOREA, December 11-13 1997.
17. E. Kindler and S. Peuker. Integrating distributed algorithms into distributed systems. In *Informatik-Berichte, No. 110: Workshop Concurrency, Specification and Programming*, pages 128–143, Berlin, September 28-30 1998.
18. Edgar Knapp. Deadlock detection in distributed databases. *ACM Computing Surveys*, 19(4):303–328, December 1987.
19. Natalija Krivokapic, Alfons Kemper, and Ehud Gudes. Deadlock detection in distributed database systems: A new algorithm and a comparative performance analysis. *The VLDB Journal*, 8(2):79–100, 1999.
20. A.D. Kshemkalyani and M. Singhal. Efficient detection and resolution of generalized distributed deadlocks. *IEEE Transactions on Software Engeneering*, 20(1):43–54, January 1994.
21. Olaf Kummer, Frank Wienberg, and Michael Duvigneau. *Renew - User Guide.* Theoretical Foundations Group. Distributed Systems Group. Department for Informatics. University of Hamburg, 2002.
22. E. Levy, H. F. Korth, and A. Silberschatz. An optimistic commit protocol for distributed transaction management. In *Proceedings of the ACM SIGMOD International Conference on Management of Data*, pages 88–97, Denver, Colorado, May, 29-31 1991.
23. Don P. Mitchel and Michael J. Merrit. A distributed algorithm for deadlock detection and resolution. In *Third Annual ACM Symposium on Principles of Distributed Computing*, pages 282–284, Vancouver, British Columbia, Canada, 1984.
24. Ron Obermarck. Distributed deadlock detection algorithm. *ACM Transactions on Database Systems*, 7(2):187–208, June 1982.
25. M.Tamer Özsu and Patrick Valduriez. *Principles of Database Systems.* Prentice Hall, second edition, 1999.

26. S. Ramsey, J. Nummenmaa, P. Thanisch, R. Pooley, and S. Gilmore. Interactive simulation of distributed transaction processing commit protocols. In P. Luker, editor, *Proceedings of the 3rd Conference of the United Kingdom Simulation Society, Keswick, U.K.*, pages 112–127, 23-25 April 1997.
27. I. Ray, E. Bertino, S. Jajodia, and L. Mancini. An advanced commit protocol for MLS distributed database systems. In *3rd ACM Conference on Computer and Communications Security*, pages 306–315, New Delhi, India, March 14-15 1996.
28. M. K. Sinha and N. Natarajan. A priority based distributed deadlock detection algorithm. *IEEE Transaction on Software Engineering*, SE-11(1):67–80, January 1985.
29. C-F Yeung and S-L Hung. A new deadlock detection algorithms for distributed real-time database systems. In *14th Symposium on Reliable Distributed Systems*, pages 146–153, Bad Neuenahr, Germany, September 13-15 1995.
30. C-F Yeung, S-L Hung, and K-Y Lam. Performance evaluation of a new distributed deadlock detection algorithm. *SIGMOD Record*, 23(3):21–26, 1994.

Time Petri Nets with Inhibitor Hyperarcs. Formal Semantics and State Space Computation

Olivier H. Roux and Didier Lime

IRCCyN (Institut de Recherche en Communication et Cybernétique de Nantes)
1, rue de la Noë B.P. 92101
44321 NANTES cedex 3, France
{Didier.Lime,Olivier-h.Roux}@irccyn.ec-nantes.fr

Abstract. In this paper, we define Time Petri Nets with Inhibitor Hyperarcs (IHTPN) as an extension of T-time Petri nets where time is associated with transitions. In this model, we consider stopwatches associated with transitions which can be reset, stopped and started by using classical arcs and branch inhibitor hyperarcs introduced by Janicki and Koutny [17]. We give a formal semantics for IHTPNs in terms of Timed Transition Systems and we position IHTPNs with regard to other classes of Petri nets in terms of timed language acceptance. We provide a method for computing the state space of IHTPNs. We first propose an exact computation using a general polyhedron representation of time constraints, then we propose an overapproximation of the polyhedra to allow a more efficient compact abstract representations of the state space based on DBM (Difference Bound Matrix).

Keywords: Time Petri nets, inhibitor hyperarc, state space, semantics, real-time systems

1 Introduction

The theory of Time Petri Nets provides a general framework to specify the behavior of reactive and real-time systems. In such applications, it is necessary to model the durations of actions in the form of a time interval and it is sometimes necessary to memorize the progress status of an action when this one is suspended then resumed.

In this class of models, some extensions of time Petri nets have been proposed to model the preemptive scheduling of tasks. Okawa and Yoneda [25] propose an approach with time Petri nets consisting of defining groups of transitions together with rates (speeds) of execution. Roux and Déplanche [30] propose an extension for time Petri nets (scheduling-TPN) that consists of mapping into the Petri net model the way the different schedulers of the system activate or suspend the tasks. For a fixed priority scheduling policy, scheduling-TPN introduce two new attributes associated to each place that respectively represent allocation (processor or resource) and priority (of the modeled task). [6] propose a similar model: Preemptive Time Petri Net (preemptive-TPN). The two new attributes are associated to transitions instead of places.

J. Cortadella and W. Reisig (Eds.): ICATPN 2004, LNCS 3099, pp. 371–390, 2004.
© Springer-Verlag Berlin Heidelberg 2004

However all these models are dedicated to the scheduling problem while stopping then resuming an action is a more general consideration. For example scheduling-TPN and preemptive-TPN are unable to model a circular priority relation which can definitively block the time flow of some transitions while no concurrent transition runs (i.e. the required resources are free). These "timed deadlock" are real and concrete problems in industrial applications, it may thus be necessary to have a model able to express them in order to detect them.

We propose to define a general T-timed extension of Petri nets with the concept of stopwatch. In this model, stopwatches are associated with transitions which can be reset, stopped and started by using classical arcs and branch inhibitor hyperarcs introduced by Janicki and Koutny [17].

We will first discuss timed extensions of Petri nets and then the concept of branch inhibitor hyperarc.

Time Petri Nets

The two main timed extensions of Petri Nets are Time Petri Nets (TPN) [24] and Timed Petri Nets [29]. While a transition can be fired within a given interval for TPN, in Timed Petri Nets, temporisation represents the minimal duration firing of transitions (or exact duration firing with an "as soon as possible" firing rule). There are also numerous way of representing time. It could be relative to places, transitions, arcs or tokens. The classes of Timed Petri Nets (time relative to place, transition...) are included in the corresponding classes of Time Petri Nets [27] TPN are mainly divided in P-TPN [18], A-TPN [10, 1] and T-TPN [4] where a time interval is relative respectively to places, arcs and transitions. Finally, Time Stream Petri Nets [11] were introduced to model multimedia applications. Since our objective is to use inhibitor arcs to stop the stopwatches associated with the transitions, we will use the T-TPN model.

It is shown in [23] that Petri nets with inhibitor arcs can be described by T-time Petri nets. Time-transitions are then used to model inhibitions. T-TPN do not make it possible to model at once a coherent flow of time and inhibitions.

Priorities, Inhibitor Arcs and Branch Inhibitor Hyperarcs

A priority net introduced by [14] is a marked net with a binary priority relation on transitions. The priority relation is assumed to be irreflexive and antisymmetric. Moreover, priority is specified only for transitions which are in a local conflict. Petri Nets with inhibitor arcs were introduced by Agerwala [2]. The standard execution rule for inhibitor arcs says that an inhibitor arc between a condition (place) p and an event (transition) t means that t can only be fired if p is unmarked. Several papers aimed at modeling priorities with inhibitor arcs. This can be achieved by using labeled nets [14] where a transition with a lower priority is replaced by a set of transitions with the same label, or by using "invisible transitions" [8]. [17] show with a counterexample that it is impossible to have a strong (not only isomorphic, but even identical reachability graphs) translation from priority nets to ordinary inhibitor nets. To cope with this problem

[17] introduced branch inhibitor hyperarcs. A branch inhibitor hyperarc between places $p_1, \ldots, p_k$ and a transition t means that t is not firable if all the places $p_1, \ldots, p_k$ are marked. However, in their example, ordinary inhibitor arcs could be used by adding a single place to the net. Branch inhibitor *hyperarcs* do not increase the expressivity of the model compared to simple inhibitor arcs but nonetheless greatly improve the convenience of the modeling.

Simultaneity and Inhibitor Arcs

There seems to be no general agreement on an exact definition of a simultaneous step for nets with inhibitor arcs : can a net fire simultaneously two transitions a and b even if the firing of the one (a) inhibits the other (b)? Both semantics can be found in the literature : models in which the simultaneous firing of $\{a, b\}$ is allowed is often called the "*a priori* concurrent semantics" [15, 17] and those which disallow $\{a, b\}$ the "*a posteriori* concurrent semantics" [26]. Both types are considered in [8, 16]. To be coherent with classical time Petri nets semantics, we will consider in this article a model which does not allow the simultaneous firing of two transitions.

Related Works

Cassez and Larsen [7] prove that stopwatch automata (SWA) with unobservable delays are as expressive as linear hybrid automata (LHA) in the sense that any timed language accepted by a LHA is also acceptable by a SWA. The reachability problem is undecidable for LHA and also undecidable for SWA. For time Petri nets as for Petri nets with inhibitor arcs, (and *a fortiori* for time Petri nets with inhibitor arcs (and hyperarcs)) boundedness of marking is undecidable [23, 14], and works on these models report undecidability results, or decidability under the assumption that the Petri net is bounded (as for reachability decidability of Time Petri Nets [28]). Boundedness and other results are obtained by computing the state space.

For dense-time models such as time Petri nets or timed automata, the state space is infinite because of the real-valued clocks; but it is possible to represent the infinite state-space by a finite partitioning in the state class graph or the region graph.

The mainstream approach to compute the state space of T-TPN is the state class graph [23, 4]. The nodes of the state class graph are sets of states (a state class is a pair consisting of a marking and a firing constraint) of the TPN and the edges are labeled with transitions names. If L is the language accepted by the state class graph and L' is the untimed language accepted by the TPN, then $L = L'$. An alternative approach has been proposed by Yoneda [32] in the form of an extension of equivalence classes which allows Computation Tree Logic (CTL) model-checking. Lilius [20] refined this approach so that it becomes possible to apply partial order reduction techniques that have been developed for untimed systems. Berthomieu and Vernadat [5] propose an alternative construction of the graph of atomic classes of Yoneda applicable to a larger class of nets. The state class method is also improved in [22] to produce a timed automaton which is timed bisimilar to the TPN, allowing TCTL model-checking. Lastly, Gardey and Roux [13] propose to extend the zone based forward algorithm of timed automata to the computation of the state space of time Petri nets. This algorithm differs from the state class graph computation by using the future of a zone instead of shifting the origin of time.

In all these approaches, temporal domains are expressed as a set of linear inequations in the form of a Difference Bound Matrix (DBM). The form of a DBM is not sufficient to express the temporal domain with the presence of stopwatches. For stopwatch automata (SWA) Cassez and Larsen [7] propose an approximate reachability algorithm obtained as a rather immediate extension of the existing reachability algorithm for timed automata. This extension consists of using DBM by the redefinition of the *future* of a zone (DBM). It yields an overapproximation of the reachable state space. In the same way, in the context of scheduling analysis, in [30, 21, 6] authors propose an overapproximation of the state space by using DBM.

Our Contribution

We propose to define a T-timed extension of Petri nets with the concept of stopwatch. The stopwatches are associated with the transitions which can be reset, stopped and started by using classical arcs and branch inhibitor hyperarcs introduced by Janicki and Koutny [17].

We give in section 2 the formal semantics of this model in the form of a timed transition system. We give in section 3 the main properties concerning this model (relation with other classes of Petri nets, decidability of boundedness and reachability problems...). We propose in section 4 an exact state space computation (based on the state class graph method) and also a more efficient computation method (at the cost of an overapproximation) that uses compact abstract representations of the state space based on DBM (Difference Bound Matrix). Finally in section 5, we show how to check timed reachability properties on this model by using the classical notion of observer.

2 Time Petri Nets with Branch Inhibitor Hyperarcs

2.1 Informal Presentation

We now present our T-time extension of branch inhibitor nets. We first recall, on examples, the models on which this extension is based *i.e.* branch inhibitor nets with *a posteriori* semantics and time Petri nets.

Two examples of branch inhibitor nets are presented in figure 1. The first net (a) has two inhibitor arcs (*i.e.* two simple hyperarcs) and the second (b) has one hyperarc. The net (a) imposes the firing of transitions t_1 and t_2 before the firing of the transition t_3 whereas the net (b) imposes only the firing of one of the transitions t_1 or t_2 before the firing of t_3. Let us recall, that with *a posteriori*

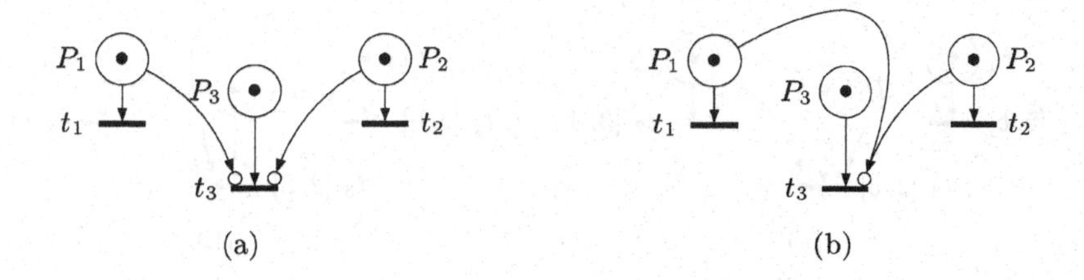

Fig. 1. Two branch inhibitor nets

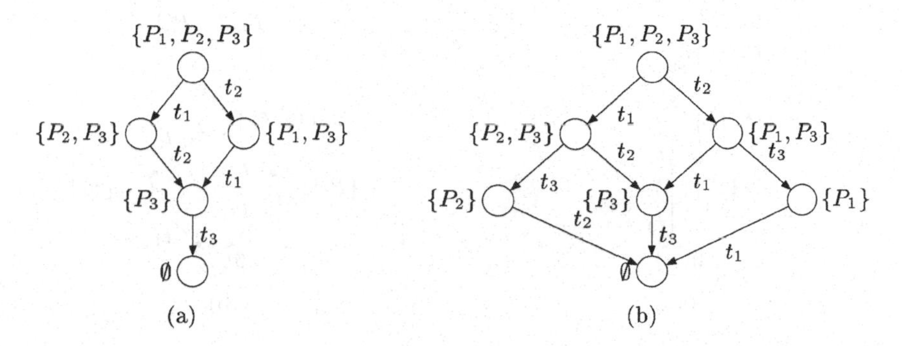

Fig. 2. Reachability graph of branch inhibitor nets of fig.1

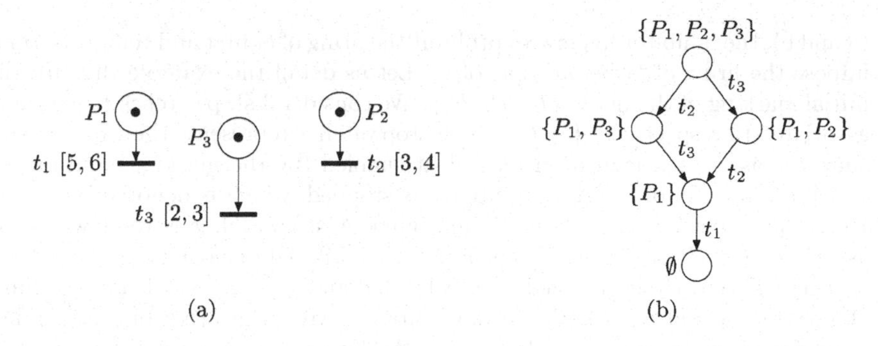

Fig. 3. A time Petri net and its state class graph

semantics, one can fire only one transition at the same time. Reachability graphs are given in figure 2.

If we remove the inhibitor arcs and add time to the underlying structure of those Petri nets, we obtain the time Petri net in figure 3a. t_1, t_2 and t_3 are enabled at the same time but as the earliest firing time of t_1 (5 time units) is strictly higher than the latest firing time of the other transitions we obtain the reachability graph (computed with the state class graph) in figure 3b.

Now let us add the branch inhibitor arcs of figure 1 to the TPN in figure 3. We obtain two T-time Petri nets with inhibitor hyperarcs in 4. In the two examples

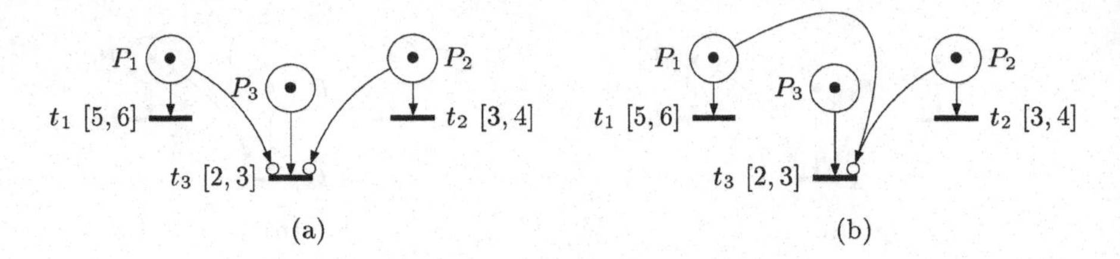

Fig. 4. Two IHTPN

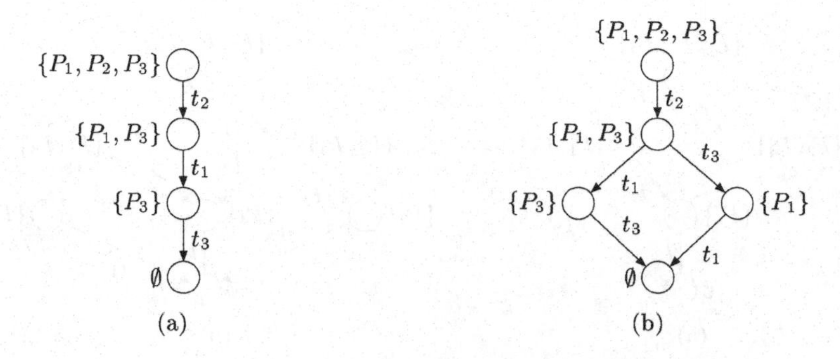

Fig. 5. Reachability graph of IHTPN of fig.4

(a and b), the inhibitor hyperarcs prohibit the firing of t_3 first and temporisations impose the firing of t_2 before that of t_1. Let us detail the example (fig.4b): the initial marking of the net is $\{P_1, P_2, P_3\}$. We consider 3 stopwatches $\{x_1, x_2, x_3\}$ associated to respectively $\{t_1, t_2, t_3\}$. A stopwatch x represents the time elapsed since the associated transition t was last enabled (by the marking) and during which t is not inhibited. A stopwatch x is stopped, which is denoted by $\dot{x} = 0$, when the associated transition t is inhibited. A stopwatch x is reset when the associated transition t is *newly enabled* (and inhibited or not).

Since P_1 and P_2 are marked, t_3 is inhibited and $\dot{x_1} = 1, \dot{x_2} = 1, \dot{x_3} = 0$. Time elapses for x_1 and x_2, and t_2 becomes firable. After the firing of t_2, t_3 is not inhibited anymore and $\dot{x_1} = 1, \dot{x_3} = 1$. We have then $x_1 \in [3, 4]$ and $x_3 = 0$ because time did not pass for t_3 which was inhibited. t_1 can fire in the interval $[1, 3]$ and t_3 must still wait between 2 and 3 time units. Reachability graphs are given in figure 5.

2.2 Formal Semantics

Definition 1 (Time Petri Net with Inhibitor Hyperarcs). *A Time Petri nets with Inhibitor Hyperarcs is a n-tuple* $\mathcal{T} = (P, T, {}^\bullet(.), (.)^\bullet, \alpha, \beta, M_0, I)$, *where*

- $P = \{p_1, p_2, \ldots, p_m\}$ *is a non-empty finite set of* places,
- $T = \{t_1, t_2, \ldots, t_n\}$ *is a non-empty finite set of* transitions,

- ${}^{\bullet}(.) \in (\mathbb{N}^P)^T$ *is the* backward incidence function,
- $(.)^{\bullet} \in (\mathbb{N}^P)^T$ *is the* forward incidence function,
- $M_0 \in \mathbb{N}^P$ *is the* initial marking *of the net,*
- $\alpha \in (\mathbb{Q}^+)^T$ *and* $\beta \in (\mathbb{Q}^+ \cup \{\infty\})^T$ *are functions giving for each transition respectively its* earliest *and* latest *firing times* $(\alpha \leq \beta)$*,*
- I *is a finite set of* branch inhibitor hyperarcs*:*
  - *each hyperarc in I is a pair (Q, t) where $t \in T$ is a transition and $Q \in \mathbb{N}^P$ is the backward inhibition valuation of the hyperarc*[1]
  - *we note $I(t)$ the number of inhibitor hyperarcs of the transition $t \in T$,*
  - *Let $i \leq I(t)$ the ith inhibitor hyperarc of the transition $t \in T$ such that $\exists Q_i \in \mathbb{N}^P, (Q_i, t) \in I$. We note ${}^{\circ}t^i = Q_i$.*

In the example presented in figure 4b : $I(t_3) = 1, {}^{\circ}t_3^1 = [1, 1, 0]$ that represents $\{P_1, P_2\}$.

A *marking* M of the net is an element of $\mathbb{N}^P$ such that $\forall p \in P, M(p)$ is the number of tokens in the place p.

A transition t is said to be *enabled* by the marking M if $M \geq {}^{\bullet}t$, (*i.e.* if the number of tokens in M in each input place of t is greater or equal to the valuation on the arc between this place and the transition). We denote it by $t \in enabled(M)$.

A transition t is said to be *inhibited* by the marking M if there are tokens in all the places of one of its inhibitor hyperarcs: $\exists i \leq I(t), {}^{\circ}t^i < M$. We denote it by $t \in inhibited(M)$.

A transition t is said to be *firable* when it has been enabled and not inhibited for at least $\alpha(t)$ time units.

A transition t_k is said to be *newly* enabled by the firing of the transition t_i from the marking M, and we denote it by $\uparrow enabled(t_k, M, t_i)$, if the transition is enabled by the new marking $M - {}^{\bullet}t_i + t_i{}^{\bullet}$ but was not by $M - {}^{\bullet}t_i$, where M is the marking of the net before the firing of t_i. Formally,

$$\begin{aligned}&\uparrow enabled(t_k, M, t_i) = ({}^{\bullet}t_k \leq M - {}^{\bullet}t_i + t_i{}^{\bullet})\\ &\wedge((t_k = t_i) \vee ({}^{\bullet}t_k > M - {}^{\bullet}t_i))\end{aligned}$$

By extension, we will denote by $\uparrow enabled(M, t_i)$ the set of transitions newly enabled by firing the transition t_i from the marking M.

Let us consider the IHTPNs in figure 6. Let us assume that the initial marking is $M(P) = 1, M(P_{inh}) = 1$ and t_1 is firable. t_2 is enabled by the marking M and by the marking $M' = M - {}^{\bullet}t_1 + t_1{}^{\bullet}$ but not by $M - {}^{\bullet}t_1$. Transitions t_1 and t_2 are newly enabled by the firing of t_1. The only difference for the example 6b is that the transition t_2 is newly enabled by the firing of t_1 but remains inhibited.

Let us now assume the initial marking is $M(P) = 2, M(P_{inh}) = 1$ and t_1 is firable. t_2 is enabled by the marking M and by the marking $M' = M - {}^{\bullet}t_1 + t_1{}^{\bullet}$ and by $M - {}^{\bullet}t_1$. t_1 is newly enabled by the firing of t_1 but not t_2: t_2 remains enabled. The only difference for the example 6b is that the transition t_2 remains enabled and inhibited.

---

[1] If valuations are equal to zero or one, then Q can be represented by a set of places.

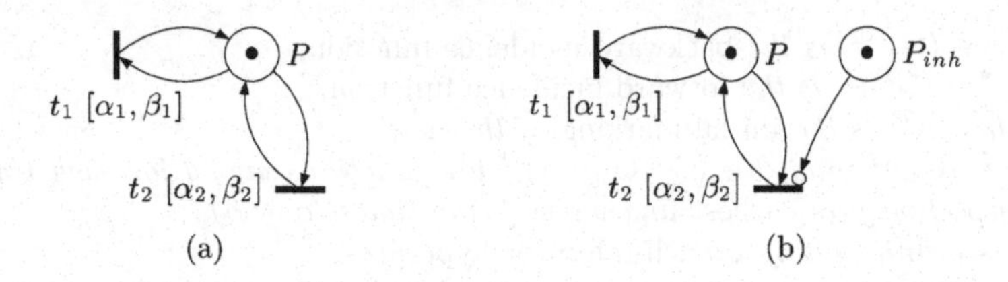

Fig. 6. Example of newly enabled transitions

We define the semantics of time Petri nets with Inhibitor Hyperarcs as *Timed Transition Systems* (TTS) [19]. In this model, two kinds of transitions may occur: *continuous* transitions when time passes and *discrete* transitions when a transition of the net fires.

A *valuation* is a mapping $\nu \in (\mathbb{R}^+)^T$ such that $\forall t \in T, \nu(t)$ is the time elapsed since t was last enabled and during which t remained not inhibited. ($\nu(t)$ represents the value of the stopwatch associated to t). Note that $\nu(t)$ is meaningful only if t is an enabled transition. $\overline{0}$ is the *null valuation* such that $\forall k, \overline{0}_k = 0$.

Definition 2 (Semantics of an IHTPN). *The semantics of a time Petri net with Inhibitor Hyperarcs $\mathcal{T}$ is defined as a TTS $\mathcal{S}_\mathcal{T} = (Q, q_0, \rightarrow)$ such that*

- $Q = \mathbb{N}^P \times (\mathbb{R}^+)^T$
- $q_0 = (M_0, \overline{0})$
- $\rightarrow \in Q \times (T \cup \mathbb{R}) \times Q$ *is the transition relation including a continuous transition relation and a discrete transition relation.*
  - *The continuous transition relation is defined* $\forall d \in \mathbb{R}^+$ *by :*

$$(M,\nu) \xrightarrow{d} (M,\nu') \text{ iff } \forall t_i \in T,$$
$$\begin{cases} \nu'(t_i) = \begin{cases} \nu(t_i) \text{ if } t_i \in enabled(M) \text{ and } t_i \in inhibited(M) \\ \nu(t_i) + d \text{ otherwise,} \end{cases} \\ M \geq {}^\bullet t_i \Rightarrow \nu'(t_i) \leq \beta(t_i) \end{cases}$$

  - *The discrete transition relation is defined* $\forall t_i \in T$ *by :*

$$(M,\nu) \xrightarrow{t_i} (M',\nu') \text{ iff },$$
$$\begin{cases} t_i \in enabled(M) \text{ and } t_i \notin inhibited(M), \\ M' = M - {}^\bullet t_i + t_i{}^\bullet, \\ \alpha(t_i) \leq \nu(t_i) \leq \beta(t_i), \\ \forall t_k \in T, \nu'(t_k) = \begin{cases} 0 \text{ if } \uparrow enabled(t_k, M, t_i), \\ \nu(t_k) \text{ otherwise} \end{cases} \end{cases}$$

Note that for transitions which are not enabled, the continuous transition relation of the semantics lets the valuations evolve. They could as well have been stopped.

Note also that, as for untimed Petri nets, branch inhibitor hyperarcs do not increase the expressivity of the model compared to simple inhibitor arcs.

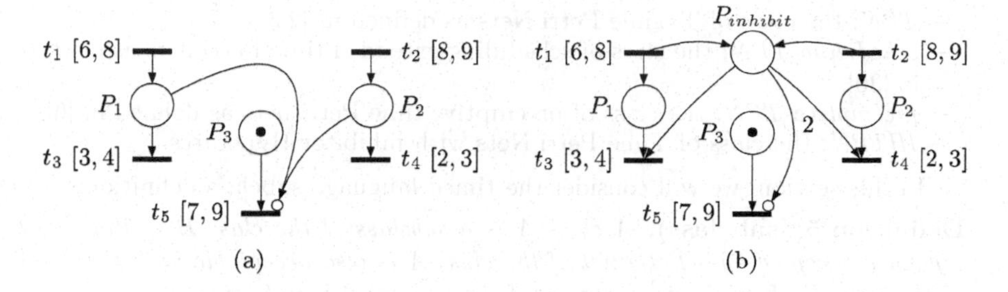

Fig. 7. Two bisimilar safe IHTPN

Figure 7 shows an example of how to translate a branch inhibitor hyperarc into simple inhibitor arcs for a safe IHTPN. The general case, for not necessarily bounded IHTPNs, is a bit more complicated and involves explicitly splitting transitions into a discrete and a continuous transitions and adding transitions not to be considered in the reachability graph. Still we are able to define a timed bisimulation between the IHTPN and its translation with simple inhibitor arcs. So, while they do not increase the expressivity of the model, branch inhibitor hyperarcs do allow a more convenient modeling. As for TPN, the behavior of an IHTPN can be defined by timed firing sequences which are sequences of pairs (t, d) where t is a transition of the IHTPN and $d \in \mathbb{R}^+$. Then a sequence $\omega = (t_1, d_1)(t_2, d_2) \cdots (t_n, d_n)$ indicates that t_1 is fired after d_1 time units, then t_2 is fired after d_2 time units and so on, so that transition t_i is fired at absolute time $\sum_{k=1}^{i} d_k$. A *marking* M is *reachable* in an IHTPN iff there is a timed firing sequence ω leading from the initial marking M_0 to M. It is common for TPN to define the *untimed* sequence from the timed one: if $\omega = (t_1, d_1)(t_2, d_2) \cdots (t_n, d_n)$ then $Untimed(\omega) = t_1 t_2 \cdots t_n$.

We can now extend the usual notion of accepted language for IHTPN.

Definition 3 (Accepted language). *The* language *accepted by an IHTPN is the set of all its timed firing sequences.*

Definition 4 (Untimed language). *The* untimed language *accepted by an IHTPN is the set of all its untimed firing sequences.*

3 Properties

3.1 Relation between Some Classes of Petri Nets and IHTPN

In this section, we will consider the following classes of Petri nets:

- *PN*: the class of classical Petri Nets,
- *priority-PN*: the class of Petri nets with priority as defined in [14],
- *bih-PN*: the class of Petri Nets with branch inhibitor hyperars as defined in [17],

- *TPN*: the class of T-Time Petri Nets as defined in [24],
- *scheduling-TPN*: the class of scheduling extended time Petri nets as defined in [30],
- *preemptive-TPN*: the class of preemptive time Petri nets as defined in [6],
- *IHTPN*: the class of Time Petri Nets with inhibitor Hyperarcs.

In this section, we will consider the timed language subclass definition:

Definition 5 (subclass). *A class A is a subclass of the class B if any timed language accepted by a Petri net of the class A is also acceptable by a Petri net of the class B. It is a strict subclass if the reciprocal is not true.*

As a consequence, classes of models are defined by the expressivity of these models as timed language acceptors.

Property 1 (Relation between TPN and IHTPN). Any TPN can be described by an IHTPN: a TPN is an IHTPN without inhibitor hyperarcs. However, it is easy to find an IHTPN which accepts the language $\forall n \in [0,1], \forall m \geq 0, (t_1, n)(t_2, m)(t_3, 1-n)$, while a TPN which accepts it does not exist. Consequently, *TPN* is a strict subclass of *IHTPN*.

Property 2 (Relation between PN and IHTPN). Any PN can be described by a TPN [23], therefore by an IHTPN. *PN* is a strict subclass of *IHTPN*.

Property 3 (Relation between bih-PN and IHTPN). Any PN with branch inhibitor hyperarcs can be described by an IHTPN. It is sufficient to take an IHTPN with the same discrete structure but in which all the firing intervals have a null earliest firing time or an infinite latest firing time. As PN with branch inhibitor hyperarcs is not a timed model, the reciprocal is not true in terms of timed language acceptance. Consequently, *bih-PN* is a strict subclass of *IHTPN*.

Property 4 (Relation between priority-PN and IHTPN). Any priority net can be described by a Petri net with inhibitor hyperarcs [17], therefore, by transitivity, by an IHTPN. *priority-PN* is a strict subclass of *IHTPN*.

Property 5 (Relation between scheduling-TPN or preemptive-TPN and IHTPN). It is easy to show that any scheduling-TPN [30, 21] and any preemptive-TPN [6] can be described by an IHTPN with an inhibitor hyperarc for each priority relation per resource. However, the reciprocal is not true: indeed, with scheduling-TPN and preemptive-TPN, it is not possible to model a circular priority relation which can definitively block the time flow of transitions while no concurrent transition runs (*i.e.* the required resources are free). Consequently, if there is an enabled transition for which time is blocked, that means that there is another one for which time is not blocked. As long as there is an enabled transition, then there will be in the future the firing of a transition. Conversely, IHTPN can model a circular relation of inhibition involving a definitive blocking of the time flow of enabled transition (see fig.8). Consequently, *scheduling-TPN* and *preemptive-TPN* are strict subclasses of *IHTPN*.

A consequence of the previous property is that IHTPNs are also adapted to the modeling of the real-time tasks scheduling.

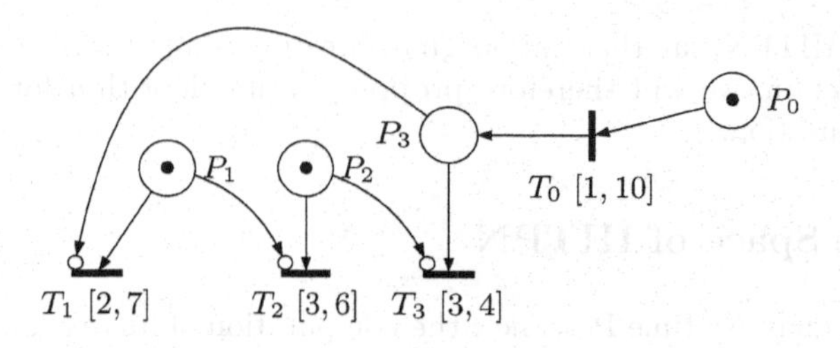

Fig. 8. IHTPN with deadlock

3.2 Boundedness and Reachability

Property 6 (Reachability and boundedness of marking problems are undecidable for IHTPN). Reachability and boundedness of marking problems are undecidable for TPN and for PN with inhibitor hyperarcs. TPNs and PNs with inhibitor hyperarcs are particular cases of IHTPNs. Consequently, reachability and boundedness of marking problems are undecidable for IHTPNs.

Remark 1. The set of reachable markings of an IHTPN is a subset of the set of reachable markings of the underlying PN. Moreover the set of reachable markings of an IHTPN is also a subset of the set of reachable markings of the underlying PN with inhibitor hyperarcs. However, the set of reachable markings of an IHTPN is not a subset of the set of reachable markings of the underlying TPN. See examples and reachability graph fig.1,2,3,4,5.

Property 7 (Sufficient and decidable condition of IHTPN marking boundedness). A not necessary but sufficient and decidable condition of IHTPN marking boundedness is: the underlying PN is bounded.

Property 8 (Sufficient and undecidable condition of IHTPN marking boundedness). A not necessary but sufficient and undecidable condition of IHTPN marking boundedness is: the underlying PN with inhibitor hyperarcs is bounded.

Definition 6 (k-boundedness). *A Petri Net is k-bounded iff for any place p of the net, no marking M such that $M(p) > k$ is reachable.*

Property 9 (k-boundedness problems for IHTPN). For classical TPN and for PN with inhibitor hyperarcs, k-boundedness is decidable because any k-bounded Petri Net has a finite set of reachable marking and for TPN, if the limits of the static firing intervals are rational numbers, it is possible to represent the infinite state space by a finite partitioning. However, for stopwatch automata, for instance, that finite partioning does not exist in the general case. While IHTPN is a model with stopwatches, it is not proved yet that IHTPNs belong to the same class as stopwatch automata. This means that k-boundedness is still an open problem for IHTPN.

The consequence of this section is that as the boundedness of marking problem is undecidable for IHTPN, we can try to compute the state space or to analyse an IHTPN but there is no guarantee for computation termination. In the next section, we will therefore propose a semi-algorithm for IHTPN state space computation.

4 State Space of IHTPN

In order to analyse a time Petri net, the computation of its reachable state space is required. However, the reachable state space of a time Petri net is obviously infinite, so a method has been proposed to partition it into a finite set of infinite *state classes* [4]. This method is briefly explained in the next subsection. The following paragraphs then describe its extension in order to compute the state space of IHTPN.

4.1 State Class Graph of a Time Petri Net

Given a bounded time Petri net, Berthomieu and Diaz have proposed a method for computing the state space as a finite set of *state classes* [4]. Basically a state class contains all the states of the net between the firing of two consecutive transitions.

Definition 7 (State class). *A* state class C*, of a time Petri net, is a pair* (M, D) *where* M *is a marking of the net and* D *a set of inequations called the* firing domain.

The inequations in D *are of two types [4]*

$$\begin{cases} \alpha_i \leq x_i \leq \beta_i \ (\forall i \text{ such that } t_i \text{ is enabled}), \\ -\gamma_{kj} \leq x_j - x_k \leq \gamma_{jk}, \ \forall j, k \text{ such that} \\ j \neq k \text{ and } (t_j, t_k) \in enabled(M)^2 \end{cases}$$

x_i *is the firing time of the enabled transition* t_i *relatively to the time when the class was entered in.*

Because of their particular form, the firing domains may be encoded using DBMs [12], which allow the use of efficient algorithms for the computation of classes.

Given a class $C = (M, D)$ and a firable transition t_f, computing the successor class $C' = (M', D')$ obtained by firing t_f consists of computing the new marking as usual and the new domain. This involves variable substitutions and eliminations and intersections, modeling the time that elapses, and computation of the canonical form of the DBM allowing efficient checks for equality. The overall complexity is $O(n^3)$ with n being the number of transitions enabled by M [3]. Computation of the state space of the TPN consists merely of the classical building of the reachability graph of state classes. That is to say, that starting from the initial state class, all the successors obtained by firing firable transitions are computed iteratively until all the produced successors have already been generated.

Example. Figure 9 shows an example of a TPN and its state class graph. The domains of the first two classes have been detailed, as well as the markings for each class. Notice that the sets of inequations of the domains fit in a DBM. This form is however insufficient to represent temporal domains with stopwatches.

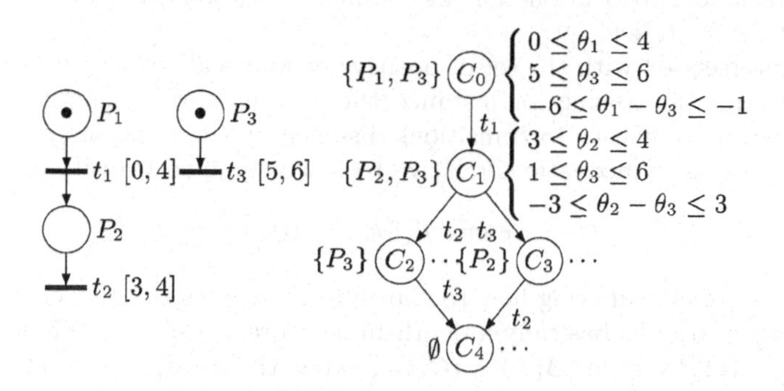

Fig. 9. A TPN and its state class graph

4.2 Exact State Space Computation of IHTPN Using Polyhedra

In the method previously presented, temporal domains are expressed as sets of linear inequations in the form of Difference Bound Matrices (DBM). The form of a DBM is not sufficient to express the temporal domain with the presence of stopwatches. Some changes are proposed to allow its construction.

Precisely, the variable substitution in the firing domain is now only done for enabled transitions that are *not inhibited.*

While we still define a state class of an IHTPN as a marking and a domain, the general form of the domain is not preserved. The new general form is that of a polyhedron with constraints involving up to n variables, with n being the number of transitions enabled by the marking of the class:

$$\begin{cases} \alpha_i \leq \theta_i \leq \beta_i, \forall t_i \in enabled(M), \\ a_{i_1}\theta_{i_1} + \cdots + a_{i_n}\theta_{i_n} \leq \gamma_{i_1 \ldots i_n}, \\ \forall(t_{i_1}, \ldots, t_{i_n}) \in enabled(M)^n \\ \text{and with } (a_{i_0}, \ldots, a_{i_n}) \in \mathbb{Z}^n. \end{cases}$$

Algorithm of Computation. We will now show the algorithm for computing the successor classes of a given class and then define the state class graph for IHTPN. But first, we need a new definition for the firability of a transition from a class:

Definition 8 (firability). *Let $C = (M, D)$ be a state class of an IHTPN. A transition t_i is said to be* firable *from C iff there exists a solution $(\theta_0^*, \ldots, \theta_{n-1}^*)$ of D, such that $\forall j \in [\![0, n-1]\!] - \{i\}$, s.t. t_j is not inhibited , $\theta_i^* \leq \theta_j^*$.*

Successors of a class. Given a class $C = (M, D)$ and a firable transition t_f, the class $C' = (M', D')$ obtained from C by the firing of t_f is given by

- $M' = M - {}^\bullet t_f + t_f{}^\bullet$
- D' is computed along the following steps, and noted $next(D, t_f)$
  1. variable substitutions for all enabled transitions that are *not* inhibited t_j: $\theta_j = \theta_f + \theta'_j$,
  2. intersection with the set of positive or null reals $\mathbb{R}^+$: $\forall i, \theta'_i \geq 0$,
  3. elimination (using for instance the Fourier-Motzkin method [9]) of all variables relative to transitions disabled by the firing of t_f,
  4. addition of inequations relative to newly enabled transitions

$$\forall t_k \in \uparrow enabled(M, t_f), \alpha(t_k) \leq \theta'_k \leq \beta(t_k).$$

State class graph. Knowing how to compute the successors of a class, the state space computation is basically a depth-first or breadth-first graph generation.

For an IHTPN $\mathcal{T}$, let $\Delta(\mathcal{T}) = (\mathcal{C}, C_0, \rightarrow)$ be the transition system such that:

- $\mathcal{C} = \mathbb{N}^P \times \mathbb{R}^T$,
- $C_0 = (M_0, D_0)$, where M_0 is the initial marking and $D_0 = \{\alpha(t_i) \leq \theta_i \leq \beta(t_i) | \, t_i \in enabled(M_0)\}$,
- $\rightarrow \in \mathcal{C} \times T \times \mathcal{C}$ is the transition relation defined by :

$$(M, D) \xrightarrow{t} (M', D') \text{ iff } \begin{cases} M' = M - {}^\bullet t + t^\bullet, \\ t \text{ is firable from } (M, D), \\ D' = next(D, t), \end{cases}$$

Then we can write the state class graph as the quotient of $\Delta(\mathcal{T})$ by a "suitable" equivalence relation. This "suitable" relation may be equality as in definition 9, in which case the state class graph has the same untimed language as its IHTPN. Or it can be inclusion as in definition 10, and then the untimed language is lost but we still have the set of reachable markings.

Given a set of inequations D, we note $[\![D]\!]$ the set of solutions of D. With this notation, we can now define equality and inclusion of classes:

Definition 9 (Equality of state classes). *Two classes $C_1 = (M_1, D_1)$ and $C_2 = (M_2, D_2)$ are* equal *if $M_1 = M_2$ and $[\![D_1]\!] = [\![D_2]\!]$.*

Definition 10 (Inclusion of state classes). *The state class $C = (M_1, D_1)$ is* included *in the state class $C_2 = (M_2, D_2)$ if $M_1 = M_2$ and $[\![D_1]\!] \subset [\![D_2]\!]$.*

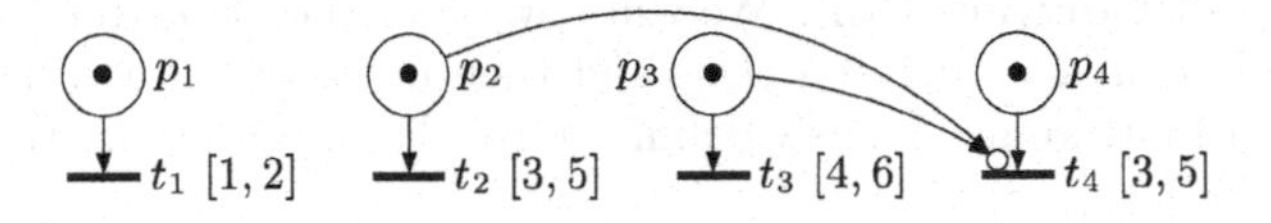

Fig. 10. A time Petri net with inhibitor hyperarcs

Example. Figure 10 shows an example of a simple IHTPN. Its initial class is $C = (M, D)$ with $M = \{p_1, p_2, p_3, p_4\}$ and

$$D = \begin{cases} 1 \leq x_1 \leq 2, \\ 3 \leq x_2 \leq 5, \\ 4 \leq x_3 \leq 6, \\ 3 \leq x_4 \leq 5, \\ -4 \leq x_1 - x_2 \leq -1, \\ -5 \leq x_1 - x_3 \leq -2, \\ -4 \leq x_1 - x_4 \leq -1, \\ -3 \leq x_2 - x_3 \leq 1, \\ -2 \leq x_2 - x_4 \leq 2, \\ -1 \leq x_3 - x_4 \leq 3 \end{cases} \tag{1}$$

t_1, t_2, t_3 and t_4 are enabled but t_4 is inhibited, hence:

$$\begin{cases} \dot{x}_1 = 1, \\ \dot{x}_2 = 1, \\ \dot{x}_3 = 1, \\ \dot{x}_4 = 0, \end{cases} \tag{2}$$

t_1 is clearly firable. Let us compute the domain of the class obtained by firing t_1. The first step is the variable substitution $x_i \leftarrow x'_i + x_1$ for all enabled transitions that are not inhibited but t_1. The domain becomes

$$\begin{cases} 1 \leq x_1 \leq 2, \\ 3 \leq x_1 + x'_2 \leq 5, \\ 4 \leq x_1 + x'_3 \leq 6, \\ 3 \leq x'_4 \leq 5, \\ -4 \leq x_1 - x_1 - x'_2 \leq -1, \\ -5 \leq x_1 - x_1 - x'_3 \leq -2, \\ -4 \leq x_1 - x_4 \leq -1, \\ -3 \leq x_1 + x'_2 - x_1 - x'_3 \leq 1, \\ -2 \leq x_1 + x'_2 - x_4 \leq 2, \\ -1 \leq x_1 + x'_3 - x_4 \leq 3 \end{cases} \tag{3}$$

The next step is the elimination of the variable x_1. We use the Fourier-Motzkin method. For that, we rewrite the inequations as follows:

$$\begin{cases} 1 \leq x_1, & x_1 \leq 2, \\ 3 - x'_2 \leq x_1, & x_1 \leq 5 - x'_2, \\ 4 - x'_3 \leq x_1, & x_1 \leq 6 - x'_3, \\ -4 + x_4, \leq x_1 & x_1 \leq -1 + x_4, \\ -2 + x_4 - x'_2 \leq x_1, & x_1 \leq 2 + x_4 - x'_2, \\ -1 + x_4 - x'_3 \leq x_1, & x_1 \leq 3 + x_4 - x'_3, \\ 3 \leq x_4 \leq 5, \\ -3 \leq x'_2 - x'_3 \leq 1, \\ -4 \leq -x'_2 \leq -1, \\ -5 \leq -x'_3 \leq -2 \end{cases} \tag{4}$$

The Fourier-Motzkin method then consists of writing that the system has solutions if and only if the lower bounds of x_1 are less or equal to the upper bounds. The obtained system is then equivalent to the initial one. After a few simplifications we obtain

$$D' = \begin{cases} 1 \leq x'_2 \leq 4, \\ 2 \leq x'_3 \leq 5, \\ 3 \leq x_4 \leq 5, \\ -3 \leq x'_2 - x'_3 \leq 1, \\ 0 \leq x'_2 - x_4 \leq 1, \\ 1 \leq x'_3 - x_4 \leq 2, \\ 4 \leq x'_2 + x_4 \leq 9, \\ 5 \leq x'_3 + x_4 \leq 10, \\ 0 \leq x'_2 + x_4 - x'_3 \leq 6, \\ 2 \leq x'_3 + x_4 - x'_2 \leq 8 \end{cases} \tag{5}$$

What we can see here is that eight inequations, given on the last four lines, are generated, which cannot be expressed with a DBM. Furthermore, we can easily see that those new inequations may give even more complex inequations (*i.e.* involving more variables) when firing another transition from the obtained class. However, in this example, those extra eight inequations are redundant with the "simple" inequations: for instance, using "simple" inequations $x'_2 \leq 4$ and $x'_4 \leq 5$ we find again $x'_2 + x'_4 \leq 9$. Formally, this means that:

$$[\![D']\!] = \left[\!\!\left[\begin{cases} 1 \leq x'_2 \leq 4, \\ 2 \leq x'_3 \leq 5, \\ 3 \leq x_4 \leq 5, \\ -3 \leq x'_2 - x'_3 \leq 1, \\ 0 \leq x'_2 - x_4 \leq 1, \\ 1 \leq x'_3 - x_4 \leq 2 \end{cases} \right]\!\!\right] \tag{6}$$

This redundancy occurs for many classes in the general case (but unfortunately not always) so it makes sense to ignore the non-DBM form inequations for a faster computation of the state class graph. This may lead, of course, to an overapproximation.

4.3 Overapproximation of the State Space

Manipulating polyhedra in the general case involves a very important computing cost. In order to be able to keep our algorithms efficient for IHTPN, we approximate the polyhedron representing the firing domain to the smallest DBM containing it. By doing this, we clearly add states in our classes that should not be reachable and thus we do an overapproximation. This is illustrated by the net in Figure 11.

After the firing sequence t_4, t_1, t_5, the transition t_6 is not firable because either t_2 or t_3 (depending on the firing time of t_4) must be fired first. The class obtained is:

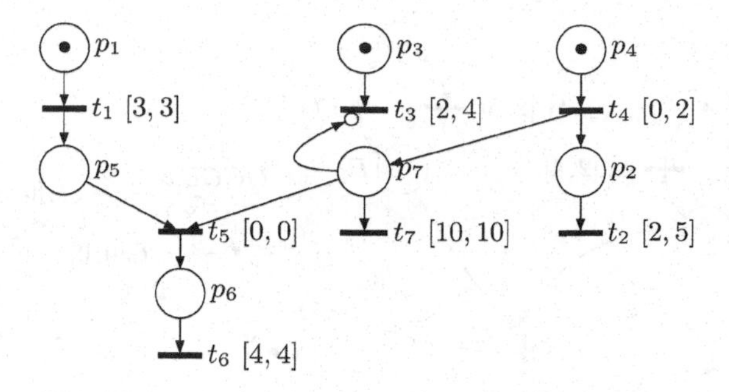

Fig. 11. A time Petri Net with Inhibitor Hyperarcs

$$\begin{cases} \{p_2, p_3, p_6\}, \\ \begin{cases} 0 \leq x_2 \leq 4, \\ 0 \leq x_3 \leq 4, \\ 4 \leq x_6 \leq 4, \\ -4 \leq x_2 - x_3 \leq 4, \\ -4 \leq x_2 - x_6 \leq 0, \\ -4 \leq x_3 - x_6 \leq 0, \\ 1 \leq x_2 + x_3 \leq 6 \end{cases} \end{cases}$$

We can easily see that t_6 is indeed not firable, for this implies $x_2 = x_3 = x_6 = 4$ and thus $x_2 + x_3 = 8$. But if we remove the $x_2 + x_3 \leq 6$ constraint in order to keep a DBM form, t_6 becomes firable. So we have here an overapproximation.

However, for the verification of safety properties the overapproximation is not a too big concern. Since we want to ensure that something "bad" never happens, we only need to check a set of states which contains the actual state space of the IHTPN. Of course, there is still a risk of being pessimistic.

5 Verification of Timed Reachability Properties

A consequence of the state space computation is that using the classical observer notion, we can verify varied timed reachability properties. An observer consists of adding places and transitions, which model the property (discrete or quantitative) to check, to the Petri net. Checking the property is transformed in checking the reachability of a given marking of the observer. In [31], the authors present generic TPN observers to model properties like absolute or relative time between the firing of transitions, causality or simultaneity. It is then possible to check properties on IHTPN with the same generic observers (thanks to property 1 in 3).

Let us consider the net in figure 12. It represents a simple IHTPN and an observer in dash lines. The observer allows to check the property: "2 successive occurrences of t_0 always happen in less than 12 time units". The property is *false* iff a marking M such that $M(FALSE) > 0$ is reachable (i.e. iff a run such that the place $FALSE$ has a token exists).

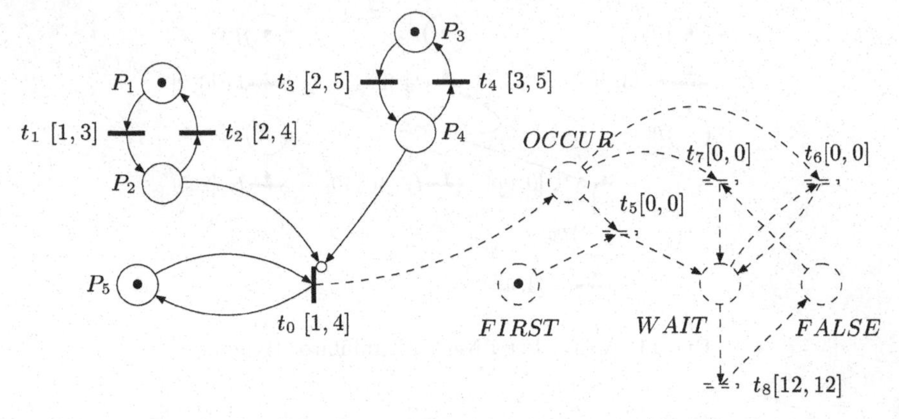

Fig. 12. Example of an IHTPN and an observer (dash lines)

Constructing the state space and looking for a run with a token in $FALSE$ allows to conclude that the property is false. Indeed, let us consider, for instance, the run[2] $s_0 \xrightarrow{2}_{t_0} s_1 \xrightarrow{0}_{t_5} s_2 \xrightarrow{0}_{t_1} s_3 \xrightarrow{0}_{t_3} s_4 \xrightarrow{4}_{t_2} s_5 \xrightarrow{0}_{t_4} s_6 \xrightarrow{2}_{t_1} s_7 \xrightarrow{0}_{t_3} s_8 \xrightarrow{4}_{t_2} s_9 \xrightarrow{1}_{t_4} s_{10} \xrightarrow{1}_{t_8} s_{11} \xrightarrow{0}_{t_0} s_{12}$.

The transition t_0 is inhibited in s_4 and s_8 ; no time elapse in s_1, s_2, s_3, s_5, s_7 and s_{11}. In this run, two successive occurrences of t_0 happen in exactly 12 time units. This run led to $M(FALSE) = 1$ in s_{11}, s_{12}.

Generally there is no need to compute the whole state space: the computation of the state space can be stopped at the first marking verifying a property.

6 Conclusion

In this paper, we have defined an extension of T-time Petri nets with inhibitor hyperarcs (IHTPN). In this model, we can consider that stopwatches are associated with transitions. A stopwatch can be reset, stopped and started by using classical arcs and inhibitor hyperarcs. It allows the memorisation of the progress status of an action that is stopped and resumed. We have also proposed a method for the computation of the state space of an IHTPN. This includes an exact computation using a general polyhedron representation of timing constraints, and an overapproximation of the polyhedra to allow more efficient compact abstract representations of the state space based on DBM (Difference Bound Matrix).

We have shown that Petri nets with inhibitor hyperarcs, time Petri nets, preemptive-TPN and scheduling-TPN are all strict subclasses of IHTPN. As a consequence, IHTPN can be used in the context of scheduling verification but IHTPN are also able to express more general problems including circular relation of inhibition involving a definitive blocking of the time flow of enabled transition. It can thus be important to have a model able to express this kind of deadlocks in order to detect them.

[2] We note $S \xrightarrow{d}_{e} S'$ to indicate a discrete step e preceded by a time flow of d time units.

Finally, we have shown how to check complex timing properties on IHTPN by using observers.

Future works include the study of the model-checking of properties expressed in temporal logics such as Timed Computation Tree Logic (TCTL). This can be carried out when the state space is computable, and when it is not the case, it can be done on the fly for some properties.

We also plan to tackle some of the problems that remain open such as the decidability of k-boundedness for IHTPN.

References

1. P. A. Abdulla and A. Nylén. Timed Petri nets and BQOs. In *22nd International Conference on Application and Theory of Petri Nets (ICATPN'01)*, volume 2075 of *Lecture Notes in Computer Science*, pages 53–72, Newcastle upon Tyne, United Kingdom, june 2001. Springer-Verlag.
2. T. Agerwala. A complete model for representing the coordination of asynchronous processes. Technical report, John Hopkins University, Baltimore, Maryland., 1974.
3. B. Berthomieu. La méthode des classes d'états pour l'analyse des réseaux temporels. In *3e congrès Modélisation des Systèmes Réactifs (MSR'2001)*, pages 275–290, Toulouse, France, october 2001. Hermes.
4. B. Berthomieu and M. Diaz. Modeling and verification of time dependent systems using time Petri nets. *IEEE transactions on software engineering*, 17(3):259–273, 1991.
5. B. Berthomieu and F. Vernadat. State class constructions for branching analysis of time Petri nets. In *9th International Conference on Tools and Algorithms for the Construction and Analysis of Systems (TACAS'2003)*, pages 442–457. Springer–Verlag, Apr 2003.
6. G. Bucci, A. Fedeli, L. Sassoli, and E. Vicario. Modeling flexible real time systems with preemptive time Petri nets. In *15th Euromicro Conference on Real-Time Systems (ECRTS'2003)*, pages 279–286, 2003.
7. Franck Cassez and Kim Guldstrand Larsen. The impressive power of stopwatches. In Catuscia Palamidesi, editor, *11th International Conference on Concurrency Theory, (CONCUR'2000)*, number 1877 in Lecture Notes in Computer Science, pages 138–152, University Park, P.A., USA, July 2000. Springer-Verlag.
8. G. Chiola, S. Donatelli, and G. Franceshinis. Priorities, inhibitor arcs and concurrency in PT nets. In *12th international conference on Application and Theory of Petri Nets*, pages 182–205, 1991.
9. G. B. Dantzig. Linear programming and extensions. *IEICE Transactions on Information and Systems*, 1963.
10. D. de Frutos Escrig, V. Valero Ruiz, and O. Marroquín Alonso. Decidability of properties of timed-arc Petri nets. In *21st International Conference on Application and Theory of Petri Nets (ICATPN'00)*, volume 1825 of *Lecture Notes in Computer Science*, pages 187–206, Aarhus, Denmark, june 2000. Springer-Verlag.
11. M. Diaz and P. Senac. Time stream Petri nets: a model for timed multimedia information. *Lecture Notes in Computer Science*, 815:219–238, 1994.
12. D. L. Dill. Timing assumptions and verification of finite-state concurrent systems. In *Workshop Automatic Verification Methods for Finite-State Systems*, volume 407, pages 197–212, 1989.

13. G. Gardey, O. H. Roux, and O. F. Roux. A zone-based method for computing the state space of a time Petri net. In *In Formal Modeling and Analysis of Timed Systems, (FORMATS'03)*, volume LNCS 2791. Springer, September 2003.
14. M. Hack. Petri net language. *Computation Structures Group Memo 127, MIT*, 1975.
15. R. Janicki and M. Koutny. Invariant semantics of nets with inhibitor arcs. In *In CONCUR'91, lncs vol.527*, pages 317–331, 1991.
16. R. Janicki and M. Koutny. Semantics of inhibitor nets. *Information and Computation*, 123(1):1–15, 1995.
17. R. Janicki and M. Koutny. On causality semantics of nets with priorities. *Fundamenta Informaticae*, 38(3):223–255, 1999.
18. Wael Khansa, J.-P. Denat, and S. Collart-Dutilleul. P-Time Petri Nets for manufacturing systems. In *International Workshop on Discrete Event Systems, WODES'96*, pages 94–102, Edinburgh (U.K.), august 1996.
19. K. G. Larsen, P. Pettersson, and W. Yi. Model-checking for real-time systems. In *Fundamentals of Computation Theory*, pages 62–88, 1995.
20. J. Lilius. Efficient state space search for time Petri nets. In *MFCS Workshop on Concurrency '98*, volume 18 of *ENTCS*. Elsevier, 1999.
21. D. Lime and O. (H.) Roux. Expressiveness and analysis of scheduling extended time Petri nets. In *5th IFAC International Conference on Fieldbus Systems and their Applications, (FET'03)*. Elsevier Science, July 2003.
22. D. Lime and O. (H.) Roux. State class timed automaton of a time Petri net. In *10th International Workshop on Petri Nets and Performance Models, (PNPM'03)*. IEEE Computer Society, September 2003.
23. M. Menasche. *Analyse des réseaux de Petri temporisés et application aux systèmes distribués*. PhD thesis, Université Paul Sabatier, Toulouse, France, 1982.
24. P. M. Merlin. *A study of the recoverability of computing systems*. PhD thesis, Department of Information and Computer Science, University of California, Irvine, CA, 1974.
25. Y. Okawa and T. Yoneda. Schedulability verification of real-time systems with extended time Petri nets. *International Journal of Mini and Microcomputers*, 18(3):148–156, 1996.
26. J.L. Peterson. Petri net theory and the modeling of systems. *Prentice-Hall, New-York*, 1981.
27. M. Pezze and M. Toung. Time Petri nets: A primer introduction. Tutorial presented at the Multi-Workshop on Formal Methods in Performance Evaluation and Applications, Zaragoza, Spain, september 1999.
28. L. Popova. On time Petri nets. *Journal on Information Processing and Cybernetics, EIK*, 27(4):227–244, 1991.
29. C. Ramchandani. *Analysis of asynchronous concurrent systems by timed Petri nets*. PhD thesis, Massachusetts Institute of Technology, Cambridge, MA, 1974. Project MAC Report MAC-TR-120.
30. O. H. Roux and A.-M. Déplanche. A t-time Petri net extension for real time-task scheduling modeling. *European Journal of Automation (JESA)*, 36(7), 2002.
31. J. Toussaint, F. Simonot-Lion, and Jean-Pierre Thomesse. Time constraint verification methods based on time Petri nets. In *6th Workshop on Future Trends in Distributed Computing Systems (FTDCS'97)*, pages 262–267, Tunis, Tunisia, 1997.
32. T. Yoneda and H. Ryuba. CTL model checking of time Petri nets using geometric regions. *IEICE Transactions on Information and Systems*, E99-D(3):297–396, march 1998.

Transit Case Study

Eric Verbeek[1] and Robert van der Toorn[2]

[1] Eindhoven University of Technology
P.O. Box 513, NL-5600 MB Eindhoven, The Netherlands
h.m.w.verbeek@tm.tue.nl
http://www.tm.tue.nl/it/staff/everbeek/

[2] Deloitte Management & ICT Consultants
The Netherlands
RvanderToorn@deloitte.nl

Abstract. One of the key issues of object-oriented modeling is inheritance. It allows for the definition of a subclass that inherits features from some superclass. When considering the dynamic behavior of objects, as captured by their life cycles, there is no general agreement on the meaning of inheritance. Basten and Van der Aalst introduced the notion of life-cycle inheritance for this purpose. Unfortunately, the search tree needed for deciding life-cycle inheritance is in general prohibitively large. This paper presents a comparative study between two possible algorithms. The first algorithm uses structural properties of both the base life cycle and the potential sub life cycle to prune the search tree, while the second is a plain exhaustive search algorithm. Test cases show that the computation times of the second algorithm can indeed be prohibitively expensive (weeks), while the computation times of the first algorithm are all within acceptable limits (seconds). An unexpected result of this case study is that it shows that we need tools for checking life-cycle inheritance.

1 Introduction

1.1 Inheritance of Behavior

One of the main goals of object-oriented design is the reuse of system components. A key concept to achieve this goal is the concept of inheritance. The inheritance mechanism allows the designer to specify a class, the subclass, that inherits features of some other class, its superclass. Thus, it is possible to specify that the subclass has the same features as the superclass, but that in addition it may have some other features.

The Unified Modeling Language (UML) [12, 7, 9] has been accepted throughout the software industry as the standard object-oriented framework for specifying, constructing, visualizing, and documenting software-intensive systems. The development of UML began in late 1994, when Booch and Rumbaugh of Rational Software Corporation began their work on unifying the OOD [6] and OMT [11] methods. In the fall of 1995, Jacobson and his Objectory company joined Rational, incorporating the OOSE method [10] in the unification effort.

J. Cortadella and W. Reisig (Eds.): ICATPN 2004, LNCS 3099, pp. 391–410, 2004.
© Springer-Verlag Berlin Heidelberg 2004

The informal definition of inheritance in UML states the following: "The mechanism by which more specific elements incorporate structure and behavior defined by more general elements." [12]. However, only the class diagrams, describing purely structural aspects of a class, are equipped with a concrete notion of inheritance. It is implicitly assumed that the behavior of the objects of a subclass, as defined by the object life cycle, is an extension of the behavior of the objects of its superclass. In the literature, several formalizations of what it means for an object life cycle to extend the behavior of another object life cycle have been studied; see [5] for an overview. Combining the usual definition of inheritance of methods and attributes with a definition of inheritance of behavior yields a complete formal definition of inheritance, thus, stimulating the reuse of life-cycle specifications during the design process. One possible formalization of behavioral inheritance is called life-cycle inheritance [5]:

> An object life cycle is a subclass of another object life cycle under life-cycle inheritance if and only if it is not possible to distinguish the external behavior of both when the new methods, that is, the methods only present in the potential subclass, are either blocked or hidden.

The notion of life-cycle inheritance has been shown to be a sound and widely applicable concept. In [5], it has been shown that it captures extensions of life cycles through common constructs such as parallelism, choices, sequencing and iteration. In [3], it is shown how life-cycle inheritance can be used to analyze the differences and the commonalities in sets of object life cycles. Furthermore, in [1], the notion of life-cycle inheritance has been successfully lifted to the various behavioral diagram techniques of UML. Also, life-cycle inheritance has been successfully applied to the workflow-management domain. There is a close correspondence between object life cycles and workflow processes. Behavioral inheritance can be used to tackle problems related to dynamic change of workflow processes [4]; furthermore, it has proven to be useful in producing correct inter-organizational workflows [2]. Finally, life-cycle inheritance has also been successfully applied in Component-Based Software Design (CBSD) [13]. In CBSD, inheritance of behavior is used, in particular, with respect to refinement and evolution of software architectures rather than the reuse of components. Different from the work we already mentioned on Petri nets and inheritance [5], in CBSD not only consider the processes of individual components, but also the interactions between several processes of different components are important. This is different from considering a process of a component in isolation. When we consider a process of a component in a software architecture, there is always an environment which is limiting the behavior of the component. These limitations should be considered carefully, otherwise they may lead to undesired behavior.

1.2 Example: A Requisition Process

We use a requisition process as an example how life-cycle inheritance can be used in the workflow domain. Through this requisition process, employees of some company are able to purchase items they need for doing their jobs. After submitting a request for requisition, the manager of the requestor is selected to approve the requisition. If the requestor has no manager, then the requisition is rejected immediately. Otherwise, two tasks are started in parallel: we record the fact that the approval of the requisition is forwarded to the selected manager, and we notify the requestor of this forward. After both tasks have been completed, we notify the selected manager that there is a requisition for him to approve. If the manager does not react in a timely manner, s/he is notified again, until s/he either approves or rejects the requisition. If s/he approves the requisition, we verify whether the managers spending limit is sufficient for the requisition. If so, the approval of the requisition is recorded and the requestor is notified. If not, the requisition is forwarded for approval to the next higher manager.

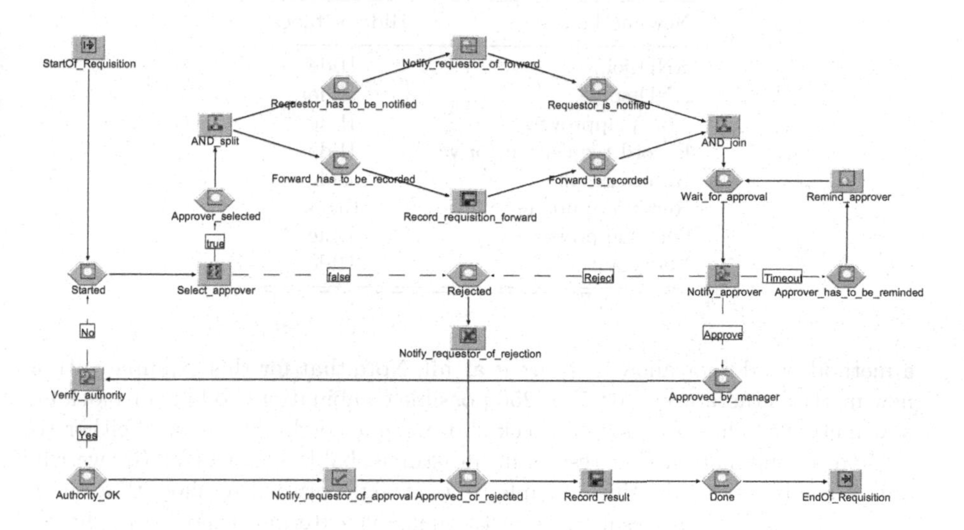

Fig. 1. A process modeled using the COSA Network editor

Figure 1 shows how this requisition process can be modeled using the COSA workflow management system (WFMS). COSA models correspond to Petri nets in a straightforward way: rectangular shapes correspond to transitions, whereas hexagonal shapes correspond to places. In contrast to this, Figure 2 models only a requestor's view on this process using the Protos business process reengineering (BPR) tool. This Protos model also corresponds to a Petri net: rectangular shapes correspond to transitions, whereas circular shapes correspond to places. Using the life-cycle inheritance relation and assuming that tasks correspond to methods, we can answer the question whether this requestor's view matches the process as modeled in COSA.

Table 1 shows a possible combination for hiding and blocking the new methods in the COSA model. If we hide a method, we allow it to occur but ignore its occurrences (it's assumed to be internal, not external), whereas if we block

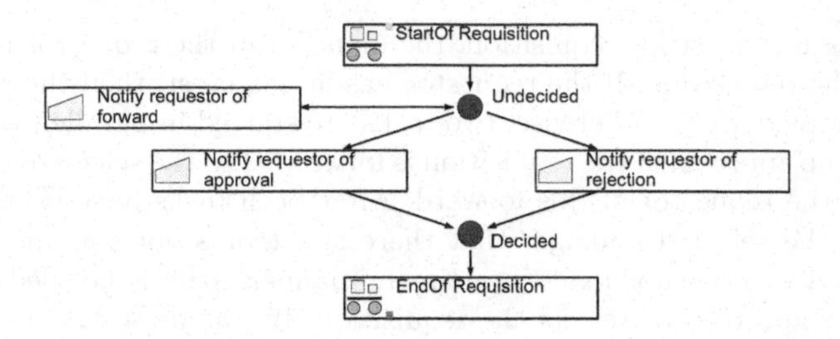

Fig. 2. A process modeled using the Protos tool

Table 1. A possible combination for hiding and blocking new methods

New method	Hide or block?
AND_join	Hide
AND_split	Hide
Notify_approver	Hide
Record_requisition_forward	Hide
Record_result	Hide
Remind_approver	Block
Select_approver	Hide
Verify_authority	Hide

a method, we do not allow it to occur at all. Note that for this example with 8 new methods there would be $2^8 = 256$ possible combinations to hide or block all new methods. Thus, we have to check 256 possible combinations until either (i) we find a combination that results in indistinguishable behavior or (ii) we run out of combinations. In this example, it is straightforward to show that there is no combination that results in indistinguishable external behavior: In the requestor's view, the requestor can get a notification of approval without having received a notification of forward, whereas this is impossible in the COSA model.

1.3 Backtracking Algorithm

The basis of deciding life-cycle inheritance is an equivalence check (to decide indistinguishable behavior), namely a branching bisimilarity check on the state spaces of both object life cycles (the base object life cycle and the potential sub object life cycle). Particularly in the workflow domain, these state spaces can be large (up to millions of states each). Therefore, such a check might be time-consuming despite the fact that efficient algorithms exist to check branching bisimilarity on state spaces [8]. As the example above has indicated, an exhaustive search algorithm for deciding life-cycle inheritance might require many equivalence checks on these state spaces: One check for every possible combination of hiding and blocking the new methods in the potential sub object life cycle, while the number of possible combinations is exponential in the number of new methods. The combination of the large state spaces and the exponential factor results in an exhaustive search algorithm that is prohibitively expensive in many cases.

In [15], a Petri-net-based backtracking algorithm has been introduced that is based on efficient pruning of the possible combinations. The main goal of this backtracking algorithm is to reduce the number of branching bisimilarity checks, using structural properties of the object life cycles at hand. The first experiments in [15] have shown that this backtracking algorithm does indeed efficiently and effectively reduce the search space. However, in [15] the backtracking algorithm was only compared to the exhaustive search algorithm using toy object life cycles. This paper presents a case study in which both algorithms are compared using a number of real-life object life cycles.

1.4 Overview

Section 2 introduces the case study. Section 3 presents it results and concludes that the life-cycle inheritance relation is absent between some object life cycles. Section 4 discusses how we corrected some of the object life cycles to obtain life-cycle inheritance. Section 5 presents the results after the corrections have been made. Section 6 concludes the paper.

2 The Transit Case

In this section we present the Transit case. The Transit system (Office for Official Publications of the European Communities 2001) is a customs system facilitating the registration and declaration of movements of goods within the Community and EFTA (European Free Trade Area) countries. Within these countries these goods may be moved under certain conditions without payment of the duties and taxes and without having to comply with any other relevant verification measures such as foreign trade requirements. Potential duties and taxes are secured by guarantees which become enforceable in the event of irregularities. The Transit process contains the following steps. A trader in a country (Trader at Destination) buys goods from trader in another country (Trader at Departure). Before the transport of the goods may start, the trader has to declare the goods at the customs office in his country (Customs Office of Departure). If the customs office accepts this declaration, then he has to arrange a financial guarantee which enforces him to fulfill his duties. Next, a customs officer may actually check the goods at the trader premises. If the results are satisfactory, then the goods may be transported. When the goods arrive at their destination in another country, the other trader has to declare the arrival of the goods at the customs office in that country (Customs Office of Arrival). Again the authorities, that are informed by their colleagues in the country of departure, have the possibility to verify the goods (either at their premises or at those of the trader) and if there are no irregularities, then the goods and the financial guarantees are released.

The second author used the Transit case to show how a set of Sequence Diagrams of a system may be used to construct an object life cycle (OLC), in

Table 2. Statistics on the six OLC's of the β case study

OLC	Places	Transitions	Reachable states
transit1	39	24	77
transit2	50	35	88
transit3	54	39	91
transit4	68	51	115
transit5	75	58	143
transit6	99	82	217

terms of a Petri net, which is an integrated system specification that comprises all possible Sequence Diagrams of the system [13]. He started with a simple object life cycle which supports a single Sequence Diagram. Then he extended this object life cycle in a number of iterations into an object life cycle which supports all Sequence Diagrams. In each iteration, he extended the functionality of an object life cycle with a number of new Sequence Diagrams. By using the life-cycle inheritance notion, he checked (manually) each extended object life cycle to make sure that the added functionality did not disturb the already existing functionality. For the remainder of this paper, we refer to this case study as the α case study.

The α case study resulted in figures of six object life cycles, transit1, ..., transit6, which were used by the first author to compare the performance of both the backtracking algorithm and the exhaustive search algorithm. For sake of completeness, we mention that the first author first had to transform these six figures into six object life cycles by hand. Table 2 shows some statistics on these six object life cycles. For the remainder of this paper, we refer to this comparison case study as the β case study. The β case study was performed on a Pentium 4 2.00 GHz computer with 256 Mb of RAM running Windows 2000 SP 3. Please note that this paper presents the results of the β case study but not of the α case study. For the results of the α case study we refer to [13].

3 Performance Results on the β Case Study

Our primary goal with the β case study was to see how the performance of our backtracking algorithm (BA) compares to the performance of an exhaustive search algorithm (ESA). For this reason, we ran both algorithms on every object life cycle (OLC) and its extension, using a prototype of the workflow verification tool Woflan [14, 16] which implements both algorithms. Table 3 shows the results. The computation times mentioned are the computation times of the algorithms

Table 3. Performance results of the β case study

potential sub OLC	Life-cycle inheritance?	Time: BA (in seconds)	(99% conf. interval)	Time: ESA (in seconds)	(99% conf. interval)
transit2	No	7.43×10^{-2}	$\pm\ 2.66\times10^{-4}$	1.69×10^{-1}	$\pm\ 3.52\times10^{-4}$
transit3	Yes	2.19×10^{-2}	$\pm\ 2.40\times10^{-4}$	1.99×10^{-2}	$\pm\ 2.45\times10^{-4}$
transit4	Yes	2.46×10^{-2}	$\pm\ 6.75\times10^{-4}$	2.15×10^{-2}	$\pm\ 2.34\times10^{-4}$
transit5	No	2.40×10^{-3}	$\pm\ 5.43\times10^{-5}$	1.35×10^{0}	$\pm\ 1.07\times10^{-2}$
transit6	No	8.28×10^{1}	$\pm\ 1.30\times10^{-1}$	1.51×10^{2}	$\pm\ 1.95\times10^{-1}$

to check life-cycle inheritance, that is, they exclude set up times and so on. For the remainder of this section, we identify each case by its potential sub object life cycle, because basically we are testing the extension object life cycle and not the extended object life cycle.

The results for cases transit3 and transit4 were to be expected: in both cases, blocking all new methods is a solution. The exhaustive search algorithm always outruns the backtracking algorithm if this is the situation, as is explained in [15]. The results of the other three cases seem normal and quite satisfactory. However, to our surprise, for none of these cases a life-cycle inheritance relation exists between the base object life cycle and the potential sub object life cycle, although this should have been the case.

At this point, it is worth mentioning that, at the moment the β case study was conducted, several researchers knowledgeable in the field of life-cycle inheritance had already took notice of the α case study, and none of them had observed that in three out of five cases the claimed life-cycle inheritance relation was absent. In our eyes, the fact that both these experts and we failed to observe this, shows that one easily underestimates the subtlety of the life-cycle inheritance relation. Thus, we conclude that we need software tools to check for this relation.

4 Diagnosing and Correcting the Processes

Of course, knowing that the α case study would be broken if these inheritance relations were not established in all five cases, we tried to diagnose the absence of these relations, even though, at this point, Woflan does not really support diagnosing errors related to the absence of life-cycle inheritance. Note that diagnosing and correcting these object life cycles was not a goal of the β case study. However, for us, it was necessary to correct the errors, and it was interesting to see to what extent Woflan could be of help.

4.1 Case transit2

Figure 3 shows the original figure[1] for the transit1 object life cycle, whereas Figure 4 shows the original figure for the transit2 object life cycle. Each object life cycle involves four parties who exchange messages. Basically, each party proceeds in the vertical plane, whereas each message proceeds in the horizontal plane. The task S_dec_dat sends the message dec_dat, the task R_dec_dat receives this message, and so on.

In the first iterative step of the α case study, the second author added the behavior related to erroneous message transfers between the two customs offices (the parties in the middle). For example, upon reception of an erroneous aar_snd message, the recipient sends a fun_nck message back to the sender, who will send the aar_snd message again upon reception of this fun_nck message. And so on, until a correct aar_snd message is received. To prevent the sender of the aar_snd message from proceeding while this message has not yet been received, he added a time_out message, although this message was not present in any of the relevant use cases. If a correct aar_snd message is received, the recipient sends a time_out message back to confirm the proper reception of the aar_snd message. Only after this confirmation can the sender of the aar_snd message proceed.

To diagnose why there exists no life-cycle inheritance relation between both object life cycles, we examined the states that were not branching bisimilar to any other state for the situation where the new methods S_time_out and R_time_out are hidden and the new methods S_fun_nck and R_fun_nck are blocked (which seems like a perfect candidate for a life-cycle inheritance relation). In the end, we discovered that the confirmation of proper reception of the arr_adv message is the cause of the absence of the life-cycle inheritance relation. In the transit1 process, the customs office that sends the arr_adv message can proceed to send the des_con message before the other customs office has received the arr_adv message. In the transit2 process, this is impossible, because the customs office that sends the arr_adv message has to wait until the other customs office has confirmed proper reception of this message.

Apparently, the solution chosen by the second author to add the behavior related to erroneous message transfers is not compatible with the life-cycle inheritance relation. This problem is clearly due to the fact that in the use cases, and thus, in practice, the time_out message does not exist. Instead, it seems safe to assume that, in practice, the sender of the arr_adv message will wait for a certain amount of time before proceeding. If the sender does not receive a fun_nck message during that period of time, s/he will proceed. Fortunately, there is a way to capture this behavior, which is shown in Figure 5: The sender will wait for a fun_nck message as long as condition p holds (that is, as long as place p is marked). After this period of time, that is, when condition p does not hold anymore, the customs office that receives the arr_adv message cannot send a fun_nck anymore. Note that this solution makes the time_out message obsolete,

[1] Please note that these figures were the only input for the β case study. For this reason, we did not retouch these figures. As a result, some text might be difficult to read. We apologize for the inconvenience.

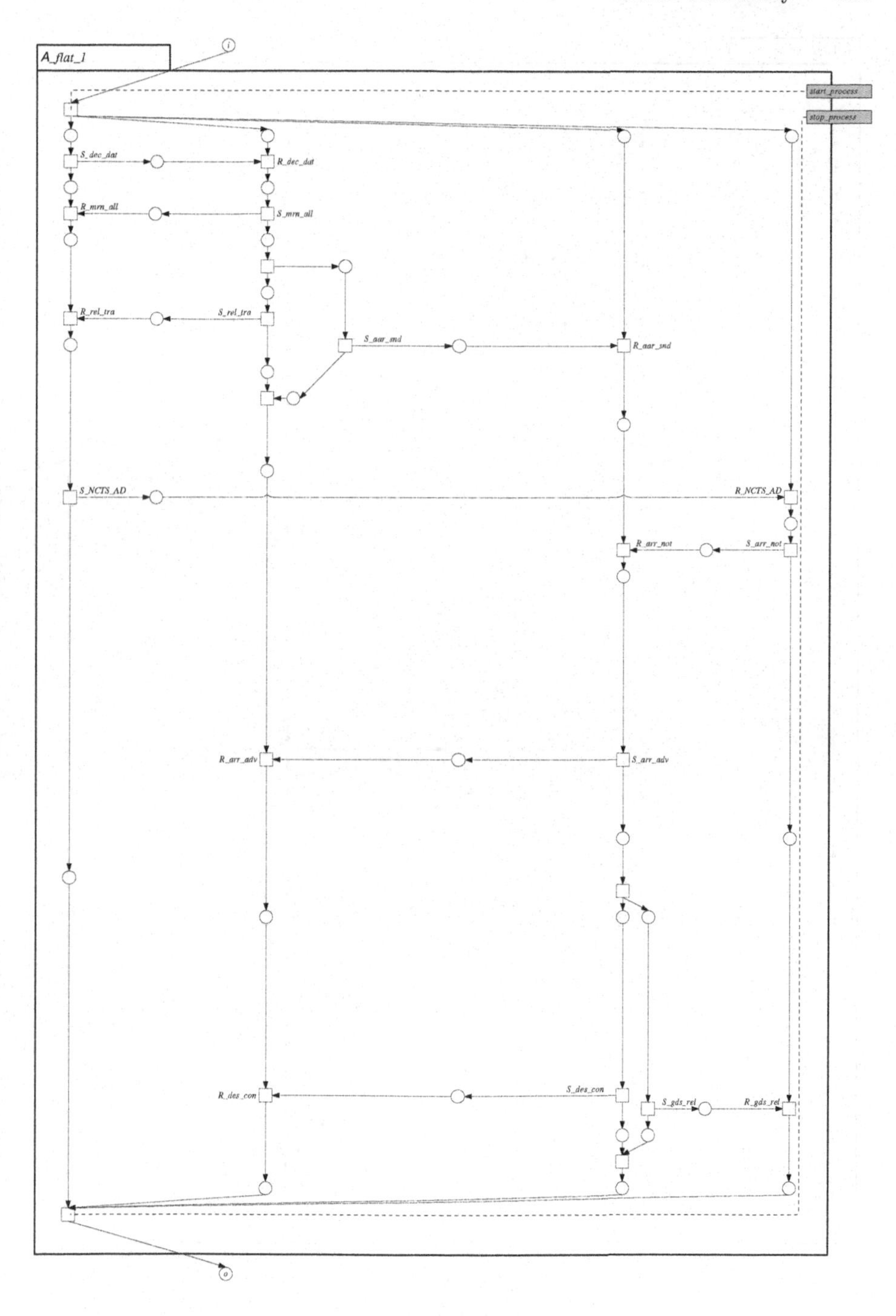

Fig. 3. transit1

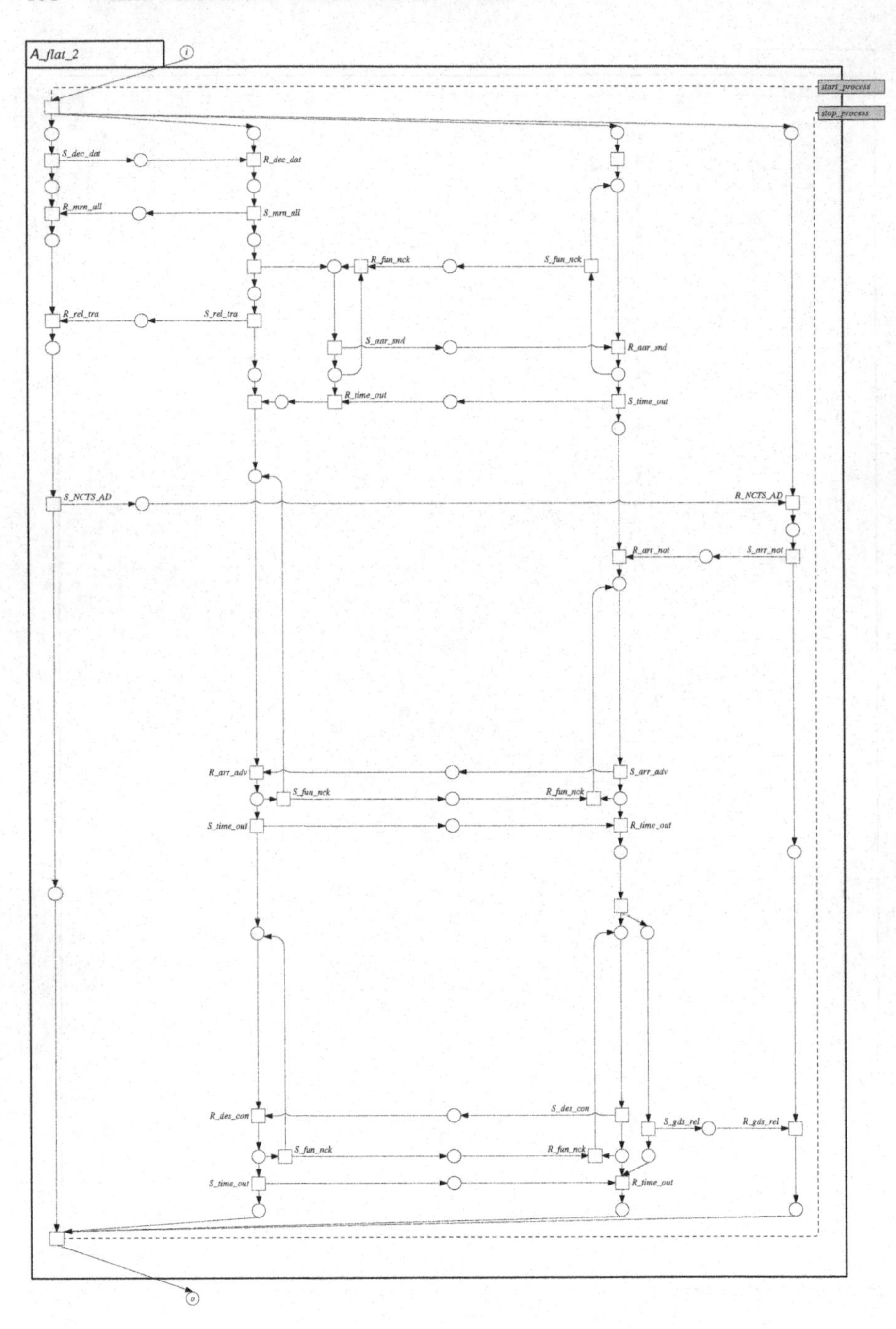

Fig. 4. transit2

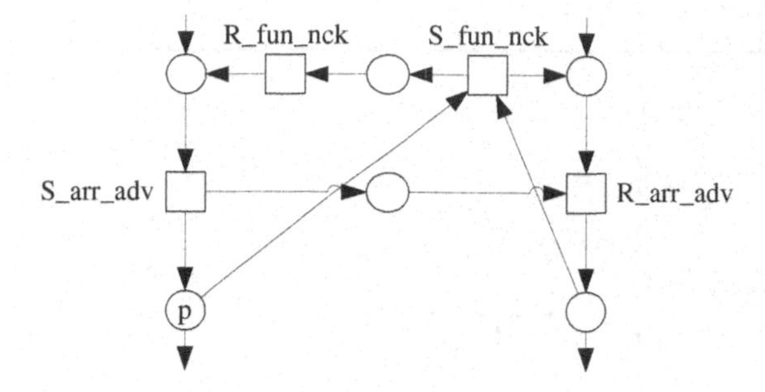

Fig. 5. fun_nck

which is a point in favor, because this message was not a part of the use cases this extension was based on. This led to our first suggestion for correcting the absence of a life-cycle inheritance relation in the transit2 case:

> Hide the methods (tasks) related to the time_out message, remove the place in-between S_time_out and R_time_out tasks, remove the connection from condition p to task R_fun_nck, and add a connection from condition p to task S_fun_nck (as is shown in Figure 5).

4.2 Case transit5

Figure 6 shows the transit4 object life cycle, whereas Figure 7 shows the transit5 object life cycle. In the transit5 object life cycle, a dummy task has been added (labeled GO), and tasks have been added for receiving the message arr_not, for sending the message time_out, and for sending and receiving the messages lar_req and lar_rsp. Note that the messages arr_not and time_out were already present in the transit4 object life cycle. As a result, we cannot hide or block the new methods that receive the message arr_not or send the message time_out. However, it is obvious that we have to hide or block the new methods labeled S_time_out: In the transit4 object life cycle we cannot send a time_out message after having send a mrn_all message, whereas in the transit5 object life cycle we can. Recall that in the previous subsection we already pointed out that the time_out message is obsolete for the extension from object life cycle transit1 to object life cycle transit2. If we would remove the tasks sending and receiving these messages in object life cycle transit2 (and, hence, from the object life cycles transit3, transit4, transit5, and transit6), the method S_time_out would not exist in the transit4 object life cycle, and, hence, we could hide or block it. Thus, applying the suggestion for correcting the previous error might also correct this one. This led to the following suggestion to correct the absence of a life-cycle inheritance relation in the transit5 case:

> Apply the suggestion for correcting the absence of a life-cycle inheritance relation in the transit2 case.

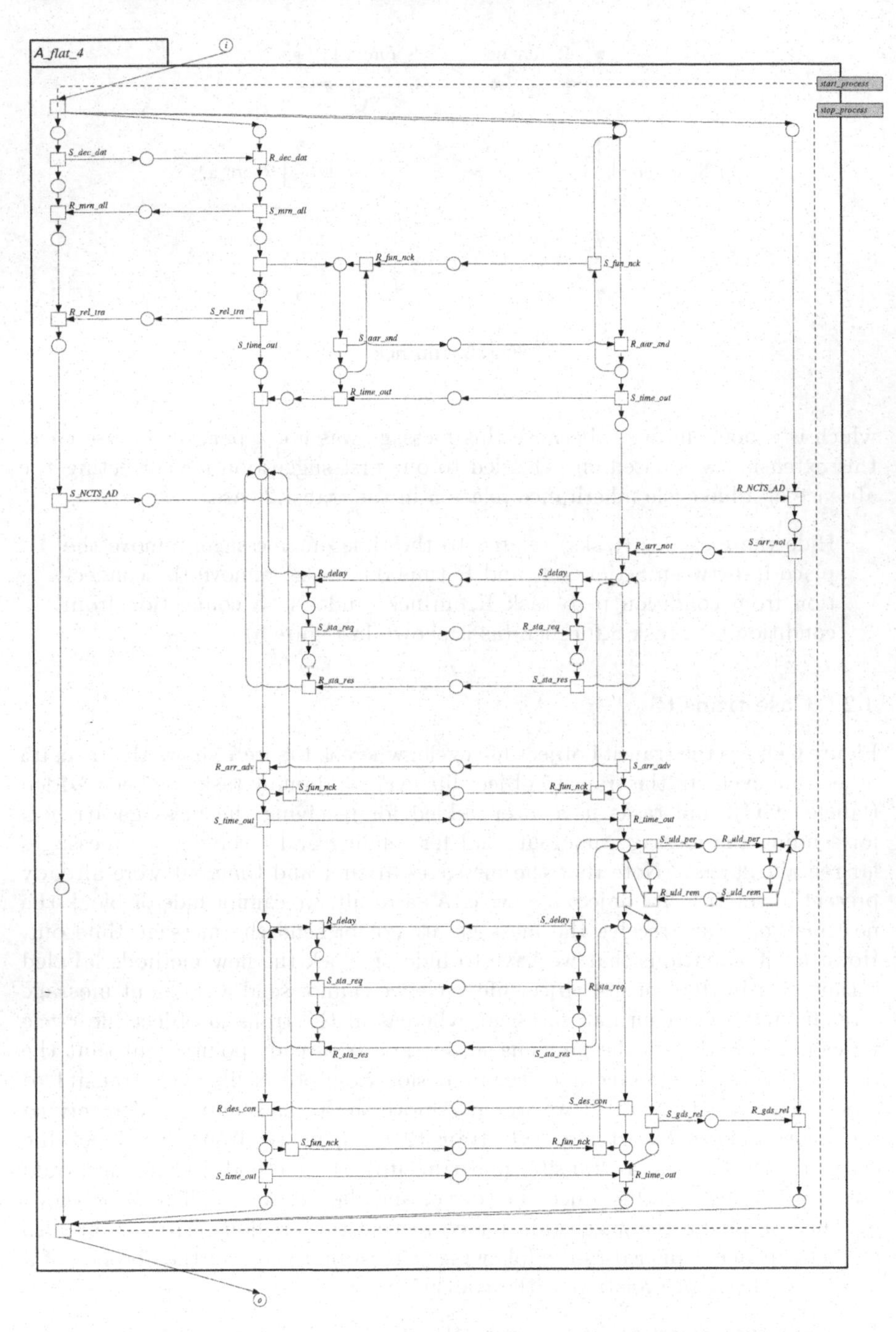

Fig. 6. transit4

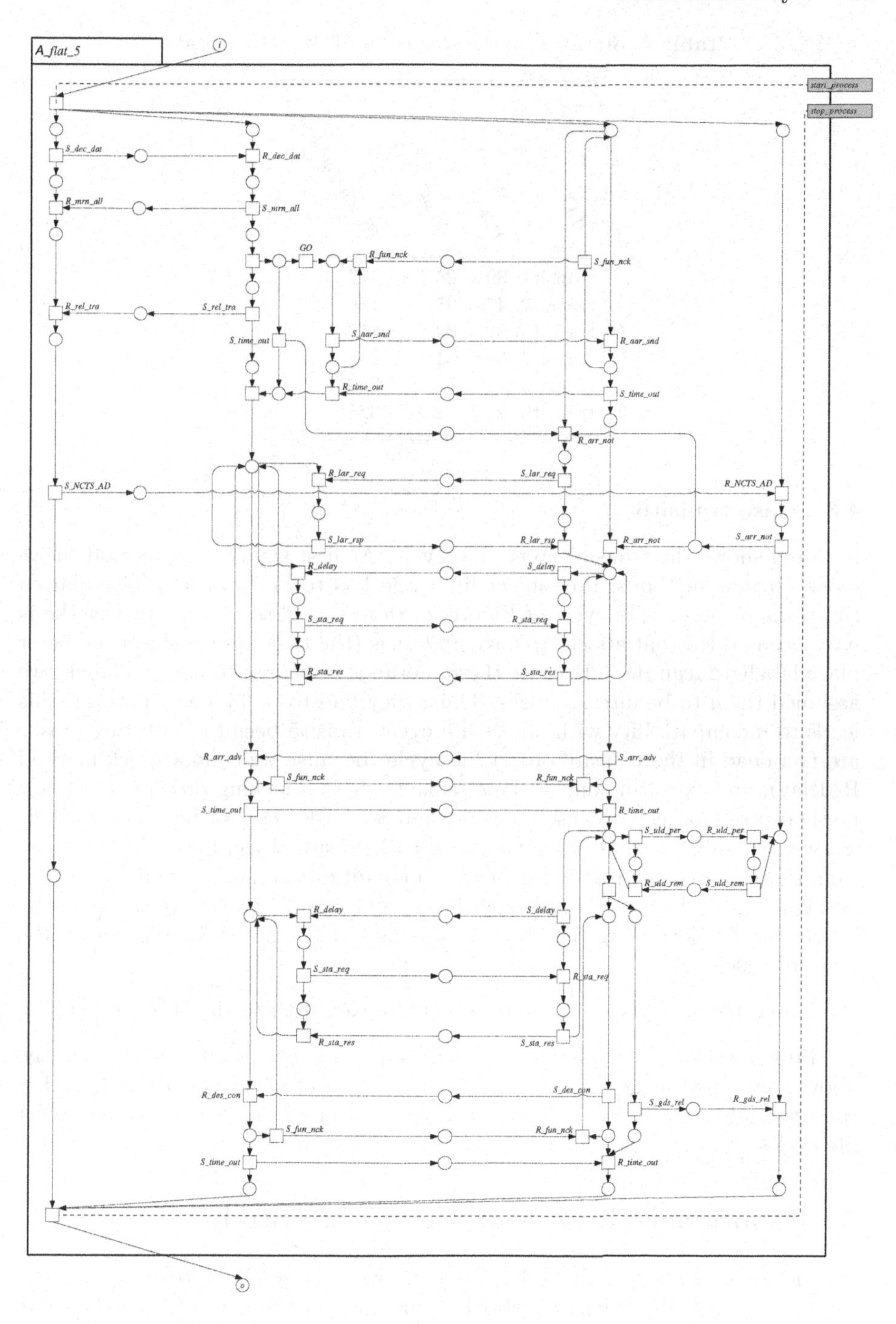

Fig. 7. transit5

Table 4. Statistics on the six OLC's of the γ case study

OLC	Places	Transitions	Reachable states
transit1	39	24	77
transit2	47	35	156
transit3	51	39	171
transit4	65	51	214
transit5	72	58	242
transit6	96	82	338

4.3 Case transit6

Figure 8 shows the transit6 object life cycle. At first sight, it seems that, if we block all new methods, this object life cycle has to be branching bisimilar to the transit5 object life cycle of Figure 7. However, observe that two methods have been added that appear to have no labels (the two input tasks of the lower place labeled S_can_dec). Because these methods appeared to have no labels, we assumed them to be internal tasks. Thus, they had to be hidden. However, this leads to incompatibility with object life cycle transit5 because both these tasks are non-dead in the transit6 object life cycle (because we cannot block method R_delay), and executing one of these tasks leads to a locking problem if all new methods are blocked (because the condition S_can_dec cannot be falsified). This led to the conviction that, when modeling the transit6 object life cycle definition, we misinterpreted the text S_can_dec for a conditions name, where it was meant as a label for both methods we believed to be unlabeled, and thus to the following suggestion for correcting the absence of a life-cycle inheritance relation in the transit6 case:

Label the methods we believed to be unlabeled with the label S_can_dec.

Table 4 shows the statistics of the six object life cycles after we applied the above mentioned suggestions and corrected the object life cycles accordingly. For the remainder of this paper, we refer to this case study on the corrected object life cycles as the γ case study.

5 Performance Results on the γ Case Study

On the corrected object life cycles, we again ran a comparison between the performance of the backtracking algorithm and the performance of the exhaustive search algorithm. Table 5 shows the results. Note that the life-cycle inheritance relation is present in all five cases. The results for the transit2, transit3, transit4,

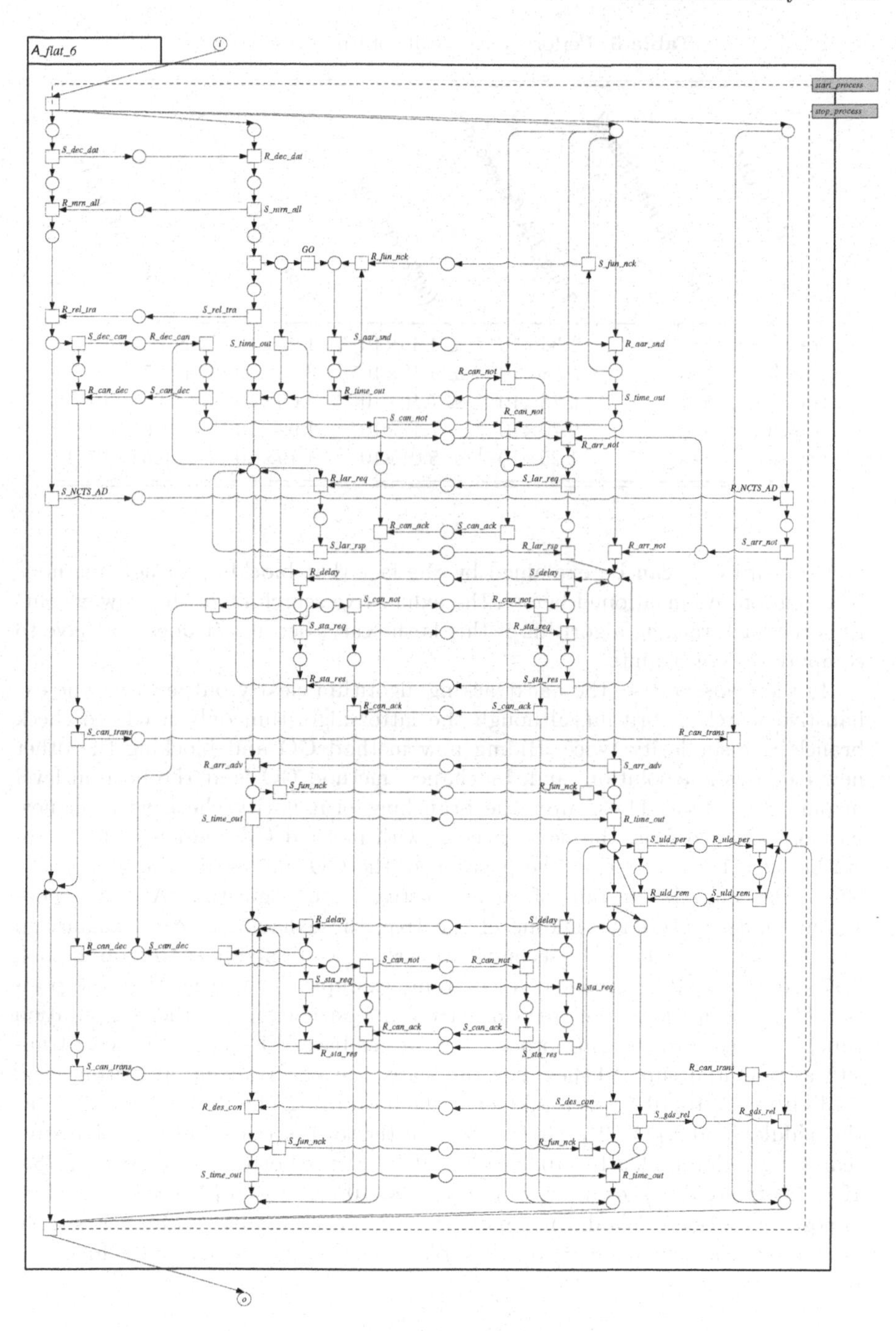

Fig. 8. transit6

Table 5. Performance results of the γ case study

potential sub OLC	Life-cycle inheritance?	Time: BA (in seconds)	(99% conf. interval)	Time: ESA (in seconds)	(99% conf. interval)
transit2	Yes	1.34×10^{-2}	$\pm\ 4.45\times10^{-4}$	1.25×10^{-2}	$\pm\ 4.50\times10^{-4}$
transit3	Yes	3.68×10^{-2}	$\pm\ 4.13\times10^{-3}$	3.44×10^{-2}	$\pm\ 1.78\times10^{-4}$
transit4	Yes	4.78×10^{-2}	$\pm\ 7.16\times10^{-4}$	4.66×10^{-2}	$\pm\ 3.41\times10^{-4}$
transit5	Yes	9.79×10^{-2}	$\pm\ 7.17\times10^{-4}$	1.92×10^{-1}	$\pm\ 1.64\times10^{-3}$
transit6	Yes	1.25×10^{-1}	$\pm\ 5.01\times10^{-4}$	1.19×10^{-1}	$\pm\ 6.54\times10^{-4}$

and transit6 case can be explained by the fact that blocking all new methods is a solution. As mentioned before, the exhaustive search algorithm always outruns the backtracking algorithm if this is the case, because it does not have to compute the constraints.

In the transit5 case, the backtracking algorithm clearly outperforms the exhaustive search algorithm, although the latter algorithm only needs to check branching bisimilarity twice: Hiding new method GO and blocking the other new methods is a solution, and, by chance, method GO is at the bottom level in our search tree. Thus, after the branching bisimilarity check with all new methods blocked fails, the next check is with method GO hidden and all others blocked. It is clear that the position of the GO method in the search tree effects the computation time of the exhaustive search algorithm: After we repositioned label GO at the middle of the tree, the exhaustive search algorithm took $4.85\times10^{-1}\pm1.05\times10^{-3}$ seconds, after we repositioned it at the top, it took $3.17\times10^{0}\pm1.17\times10^{-2}$ seconds. An extreme example illustrating that the position of labels in the search tree can have a dramatic effect on the computation time, is the case when using process transit1 as base WF-net and process transit6 as potential sub WF-net. For this case, the backtracking algorithm took $2.73\times10^{-2}\pm2.10\times10^{-4}$ (method GO at the bottom), $2.71\times10^{-4}\pm2.01\times10^{-4}$ (at the middle), and $2.71\times10^{-2}\pm3.25\times10^{-4}$ (at the top) seconds, but the exhaustive search algorithm took $3.19\times10^{-2}\pm1.78\times10^{-4}$ (at the bottom), $1.45\times10^{2}\pm2.78\times10^{-1}$ (at the middle), or approximately 2.38×10^{6} (at the top) seconds (that is, almost four weeks). We did not measure the latter number, because it simply takes too much time; instead, we extrapolated the previous result in the following way:

- The search tree contains 28 levels (28 new methods), where level 1 is the top level, level 15 is the middle level, and level 28 is the bottom level;

– Thus, finding the solution when method GO is positioned at the middle takes 8193 ($2^{28-15}+1$) branching bisimilarity checks, which took approximately 145 seconds;
– Because finding a solution when method GO is positioned at the top takes 134,217,729 ($2^{28}+1$) branching bisimilarity checks, this will take approximately 2.38×10^6 ($((2^{27}+1)/(2^{13}+1))\times145$) seconds.

Note that we assume that the branching bisimilarity check is the dominant factor with regard to the computation time and that the branching bisimilarity checks, when method GO is positioned in the middle, are representative for all these checks.

6 Conclusion

The main first conclusion of the β and γ case studies is that the backtracking algorithm can outrun the exhaustive search algorithm by orders of magnitudes (for example, a fraction of a second instead of almost four weeks). In contrast with this, the exhaustive search algorithm might outrun the backtracking algorithms too, but usually this is only the case when blocking all new methods yields a life-cycle inheritance relation, and the difference is clearly within acceptable limits.

We can also try to control the exhaustive search algorithm by using intermediate steps, as both case studies shows. As mentioned, when trying to check for life-cycle inheritance between object life cycles transit1 and transit6 in case method GO is at the top level of the search tree, the exhaustive search algorithm takes about four weeks to compute. However, using four intermediate steps, the combination of five exhaustive search algorithms takes approximately 3.29 (0.015 + 0.0344 + 0.0466 + 3.17 + 0.119) seconds. Thus, divide and conquer might also be a good technique to lessen performance problems with the exhaustive search algorithm. However, this might not always be possible.

Another conclusion is that, although the authors of [15] did it for different reasons, it seemed to be a wise decision to try to block a new method before we try to hide it: It seems that, in general, the majority of the new methods needs to be blocked. (Note that the case study performed in the ATPN 2003 [15] also supports this conclusion).

A third conclusion is that we really need a software tool to check whether a life-cycle inheritance relation exists between two object life cycles. In the α case study, a number of knowledgeable people did not detect that in certain cases the claimed life-cycle inheritance relation was absent. Only after we checked this with Woflan, this became apparent.

At the moment, Woflan does not provide any diagnostic information related to life-cycle inheritance, except perhaps for a branching bisimulation relation in case of a seemingly perfect hiding and blocking scheme. Using such a scheme, we were able to correct one absent life-cycle inheritance relation, but it took considerable effort. The other two cases in which the life-cycle relation was absent were diagnosed by simply accepting the fact that this relation was absent and

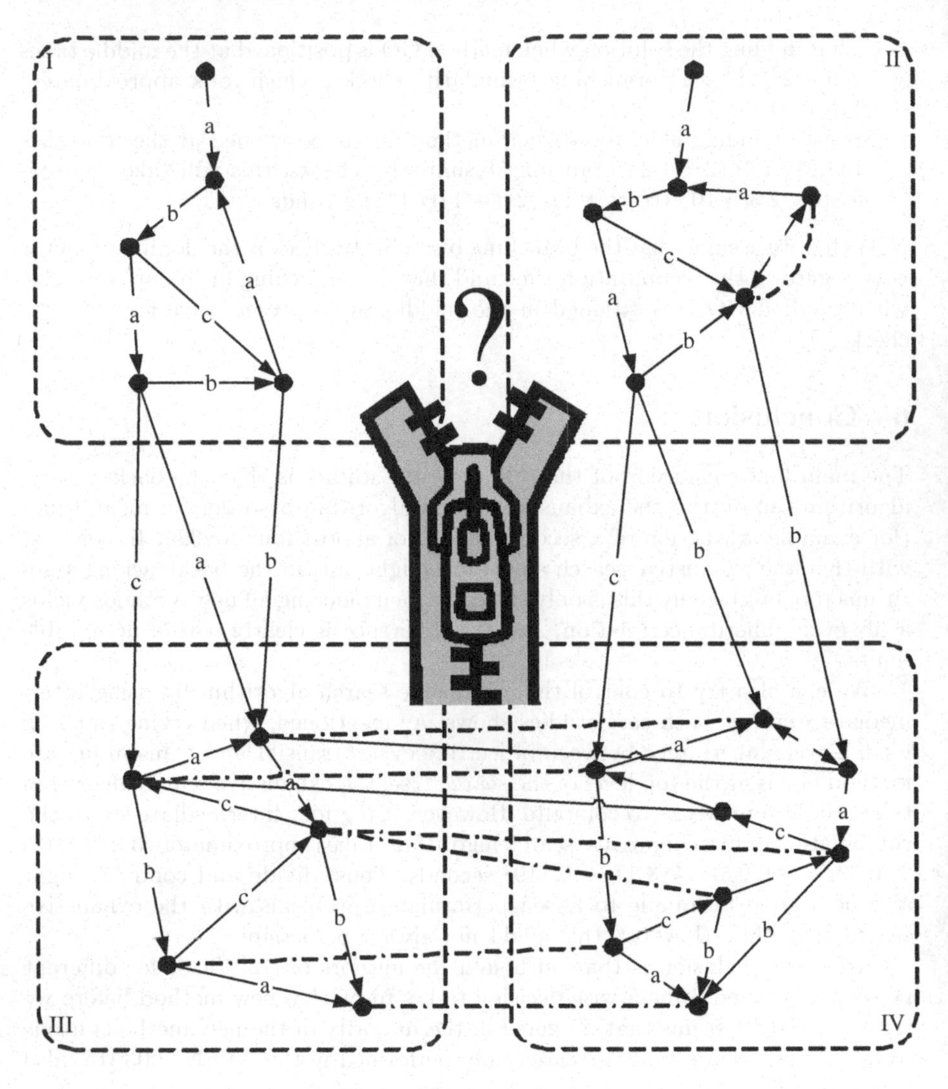

Fig. 9. The 'zipper' scheme

to look for possible causes. However, it would be nice if Woflan could give some guidance when trying to diagnose the absence of a life-cycle inheritance relation. A possible idea would be to have the designer of the process definitions specify a hiding and blocking scheme, and to report back (in some way) the boundary between branching bisimilar states and non-branching bisimilar states. By definition, the successful terminal states are branching bisimilar, and, obviously, the initial states are not. Somewhere between the initial states and the successful terminal states a boundary exists that separates branching bisimilar states from non-branching bisimilar states. Note that, by definition, arcs can only cross this boundary from non-branching bisimilar states to branching bisimilar states. Apparently, for some reason, on this boundary the branching bisimulation gets lost, which makes it very interesting from a diagnosis point of view. Figure 9 visualizes this 'zipper' scheme. We conclude with a remark concerning a possible positive relation between the computation time needed to check a property and the existence of violations of these properties (errors), that is, in case of errors, the computation time increases. If we rank the computation times for the exhaustive search algorithm as presented by tables 3 and 5 from high to low, then we observe that the cases where no life-cycle inheritance relation is present rank first, second, and fourth. If we do the same for the backtracking algorithm, then they rank first, fourth, and tenth (last). Although the transit5 case of the β case study ranks last, the other cases seem to suggest that the computation time increases in the presence of errors. Our experience is that such a correlation also exists for the soundness property (see [4]). From [4, 5] it is clear that an object life cycle corresponds to a sound workflow net. As a result, life-cycle inheritance is only defined on sound workflow nets. Thus, before checking life-cycle inheritance, we need to check soundness. We used Woflan [16] to check whether the object life cycles where sound. Empirical data on these soundness checks are not presented in this paper, but these data suggest that there exists a positive correlation between the existence of soundness errors and the computation time needed to check soundness.

References

1. W.M.P. van der Aalst. Inheritance of dynamic behaviour in UML. In D. Moldt, editor, *MOCA'02, Second Workshop on Modelling of Objects, Components, and Agents*, pages 105–120, Aarhus, Denmark, August 2002. University of Aarhus, Report DAIMI PB - 561. http://www.daimi.au.dk/CPnets/workshop02/moca/papers/.
2. W.M.P. van der Aalst. Inheritance of interorganizational workflows to enable business-to-business E-commerce. *Electronic Commerce Research*, 2(3):195–231, 2002.
3. W.M.P. van der Aalst and T. Basten. Identifying commonalities and differences in object life cycles using behavioral inheritance. In J.-M. Colom and M. Koutny, editors, *Applications and Theory of Petri Nets 2001*, volume 2075 of *Lecture Notes in Computer Science*, pages 32–52, Newcastle, UK, 2001. Springer, Berlin, Germany.
4. W.M.P. van der Aalst and T. Basten. Inheritance of workflows: An approach to tackling problems related to change. *Theoretical Computer Science*, 270(1-2):125–203, 2002.
5. T. Basten and W.M.P. van der Aalst. Inheritance of behavior. *Journal of Logic and Algebraic Programming*, 47(2):47–145, 2001.
6. G. Booch. *Object-Oriented Analysis and Design: With Applications.* Benjamin/Cunnings, Redwood City, California, USA, 1994.
7. G. Booch, J. Rumbaugh, and I. Jacobson. *The Unified Modeling Language User Guide.* Addison-Wesley, Reading, Massachusetts, USA, 1998.

8. J.F. Groote and F.W. Vaandrager. An efficient algorithm for branching bisimulation and stuttering equivalence. In M.S. Paterson, editor, *Automata, Languages and Programming*, volume 443 of *Lecture Notes in Computer Science*, pages 626–638, Warwick University, England, July 1990. Springer, Berlin, Germany.
9. Object Management Group. Omg unified modeling language. http://www.omg.com/uml/.
10. I. Jacobson, M. Ericsson, and A. Jacobson. *The Object Advantage: Business Process Reengineering with Object Technology.* Addison-Wesley, Reading, Massachusetts, USA, 1991.
11. J. Rumbaugh, M. Blaha, W. Premerlani, F. Eddy, and W. Lorensen. *Object-Oriented Modeling and Design.* Prentice-Hall, Englewoord Cliffs, New Jersey, USA, 1991.
12. J. Rumbaugh, I. Jacobson, and G. Booch. *The Unified Modeling Language Reference Manual.* Addison-Wesley, Reading, Massachusetts, USA, 1998.
13. R.A. van de Toorn. *Component-Based Software Design with Petri Nets: an Approach Based on Inheritance of Behavior.* PhD thesis, Eindhoven University of Technology, Eindhoven, The Netherlands, January 2004. (forthcoming).
14. H.M.W. Verbeek and W.M.P. van der Aalst. Woflan 2.0: A Petri-net-based workflow diagnosis tool. In M. Nielsen and D. Simpson, editors, *Application and Theory of Petri Nets 2000*, volume 1825 of *Lecture Notes in Computer Science*, pages 475–484. Springer, Berlin, Germany, 2000.
15. H.M.W. Verbeek and T. Basten. Deciding life-cycle inheritance on Petri nets. In W.M.P. van der Aalst and E. Best, editors, *24th International Conference on Application and Theory of Petri Nets (ICATPN 2003)*, volume 2679 of *Lecture Notes in Computer Science*, pages 44–63, Eindhoven, The Netherlands, June 2003. Springer, Berlin, Germany.
16. H.M.W. Verbeek, T. Basten, and W.M.P. van der Aalst. Diagnozing workflow processes using Woflan. *The Computer Journal*, 44(4):246–279, 2001.

Eliminating Internal Behaviour in Petri Nets

Harro Wimmel

Parallel Systems, Faculty of Computing Science
Carl von Ossietzky Universität Oldenburg, Germany

Abstract. A safe labelled marked graph (Petri Net) with internal transitions can be transformed into a pomset-equivalent safe labelled marked graph without internal transitions.

Keywords: Partial order theory of concurrency, Petri nets, Pomsets.

1 Introduction

It is often desirable to transform a system with internal (silent, unobservable) behaviour, under invariance of certain desirable behavioural properties, into one which has no unobservable behaviour. In this paper, we show that for a class of Petri nets [9], a transformation preserving pomset [8] behaviour can be found.

We will focus our attention on the class of transition-labelled safe marked graphs [5, 6], where a label τ indicates a silent transition. Our main result states that such a marked graph can be systematically transformed into another one which is also safe and has the same pomsets as the original one, but has no τ-labelled transitions.

Based on the work in [10, 12], the author has conjectured for some time that the restriction to marked graphs can be lifted in this result, i.e., that *every* safe Petri net can be transformed into a pomset-equivalent safe τ-free Petri net. From [4] it follows then that the result would even hold for bounded Petri nets. The present paper can, hopefully, serve as a first step in proving the (more general) conjecture.

In Section 2, we define nets, labelled marked graphs, pomsets, and a few auxiliary notions. Section 3 contains the proof of the main result. Concluding remarks can be found in Section 4.

2 Definitions

Let Σ denote some fixed alphabet of (observable) actions. We use a, b, ... as names for actions (typical elements of Σ). The symbol $\tau \notin \Sigma$ will be used for unobservable actions.

We call a relation $R \subseteq X \times X$ a partial order on X if it is irreflexive and transitive. Given a partial order R on X, we call $x, y \in X$ *concurrent*, $x \; co_R \; y$, if neither xRy nor yRx. We call a set $c \subseteq X$ an *antichain* if all $x, y \in c$ are concurrent, and a *cut* if additionally for all $x \in X \backslash c$ there is some $y \in c$ with xRy or yRx; i.e., a cut is a maximal antichain. $c \subseteq X$ is a B-cut for some $B \subseteq X$ if c is a cut in X and $c \subseteq B$.

J. Cortadella and W. Reisig (Eds.): ICATPN 2004, LNCS 3099, pp. 411–425, 2004.
© Springer-Verlag Berlin Heidelberg 2004

Definition 1. *Posets, causality structures, pomsets*

A *partially ordered set (poset)* is a pair (X, R) of a set X and a partial order $R \subseteq X \times X$. We call x_1 a *predecessor* of x_2 (and x_2 a *successor* of x_1) if $x_1 R x_2$. Further, x_1 is a *direct predecessor* of x_2 (and x_2 is a *direct successor* of x_1) if additionally there is no x with $x_1 R x$ and $x R x_2$. (X, R) has the *finite precedence property* if for every $x \in X$, the set of predecessors of x is finite.

A *causality structure* κ *over* $\Sigma \cup \{\tau\}$ is a triple $\kappa = (E, R, l)$ consisting of a poset (E, R) (the letter E reminds of events), and a labelling function $l \colon E \to \Sigma \cup \{\tau\}$. Two causality structures $\kappa_1 = (E_1, R_1, l_1)$ and $\kappa_2 = (E_2, R_2, l_2)$ are *isomorphic* if there is a bijection $\beta \colon E_1 \to E_2$ with $\beta \circ R_2 \circ \beta^{-1} = R_1$ and $\beta \circ l_2 = l_1$ [1]. A *pomset (partially ordered multiset)* is an isomorphism class of causality structures. By $[\kappa]$ we denote the pomset containing the causality structure κ as a representative. A pomset $[(E, R, l)]$ is finite if E is finite. ■

Graphically, a finite pomset $[(E, R, l)]$ will be presented as a directed graph whose nodes are labelled by elements of the multiset $l(E)$ and whose directed arcs represent the direct predecessor relation.

In this paper we will use Petri nets without arc weights, whose transitions may be labelled by symbols from the fixed set $\Sigma \cup \{\tau\}$.

Definition 2. *Petri net*

An *(unlabelled, unmarked) net* is a triple (S, T, F) where S and T are sets with $S \cap T = \emptyset$, and F is a relation $F \subseteq (S \times T) \cup (T \times S)$. Let $F(x) = \{y \mid (x, y) \in F\}$ and $F^{-1}(y) = \{x \mid (x, y) \in F\}$.

A *marking* of (S, T, F) is a function $M \colon S \to \mathbb{N}$. We say that there are k tokens on a place s under M iff $M(s) = k$. A *marked net* is a tuple (S, T, F, M_0), where (S, T, F) is a net and M_0 is a marking.

A net is *T-restricted* if $\forall t \in T \colon F^{-1}(t) \neq \emptyset \neq F(t)$. A net (S, T, F) is a *T-net* if $\forall s \in S \colon |F^{-1}(s)| \leq 1 \wedge |F(s)| \leq 1$, and a *marked graph* if it is a marked T-net.

A net (B, E, G) is an *occurrence net* if it is a T-net, the transitive closure of G, G^+, is irreflexive (and hence, $(B \cup E, G^+)$ is a partial order), and $(B \cup E, G^+)$ has the finite precedence property. Elements of B and E are called *conditions* and *events*, respectively.

A *(labelled, marked) net* is a 5-tuple (S, T, F, M_0, λ), where (S, T, F, M_0) is a marked net and $\lambda \colon T \to \Sigma \cup \{\tau\}$ is a *(transition) labelling*. We call the labelling λ *injective on observables* if $\lambda|_\Sigma$ (i.e., λ restricted to observable actions) is injective. ■

We assume all nets to be T-restricted. Except for (possibly) occurrence nets, we also assume all nets to be finite, i.e. S and T are finite.

The token game for marked nets is defined as follows. A marking M *activates* a transition t (in symbols: $M[t\rangle$) iff $\forall s \in F^{-1}(t) \colon M(s) \geq 1$. Moreover, M leads to a marking M' through the *occurrence* of t (also called the *firing of* t, in symbols: $M[t\rangle M'$) iff $M[t\rangle$ and $\forall s \in S \colon M'(s) = M(s) - F(s, t) + F(t, s)$ (here, we have identified F with its characteristic function). We define inductively that an *occurrence sequence* $\varrho \in T^*$ leads from M to M':

[1] We interpret $\circ$ 'relationally', i.e. $(x, y) \in Q \circ R \iff \exists z \colon (x, z) \in Q \wedge (z, y) \in R$.

$$M[\varepsilon\rangle M' \iff M = M'$$
$$M[\varrho t\rangle M' \iff \exists M'' \colon M[\varrho\rangle M'' \wedge M''[t\rangle M',$$

with ε being the empty occurrence sequence. By $OS(N)$, we denote the set of all occurrence sequences starting from the initial marking M_0 of N [2]. A marking M' is reachable from a marking M iff there exists an occurrence sequence $\varrho \in T^*$ with $M[\varrho\rangle M'$. A marked net $N = (S, T, F, M_0)$ is *k-bounded* (for $k \in \mathbb{N}\backslash\{0\}$) iff in every marking M reachable from M_0 all places $s \in S$ contain at most k tokens, and *safe* iff it is 1-bounded. It is *live* iff for every marking M reachable from M_0 and for every transition $t \in T$, there is a marking M' which is reachable from M and activates t.

We formalise the pomset behaviour of a labelled, marked net by first defining its processes[3].

Definition 3. *Processes*
A *process* π of a net $N = (S, T, F, M_0, \lambda)$ is a tuple $\pi = (B, E, G, r, \lambda')$, where $r \colon (B \cup E) \to (S \cup T)$, $\lambda' \colon E \to \Sigma \cup \{\tau\}$, and:

- (B, E, G) is an occurrence net.
- $r(B) \subseteq S$ and $r(E) \subseteq T$, i.e. r 'folds' conditions onto places and events onto transitions.
- interpreted as a multiset mapping, r satisfies an initiality condition:

  $$r(Min_{G^+}(B \cup E)) = M_0,$$

  and a progress condition:

  $$\forall e \in E \colon r(G^{-1}(e)) = F^{-1}(r(e)) \wedge r(G(e)) = F(r(e)).$$

  For safe nets (to which we will be limited here), it is sufficient to interpret r as a standard mapping on sets.
- The labelling is inherited from transitions to events: $\forall e \in E \colon \lambda'(e) = \lambda(r(e))$.

■

We call two processes *isomorphic* if one of them results from the other by renaming conditions and events, and we do not distinguish between two isomorphic processes. Given a finite B-cut c in a (possibly infinite) process π, the part restricted between $Min_{G^+}(B \cup E)$ (the minimal elements of $B \cup E$ regarding G^+) and c is again a process, which we call a *prefix* of π; due to the finite precedence property and the finiteness of c, prefixes are always finite.

Possible observable behaviour is specified by the labelling of transitions and events, respectively. As a pomset describes such a behaviour, we consider only the observable events and their precedence relations in a process.

[2] Such sequences are the 'interleaving behaviour' of N.

[3] A process reflects a simulation of the token game of a net in a 'truly concurrent' way.

Definition 4. *Pomsets of a Petri net*

Let $N = (S, T, F, M_0, \lambda)$ and let $\pi = (B, E, G, r, \lambda')$ be a process of N. Let $E' := \{e \in E \mid \lambda'(e) \neq \tau\}$ be the set of observable events in π. The pomset $\Phi(\pi)$ is defined as $[(E', G^+ \cap (E' \times E'), \lambda'|_{E'})]$.

The pomsets of a labelled, marked Petri net $N = (S, T, F, M_0, \lambda)$ are defined as

$$Pomsets(N) \quad = \quad \{\Phi(\pi) \mid \pi \text{ is a finite process of } N\}.$$

■

Thus, the pomset $\Phi(\pi)$ is an abstraction of the process π. We allow for autoconcurrency here, i.e., multiple instances of the same action may occur in a pomset $\Phi(\pi)$ without any causal ordering between them. Note that this is possible even if the Petri net is safe.

Prefixes of pomsets can be defined via processes, if π is a prefix of π', $\Phi(\pi) = [(E, R, l)]$ is a prefix of $\Phi(\pi') = [(E', R', l')]$. This naturally means that $E \subseteq E'$, $e \in E \wedge e'Re \Rightarrow e' \in E$, $R = R'|_{E \times E}$, and $l = l'|_E$.

3 Marked Graph Transformation

We start with a live and safe, strongly connected, marked graph $N = (S, T, F, M_0, \lambda)$ where λ is injective on observables. Our objective is to transform this net into a pomset-equivalent one, N' (i.e., a live and safe marked graph with the same pomsets as N), which however does not have any τ as a transition label.

Despite the fact that the theory on live and safe marked graphs has been known for a long time [5, 6], its known consequences are of only partial use for the problem considered here, since the target net N' may (and in general, must) have a non-injective labelling. In fact, the pomset theory of marked graphs with arbitrary transition labellings is not as well studied as the theory of unlabelled marked graphs.

The example shown in Figure 1 exhibits some of the problems.

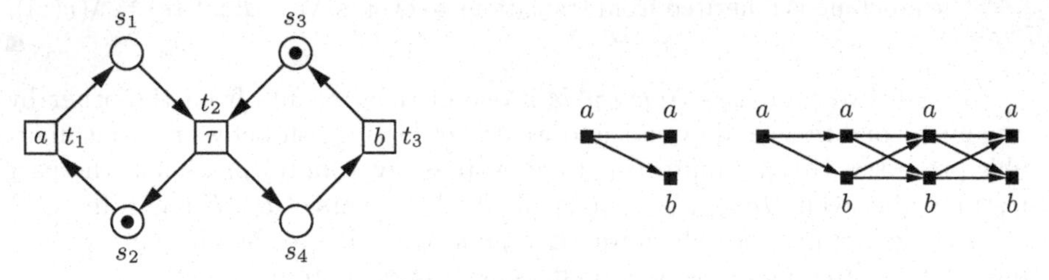

Fig. 1. A marked graph (left) and two of its pomsets (right)

Eliminating the τ-labelled transition t_2 in the net itself (without recourse to processes) seems difficult. For instance, one might be tempted to replace the τ-transition t_2 by paths between the remaining transitions, which carry the minimum amount of tokens of paths in the original system. Then one arrives at the net shown in Figure 2. This net is not safe (albeit 2-bounded), and, more significantly, has different pomset behaviour. The pomset shown on the right-hand side of Figure 2, which arises when the first occurrence of b uses the token produced by the second occurrence of a, cannot arise in Figure 1, since there the second a and the first b are necessarily concurrent.

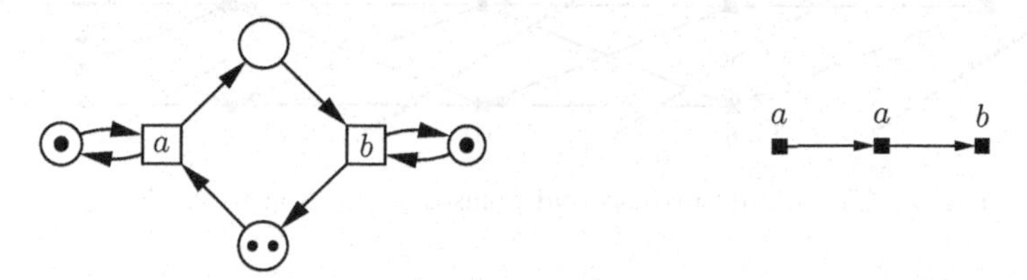

Fig. 2. An incorrect translation (left) and one of its pomsets (right)

The main idea of our construction is the following:

- Use standard theory to deduce that N has a unique (up to isomorphism) infinite process and infinite pomset.
- Carefully choose a prefix of that (τ-free!) infinite pomset.
- Transform the prefix into a live and safe marked graph N'.
- Prove that N and N' have the same set of pomsets.

To see how this works on the example, see Figure 3 which shows the (unique) infinite process and infinite pomset of the net on the left-hand side of Figure 1. From the theory of live and safe marked graphs, it is known [5, 6] that the process is divided into segments in which every transition occurs exactly once. These segments are separated by B-cuts, shown in the figure by dashed lines, each of which corresponds to the initial marking, and the process can be viewed as a concatenation ('glueing' corresponding initial and final conditions) of such segments. We also show where these cuts would go through the infinite pomset. Clearly, the pomsets shown on the right-hand side of Figure 1 correspond to certain prefixes of this infinite pomset, and there is no prefix which corresponds to the pomset shown on the right-hand side of Figure 2.

Suppose that we cut the process at any one of the dashed lines and consider only the prefix to the left of that line. Then the final conditions of this prefix can be folded back to the initial cut, yielding (when this initial cut is marked by a token on each place) a marked graph which is live and safe and has the same pomset behaviour as the original net. However, it still has τ-labelled transitions, so this does not help. Luckily, we also have the corresponding prefix of the infinite pomset, and this prefix does not contain τ-labelled nodes. We can obtain a net from this prefix by replacing each edge by an edge-place-edge combination. The edges pointing into nirvana are then folded back to some "earlier" a or b node. Places in front of such an edge are marked by a token. The result for our example can be seen in figure 4. While the net on the left hand side (using only the segment of the pomset between the first two dashed lines) looks very much

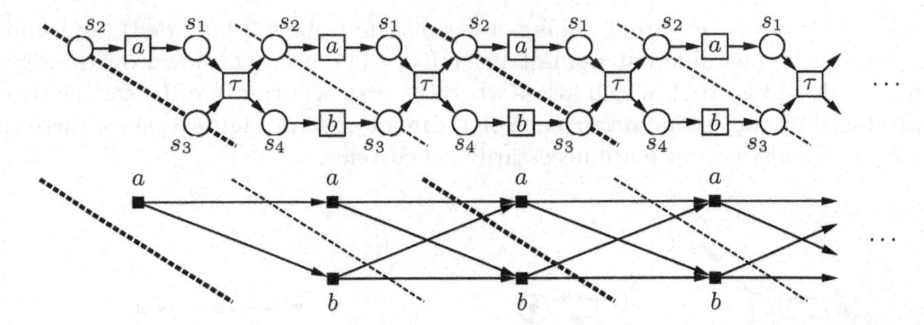

Fig. 3. The infinite process and pomset of the first example

like figure 2 (and is wrong for almost the same reasons), the net on the right has indeed the same pomset behaviour as the original net with unobservable transitions. For this net we have used the first two segments of the infinite pomset, i.e. those parts that lie between the first and the third cut (drawn as bold dashed lines).

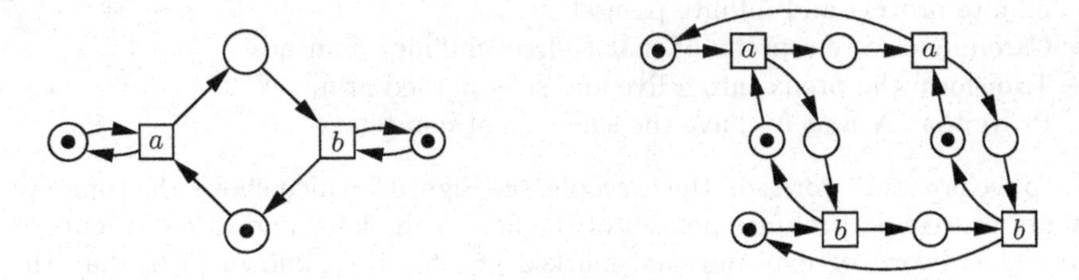

Fig. 4. Nets constructed from the first resp. the first two segments of the pomset

Each segment of the infinite pomset contains exactly one instance of every action, since our labelling is injective on observables. Therefore, it is quite easy to define segments of the infinite pomset.

Definition 5. *Segments of the infinite pomset*

Let $N = (S, T, F, M_0, \lambda)$ be a live and safe marked graph. A pomset $\varphi = [(E, R, l)] \in \mathit{Pomsets}(N)$ consists of the first n segments (of the infinite pomset of N) iff $|l^{-1}(a)| = n$ for every $a \in \lambda(T) \setminus \{\tau\}$. ■

Note, that the pomset φ above is obviously unique for each n. For some number n the pomset φ will hopefully be sufficient for constructing N'.

Suppose, we would use n segments for the construction of N'. The infinite pomset can then be divided into isomorphic blocks of n segments each, where the first block corresponds to the first n segments, the second block to the next n segments, and so on. Each node of the infinite pomset is in a unique block and has – by the isomorphism – a corresponding node in the first block, from which N' is constructed. If there is a node leading to a nirvana node, it is folded back to the corresponding node in the first block. This is shown in Fig. 4 for $n = 1$ and $n = 2$. If all edges leaving the first block just lead to the second block (as it is the case for $n = 2$ in the example) everything will be fine. If an edge leads from the first block to a later block (e.g. the third) things are different. At the border from the second to the third block there will now be two edges representing the same connection via isomorphism: One edge from the first to the third block, one edge from the second to the fourth block. In the construction of N', both edges will be represented by the same place. Since the edges occur concurrently (both crossing the cut at the end of block two), the corresponding place in N' must contain more than one token at some time. Even if there weren't safety concerns, this would be a bad idea: We cannot distinguish which token belongs to which path. Tokens could overtake each other and would ruin the partial order relation of the pomsets of N'. Therefore, n must be chosen large enough such that all edges leaving the first block end in the second block. If a can fire twice before b fires the first time – as in the example – the third firing of a depends on the first firing of b. Since the first b is in the first block, the third a must be in block one or two. This suggests that the "degree of independence" between two transitions could determine how many segments we need for block one and thus N'.

Definition 6. *Synchronic distances*

Let $N = (S, T, F, M_0, \lambda)$ be a marked, labelled net. For all transitions t_1, $t_2 \in T$ the *asymmetric synchronic distance* $asd(t_1, t_2)$ is defined as

$$asd(t_1, t_2) := Max_{\sigma \in OS(N)}(\#_{t_1}(\sigma) - \#_{t_2}(\sigma)).^4$$

The *symmetric synchronic distance* $ssd(t_1, t_2)$ of t_1, t_2 is $ssd(t_1, t_2) := asd(t_1, t_2) + asd(t_2, t_1)$.[5] The *visible synchronic distance* $vsd(N)$ *in* N is

$$vsd(N) := Max_{t_1, t_2 \in T, \lambda(t_1) \neq \tau \neq \lambda(t_2)} ssd(t_1, t_2). \qquad \blacksquare$$

Note, that the definition could also have been based on prefixes of the infinite pomset instead of the set of occurrence sequences $OS(N)$, since occurrence sequences are just linearisations of such prefixes. Both approaches are equivalent and yield the same results for $asd(t_1, t_2)$.

In our example, we find $asd(t_1, t_3) = 2$ and $asd(t_3, t_1) = 0$, therefore $ssd(t_1, t_3) = 2$ and $vsd(N) = 2$. We propose that $vsd(N)$ is the number of segments needed in the construction of N'. Note, that $vsd(N)$ is finite for every strongly connected marked graph N: For every pair of transitions x, y we find a path from x to y and if there are m tokens on such a path, y can fire at most m times without having to fire x, i.e. $asd(y, x) \leq m$.

Let from this point on $N = (S, T, F, M_0, \lambda)$ be a strongly connected, live, and safe marked graph which is labelled injectively on observables. We will now prove some properties about the pomsets of N, then present the algorithm which

[4] $\#_t(\sigma)$ denotes the number of occurrences of t in σ.

[5] Usually, the synchronic distance is defined as $Max\{asd(t_1, t_2), asd(t_2, t_1)\}$. Note that $ssd(t_1, t_2) > 0$ if $t_1 \neq t_2$ and N is live. For $t_1 = t_2$ we always have $ssd(t_1, t_2) = 0$.

will construct a live, safe, and τ-free marked graph N' with $Pomsets(N) = Pomsets(N')$, and finally prove the correctness of this algorithm.

So, let us have a look at the behaviour of N. Since we have injectivity on observables and safety, two labels a in a pomset $[(E, R, l)]$ of N belong to non-concurrent events, i.e. all events with the same label are (totally) ordered. From this point on, let the events with some label $a \in l(E)$ be named a_0, a_1, a_2, ... with $a_i R a_{i+1}$ for $0 \leq i < |l^{-1}(a)| - 1$.

It is known from the theory of marked graphs [5, 6] that there is an occurrence sequence from a marking M to itself containing each transition exactly once if M is a live marking. Consider now the consequences for the pomsets of N.

Lemma 1. *Pomsets have no backward arcs*

Let $\varphi = [(E, R, l)] \in Pomsets(N)$ be a pomset of N. For any labels $a, b \in l(E)$ and $i, j \in \mathbb{N}$ with $a_i, b_j \in E$, if $i > j$ then $\neg(a_i R b_j)$.

Proof. For $a = b$ this is obvious, so let $a \neq b$ now. Take some a_i and b_j with $i > j$. Since M_0 is a live marking, every transition can be fired exactly j times. The corresponding process contains an event b_j but no event a_i, i.e. b_j does not depend on a_i. This relation also holds in the infinite process and all its prefixes containing b_j. Therefore, in every pomset containing a_i and b_j it holds that $\neg(a_i R b_j)$. ■

Lemma 2. *Pomset relation does not depend on absolute index*

Let $\varphi = [(E, R, l)] \in Pomsets(N)$ be a pomset of N. For any two labels $a, b \in l(E)$ and $i, j \in \mathbb{N}$ with $i \leq j$ such that $a_{i+1}, b_{j+1} \in E$ it holds that $(a_i R b_j) \iff (a_{i+1} R b_{j+1})$.

Proof. This follows from the fact that the initial marking M_0 can be reproduced by firing each transition exactly once. Let c be the initial B-cut of the infinite process π (the first thick dashed line in Fig. 3), and let c' be the B-cut at which the marking M_0 is reproduced for the first time (the first thin dashed line in Fig. 3). From both points of view, c and c', the future of the process looks exactly the same, due to lack of conflicts. I.e., if a_i relates to b_j viewed from c, then a_{i+1} relates to b_{j+1} viewed from c', and vice versa. As long as all these events are in the prefix π' of π with $\Phi(\pi') = \varphi$, the relation $(a_i R b_j) \iff (a_{i+1} R b_{j+1})$ holds also in φ. ■

Lemma 3. *Concurrency is limited to index difference asd*

Let $\varphi = [(E, R, l)] \in Pomsets(N)$ be a pomset of N. For any two labels $a, b \in l(E)$ and $i, j \in \mathbb{N}$ with $a_i, b_j \in E$, $i + asd(b, a) \leq j \iff a_i R b_j$ holds.

Proof. If $i + asd(b, a) \leq j$, the j-th b cannot fire before the i-th a, i.e. a_i and b_j must be in relation R in the infinite process. Since both appear also in the prefix π with $\Phi(\pi) = \varphi$, they are also in relation R in π and φ, i.e. $a_i R b_j$.

Now assume $a_i R b_j$ and let σ be an arbitrary occurrence sequence with $\#_b(\sigma) = k \in \mathbb{N}$. We have to show that $i + (k - \#_a(\sigma)) \leq j$. Since this clearly holds if $k \leq j - i$, assume that $k > j - i$. From $a_i R b_j$ it follows by Lemma 2 that $a_{k-(j-i)} R b_k$. Hence $\#_a(\sigma) \geq k - (j - i)$. ■

These three lemmas tell us about the differences of two pomsets of two marked graphs (strongly connected, live, safe, and injectively labelled on observables) with a known upper bound on the *vsd* values. If the two pomsets have the same set of events with the same labelling, differences can always be found by checking the relation of all pairs of events (a_0, b_j) with $0 \leq j < n$ for some upper bound n on the *vsd* values of both nets.

We will now construct a net N' for which we show that it is a strongly connected, live, and safe marked graph without τ-labels. As a last step, it remains to be shown that $Pomsets(N) = Pomsets(N')$.

Definition 7. *Construction algorithm for* N'

Let $n := vsd(N)$. Then, $N' := (S', T', F', M'_0, \lambda')$ is defined as follows.

1. For each $x \in T$ there are transitions t_{x0} to $t_{x(n-1)}$ in T'. Every transition $t_{xi} \in T'$ is labelled $\lambda'(t_{xi}) := \lambda(x)$. We say t_{xi} *is of origin* x.
2. For each $x \in T$ and $i \in \{0, \ldots, n-1\}$ there is a place $p_{xi,x(i+1 \bmod n)}$ in P'. There are arcs
$$t_{xi}\, F'\, p_{xi,x(i+1 \bmod n)} \quad \text{and} \quad p_{xi,x(i+1 \bmod n)}\, F'\, t_{x(i+1 \bmod n)}$$
and a token lies on $p_{x(n-1),x0}$.
3. For $x, y \in T$ with $x \neq y$ and $i, j \in \{0, \ldots, n-1\}$ there is a place $p_{xi,yj}$ in P' with
$$t_{xi}\, F'\, p_{xi,yj}\, F'\, t_{yj} \qquad \text{iff } i + asd(y,x) \equiv j \bmod n.$$
A token lies on $p_{xi,yj}$ iff $i + asd(y,x) = j + n$.
4. There are no other places, transitions, arcs, or tokens in N'.

■

Step 1 and 2 of this algorithm provide cycles (i.e. closed paths, beginning and ending at the same node) of n transitions, all with the same label. Thus, all transitions with the same origin in N are totally ordered in every pomset of N'. In the net on the right-hand side of figure 4, there are the cycle through both a's in the top and the cycle through both b's in the bottom of the picture. Step 3 creates the dependencies between transitions of different origins. If the j-th instance of y but not the $j-1$-st instance of y depends on the i-th instance of x (in the infinite pomset of N), we find that $j-i$ y's can fire before an x must fire. Thus, $asd(y,x) = j - i$, or $i + asd(y,x) = j$. In case $j \geq n$ we would leave the net. Since the net is cyclic, we subtract n from j, i.e. we calculate modulo n, starting again from the front.

Interestingly, Step 2 is superfluous (and becomes a part of Step 3) if we define $asd(x,x) := 1$ for any transition x in N. Since the cycles created in Step 2 play a special role in some of our proofs we will not follow this line.

It is rather easy to see that N' as constructed in the algorithm is not only a marked graph but also strongly connected, live and safe.

Lemma 4. N' *is a marked graph*

The net N' as constructed in Definition 7 is a marked graph.

Proof. Each place $p_{xi,yj}$ has exactly t_{xi} as its only input and t_{yj} as its only output transition. ■

Lemma 5. *N' is strongly connected*

The marked graph N' as constructed in Definition 7 is strongly connected.

Proof. Note first, that all transitions t_{xi} in N' with the same origin x are connected by the cycles constructed in Step 2 of the construction, i.e. they form strongly connected components. Since for transitions $x, y \in T$ there is some $m \in \mathbb{N}$ with $asd(y,x) \leq m$, the place $p_{x0,y(asd(y,x))}$ exists and connects the component with origin x to the component with origin y. Since this holds for any two transitions, all strongly connected components are pairwise connected in both directions. ∎

Lemma 6. *N' is live*

The strongly connected marked graph N' built in Definition 7 is live.

Proof. It is known from the theory of strongly connected marked graphs that a marked graph is live iff every cycle contains at least one token [11]. Note, that transitions in a path in N' usually have an (not strictly) increasing index (e.g. from t_{x4} to t_{x5} to $t_{y(5+asd(y,x))}$ and so on). To be a cycle, a path must fulfill one of two conditions:

1. One section of the path has a decreasing index. By construction, whenever the index decreases on a path from one transition to the next, there will be a token on the place in between.
2. All transitions on the path have the same index. Suppose, such a path forms a cycle and there is no token on it. The construction does not create cycles containing only a single transition t_{xi} unless $vsd(N) = 1$. In this case, $i = 0$ and the cycle is $t_{x0}\ F\ p_{x0,x0}\ F\ t_{x0}$ where $p_{x0,x0}$ holds a token, a contradiction. So let now the path contain at least two transitions. For each consecutive pair of transitions t_{xi}, t_{yi} we know from the construction that either $asd(y,x) = 0$ or $asd(y,x) = vsd(N)$ must hold, otherwise these two transitions would not be directly connected. In the latter case, there would be a token on $p_{xi,yi}$, so the former case must hold. This means, in N y cannot fire before x. As such a condition holds for each consecutive pair on the cycle, none of the corresponding transitions in N can ever fire. This is a contradiction to the liveness of N. ∎

Lemma 7. *N' is safe*

The strongly connected, live marked graph N' built in Definition 7 is safe.

Proof. Since N' is live, it suffices to show that each place of N' lies on a cycle with at most one token on it [11]. Let $n := vsd(N)$. The places $p_{xi,x(i+1)}$ are contained in the cycle

$$p_{x(n-1),x0}\, F'\, t_{x0}\, F'\, p_{x0,x1}\, F'\, t_{x1}\, F'\, p_{x1,x2}\, F'\, t_{x2}\, F' \ldots F'\, t_{x(n-1)}\, F'\, p_{x(n-1),x0},$$

where the only token is on the place $p_{x(n-1),x0}$. For a place $p_{xi,yj}$ with $x \neq y$ we distinguish two cases:

1. $i + asd(y,x) = j + n$, i.e. there is a token on $p_{xi,yj}$. We find a path

$$t_{xi}\, F'\, p_{xi,yj}\, F'\, t_{yj}\, F'\, p_{yj,x(j+asd(x,y))}\, F'\, t_{x(j+asd(x,y))}$$

where $j + asd(x,y) = i + asd(y,x) + asd(x,y) - n \le i$. If $j = i$ we know $asd(y,x) = n$ and thus $asd(x,y) = 0$. The cycle is complete and only $p_{xi,yj}$ contains a token. Otherwise we complete the path to a cycle with

$$t_{x(j+asd(x,y))}\, F'\, p_{x(j+asd(x,y)),x(j+asd(x,y)+1)}\, F' \ldots F'\, p_{x(i-1),xi}\, F'\, t_{xi}.$$

The dots ... represent a section of the cycle from Step 2 of the construction of N', which leads through transitions of origin x only. Since $j+asd(x,y) \le i$, none of the intermediate places on this path contains a token. So, the cycle altogether has exactly one token, which lies on $p_{xi,yj}$.
2. $i + asd(y,x) = j$. There are two subcases: either $j + asd(x,y) \ge n$ or not. In the former case, we obtain the path

$$t_{xi}\, F'\, p_{xi,yj}\, F'\, t_{yj}\, F'\, p_{yj,x(j+asd(x,y)-n)}\, F'\, t_{x(j+asd(x,y)-n)}$$

with $j + asd(x,y) - n = i + asd(y,x) + asd(x,y) - n \le i$. We can complete the cycle in the same way as in case 1, except that the only token now lies on $p_{yj,x(j+asd(x,y)-n)}$. In the latter case, we obtain the path

$$t_{xi}\, F'\, p_{xi,yj}\, F'\, t_{yj}\, F'\, p_{yj,x(j+asd(x,y))}\, F'\, t_{x(j+asd(x,y))}$$

as in case 1, but since $j + asd(x,y) = i + asd(y,x) + asd(x,y) > i$, the completion of the cycle runs through $p_{x(n-1),x0}$, which contains a token. Neither $p_{xi,yj}$ nor $p_{yj,x(j+asd(x,y))}$ contain a token, so this is the only token on the cycle.

Now we know that each place in N' lies on some cycle containing exactly one token. Therefore, N' must be safe. ■

For the final proof of the equivalence of the pomset behaviours, two more properties of N' are important: first, the order of firing in N' is according to the indices, and second, if there is any token-free path from t_{x0} to t_{yj}, then there is also a token-free path from transition t_{x0} to t_{yj} not going through any t_{zk} with $x \ne z \ne y$. This path will not have indices greater than j. In other words, $asd(y,x) \le j$.

Lemma 8. *Enabledness in N'*

For every transition $x \in T$ and every occurrence sequence σt_{xi} of N' holds that $M'_0[\sigma t_{xi}\rangle$ implies $i = \#_x(origin(\sigma)) \mod vsd(N)$ (where *origin* maps transitions t_{yj} to their origin y).

Proof. The cycle created in Step 2 of Definition 7 contains exactly one token which lies before t_{x0}. Everything else is obvious. ■

Lemma 9. *Detours in N'*

Let t_{x0} and t_{yj} be two transitions in N'. If there is a token-free path in N' from t_{x0} to t_{yj} then $asd(y,x) \leq j$.

Proof. Note first, that "token-free" implies that the path does not contain places $p_{xi,yj}$ with $j < i$ since these contain tokens. Indeed every cycle contains a token, i.e. the path is also cycle-free. We make an induction over the number i of intermediate transitions on the path. If $i = 0$, there are no intermediate transitions, so the lemma holds because $asd(y,x) = j$ by point 3 of Definition 7. Let $i > 0$ now and assume the lemma holds up to $i-1$ intermediate transitions. Suppose the path looks like

$$t_{x0}\, F' \ldots F'\, t_{zm}\, F' \ldots F'\, t_{yj}$$

where t_{zm} is the last non-y transition on the path. The last part of the path then looks like $t_{zm}\, F'\, p_{zm,yq}\, F'\, t_{yq}\, F' \ldots F'\, t_{yj}$. Since $p_{zm,yq}$ exists, $asd(y,z) = q-m \leq j-m$. By induction, $asd(z,x) \leq m$. Obviously, a triangle inequality holds for asd, so we can calculate $asd(y,x) \leq asd(y,z) + asd(z,x) \leq (j-m) + m = j$. ∎

Lemma 10. *Relation R in pomsets is identical for N and N'*

Let $\varphi = [(E,R,l)] \in Pomsets(N)$ and $\varphi' = [(E,R',l)] \in Pomsets(N')$ be pomsets of N resp. N' with the same multiset of labels (events). Then, $R = R'$.

Proof. Note first, that Lemma 1 and Lemma 2 hold for N' despite the fact that N' is not labelled injectively on observables. The only difference is that firing each transition once will not reproduce the initial marking M_0'. Instead of firing every transition once, in N' one transition of each origin can be fired. Due to the symmetry of N' ("turning N' by $360/vsd(N)$ degrees" and firing one transition of each origin leads to a net isomorphic to N') the proofs remain valid. Take now some arbitrary pair (a_0, b_j) with $a \neq b$ and $j \geq 0$, since other pairs are not of interest (see Lemmata 1 and 2).

Suppose first, that $a_0 R b_j$ holds. As a consequence, $asd(b,a) \leq j$ follows from Lemma 3. In N' there is a path $t_{a0}\; F'\; p_{a0,bj}\; F'\; t_{bj}$ if $asd(b,a) = j$, and a path $t_{a0}\; F'\; p_{a0,b(asd(b,a))}\; F'\; t_{b(asd(b,a))}\; F' \;\ldots\; F'\; t_{bj}$ if $asd(b,a) < j$. By Lemma 8, t_{a0} and $t_{b(j \bmod vsd(N))}$ are the transitions that are responsible for creating a_0 and b_j in the infinite process π of N'. If there is a token-free path from t_{a0} to $t_{b(j \bmod vsd(N))}$, $j < vsd(N)$ and a_0 and b_j are ordered in π. They are also ordered in the prefix π' of π with $\Phi(\pi') = \varphi'$ and thus in φ', i.e. $a_0 R' b_j$ for $j < vsd(N)$. Otherwise, $j \geq vsd(N)$, and the tokens on paths from t_{a0} to $t_{b(j \bmod vsd(N))}$ are used for $b_{(j \bmod vsd(N))}$ instead of b_j. Since $a_0 R' b_{(j \bmod vsd(N))}$ is already proved and $b_i R' b_j$ for all $i < j$ by construction it is clear that $a_0 R' b_j$.

Suppose now, that $a_0 R' b_j$ holds. If $j \geq vsd(N)$, $a_0 R b_j$ holds by Lemma 3 as $0 + asd(b,a) \leq 0 + vsd(N) \leq j$. Otherwise, we find a path from t_{a0} to t_{bj} in N'. By Lemma 1 there are no events c_k with $k \geq vsd(N) > j$ on the corresponding path in φ', i.e. the path is token free. By Lemma 9 $asd(b,a) \leq j$, and this implies $a_0 R b_j$ by Lemma 3. ∎

Lemma 11. *N and N' have the same pomsets*
Then, $Pomsets(N) = Pomsets(N')$.

Proof. Let $\varphi = [(E, R, l)] \in Pomsets(N)$ be some pomset. Let $i := \max\{|l^{-1}(a)| \mid a \in \Sigma\}$. Since M_0 is reproduced by firing each transition exactly once, we can find a pomset $\varphi' = [(E', R', l')] \in Pomsets(N)$ where for each label $a \in \Sigma$ there are exactly i events e with $l(e) = a$; φ is a prefix of φ'. The analogous argument for N' yields a pomset $\varrho' = [(E', R'', l')] \in Pomsets(N')$. By Lemma 10, $R' = R''$, i.e. φ is a prefix of ϱ'. Thus, $\varphi \in Pomsets(N')$. The same proof works for showing that pomsets in $Pomsets(N')$ are also in $Pomsets(N)$. ∎

Theorem 1. *N' is a τ-free version of N*
Let $N = (S, T, F, M_0, \lambda)$ be a strongly connected, live, and safe marked graph with a labelling that is injective on observables. Then, N' — as constructed in Definition 7 — is a strongly connected, live, and safe marked graph without τ-transitions that has the same pomset behaviour as N, i.e. $Pomsets(N) = Pomsets(N')$.

Proof. Follows from Lemma 4, Lemma 5, Lemma 6, Lemma 7, Lemma 11, and the fact that no τ labels are introduced in Definition 7. ∎

Theorem 2. *τ-Elimination in arbitrary safe marked graphs*
Let $N = (S, T, F, M_0, \lambda)$ be a safe marked graph. There is a safe marked graph $N' = (S', T', F', M_0', \lambda')$ with $\tau \notin \lambda'(T')$ and $Pomsets(N) = Pomsets(N')$.

Proof. If N is not labelled injectively on observables, we replace the labelling by one that is injective on observables (but with the same τ-labels). Obviously, there is a map h going from the latter to the former labelling. Apply Theorem 2 to the newly labelled marked graph and then h to each transition label of the resulting marked graph. We obtain the desired marked graph if the second application of Theorem 2 works correctly. This can be shown as follows.

Observe first, that in a safe and strongly connected marked graph either all transitions are live, or none is. If some transition is not live, the places in its postset cannot be marked arbitrarily often. Thus, the transitions in the postsets of these places are not live either. Due to strong connectedness, this propagates through the marked graph, and no transition at all is live.

If N does not contain live transitions, its maximal process is finite. If this process $\pi = (B, E, G, r, \lambda'')$ contains a τ-labelled event e, we may replace e and its pre- and postset by the set of conditions $G^{-1}(e) \times G(e)$ and connect them accordingly, i.e. (b, b') lies in the preset of e' if $e'Gb$ and in the postset of e'' if $b'Ge''$. The pomsets remain unchanged and the resulting "process" is the marked graph N'.

Without loss of generality we may now assume that N is weakly connected, otherwise we can handle each weakly connected component of N separately. So, let N consist of exactly one weakly connected component. If N has only one strongly connected component, N is either live and Theorem 1 is applicable, or N is non-live and the construction of the previous paragraph can be used. In both cases, we are done. So assume now, that N has at least two different strongly connected components A and B with a path from A to B, but not from B to A. Since marked graphs do not branch at places, the path must lead from a transition of A to a transition of B, and it must contain at least one place. If A were live, this place would not be safe. Thus, A is not live. Since B consumes tokens from this place (directly or indirectly), it cannot be live either. As this holds for any two different strongly connected components in N, there cannot be any live transitions in N. Again, the marked graph without τ-labels can be obtained according to the previous paragraph. ∎

The assumption of safeness can, however, not be easily lifted, since two tokens on the same place lead to a (dynamic) conflict. Take a transition and two places with arcs to the transition back and forth. Each place contains two tokens. Pomsets of this marked graph consist of two chains of events which are interlinked at arbitrary points. The links depend on which two tokens are chosen when firing the transition. Accordingly, there is no unique maximal process either, which is the main prerequisite for our proof.

4 Conclusions

We have presented a technique to remove unobservable transitions from some safe marked graph without changing the pomset behaviour. Such an elimination algorithm is well-known e.g. for S-systems, which can be interpreted as a set of independent (and isomorphic) finite automata. For the latter, τ-elimination is a standard procedure. The algorithm for safe marked graphs and its proof are clearly more involved. Neither of these requirements – being an S-system or a safe marked graph – can be lifted easily. In the case of S-systems, the removal of τ's according to the theory of finite automata will not be possible if we lift the "no branching at transitions" condition. In the case of marked graphs, allowing conflicts will invalidate our construction, as there is no unique infinite process and pomset anymore on which the whole proof is based. Turning to event structures [7], which incorporate a conflict relation, might seem to be a good idea, but it is not clear if and how the backfolding (i.e. completion of the cycles in the construction of the net N') will work as the conflict relation can become rather complex when unobservable transitions are involved.

Acknowledgment

I would like to thank Javier Esparza for drawing my attention to the fact that the sub-problem solved in this paper is interesting in itself, and Eike Best and Philippe Darondeau for some interesting discussions on this topic and on the more general conjecture.

References

1. E. Best, C. Fernández: *Nonsequential Processes: A Petri Net View.* EATCS Monographs on Theoretical Computer Science No. 13, Springer Verlag, 1988.
2. E. Best: Semantics of Sequential and Nonsequential Programs. Prentice Hall, 1996.
3. E. Best, R. Devillers: Sequential and Concurrent Behaviour in Petri Net Theory, *Theoretical Computer Science* **55**, pp. 87–136, 1988.
4. E. Best, H. Wimmel: Reducing k-Safe Petri Nets to Pomset-Equivalent 1-Safe Petri Nets, 21st International Conference on Application and Theory of Petri Nets, *Lecture Notes in Computer Science* **1825**, pp. 63–82, 2000.
5. F. Commoner, A. Holt, S. Even, A. Pnueli: Marked Directed Graphs. *Journal of Computer and System Sciences* **5**, pp. 511–523, 1971.
6. H.J. Genrich, K. Lautenbach: Synchronisationsgraphen. *Acta Informatica* **2**, pp. 143–161, 1973.
7. M. Nielsen, P.S. Thiagarajan: Regular Event Structures and Finite Petri Nets: The Conflict-Free Case. *Lecture Notes in Computer Science* **2360**, p. 335, 2002.
8. V. Pratt: Modelling Concurrency with Partial Orders. *International Journal of Parallel Programming* **15**, pp. 33–71, 1986.
9. L. Priese, H. Wimmel: *Petri-Netze.* 376 pages, ISBN 3-540-44289-8, Springer-Verlag, 2003.
10. L. Priese, H. Wimmel: A Uniform Approach to True-Concurrency and Interleaving Semantics for Petri-Nets. *Theoretical Computer Science* **206**, pp. 219–256, 1998.
11. P.H. Starke: Analyse von Petri-Netz-Modellen, Teubner, 1990.
12. H. Wimmel: Algebraische Semantiken für Petri-Netze (Dissertation). Universität Koblenz-Landau, 2000.

Infinity of Intermediate States Is Decidable for Petri Nets

Harro Wimmel

Parallel Systems, Department of Computing Science
Carl von Ossietzky Universität Oldenburg, Germany

Abstract. Based on the algorithms to decide reachability for Petri nets, we show how to decide whether the number of markings reachable on paths between two given markings is finite or infinite.

Keywords: Petri nets, reachability problem, decidability.

1 Introduction

Decidability problems come in different flavours. In formal language theory e.g., common problems are the word problem, the emptiness problem, and the finiteness problem, asking whether a word can be accepted by an automaton or if the language of an automaton is empty or finite, respectively. In Petri net theory, the word problem is (at least for nets without any unobservable transitions) decidable in exponential time, by simply using brute force. The emptiness problem can be formulated as the question "Is there or is there not a firing sequence leading from one specified marking to another?", which is better known as the reachability problem. The reachability problem is also decidable (see [6], [7], [2], [3], [9], or [10]), but only a lower bound for its complexity is known: it takes at least exponential space to solve [5]. The finiteness problem would be a question such as "Given markings m and m', can we reach finitely many or infinitely many different markings on paths from m to m'?" As far as we know, this kind of question has not yet been addressed for Petri nets. We will call this the *intermediate markings problem* IMP.

A seemingly similar problem is the boundedness problem: Given a marking m, is the number of markings reachable from m finite or infinite? It is well-known that this problem is decidable by constructing the coverability graph [1] which shows us if we can obtain arbitrarily large markings. Constructing coverability graphs from m and "backwards" from m' would seem a good idea to solve IMP, but unluckily we cannot find out from these coverability graphs at which markings they connect, i.e. if and how many markings are reachable from m as well as backwards from m'. Reachability graphs cannot help either since we would need to search for infinitely many paths then. It seems that IMP cannot be solved that easily.

Solving the problem will have an impact in other fields as well. When dealing e.g. with Netcharts [8], an extension of Message Sequence Charts, the following

J. Cortadella and W. Reisig (Eds.): ICATPN 2004, LNCS 3099, pp. 426–434, 2004.
© Springer-Verlag Berlin Heidelberg 2004

problem arises: A set of sequential processes communicates via buffers. This system terminates when all sequential processes terminate and all buffers are empty. If there is an upper bound n for the maximal size of the buffers, the system can be realized in one way or another with buffers of length n and we obtain a finite implementation. We will not deal with such an application here, but it is quite clear that the number of states reachable in a terminating system (determined by the control flow and the number of elements in each buffer) is finite iff the upper bound n exists. If we construct a Petri net model for this system, it suffices to decide whether we can reach only a finite number of markings between the initial and the terminating marking, i.e. we need to decide IMP.

In this paper, we will show that IMP is indeed decidable. In section 2 we present the basic definitions necessary to understand the problem. Section 3 presents those parts of the proof of decidability of the reachability problem that are necessary to deal with the intermediate markings problem IMP. In section 4 we finally show that the latter problem is also decidable.

Acknowledgment. I'd like to thank Javier Esparza for pointing me to this question.

2 Basic Definitions

In this section, we present the basic definitions necessary for solving the intermediate markings problem. These definitions include Petri nets and reachability and coverability graphs.

Definition 1. *Petri net*

An *(unmarked) net* is a triple (S,T,F) where S and T are finite sets with $S \cap T = \emptyset$, and F is a mapping $F\colon (S \times T) \cup (T \times S) \to \mathbb{N}$.

A *marking* of (S,T,F) is a function $m\colon S \to \mathbb{N}$. We say that a place s *has* k *tokens under* m if $m(s) = k$.

The transition rule for Petri nets is defined as follows. A marking m *activates* a transition t (in symbols: $m[t\rangle$) iff $\forall s \in S\colon m(s) \geq F(s,t)$. Moreover, m leads to a marking m' through the *occurrence* of t (in symbols: $m[t\rangle m'$, we also say that t *fires under* m and yields m' as a result) iff $m[t\rangle$ and $\forall s \in S\colon m'(s) = m(s) - F(s,t) + F(t,s)$. We define inductively that an *occurrence sequence* $\sigma \in T^*$ leads from m to m':

$$\begin{aligned} m[\varepsilon\rangle m' &\iff m = m' \\ m[\sigma t\rangle m' &\iff \exists m''\colon m[\sigma\rangle m'' \wedge m''[t\rangle m', \end{aligned}$$

where ε is the empty word. ∎

In the following, let ω be a symbol representing an unbounded number. We abbreviate $\mathbb{N} \cup \{\omega\}$ by $\mathbb{N}_\omega$. The usual order $\leq$ on natural numbers extends to $\mathbb{N}_\omega$ by defining $\forall n \in \mathbb{N}\colon n \leq \omega$. It is quite easy to extend the definition of markings to ω-markings $m: S \to \mathbb{N}_\omega$. We introduce a special ordering $\preceq_\omega$ for ω-markings, where $m \preceq_\omega m'$ iff for all places s either $m(s) = m'(s)$ or $m'(s) = \omega$.

The firing of transitions can also be easily adapted by just defining that $\omega - i$ and $\omega + i$ both equal ω for all $i \in \mathbb{N}$. The usual firing rule can now be applied, and firing a transition will never increase or decrease the number of tokens for a place s with $m(s) = \omega$.

An ω-marking m' is reachable from an ω-marking m in a net N iff there exists an occurrence sequence leading from m to m'. The set of ω-markings reachable from m will be denoted by $\mathcal{R}(m)$.

Definition 2. *Reachability and Intermediate Markings Problem*

The *reachability problem* RP is the set

$$\mathsf{RP} := \{(N, m, m') \mid N = (S, T, F) \text{ is a net, } m, m' \in \mathbb{N}^S, \exists \tau : m[\tau\rangle m'\}$$

and the *intermediate markings problem* IMP is the set

$$\mathsf{IMP} := \{(N, m, m') \mid N = (S, T, F) \text{ is a net, } m, m' \in \mathbb{N}^S, \text{ and the set } I = \{m'' \mid \exists \tau, \tau' : m[\tau\rangle m''[\tau'\rangle m'\} \text{ is infinite}\}.$$

■

The reachability problem RP, i.e. the problem if, given some net and two markings, we can reach one marking from the other, is known to be decidable. This was first proved by Mayr [6, 7] with an alternative proof by Kosaraju [2] only a bit later. In the 90's, Lambert [3] presented another proof, which also found entrance to Priese and Wimmel [9].

The intermediate markings problem IMP is the problem we deal with in this paper. IMP is the question whether there are infinitely many different markings reachable on paths from one given marking to another or not. To show its decidability we need the proof of decidability of RP. Following the concepts of this proof, we will present the ideas necessary to prove the decidability of IMP.

To deal with reachability, reachability graphs and coverability graphs were invented [1].

Definition 3. *Reachability and Coverability Graphs*

Let $N = (S, T, F)$ be a net and $m_0: S \to \mathbb{N}_\omega$ an ω-marking of N. The *reachability graph* R of (N, m_0) is the edge-labelled graph $R = (\mathcal{R}(m_0), E, T)$, where an edge $e = (m_1, t, m_2)$ with labelling t is in E iff $m_1[t\rangle m_2$.

A *coverability graph* $C = (V, E, T)$ of (N, m_0) is defined inductively. First, m_0 is in V. Then, if $m_1 \in V$ and $m_1[t\rangle m_2$, check for every m on a path from m_0 to m_1 if $m \leq m_2$. If the latter holds, change $m_2(s)$ to ω whenever $m_2(s) > m(s)$. Now, add (the changed) m_2 to V (if m_2 is not yet in V) and (m_1, t, m_2) to E. Repeat, until no new nodes and edges can be added anymore. ■

While the reachability graph can be (and usually is) of infinite size, a coverability graph is always finite, but cannot be used to decide reachability due to its inexactness (an ω can stand for any natural number, but it is possible that a certain specific number of tokens on an ω-place can't be reached). Still, the concept of a coverability graph is valuable in dealing with reachability, as it allows for a partial solution to the reachability problem: A marking m is not reachable if there is no m' with $m' \geq m$ in the coverability graph. For a complete solution of the reachability problem, the concept of the coverability graph needs to be extended.

3 A Look at the Decidability of RP

We present here some main ideas behind the proof of decidability of RP according to Lambert [3] (see also [9]). This proof is based on marked graph transition sequences (MGTS), which are sequences of special instances of coverability graphs $C_1, \ldots, C_n$ alternating with transitions $t_1, \ldots, t_{n-1}$. These special instances of coverability graphs are called precovering graphs in [3] and have additional structure that will not be of any use for our purposes. In this paper, each coverability graph C_i is augmented only by two additional ω-markings $m_{i,in}$, $m_{i,out}$. If C_i is constructed from an initial marking M_i then $m_{i,in} \preceq_\omega M_i$ and $m_{i,out} \preceq_\omega M_i$ must hold, i.e. for all places s $M_i(s) < \omega$ implies $M_i(s) = m_{i,in}(s) = m_{i,out}(s)$. The transitions $t_1, \ldots, t_{n-1}$ in an MGTS connect the output $m_{i,out}$ of one coverability graph to the input $m_{i+1,in}$ of the next, see Fig. 1 with $n = 4$, where the circles represent the coverability graphs with their initial markings in the center.

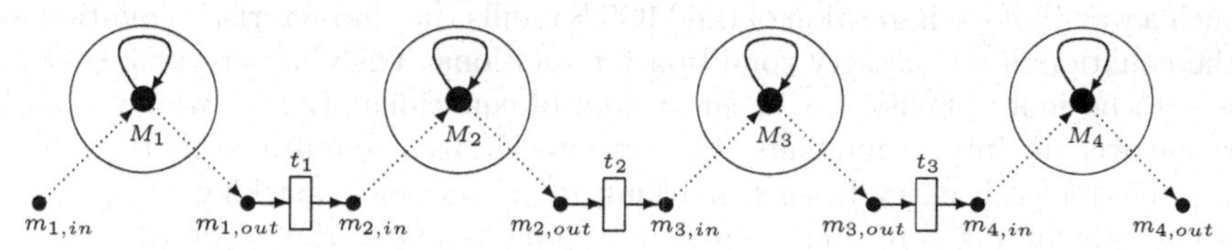

Fig. 1. A marked graph transition sequence (MGTS). Dots represent markings, the circles represent coverability graphs (with more than one node in general), solid lines inside these circles are occurrence sequences (which must be firable in the Petri net). The dotted lines show the entry to and exit from coverability graphs, which do not change the actual marking in the Petri net. Both $m_{i,in} \preceq_\omega M_i$ and $m_{i,out} \preceq_\omega M_i$ hold for every i

A solution of an MGTS is by definition an occurrence sequence leading through the MGTS. In the above figure it begins with the marking $m_{1,in}$, leads in cycles through the first coverability graph until the marking $m_{1,out}$ is reached, then t_1 can fire to reach $m_{2,in}$, from which the second coverability graph is entered, and so on, until the MGTS ends. Whenever the marking of some node has a finite value for some place, this value must be reached exactly by the occurrence sequence when the node is traversed. If the value is ω, there are no such conditions. The set of solutions of an MGTS G is denoted $L(G)$ (see also [3], page 90). It is quite obvious to see that an instance of the reachability problem can be formulated as the problem of finding a solution for a special MGTS, which is depicted in Fig 2. The node $\boldsymbol{\omega}$ (with all entries ω) is the only node of the coverability graph, i.e. we allow arbitrary ω-markings and firings of transitions between $m_{1,in}$ and $m_{1,out}$, but the sequence must begin exactly at the (finite) marking $m_{1,in}$ and end exactly at the marking $m_{1,out}$. If we can calculate the solutions for this MGTS, we have solved the reachability problem for $(N, m_{1,in}, m_{1,out})$ at the same time. We get a positive answer if there is a solution for the MGTS, otherwise a negative answer.

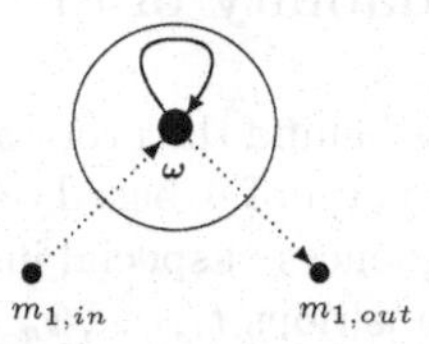

Fig. 2. The MGTS representation of an instance $(N, m_{1,in}, m_{1,out})$ of the reachability problem. The MGTS consists of one coverability graph with just one node $\boldsymbol{\omega}$ which represents the ω-marking where all places have an unbounded number of tokens and from which every transition can fire. A solution for this MGTS is an occurrence sequence from $m_{1,in}$ to $m_{1,out}$

So, the "only" problem that remains to decide RP is to solve arbitrary MGTS. Lambert shows that for each MGTS a characteristic equation can be defined in such a way that each solution of the MGTS fulfills the characteristic equation, i.e. the equation is a necessary condition for solutions. The characteristic equation is – to be more precise – a linear system of equations $Ax = b$ where A and b range over the integer numbers. The system can become quite large: there is one variable for each entry in each marking $m_{i,in}$ and each marking $m_{i,out}$ in the MGTS (including zero and ω entries) as well as one variable for each edge in each coverability graph of the MGTS. Since markings must not become negative and transitions cannot fire backwards, solutions sought must be semi-positive. The (possibly empty) set of semi-positive solutions of such a system can always be computed according to a result in [4].

If the characteristic equation were a sufficient and necessary condition for solutions of an MGTS, RP would be solved immediately. This is not the case in general, but Lambert shows under which circumstances the characteristic equation is also a sufficient condition for solutions of an MGTS. Generally speaking, this holds iff the variables for the edges and the variables for all ω-entries of the markings are unbounded in the set of solutions of the equation system and $m_{i,out}[t_i\rangle$ holds for all i. An MGTS to which such a sufficient characteristic equation belongs is called *perfect*. The unboundedness of the variables can be tested effectively according to [4].

Since not all MGTS are perfect, a decomposition theorem for MGTS is presented in [3]. This theorem produces from one MGTS G a new set of MGTS's, all of which are to a greater degree perfect, such that the new set as a whole has the same solutions as the original MGTS. This means that each occurrence sequence leading through the original MGTS and solving it will also lead through at least one of the derived MGTS's and solve it, and vice versa. The degree of perfectness is discrete and cannot be increased indefinitely, therefore at some point after repeatedly applying the decomposition theorem we arrive at a finite set Γ of perfect MGTS's. With the assumption that $m_{1,in}$ and $m_{n,out}$ are ω-free the decomposition theorem (cf. [3], page 94, or [9]) is simplified to the following form.

Theorem 1. *Decomposition*

Let G be an MGTS. G can be decomposed into a finite (possibly empty) computable set Γ of perfect MGTS's with the same overall set of solutions as G, i.e. $L(G) = \bigcup_{\gamma \in \Gamma} L(\gamma)$, and the same first input marking $m_{1,in}$ and last output marking $m_{n,out}$.

By applying the decomposition theorem to the MGTS of Fig. 2 (for the instance $(N, m_{1,in}, m_{1,out})$ of the reachability problem) we obtain a set Γ of perfect MGTS's. For each of these perfect MGTS γ we can decide whether it has solutions, i.e. whether $L(\gamma) \neq \emptyset$. If at least one has a solution, we obtain a positive answer to the reachability problem, otherwise a negative answer. I.e., the algorithm for deciding RP looks like this:

input $(N, m_{1,in}, m_{1,out})$
`create` G `according to Fig. 2`
`decompose` G `into` Γ `with` γ `perfect for all` $\gamma \in \Gamma$
if $\exists \gamma \in \Gamma$ `with` $L(\gamma) \neq \emptyset$ **answer yes else answer no**

The reachability problem is not only decidable, it is also possible to calculate an occurrence sequence solving it, if one exists. Lambert's iteration lemma (see [3] or [9]) states the following:

Lemma 1. *Iteration lemma*

Let G be a perfect MGTS (consisting of n coverability graphs and intermediate transitions t_1 to t_{n-1}) with at least one solution. Then, for each sequence of pairs of *covering sequences* $((u_i, v_i))_{1 \leq i \leq n}$ (see below) we can compute $k_0 \in \mathbb{N}$ and occurrence sequences β_i, w_i from M_i to M_i such that *for every* $k \geq k_0$

$$(u_1)^k \beta_1 (w_1)^k (v_1)^k t_1 (u_2)^k \beta_2 (w_2)^k (v_2)^k t_2 \dots t_{n-1} (u_n)^k \beta_n (w_n)^k (v_n)^k \qquad (1)$$

is a solution of G. ■

A covering sequence u_i is an occurrence sequence from M_i to M_i with the following special properties:

- The sequence u_i can fire under the marking $m_{i,in}$.
- If $M_i(s) = \omega > m_{i,in}(s)$ for some place s, u_i will add tokens to s when firing.
- If $M_i(s) < \omega$ (and thus $m_{i,in}(s) = M_i(s)$) for some place s, u_i will not change the number of tokens on s when firing (obvious, since it leads from M_i to M_i).
- If $M_i(s) = \omega = m_{i,in}(s)$ for some place s, then it is not defined what firing u_i does with respect to s.

For v_i, analogous properties are required, except that the reverse of v_i must be able to fire backwards from $m_{i,out}$. Lambert ([3], page 86) has proved that such covering sequences always exist and at least one can be computed. Therefore, each part $(u_i)^k\beta_i(w_i)^k(v_i)^k$ leads from the center node M_i (resp. $m_{i,in}$) of the ith coverability graph of the MGTS back to itself (resp. $m_{i,out}$), and the next intermediate transition t_i leads from there to the next coverability graph in the sequence. Firing the covering sequence u_i k_0 times will pump up the marking to the level necessary such that β_i can fire and v_i will pump down the marking later to be able to reach $m_{i,out}$ exactly (for the non-ω entries, of course).

4 Deciding IMP

We now use the concepts from the proof of decidability of RP and draw some additional conclusions, which lead to the following theorem.

Theorem 2. *Decidability of* IMP

IMP is decidable.

To prove this we first observe that it is sufficient to analyse only perfect MGTS. Consider the following algorithm for IMP:

input $(N, m_{1,in}, m_{1,out})$
`create` G `according to Fig. 2`
`decompose` G `into` Γ `with` γ `perfect for all` $\gamma \in \Gamma$
if $\exists \gamma \in \Gamma$ `with` $L(\gamma) \neq \emptyset$ `and there are infinitely many different markings on solution paths from` $m_{1,in}$ `to` $m_{n,out}$ `in` γ **answer** `yes`
else answer `no`

This algorithm matches with the one for RP except for one criterion: checking whether there are infinitely many different markings on solution paths from $m_{1,in}$ to $m_{n,out}$ in some perfect MGTS γ. If this criterion is decidable, we can decide IMP altogether. The center nodes M_i of coverability graphs in a perfect MGTS now play an important role.

Theorem 3. *ω-entries in the M_i*

Let G be a perfect MGTS with at least one solution, i.e. $L(G) \neq \emptyset$. The set of markings traversed in solutions of G is infinite iff there is an M_i containing an ω in one of the coverability graphs of G (with $1 \leq i \leq n$).

Proof. Assume first, that $M_i(s) < \omega$ for all places s and $1 \leq i \leq n$. Every coverability graph in an MGTS is strongly connected by definition (see [3], page 86). Therefore, all nodes in any of the coverability graphs (plus, due to $m_{i,in} \preceq_\omega M_i$ and $m_{i,out} \preceq_\omega M_i$, the input and output markings), are finite, i.e. do not contain any ω. We conclude that exactly the markings (nodes) in the MGTS G can be reached when traversing it. Since G is finite, the number of reachable markings between overall input and output of G is also finite. In a solution such as (1), a covering sequence (such as u_1, v_1, ...) does not change the number of tokens for any place.

Suppose now that $M_i(s) < \omega$ does not hold for all M_i and places s. Let then i be the index of the leftmost coverability graph in G such that $M_i(s) = \omega$ holds for some place s, i.e. i is minimal with this property. Then, $m_{i,in}$ does not contain any ω, which can be seen as follows: Assume $m_{i,in}(s') = \omega$ for some place s'. This can only be the case if $i > 1$: Since $m_{1,in}$ is the input marking for our test of membership in IMP, it must be finite. According to [3] (page 92) perfectness means that if $m_{i,in}(s) = \omega$ then in solutions of the characteristic equation the number of tokens on s at this point can be arbitrarily high. On the other hand, the definition of the characteristic equation itself ([3], page 91) states that if $m_{i-1,out}(s) < \omega$, all solutions must reach this value exactly. Thus, if we make the step $m_{i-1,out}[t_{i-1}\rangle m$ in a solution, then $m(s)$ can take only one value, which is a contradiction to $m_{i,in}(s) = \omega$ and perfectness.

It follows that $m_{i,in}(s) < \omega = M_i(s)$. Let now (1) be a solution for the perfect MGTS G. Then u_i increases the number of tokens on s when firing (as this is one of the covering sequence properties of u_i, see the previous section). Since $m_{i,in}$ is finite, it is reached exactly (independent from the choice of k in the solution), and from this point on, the part of the solution beginning with $(u_i)^k \ldots$ can fire for any $k \geq k_0$. Consequently, we reach different markings for different k after $(u_i)^k$. Since there are an infinite number of possibilities for k, we will encounter an infinite number of different markings while traversing solutions. This is shown graphically in Fig 3. ∎

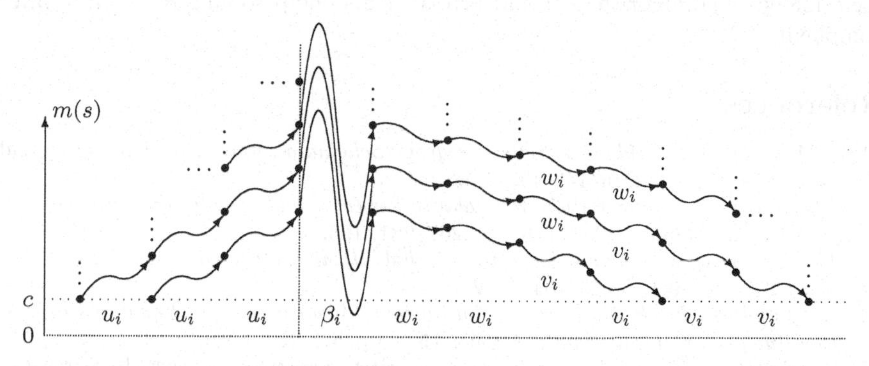

Fig. 3. Traversing solutions with different k (2, 3, 4) in the ω-case. Firing u_i increases, firing v_i usually decreases the number of tokens on some place s. The long, vertical, dotted line contains an infinite set of pairwise different reachable markings

Since we can decide for any perfect MGTS if it contains an ω somewhere, Theorem 2 is correct and we have decided IMP.

There is another problem looking quite similar to IMP: Given two markings $m_{1,in}$, $m_{1,out}$, are there infinitely many occurrence sequences leading from $m_{1,in}$ to $m_{1,out}$, i.e. is $L(G)$ (with G from Fig. 2) infinite? This problem can be solved using a slightly extended method. First, check if the problem is in IMP, answer "yes" if this is the case. If not, all perfect MGTS's constructed with the decomposition theorem will be ω-free. Remove all those perfect MGTS's which have no solution. If any of the remaining perfect MGTS's contains a non-trivial coverability graph C_i (where non-trivial means with at least one edge) we can find a non-empty covering sequence u_i from M_i to M_i (due to strong connectedness). Using this covering sequence, the iteration lemma yields infinitely many solutions. The answer will also be "yes". If all the perfect MGTS's contain only trivial coverability graphs, each of the MGTS's represents just one solution, namely the sequence $t_1 t_2 \ldots t_{n-1}$ of transitions in the MGTS. Since there are only finitely many of them, the answer is "no".

5 Conclusions

It was shown in this paper that, besides the word and reachability problem for Petri nets, also the finiteness problem, called the intermediate markings problem here, is decidable. The lower bound for the complexity of the algorithm for deciding IMP presented in this paper can be expected to be of at least exponential space, as it exploits the whole proof for the decidability of the reachability problem. A proof would of course need to employ some kind of reduction and will not be done here. Due to the fact that coverability graphs can become arbitrarily large, the complexity of our algorithm has no known upper bound. Whether better algorithms for deciding IMP can be found is unclear, though it does not seem likely: Such algorithms would probably also help solving RP with a lower complexity.

References

1. R.M. Karp, R.E. Miller: *Parallel Program Schemata*, Journal of Computer and System Sciences **3**:2, pp.147–195, 1969.
2. S.R. Kosaraju: *Decidability of reachability in vector addition systems*, Proceedings of the 14th Annual ACM STOC, pp.267–281, 1982.
3. J.L. Lambert: *A structure to decide reachability in Petri nets*, Theoretical Computer Science **99**:1, pp.79–104, 1992.
4. J.L. Lambert: *Finding a Partial Solution to a Linear System of Equations in Positive Integers*, Comput. Math. Applic. **15**:3, pp.209–212, 1988.
5. R.J. Lipton: *The Reachability Problem Requires Exponential Space*, Research Report 62, Yale University, 1976.
6. E.W. Mayr: *Ein Algorithmus für das allgemeine Erreichbarkeitsproblem bei Petrinetzen und damit zusammenhängende Probleme*, Technischer Bericht der Technischen Universität München TUM-I8010 (Dissertation), Institut für Informatik, 1980.
7. E.W. Mayr: An algorithm for the general Petri net reachability problem, *SIAM Journal of Computing* **13**:3, pp.441–460, 1984.
8. M. Mukund, K. Narayan Kumar, P.S. Thiagarajan: *Netcharts: Bridging the gap between HMSCs and executable specifications*, Proceedings of CONCUR '03, Lecture Notes in Computer Science **2761**, pp.296–310, 2003.
9. L. Priese, H. Wimmel: *Petri-Netze.* 376 pages, ISBN 3-540-44289-8, Springer-Verlag, 2003.
10. C. Reutenauer: *The Mathematics of Petri nets*, Prentice Hall, 1990.

Operation Net System: A Formal Design Representation Model for High-Level Synthesis of Asynchronous Systems Based on Transformations

Dong-Hoon Yoo[1], Dong-Ik Lee[1,*], and Jeong-A Lee[2]

[1] Department of Information and Communications,
Kwang-Ju Institute of Science and Technology, Republic of Korea
dhyoo@kjist.ac.kr

[2] Department of Computer Engineering,
Chosun University, Republic of Korea
jalee@chosun.ac.kr

Abstract. This paper proposes a formal design representation model, called *Operation Net System*, for high-level synthesis of asynchronous systems which is based on *transformational approaches*. Operation Net System consists of *Operation Net*, which is based on hierarchical timed Petri nets, and a *module graph* to *unify* the description of control and data parts of an asynchronous system. It is used as an *intermediate* design representation during transformations. Several *semantic-preserving* basic transformations are defined and used to successively transform an initial design representation into an optimized implementation satisfying designer's requirements. The hierarchical concept of Operation Net reduces the complexity of net operations such as finding the relation of two operations by hiding sub-hierarchical blocks. The selection of the basic transformations is guided by an exploration strategy which performs operation scheduling, resource allocation, and module binding *simultaneously*. To deal with the complexity of the exploration, we use an iterative algorithm such as *Tabu search*. This *integration* of high-level synthesis sub-tasks enables designers to get a better chance to reach a globally optimized solution.

1 Introduction

Power consumption, electromagnetic compatibility, clock skew, robustness and scalability have become increasingly critical issues in modern VLSI systems. This draws more interest in asynchronous designs because the asynchronous designs compare favourably with synchronous designs with respect to the above criteria. Although asynchronous techniques have many advantages over the synchronous

* Professor Dong-Ik Lee who was the Ph.D. thesis adviser of the first author has passed away last October 5. This paper was co-authored with Prof. Dong-Ik Lee before he died.

J. Cortadella and W. Reisig (Eds.): ICATPN 2004, LNCS 3099, pp. 435–453, 2004.
© Springer-Verlag Berlin Heidelberg 2004

counterpart, asynchronous designs are not prevailing due to the lack of high-quality CAD tools. That is, there is a lack of approaches providing the systematic design-space exploration and optimization of large-scale asynchronous systems.

There have been many approaches to design large-scale asynchronous systems [1–7]. A conventional approach is to use the combination of several design methodologies supported by manual optimizations or academic automation tools [1]. This approach allows a number of aggressive optimizations. However, it is an error-prone and time consuming job. Also, it does not provide systematic design-space exploration. On the other hand, syntax-directed methods are successfully adopted for commercial products [8]. These methods start from an high-level abstraction, such as a concurrent program, and generate a circuit by translating each program construct into a corresponding sub-circuit, such as *handshake circuits* [4] or *macro-modules* [3]. Although they are well supported by automated tools such as `Tangram` [5], they do not provide a chance to explore design-space other than peephole optimization techniques. There are approaches to use Petri nets as an intermediate format [6, 7, 9]. These approaches translate the given hardware description languages to intermediate Petri nets. The Petri nets are used for partitioning, performance analysis, and control circuits generation. In [9], we proposed a global optimization method, which translates a concurrent program into signal-level Petri nets, for syntax-directed approaches. However, all these frameworks do not provide optimization techniques such as high-level synthesis including operation scheduling, resource allocation, and binding.

The contribution of this paper is to propose a new high-level synthesis method with defining a *formal design representation model* and its *transformation patterns* which are suitable for the synthesis of large-scale asynchronous systems. The representation model consists of a hierarchical timed Petri net and a data path, called *Operation Net* and a *module graph* respectively, to *unify* the description of control and data parts of asynchronous systems. The model is used as an *intermediate* design representation during transformations. Various *semantic-preserving* basic transformations which are suited to asynchronous systems are defined. And they are used to successively transform an initial design to an optimized implementation. The selection of the basic transformations is guided by an optimization strategy which performs the high-level sub-tasks *simultaneously*. To deal with the complexity of the method, we use an iterative algorithm such as *Tabu search* [10, 11].

In section 2, we present an introduction to asynchronous high-level synthesis and transformational approaches. In section 3 and 4, we introduce the proposed design representation and basic transformation patterns. In section 5, we present an overview of optimization algorithm. And we conclude and present future work in section 6.

2 High-Level Synthesis: An Overview

High-level synthesis, sometimes called *behavioral synthesis*, is the design task of mapping an abstract behavioral description of a digital system onto a register-transfer level design to implement that behavior. High-level synthesis has the potential to greatly improve both designer productivity and design-space exploration.

In convention, high-level synthesis consist of three sub-tasks; operation scheduling, resource allocation, and module binding. Scheduling in a synchronous design deals with the assignment of each operation to a control step corresponding to a clock cycle or time interval which is determined by the *worst-case delay* of all the operations executed in the control step. Allocation and binding carry out selection and assignment of hardware resources for the given design.

Asynchronous systems have significant differences against their synchronous counterparts from the viewpoint of a timing model [12]. Firstly, asynchronous systems deal with *continuous* time domain, therefore initiation and completion of each operation are considered as *events* which can occur at any time. Secondly, each operation has a *variable* and *data-dependent* delay. Therefore scheduling in an asynchronous design is not determining control steps of operations but defining the *partial ordering* of operations.

High-level synthesis is complex and computationally expensive. To reduce the complexity, a synthesis approach either

1. partitions the synthesis task into several sub-tasks and performs one sub-task at a time, or
2. partitions the task into a sequence of transformation steps each of which makes a small change to the intermediate result of the earlier steps.

The latter approach is called a *transformational approach.* We can integrate the sub-tasks of high-level synthesis and get a globally optimized solution using the transformational approach. Also, we can employ optimization heuristics such as *Tabu search* [10, 11] because the transformational approach can be viewed as a *neighborhood search* optimization method. The conceptual comparison of the traditional and the transformational approaches is illustrated in Figure 1.

One of the important issues of the transformational approach is a *design representation model* which captures intermediate results during synthesis. Petri nets are the one of best candidates for the design representation of asynchronous systems because they can easily express concurrent and asynchronous aspects of asynchronous systems. Another important issue is *basic transformation rules* which are correctness-preserving. Each transformation should preserve behavioral correctness such as data dependency.

Several *high-level synthesis* methods for asynchronous systems have been proposed using the traditional approach in [12, 13]. Badia and Cortadella [12] were the first to propose asynchronous operation scheduling. The method uses *event-list* based scheduling on a design representation model, called a data-flow graph(DFG). Bachman developed a CAD tool called `Mercury` [13] for scheduling and resource allocation using *resource edges* which constrain the orders of operations on a DFG. These approaches are only applicable to each DFG segment of an overall control/data-flow representation and do not try to get a globally optimized solution on the control/data-flow representation. Also, they do not consider other high-level sub-tasks such as resource allocation or module binding.

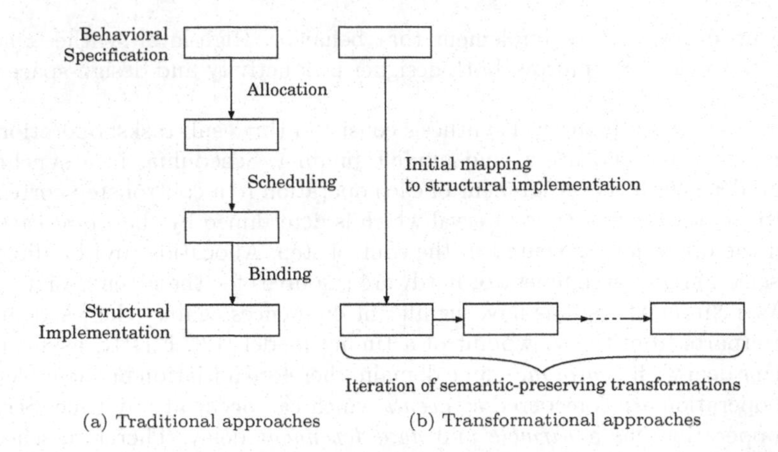

Fig. 1. Comparison between traditional and transformational approaches

A *transformational approach* for high-level synthesis of synchronous systems is proposed by Peng at al. in [14, 15]. This method takes account of the three high-level sub-tasks simultaneously. They defined a design representation model which consists of an extended timed Petri net (ETPN) and a data flow graph to unify the description of control and data parts of synchronous systems. Various transformation patterns are defined and they are successively applied to the design representation to get efficient implementation. It is noticeable that these transformation patterns are based on a clock cycle and most transformations can be used only for synchronous designs. Also, there is lack of transformation patterns for changing order of operations which play an important role in high-level synthesis of asynchronous systems. Moreover, the ETPN does not support an efficient algorithm for finding the relation of two operations, although finding relation of two operations is a key net operation for reordering of operations. Therefore, the ETPN and these transformation patterns are not applicable to high-level synthesis of asynchronous systems.

In this paper, we propose a *design representation model* for high-level synthesis of asynchronous systems. The model consists of *Operation Net* and a *module graph* to unify the description of control and data parts of asynchronous systems. The Operation Net is based on hierarchical timed Petri nets and the module graph is based on Peng's data flow graph. Also, we defined various transformation patterns for reordering of operations on Operation Net. Important characteristics of the proposed method are as follows.

- *The method integrates sub-tasks of high-level synthesis* such as operation scheduling, resource allocation, and module binding. This integration enables designers to get a better chance to reach a globally optimized solution.
- *An extended Petri net is used as a design representation.* Petri nets are widely used as a specification [16] or an intermediate design representation [6, 7] in asynchronous design since they can easily express concurrent and

asynchronous aspects of asynchronous digital systems. The proposed method can be adopted to any approaches which use Petri nets as an intermediate design representation to improve the quality of the design.

- *The hierarchical concept of Operation Net reduces the complexity of net operations by hiding sub-hierarchical blocks.* For example, finding the relation of two operations, which can be serial or parallel, plays an important role in the optimization of a net. It requires, however, high computational complexity in ordinary Petri nets. Since the hierarchical abstraction simplifies the net structure, it can reduce the complexity of net operations.

Once a behavioral specification is translated into an initial design representation, it can be viewed as a primitive implementation. We assume that the primitive implementation has been already generated from hardware description languages (HDLs) such as VHDL, Verilog, or concurrent programming languages. And then the primitive implementation is transformed into an optimized implementation using the basic transformations which are selected by an optimization strategy. Methods translating HDLs into Petri nets can be found in [6, 15, 7, 9].

3 Design Representation

The design representation consists of *Operation Net* (OPN) and a *module graph* which describe control and a data path respectively. Our design representation is based on the model of [14]. However, we renovate the model to be suitable for asynchronous design.

OPN is a timed Petri net equipped with hierarchies, block-structures, and conditional choices. OPN is similar to a Control-Data Flow Graph (CDFG) [17] but more suitable for asynchronous systems which have concurrency and asynchrony inherently. The module graph describes the allocated and bound hardware resource modules and flow relations between the modules.

Examples of OPN and the module graph are illustrated in Figure 2. The OPN in Figure 2:(a) represents the control flow of a differential equation solver while the module graph in Figure 2:(b) represents corresponding *allocated* and *bound* data modules for the OPN. A place controls arcs of the module graph which have the same number as the place. A data module, which is represented by a rectangle in the module graph, is allocated and bound to the place which controls input arcs of the data module. The module is called the *associated* data module of the place, denoted as $AM(p)$ where p is the place. If control flow arrives at a place p, the associated module $AM(p)$ performs the corresponding operation.

3.1 Basic Definition

Using OPN, we describe control constructs of HDLs such as loops and conditional branches. For example, figure 2:(a) expresses a loop construct while the place

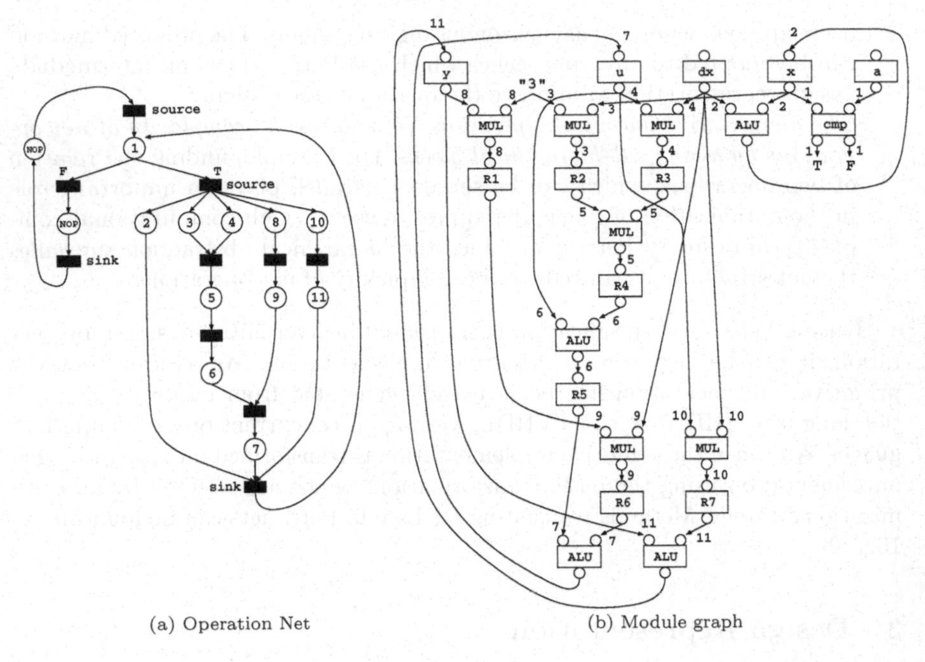

(a) Operation Net (b) Module graph

Fig. 2. A data representation for a differential equation solver

"1" checks a loop condition by controlling the associated comparator module in figure 2:(b).

OPN has three major properties. First, it has *conditional choices.* A choice is introduced when a place has more than two output transitions. In OPN, a choice can select a *guarded* transition whose guard is evaluated to *true.* For example, the choice on the place "1" of figure 2:(a) can select one of the transitions labeled "T" or "F" when one of their guards is evaluated to true. Second, OPN has *hierarchies.* Loop and branching constructs are modelled in an OPN through the use of hierarchies. Each loop or branching body forms a new sub-hierarchical OPN. And the hierarchy is represented by labeling places with a hierarchy level. The entrance of a loop or branching body is the place on a conditional choice. For example, in Figure 2:(a), a loop construct forms a sub-hierarchical OPN including control places "2", "3", "4", "5", "6", "7", "8", "9", "10", and "11". And the entrance of the sub-hierarchical OPN is the place "1" on the conditional choice. Third, OPN is *block-structured.* That is, each sub-hierarchical OPN always begins with a *source* transition and ends with a *sink* transition as shown in Figure 2:(a).

We introduce the formal definition of Operation Net as follows.

Definition 1. ***Operation Net (OPN)*** *is a 8-tuple,* $N = (\Sigma, P, T, F, H, E, K, M_0)$*, where:*

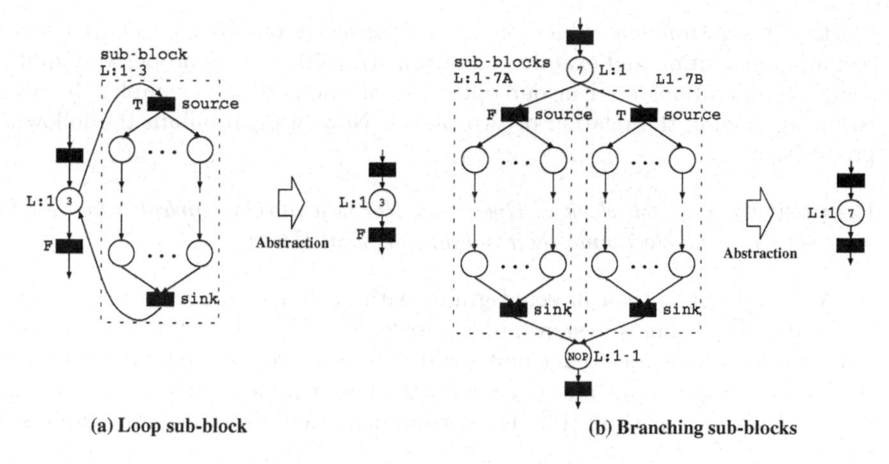

Fig. 3. An example of blocks

- Σ *is a finite set of* ***block names****, or* ***hierarchy levels****.*
- $P = \{p_1, p_2, ..., p_m\}$ *is a finite set of* ***control places*** *each of which represents the operation control of a corresponding data module.*
- $T = \{t_1, t_2, ..., t_n\}$ *is a finite set of* ***transitions****.*
- $F \subseteq (P \times T) \cup (T \times P)$ *is a set of* ***control flow relation****.*
- $H : P \rightarrow \Sigma$ *is a* ***hierarchy function*** *which is a mapping from places to block names.*
- $E \subseteq T$ *is a finite set of* ***source transitions****.*
- $K \subseteq T$ *is a finite set of* ***sink transitions****.*
- $M_0 : P \rightarrow \{0, 1\}$ *is an* ***initial marking****.*

The behavior of OPN is the same as ordinary Petri nets [18] except for conditional choices.

The block name of a place contains the information of the hierarchy and the block where the place is laid. For example, the place "3" in Figure 3:(a) has the block name "L:1" and it is an entrance place of the sub-hierarchical block, i.e., the loop body. Therefore, all places of the sub-block have the block name "L:1-3". This block name means that the places are on the sub-hierarchical block of the place "3" which is on the block "L:1". Since sub-hierarchical blocks of a branching construct are executed mutual exclusively, the block names have special tags to distinguish alternative blocks such as "L:1-7A" and "L:1-7B" as shown in Figure 3:(b). The NOP place in Figure 3:(b) is for merging several flows of the alternative sub-hierarchical blocks.

The abstraction of hierarchies hides all conditional choices in a block as shown in Figure 3. For example, the sub-hierarchical blocks and a merge place in Figure 3 are hidden as places "3" and "7". These entrance places which are on the conditional choices are called *fusion places*. And all choices of OPN are introduced on the fusion places. Therefore, we can imagine that there are no choices in a block, if all fusion places of the block hide their sub-hierarchical blocks. This *abstraction* of hierarchies enables places on a block to have exactly one input transition and exactly one output transition. This property simplifies some net operations used in our optimization method, for example, the algorithm for finding the relation of two places. Now, we can obtain the following proposition.

Proposition 1. *Each block of Operation Net is a* **marked graph** *when all fusion places of the block hide their sub-hierarchical blocks.*

A module graph is a directed graph with each vertex having one or more input and output ports as shown in Figure 2:(b). The vertices are used to model data manipulation and storage units, while the arcs are used to model data flow. Bubbles represent input and output ports. The module graph is based on the model of the data path in [14]. The formal definition of the module graph is as follows.

Definition 2. *A* **module graph** *is a 5-tuple* $D = (M, I, O, A, B)$*, where:*

- $M = \{m_1, m_2, \ldots, m_n\}$ *is a finite set of* **modules** *each of which represents a data manipulation or storage unit.*
- $I = I(m_1) \cup I(m_2) \cup \ldots \cup I(m_n)$ *is a finite set of* **input ports** *where* $I(m_j)$ *is the set of input ports associated with the module* m_j*.*
- $O = O(m_1) \cup O(m_2) \cup \ldots \cup O(m_n)$ *is a finite set of* **output ports** *where* $O(m_j)$ *is the set of output ports associated with the module* m_j*.*
- $A \subseteq O \times I = \{\langle o, i \rangle \mid o \in O, i \in I\}$ *is a finite set of* **arcs** *each of which represents a connection from an output port of a module to an input port of another module or the same module.*
- $B : O \rightarrow 2^{OP}$ *is a mapping from output ports to sets of operations.* $OP = \{op_1, op_2, \ldots, op_m\}$ *is a finite set of* **operations** *which consists of sequential and combinatorial operations.*

Note that the operations are not associated with the modules but the output ports. Therefore, a module can perform several different operations on its multiple output ports at the same time. Also, an output port can perform different operations at different times.

We introduce a *unified* design representation, called *Operation Net System*, as follows.

Definition 3. ***Operation Net System (ONS)*** *is a 4-tuple* $S = (N, D, C, G)$*, where:*

- N *is Operation Net (OPN).*
- D *is a module graph.*
- $C : P \rightarrow 2^A$ *is a* **control function** *which is a mapping from places to sets of arcs* A *of the module graph.*
- $G : O \rightarrow 2^T$ *is a* **guard function** *which is a mapping from output ports of modules* O *in the module graph to sets of transitions.*

If a control place holds a token, a control signal is sent to the corresponding arcs of the module graph which are specified by the control function C. The control signal indicates that valid data flows though the controlled arcs. Then the *associated* data modules are activated and the corresponding operation is performed with the data. A transition may be guarded by conditions produced from the data module which is described by the guard function G. Using this guard function, the results of some data operations influence the control flow of conditional choices by sending control signals. For example, the conditional choice on the place "1" of Figure 2:(a) is controlled by the comparator in Figure 2:(b) which generates T or F control signals.

Now, we define the relations of two places on OPN. If two places are in *serial relation*, the places cannot hold tokens at the same time. Therefore, the associated data modules of the places cannot be activated at the same time. If two places are in *parallel relation*, the places can hold tokens at the same time. Therefore, the associated data modules of the places can be activated at the same time. Let the **pre-path** (**post-path**) of a place p in a block of OPN, denoted as $W_{pre}(p)$ ($W_{post}(p)$), be a set of places and transitions on all paths from p to the source (sink) transition of the block without stepping down to sub-hierarchical blocks. The **abstraction** of a place p, denoted as $ABS(p)$, returns the fusion place of p. And the **depth** of a place p, denoted as $H_{depth}(p)$, returns the hierarchy depth of p from a top-level hierarchy. Since each block of OPN is a marked graph (see Proposition 1), we can easily define, or figure out, the two relation as follows.

Definition 4. *Two places p_i and p_j are in* **serial relation**, *denoted as $p_i;p_j$, or $p_j;p_i$, if any of followings is true:*

1. *If the two places are in the same block, $p_i \in W_{pre}(p_j) \wedge p_j \in W_{post}(p_i)$, or $p_i \in W_{post}(p_j) \wedge p_j \in W_{pre}(p_i)$.*
2. *The two places are in two different mutually exclusive sub-blocks of a conditional branch*[1].
3. *Otherwise, i.e., if $H_{depth}(p_i) \neq H_{depth}(p_j)$, abstractions of the two places satisfy the condition 1 or 2 in the following hierarchy levels where p_k is the common fusion place, which can be get by iteratively applying ABS on p_i and p_j until these abstractions are up to the same fusion place, i.e., p_k,*
   (a) *the sub-hierarchical block H_k of the fusion place p_k, which includes both abstraction places of p_i and p_j, in the case of condition 1.*
   (b) *two mutually exclusive sub-blocks H_k and H_l of the fusion place p_k, which include the abstraction places of p_i and p_j respectively, in the case of condition 2.*

[1] Strictly speaking, the condition 2 of Definition 4 is not the serial relation. If two places are in different mutually exclusive sub-blocks of a conditional branch, there is no precedence relation between the two places. However, the places cannot hold tokens at the same time. Therefore, we consider this case as the serial relation for convenience.

Definition 5. *Two places p_i and p_j are in* ***parallel relation****, denoted as $p_i \| p_j$, or $p_j \| p_i$, if the two places are not in the serial relation.*

We introduce some restrictions on Operation Net System to prevent ambiguous behaviors of a system such as non-deterministic properties or undefined values.

Definition 6. *Operation Net System $S = (N, D, C, G)$ is well designed if:*

- $\forall p_i \in P, M(p_i) \leq 1$ *for each $M \in R$ where R is a set of reachable markings. That is, Operation Net must be* ***safe****. Therefore, two operations cannot be executed on the same data module at any moment.*
- t_i *and t_j must be guarded and these guards, or conditions, must not be true at the same time where $\langle p, t_i \rangle, \langle p, t_j \rangle \in F$. That is, Operation Net does not allow non-deterministic choices.*
- $\forall M \in R, M(p_i) + M(p_j) \leq 1$ *when $AM(p_i) = AM(p_j)$. That is, if a data module is associated with two places, the places cannot hold tokens at the same time. In other words, if two places shares a data module, the places must be in serial relation.*
- $AM(p_i) \neq AM(p_j)$ *when $\exists M \in R, M(p_i) = 1 \wedge M(p_j) = 1$. That is, if two places are allowed to hold tokens at the same time, their associated modules must be different. In other words, if two places are in the parallel relation, they cannot share a data module.*

From now on, we only consider the *well designed* Operation Net System. The basic idea of optimization starts from the relation of two control places. If two places are in the serial relation but they are data-independent operations, the places can be changed to be in parallel relation. If two places are in parallel relation, the places can be changed to be serial relation but share a data module if they are the same operation type. Therefore, parallelization can improve the performance of a system while serialization can reduce the area of the system. These transformations must be *data-invariant* which means that all data dependent operations must keep their orders during transformations. We say that a transformation is *semantic-preserving* if the transformation is data-invariant.

If two places are data dependent, the associated modules cannot be executed at the same time, that is, two places must be in serial relation. Let the **pre-module set** of a control place p, denoted as $MS_{pre}(p)$, be a set of storage modules which have some output ports connected to an arc controlled by p. And let the **post-module set** of p, denoted as $MS_{post}(p)$, is a set of storage modules which have some input ports connected to an arc controlled by p. Then, we define the data dependent relation as follows.

Definition 7. *Control places p_i and p_j are* ***data dependent****, denoted as $p_i \leftrightarrow p_j$, if any of the following is true:*

- $MS_{post}(p_i) \cap MS_{pre}(p_j) \neq \emptyset$.
- $MS_{pre}(p_i) \cap MS_{post}(p_j) \neq \emptyset$.
- $MS_{post}(p_i) \cap MS_{post}(p_j) \neq \emptyset$.

If all data dependent operations are executed in the predefined order specified in a behavioral description, the semantics of the given system is not changed. However, data independent operations can be reordered without changing the semantics for optimization. This means that we can reconstruct the control structure, i.e., Operation Net, without changing data dependency to optimize the modelled system. For example, we can reconstruct OPN to improve performance by carrying out as many operations in parallel as possible.

3.2 Timing

In the previous section, we introduced Operation Net System (ONS) with a *non-timed* Petri net. Since the normal Petri net cannot model the notion of execution time, it is impossible to introduce any performance measure of the modelled system. Therefore, we introduce timing into the ONS by requesting a token of each control place to remain during a delay range which is assigned to the associated data module of the place. The delay range consists of minimum and maximum execution delays of the data module.

Definition 8. ***Timed Operation Net System (TONS)*** *is a two-tuple,* $\langle S, R\rangle$*, where:*

- $S = (N, D, C, G)$ *is Operation Net System.*
- $R = \{[l_1, u_1], [l_2, u_2], \ldots, [l_n, u_n]\}$ *is a finite set of* ***bounded timing constraints*** *each of which is assigned to a control place. Each timing constraint consists of* ***lower bound*** l *and* ***upper bound*** u*.*

The behavior of timed Operation Net System is the same as its corresponding Operation Net System except that:

1. *A place is* ***readied*** *when it gets to have a token. A place is* ***satisfied*** *if it has been readied at least* l *time units. A place becomes* ***expired*** *when it has been readied* u *time units.*
2. *A transition cannot fire until its every input place is satisfied, and it must fire before its every input place has expired.*
3. *A guarded transition will be fired when it satisfies above condition and its guarding condition is true.*

Note that every conflict, or choice, of OPN is conditional, i.e., every transition involved in a conflict has a guarding condition, and only one of the conditions can be evaluated to true by the result of operation in the corresponding data module(See the second condition of Definition 6). Therefore, ONS and TONS have no non-deterministic choices or arbitration.

4 Basic Transformations

In this section, we introduce basic transformation patterns which are used to optimize ONS. These transformations consist of data-oriented transformations and control-oriented transformations for a module graph and OPN, respectively.

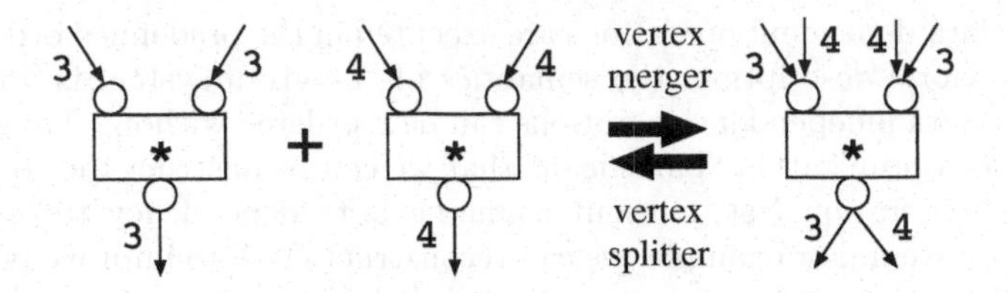

Fig. 4. An example of vertex merger and splitter

4.1 Data-Oriented Transformations

The transformations on a module graph must not change the predefined orders of operations. That is, it must keep the data dependent relation of all operations to preserve the semantics of a design. The *vertex merger* and *vertex splitter* transformations of [14] are the transformations for this purpose. The vertex merger is a simple transformation which merges two data modules into a data module. It is used to share hardware resources for optimization of the design from the point of view of area costs. Before performing the vertex merger, the associated control places of the vertices must be in serial relation, i.e., the places cannot hold tokens at the same time. Also, two operations must be the same operation type.

The vertex splitter transformation is an inverse transformation to the vertex merger. A data module shared by two control places are split to two modules. The arcs together with the controlling places follow their related vertices. Although the vertex splitter increases the area of the design, it can improve the performance with parallelization which will be introduced in following section.

The example of the vertex merger and splitter is presented in Figure 4. The vertices of Figure 4 is from Figure 2:(b). Note that the vertex merger must be initiated after the associated places "3" and "4" are serialized.

4.2 Control-Oriented Transformations

Transformations on OPN are performed in each block of the OPN. For any two control places, the transformations serialize or parallelize the places for each case of possible optimization. That is, serialization makes a chance to share a hardware resource with the vertex merger. And parallelization makes a chance to improve the performance with the vertex splitter.

Serialization. Let $first(p)$ and $last(p)$ be the input and the output transition of a control place p. And let the **join-point** of p_j with p_i, denoted as $JP_{p_i}(p_j)$, be a set of join transitions where **post-path** of p_j, $W_{post}(p_j)$, meets **post-path** of p_i, $W_{post}(p_i)$. We introduce the definition of serialization as follows.

Definition 9. *Operation Net System* $S = (N, D, C, G)$ *is transformed into* S' $= (N', D', C', G')$ *by serialization transformations. The serialization of two control places* p_i *and* p_j *is defined as follows:*

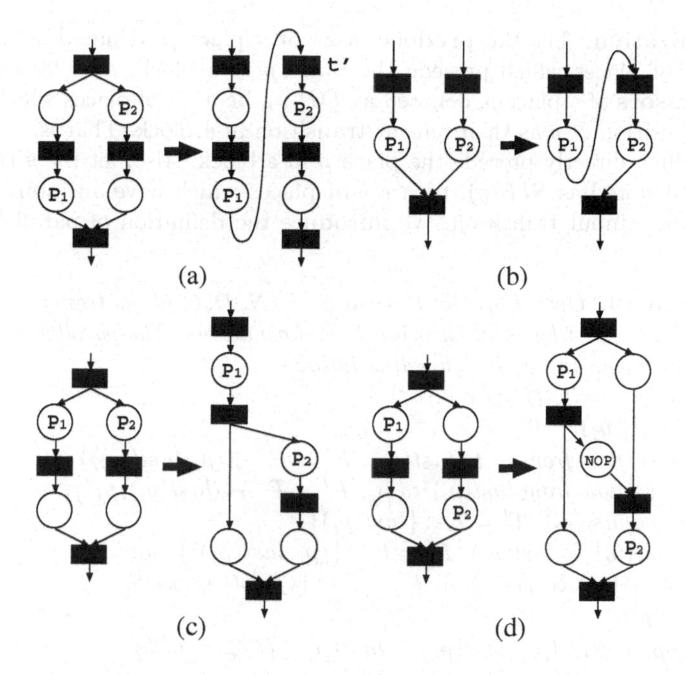

Fig. 5. An example for serialization of p_1 and p_2

1. *if* $last(p_i) \in JP_{p_i}(p_j)$,
   (1) *if* $first(p_j) \in W_{pre}(p_i)$, *(Figure 5:(a))*
   · *make a new transition* t', $T' = T \cup \{t'\}$,
   · *change* $last(p_i)$ *to* t', $F' = (F - \{\langle p_i, last(p_i)\rangle\}) \cup \{\langle p_i, t'\rangle\}$, *and*
   · *change* $first(p_j)$ *to* t', $F' = (F - \{\langle first(p_j), p_j\rangle\}) \cup \{\langle t', p_j\rangle\}$.
   (2) *if* $first(p_j) \notin W_{pre}(p_i)$, *(Figure 5:(b))*
   · *change* $last(p_i)$ *to* $first(p_j)$, $F' = (F - \{\langle p_i, last(p_i)\rangle\}) \cup \{\langle p_i, first(p_j)\rangle\}$.
2. *if* $last(p_i) \notin JP_{p_i}(p_j)$,
   (1) *if* $first(p_j) \in W_{pre}(p_i)$, *(Figure 5:(c))*
   · *change* $first(p_j)$ *to* $last(p_i)$, $F' = (F - \{\langle first(p_j), p_j\rangle\}) \cup \{\langle last(p_i), p_j\rangle\}$.
   (2) *if* $first(p_j) \notin W_{pre}(p_i)$, *(Figure 5:(d))*
   · *make a new place NOP (no operation),* $P' = P \cup \{NOP\}$,
   · *let* $first(NOP)$ *be* $last(p_i)$, $F' = F \cup \{\langle last(p_i), NOP\rangle\}$, *and*
   · *let* $last(NOP)$ *be* $first(p_j)$, $F' = F \cup \{\langle NOP, first(p_j)\rangle\}$.

Intuitively, the transformations of Definition 9 do not change the data dependent relation of operations of the module graph. Also, the transformations satisfy the well designed properties of ONS because they keep the Operation Net safe and do not touch guarded transitions. Therefore, we can obtain the following proposition.

Proposition 2. *All serialization transformations in Definition 9 are semantics-preserving.*

Parallelization. Let the **predecessors** of a place p, denoted as $PRED(p)$, be a set of places which precede the place p in a block. And let the **direct-predecessors** of a place p, denoted as $DP(p)$, be a set of places which have the input transition of p as their output transition in a block. That is, it is a set of places which directly precede the place p in a block. Also, let the **siblings** of a place p, denoted as $SIB(p)$, be a set of places which have an input transition of p as their input transition. We introduce the definition of parallelization as follows.

Definition 10. *Operation Net System $S = (N, D, C, G)$ is transformed into $S' = (N', D', C', G')$ by parallelization transformations. The parallelization of two control places p_i and p_j is defined as follows:*

1. if $|DP(p_j)| = 1$, (Figure 6:(a))

(1) if $|SIB(p_j)| = 0$,

· *remove flow from p_i to $last(p_i)$, $F' = F - \{\langle p_i, last(p_i)\rangle\}$,*
· *remove flow from $last(p_i)$ to p_j, $F' = F - \{\langle last(p_i), p_j\rangle\}$,*
· *remove $last(p_i)$, $T' = T - \{last(p_i)\}$,*
· *let $last(p_i)$ be $last(p_j)$, $F' = F \cup \{\langle p_i, last(p_j)\rangle\}$, and*
· *let $first(p_j)$ be $first(p_i)$, $F' = F \cup \{\langle first(p_i), p_j\rangle\}$.*

(2) if $|SIB(p_j)| \geq 1$,

a. if $\exists p_k \in SIB(p_j), last(p_k) = last(p_j)$, (Figure 6:(b))
· *change $first(p_j)$ to $first(p_i)$, $F' = (F - \{\langle last(p_i), p_j\rangle\}) \cup \{\langle first(p_i), p_j\rangle\}$,*

b. otherwise, (Figure 6:(c))
· *make a new place NOP, $P' = P \cup \{NOP\}$,*
· *change $first(p_j)$ to $first(p_i)$, $F' = (F - \{\langle first(p_j), p_j\rangle\}) \cup \{\langle first(p_i), p_j\rangle\}$,*
· *let $first(NOP)$ be $last(p_i)$, $F' = F \cup \{\langle last(p_i), NOP\rangle\}$, and*
· *let $last(NOP)$ be $last(p_j)$, $F' = F \cup \{\langle NOP, last(p_j)\rangle\}$.*

2. if $|DP(p_j)| > 1$ and $NOP \notin DP(p_j)$,

(1) if $|SIB(p_j)| = 0$,

a. $\exists p_k \in DP(p_j), first(p_k) = first(p_i)$, (Figure 6:(d))
· *change $last(p_i)$ to $last(p_j)$, $F' = (F - \{\langle p_i, last(p_i)\rangle\}) \cup \{\langle p_i, last(p_j)\rangle\}$.*

b. otherwise, (Figure 6:(e))
· *make a new place NOP, $P' = P \cup \{NOP\}$,*
· *change $last(p_i)$ to $last(p_j)$, $F' = (F - \{\langle p_i, last(p_i)\rangle\}) \cup \{\langle p_i, last(p_j)\rangle\}$,*
· *let $first(NOP)$ be $first(p_i)$, $F' = F \cup \{\langle first(p_i), NOP\rangle\}$, and*
· *let $last(NOP)$ be $first(p_j)$, $F' = F \cup \{\langle NOP, first(p_j)\rangle\}$.*

3. if $|DP(p_j)| > 1$ and $NOP \in DP(p_j)$,

(1) if $p_i \in DP(p_j)$ and $p_i \neq NOP$,
· *the transformation is the same as procedure 2.*

(2) if $|SIB(p_j)| = 0$,

a. $\exists p_k \in DP(p_j), first(p_k) = first(p_i)$, (Figure 6:(f))
· *check data dependency between all places in $\{p | p \in (PRED(NOP) - \{p_i\}) \wedge p \| p_i\}$ and p_j, and if there is no dependency,*
· *change $last(NOP)$ to $last(p_j)$, $F' = (F - \{\langle NOP, last(NOP)\rangle\}) \cup \{\langle NOP, last(p_j)\rangle\}$.*

b. otherwise, (Figure 6:(g))

· *check data dependency between all places in* $\{p|p \in (PRED(NOP) - \{p_i\}) \wedge p||p_i\}$ *and* p_j*, and if there is no dependency,*

· *change* $first(NOP)$ *to* $first(p_i)$, $F' = (F - \{\langle first(NOP), NOP\rangle\}) \cup \{\langle first(p_i), NOP\rangle\}$.

$\{p|p \in (PRED(NOP) - \{p_i\}) \wedge p||p_i\}$ of the condition 3.(2).a in Definition 10 is a set of places which are concurrent with the place p_i. It is easily calculated by $PRED(NOP) - PRED(p_i) - p_i$ using Proposition 1. The transformations of Definition 10 are also semantics-preserving by the same reason of the serialization. Therefore, we can obtain the follow proposition.

Proposition 3. *All parallelization transformations in Definition 10 are semantics-preserving.*

Removal of Redundant NOP. Several NOP places can be generated during the transformations as shown in Definition 9 and 10. These NOP places are not for controlling data modules but just for keeping the precedence relation of control places. Some of the NOP places are redundant, i.e., there exists another path from the input transition of a NOP to the output transition of the NOP except NOP itself. Therefore, we can prevent unnecessary transformations caused by NOP by removing the redundant NOP places.

5 Exploration Procedure

In this section, we explain the overall design-space exploration procedure of the proposed method. As previously mentioned, the proposed method starts with an initial Operation Net System (ONS) which is a primitive implementation translated from a behavioral description. ONS represents the control and data parts which are allocated and bound to each other. Usually in the initial ONS, each control place is bound its own data module. That is, there is no hardware resource sharing but there is a chance to improve the performance by arranging as many control places in parallel as possible.

The goal of the exploration is to satisfy the design objectives given by designers in terms of the performance and the area of a design. We can reduce the area of the design by carrying out the vertex merger with the serialization transformation. And we can improve the performance by carrying out the parallelization transformation with the vertex splitter. The selection of the transformations is guided by an optimization algorithm.

The exploration algorithm performs serialization and parallelization transformations on each block of OPN. Before and after each transformation, the vertex merger and splitter are performed together to preserve the well designed properties. Also, the vertex merger and splitter can be performed between the blocks to share hardware resources. The initial ONS is iteratively improved by a sequence of transformations until the intermediate ONS meets the design objectives.

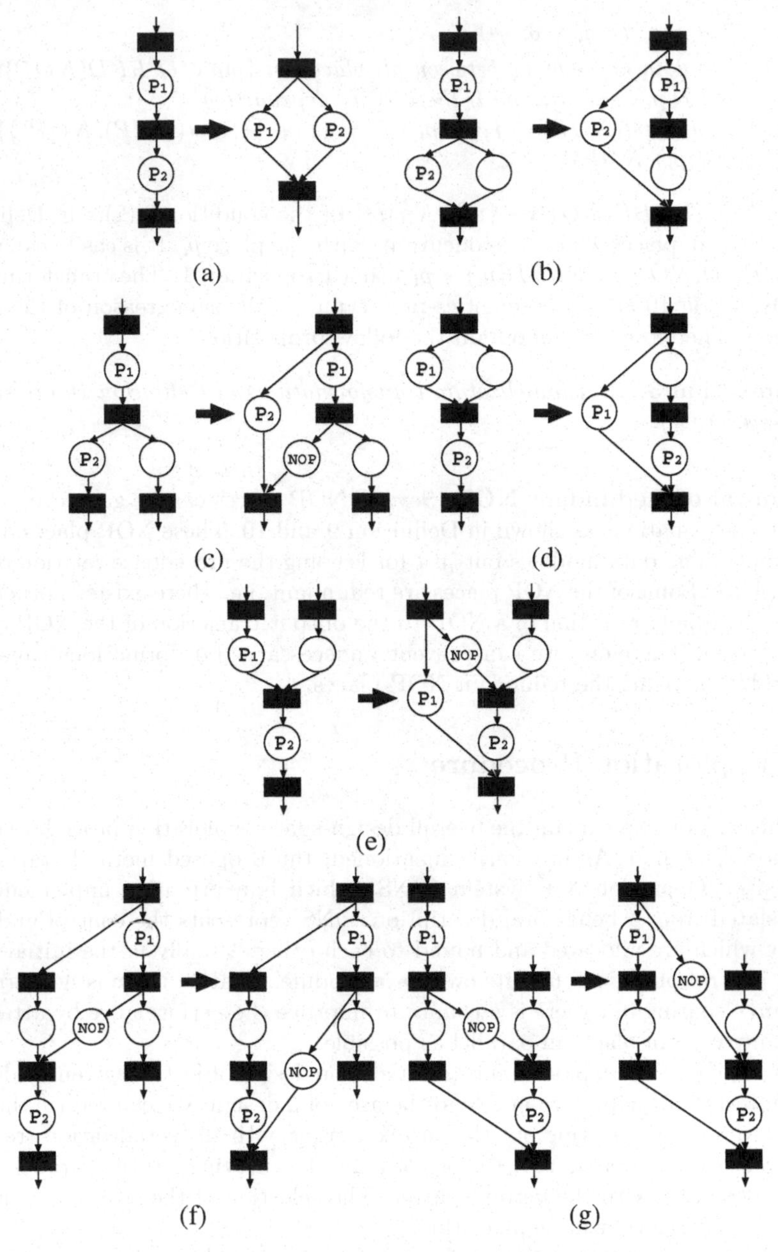

Fig. 6. An example for parallelization of p_1 and p_2

The stepwise improvement property of the transformational approach enables us to use an iterative algorithm such as *Tabu search* [10]. The Tabu search is a kind of a *neighborhood search* which searches only a small area around the current

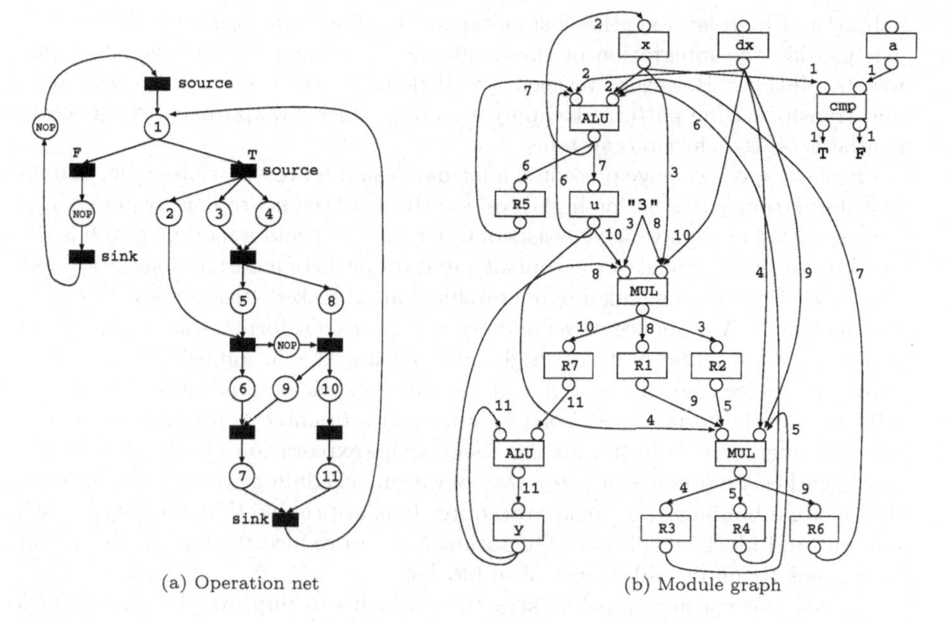

(a) Operation net (b) Module graph

Fig. 7. Transformed Operation Net and module graph

solution in each step. The integration of allocation, operation scheduling, and binding will result in a better chance to reach the globally optimal solution using the iterative algorithm. We are developing the Tabu search algorithm which is suitable for our approach.

The transformed OPN and the module graph of Figure 2 are shown in Figure 7. The number of data modules of Figure 2 is total 11 (6 multipliers, 4 ALUs, and 1 comparator). And latency is calculated by the sum of delays of 1 comparator, 2 multipliers, and 2 ALUs which are on the critical path of the Operation Net. In Figure 7, the number of data modules is 5 (1 comparator, 2 multipliers, and 2 ALUs) and latency is the sum of delays of 1 comparator, 3 multipliers, and 1 ALU. Therefore, the optimized design uses less than the half number of data resources compared with the original design while it loses performance with the small additional delay which is the difference of the delays of a multiplier and an ALU.

6 Conclusions and Future Work

Although high-level synthesis provides systematic design-space exploration and optimization of large-scale systems in the early design phase, there were few researches on high-level synthesis of asynchronous systems. Previous researches have focused on operation scheduling of asynchronous systems on a DFG segment of a control/data flow graph without considering globally optimized solutions. Also, there is lack of high-level synthesis method considering the three subtasks of high-level synthesis simultaneously. The transformational approach can provide the integration of those sub-tasks and help to find globally optimized solutions. However, we need a well defined design representation model and transformation patterns to apply a transformational approach to high-level synthesis of asynchronous systems.

In this paper, we have presented a formal design representation model and its transformation patterns for high-level synthesis of asynchronous systems. The design representation which consists of Operation Net and a module graph is defined to *unify* the model of the control and data parts of a design. Operation Net is a timed Petri net extended by hierarchical and blocked structures with conditional choices. A sequence of *semantic-preserving* transformations is applied on the design representation to get a globally optimized solution using an iterative algorithm. Since control oriented transformations such as serialization and parallelization are performed with corresponding data oriented transformations such as vertex merger and splitter during design space exploration, the proposed high-level synthesis method can *integrate* operation scheduling, resource allocation, and module binding in a single procedure. It is noticeable that the hierarchical concept and block structure of Operation Net can reduce the complexity of net operations by hiding sub-hierarchical blocks.

We do not consider pipeline structure, which can improve throughput of a system, yet. Pipeline transformations on the proposed model will be studied for asynchronous systems. Also, we plan to develop the automatic generation of control circuits for Operation Net. The *process oriented control circuit generation* method [19] presented by the authors and *automatic distributed control generation* method [20] can be used for our approach.

Acknowledgements

This work has been supported in part by the IT SoC of Korea IT Industry Promotion Agency(KIPA).

References

1. Rotem, S., Stevens, K., Ginosar, R., Beerel, P., Myers, C., Yun, K., Kol, R., Dike, C., Roncken, M., Agapiev, B.: RAPPID: An asynchronous instruction length decoder. In: Proc. International Symposium on Advanced Research in Asynchronous Circuits and Systems. (1999) 60–70
2. Martin, A.J.: Programming in VLSI: From communicating processes to delay-insensitive circuits. In Hoare, C.A.R., ed.: Developments in Concurrency and Communication. UT Year of Programming Series, Addison-Wesley (1990) 1–64
3. Brunvand, E.: Translating Concurrent Communicating Programs into Asynchronous Circuits. PhD thesis, Carnegie Mellon University (1991)
4. Berkel, K.v.: Handshake Circuits: an Asynchronous Architecture for VLSI Programming. Volume 5 of International Series on Parallel Computation. Cambridge University Press (1993)

5. Peeters, A.M.G.: Single-Rail Handshake Circuits. PhD thesis, Eindhoven University of Technology (1996)
6. Akella, V., Gopalakrishnan, G.: SHILPA: A high-level synthesis system for self-timed circuits. In: Proc. International Conf. Computer-Aided Design (ICCAD), IEEE Computer Society Press (1992) 587–591
7. Jacobson, H., Brunvand, E., Gopalakrishnan, G., Kudva, P.: High-level asynchronous system design using the ACK framework. In: Proc. International Symposium on Advanced Research in Asynchronous Circuits and Systems, IEEE Computer Society Press (2000) 93–103
8. Kessels, J., Marston, P.: Designing asynchronous standby circuits for a low-power pager. Proceedings of the IEEE **87** (1999) 257–267
9. Yoo, D.H., Lee, D.I.: Translating concurrent programs into speed-independent circuits through petri net transformations. IEICE Transactions on Fundamentals **E83-A** (2000) 2203–2211
10. Glover, F.: Tabu search - Part I. ORSA Journal on Computing **1** (1989) 190–206
11. Glover, F.: Tabu search - Part II. ORSA Journal on Computing **2** (1989) 4–32
12. Badia, R.M., Cortadella, J.: High-level synthesis of asynchronous systems: Scheduling and process synchronization. In: Proc. European Conference on Design Automation (EDAC), IEEE Computer Society Press (1993) 70–74
13. Bachman, B.M.: Architectural-level synthesis of asynchronous systems. Master's thesis, The University of Utah (1998)
14. Peng, Z., Kuchcinski, K.: Automated transformation of algorithms into register–transfer level implementations. IEEE Transactions on Computer-Aided Design **13** (1994)
15. Eles, P., Kuchcinski, K., Peng, Z.: System Synthesis with VHDL. Kluwer Academic Publishers (1999)
16. Chu, T.A.: Synthesis of Self-timed VLSI Circuits from Graph-theoretic Specifications. PhD thesis, MIT (1987)
17. Micheli, G.D.: Synthesis and Optimization of Digital Circuits. McGraw-Hill (1994)
18. Murata, T.: Petri nets: Properties, analysis, applications. Proceedings of the IEEE **77** (1989) 541–580
19. Kim, E., Lee, J.G., Lee, D.I.: Automatic process-oriented control circuit generation for asynchronous high-level synthesis. In: Proc. International Symposium on Advanced Research in Asynchronous Circuits and Systems, IEEE Computer Society Press (2000) 104–113
20. Theobald, M., Nowick, S.M.: Transformations for the synthesis and optimization of asynchronous distributed control. In: Proc. ACM/IEEE Design Automation Conference. (2001)

EMiT: A Process Mining Tool

Boudewijn F. van Dongen and Wil M.P. van der Aalst

Department of Technology Management, Eindhoven University of Technology
P.O. Box 513, NL-5600 MB, Eindhoven, The Netherlands
b.f.v.dongen@tue.nl

Abstract. Process mining offers a way to distill process models from event logs originating from transactional systems in logistics, banking, e-business, health-care, etc. The algorithms used for process mining are complex and in practise large logs are needed to derive a high-quality process model. To support these efforts, the process mining tool *EMiT* has been built. *EMiT* is a tool that imports event logs using a standard XML format as input. Using an extended version of the α-algorithm [3, 8] it can discover the underlying process model and represent it in terms of a Petri net. This Petri net is then visualized by the program, automatically generating a "smart" layout of the model. To support the practical application of the tool, various adapters have been developed that allow for the translation of system-specific logs to the standard XML format. As a running example, we use an event log generated by the workflow management system Staffware.

1 Introduction

During the last decade workflow management concepts and technology [2, 4, 10, 11] have been applied in many enterprise information systems. Workflow management systems such as Staffware, IBM MQSeries, COSA, etc. offer generic modeling and enactment capabilities for structured business processes. By making graphical process definitions, i.e., models describing the life-cycle of a typical case (workflow instance) in isolation, one can configure these systems to support business processes. Besides pure workflow management systems many other software systems have adopted workflow technology. Consider for example ERP (Enterprise Resource Planning) systems such as SAP, PeopleSoft, Baan and Oracle, CRM (Customer Relationship Management) software, etc. Despite its promise, many problems are encountered when applying workflow technology. One of the problems is that these systems require a workflow design, i.e., a designer has to construct a detailed model accurately describing the routing of work. Modeling a workflow is far from trivial: It requires deep knowledge of the workflow language and lengthy discussions with the workers and management involved.

Instead of starting with a workflow design, one could also start by gathering information about the workflow processes as they take place. In this paper it is assumed that it is possible to record events such that (i) each event refers to a task (i.e., a well-defined step in the workflow), (ii) each event refers to a case

J. Cortadella and W. Reisig (Eds.): ICATPN 2004, LNCS 3099, pp. 454–463, 2004.
© Springer-Verlag Berlin Heidelberg 2004

(i.e., a workflow instance), and (iii) events are totally ordered. Most information system will offer this information in some form. Note that (transactional) systems such as ERP, CRM, or workflow management systems indeed provide event logs. It is also important to note that the applicability of process mining is not limited to workflow management systems. The only requirement is that it is possible to collect logs with event data. These event logs are used to construct a process specification which adequately models the behavior registered.

The term *process mining* is used for the method of distilling a structured process description from a set of real executions. In this paper we do not give an overview of related work in this area. Instead we refer to the survey paper [6] and a special issue of *Computers in Industry* [7].

In this paper, the process mining tool *EMiT* is presented. Using an example log generated by Staffware, we illustrate the various aspects of the tool. In Section 3, the XML format used to store logs is described. Section 4 discusses the three main steps of the mining process. Section 5 discusses the export and visualization of the Petri nets discovered in the mining process. Finally, we conclude the paper.

2 Running Example

To illustrate the functionality of *EMiT*, an example log generated by the Staffware [12] is used. Since Staffware is one of the leading workflow management systems, it is a nice illustration of the practical applicability of *EMiT*. For presentation purposes we consider a log holding only six cases, as shown in Table 1.

Although the log presented here is rather small (only six cases) it is already hard to find the structure of the underlying workflow net by just examining the log. Therefore, a tool like *EMiT* is needed. However, before *EMiT* can be used, the log is translated into a generic input format that is tool-independent, i.e., *EMiT* does not rely on the specific format used by a system like Staffware. Instead there is a "Staffware adapter" translating Staffware logs to the XML format described in the next section.

3 A Common XML Log Format

Every log file contains detailed information about the events as they take place. However, commercial workflow system use propriety logging formats. Therefore, *EMiT* uses a tool-independent XML format. Since events in the log refer to state changes a first step is to describe the states in which each specific task can be. For this purpose we use a transactional model.

3.1 A Transactional Model

To describe the state of a task the Finite State Machine (FSM) shown in Figure 1 is used. The FSM describes all possible states of a task from creation to completion. The arrows in this figure describe all possible transitions between

Table 1. A Staffware workflow log.

Case 1

Step description	Event	User	yyyy/mm/dd hh:mm
	Start	bvdongen@staffw_	2002/04/18 09:05
A	Processed To	bvdongen@staffw_	2002/04/18 09:05
A	Released By	bvdongen@staffw_	2002/04/18 09:05
B	Processed To	bvdongen@staffw_	2002/04/18 09:05
B	Released By	bvdongen@staffw_	2002/04/18 09:05
D	Processed To	bvdongen@staffw_	2002/04/18 09:05
E	Processed To	bvdongen@staffw_	2002/04/18 09:05
D	Released By	bvdongen@staffw_	2002/04/18 09:06
E	Released By	bvdongen@staffw_	2002/04/18 09:06
G	Processed To	bvdongen@staffw_	2002/04/18 09:06
G	Released By	bvdongen@staffw_	2002/04/18 09:06
I	Processed To	bvdongen@staffw_	2002/04/18 09:06
I	Released By	bvdongen@staffw_	2002/04/18 09:06
	Terminated		2002/04/18 09:06

Case 2

Step description	Event	User	yyyy/mm/dd hh:mm
	Start	bvdongen@staffw_	2002/04/18 09:05
A	Processed To	bvdongen@staffw_	2002/04/18 09:05
A	Released By	bvdongen@staffw_	2002/04/18 09:05
B	Processed To	bvdongen@staffw_	2002/04/18 09:05
B	Released By	bvdongen@staffw_	2002/04/18 09:05
D	Processed To	bvdongen@staffw_	2002/04/18 09:05
E	Processed To	bvdongen@staffw_	2002/04/18 09:05
E	Released By	bvdongen@staffw_	2002/04/18 09:06
D	Released By	bvdongen@staffw_	2002/04/18 09:06
G	Processed To	bvdongen@staffw_	2002/04/18 09:06
G	Released By	bvdongen@staffw_	2002/04/18 09:06
I	Processed To	bvdongen@staffw_	2002/04/18 09:06
I	Released By	bvdongen@staffw_	2002/04/18 09:06
	Terminated		2002/04/18 09:07

Case 3

Step description	Event	User	yyyy/mm/dd hh:mm
	Start	bvdongen@staffw_	2002/04/18 09:05
A	Processed To	bvdongen@staffw_	2002/04/18 09:05
A	Released By	bvdongen@staffw_	2002/04/18 09:05
C	Processed To	bvdongen@staffw_	2002/04/18 09:05
C	Released By	bvdongen@staffw_	2002/04/18 09:05
F	Processed To	bvdongen@staffw_	2002/04/18 09:05
E	Processed To	bvdongen@staffw_	2002/04/18 09:05
E	Released By	bvdongen@staffw_	2002/04/18 09:06
F	Released By	bvdongen@staffw_	2002/04/18 09:06
H	Processed To	bvdongen@staffw_	2002/04/18 09:06
H	Released By	bvdongen@staffw_	2002/04/18 09:06
I	Processed To	bvdongen@staffw_	2002/04/18 09:06
I	Released By	bvdongen@staffw_	2002/04/18 09:06
	Terminated		2002/04/18 09:07

Case 4

Step description	Event	User	yyyy/mm/dd hh:mm
	Start	bvdongen@staffw_	2002/04/18 09:05
A	Processed To	bvdongen@staffw_	2002/04/18 09:05
A	Released By	bvdongen@staffw_	2002/04/18 09:05
C	Processed To	bvdongen@staffw_	2002/04/18 09:05
C	Released By	bvdongen@staffw_	2002/04/18 09:05
F	Processed To	bvdongen@staffw_	2002/04/18 09:05
E	Processed To	bvdongen@staffw_	2002/04/18 09:05
F	Released By	bvdongen@staffw_	2002/04/18 09:06
E	Released By	bvdongen@staffw_	2002/04/18 09:06
H	Processed To	bvdongen@staffw_	2002/04/18 09:06
H	Released By	bvdongen@staffw_	2002/04/18 09:06
I	Processed To	bvdongen@staffw_	2002/04/18 09:06
I	Released By	bvdongen@staffw_	2002/04/18 09:07
	Terminated		2002/04/18 09:07

Case 5

Step description	Event	User	yyyy/mm/dd hh:mm
	Start	bvdongen@staffw_	2002/04/18 13:47
A	Processed To	bvdongen@staffw_	2002/04/18 13:47
A	Released By	bvdongen@staffw_	2002/04/18 13:49
C	Processed To	bvdongen@staffw_	2002/04/18 13:49
C	Released By	bvdongen@staffw_	2002/04/18 13:53
F	Processed To	bvdongen@staffw_	2002/04/18 13:53
E	Processed To	bvdongen@staffw_	2002/04/18 13:53
E	Released By	bvdongen@staffw_	2002/04/18 13:56
F	Released By	bvdongen@staffw_	2002/04/18 13:57
H	Processed To	bvdongen@staffw_	2002/04/18 13:57
H	Released By	bvdongen@staffw_	2002/04/18 13:59
I	Processed To	bvdongen@staffw_	2002/04/18 13:59
I	Released By	bvdongen@staffw_	2002/04/18 14:04
	Terminated		2002/04/18 09:07

Case 6

Step description	Event	User	yyyy/mm/dd hh:mm
	Start	bvdongen@staffw_	2002/04/18 13:48
A	Processed To	bvdongen@staffw_	2002/04/18 13:48
A	Released By	bvdongen@staffw_	2002/04/18 13:48
B	Processed To	bvdongen@staffw_	2002/04/18 13:48
B	Released By	bvdongen@staffw_	2002/04/18 13:53
D	Processed To	bvdongen@staffw_	2002/04/18 13:53
E	Processed To	bvdongen@staffw_	2002/04/18 13:53
D	Released By	bvdongen@staffw_	2002/04/18 13:56
E	Released By	bvdongen@staffw_	2002/04/18 14:10
G	Processed To	bvdongen@staffw_	2002/04/18 14:10
G	Released By	bvdongen@staffw_	2002/04/18 14:13
I	Processed To	bvdongen@staffw_	2002/04/18 14:13
I	Released By	bvdongen@staffw_	2002/04/18 14:15
	Terminated		2002/04/18 14:15

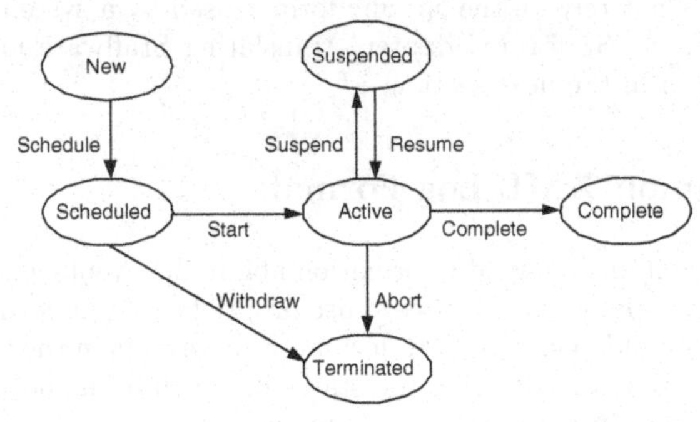

Fig. 1. An FMS showing the states and transitions of a task.

states and it is assumed that these transitions are atomic events (i.e. events that take no time) that appear in the log. All states and the transitions between those states will be discussed below.

- *New*: This is the state in which a task starts. After creation of the task this is the initial state. From this state the task can be scheduled (i.e. it will appear as a work-item in the worklist). In the Staffware example, this schedule event is shown as the "processed to" line.
- *Scheduled*: When a task is sent to one or more users, it becomes scheduled. After that, two things can happen. Either the task is picked up by a user who starts working on it, or it is withdrawn from the worklist. The "start" event is not logged by Staffware, but most other systems do, e.g., InConcert logs this event as "TASK_ACQUIRE". Staffware logs the "withdraw" event. Again Staffware uses a different term ("withdrawn" rather than "withdraw").
- *Active*: This state describes the state of the task while a user is actually working on it. The user is for example filling out the form that belongs to this task. Now three things can happen. First, a task can be suspended (for example when the user goes home at night while the work is not completely finished). Second, the task can be completed successfully and third, the task can be aborted. In the Staffware log, only the "complete" event appears as "Released by". The "abort" and "suspend" events are not logged in Staffware.
- *Completed*: Now the task is successfully completed.
- *Suspended*: When a task is suspended, the only thing that can happen is that it becomes *active* again by resuming the task. Note that it is not possible for a task to be aborted or completed from this state. Before that, it has been resumed again.
- *Terminated*: Now the task is not successfully completed, but it cannot be restarted again.

It can be seen that the Staffware example does not contain all the possible information. For example, it cannot be shown at which moment an employee actually started working on a task. On the other hand, the Staffware log contains information that does not fit into our FSM. Some of that information however can be very useful. Therefore, another kind of event is defined, namely *normal*. In our example the "Start" and "Terminated" events that refer to the start and end of a case have *normal* as event type.

Note that Staffware is just used as an example. For other systems, e.g., ERP systems, CRM systems, or other workflow management systems, alternative mappings are used. However, the resulting mapping is always made onto the Finite State Machine described in Figure 1.

3.2 XML Format

After defining the possible states of a task, a standard format for storing logs can be defined. This format is described by the following DTD[1]:

[1] The DTD describes the current format of EMiT. A new and extended format specified in terms of an XML schema has been defined. Future generations of *EMiT* will be based on this format. See http://www.processmining.org for more details.

```
<!ELEMENT WorkFlow_log (source?,process+)>
  <!ELEMENT source EMPTY>
    <!ATTLIST source program CDATA #REQUIRED>
  <!ELEMENT process (case*)>
    <!ATTLIST process id ID #REQUIRED>
    <!ATTLIST process description CDATA "none">
    <!ELEMENT case (log_line*)>
      <!ATTLIST case id ID #REQUIRED>
      <!ATTLIST case description CDATA "none">
      <!ELEMENT log_line (task_name, event?, date?, time?, originator?)>
        <!ELEMENT task_name (#PCDATA)>
        <!ELEMENT event EMPTY>
          <!ATTLIST event kind (normal|schedule|start|withdraw|
                               suspend|resume|abort|complete) #REQUIRED>
        <!ELEMENT date (#PCDATA)>
        <!ELEMENT time (#PCDATA)>
        <!ELEMENT originator (#PCDATA)>
```

This XML format describes a workflow log in the following way. First, the process that is logged needs to be specified. In the Staffware example as shown in Table 1 only one process is shown. Then, for each process a number of cases is specified. The Staffware example contains six cases. Each of these cases consist of a number of lines in the log. For each line, a "log_line" element is used. This element typically contains the name of the task that is present in the log. Further, it contains information about the state transition of the task in the transactional model, a timestamp and the originator of the task. Except for the originator of the task, all this information is used by *EMiT* in the mining process. (The originator information is added for future extensions.) *EMiT* can be used to convert log files from different systems into the common XML format. For this purpose, external programs are called by *EMiT*. The list of programs available can be extended by anyone who uses the tool to suit their own needs.

4 The Mining Process

The mining process consists of a number of steps, namely the pre-processing, the processing and the post-processing. The core algorithm used in the processing phase that is implemented in *EMiT* is the α-algorithm. This algorithm is not presented in this paper. For more information the reader is referred to [3, 8, 14] or http://www.processmining.org.

4.1 Pre-processing

In the pre-processing phase, the log is read into *EMiT* and the log based ordering relations are inferred based on that log. In order to build these relations, the assumption is made that the events in the log are totally ordered. However, several refinements can be made. First of all, it is possible to specify which events should be used in the rediscovering process. When using the default settings, only the "complete" event of a task is taken into account. Basically, the log on which the ordering relations will be built is the original log where all entries (element type "log_line") corresponding to event types not selected are removed. However, if a case contains a "withdraw" or "abort" event and that event is not specified then the whole case is excluded. The second option is that parallel relations can be inferred based on time-information present in the log. As described in [8], two tasks A and B are considered to be in parallel if and only if in the log, A is directly followed by B at least once and B is directly followed by A at least once. However, here the assumption is made that A and B are atomic actions, while in real situations tasks span a certain time interval. The beginning and the end of such a task however can be considered atomic. *EMiT* is capable of inferring parallelism for all events of a certain task, if first task A is started and then task B is started before task A has ended. In order to let *EMiT* do this, the user can give the boundaries of the intervals in terms of event types. For the Staffware example, one profile is set to "schedule" and "complete". For Staffware this is a logical definition since tasks are always scheduled in the same order by the system, i.e., although things can be executed in parallel, they are scheduled sequentially. By setting the profile to "schedule" and "complete", *EMiT* is still able to detect parallelism.

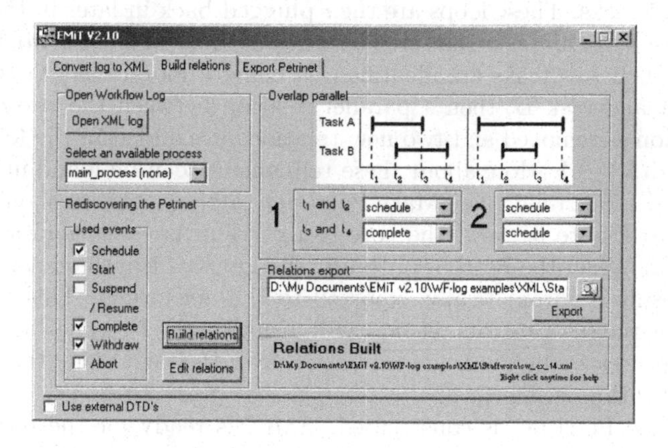

Fig. 2. *EMiT* pre-processing.

When all options are set, the pre-processing phase can start by clicking the "Build relations" button as shown in Figure 2. Now, the relations are built as described above, but also the loops of length one and two are identified. By clicking the "Edit relations" button, the program advances to the processing phase.

4.2 Processing

In the next phase, the core α-algorithm is called. However, since the α-algorithm cannot deal with loops of length one and two, some refinements are made. First

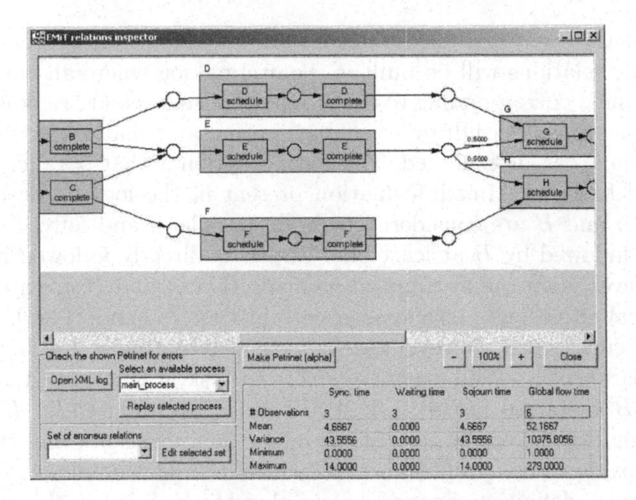

Fig. 3. *EMiT* processing.

of all, all tasks that are identified as a loop of length one (cf. [3, 8]) are taken out of the set of tasks. These loops are then plugged back in later in the processing phase. Second, for all tasks that are identified as loops of length two, the relations are changed. If two tasks are identified as being a loop of length two together, say task A and task B, then a parallel relation between the two exists, $A \| B$. This relation is removed and two new relations are added in the following way: $A \rightarrow B$ and $B \rightarrow A$. More about these refinements can be found in [1].

To build a Petri net, the "Make Petri net (alpha)" button should be clicked as shown in Figure 3. Now, the core α-algorithm is called and a Petri net is constructed. When the Petri net is built, the loops of length one are added to it and the result is automatically exported to the *dot* format. This format serves as input for the *dot* program [9] that will visualize the Petri net in a smart way. The output created by *dot* is loaded back into *EMiT* again and the result of the mining algorithm is shown.

When the Petri net is constructed, *EMiT* is ready for the post-processing phase.

4.3 Post-processing

In this phase, the original log is loaded into the program again. Using the original log and the Petri net generated in the processing step, additional information can be derived. Besides, the relations inferred in the pre-processing phase can be altered. To calculate additional information for the Petri net, the original log is used. Since the Petri net that is rediscovered should be able to generate the same log traces as were given as input, a simple algorithm is used. First, a token is placed in the source place of the Petri net. Now, for one case, the tasks appearing in the log are fired one by one. Of course, only the tasks that are actually used in the pre-processing are taken into account. Every time a token enters a place by firing a transition, a timestamp is logged for this token. When the token is consumed again, timing information is added to the place. This timing information consist of three parts. First, there is the *waiting time*. This is the time a token has spent in a place while the transition consuming the token was enabled. Second, there is the *synchronization* time. This is the time a token spent in a place while the transition consuming the token was not yet enabled. This typically happens if that transition needs multiple tokens to fire. Finally there is the sojourn time, which is the sum of the two. An example of these times for the place connecting *D-complete* with *G-schedule* can be found in the lower right part of Figure 3.

This process is repeated for all cases. If everything goes well, then a message is given and the picture showing the Petri net is updated. This update is done to show the probabilities for each choice in the Petri net. If a place has multiple outgoing arcs, a choice has to be made. The probability that each arc is chosen is basically the number of occurrences of the transition on that arc in the log divided by the sum of all occurrences of all transitions with an incoming arc from that place. The timing information for each place can be made visible in the lower right corner of the window by just clicking on a specific place in the picture. Some metrics for the waiting time, the synchronization time and the sojourn time are shown.

If, when replaying the original log, a transition is unable to fire, a message is generated stating the name of the transition. Such a message is only given for the first error. The second error message that can be generated is that after completing a case there are still tokens remaining in the net. In order to solve problems generated by the program, it is possible to change the set of ordering relations that are inferred in the pre-processing phase. By clicking on a transition in the Petri net, all relations involving that transition will be shown (except the # relation, which represents the fact that two tasks have no causal or parallel relation). These relations can then be altered.

5 Exporting the Petri Net

After the Petri net has been rediscovered, it can be exported. By clicking the "Export Petri net" button the Petri net is saved in three different formats. Each format is saved in a separate file(s):

- *low detail dot*: This output format is used to export only the basic Petri net with probabilities added to all arcs (if available). No timing information will be shown in this output file. Together with the *dot* file, a number of HTML files are created. These files can be used together with the *dot* "jpg" and "imap" export to make a web page. Each of these HTML files contain timing information for a specific place.
- *high detail dot*: This output format is used to export the basic Petri net with probabilities added to all arc (if available) and timing information.

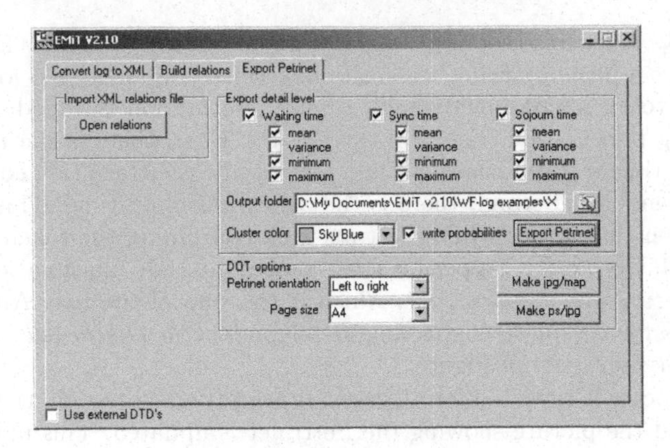

Fig. 4. *EMiT* post-processing.

– *Woflan*: This output format can be imported by Woflan for further analysis [13]. Woflan supports verification of the discovered model.

After exporting the discovered model in various formats, some options for *dot* can be set. The "P-net orientation" setting specifies whether the Petri net should be draw from left to right or from top to bottom. The "Page size" setting can be set to "A4" or "Letter". If these options are set, *dot* can be used to create pictures and other exports. The buttons "Make jpg/map" and "Make ps/jpg" can be used to generate the graphics files. This way *EMiT* provides several ways to visualize the mining result.

6 Conclusion

In this paper *EMiT* is presented as a tool to mine process models from timed logs. The whole conversion process from a log file in some unspecified format towards a Petri net representation of a process model is described in a number of different steps. Along the way, an example is used to illustrate the use of the tool. We invite the reader to use the tool. The tool is available for download from http://www.processmining.org. This download includes some 30 examples, both artificial and practical. The example presented in this paper is called "sw_ex_14.log" and it is converted into "sw_ex_14.xml". Both files are included in the download.

In the future we plan to extend the tool in various ways. First of all, we plan to enhance the mining algorithm to be able to deal with complex process patterns, invisible/duplicate tasks, noise, etc. [1]. Second, we are developing additional adaptors to extract information from more real-life systems (e.g., SAP and FLOWer). Third, we are working on embedding parts of the system in the ARIS Process Performance Monitoring (PPM) tool and exporting the result to BPR tools like Protos. We are extending the scope of the mining process to include organizational data, data flow, social network analysis, etc. Last but not least, we are merging the tools EMiT, Thumb [14], and MinSoN [5] into an integrated tool.

References

1. A.K.A. de Medeiros and W.M.P. van der Aalst and A.J.M.M. Weijters. Workflow Mining: Current Status and Future Directions. In R. Meersman, Z. Tari, and D.C. Schmidt, editors, *On The Move to Meaningful Internet Systems 2003: CoopIS, DOA, and ODBASE*, volume 2888 of *Lecture Notes in Computer Science*, pages 389–406. Springer-Verlag, Berlin, 2003.
2. W.M.P. van der Aalst, J. Desel, and A. Oberweis, editors. *Business Process Management: Models, Techniques, and Empirical Studies*, volume 1806 of *Lecture Notes in Computer Science*. Springer-Verlag, Berlin, 2000.
3. W.M.P. van der Aalst and B.F. van Dongen. Discovering Workflow Performance Models from Timed Logs. In Y. Han, S. Tai, and D. Wikarski, editors, *International Conference on Engineering and Deployment of Cooperative Information Systems (EDCIS 2002)*, volume 2480 of *Lecture Notes in Computer Science*, pages 45–63. Springer-Verlag, Berlin, 2002.
4. W.M.P. van der Aalst and K.M. van Hee. *Workflow Management: Models, Methods, and Systems*. MIT press, Cambridge, MA, 2002.
5. W.M.P. van der Aalst and M. Song. Mining Social Networks: Uncovering interaction patterns in business processes. In M. Weske, B. Pernici, and J. Desel, editors, *International Conference on Business Process Management (BPM 2004)*, Lecture Notes in Computer Science, Springer-Verlag, Berlin, 2004.
6. W.M.P. van der Aalst, B.F. van Dongen, J. Herbst, L. Maruster, G. Schimm, and A.J.M.M. Weijters. Workflow Mining: A Survey of Issues and Approaches. *Data and Knowledge Engineering*, 47(2):237–267, 2003.
7. W.M.P. van der Aalst and A.J.M.M. Weijters, editors. *Process Mining*, Special Issue of Computers in Industry, Volume 53, Number 3. Elsevier Science Publishers, Amsterdam, 2004.
8. W.M.P. van der Aalst, A.J.M.M. Weijters, and L. Maruster. Workflow Mining: Discovering Process Models from Event Logs. QUT Technical report, FIT-TR-2003-03, Queensland University of Technology, Brisbane, 2003. (Accepted for publication in IEEE Transactions on Knowledge and Data Engineering.).
9. AT&T. Graphviz - Open Source Graph Drawing Software (including DOT). http://www.research.att.com/sw/tools/graphviz/.
10. S. Jablonski and C. Bussler. *Workflow Management: Modeling Concepts, Architecture, and Implementation*. International Thomson Computer Press, London, UK, 1996.
11. F. Leymann and D. Roller. *Production Workflow: Concepts and Techniques*. Prentice-Hall PTR, Upper Saddle River, New Jersey, USA, 1999.
12. Staffware. *Staffware 2000 / GWD User Manual*. Staffware plc, Berkshire, United Kingdom, 2000.
13. H.M.W. Verbeek, T. Basten, and W.M.P. van der Aalst. Diagnosing Workflow Processes using Woflan. *The Computer Journal*, 44(4):246–279, 2001.
14. A.J.M.M. Weijters and W.M.P. van der Aalst. Rediscovering Workflow Models from Event-Based Data using Little Thumb. *Integrated Computer-Aided Engineering*, 10(2):151–162, 2003.

3D-Visualization of Petri Net Models: Concept and Realization

Ekkart Kindler and Csaba Páles

Department of Computer Science, University of Paderborn
{kindler,cpales}@upb.de

Abstract. We present a simple concept for the 3D-visualization of systems that are modelled as a Petri net. To this end, the Petri net is equipped with some information on the physical objects corresponding to the tokens on the places. Moreover, we discuss a prototype of a tool implementing this concept: PNVis version 0.8.0.

Keywords: Petri nets, 3D-visualization, animation.

1 Introduction

Petri nets are a well-accepted formalism for modelling concurrent and distributed systems in various application areas: Workflow management, embedded systems, production systems, and traffic control are but a few examples. The main advantages of Petri nets are their graphical notation, their simple semantics, and the rich theory for analyzing their behaviour.

In spite of their graphical nature, getting an understanding of a complex system just from studying the Petri net model itself is quite hard – if not impossible. In particular, this applies to experts from some application area who, typically, are not experts in Petri nets. 'Playing the token-game' is not enough for understanding the behaviour of a complex system. Using suggestive icons for transitions and places of the Petri net in order to indicate the corresponding action or document in the application area is only a first step.

Therefore, there have been different approaches that try to visualize the behaviour of a Petri net in a way understandable for experts in the application area. At best, there will be an animation of the model using icons and graphical features from the corresponding application area. *ExSpect* [14], for example, uses the concept of a *dashboard* in order to visualize the dynamic behaviour of a system in a way that is familiar to the experts in the application area (e. g. by using flow meters, flashing lights, etc. as used in typical control panels). In ExSpect, it is even possible to interact with the simulation via this dashboard. Another example is the Mimic library of Design/CPN [4, 13], which allows a Design/CPN simulation to manipulate graphical objects, and the user can interact with the simulation via these graphical elements. This way, one can get a good impression of the 'look and feel' of the final product. A good example is the model of a mobile phone [10]. Another approach for visualizing Petri nets is based on graph transformations and their animation: GenGED [1, 2].

J. Cortadella and W. Reisig (Eds.): ICATPN 2004, LNCS 3099, pp. 464–473, 2004.
© Springer-Verlag Berlin Heidelberg 2004

In the *PNVis* project, we take the next step: The graphical objects manipulated by the simulation are no longer considered to be artifacts for visualizing information; rather we consider them as a part of the system model. Actually, they are considered as the *physical part* of the system. Though simple, this step has several benefits: First, it makes the interaction between the control system and the physical world explicit. Second, the physical part can be used for a realistic 3D-visualization of the dynamic behaviour by using the shape and the dynamic properties of the physical components. Third, the properties of the physical objects can be used for analysis and verification purposes. For example, we can exploit the fact that two physical components cannot be at the same place at a time.

In this paper, we show how a Petri net can be equipped with the information on the physical objects. Moreover, we discuss the concepts and a prototype of a tool that uses this information for a 3D-visualization of the system: PNVis version 0.8.0. Here, the focus is on those aspects of the physical objects that are necessary for visualization: basically the shape of the objects. The prototype of PNVis is restricted to low-level Petri nets. The PNVis project, however, has a much wider scope. For example, we would like to use some physical properties of the objects such as their weight for analysis purposes. Moreover, PNVis will support high-level Petri nets, and it will provide concepts for constructing a system from components in a hierarchical way.

PEP [11] was one of the first Petri Net tools that came up with a 3D-visualization of Petri net models: *SimPep* [6]. Basically, *SimPep* triggers animations in a *VRML model* [5] while simulating the underlying Petri net. But, this simulation imposes a sever restriction on the animations: There is only one animation at a time; concurrent animations of independent objects are impossible. In this paper, we will present concepts that allow us to have concurrent animations of independent objects. The trick is to associate animations with places rather than with transitions.

2 Concepts

In order to animate the behaviour of a Petri net in a 3D-visualization, the net must be equipped with some information on the physical objects. Moreover, the behaviour of the objects must be related to the dynamic behaviour of the Petri net. In the following, we will discuss how to add this information to a net.

Geometry, shapes and animation functions. In a first step, we distinguish those places of a Petri net that correspond to physical objects. We call these *animation* places. The idea is that each token on an animation place corresponds to a physical object with its individual appearance and behaviour. In order to animate a physical object, we need two pieces of information: its shape and its behaviour.

Defining the *shape* of the object is easy: Each animation place is associated with a *3D-model* (e.g. a VRML model [5]) that defines the shape of all tokens on this place. Defining the *behaviour* of an object is similar: Each animation

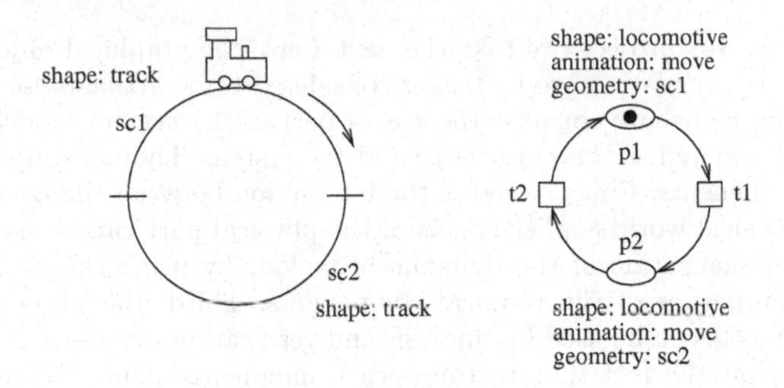

Fig. 1. A simple toy-train model

place is associated with an *animation function*, where the animation function is composed from some predefined animation functions. When a token is produced on an animation place, an object with the corresponding shape appears and behaves according to the animation function. For example, the object could *move* along a predefined line or the object could *appear* at some point.

In order to illustrate these concepts, let us consider a simple example: a model of a toy-train. Figure 1 shows the layout of a toy-train, which consists of two semicircle tracks *sc1* and *sc2*, which are composed to a full circle. We call this layout the underlying *geometry*. For defining such a geometry, there is a set of predefined geometrical objects such as *lines*, *circle* segments, and *points*. In our example, there is one toy-locomotive moving clockwise on this circle. The right-hand side of Fig. 1 shows the corresponding Petri net model, where both places *p1* and *p2* are animation places. In this example, the correspondence between the Petri net model and the physical model is clear from the layout. Formally, this correspondence is defined by annotating each place with a reference to the corresponding element in the geometry. Place *p1* corresponds to the upper semicircle of the geometry *sc1* and place *p2* corresponds to the lower semicircle of the geometry *sc2*. The annotation *shape*[1] defines the shape of the physical objects corresponding to the tokens on this place. In our example, it is a locomotive for both places, where the details of the definition of the shape will be discussed in Sect. 3. For now, you can think of it as a reference to some VRML model of a toy locomotive. The annotation *animation* defines the behaviour of the object corresponding to a token on a place. This behaviour will be started when a token is added to the place. In our example, the behaviour is a *move* animation. Note that, without additional parameters, the animation function refers to the geometry object corresponding to that place. So, a locomotive corresponding to a token on place *p1* will move on semicircle *sc1*, and a locomotive corresponding to a token on place *p2* will move on semicircle *sc2*.

In order to make our example complete, we must also give some information on how to visualize the geometry objects themselves. To this end, each geometry

[1] Note that, in the implementation of the net type for the PNK, we call this extension *dynamic shape*.

Fig. 2. Screenshot of the visualization

shape: locomotive
animation: move
geometry: sc1
id:train
id:train
p1
t2
t1
p2
id:train
id:train
shape: locomotive
animation: move
geometry: sc2

Fig. 3. The model with identities

object is equipped with an annotation *shape*, which defines the graphical appearance of the geometry object. In our example, the semicircles will be visualized as tracks. The precise definition of these tracks and their appearance will be discussed in Sect. 3. Once we have provided all this information, we can start the 3D-visualization of this system. Figure 2 shows a screenshot of the animation of our example, where the locomotive on place *p1* has almost reached the end of its move animation on *sc1*.

Object identities. Up to now, the objects corresponding to the tokens on the two places *p1* and *p2* are completely independent of each other. When transition *t1* fires, the object corresponding to the token on place *p1* is deleted and a new object corresponding to the token on place *p2* is created and the move animation is started. Apart from the fact that this constant deletion and new creation of 3D-objects would be quite inefficient, this behaviour is not what happens in reality. In reality, the same physical object, the locomotive, moves from track *sc1* to track *sc2*. In order to keep the identity of a physical object when a 'token is moved from one place to another', we equip the arcs of the Petri net with an annotation *id*, which is some identifier *n*. We call *n* the *identity* of that arc. By assigning the same identity to an in-coming arc and an out-going arc of a transition, we express that the corresponding object is moved between those two places. In order not to clone a physical object, we require that there is a one-to-one *correspondence* between the identities of the in-coming and out-going arcs of a transition; i. e. each identifier occurs exactly once in all in-coming arcs and exactly once in all out-going arcs. Figure 3 shows the toy-train example equipped with such identities. Of course, we may have arcs without identity annotations. For an in-coming arc of a transition, this means that the corresponding object will be deleted. For an out-going arc of a transition, this means that a corresponding object will be created, where the shape of the newly created object is defined by the shape annotation of the place.

Finished and unfinished animations. Next, we consider the relation of the behaviour of the Petri net and the animations of the objects corresponding to the

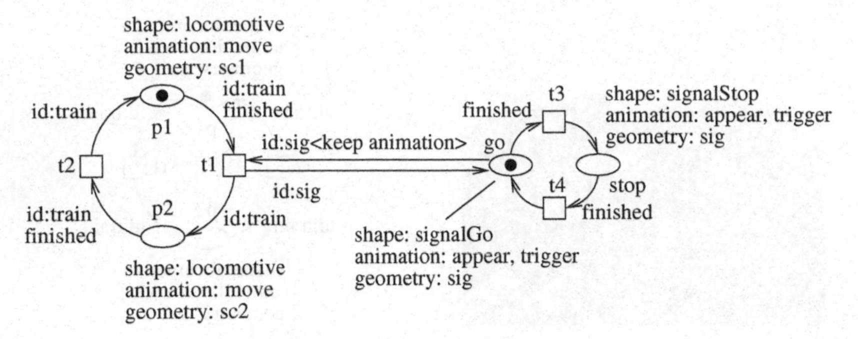

Fig. 4. A toy-train with a signal

tokens in more detail. When a token is added to an animation place by firing a transition, the animation for the corresponding object is started. But, what will happen, if a token is removed from a place before the animation on its corresponding object is terminated? One idea would be to immediately stop the animation. In our example, this would not make much sense, because the locomotive would appear to jump from some intermediate position of the track to the start of the next track. Assuming that transition firing does not take any time, this behaviour is physically impossible. In our example, we would like to remove a token from a place only when the animation of the corresponding object is finished. On the other hand, there are examples in which it makes sense to remove a token from a place while an animation on the corresponding object is running. Whether a transition may remove or must not remove a token with a corresponding animation running must be explicitly defined in the Petri net model. When the animation must be finished before the token may be removed, we add a label *finished* to the corresponding arc. If there is no such annotation, the transition need not wait until the animation of the corresponding object is terminated. In that case, there are two possibilities to proceed: Either the animation of the object is stopped or the animation is continued on the new place. When the animation should be continued for the token on the new place, the corresponding arc has an id with an additionally tag <keep animation>. When there is no such tag, the running animation is stopped and a new animation is started on the new place.

In order to illustrate these new concepts, we extend our example: We assume that there is a signal at the end of track *sc1* for which we add a position *sig* in the geometry somewhere at the end of track *sc1*. The idea is that the locomotive should stop at the end of track *sc1*, when the signal is in state *stop*; when the signal is in state *go*, the locomotive may enter track *sc2*. The Petri net in Fig. 4 models this behaviour. The two places *p1* and *p2* as well as the transitions *t1* and *t2* are the same as before. The arcs are equipped with identities in order to keep the same object, i. e. the locomotive, on the tracks. The annotation *finished* guarantees that the transitions wait until the move animation of the locomotive has come to an end (i. e. the locomotive has reached the end of the track). The two states of the signal are represented by the places *stop* and *go*.

The object corresponding to a token on place *stop* is a signal with its red light on: *signalStop*. The object corresponding to a token on place *go* is a signal with its green light on: *signalGo*. These objects will appear at the point *sig* of the geometry (at the end of *sc1*). Due to the loop between place *go* and transition *t1*, transition *t1* can fire only when the signal is in state *go*. The interesting parts of this model are the identities of transition *t1*; when transition *t1* is fired, the object of the signal from place *go* stays on this place. Moreover, the animation is not restarted, because the identity is equipped with the *keep animation* tag.

Another interesting issue is the animation of the signal. The animation function is composed from two predefined animation functions: *appear, trigger*. The meaning is that these animations are started sequentially. When the first animation function has finished, the second starts. So, in both cases the signal appears at position sig; then, it behaves as a trigger. A *trigger* is an animation function that simply waits for a user to click on that object in the 3D-visualization. When this happens, the animation terminates. In combination with the annotations *finished* at the in-coming arcs of transitions *t3* and *t4*, the user can toggle the state of the signal by clicking on the signal. A user's click on the signal object will finish the *trigger* animation running for this object; once the animation function is finished transition *t3* resp. *t4* will fire.

Animation results. In order to allow us more complex interactions between the Petri net model and the animations, the animation functions are equipped with a result value. The result of an animation could depend on the outcome of the animation function. For example, the outcome of the trigger animation, could depend on the part of the object the user clicked on. In some cases, we would like a transition to fire only when the animation function returns a particular result *n*. To this end, we annotate the corresponding arc with *result:*{*n*}. Actually, the annotation result may give a range of values *result:*{*0..3*} or *result:*{*0..*} or *result:*{*..10*}, where the last two annotations denote ranges that are open in one direction. The particular annotation *result:*{*..*} represents the full range of possible return values, which means that the corresponding animation must have terminated, but its value does not matter at all. Therefore, the notation *finished* introduced earlier is just a shorthand for *result:*{*..*}.

Collisions. In our previous examples, there was only one locomotive. Figure 5 shows a screenshot of a more complex example, where there are two locomotives, two signals, and two switches. All objects are animated independently of each other. In particular, the signals as well as the switches can be toggled by the user by clicking on the corresponding objects. This way, the user can control the route of the locomotives. In this scenario, it could well happen that two locomotives move on the same track. In principle, the animations of the different tokens in a Petri net are completely independent of each other. But, they may interfere, when two objects approach each other. The reason is that objects are considered to be solid. And solid objects cannot be at the same position at the same time. Consider the situation shown in Fig. 5 again. Suppose that the first locomotive stops in front of the stop signal. Eventually, the second locomotive will approach the first locomotive. Then, the move animation of the second locomotive will be

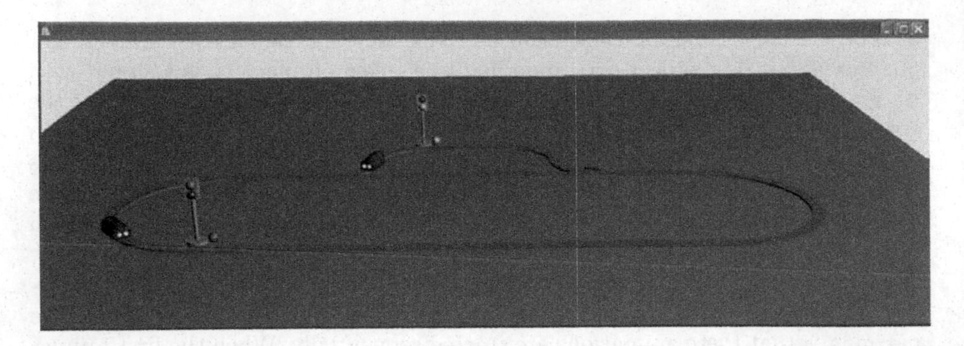

Fig. 5. Screenshot of a more complex toy-train

suspended (but not finished). So the second locomotive will stop right behind the first locomotive without finishing its animation. When the user clicks on the signal again, it is switched to go. Then, the first locomotive will be moved to the next track and a move animation on the next track will be started for this locomotive. When the first locomotive has moved a little bit, the second locomotive will resume its movement again and, eventually, will finish its animation. This way, the animation reflects the fact that objects are solid. Currently, we avoid collisions of solid objects, by suspending the corresponding animations when there is another object in front of it. This behaviour was inspired by material flow systems in which collisions of shuttles are avoided by infra-red detectors. But, we could also model other behaviour; for example, we could also stop the animation of objects, when they collide and return a special result value. This way, the Petri net model can be aware of collisions. More detailed concepts for reacting on collisions, however, need further investigations.

Extensions. Here we have discussed those concepts only, that are already implemented in PNVis version 0.8.0. Future versions will support high-level Petri nets and parameterized animation functions and parameterized 3D-models [7].

3 Realization

The above concepts have been implemented in a prototype tool called PNVis, which is based on the *Petri Net Kernel* (PNK) [15] and uses Java3D for implementing the 3D-visualization. In the following, we discuss how the additional information is provided to PNVis.

There are three types of information that must be provided to the tool: the annotations of the Petri net, the geometry, and the 3D-models for the animated objects and the geometry objects. The annotations for the Petri net can be easily added as extensions to the Petri, by defining a new Petri net type. Here, we do not discuss the definition of such a Petri net type. Basically, there is a list of new annotations for each element of a Petri net, which is similar to the concept of annotations in PNML [3]. Moreover, the Petri net will have two global annotations: a reference to a *geometry file* and to a *models file*.

Geometry file. The *geometry file* defines the underlying geometry of the system, i. e. it lists all the geometry objects in some XML syntax. For our toy-train with one signal, the geometry file looks as follows:

```
<geometry>
   <circle id="sc1" shape="track" angle="180"
           cx="0" cy="0" cz="0" sx="-10" sy="0" sz="0" />
   <circle id="sc2" shape="track" angle="180"
           cx="0" cy="0" cz="0" sx="10" sy="0" sz="0" />
   <point  id="sig" x="13" y="0" z="0" />
</geometry>
```

Basically, the XML file consist of a list of predefined geometry objects, which are *points*, *lines*, *circles*, and *Bezièr curves.* Moreover, a geometry object could be composed from many predefined geometry objects. We call such a geometry object a *compound object*[2]. The attribute *id* is the unique identifier of the corresponding geometry object. This identifier will be used in the geometry annotations of the places of the Petri net in order to establish the correspondence between the Petri net and the geometry. The attribute *shape* defines the graphical appearance of the geometry object, which is a reference to a definition in the models file. The other attributes depend on the chosen geometry object. For example, attributes *cx*, *cy*, and *cz* define the center of a circle segment, attributes *sx*, *sy*, and *sz* define the start point of a circle segment, and attribute *angle* defines the angle of the circle segment (in clockwise orientation)[3].

Models file. The *models file* defines the graphical appearance of the shapes used in the geometry file and the Petri net model. We call the shapes for the geometry file *static models*, and we call the shapes for the places of the Petri net *dynamic models.* For our toy-train, the models file looks as follows:

```
<models>
   <static>
      <model id="track">
          <profile> <rectangle height="1.5" width="3" /> </profile>
          <texture name="track.jpg" />
      </model>
   </static>
   <dynamic>
      <model id="locomotive"><file name="locomotive.wrl" /></model>
      <model id="signalGo"><file name="lampRed.wrl" /></model>
      <model id="signalStop"><file name="lampGreen.wrl" /></model>
   </dynamic>
</models>
```

[2] PNVis version 0.8.0 does not support Bezièr curves and compound geometry objects.

[3] Note that PNVis version 0.8.0 ignores the z-coordinates.

In the *static* section, we have the definition of tracks, which define the graphical appearance of the geometry objects in the visualization. It is defined by giving a profile and a texture. The idea is that the profile will be moved along the geometry object in order to defines its outline. The texture will be placed on this outline. This way, we need only one definition of a static shape for all types of geometry objects. In our example, the profile is a rectangle and the texture is some JPEG file[4].

In the *dynamic* section, we define several 3D-models (one for each model referred to in the Petri net). Here, we refer to some VRML models.

In fact, the Petri net from Fig. 4 along with the above geometry file, the model file, and the VRML files are sufficient for visualizing the Petri net model with our tool. The separation of the geometry file and the model file allows us to easily exchange the underlying layout as well as the graphical appearance of a model. This way, we have a clear separation between the dynamic behaviour which is modelled in the Petri net, the underlying layout, which is defined in the geometry file, and the graphical appearance, which is defined in the models file.

4 Conclusion

In this paper, we have introduced concepts that allow us to easily equip a Petri net with a 3D-visualization. What is more, for obtaining a visualization, no programming is necessary. We only need to provide some 3D-models, a geometry, and some animation functions from a set of predefined animation functions.

One of the principles underlying these concepts is *separation of concerns*. The 3D-visualization part is quite independent from the Petri net itself. This way, the concept provides an abstraction mechanism, and it is possible to analyze the behaviour of the system without considering the details of the physical model. But, this is not always possible. For example, collisions of objects could result in deadlocks that are not present in the Petri net model alone. The investigation of such problems and the definition of sufficient conditions for the independence of the Petri net properties from the physical properties is one of the future research directions.

The implementation of *PNVis version 0.8.0* is now freely available and demonstrates that the concepts are feasible. PNVis runs on all systems on which Java and Java3D are installed. More detailed information on PNVis and its code can be found at [12]. Clearly, there could be much more features for obtaining more realistic animations. Such features will be added in a future version of PNVis; in particular, PNVis will also support high-level Petri nets. Which other features are necessary and appropriate is another direction of future research.

Acknowledgments

We would like to thank some anonymous reviewers for their comments on earlier versions of this paper.

[4] Note that PNVis version 0.8.0 supports the profile rectangle only, and it completely ignores textures.

References

1. R. Bardohl, C. Ermel, and L. Ribeiro. Towards visual specification and animation of Petri net based models. In *Workshop on Graph Transformation Systems (GRATRA '00)*, pages 22–31, March 2000.
2. Roswitha Bardohl, Claudia Ermel, and Julia Padberg. Formal relationship between Petri nets and graph grammars as basis for animation views in GenGED. In *Integrated Design and Process Technology IDPT 2002*, Society for Design and Process Science, June 2002.
3. Jonathan Billington, Søren Christensen, Kees van Hee, Ekkart Kindler, Olaf Kummer, Laure Petrucci, Reinier Post, Christian Stehno, and Michael Weber. The Petri Net Markup Language: Concepts, technology, and tools. In W. van der Aalst and E. Best, editors, *Application and Theory of Petri Nets 2003, 24th International Conference, LNCS* 2679 , pages 483–505. Springer, June 2003.
4. Design/CPN. `http://www.daimi.au.dk/designCPN/`. 2004/03/12.
5. ISO/IEC International Standard. Information technology – Computer graphics and image processing – The Virtual Reality Modeling Language (VRML) – Part 1: Functional specification and UTF-8 encoding. ISO/IEC 14772-1, 1997.
6. Michael Kater. SimPEP: 3D-Visualisierung und Animation paralleler Prozesse. Masters thesis (in German), Universität Hildesheim, April 1998.
7. Ekkart Kindler and Csaba Páles. PNVis: Documentation of version 0.8.0. PNVis homepage [12], March 2004 (evolving document).
8. Ekkart Kindler and Wolfgang Reisig. Algebraic system nets for modelling distributed algorithms. *Petri Net Newsletter*, 51:16–31, December 1996.
9. Ekkart Kindler and Hagen Völzer. Algebraic nets with flexible arcs. *Theoretical Computer Science*, 262:285–310, July 2001.
10. Louise Lorentsen, Antti-Pekka Tuovinen, and Jianli Xu. Modelling of features and feature interactions in Nokia mobile phones using coloured Petri nets. In J. Esparza and C. Lakos, editors, *Application and Theory of Petri Nets 2002, 23rd International Conference, LNCS* 2360, pages 294–313. Springer, June 2002.
11. The PEP Tool. `http://parsys.informatik.uni-oldenburg.de/~pep`. 2004/03/12.
12. PNVis homepage. `http://www.upb.de/cs/kindler/research/PNVis`. 2004/03/12.
13. Jens Linneberg Rasmusen and Mejar Singh. *Mimic/CPN: A Graphical Animation Utility for Design/CPN.* Computer Science Department, Aarhus University, Aarhus Denmark, December 1995.
14. Eric Verbeek. ExSpect 6.4x product infromation. In K. H. Mortensen, editor, *Petri Nets 2000: Tool Demonstrations*, pages 39–41, June 2000.
15. Michael Weber and Ekkart Kindler. The Petri Net Kernel. In H. Ehrig, W. Reisig, G. Rozenberg, and H. Weber, editors, *Petri Net Technologies for Modeling Communication Based Systems, LNCS* 2472, pages 109–123. Springer, 2003.

An Approach to Distributed State Space Exploration for Coloured Petri Nets*

Lars M. Kristensen[1,**] and Laure Petrucci[2]

[1] Department of Computer Science, University of Aarhus
IT-parken, Aabogade 34, DK-8200 Aarhus N, Denmark
lmkristensen@daimi.au.dk

[2] LIPN, CNRS UMR 7030, Université Paris XIII
99, avenue Jean-Baptiste Clément
F-93430 Villetaneuse, France
petrucci@lipn.univ-paris13.fr

Abstract. We present an approach and associated computer tool support for conducting distributed state space exploration for Coloured Petri Nets (CPNs). The distributed state space exploration is based on the introduction of a coordinating process and a number of worker processes. The worker processes are responsible for the storage of states and the computation of successor states. The coordinator process is responsible for the distribution of states and termination detection. A main virtue of our approach is that it can be directly implemented in the existing single-threaded framework of Design/CPN and CPN Tools. This makes the distributed state space exploration and analysis largely transparent to the analyst. We illustrate the use of the developed tool on an example.

1 Introduction

State space exploration is one of the main approaches to computer-aided validation and verification [4, 7]. The basic idea of state space exploration is to compute all reachable states and state changes of the system and representing these as a directed graph. The main advantage of such exploration methods is that they are highly automatic to use and allow for investigating of many properties of the system under consideration. The main disadvantage of state space exploration methods is the *state space explosion problem* [17].

A wide variety of methods (see [17] for a survey) have been suggested in the literature to alleviate the state space explosion problem. Recently [5, 6], there has also been increased interest in exploiting the memory and computing power of several machines to conduct distributed state space exploration. Distributed exploration does not alleviate the state space explosion problem, but it increases the memory available for storage of the state space, and has the potential for a

* This work was started when both authors were at LSV, CNRS UMR 8643, ENS de Cachan, France.

** Supported by the Danish Natural Science Research Council.

J. Cortadella and W. Reisig (Eds.): ICATPN 2004, LNCS 3099, pp. 474–483, 2004.
© Springer-Verlag Berlin Heidelberg 2004

linear speed-up in time. Distributed state space exploration has been developed, e.g., for the SPIN [1], UPPAAL [2], and Murϕ [16] tools.

In this paper we consider distributed state space exploration for Coloured Petri Nets (CPNs) [14]. Modelling and analysis of CPN models are supported by Design/CPN [9] and CPN Tools [8]. Until now, only very limited investigations have been conducted on distributed state space exploration for CPNs [11]. The contribution of this paper is to explore the use of distributed state space exploration in the context of CPNs and their associated computer tools Design/CPN and CPN Tools. A main requirement in the developent of our approach has been to exploit as much as possible the existing support for state spaces, and to make the distributed state space exploration largely transparent to the analyst.

The rest of this paper is organised as follows. Section 2 introduces some basic notations for state spaces. Section 3 presents our algorithm for distributed state space exploration and section 4 gives some experimental results with this algorithm on an example. Finally, section 5 contains the conclusions.

2 Background

Figure 1 lists the standard algorithm for sequential explicit state space exploration. The algorithm operates on two sets: UNPROCESSED which is a set of states for which successor states have not yet been calculated and NODES which is the set of already visited states. The algorithm starts from the initial state M_0 and conducts a loop until the set of unprocessed states is empty. In each iteration of the loop (lines 3–11), a state M is selected from the set of unprocessed states and the successor states M' of M are examined in turn. Successor states that have not been previously visited are inserted into the set of unprocessed states.

```
 1: UNPROCESSED ← {M0}
 2: NODES ← {M0}
 3: while ¬ UNPROCESSED.EMPTY() do
 4:   M ← UNPROCESSED.GETNEXTELEMENT()
 5:   for all ((t,b),M') such that M[(t,b)⟩M' do
 6:     if ¬(NODES.CONTAINS(M')) then
 7:       NODES.ADD(M')
 8:       UNPROCESSED.ADD(M')
 9:     end if
10:   end for
11: end while
```

Fig. 1. Sequential state space exploration algorithm.

The time taken to explore the state space of a CPN is determined by the computation of enabled binding elements and corresponding successor states in line 5, and the time used to determine whether a newly generated state has already been explored before in line 6. This computation can be costly for large CPN models having many tokens and complex arc expressions. The space used to store the state space is another main factor in the algorithm. A distributed computation of the state space could potentially achieve the following: 1) Computation of successor states could be done in parallel for several states at a time, 2) Determining whether a state has already been explored could be done in parallel for several states at a time, and 3) the total amount of memory available for the state space exploration would be increased by having several machines.

3 Distributed State Space Exploration

Based on previous work [3, 15, 16] on distributed state space exploration, we choose to distribute both storage and calculation of successor states to a number of *worker processes*. The alternative would have been to only distribute the storage of states and compute sucessor states centrally. Distribution of successor states was considered of particular importance for CPNs, where the computation of successor states can be costly in case of many tokens on places and/or complex arc inscriptions. Unlike the approaches reported in [15], we introduce a central *coordinator process*. The purpose of the coordinator process is to distribute states. The introduction of the coordinator process makes the implementation of distributed state space exploration simpler given the single-threaded nature of Design/CPN and CPN Tools. Furthermore, the introduction of the coordinator process simplifies the termination detection. A third advantage of this architecture is that the coordinator can also be used later to control the verification process while the user interacts with the coordinator process only.

The basic idea is that when a worker computes a successsor state, it first checks if it itself is responsible for the state. Determing this is based on the use of an external hash function known to all workers and the coordinator process. If so, it checks locally whether the state is a new one. Otherwise, it sends the state to the coordinator. The coordinator then sends the state to the worker responsible for the state. The main disadvantage of this architecture is that the central coordinator node could become a bottleneck. On the other hand, the computation that needs to be done by the coordinator is very limited since it only consists in relaying states to the appropriate worker process.

The worker and the coordinator processes are all SML/NJ processes (Standard ML of New Jersey), and the communication infrastructure between the nodes is based on the Comms/CPN library [12]. This library supports communication between SML/NJ processes and external applications. In this particular case, all communications will be between SML/NJ processes. A main problem with the SML/NJ processes is that they are single-threaded and Comms/CPN only supports a blocking receive primitive. We have therefore added a CANRECEIVE primitive to Comms/CPN to support a polling receive. We assume that each of the workers will run on machines with equal computing power and communication between workers and coordinator will be on a local area network.

Figure 2 gives the algorithm executed by the coordinator process during the distributed state space exploration. The coordinator starts by sending the initial

```
 1: computed ← false
 2:
 3: SEND(STATE M0,worker(h_ext(M0)))
 4: nextprobe ← h_ext(M0)
 5: SEND(PROBE,worker(nextprobe))
 6: nextprobe ← nextprobe +1
 7:
 8: while ¬ computed do
 9:   for all i ∈ {1,...,n} do
10:     if CANRECEIVE(worker(i)) then
11:       RECEIVE(message,worker(i))
12:       if message == STATE M then
13:         nextprobe ← MIN(nextprobe,h_ext(M))
14:         SEND(STATE M,worker(h_ext(M)))
15:       else
16:         {Probe was returned}
17:         if nextprobe > n then
18:           computed ← true
19:         else
20:           SEND(PROBE,worker(nextprobe))
21:           nextprobe ← nextprobe +1
22:         end if
23:       end if
24:     end if
25:   end for
26: end while
27:
28: for all i ∈ {1,...,n} do
29:   SEND(STOP,worker(i))
30: end for
```

Fig. 2. State space exploration algorithm for the coordinator.

state M_0 to the worker determined by the external hash function $h_{\text{ext}} : \mathbb{M} \rightarrow \{1, 2, \ldots, n\}$ used to distribute states between the n workers. It then sends a PROBE *message* to this process to be able to detect when the process has finished processing the state just sent. The PROBE messages and variable *nextprobe* are used to detect termination. We will explain how the termination detection works after presenting the algorithm for the workers. The coordinator then runs a loop where each of the worker processes is polled for messages using the CANRECEIVE primitive. If the received message is a STATE, this state is sent to the appropriate worker process and *nextprobe* is updated accordingly. If a PROBE message is received from a worker process, then it is passed on to the next worker and *nextprobe* updated accordingly.

Figure 3 lists the algorithm executed by each of the workers. The workers run in a loop exploring states received from the coordinator. Each worker will terminate once a STOP message is received from the coordinator. Whenever a state M is received, it is first checked if the state is already stored by the worker.

```
 1: stop ← false
 2:
 3: while ¬ stop do
 4:    RECEIVE(message)
 5:    if message == STATE M then
 6:       if ¬NODES.CONTAINS(M) then
 7:          NODES.ADD(M)
 8:          UNPROCESSED ← {M}
 9:
10:          while ¬ UNPROCESSED.EMPTY() do
11:             M ← UNPROCESSED.GETNEXTELEMENT()
12:             for all ((t,b), M') such that M[(t,b)⟩M' do
13:                if h_ext(M') ≠ i then
14:                   SEND(STATE M')
15:                else if ¬NODES.CONTAINS(M') then
16:                   NODES.ADD(M')
17:                   UNPROCESSED.ADD(M')
18:                end if
19:             end for
20:          end while
21:       end if
22:    else if message == PROBE then
23:       SEND(PROBE)
24:    else
25:       stop ← true
26:    end if
27: end while
```

Fig. 3. State space exploration algorithm for worker i.

If not, then it is added to UNPROCESSED and an exploration starting in M is conducted. States encountered in this exploration which belong to other workers are transmitted to the coordinator process, whereas encountered states that belong to the worker but are not currently stored are added to UNPROCESSED. When the exploration of the state space from the received state terminates, the worker goes back waiting for the next message. If a PROBE message is read, this message is sent back to the coordinator, and in case of a STOP message from the coordinator, the worker will stop its exploration.

The basic idea to detect termination of the state space exploration is that the coordinator passes a PROBE message among the workers. This solution is inspired by the distributed deadlock detection in [10]. The coordinator keeps track, in the *nextprobe* variable, of the next worker to send the probe to. The idea is that workers with an identity strictly smaller than *nextprobe* are known to be blocked, i.e., they are waiting for an incoming message in line 4 of the algorithm in figure 3. Hence, when the coordinator sends a state to a worker with an identity which is smaller than the current *nextprobe*, this worker will be become active and thus the coordinator decreases the value of *nextprobe* accordingly (see line 13 in figure 2). When the coordinator sends the probe to a worker (see line 20 in figure 2), it updates the *nextprobe* to be the next worker. Hence, if states are not given to workers with a lower identity before the probe is returned, the next worker will then be probed. When the coordinator receives the probe from the last worker (line 17 of the algorithm in figure 2), then all workers are known to be in the blocked state and hence the state space exploration has been completed.

The distributed state space exploration requires a hash function h_{ext} mapping states onto workers. Assuming that the workers are running on machines of equal computing power in terms of memory and CPU, then this function should achieve two goals. Firstly, it should distribute the states uniformly across the n workers. Secondly, it should ensure a certain degree of *locality*, i.e., to reduce the communication overhead we would like as many successors states of a given state to reside on the same machine. To some extent these are conflicting goals.

To achieve a degree of locality we can select a small subset of the places P_h of the CPN and let the hash function depend on the marking of these places. All transitions in the CPN model which do not have a place in P_h as input or output will hence not affect the hash value and when such a transition is enabled from a state M, the successor state M' will belong to the same machine as M. Intuitively, we can thus control the degree of locality by the size of the set P_h. On the other hand, the set of reachable markings of the set P_h must accommodate a preferably even distribution of states onto the n workers.

Another issue is the relationship between the external hash function h_{ext} and the internal hash function h_{int} used in the internal hash table on each of the workers. Here we assume that all workers will be using the same internal hash function. We need to ensure that these hash functions are independent, i.e., that the external hash function h_{ext} does not cause the internal hash function h_{int} to map the states stored on that worker into a very small set of values. One approach to avoid this is to let the internal hash function h_{int} depend on the marking of places in $P \setminus P_h$. It should however be mentioned that the marking of places in $P \setminus P_h$ could be related via place invariants to the marking of places in P_h, and hence the markings may not be totally independent. The issue of the external hash function will be investigated further in section 4.

4 An Example

We now present a set of initial results obtained with the basic state exploration algorithm presented in the previous sections. We consider the distributed database system from [14] in figure 4. The CPN model describes the communication between a set of database managers $D = \{d_1, d_2, \ldots, d_N\}$ for maintaining consistent copies of a database in a distributed system. The idea of the protocol is that when a database manager updates its local copy of the database, requests are sent to the other database managers for updating their copy of the database. When each database manager has updated its copy, it sends an acknowledgement back to the initiating database manager to confirm that the update has now been performed. A database manager in the protocol can either be in a state Waiting

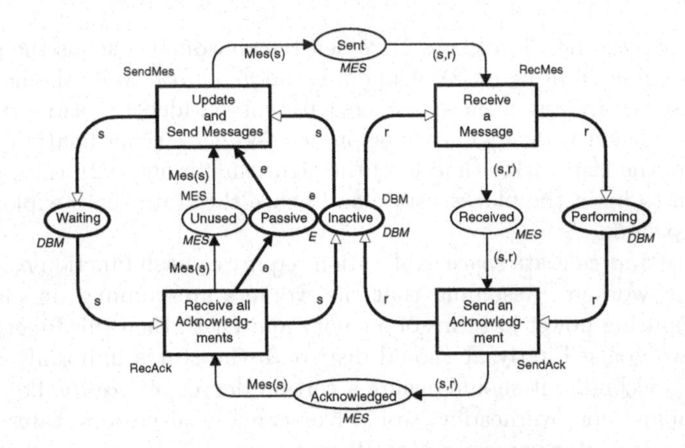

Fig. 4. The distributed database example.

(for acknowledgement), Performing (an update requested by another database manager) or Inactive after having sent the acknowledgement back. All database managers are initially Inactive.

Table 1 gives some statistics for the distributed data base system where we have used 1–10 workers. For these experiments, the external hash function h_{ext} is based on the standard internal hash function h_{int} used by Design/CPN:

$$h_{\text{ext}}(M) = (h_{\text{int}}(M) \bmod n) + 1 \tag{1}$$

The $|DBM|$ column gives the number of database managers considered and the n column specifies the number of workers. *States* gives the total number of states in the state space and *Time* specifies the total time used for the exploration as measured in the coordinator process. The exploration time is written on the form *mm:ss* where *mm* is minutes and *ss* is seconds. The *Transmit* column specifies the total number of states transmitted. *External* indicates the number of successor states computed that were external while *Internal* gives the number of successor states computed that were internal. This is measured across all the workers. Column *Stored* gives the total number of states received by the worker which were already stored by the worker when received.

Table 2 gives detailed statistics for the workers. The $|DBM|$ column gives the number of managers considered and the n column specifies the number of workers. For each worker $W1$–$W10$, we list the number of states stored on the worker and in Table 3 the total time that the worker spent in the blocking state.

To reduce the number of states transmitted, a small cache is introduced on each process. This cache contains a set of recently sent states and is consulted before transmitting a state. The best results obtained in our experiments with the database example uses a cache with 200 states. Hence, this configuration is used hereafter.

Until now we have used the internal hash function in each worker as a basis for the external hash function. For the DBM system it is however possible to

Table 1. Initial experimental results for the database system.

|DBM|	n	States	Time	Transmit	External	Internal	Stored
10	1	196,831	112:49	1	0	1,181,000	0
10	2	196,831	58:25	590,501	590,500	590,500	585,380
10	3	196,831	63:00	1,180,981	1,180,980	20	984,160
10	4	196,831	83:03	1,180,991	1,180,990	10	984,160
10	5	196,831	108:59	1,181,001	1,181,000	0	984,170
10	6	196,831	47:38	1,180,991	1,180,990	10	984,160
10	7	196,831	40:48	590,511	590,510	590,490	492,090
10	8	196,831	27:12	1,180,991	1,180,990	10	984,160
10	9	196,831	57:08	1,181,001	1,181,000	0	984,170
10	10	196,831	77:07	1,181,001	1,181,000	0	984,170

Table 2. Number of nodes on workers for the database system.

|DBM|	n	W1	W2	W3	W4	W5	W6	W7	W8	W9	W10
10	1	196,831									
10	2	98,410	98,421								
10	3	65,610	65,610	65,611							
10	4	49,210	49,211	49,200	49,210						
10	5	40,330	47,521	25,280	53,940	29,760					
10	6	31,950	33,660	34,510	33,660	31,950	31,101				
10	7	6,720	46,090	20,160	23,221	40,320	6,560	53,760			
10	8	24,130	24,240	24,600	24,240	25,080	24,971	24,600	24,970		
10	9	12,960	29,160	17,101	23,490	23,490	17,100	29,160	12,960	31,410	
10	10	20,160	23,761	12,640	26,970	14,880	20,170	23,760	12,640	26,970	14,880

come up with a tailored hash function based on the observation that except for the initial marking only one of the database managers is active at a time. If we let $Waiting(M)$ denote the database manager which is waiting (if any) in the state M and 1 if there is none, then a possible external hash function for the DBM system would be:

$$h_{\mathsf{ext}}(M) = (Waiting(M) \bmod N) + 1 \tag{2}$$

With this hash function it only makes sense to have at most N workers. The experiments conducted with 10 database managers and a cache size of 200 states are presented in Table 4. The hash function splits the state space into 9 sets of 19,683 states each, plus one containing also the initial marking (19,684 states). These sets are evenly distributed among the computation nodes. In these experiments 18 of the new states computed are external while the 1,180,982 others are internal. Hence, there is very little communication between the processes and they can perform their local computations without waiting for other workers to provide new states to handle. The results obtained are thus considerably better than with the original hash function. For 9 and 8 workers, some computation nodes are handling 2 groups of states, which explains the longer computation time, and the longer blocking time for the other workers.

Table 3. Waiting time of workers for the database system.

|DBM|	n	W1	W2	W3	W4	W5	W6	W7	W8	W9	W10
10	1	0:00									
10	2	4:22	8:33								
10	3	23:43	23:38	18:40							
10	4	43:52	21:57	29:08	55:24						
10	5	27:22	40:50	80:46	64:01	99:50					
10	6	13:30	10:34	10:15	15:49	33:46	32:39				
10	7	40:03	7:10	35:10	35:12	25:10	40:16	1:07			
10	8	12:43	5:36	7:20	15:07	7:11	11:57	6:57	10:44		
10	9	54:21	37:40	51:23	48:07	49:24	52:04	39:44	54:23	9:53	
10	10	42:23	51:36	67:18	60:19	73:56	52:49	51:33	68:36	14:01	64:17

Table 4. Experimental results using the tailored hash function.

n	Time	Block. Time
10	3:39	2:14
9	5:48	4:22
8	5:58	4:33

5 Conclusions and Future Work

We have described an approach to conducting distributed state space exploration for CPNs and presented some experimental results obatined with an implementation of our approach within Design/CPN. The experimental results are encouraging and indicate that the art of distributed state space exploration is in the choice of a good external hash function mapping states onto workers. In the general case, we would like to automatically derive an external hash function without relying on the user knowledge of the system to specify a good hash function. Hence, we need a way of determining a set of places that can be used to obtain a good hash function. A possible approach to this is to conduct an initial partial state space exploration (depth-first and breadth-first) and from the markings encountered attempt to derive a hash function. This hash function can then be based on counting the tokens on a selected set of places, or eventually, when dealing with places having simple colour sets such as integers or enumerations, on the rank of the colours in the place. Future work includes exploring more elaborated approaches to perform model-checking in our distributed framework. A possible starting point for this work would be [13, 15].

References

1. J. Barnat, L. Brim, and J. Stříbrná. Distrubuted LTL model checking in SPIN. In *Proc. of SPIN 2001*, volume 2057 of *LNCS*, pages 200–216. Springer-Verlag, 2001.
2. G. Behrmann. A Performance Study of Distributed Timed Automata Reachability Analysis. In Lubos Brim and Orna Grumberg, editors, *Electronic Notes in Theoretical Computer Science*, volume 68. Elsevier, 2002.

3. G. Behrmann, T. Hune, and F. Vaandrager. Distributed Timed Model Checking - How the Search Order Matters. In *Proc. of CAV'00*, volume 1855 of *LNCS*, pages 216–231. Springer-Verlag, 2000.
4. B. Bérard, M. Bidoit, A. Finkel, F. Laroussinie, A. Petit, L. Petrucci, and Ph. Schnoebelen. *Systems and Software Verification. Model-Checking Techniques and Tools.* Springer-Verlag, 2001.
5. L. Brim and O. Grumberg, editors. *Proc. of 1st Workshop on Parallel and Distributed Model Checking*, volume 68 of *Electronic Notes in Theoretical Computer Science*, October 2002.
6. L. Brim and O. Grumberg, editors. *Proc. of 2nd Workshop on Parallel and Distributed Model Checking*, volume 89 of *Electronic Notes in Theoretical Computer Science*, September 2003.
7. E. Clarke, O. Grumberg, and D. Peled. *Model Checking.* The MIT Press, 1999.
8. The CPN Tools Homepage. `http://www.daimi.au.dk/CPNtools`.
9. The Design/CPN Homepage. `http://www.daimi.au.dk/designCPN`.
10. E.W. Dijkstra. Derivation of a Termination Detection Algorithm for Distributed Computations. *Information Processing Letters*, 16:217–219, 1983.
11. G. Farret and G. Carré. Distributed Methods for Computation and Analysis of a Coloured Petri Net State Space. Master's thesis, Department of Computer Science, University of Aarhus, August 2002.
12. G. Gallasch and L. M. Kristensen. Comms/CPN: A Communication Infrastructure for External Communication with Design/CPN. In *Proc. of the 3rd Workshop on Practical Use of Coloured Petri Nets and the CPN Tools (CPN'01)*, pages 79–93. Department of Computer Science, University of Aarhus, 2001. DAIMI PB-554, ISSN 0105-8517.
13. H. Garavel, R. Mateescu, and I. Smarandache. Parallel State Space Construction for Model-Checking. In *Proc. of SPIN 2001*, volume 2057 of *LNCS*, pages 217–234. Springer-Verlag, 2001.
14. K. Jensen. *Coloured Petri Nets - Basic Concepts, Analysis Methods and Practical Use. - Volume 1: Basic Concepts.* Springer-Verlag, 1992.
15. F. Lerda and R. Sisto. Distributed-Memory Model-Checking with SPIN. In *Proc. of SPIN 1999*, volume 1680 of *LNCS*, pages 22–39. Springer-Verlag, 1999.
16. U. Stern and D.L. Dill. Parallelizing the Murϕ Verifier. In *Prooceedings of CAV'97*, volume 1254 of *LNCS*, pages 256–278. Springer-Verlag, 1997.
17. A. Valmari. The State Explosion Problem. In *Lectures on Petri Nets I: Basic Models*, volume 1491 of *LNCS*, pages 429–528. Springer-Verlag, 1998.

An Extensible Editor and Simulation Engine for Petri Nets: RENEW

Olaf Kummer, Frank Wienberg, Michael Duvigneau, Jörn Schumacher, Michael Köhler, Daniel Moldt, Heiko Rölke, and Rüdiger Valk

University of Hamburg, Department of Computer Science
Vogt-Kölln-Str. 30, D-22527 Hamburg,
{kummer,wienberg,duvigneau,6schumac,koehler,moldt,roelke,valk}
@informatik.uni-hamburg.de

Abstract. Renew is a computer tool that supports the development and execution of object-oriented Petri nets, which include net instances, synchronous channels, and seamless Java integration for easy modelling. Renew is available free of charge including the Java source code.
Due to the growing application area more and more requirements had to be fulfilled by the tool set. Therefore, the architecture of the tool has been refactored to gain more flexibility. Now new features allow for plug-ins on the level of concepts (net formalisms) and on the level of applications (e.g. workflow or agents).

Keywords: Reference nets, RENEW, plug-in, architecture, high-level Petri nets, nets-within-nets, tool, integrated development environment

1 Introduction

Petri nets are a well established means to describe concurrent systems. Object-oriented analysis and programming techniques are currently the de-facto standard of software development. This has lead to the invention of a variety of object-oriented Petri net formalisms. One of the most widely used object-oriented languages is Java, which features a relatively clean language design, good portability, and a powerful set of system libraries.

RENEW [13] is a Java-based high-level Petri net simulator that provides a flexible modelling approach based on reference nets. The publicly available version 1.6 has undergone several improvements and contains many features. However, different application areas require different functionality of the tool set. Therefore, we made a major redesign of the tool which will be called RENEW 2.0 or for short RENEW in the following and is the topic of this paper. Before it is released, RENEW is undergoing a final comprehensive test in a larger project to ensure the same high quality as the previous versions.

RENEW is an integrated environment that gives the user access to all required tools including an editor. Underlying design principles of the editor are: *easy to use* interface, minimal input for the user, direct relation to the functionality and provision of a high-level formalism. Fig. 1 shows a screen shot where the main tool bars and a net drawing are visible.

J. Cortadella and W. Reisig (Eds.): ICATPN 2004, LNCS 3099, pp. 484–493, 2004.
© Springer-Verlag Berlin Heidelberg 2004

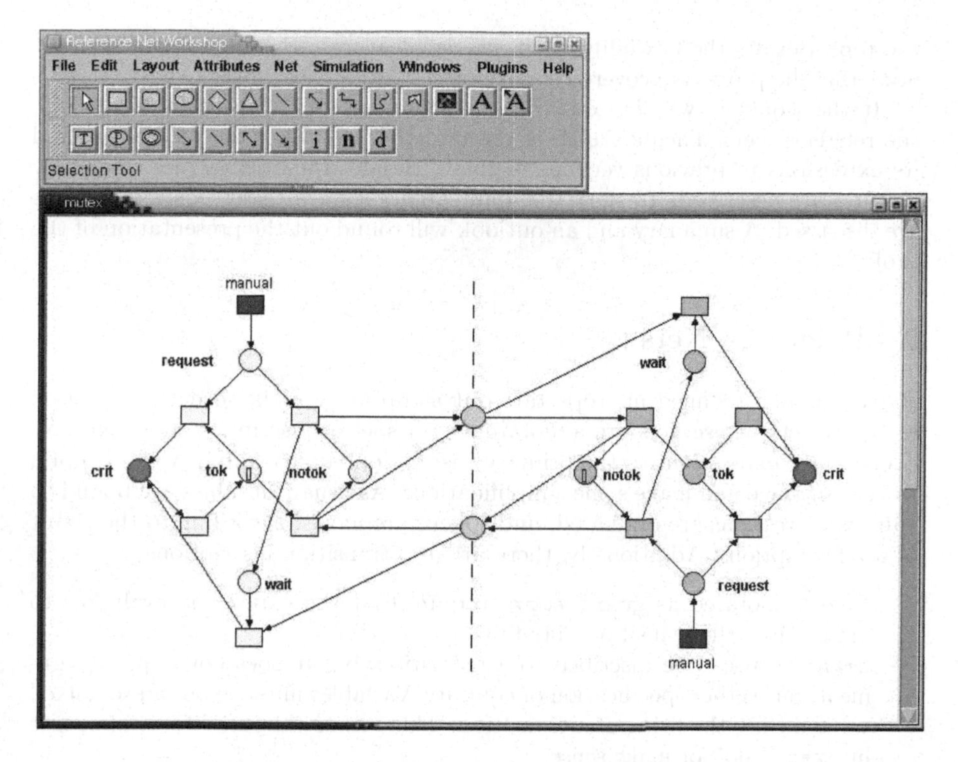

Fig. 1. A screen shot of the RENEW application

The buttons in the visible tool bars mostly deal with the graphical creation of net diagrams. They help the user to create and layout new nets and to illustrate them with additional graphics, as well as to edit existing nets. The creation of new nets is supported by functions that allow the easy creation of new places, transitions, arcs, and textual inscriptions. With a simple mouse drag you can create a new node *and* an arc that connects it to an existing node.

Besides the graphical editor, the tool allows for the use of Petri nets in server-side processes without the graphically animated token game. Based on customised compilers different formalisms were supported in previous versions of RENEW. Several extensions of the Petri net formalism had been integrated that included clear arcs, flexible arcs and inhibitor arcs. The expressiveness of timed Petri nets, where time stamps are attached to tokens and to input and output arcs, was provided by the tool, too.

Our goal for the new Version 2.0 is to add flexibility and extensibility as a key feature at the level of supported formalisms and at the user level to the tool. The desire for more flexibility comes from the different applications of RENEW. Especially its use in the areas of workflow and agent systems required various extensions to the tool set. The major refactoring of RENEW (see [14]) allows such extensions to be plugged in and out of the application without overloading the tool. Besides the flexibility itself, no new features are added to the tool for now. But the process uncovered some existing hidden features.

In the sequel we will first describe the characteristics of the main formalism, the reference nets. Then a sketch of the underlying architecture of the tool and its extensions to previous versions is given. Besides the architecture some new features are described. To show the applicability some already existing plug-ins are discussed. A summary and an outlook will round out the presentation of the tool.

2 Reference Nets

Reference nets (defined in [10]) start out as ordinary higher-order nets, based on Petri nets whose arcs are annotated by a special inscription language. We choose Java expressions as the primary inscription language, but we add tuples to the language and make some simplifications. As usual, variables are bound to values, expressions are evaluated, and tokens are moved according to the result of arc inscriptions. Additionally, there are also transition inscriptions.

- Guards, notated as `guard` *expr*, require that the expression evaluates to `true` before the transition may fire.
- *expr*=*expr* can be inscribed to a transition, but it does not imply assignment, but rather specification of equality. Variables must be bound to a fixed value during the firing of a transition. This means that modify assignments like `x=x+1` do not make sense.
- Java expressions might be evaluated, even when it turns out that the transition is not enabled and cannot fire. This causes problems for some Java method calls, therefore the notation `action` *expr* is provided. It guarantees that *expr* will be evaluated exactly once, a feature that is needed when side effects (e.g. changes to Java objects) come into play.

When a net is constructed, it is merely a net template without any marking. But it is then possible to create a net instance from the template. In fact, an arbitrary number of net instances can be created dynamically during a simulation. Only net instances, not the templates, have got a marking that can change over time.

- A net instance is created by a transition inscription of the form *var*`:new` *netname*. It means that the variable *var* will be assigned a new net instance of the template *netname*. The net name must be uniquely chosen for each template that we specify.

It should be noted that RENEW supports the concepts of *Nets-within-Nets* (see [16]) which is a major research topic of our group. In order to exchange information between different net instances, synchronous channels were implemented. They provide greater expressiveness compared to message passing. Unlike the synchronous channels from [2], which are completely symmetric, we will impose a direction of invocation. The invoking side of a channel will be known as the downlink, the invoked side is called the uplink.

- An uplink is specified as a transition inscription *:channelname(expr,...)*. It provides a name for the channel and an arbitrary number of parameter expressions.
- A downlink looks like *netexpr:channelname(expr,...)* where *netexpr* is an expression that must evaluate to a net reference. The syntactic difference reflects the semantic difference that the invoked object must be known before the synchronisation starts.

To fire a transition that has a downlink, the referenced net instance must provide an uplink with the same name and parameter count and it must be possible to bind the variables suitably so that the channel expressions evaluate to the same values on both sides. The transitions can then fire simultaneously. Note that channels are bidirectional for all parameters except the downlink's *netexpr*. E.g., a net might pass a value through the first parameter of a downlink and the called net might return a result through the second parameter of the same channel. This is similar to the undirected parameters of Prolog predicates, but different from the invocations of Cooperative Nets by Sibertin-Blanc (see [15]).

A transition may have an arbitrary number of downlinks, but at most one uplink. (Again, the similarity with horn clauses in Prolog does not occur by chance.) A transition without uplinks will be called a spontaneous transition, because it may fire without being invoked by another transition. A transition may have an uplink and downlinks at the same time. A transition may synchronise multiple times with the same net and even with the same transition.

3 Architecture and Extensibility

In the last year, the tool has undergone a large architectural refactoring. The application has been decomposed into several components, each component is seen as a "plug-in". The intention of this decomposition is an increased flexibility and extensibility of the tool so that new features can be added in an easy way. However, in the first step, the architectural changes do not add new features to the tool. But some existing features that were hidden in the system can now be accessed more easily.

All components of the application are now managed by a central plug-in system. The system allows for addition and removal of plug-ins at runtime. Many plug-ins provide extension interfaces where additional features of other plug-ins can be registered.

The basic functionality of RENEW is provided by the plug-ins depicted in Fig. 2. To run a simulation of some reference nets without graphical feedback, the plug-ins Util, Core, Simulator and Formalism are needed. The configuration of the simulation can then be done by setting several properties on the command line.

The main component is the Simulator plug-in. It comprises the simulation engine and some packages providing in- and output abstractions for the simulation. Nets can be fed to the simulation engine by using the so-called `shadow`-API, which abstracts from the layout information and retains only the topological net

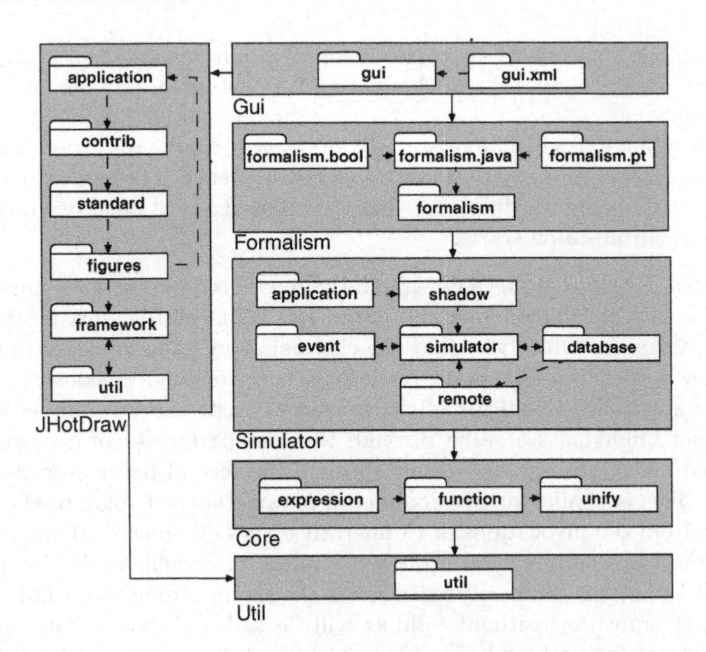

Fig. 2. Basic plug-ins of RENEW: packages and dependencies (based on [14, p. 91])

structure. This provides a convenient way to create non-graphical nets algorithmically.

Other packages provide interfaces to observe and influence the state of a running simulation. Also persistent storage of the simulation state in a relational database engine like mySQL or Oracle is provided, so that the simulation can continue from the last consistent state after a power failure (see [7]).

If graphical feedback is desired, the plug-ins Gui and JHotDraw can be added. This can happen while a simulation is running, the user interface will attach itself to the running simulation engine. It is also possible to quit the user interface without terminating the simulation.

The simulation engine can be enabled for remote access, so that the graphical user interfaces of independent RENEW instances on other hosts can inspect and influence the simulation state. This separation of simulation engine and user interface has already been introduced in RENEW 1.5, but now its use has become easier. If a simulation is started from within the user interface, a configuration dialog (see Fig. 3) provides a convenient way to set the properties of the next simulation run.

In the design of RENEW it has been clear from the very beginning that the tool must be able to handle different formalisms by just exchanging a compiler component. A compiler converts a shadow net into an internal format by parsing the textual net inscriptions. A custom compiler can reuse the existing classes of the Core plug-in which provides predefined building blocks for expressions, places, arcs, and so on. The compiler just needs to compose the compiled

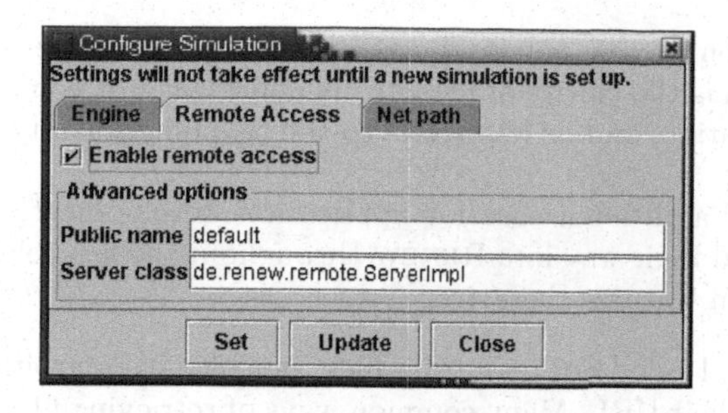

Fig. 3. Simulation configuration dialog

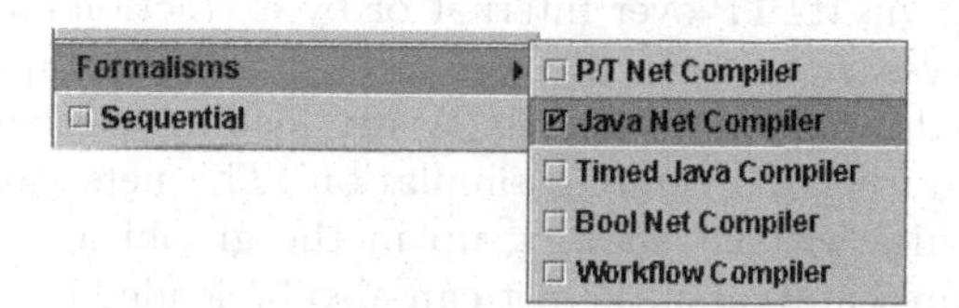

Fig. 4. Compiler selection menu

nets out of these classes. Custom classes are only needed for non-standard net components.

Some exemplary formalisms existed in previous versions of the tool, but without much help for their configuration. Now the Formalism plug-in provides a registry for known compilers. All registered compilers are presented to the user by a menu in the user interface, if the Gui plug-in is loaded (see Fig. 4). As a formalism can be accompanied by new graphical elements to be used in net drawings, the plug-in can add tool bars or menu entries to the application's user interface. New formalisms can be easily included in the system by creating a new plug-in that comprises the compiler and related classes.

A refactoring of the simulation engine has clarified the separation of formalisms (compilers), common Petri net abstractions (like transitions and places), and the core engine, which in fact is just a generic binding search algorithm (see [10]). With some changes to the user interface, we will soon have a real multi-formalism tool that supports a different formalism for each net template within one simulation. This means that some parts of a system can be modelled as simple P/T-nets and some parts as coloured Petri nets.

4 Development Support

RENEW is used in our group to develop and run medium-sized applications implemented in a mixture of reference nets and Java code (e.g. roundabout 100 nets and the same number of Java classes). The mix of reference nets and Java is well supported because any Java object can be used as a token in nets and because any net can be wrapped by a so-called "stub" to make it appear as a

Java class. When mixing nets and classic code, the developer can benefit from both sides: he has the clarity of nets at his hand, when it comes to concurrency and synchronisation, and he has access to the rich functionality of the Java class libraries.

Pursuant of additional ease for application development, many other features have found their way into RENEW throughout the last releases. Among the developer-driven features there are:

- Nets can be loaded not only from files, but also from arbitrary locations by specifying their URL. Many common ways of retrieving files can be used this way, as long as the Java runtime system knows a way to access the specified resource (e.g. via HTTP over Internet or by extraction out of `.jar` files).
- It is not necessary to open all nets belonging to an application prior to the start of the simulation engine. A net loading mechanism fetches nets on demand during the running simulation. The nets can be loaded from shadow net files without showing up in the graphical user interface. The graphical representation of the net can also be loaded on demand when the user inspects a net instance during the simulation.
- An implementation of last year's PNML-standard draft (*P*etri *N*et *M*arkup *L*anguage, see [8]) is provided, allowing import and export of nets from or to other tools.
- Breakpoints can be attached to transitions or places, causing the simulation to halt when a transition fires or the marking of a place changes.
- Transitions can be marked as "manual". The simulation engine will not fire these transitions during automatic runs. But such a transition can be fired explicitly from the graphical user interface at any time (given it is activated).
- The interactive debugging of complex mixed systems using nets and Java code is supported by an in-depth inspection of Java token objects.

Of course, the database backing and remote access features mentioned in the previous section also have their origins in application development. All these features are of great value for our existing plug-ins, like those presented in the next section.

5 Plug-in Examples

This section describes some current plug-ins that have been built for RENEW. The plug-ins presented in the following are not all available to the general public, some being experimental or highly application specific. But they can serve as examples for the extensibility of the tool.

Workflow. Workflow support requires a workflow management system. A standard has been defined by the WfMC (Workflow management coalition, see [17]). The Workflow plug-in extends RENEW in two ways: It uses an interface provided by the engine to monitor and control the activation of transitions in general, and it provides a specialised workflow compiler to the Formalism plug-in. Some sketches of the resulting workflow engine have been given in [6].

With the plug-in, RENEW can serve as a development environment and execution engine for workflow systems, where the firing of transitions is coupled with the execution of workflow tasks. Overall, the suitability of RENEW for workflow support and the tool's extensibility is demonstrated by the Workflow plug-in.

Multi-agent Systems. Since multi-agent systems (MAS) are inherently concurrent, Petri nets recommend themselves to be used in that area. In our group, we have designed a MAS architecture called MULAN (short for *Mul*ti *A*gent *N*ets, see [9]) with reference nets. This architecture is implemented with a mixture of nets and Java code by the CAPA plug-in (*C*oncurrent *A*gent *P*latform *A*rchitecture, presented in [4]). This plug-in mostly runs within the RENEW simulation engine, but in addition provides customised graphical representations for special token objects.

The MulanViewer plug-in reflects the high potential of the whole architecture. It enhances the user interface of RENEW by an overview window that summarises the state of certain places in certain nets of the Capa plug-in. The viewer also provides fast navigation through the important net instances of the MAS to inspect the token game.

A different kind of plug-in is the Diagram plug-in: It extends the editor component of RENEW by a special mode to draw Agent Interaction Protocols (AIPs) as they are proposed by AUML (see [12]). These protocol diagrams are not Petri nets, but the plug-in prototype can generate net structures reflecting the control flow corresponding to the AIP for each participating agent.

Modelling Support. The task of modelling benefits enormously from elaborated tool support. Possible features are the direct input of model elements or flexible menu bars that adapt to the current task. The menus of the former RENEW were static and could only be extended by releasing a new version of the tool. By making the menu bar flexible, we can add any tool bar the installed plug-ins provide.

An example is the NetComponents plug-in, as presented in [1]. It allows to add new tool bars to the user interface at runtime, where each button creates a small component of net structure that is often used during a development process. Different component sets are available, depending on the application area: the tool bar from [1] comprises patterns for creating MULAN/CAPA protocol nets. A different tool bar covers some workflow patterns, as presented in [11].

6 Conclusions

RENEW, based on its new architecture, provides all major features of the old versions up to now. This is the result of a refactoring process. We concentrated on the aspect of flexibility to provide a better tool set to the RENEW-users without introducing new features besides the plug-in mechanism. However, as a result of the new architecture as a side effect the tool can now easily be extended based on the plug-in concept.

The underlying algorithms and the support of concurrency, which is even present inside the tool itself, has not been tackled in this paper. However, the basic architecture and important newly added features with respect to usability have been presented.

The plug-in architecture now allows for the easy use of multiple formalisms and net variants. Basis of this is the clean separation between the underlying simulation engine and the formalisms on top. The former versions allowed the use of customised compilers that had to be specified when starting the tool. Now the refactored tool allows for an easy formalism setup between simulation runs. In the near future, we will be able to combine different formalisms within one simulation environment.

The multi-formalism simulation can be useful within our agent-oriented modelling approach, where we consider open systems. Each agent can come along to an agent platform, which we consider to be a RENEW-simulator, with its own formalism. This depends on the level of abstraction when building the models. The idea is to use the weakest formalism (with respect to the expressibility) to reach a certain task during modelling. In combination with the support for the PNML-standard this allows the modeller to exchange the nets with other tools adequate for the chosen formalisms. Especially in the area of verification a weaker formalism commonly allows the use of more powerful tools.

An important outcome of the work done for the new architecture is that we are able to build a powerful agent development and simulation tool set. The integration into the FIPA-standard environment (see [5]) has been functionally reached with [4], followed by an integration into the Agentcities context (see [3]). The functionality built into plug-ins will enhance the flexibility of our tool set to become a full agent execution environment which is autonomous, open, concurrent, adaptive and mobile.

In the future the number of plug-ins will increase. Any programmer may build those plug-ins that support his own needs in the directions of a development environment that is based on high-level Petri nets, as we do. Furthermore, others might realize that they need their own special plug-in to support a certain net formalism or even other formalisms. They can then easily add their special drawing tools and compilers. The main restriction is that they have to stick to our architecture and the interfaces provided so far. Due to the underlying design principles the coupling, however, will be loose.

References

1. L. Cabac, D. Moldt, and H. Rölke. A Proposal for Structuring Petri Net-Based Agent Interaction Protocols. In W. v.d. Aalst and E. Best, editors, *Applications and Theory of Petri Nets. ICATPN 2003, Eindhoven. Proceedings*, number 2679 in LNCS, pages 102–120. Springer, 2003.

2. S. Christensen and N. Damgaard Hansen. Coloured Petri Nets Extended with Channels for Synchronous Communication. Report DAIMI PB–390, Aarhus University, 1992.
3. M. Duvigneau, M. Köhler, D. Moldt, C. Reese, and H. Rölke. Agent-based Settler Game. In *Agentcities Agent Technology Competition, Barcelona, Spain. Proceedings*, February 2003.
4. M. Duvigneau, D. Moldt, and H. Rölke. Concurrent Architecture for a Multi-agent Platform. In F. Giunchiglia, J. Odell, and G. Weiss, editors, *Agent-Oriented Software Engineering III. Workshop, AOSE 2002, Bologna.*, number 1420 in LNCS, pages 59–72. Springer, 2003.
5. Foundation for Intelligent Physical Agents (FIPA). `http://www.fipa.org`.
6. T. Jacob, O. Kummer, D. Moldt, and U. Ultes-Nitsche. Implementation of Workflow Systems using Reference Nets – Security and Operability Aspects. In K. Jensen, editor, *4th Workshop and Tutorial on Practical Use of Coloured Petri Nets and the CPN Tools.* University of Aarhus, 2002. Report DAIMI PB-560.
7. T. Jacob, O. Kummer, and D. Moldt. Persistent Petri Net Execution. *Petri Net Newsletter*, 61:18–26, October 2001.
8. J. Billington et. al. The Petri Net Markup Language: Concepts, Technology, and Tools. In W. v.d. Aalst and E. Best, editors, *Applications and Theory of Petri Nets. ICATPN 2003, Eindhoven. Proceedings*, number 2679 in LNCS, pages 483–505. Springer, 2003.
9. M. Köhler, D. Moldt, and H. Rölke. Modelling the Structure and Behaviour of Petri Net Agents. In J.-M. Colom and M. Koutny, editors, *Application and Theory of Petri Nets. ICATPN 2001, Newcastle upon Tyne. Proceedings*, number 2075 in LNCS, pages 224–241. Springer, 2001.
10. O. Kummer. *Referenznetze.* Logos-Verlag, Berlin, 2002.
11. D. Moldt and H. Rölke. Pattern Based Workflow Design Using Reference Nets. In W. v.d. Aalst, A. ter Hofstede, and M. Weske, editors, *Business Process Management. BPM 2003, Eindhoven. Proceedings*, number 2678 in LNCS, pages 246–260. Springer, 2003.
12. J. Odell, H. Van Dyke Parunak, and B. Bauer. Extending UML for Agents. In G. Wagner, Y. Lesperance, and E. Yu, editors, *Agent-Oriented Information Systems. Workshop at the 17th National Conference on Artificial Intelligence (AAAI), AOIS 2000, Austin. Proceedings*, pages 3–17, Austin, TX, 2000.
13. Renew – The Reference Net Workshop. WWW page at `http://renew.de/`. Contains the documentation for Renew and an introduction to reference nets.
14. J. Schumacher. Eine Plugin-Architektur für Renew. Konzepte, Methoden, Umsetzung. Diplomarbeit, Universität Hamburg, October 2003.
15. C. Sibertin-Blanc. Cooperative Nets. In R. Valette, editor, *Application and Theory of Petri Nets. ICATPN '94, Zaragoza. Proceedings*, volume 815 of *LNCS*, pages 471–490. Springer, 1994.
16. R. Valk. Petri nets as token objects: An introduction to elementary object nets. In J. Desel and M. Silva, editors, *Application and Theory of Petri Nets. ICATPN '98, Lisbon. Proceedings*, number 1420 in LNCS, pages 1–25. Springer, June 1998.
17. Workflow Management Coalition Homepage. URL: `http://www.wfmc.org/`, 2003.

Web Supported Enactment of Petri-Net Based Workflows with XRL/Flower

Alexander Norta

Eindhoven University of Technology, Faculty of Technology and Management,
Department of Information and Technology,
P.O. Box 513, NL-5600 MB, Eindhoven, The Netherlands
a.norta@tm.tue.nl

Abstract. This paper describes concepts and features of a Web-based system called XRL/flower for carrying out Petri-net based workflows described with XRL (eXchangeable Routing Language). XRL/flower uses XML technology and is implemented in Java on top of the Petri-net Kernel PNK. Standard XML tools can be deployed to parse, check, and handle XRL documents. The XRL enactment application is complemented with a Web server, allowing actors to interact with the system through the internet. A database allows the enactment engine and the Web server to exchange information with each other. Since XRL is instance based, a modelled workflow serves as a template that needs to be copied and may be possibly refined for enactment. For that purpose XRL constructs are automatically translated into Petri-net constructs. As a result, the system is easy to extend: For supporting a new control flow primitive, the engine itself does not need to change. Furthermore, the Petri net representation can be analyzed using state-of-the-art analysis techniques and tools.

1 Introduction

XRL/flower [25] is a system of software tools that is developed at the Department of Technology Management at the TU-Eindhoven. It is intended to serve as middleware between organizations for Web-based enactment of processes modelled with XRL (eXchangeable Routing Language). XRL [5, 19] is an instance-based workflow language that uses XML for the representation of process definitions and Petri nets for its semantics resulting in an unambiguous understanding of XRL.

Carefully extracted control flow patterns [3, 4, 13, 14] are contained in the definition of XRL resulting in superior control flow expressive power. As shown in [6], the semantics of XRL is expressed in terms of *Work-Flow nets* (WF-nets), which permits the use of theoretical results and standard tools such as Woflan [2, 24] for checking the appealing notion of soundness. It should be noted that a WF-net specifies the dynamic behavior of a single case in isolation. An informal description of WF-nets and soundness is given in the following paragraphs.

A WF-net is a special subclass of Petri nets that has one input place (i) and one output place (o) as any case handled by the procedure represented by the WF-net is created when it enters the workflow management system and is deleted once it is completely handled by the workflow management system, i.e., the WF-net specifies the life-cycle

J. Cortadella and W. Reisig (Eds.): ICATPN 2004, LNCS 3099, pp. 494–503, 2004.
© Springer-Verlag Berlin Heidelberg 2004

of a case. Furthermore, there may be no 'dangling tasks and/or conditions', i.e., tasks and conditions that do not contribute to the processing of cases.

Additionally, the requirement should be verified that for any case, the procedure will eventually terminate and the moment the procedure terminates there is a token in place o and all other places are empty. Moreover, there should be no dead tasks, i.e., it should be possible to execute an arbitrary task by following the appropriate route through the WF-net. These two additional requirements correspond to the so-called *soundness property* [1].

The Petri-net semantics of XRL is realized by mapping to PNML [15, 17, 26], an XML-based interchange format that permits the definition of *Petri-net types*. For that purpose a stylesheet translator is employed that contains mapping rules to PNML for every XRL control-flow construct that are described in [6].

The resulting WF-net represented in PNML can be loaded into the so-called Petri-net kernel PNK [16, 18], which provides an infrastructure offering methods for the administration and modification of Petri nets. The PNK supports a quick, modular, object oriented implementation and integration of Petri-net algorithms and applications. A simple interface allows access to basic net information.

The development of XRL and subsequently XRL/flower can be seen as a reaction to several XML-based standards for business process modelling that have emerged in recent years. Some examples of relevant past and present acronyms are BPML [9], WSFL [12], WSCI [7], BPEL4WS [11], XPDL [10], XLANG [21], and so forth. However, they all share the commonality of lacking precise semantics, which can result in different interpretations of how to support the mentioned standards with enactment technology. In contrast, XRL is equipped with very clear Petri-net semantics. Differently to the mentioned XML-standards one can determine *before* enactment whether an XRL modelled workflow is sound or not. Such analysis power is crucial for avoiding the occurrence of abnormalities such as deadlocks during carrying out business transactions.

The remainder of this paper is structured as follows: Section 2 describes the toolset architecture of XRL/flower. Furthermore, the life cycle of XRL instances is depicted. Next, Section 3 shows how implemented modules of XRL/flower exchange information through a database whose model is presented. A conclusion follows in Section 4.

2 XRL/Flower Architecture

The workflow management system XRL/flower is the result of XRL features. Since XRL is based on both XML for syntax and Petri nets for semantics, standard XML tools can be deployed to parse, check, and handle XRL documents. The Petri-net representation allows for a straightforward implementation of the workflow enactment engine. XRL constructs are automatically transformed to Petri-net constructs. This allows for an efficient implementation and the system is easy to extend by employing an XSL translator for mapping routing elements to PNML. Thus, for supporting a new control flow primitive, only a transformation to the Petri-net format needs to be added and the engine itself does not need to change.

Figure 1 shows the toolset architecture of XRL/flower where grey shaded elements are largely implemented. Using both the control-flow data for the workflow case and

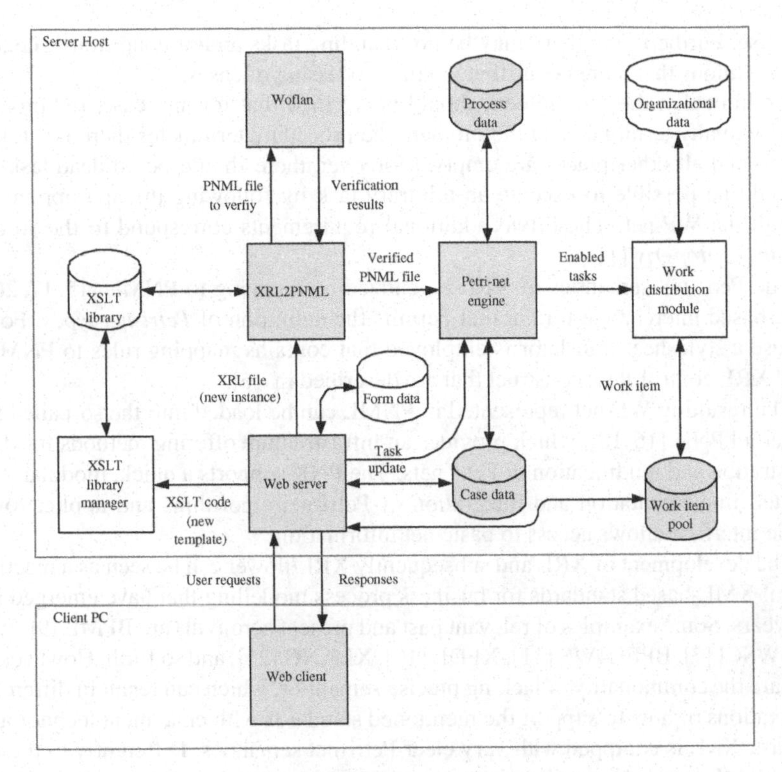

Fig. 1. XRL/flower architecture

case specific data, the Petri-net engine computes the set of enabled tasks, that is, the set of work items that are ready. The engine sends this set to the work distribution module. Based on information of organizational roles and actors, the work distribution module fills the work item pool. Resources that may carry out those ready worklist items can log into XRL/flower through the Web server. If the actor has registered beforehand, an online worklist manager displays the ready items in the Web client that are assigned by the work distribution module. By accepting a chosen worklist item, its content is displayed.

In order to enable an actor to perform an activity, the Web server fills the appropriate form template with case specific data for the activity. The Web server stores updated case data and signals the Petri-net engine the activity has been completed. The Petri-net engine then recomputes a new set of work items that are ready. The actor can also start an XRL instance by sending the corresponding XRL file to the Web server. The Web server forwards the XRL file to the XRL2PNML module that transforms XRL to PNML (Petri-Net Markup Language), which is a standard representation language for a Petri net in XML format [15].

The activity diagram depicted in Figure 2 shows the life cycle of an XRL modelled workflow. First, the XRL2PNML module performs a transformation from XRL to two PNML files: one for verification and one for enactment sharing similar soundness char-

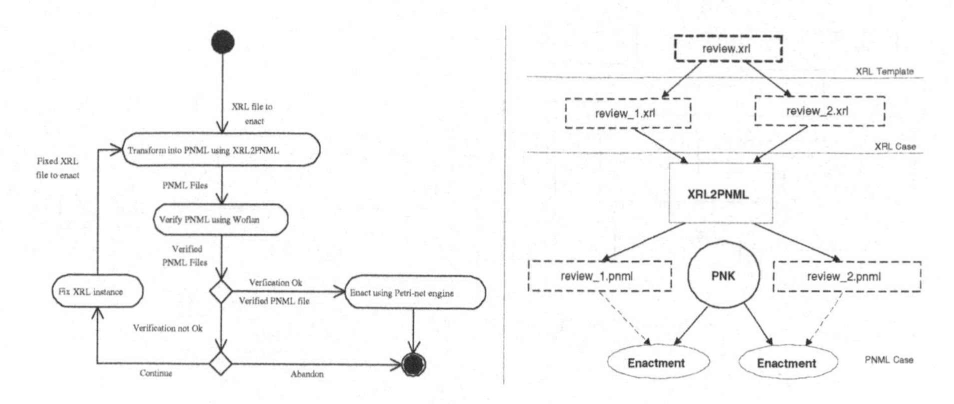

Fig. 2. XRL case life-cycle

acteristics. The first PNML file is verified using the Woflan tool. Based on the result either the second PNML file is sent to the Petri-net engine for enactment, or the actor is informed about the XRL instance containing flaws. In the latter case, the actor may either abandon the new instance, or modify it to fix the errors. Of course, the fixed instance is also verified before it is enacted.

The right hand visualization of Figure 2 serves to complete the understanding of XRL/ flower instance handling. Dashed rectangles represent workflow files, rectangles stand for XRL/flower modules, and circles depict component processes. An XRL workflow may serve as a template of which instance files are created. Every instance of the workflow is uniquely identifyable by an extra underscore and instance number at the end of its name. After the XRL2PNML module translates every instance to PNML and assuming soundness evaluation succeeds, instances are loaded into the Petri-net engine module. The latter consists of the Petri-net kernel PNK and an enactment application built on top of PNK. For every workflow instance that needs to be carried out, a new enactment application process is created. Thus, multiple instance enactments can be handled concurrently by XRL/flower's Petri-net enactment module.

3 Component Description

Several parts of the XRL/flower toolset in Figure 1 are grey shaded, which means they are largely implemented in the currently available prototype [20]. This section describes the existing functionality of tool modules depicted in Figure 1 and how they interact with each other.

A relevant preliminary for describing the component's way of interacting is the database of which a model is depicted in Figure 3. The central entity of the model is `WFCase` that contains attributes about every enacted case instance's unique identifier, starting and ending timestamp. A case file can contain variables with optional values present at case start time. Likewise documents can be contained in the case file in combination with a `uri`.

For attaining increased flexibility with respect to organizational fluctuation, a split between organizational roles and actors is sensible. Attributes can be attached to roles

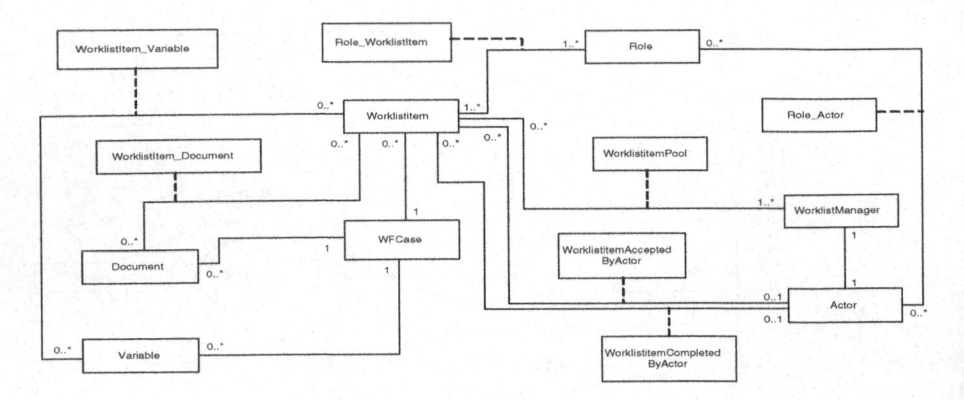

Fig. 3. Database model of XRL/flower

into which actors can slip. The database model in Figure 3 shows that an actor may slip into multiple roles and a role may be held by many actors. When an actor registers for the first time with XRL/flower, a worklistitem manager entry is inserted in the entity `WorklistManager`.

Ready worklist items are inserted with a start timestamp in the `Worklistitem` entity and a worklist-item-manager identifier is instantaneously added to the worklistitem entry. Variable and document links can be contained in a XRL task definition. Thus, further tuples need to be inserted for assigning the appropriate variables and documents to a particular worklist item entry. The database model of Figure 3 also shows entities for the acceptance of a worklist item and completion of an activity that are committed by an actor together with time stamps through the Web client.

3.1 XRL2PNML

The XRL2PNML module consist of a stylesheet translator containing mapping semantics from XRL to PNML described in [6]. As a result carefully extracted control-flow patterns contained in the XRL modelled workflow are converted into a WF-net represented in PNML format.

3.2 Woflan

The soundness property of WF-net mapped workflows can be verified with the analysis tool Woflan [22–24]. This way it is possible to eliminate modelling abnormalities, e.g. deadlocks, before workflow enactment, which leads to the avoidance of costly run-time failures.

3.3 Petri-Net Enactment Engine

The core of this module is the Petri-net kernel PNK on top of which an enactment application is implemented. PNML files representing case instances can be loaded into the PNK that translates workflow node tags into an object instance net. Next, the case name is detected in the PNML file and inserted to the `WFCase` entity. Parsers check for variable and document tags that are automatically inserted to the database together with their optionally present values.

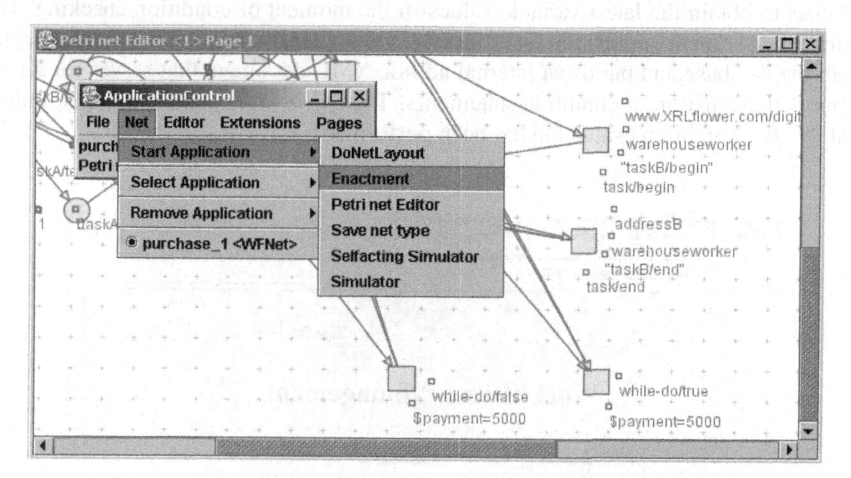

Fig. 4. Enactment application of the Petri-net enactment module

Since the PNML file is marked as a WF-net, the Petri-net module offers permitted applications in a pull-down menu that may be carried out. Figure 4 shows the enactment application selection in front of an editor displaying the PNK loaded workflow instance. After choosing the enactment application, a simple Petri-net firing rule can be started that allocates all enabled transitions and fires them randomly in a loop until no enabled transitions can be detected [25].

Several firing transitions carry labels that trigger particular enactment application methods. Tasks mapped from XRL to PNML have a starting and ending transition labelled `task/begin` and `task/end` respectively [6]. A `task/begin` label triggers the Petri-net engine module to look for the role allowed to perform the task. After allocating the set of concrete actors who fill that role, corresponding entries are inserted to the database of the actor's worklist managers. The `task/begin` labelled transition waits with firing until the Petri-net engine module reads from the database that the worklist item has been accepted by an actor. A `task/end` labelled transition means the enactment process is waiting for the actor to complete the task. In such a case the firing rule commits a task completion time stamp to the database.

Some XRL elements require condition defintions that are mapped to PNML as the editor in the background of Figure 4 shows with the example of `while-do/true` and `while-do/false` labelled transitions. Both labels are the PNML mapping result of the XRL `while-do` element [6] where the condition statement is mentioned once. In the first PNML-label case the labelled transition evaluates to true and in the latter case transition firing takes place when the condition is false. The Petri-net engine uses XPath [8] for condition evaluation.

The condition element of XRL is equally split into a condition/tbegin and condition/fbegin transitions [6] after XSL translation to PNML. Again condition/tbegin labelled transitions fire when the attached condition evaluates to true and condition/fbegin transitions need the condition to be false for firing.

Variables used in conditions can change during case enactment. Thus, the firing rule has to obtain the latest variable values in the moment of condition checking. The Petri-net enactment application reads all case variables and their corresponding values from the database and builds an internal ad hoc XML-document that serves for XPath supported evaluation of condition statements. This XML-document is dropped immediately after condition evaluation has been performed.

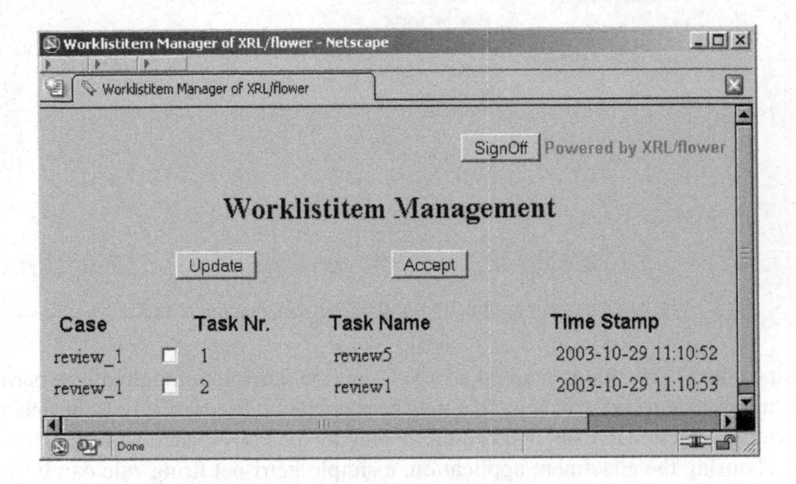

Fig. 5. Worklistitem manager created by the Web server

Finally, wait_all and wait_any of XRL are elements that either wait for the occurrence of all or any associated events respectively [6]. Both XRL elements wait for the event occurrences until a timeout has expired. Mapped to PNML, the Petri-net engine firing rule picks up the timeout/begin labelled transition's timestamp and waits till the system timestamp is later than the transition timestamp. Once the system timestamp is late enough the transition fires. When the workflow terminates, i.e., one token has reached the sink place and no tokens are left over in the Petri-net, the enactment application commits an end timestamp to the database entry of the WFCase.

3.4 Web Server/Web Client

An actor who wants to interact with the XRL/flower system must first log in through the Web client. If the actor can not be detected in the database, the Web server sends a form to the client for entering assorted data that is inserted through the Web server to the database.

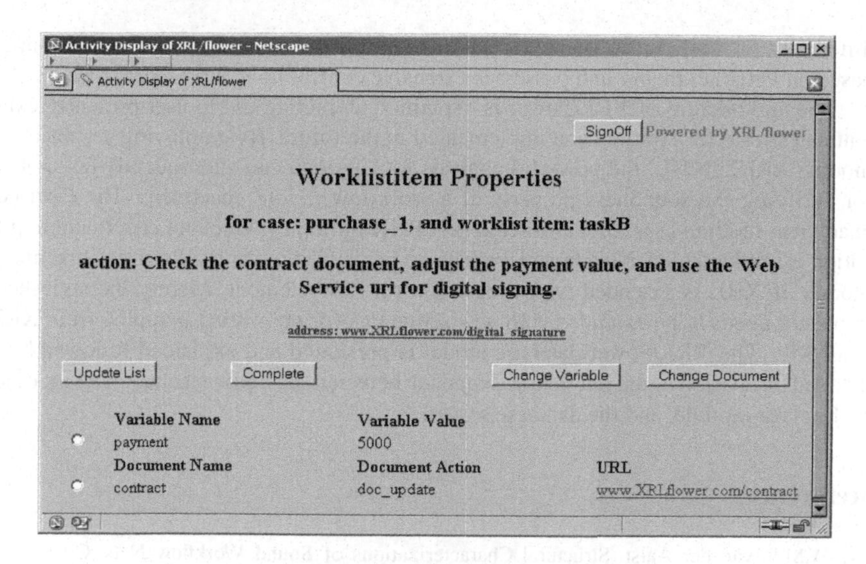

Fig. 6. Display of an activity

If the actor logging into XRL/flower can be detected in the database, the Web server checks in the `WorklistitemPool` entity for ready worklist items. The detected set is displayed in the Web client as can be seen in Figure 5. Clicking the `Accept` button triggers the Web server to perform a corresponding insertion in the database (see Figure 3).

In a next step the Web server reads all variable and document entries associated with the activity and displays them in the Web client for the actor. A chosen worklist item can be accepted through the Web interface, which results in a display of all activity properties as Figure 6 shows. An address attribute is displayed delivering a Web Service `uri` the actor can use for carrying out an activity operation. The Web client allows the actor to choose particular variables and documents for changing the delivered values and commits them to the database. Such changes are relevant for the evaluation result of XPath conditions used in the workflow case.

Figure 6 shows a `Complete` button the actor may click after having carried out the activity instruction. Such action commits a corresponding entry in the database containing a timestamp. As a result the Web server collects the new set of ready worklist items the Petri-net engine module has generated and inserted to the database and displays them again in the Web client. Once the actor has completed carrying out activities with XRL/flower support, clicking the `SignOff` button results in a log off.

4 Conclusion

This paper presents the Web-based workflow management system XRL/flower for enactment of XRL defined workflows. XRL significantly differentiates itself from emerging standards like BPEL4WS by being extensible and instance based. Furthermore, differently to XML-based standards XRL is equipped with clearly defined semantics based on Petri-net theory and possesses extensive control flow expressive power.

The architecture of XRL/flower is explained, depicting the toolset presently existent and still to be explored and implemented in the future. By employing a translation module XRL2PNML, the powerful analysis tool Woflan can subsequently be applied for verifying the soundness property of a workflow *before* enactment. The Petri-net enactment module uses the Petri-net kernel PNK on top of which an enactment application is implemented. No adaptations have to be carried out at the Petri-net enactment module if XRL is extended with a new control-flow element. Merely the stylesheet translator needs to be extended with new semantics for permitting mapping from XRL to PNML. The XRL/flower database model is presented and explained followed by a detailed discussion of the interaction sequence between the Petri-net enactment module, Web server module, and the database server.

References

1. W.M.P. van der Aalst. Structural Characterizations of Sound Workflow Nets. Computing Science Reports 96/23, Eindhoven University of Technology, Eindhoven, 1996.
2. W.M.P. van der Aalst. The Application of Petri Nets to Workflow Management. *The Journal of Circuits, Systems and Computers*, 8(1):21–66, 1998.
3. W.M.P. van der Aalst, A.H.M. ter Hofstede, B. Kiepuszewski, and A.P. Barros. Workflow Patterns Home Page. http://www.tm.tue.nl/it/research/patterns/.
4. W.M.P. van der Aalst, A.H.M. ter Hofstede, B. Kiepuszewski, and A.P. Barros. Advanced Workflow Patterns. In O. Etzion and P. Scheuermann, editors, *7th International Conference on Cooperative Information Systems (CoopIS 2000)*, volume 1901 of *Lecture Notes in Computer Science*, pages 18–29. Springer-Verlag, Berlin, 2000.
5. W.M.P. van der Aalst and A. Kumar. Xml based schema definition for support of inter-organizational workflow. *Information Systems Research*, 14(1):23–47, March 2003.
6. W.M.P. van der Aalst, H.M.W. Verbeek, and A. Kumar. XRL/Woflan: Verification of an XML/Petri-net based language for inter-organizational workflows (Best paper award). In K. Altinkemer and K. Chari, editors, *Proceedings of the 6th Informs Conference on Information Systems and Technology (CIST-2001)*, pages 30–45. Informs, Linthicum, MD, 2001.
7. BEA Systems, Intalio, SAP AG , Sun Microsystems. *Web Service Choreography Interface (WSCI) 1.0 Specification.* http://wwws.sun.com/software/xml/developers/wsci/, 2003.
8. A. Berglund, S. Boag, and D. Chamberlin et al. *XML Path Language (XPath) 2.0.* http://www.w3.org/TR/2003/WD-xpath20-20030822, 2003.
9. BPML.org. *Business Process Modeling Language (BPML) version 1.0.* Accessed August 2003 from www.bpmi.org, 2003.
10. Workflow Management Coalition. *XML Process Definition Language.* http://www.wfmc.org/standards/docs/TC-1025_10_xpdl_102502.pdf, 2002.
11. F. Curbera, Y. Goland, J. Klein, F. Leymann, D. Roller, S. Thatte, and S. Weerawarana. *Business Process Execution Language for Web-Services.* http://www-106.ibm.com/developerworks/library/ws-bpel/, 2003.
12. IBM. *Web Service Flow Language (WSFL) 1.0 Specification.* http://www-3.ibm.com/software/solutions/webservices/pdf/WSFL.pdf, 2003.
13. B. Kiepuszewski. *Expressiveness and Suitability of Languages for Control Flow Modelling in Workflows.* PhD thesis, Queensland University of Technology, Queensland University of Technology, Brisbane, Australia, 2002.

14. B. Kiepuszewski, A.H.M. ter Hofstede, and W.M.P. van der Aalst. Fundamentals of Control Flow in Workflows. *Acta Informatica*, 39(3):143–209, March 2003.
15. E. Kindler, J. Billington, and S. Christensen et al. The petri net markup language: Concepts, technology, and tools. In W.M.P. van der Aalst and E. Best, editors, *Proceedings of the 24th AInternational Conference,ICATPN 2003*, number 2679 in Lecture Notes in Computer Science, pages 483–505, Eindhoven, The Netherlands, 2003. Springer Verlag, Berlin.
16. E. Kindler and M. Weber et al. *Petri Net Kernel (PNK) Home Page.* http://www.informatik.hu-berlin.de/top/pnk/, 2003.
17. E. Kindler and M. Weber et al. *Petri Net Markup Language (PNML) Home Page.* http://www.informatik.hu-berlin.de/top/pnml/, 2003.
18. E. Kindler and M. Weber. The petri net kernel - an infrastructure for building petri net tools. *International Journal on Software Tools for Technology Transfer*, 3(4):486–497, 2001.
19. A. Norta. XRL Home Page. http://www.tm.tue.nl/it/research/xrl/.
20. A. Norta. XRL/flower Home Page. http://www.tm.tue.nl/it/research/xrl/flower.
21. S. Thatte. *XLANG: Web Service for Business Process Design*, 2003.
22. H.M.W. Verbeek and W.M.P. van der Aalst. Woflan Home Page, Eindhoven University of Technology, Eindhoven, The Netherlands. http://www.tm.tue.nl/it/woflan.
23. H.M.W. Verbeek and W.M.P. van der Aalst. Woflan 2.0: A Petri-net-based Workflow Diagnosis Tool. In M. Nielsen and D. Simpson, editors, *Application and Theory of Petri Nets 2000*, volume 1825 of *Lecture Notes in Computer Science*, pages 475–484. Springer-Verlag, Berlin, 2000.
24. H.M.W. Verbeek, T. Basten, and W.M.P. van der Aalst. Diagnosing Workflow Processes Using Woflan. *The Computer Journal, British Computer Society*, 44(4):246–279, 2001.
25. H.M.W Verbeek, A. Hirnschall, and W.M.P. van der Aalst. XRL/Flower: Supporting inter-organizational workflows using XML/Petri-net technology. In C. Bussler, R. Hull, S. McIlraith, M.E. Orlowska, B. Pernici, and J. Yang, editors, *Web Services, E-Business, and the Semantic Web*, CAiSE 2002 International Workshop, WES 2002, Toronto, Canada, pages 93–109. LNCS Springer, May 2002.
26. M. Weber and E. Kindler. The petri net markup language. In H. Ehrig, W. Reisig, G. Rozenberg, and H. Weber, editors, *Petri Net Technology for Communication-Based Systems Advances in Petri Nets*, number 2472 in Lecture Notes in Computer Science, page 455 p. Springer Verlag, Berlin, 2003.

Author Index

Lecture Notes in Computer Science

For information about Vols. 1–3005

please contact your bookseller or Springer-Verlag

Vol. 3120: J. Shawe-Taylor, Y. Singer (Eds.), Learning Theory. X, 648 pages. 2004.

Vol. 3109: S.C. Sahinalp, S. Muthukrishnan, U. Dogrusoz (Eds.), Combinatorial Pattern Matching. XII, 486 pages. 2004.

Vol. 3105: S. Göbel, U. Spierling, A. Hoffmann, I. Iurgel, O. Schneider, J. Dechau, A. Feix (Eds.), Technologies for Interactive Digital Storytelling and Entertainment. XVI, 304 pages. 2004.

Vol. 3099: J. Cortadella, W. Reisig (Eds.), Applications and Theory of Petri Nets 2004. XI, 505 pages. 2004.

Vol. 3098: J. Desel, W. Reisig, G. Rozenberg (Eds.), Lectures on Concurrency and Petri Nets. VIII, 849 pages. 2004.

Vol. 3096: G. Melnik, H. Holz (Eds.), Advances in Learning Software Organizations. X, 173 pages. 2004.

Vol. 3094: A. Nürnberger, M. Detyniecki (Eds.), Adaptive Multimedia Retrieval. VIII, 229 pages. 2004.

Vol. 3093: S.K. Katsikas, S. Gritzalis, J. Lopez (Eds.), Public Key Infrastructure. XIII, 380 pages. 2004.

Vol. 3092: J. Eckstein, H. Baumeister (Eds.), Extreme Programming and Agile Processes in Software Engineering. XVI, 358 pages. 2004.

Vol. 3091: V. van Oostrom (Ed.), Rewriting Techniques and Applications. X, 313 pages. 2004.

Vol. 3089: M. Jakobsson, M. Yung, J. Zhou (Eds.), Applied Cryptography and Network Security. XIV, 510 pages. 2004.

Vol. 3086: M. Odersky (Ed.), ECOOP 2004 – Object-Oriented Programming. XIII, 611 pages. 2004.

Vol. 3085: S. Berardi, M. Coppo, F. Damiani (Eds.), Types for Proofs and Programs. X, 409 pages. 2004.

Vol. 3084: A. Persson, J. Stirna (Eds.), Advanced Information Systems Engineering. XIV, 596 pages. 2004.

Vol. 3083: W. Emmerich, A.L. Wolf (Eds.), Component Deployment. X, 249 pages. 2004.

Vol. 3079: Z. Mammeri, P. Lorenz (Eds.), High Speed Networks and Multimedia Communications. XVIII, 1103 pages. 2004.

Vol. 3078: S. Cotin, D.N. Metaxas (Eds.), Medical Simulation. XVI, 296 pages. 2004.

Vol. 3077: F. Roli, J. Kittler, T. Windeatt (Eds.), Multiple Classifier Systems. XII, 386 pages. 2004.

Vol. 3076: D. Buell (Ed.), Algorithmic Number Theory. XI, 451 pages. 2004.

Vol. 3074: B. Kuijpers, P. Revesz (Eds.), Constraint Databases and Applications. XII, 181 pages. 2004.

Vol. 3073: H. Chen, R. Moore, D.D. Zeng, J. Leavitt (Eds.), Intelligence and Security Informatics. XV, 536 pages. 2004.

Vol. 3072: D. Zhang, A.K. Jain (Eds.), Biometric Authentication. XVII, 800 pages. 2004.

Vol. 3070: L. Rutkowski, J. Siekmann, R. Tadeusiewicz, L.A. Zadeh (Eds.), Artificial Intelligence and Soft Computing - ICAISC 2004. XXV, 1208 pages. 2004. (Subseries LNAI).

Vol. 3068: E. André, L. Dybkj{\}ae r, W. Minker, P. Heisterkamp (Eds.), Affective Dialogue Systems. XII, 324 pages. 2004. (Subseries LNAI).

Vol. 3067: M. Dastani, J. Dix, A. El Fallah-Seghrouchni (Eds.), Programming Multi-Agent Systems. X, 221 pages. 2004. (Subseries LNAI).

Vol. 3066: S. Tsumoto, R. S lowiński, J. Komorowski, J.W. Grzymala-Busse (Eds.), Rough Sets and Current Trends in Computing. XX, 853 pages. 2004. (Subseries LNAI).

Vol. 3065: A. Lomuscio, D. Nute (Eds.), Deontic Logic in Computer Science. X, 275 pages. 2004. (Subseries LNAI).

Vol. 3064: D. Bienstock, G. Nemhauser (Eds.), Integer Programming and Combinatorial Optimization. XI, 445 pages. 2004.

Vol. 3063: A. Llamosí, A. Strohmeier (Eds.), Reliable Software Technologies - Ada-Europe 2004. XIII, 333 pages. 2004.

Vol. 3062: J.L. Pfaltz, M. Nagl, B. Böhlen (Eds.), Applications of Graph Transformations with Industrial Relevance. XV, 500 pages. 2004.

Vol. 3061: F.F. Ramas, H. Unger, V. Larios (Eds.), Advanced Distributed Systems. VIII, 285 pages. 2004.

Vol. 3060: A.Y. Tawfik, S.D. Goodwin (Eds.), Advances in Artificial Intelligence. XIII, 582 pages. 2004. (Subseries LNAI).

Vol. 3059: C.C. Ribeiro, S.L. Martins (Eds.), Experimental and Efficient Algorithms. X, 586 pages. 2004.

Vol. 3058: N. Sebe, M.S. Lew, T.S. Huang (Eds.), Computer Vision in Human-Computer Interaction. X, 233 pages. 2004.

Vol. 3057: B. Jayaraman (Ed.), Practical Aspects of Declarative Languages. VIII, 255 pages. 2004.

Vol. 3056: H. Dai, R. Srikant, C. Zhang (Eds.), Advances in Knowledge Discovery and Data Mining. XIX, 713 pages. 2004. (Subseries LNAI).

Vol. 3055: H. Christiansen, M.-S. Hacid, T. Andreasen, H.L. Larsen (Eds.), Flexible Query Answering Systems. X, 500 pages. 2004. (Subseries LNAI).

Vol. 3054: I. Crnkovic, J.A. Stafford, H.W. Schmidt, K. Wallnau (Eds.), Component-Based Software Engineering. XI, 311 pages. 2004.

Vol. 3052: W. Zimmermann, B. Thalheim (Eds.), Abstract State Machines 2004. Advances in Theory and Practice. XII, 235 pages. 2004.

Vol. 3051: R. Berghammer, B. Möller, G. Struth (Eds.), Relational and Kleene-Algebraic Methods in Computer Science. X, 279 pages. 2004.

Vol. 3050: J. Domingo-Ferrer, V. Torra (Eds.), Privacy in Statistical Databases. IX, 367 pages. 2004.

Vol. 3049: M. Bruynooghe, K.-K. Lau (Eds.), Program Development in Computational Logic. VIII, 539 pages. 2004.

Vol. 3047: F. Oquendo, B. Warboys, R. Morrison (Eds.), Software Architecture. X, 279 pages. 2004.

Vol. 3046: A. Laganà, M.L. Gavrilova, V. Kumar, Y. Mun, C.K. Tan, O. Gervasi (Eds.), Computational Science and Its Applications – ICCSA 2004. LIII, 1016 pages. 2004.

Vol. 3045: A. Laganà, M.L. Gavrilova, V. Kumar, Y. Mun, C.K. Tan, O. Gervasi (Eds.), Computational Science and Its Applications – ICCSA 2004. LIII, 1040 pages. 2004.

Vol. 3044: A. Laganà, M.L. Gavrilova, V. Kumar, Y. Mun, C.K. Tan, O. Gervasi (Eds.), Computational Science and Its Applications – ICCSA 2004. LIII, 1140 pages. 2004.

Vol. 3043: A. Laganà, M.L. Gavrilova, V. Kumar, Y. Mun, C.K. Tan, O. Gervasi (Eds.), Computational Science and Its Applications – ICCSA 2004. LIII, 1180 pages. 2004.

Vol. 3042: N. Mitrou, K. Kontovasilis, G.N. Rouskas, I. Iliadis, L. Merakos (Eds.), NETWORKING 2004, Networking Technologies, Services, and Protocols; Performance of Computer and Communication Networks; Mobile and Wireless Communications. XXXIII, 1519 pages. 2004.

Vol. 3040: R. Conejo, M. Urretavizcaya, J.-L. Pérez-de-la-Cruz (Eds.), Current Topics in Artificial Intelligence. XIV, 689 pages. 2004. (Subseries LNAI).

Vol. 3039: M. Bubak, G.D.v. Albada, P.M. Sloot, J.J. Dongarra (Eds.), Computational Science - ICCS 2004. LXVI, 1271 pages. 2004.

Vol. 3038: M. Bubak, G.D.v. Albada, P.M. Sloot, J.J. Dongarra (Eds.), Computational Science - ICCS 2004. LXVI, 1311 pages. 2004.

Vol. 3037: M. Bubak, G.D.v. Albada, P.M. Sloot, J.J. Dongarra (Eds.), Computational Science - ICCS 2004. LXVI, 745 pages. 2004.

Vol. 3036: M. Bubak, G.D.v. Albada, P.M. Sloot, J.J. Dongarra (Eds.), Computational Science - ICCS 2004. LXVI, 713 pages. 2004.

Vol. 3035: M.A. Wimmer (Ed.), Knowledge Management in Electronic Government. XII, 326 pages. 2004. (Subseries LNAI).

Vol. 3034: J. Favela, E. Menasalvas, E. Chávez (Eds.), Advances in Web Intelligence. XIII, 227 pages. 2004. (Subseries LNAI).

Vol. 3033: M. Li, X.-H. Sun, Q. Deng, J. Ni (Eds.), Grid and Cooperative Computing. XXXVIII, 1076 pages. 2004.

Vol. 3032: M. Li, X.-H. Sun, Q. Deng, J. Ni (Eds.), Grid and Cooperative Computing. XXXVII, 1112 pages. 2004.

Vol. 3031: A. Butz, A. Krüger, P. Olivier (Eds.), Smart Graphics. X, 165 pages. 2004.

Vol. 3030: P. Giorgini, B. Henderson-Sellers, M. Winikoff (Eds.), Agent-Oriented Information Systems. XIV, 207 pages. 2004. (Subseries LNAI).

Vol. 3029: B. Orchard, C. Yang, M. Ali (Eds.), Innovations in Applied Artificial Intelligence. XXI, 1272 pages. 2004. (Subseries LNAI).

Vol. 3028: D. Neuenschwander, Probabilistic and Statistical Methods in Cryptology. X, 158 pages. 2004.

Vol. 3027: C. Cachin, J. Camenisch (Eds.), Advances in Cryptology - EUROCRYPT 2004. XI, 628 pages. 2004.

Vol. 3026: C. Ramamoorthy, R. Lee, K.W. Lee (Eds.), Software Engineering Research and Applications. XV, 377 pages. 2004.

Vol. 3025: G.A. Vouros, T. Panayiotopoulos (Eds.), Methods and Applications of Artificial Intelligence. XV, 546 pages. 2004. (Subseries LNAI).

Vol. 3024: T. Pajdla, J. Matas (Eds.), Computer Vision - ECCV 2004. XXVIII, 621 pages. 2004.

Vol. 3023: T. Pajdla, J. Matas (Eds.), Computer Vision - ECCV 2004. XXVIII, 611 pages. 2004.

Vol. 3022: T. Pajdla, J. Matas (Eds.), Computer Vision - ECCV 2004. XXVIII, 621 pages. 2004.

Vol. 3021: T. Pajdla, J. Matas (Eds.), Computer Vision - ECCV 2004. XXVIII, 633 pages. 2004.

Vol. 3019: R. Wyrzykowski, J.J. Dongarra, M. Paprzycki, J. Wasniewski (Eds.), Parallel Processing and Applied Mathematics. XIX, 1174 pages. 2004.

Vol. 3018: M. Bruynooghe (Ed.), Logic Based Program Synthesis and Transformation. X, 233 pages. 2004.

Vol. 3017: B. Roy, W. Meier (Eds.), Fast Software Encryption. XI, 485 pages. 2004.

Vol. 3016: C. Lengauer, D. Batory, C. Consel, M. Odersky (Eds.), Domain-Specific Program Generation. XII, 325 pages. 2004.

Vol. 3015: C. Barakat, I. Pratt (Eds.), Passive and Active Network Measurement. XI, 300 pages. 2004.

Vol. 3014: F. van der Linden (Ed.), Software Product-Family Engineering. IX, 486 pages. 2004.

Vol. 3012: K. Kurumatani, S.-H. Chen, A. Ohuchi (Eds.), Multi-Agnets for Mass User Support. X, 217 pages. 2004. (Subseries LNAI).

Vol. 3011: J.-C. Régin, M. Rueher (Eds.), Integration of AI and OR Techniques in Constraint Programming for Combinatorial Optimization Problems. XI, 415 pages. 2004.

Vol. 3010: K.R. Apt, F. Fages, F. Rossi, P. Szeredi, J. Váncza (Eds.), Recent Advances in Constraints. VIII, 285 pages. 2004. (Subseries LNAI).

Vol. 3009: F. Bomarius, H. Iida (Eds.), Product Focused Software Process Improvement. XIV, 584 pages. 2004.

Vol. 3008: S. Heuel, Uncertain Projective Geometry. XVII, 205 pages. 2004.

Vol. 3007: J.X. Yu, X. Lin, H. Lu, Y. Zhang (Eds.), Advanced Web Technologies and Applications. XXII, 936 pages. 2004.

Vol. 3006: M. Matsui, R. Zuccherato (Eds.), Selected Areas in Cryptography. XI, 361 pages. 2004.